Springer Series in Computational Neuroscience

Computational Neuroscience gathers monographs and edited volumes on all aspects of computational neuroscience, including theoretical and mathematical neuroscience, biophysics of the brain, models of neurons and neural networks, and methods of data analysis (e.g. information theory). The editors welcome suggestions and projects for inclusion in the series.

About the Editors:

Alain Destexhe is Research Director at the Centre National de la Recherche Scientifique (CNRS), France, and Romain Brette is Research Director at the Institut National de la Santé et de la Recherche Médicale (INSERM), France.

Ad Aertsen • Sonja Grün • Pedro E. Maldonado •
Günther Palm
Editors

Introducing Computation to Neuroscience

Selected Papers of George Gerstein

Springer

Editors
Ad Aertsen
Faculty of Biology
University of Freiburg
Freiburg, Germany

Sonja Grün
Neuroscience & Medicine
Research Centre Jülich
Jülich, Germany

Pedro E. Maldonado
Department of Neuroscience
Universidad de Chile
Santiago, Chile

Günther Palm
Neural Information Processing
University of Ulm
Ulm, Germany

ISSN 2197-1900 ISSN 2197-1919 (electronic)
Springer Series in Computational Neuroscience
ISBN 978-3-030-87449-0 ISBN 978-3-030-87447-6 (eBook)
https://doi.org/10.1007/978-3-030-87447-6

This Springer imprint is published by the registered company Springer Nature Switzerland AG
The registered company address is: Gewerbestrasse 11, 6330 Cham, Switzerland

Preface

This book brings together a selection of papers by George Gerstein, representing his long-term endeavor of making neuroscience into a real science inspired by physics, where he had his roots. In this, George was many years ahead in the field, consistently striving for quantitative analyses, mechanistic models, conceptual clarity, and, foremost, true understanding. In doing so, George pioneered Computational Neuroscience, many years before the term was even born.

The overarching goal of George Gerstein's research program throughout his productive career was to understand the functional organization of neuronal networks in the brain. How do neurons in these networks interact to make the network accomplish a certain function, e.g., in sensory perception, while providing the neurons with the stimulus–response properties they reveal when recorded in experiments? To quote from his own writings (Gerstein et al. 1989): *"Long before emergence of all the current complexities of single neuron anatomy and physiology, it was recognized that the computational processes constantly carried out by the nervous system could not be explained by the properties of single neurons alone. Sherrington, in the 1930's, verbalized the idea that neurons must cooperate in fulfilling a complex task. This concept has been completely assimilated into neurobiological thought and explanation, but only in the most general way. Experimental access to the properties of real neuronal assemblies has largely been impossible until quite recently. Thus, the development of the assembly concept has been somewhat unfettered."*

In this research effort, George Gerstein played an active role, both as an experimenter and as a theoretician. Thus, he was quite aware of the pros and cons of both approaches and of the limitations of either one. To quote from his own writings once more (Gerstein and Turner 1993): *"Over the years, the relationship between experimenters and theoreticians in the study of brain function has been at best uneasy. Fault can be found on both sides; in part there is a communication difficulty. The principal problem is the lack of direct confrontation: theories are rarely couched in terms amenable to experimental testing, while experiments are rarely aimed at theoretical issues. Theoreticians often use large populations of elements and implied observation times that are wildly unrealistic for the measurement possibilities of the laboratory. Conversely, even those experimenters who are observing a population of neurons are all too often thinking in terms similar to single neuron stimulus designs."*

Yet, George, being a staunch pragmatist and optimist, consistently endeavored to connect the two approaches, in a constructive effort to join forces and, thereby, optimizing the chances of success (Gerstein and Turner 1993): *"At this point in time, the confrontation we are seeking might therefore come at an intermediate, less global level, involving modest numbers of elements. Here we may be able to seek emergent properties of neuronal assemblies from both the experimental and the theoretical point of view."*

George never liked to write a book. Despite many requests, he consistently refused. His response invariably was: *"Books are about old papers. I prefer to write new ones."* And so he did, until the very end. Thus, we found ourselves with the task of collecting a selection of George Gerstein's many seminal contributions to neuroscience, be they experimental, theoretical, or computational, into a single, comprehensive volume. Our aim, in doing so, was also to provide the readers, especially the younger ones who may not even be aware that there was ever a time in neuroscience without the PSTH, the cross-correlation approach, or stochastic neuron models, with the fresh introduction of these various concepts in the original literature.

We organized the material in a series of chapters by subject, ordered in time, each one containing one or more of George Gerstein's papers. References to the included papers throughout this volume are highlighted in boldface type. Several of George's former students and collaborators kindly contributed an Introduction to these various chapters, intending to put the reprinted papers in their historical and present context and to add the personal touch of their involvement in generating these papers. Thereby, each of these chapters introduces one (or more) of George Gerstein's seminal contributions to the Neurosciences, be it in the domain of neuron and network modeling or in the area of neuronal data analysis techniques, ordered by their publication date. Specifically, the various chapters highlight the following innovative computational approaches and their applications in neurophysiological experiments first introduced in George Gerstein's papers:

Chapter 1: the first stochastic model of a neuron's spiking activity

Chapter 2: the introduction of statistical techniques from stochastic point process theory (e.g., inter-spike interval histogram, serial correlation, Peri-Stimulus Time Histogram (PSTH), auto- and cross-correlation) to the analysis of neuronal spike train data

Chapter 3: setting up a lab for single neuron recordings in the inferior temporal cortex of the awake behaving monkey and demonstrating the effect of attention on neuronal responses to visual stimuli

Chapter 4: demonstrating plasticity of neuronal interactions in small neuronal networks upon stimulation

Chapter 5: designing novel methods for finding repeating spike patterns in single- and multi-neuron spike recordings and for identifying cooperating groups of neurons (gravitational clustering)

Chapter 6: using gravitational clustering analysis to identify neuronal assemblies in the respiratory brain stem and their reconfiguration with breathing phase
Chapter 7: characterizing the dynamics of spike correlations and neuronal interactions (Joint-PSTH) and their dependence on sensory stimuli and behavior
Chapter 8: introducing different measures for statistical significance of spike train correlations
Chapter 9: detecting recurrent spatiotemporal firing patterns in multi-neuron spike train recordings
Chapter 10: interpreting cross-correlations between unresolved multi-neuron recordings
Chapter 11: demonstrating plasticity in receptive field properties and neuronal interactions in auditory cortex upon intracortical microstimulation
Chapter 12: demonstrating appetitive conditioning-induced plasticity of receptive fields and state-dependent modulation of stimulus-evoked responses and their across-trial variability in auditory cortex
Chapter 13: improving the signal-to-noise ratio in multi-neuron recordings and the sensitivity of gravitational clustering in measuring neuronal interactions
Chapter 14: detecting higher-order correlations and other signs of neuronal assembly activity in network model simulations and experimental data from large-scale multi-neuron recordings

We hope that, taken together, these various chapters and their Introductions will help the reader appreciate how much our current thinking on brain function owes to George Gerstein's research and the insights gained from it. George was not only an innovative thinker in theory, experiment, and analysis methodology, he also inspired by his mode of thinking science. Thus, in a way, this book is both a product of this inspiration and a tribute to it.

The volume closes with two separate contributions: a commemoration of the first Nencki Award being awarded in 2008 to George Gerstein, including a summary **CV of his education and scientific career**, written and published by Andrzej Wróbel, and an **Epilogue**, presenting an Obituary to George Gerstein, written by his colleagues Ad Aertsen and Moshe Abeles.

Acknowledgment

We thank Petra O'Brien and Maren Maus for their support in getting this book together, and Markus Diesmann for his TeXnical advice.

References

Gerstein GL, Bedenbaugh P, Aertsen AMHJ (1989) Neuronal assemblies. IEEE Trans Biomed Eng 36(1):4–14. https://doi.org/10.1109/10.16444

Gerstein GL, Turner MR (1993) Neural assemblies as building blocks of cortical computation. In: Schwartz EL (ed) Computational neuroscience. MIT Press, Cambridge, pp 179–191

Freiburg, Germany — Ad Aertsen
Jülich, Germany — Sonja Grün
Santiago, Chile — Pedro E. Maldonado
Ulm, Germany — Günther Palm
February, 2020

Contents

About the Editors

Ad Aertsen is an Emeritus Professor of Neurobiology and Biophysics of the Faculty of Biology at the University of Freiburg, Germany. He is a Founding Director of the Bernstein Center Freiburg. He studied Physics and Applied Mathematics at the University of Utrecht, the Netherlands, where he received his Diploma (comp. to M.Sc.) in Physics in 1973. He gained a Ph.D. in Physics in 1981 under Peter Johannesma and Jos Eggermont from the Neurophysics Laboratory at the University of Nijmegen, the Netherlands. He pursued his postdoctoral studies in Physiology with George Gerstein at the University of Pennsylvania, Philadelphia, USA. He was a research group leader with Valentino Braitenberg at the Max-Planck-Institute for Biological Cybernetics in Tübingen, Germany, a visiting professor with Eilon Vaadia and Moshe Abeles at the Hebrew University in Jerusalem, Israel, a group leader with Werner von Seelen at the Ruhr-University in Bochum, Germany, and an Associate Professor at the Weizmann Institute of Science in Rehovot, Israel, working with Amos Arieli and Amiram Grinvald. He moved to the Freiburg University, Germany, in 1996. There, his research focused on the recording, analysis, and modeling of neuronal assembly activity in cortical and subcortical networks and on the development of brain–machine interfaces, including neuronal motor prostheses.

Sonja Grün is a Professor of Theoretical Systems Neurobiology at the RWTH Aachen University, Germany, and Director of the Institutes of Neuroscience and Medicine INM-6 and INM-10 at Research Centre Jülich, Germany. She studied Physics at the Eberhard Karls University in Tübingen and performed her diploma thesis in Physics (comp. to M.Sc.) at the Max-Planck Institute for Biological Cybernetics in Tübingen in 1991. Under the supervision of Ad Aertsen and Christoph von der Malsburg, she pursued her doctoral thesis in Theoretical Neuroscience and got her Dr. rer. nat. in Physics from the Ruhr-University Bochum in 1996. For her postdoctoral studies, she joined the lab of Moshe Abeles at the Hebrew University in Jerusalem and worked in experimental neuroscience. From 1998 to 2002, she worked as a senior fellow in the Department of Wolf Singer at the

Max-Planck Institute for Brain Research in Frankfurt/Main. In 2003, she got her habilitation and venia legendi in Neurobiology and Biophysics from the Albert Ludwigs University, Freiburg, Germany. From 2003 to 2006, she was Assistant Professor for Neuroinformatics/Theoretical Neuroscience at the Free University in Berlin. In 2006, she moved to Japan to fill a position as a Unit Leader and later as a Team Leader at the RIKEN Brain Science Institute, Wako City, Japan. Since 2011, Sonja is at the Research Centre Jülich and a Full Professor at the RWTH Aachen University. From 2014 to 2016, she was a guest professor at the Osaka University, Japan. Her research interests are (a) dynamical interactions in the brain network relevant for behavior and cognition, (b) development of statistical analysis tools for activity data from behaving animals, (c) development of collaborative, reproducible digital workflows, and (d) closing the loop between neural network models and experimental data.

Pedro E. Maldonado is a Full Professor at the Faculty of Medicine, Universidad de Chile. He is currently the Chairman of the Department of Neuroscience. He is also an Associate Researcher at the Millennium Institute of Biomedical Neuroscience (BNI) and also a visiting scientist at the Institute of Neuroscience and Medicine at the Jülich Research Center in Germany. He earned a bachelor's degree in Biological Sciences (1984) and a master's degree in Physiology (1986) at the Faculty of Sciences of the Universidad de Chile, under the supervision of Francisco Varela and Humberto Maturana. He then worked as an instructor at the Faculty of Medicine, in the University of Miami for two years, to later pursue a Ph.D. degree in Physiology in 1993 as a student of George L. Gerstein, in the University of Pennsylvania. He completed a postdoctoral period with Charles Gray at the Neuroscience Center of the University of California, Davis, USA. He returned to Chile in 1998 as a faculty member at the Faculty of Medicine, Universidad de Chile. His research interest relates to understanding how higher cognitive functions such as perception, memory, and complex motor acts emerge from the integration and coordination of large populations of neurons in the brain. He works in visual perception, with a focus on ecological paradigms and active sensing, as closed-loop models.

Günther Palm began his studies of mathematics at the University of Hamburg and graduated at the Eberhard-Karls-University of Tübingen with a PhD thesis on *"Entropie und Generatoren in dynamischen Verbänden"* supervised by Prof. Dr. Rainer Nagel in 1975. He then worked as a research assistant at the Max Planck Institute for Biological Cybernetics, Tübingen, on topics of quantitative neuroanatomy, information theory, nonlinear systems theory, associative memory, and brain theory from 1975 to 1988. During that time, he spent one year (1983/1984) in Berlin as a fellow of the Wissenschaftskolleg. In 1988, he became professor for theoretical brain research at the University of Düsseldorf. Since 1991, he is director of the Institute of Neural Information Processing at Ulm University. He retired in 2016 and is working part time on neural data analysis at the Forschungszentrum Jülich since 2017. Professor Palm's research focus is on information theory, neural networks, associative memory, and specifically on Hebbian cell assemblies. By 2015, he has published more than 300 peer-reviewed articles in international journals and 60 invited contributions and (co-)edited 8 books. He is author of the monographs *Neural Assemblies: An Alternative Approach to Artificial Intelligence* (1982) and *Novelty, Information and Surprise* (2012).

Chapter 1
Leaky Integrate and Fire Neurons

RC Circuits to Model Neurons and Reward Monkeys

Eberhard Fetz

George Gerstein and I first met in 1962 when George was transitioning from postdoctoral fellow to junior faculty at the MIT and I was a graduate student in the Physics Department. We were both converts from physics to neuroscience and understood the perspectives of a hardcore science, in which mathematical descriptions could be applied and principles derived. George was fruitfully exploring Vincent Hakima range of sophisticated analytical strategies to capture and quantify neural dynamics. I had just had the epiphany (on LSD) that the brain was the most interesting physical system in the universe—the source of all conscious experience and behavior, and not least, where I lived. I was intrigued by the riddle of how neural activity generates all this endless experience. I thought that neural explanations of behavior would probably be like statistical mechanical explanations of thermodynamics. Of course the fatal flaw in that notion was the fact that statistical mechanics succeeded because all particles could be assumed to behave identically, whereas in the brain each neuron behaves differently.

I was looking for an appropriate advisor to guide my physics thesis and still somehow begin working on the brain. With his two physics degrees from Harvard, George was the perfect candidate. He would probably have been my dissertation advisor had he stayed at the MIT rather than take off for the University of Pennsylvania. But before he left, we did collaborate on a project to describe the firing properties of a simple RC model of spiking neurons. This leaky integrator summed exponentially decaying voltage pulses representing post-synaptic potentials and

E. Fetz (✉)
Department of Physiology and Biophysics, University of Washington, Seattle, WA, USA
e-mail: fetz@uw.edu

A. Aertsen et al. (eds.), *Introducing Computation to Neuroscience*, Springer Series in Computational Neuroscience, https://doi.org/10.1007/978-3-030-87447-6_1

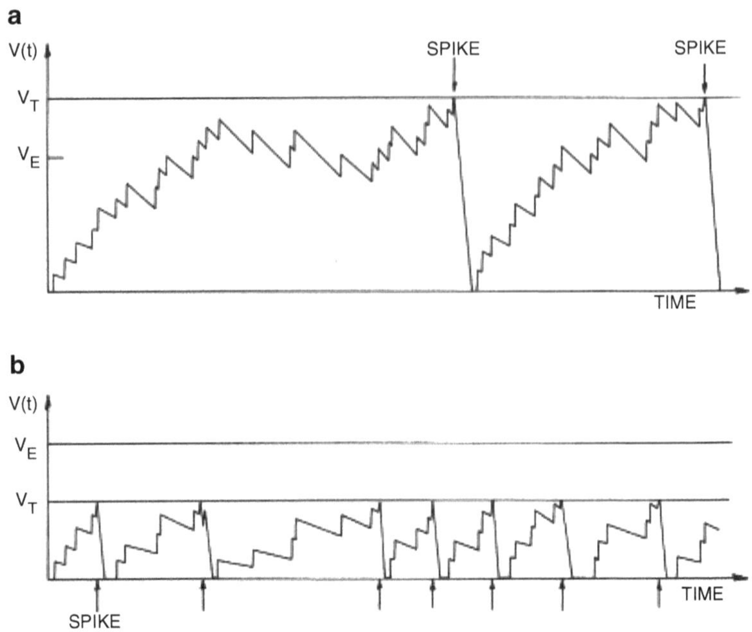

Fig. 1.1 RC neuron potential as function of time with Poisson distributed EPSP inputs. Threshold (V_T) could be above (**a**) or below (**b**) equilibrium potential (V_E)

generated spikes when the sum reached threshold. The resultant spike trains were grist for the statistical algorithms that George and Nelson Kiang had developed to describe the firing properties of biological neurons. The interspike and joint interval histograms generated by the RC model resembled those seen in cerebral cortex. The results could be summarized in terms of two regimes, depending on whether the model firing threshold was above or below the mean asymptotic equilibrium level of the "membrane potential" (Fig. 1.1). The model was also used to compare the dynamics produced with purely excitatory and with mixed excitatory and inhibitory pulses. These results were written up in a short Quarterly Progress Report for the Research Laboratory of Electronics (Fetz and Gerstein 1963).

The closest that our RLE report came to being recognized in the literature was a citation in Gerstein and Mandelbrot (1964). That paper explored various random walk models that took discrete steps toward or away from threshold. Those crossings also generated pulse trains for analysis. In contrast, our RC model involved exponentially decaying PSPs, but the two approaches could be made to exhibit many similar properties. The random walk can also be used to model many other situations. More recently, the random walk model has proven highly fruitful in capturing the neural processes underlying decision making based on accumulating sensory evidence (e.g., Gold and Shadlen (2007)).

The last challenge that I tried to solve with George was a closed analytical expression for the interspike interval distribution of the RC neuron model in terms of model parameters. I finally abandoned this effort after George's departure, turning to recording actual physiological neurons in the cat spinal cord in the lab of Patrick Wall, who became my thesis advisor. Pat provided a proper introduction to neurophysiology and supervised my thesis on "Pyramidal tract effects on dorsal horn interneurons."

(When my physics graduate committee got wind of the title of my thesis, they wondered whether this was appropriate for a physics degree. They called a meeting for me to present the proposed research, and I emphasized source-sink mapping of spinal cord field potentials, using Laplacians and other Maxwellian concepts. Afterward, I was dismissed while they deliberated; probably thanks to Pat's input, they decided that I had just passed my thesis defense. They obviously did not want to hear any more of it.)

After graduating, I crossed the country to do postdoctoral work at the University of Washington and learn more neurophysiology. I intended to work with Arnie Towe on pyramidal tract neurons of chloralosed cats. But being more interested in the activity of neurons in awake behaving monkeys, I joined Eric Luschei, a brilliant graduate student in Mitchell Glickstein's lab, who had developed the chronic unit recording techniques at the same time as Ed Evarts. When Mitch and Eric left Seattle for Brown University, I was free to pursue my own interests and began to investigate operant conditioning of cortical cell activity. These experiments used the RC circuit that George and I had explored, but now to reward monkeys for generating changes in cortical cell activity. The initial version was a simple leaky integrator that summed pulses from spikes of a motor cortex neuron. When the monkey increased the firing rate sufficiently to reach integrator threshold, he triggered a feeder that delivered applesauce (Fig. 1.2, from Fetz and Baker (1973) [this paper and others from our lab can be accessed at: https://depts.washington.edu/fetzweb]). A more sophisticated version integrated multiple inputs, requiring the monkey to control patterns of simultaneous cell and muscle activity to reach threshold (Fetz and Finocchio 1975). The remarkable finding was that the monkeys could quickly generate several new and orthogonal response patterns during the course of a single session.

In order to provide the monkey with continuous feedback about how close his neural activity was getting to reinforcement threshold, the integrator voltage was displayed on a meter facing the monkey. The rightward position of the meter arm corresponded to the feeder trigger, and monkeys quickly learned to do what it took to drive that meter to the right and get applesauce. At that time, I saw this meter as a device for providing biofeedback about neural activity. Had I realized that neural control of the meter arm was a first step toward brain-controlled prosthetic arms, my career might have been different. In retrospect it would probably have been more exciting to pursue further experiments in volitional control of neural activity to drive external devices. Instead, my physics training diverted me to focus on proving causality between premotor neurons and muscles in behaving monkeys, using spike-triggered averages of EMG activity.

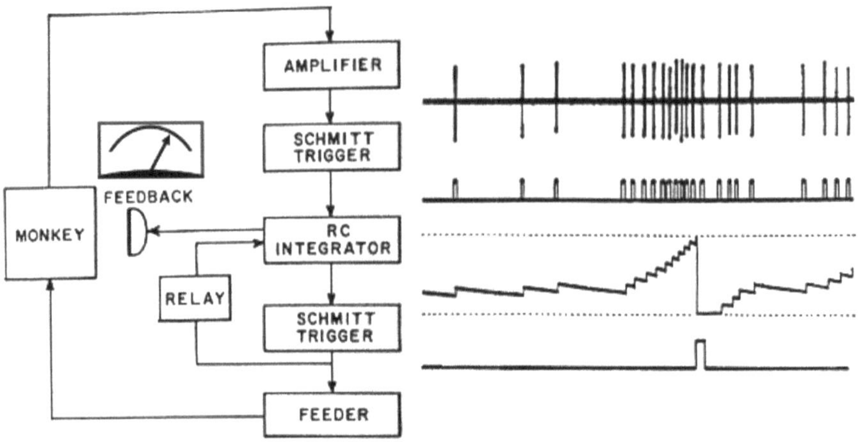

Fig. 1.2 RC integrator used to reward monkey for bursts of activity in neurons. Integrator voltage deflected neurofeedback meter arm

The basic RC circuit that George and I initially investigated has been extended to artificial neural networks of multiple "integrate-and-fire" (IAF) units. The IAF units sum more realistically shaped post-synaptic potentials to firing threshold and affect their targets via weighted connections. The connection weights can be derived from networks of continuous units trained with back-propagation to perform particular spatiotemporal transforms (Fetz 1993). The conversion involves replacing individual continuous units with a group of IAF units connected with the same relative weight (Maier et al. 2003). More recently we have found that networks of IAF units endowed with spike-timing-dependent plasticity can be used to simulate cortical plasticity induced with closed-loop activity-dependent stimulation (Shupe and Fetz, unpublished).

Our studies of motor cortex neurons and their relation to limb movement have provided many occasions to apply the analytic techniques pioneered by George and his colleagues. The basic principles of spike train correlations first described in Perkel et al. (1967) were fundamental to investigating synaptic interactions between cortical neurons (Fetz et al. 1991; Smith and Fetz 2009). We also investigated the synaptic mechanisms underlying features in correlated spike trains with intracellular recordings of post-synaptic potentials (Cope et al. 1987; Matsumura et al. 1996).

Another useful technique pioneered by George was the Joint Peri-Stimulus Time Histogram (JPSTH) analysis of Aertsen et al. (1989). We used the JPSTH to dissect the time-varying correlations between motor cortex neurons during movement. Figure 1.3 is an unpublished figure from the thesis of Wade Smith 1989. The histograms along the x- and y-axes at left show the response averages of two simultaneously recorded neurons during performance of a ramp-and-hold wrist flexion-extension task. The overall cross-correlogram between the two neurons (upper-right histogram) shows a broad central common input peak, but notably, the

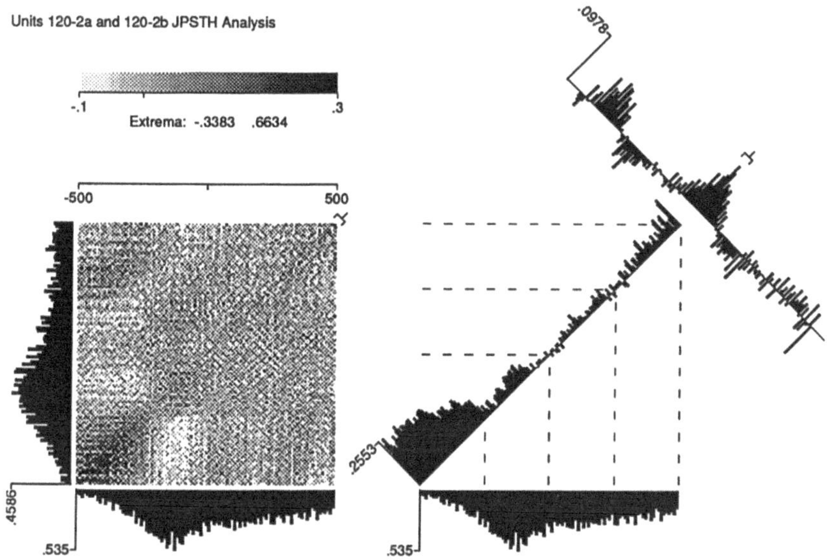

Fig. 1.3 JPSTH of two motor cortex neurons during movement task. From Smith, Doctoral Dissertation, University of Washington, 1989

time resolution of the central slice plotted along the diagonal reveals that most of the above-chance common input occurs well before the maximum firing rates of the neurons.

I last saw George on the happy occasion of his retirement symposium at the University of Pennsylvania in 2003. Many of his friends and collaborators convened to look back at his full and productive life, pushing the envelope with innovative computational approaches and wide range of experimental achievements. George was still full of enthusiasm for what struck him as interesting issues and even seemed a bit more tolerant of what he used to dismiss with "Bah, Humbug." We are all indebted to George for his many seminal contributions and will miss his steadfast friendship.

References

Aertsen AM, Gerstein GL, Habib MK, Palm G (1989) Dynamics of neuronal firing correlation: modulation of "effective connectivity". J Neurophysiol 61(5):900–917. https://doi.org/10.1152/jn.1989.61.5.900

Cope TC, Fetz EE, Matsumara M (1987) Cross' correlation assessment of synaptic strength of single Ia fibre connections with triceps surae motoneurones in cats. Am J Physiol 390:161–188. https://doi.org/10.1113/jphysiol.1987.sp016692

Fetz E (1993) Dynamic recurrent neural network models of sensorimotor behavior. In: Gardner D (ed) The neurobiology of neural networks. MIT Press, Cambridge, pp 165–190

Fetz EE, Baker MA (1973) Operantly conditioned patterns of precentral unit activity and correlated responses in adjacent cells and contralateral muscles. J Neurophysiol 36(2):179–204. https://doi.org/10.1152/jn.1973.36.2.179

Fetz EE, Finocchio DV (1975) Correlations between activity of motor cortex cells and arm muscles during operantly conditioned response patterns. Exp Brain Res 23(3):217–240. https://doi.org/10.1007/BF00239736

Fetz EE, Gerstein G (1963) An RC model for spontaneous activity of single neurons. Quarterly Progress Report #71, Research Laboratory of Electronics, MIT

Fetz E, Toyama K, Smith W (1991) Synaptic interactions between cortical neurons. In: Peters A, Jones E (eds) Cerebral cortex, vol 9. Springer, Boston, pp 1–47. https://doi.org/10.1007/978-1-4615-6622-9_1

Gerstein GL, Mandelbrot B (1964) Random walk models for the spike activity of a single neuron. Biophys J 4(1, Part 1):41–68. https://doi.org/10.1016/S0006-3495(64)86768-0

Gold J, Shadlen MN (2007) The neural basis of decision making. Annu Rev Neurosci 30:535–574. https://doi.org/10.1146/annurev.neuro.29.051605.113038

Maier MA, Shupe LE, Fetz EE (2003) Recurrent neural networks of integrate-and-fire cells simulating short-term memory and wrist movement tasks derived from continuous dynamic networks. J Physiol Paris 97(4–6):601–612. https://doi.org/10.1016/j.jphysparis.2004.01.017

Matsumura M, Chen M, Sawaguchi T, Kubota K, Fetz EE (1996) Synaptic interactions between primate precentral cortex neurons revealed by spike-triggered averaging of intracellular membrane potentials in vivo. J Neurosci 16(23):7757–7767. https://doi.org/10.1523/JNEUROSCI.16-23-07757.1996

Perkel DH, Gerstein GL, Moore GP (1967) Neuronal spike trains and stochastic point processes. II. Simultaneous spike trains. Biophys J 7(4):419–440. https://doi.org/10.1016/s0006-3495(67)86597-4

Smith W (1989) Synaptic interactions between identified motor cortex neurons in the active primate. Doctoral Dissertation, University of Washington

Smith WS, Fetz EE (2009) Synaptic interactions between forelimb-related motor cortex neurons in behaving primates. J Neurophysiol 102(2):1026–1039. https://doi.org/10.1152/jn.91051.2008

C. AN RC MODEL FOR SPONTANEOUS ACTIVITY OF SINGLE NEURONS

Spontaneously active single neurons produce spike trains whose temporal characteristics can be conveniently described in terms of: (a) an interval histogram[1] of the time intervals between successive spikes, and (b) a two-dimensional joint-interval histogram of two successive spike intervals.[2] These measures can be used to differentiate and characterize various modes of spontaneous activity in actual neurons. Rodieck, Kiang, and Gerstein[2] computed the histograms of neurons in cochlear nucleus of anesthetized cats, and found typical interval histograms to be: (a) exponential, with a small dead time, (b) approximately Gaussian, or (c) long-tailed, that is, unimodal and asymmetric, with tails longer than exponential.

Various mathematical models have been proposed to explain the observed spontaneous activity of neurons. Kuffler, Fitzhugh, and Barlow[3] have suggested that interval histograms of spontaneous spike activity might be described by incomplete Gamma functions. Viernstein and Grossman,[4] Verveen,[5] and, more recently, Weiss[6,7] have examined models based on threshold crossings of a random noise with Gaussian amplitude distribution. Gerstein and Mandelbrot[8] have discussed a two-parameter model based on a random walk "toward" a firing threshold. Unfortunately, none of these models is able to describe all of the three types of spontaneous activity mentioned above.

In the present report we shall examine another type of model which, in spite of its simplicity, has not been discussed in published work on this subject. This model incorporates the following experimental facts: In its resting state the neuron maintains a constant potential difference across its cell membrane, the internal potential being approximately 70 mv below the external potential. The membrane polarization can be changed by virtue of certain chemicals released by afferent neurons. Such chemicals may either depolarize the membrane – that is, reduce the potential difference – thereby giving rise to what Eccles[9] terms an excitatory postsynaptic potential (EPSP), or they may hyperpolarize the membrane to produce an inhibitory postsynaptic potential (IPSP). These chemicals are found to be released at the synaptic knobs in quantized amounts, this action causing depolarization and hyperpolarization "pulses" that decay approximately exponentially in time and along the membrane surface.[10] Under certain circumstances these pulses have been observed with microelectrodes.[11] The effects of successive pulses are cumulative; if the membrane is depolarized beyond a certain critical value called threshold, a large all-or-nothing action spike which propagates

(XVI. COMMUNICATIONS BIOPHYSICS)

without decrement along the neuron's axon, is produced.

These properties of active neurons suggest the following model: The neuron's membrane polarization is assumed to be represented by a single variable called the potential. The potential may change discontinuously by positive or negative steps of size h, representing the EPSP and IPSP pulses. The potential also varies continuously by decaying exponentially toward a fixed value V_o, called the resting potential. If the potential reaches a positive value V_T, called threshold, the neuron is said to "fire," that is, to produce an action spike, and the potential returns to V_o.

Under conditions of spontaneous activity, it seems reasonable to assume that the positive and negative pulses are generated by Poisson processes. Hence, the probability density $P_{\pm}(t)$ of time intervals of length t between positive or negative pulses would be exponential:

$$P_{\pm}(t) = R_{\pm} e^{-R_{\pm} t},$$

where $R_{\pm}$ represents the average rate of positive or negative pulses. For successive pulses to be summed, the pulse rates $R_{\pm}$ must exceed the decay rate of the potential

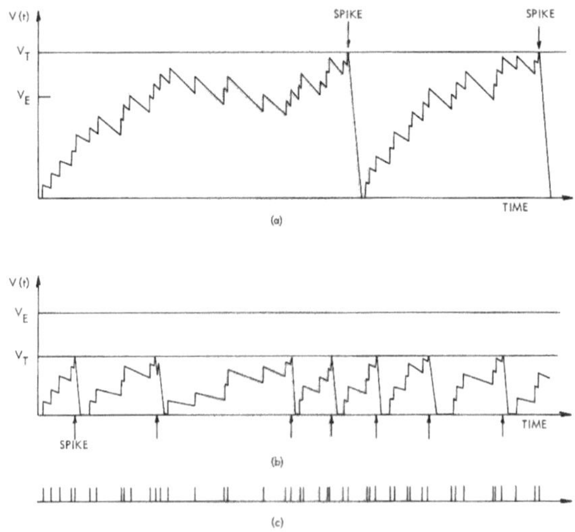

Fig. XVI-2. Neuron potential as a function of time. Neuron time constant $\tau = 7\tau_+$, $\tau_- = \infty$. (a) $V_T > V_E$ mode. (b) $V_T < V_E$ mode. (c) Train of positive pulses producing the potential.

toward its resting state. Figure XVI-2 illustrates a typical path of the potential as a function of time when all pulses are positive (cf. the depolarization of muscle membrane under successive EPSP's observed by Del Castillo and Katz[12]).

This model produces a spike train whose interval histogram is difficult to compute explicitly. The histogram is essentially the probability density of first-passage times from resting potential to threshold.

It is easy to simulate the model with a simple integrating RC circuit. The capacitor voltage into which the positive and negative pulses are fed represents the membrane potential. The incoming pulses are generated by a discriminator whose threshold level is set far above the mean of a random-noise signal; thus the density of pulse intervals will be exponential. These pulses are summed when the RC time constant τ exceeds the mean pulse interval $\tau_{\pm} = 1/R_{\pm}$. When the potential reaches the threshold value V_T, a discriminator produces a neuron "spike," and triggers a large negative pulse that reduces the potential to V_o. The spike train generated in this way is processed by the same TX-0 program that computes histograms for actual neurons.[13]

We consider first the case in which all incoming pulses are positive (Fig. XVI-2). With the neuron threshold set sufficiently high the potential will increase to an "equilibrium" value V_E, at which the average increase in potential because of incoming pulses equals the average decay – that is, $V_E = \tau/\tau_+$. The potential will fluctuate about V_E because of the fluctuations of pulse intervals about their mean, τ_+. The neuron threshold can be set above or below this equilibrium level, thereby giving rise to two distinct modes of operation:

(i) Threshold above equilibrium ($V_T > V_E$). The potential rises to V_E and fluctuates about this value until it reaches threshold. This mode would be expected to give rise to rather long intervals between firings (Fig. XVI-2a).

(ii) Threshold below equilibrium ($V_T < V_E$). The potential crosses threshold before reaching V_T. This mode produces more rapid and more regular firings (Fig. XVI-2b).

The interval histograms obtained from the model under these conditions are shown in Figs. XVI-3 and XVI-4, together with their semi-log plots. The circuit's time constant was directly measured to be $\tau = 180$ msec, while the mean interval between incoming pulses was 1.5 msec. With threshold above equilibrium the spike interval histogram is asymmetric, with an exponential tail for long times. With threshold below equilibrium the histogram approximates a Gaussian shape, but remains asymmetric owing to an excess of long intervals.

Figures XVI-5 and XVI-6 show joint-interval histograms for both modes; the intensity of the point (t_1, t_2) is proportional to the number of times interval t_1 is followed by interval t_2. As would be expected from the fact that successive intervals are statistically independent (since after each firing the potential is reset to the same V_o), both

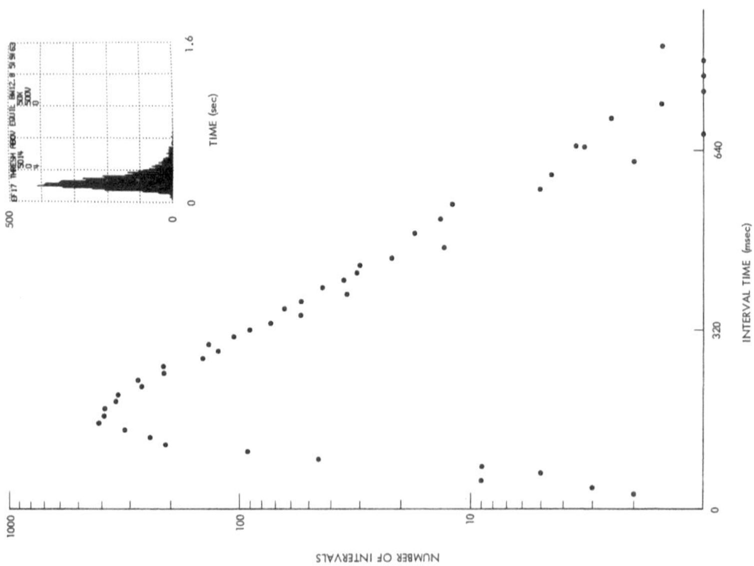

Fig. XVI-3. Interval histogram for $V_T > V_E$. Positive pulses only. $\tau = 180$ msec; $\tau_+ = 1.5$ msec; $\tau_- = \infty$.

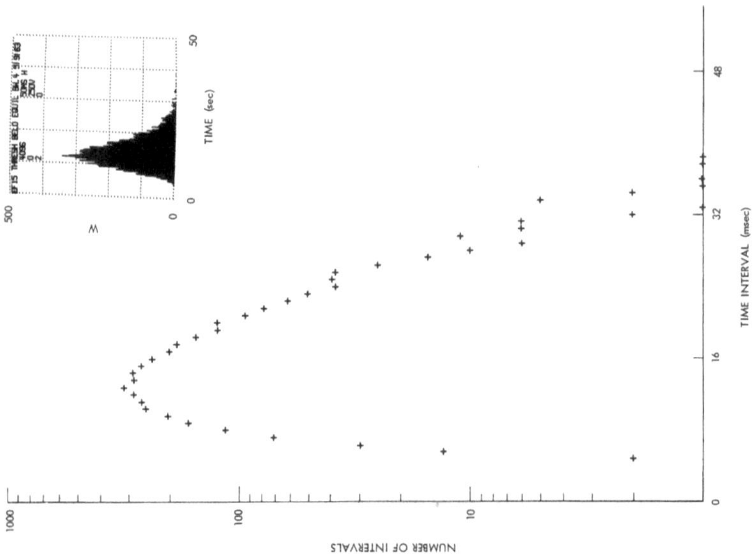

Fig. XVI-4. Interval histogram for $V_T < V_E$. Positive pulses only. $\tau = 180$ msec; $\tau_+ = 1.5$ msec; $\tau_- = \infty$.

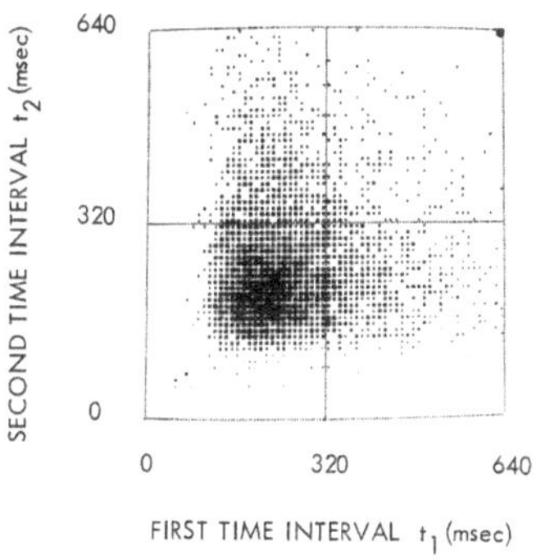

Fig. XVI-5. Joint-interval histogram for $V_T > V_E$. (Same time constants as in Fig. XVI-3.)

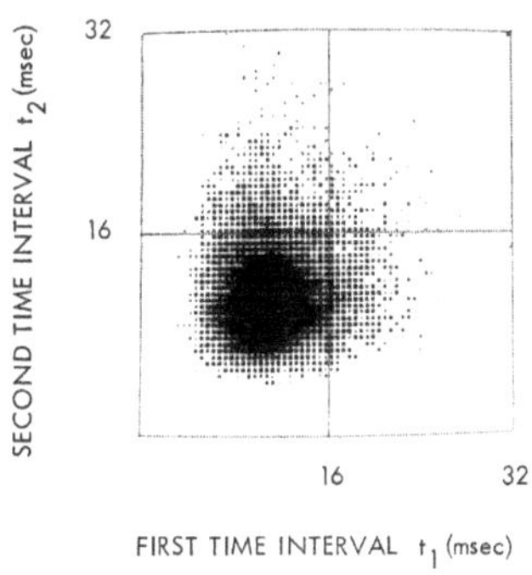

Fig. XVI-6. Joint-interval histogram for $V_T < V_E$. (Same time constants as in Fig. XVI-4.)

joint-interval histograms are symmetric about 45°. The $V_T > V_E$ plot shows that for intervals exceeding the mean, lines of equal joint probability density are approximately straight lines parallel to $t_1 + t_2 = 0$ – as would be expected for Poisson events. The $V_T < V_E$ mode has almost circular lines of equal probability density, which is characteristic of Gaussian interval densities. Thus with positive pulses the model gives exponential and almost Gaussian interval histograms, but not those with longer-than-exponential tails.

A more general situation involves both positive and negative pulses of equal magnitude h (Fig. XVI-7). We consider first the situation when the mean rate of positive

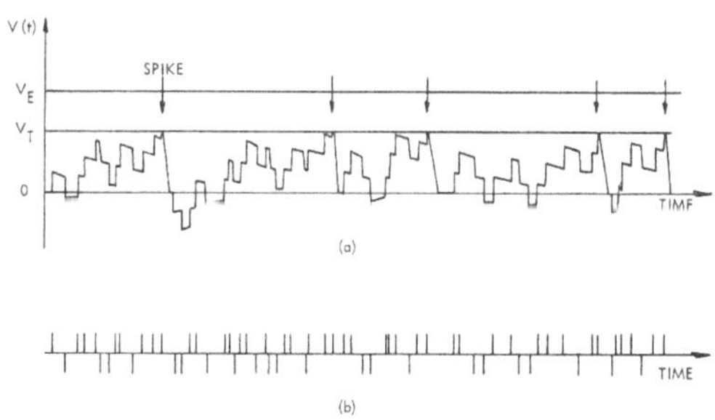

Fig. XVI-7. (a) Neuron potential as a function of time for positive and negative pulses. $R_+ > R_-$. (b) Train of positive and negative pulses producing the potential.

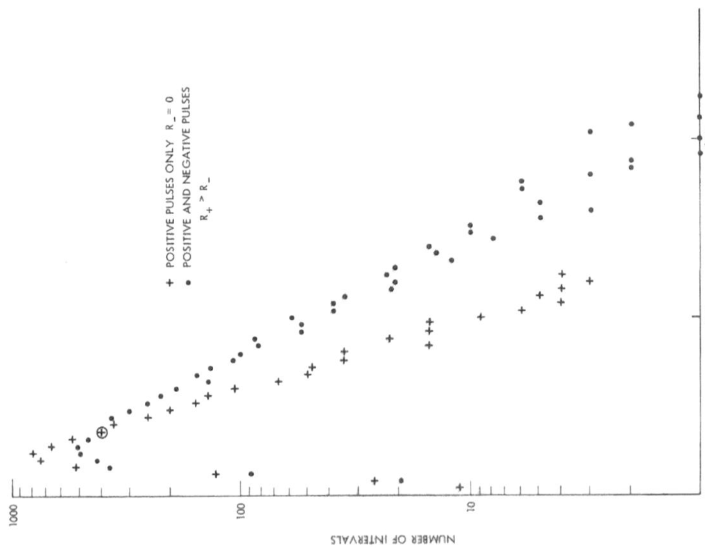

Fig. XVI-8. Interval histogram for $V_T > V_E$, with and without negative pulses.

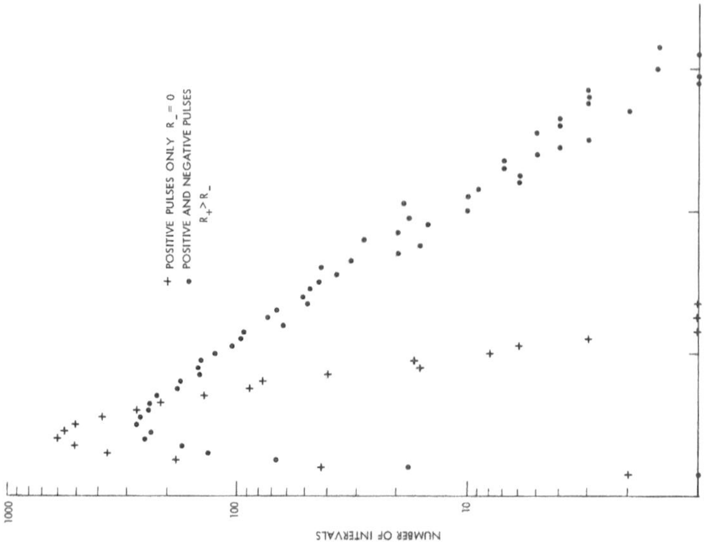

Fig. XVI-9. Interval histogram for $V_T < V_E$, with and without negative pulses.

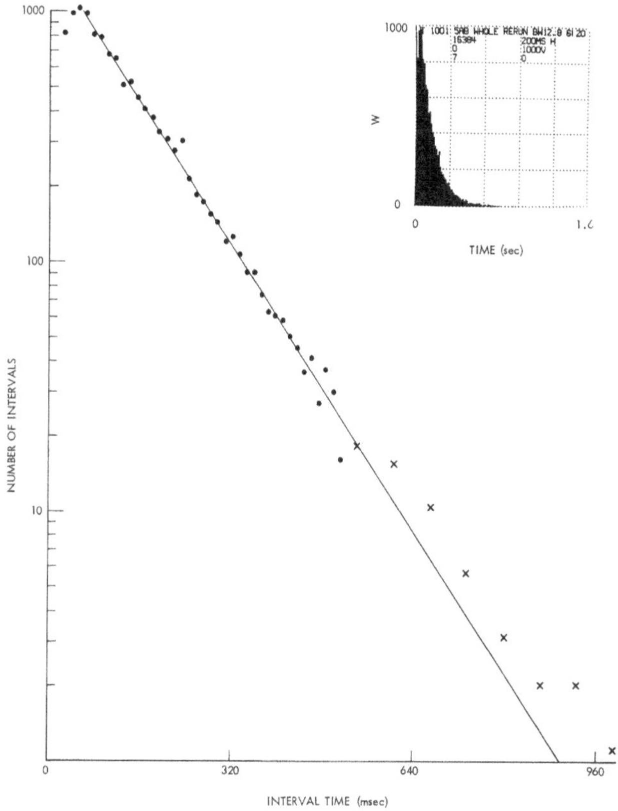

Fig. XVI-10. Interval histogram for approximately equal positive and negative pulse rates ($V_T \approx 0$). $\tau = 50$ msec; $\tau_+ = \tau_- = 1.2$ msec. (X signifies lumped data.)

pulses, R_+, exceeds that of negative pulses R_-; in this case the potential will again have a positive equilibrium value, $V_T = h\frac{(R_+-R_-)}{1/\tau}$. Figure XVI-8 shows a comparison of the histograms for $V_T > V_E$ with and without negative pulses; the negative pulses are seen to extend the histogram but not to alter its shape; the tail remains exponential. Similarly, Fig. XVI-9 shows a histogram for $V_T < V_E$ before and after negative pulses are introduced; with the addition of negative pulses the histogram's tail approaches an exponential – probably due to the fact that the negative pulses reduce the equilibrium level and bring threshold effectively closer to equilibrium. Thus, when R_+ is considerably larger than R_-, introducing negative pulses does not result in any new histograms.

The model can also be operated with the rates of positive and negative pulses approximately equal; in this case the equilibrium potential $V_T = 0$. Figure XVI-10 shows an interval histogram obtained for $\tau_+ = \tau_- = 1\text{-}2$ msec and $\tau = 50$ msec. The semi-log plot shows a slight deviation from exponential, in the direction that is characteristic of long-tailed histograms. This deviation could probably be made more pronounced by further adjustments of the time constants and perhaps by revision of certain assumptions (as stated below).

Thus, under the various conditions that we have investigated, the RC model seems capable of simulating most of the spontaneous spike trains observed in actual neurons. Within these limitations the results suggest that neurons with quasi-Gaussian histograms have a higher rate of incoming EPSP's – or a greater predominance of EPSP's over IPSP's – than those with exponential histograms (if equal threshold levels are assumed).

Similarly, a neuron with a long-tailed interval density would have approximately equal rates of incoming EPSP and IPSP.

The RC model may be able to generate other histogram shapes – and perhaps produce more pronounced long-tailed densities – if some of the following assumptions are revised.

(a) A crucial assumption thus far has been that the interval density of incoming pulses will be exponential. Other density shapes such as Gaussian have not yet been investigated, and, when they are, they may yield different histograms.

(b) The sizes of the positive and negative pulses have been assumed to be constant and equal. Neurophysiological experiments indicate that the effect of a single IPSP depends strongly on the degree of membrane polarization.[13]

(c) The assumption of independence of successive intervals – implicit in the automatic return to V_o after firing – might be relaxed in order to produce different interval and joint-interval histograms.

E. E. Fetz, G. L. Gerstein

References

1. G. L. Gerstein and N. Y-S. Kiang, An approach to the quantitative analysis of electrophysiological data from single neurons, Biophys. J. 1, 15 (1960).

2. R. W. Rodieck, N. Y-S. Kiang, and G. L. Gerstein, Some quantitative methods for the study of spontaneous activity of single neurons, Biophys. J. 2, 351 (1962).

3. S. W. Kuffler, R. Fitzhugh, and H. B. Barlow, Maintained activity in the cat's retina in light and in darkness, J. Gen. Physiol. 40, 683 (1957).

4. L. J. Viernstein and R. G. Grossman, Neural discharge patterns in the transmission of sensory information, Fourth Symposium on Information Theory, edited by Colin Cherry (Butterworth and Co., London, 1961).

5. A. A. Verveen, Fluctuation in Excitability, Ph.D. Thesis, University of Amsterdam (Drukery Holland, N.V., 1961).

6. T. F. Weiss, A model for firing patterns of auditory nerve fibers, Quarterly Progress Report No. 69, Research Laboratory of Electronics, M.I.T., April 15, 1963, p. 217.

7. T. F. Weiss, A Model for Firing Patterns of Auditory Nerve Fibers, Ph.D. Thesis, Department of Electrical Engineering, M.I.T., June 1963.

8. G. L. Gerstein and B. Mandelbrot, Random walk models for the spike activity of a single neuron, Biophys. J. (in press).

9. J. C. Eccles, The Physiology of Nerve Cells (The Johns Hopkins Press, Baltimore, 1957).

10. M. A. B. Brazier, The Electrical Activity of the Nervous System (Macmillan Company, New York, 1960).

11. J. Del Castillo and B. Katz, Statistical factors involved in neuromuscular facilitation and depression, J. Physiol. 124, 574 (1954).

12. Ibid., p. 578.

13. G. L. Gerstein, Analysis of firing patterns in single neurons, Science 131, 1811 (1960).

RANDOM WALK MODELS FOR THE SPIKE ACTIVITY OF A SINGLE NEURON

GEORGE L. GERSTEIN *and* BENOIT MANDELBROT
From the Center for Communication Sciences, Research Laboratory of Electronics, Massachusetts Institute of Technology, Cambridge, Massachusetts; Harvard University, Cambridge, Massachusetts; and I. B. M. Thomas J. Watson Research Center, Yorktown Heights, New York

ABSTRACT Quantitative methods for the study of the statistical properties of spontaneously occurring spike trains from single neurons have recently been presented. Such measurements suggest a number of descriptive mathematical models. One of these, based on a random walk towards an absorbing barrier, can describe a wide range of neuronal activity in terms of two parameters. These parameters are readily associated with known physiological mechanisms.

INTRODUCTION

A prominent characteristic of the electrical activity in the nervous system is the prevalence of spontaneous activity, both at the level of the spatially summed potentials observed with a gross electrode and at the level of the action potentials of a single neuron. In a recent paper Rodieck, Kiang, and Gerstein (1962), using examples recorded in the cochlear nucleus of anesthetized cats, set forth some quantitative methods that may be used to study the statistical properties of the train of action potentials associated with spontaneous activity of a single neuron. It was shown in that paper that three types of measurements are particularly useful in order to establish a statistical description of the spike train:[1]

(*a*) The interval histogram—a histogram of the distribution of time intervals between successive spikes. This measurement estimates the probability density of intervals.

(*b*) The joint interval histogram—a histogram of the joint distribution of two successive interspike intervals. This measurement estimates the joint probability density of successive intervals.

(*c*) The scaled interval histogram—a histogram of the intervals between every $2^{m\text{th}}$ spike when m is a positive integer. (For $m = 0$ this description is identical with the interval histogram.) This measurement estimates the probability density

[1] We shall use spikes and action potentials interchangeably.

of particular intervals that are themselves the sum of 2, 4, 8, . . . successive interspike intervals.

In the present paper we shall use some of the data and measurements of Rodieck *et al.* (1962) as a basis for several mathematical models of the neuron. We shall show that the parameters in these models can be given significance in physiological terms, and that these models can generate many of the observed characteristics of spike trains from single neurons.

SOME BASIC OBSERVATIONS

The Poisson Model

Whenever one deals with a sequence of events that seem randomly spaced in time, it is usual to assume as a first approximation that the probability for the event to occur during a time increment dt is equal to $R\,dt$, independently of the past history of the process. Here, R is a constant that defines the average time rate of occurrence of the events. If the events are action potentials, these conditions cannot be met fully because the interspike interval has a lower bound, D, resulting from the refractory time of the neuron. If there are no other constraints, the interspike intervals will be distributed with a probability density

$$\begin{aligned} I(\tau) &= R \exp\left[-R(\tau - D)\right] \quad \text{for} \quad \tau \geq D \\ &= 0 \qquad\qquad\qquad\qquad\quad \text{for} \quad \tau < D, \end{aligned} \tag{1}$$

where τ is the duration of the interspike interval.

In this type of model successive interspike intervals are independent random variables, so that the joint probability density for interval τ_1 followed immediately by interval τ_2 can be written

$$\begin{aligned} I(\tau_1, \tau_2) &= R^2 \exp\left[-R(\tau_1 - D + \tau_2 - D)\right] \\ &= R^2 \exp\left[-R(\tau_1 + \tau_2 - 2D)\right] \quad \text{for} \quad \tau_1 \geq D, \quad \tau_2 \geq D \end{aligned} \tag{2}$$

It follows that the lines of equal probability density in the joint distribution of τ_1 and τ_2 will be straight lines that are parallel to the second bisectrix $\tau_1 + \tau_2 = 0$.

Consider also the probability density $I_k(\tau)$ for the interval between a spike and the kth spike that follows it. (This represents a slight generalization of the measurement defined by the scaled interval histogram, in which k was restricted to be of the form 2^m, when m is a positive integer.) In the present case of an exponential distribution, each such interval is the sum of k independent exponential random variables. The resulting probability density is called the gamma distribution of order k:

$$I_k(\tau) = R^k \tau^{k-1} \exp(-R\tau)[\Gamma(k)]^{-1}, \tag{3}$$

where Γ represents the gamma function. As k increases, this distribution rapidly becomes peaked, symmetric, and tends towards the Gaussian.

Thus, in order to test empirically the conjecture that a spike train may be described as a Poisson process, one must examine the basic distribution of interspike intervals $I(\tau)$, the joint distribution of successive intervals $I(\tau_1, \tau_2)$, and, finally, the distribution of scaled intervals $I_k(\tau)$. It is clear that there are indeed cases in which all three theoretical predictions of the Poisson process are verified. For example, consider the data given by Rodieck *et al.* (1962) for Unit 259-2 which show an exponential distribution of intervals.[2] We show in Fig. 1*a* that for this unit the lines of constant probability density in the joint interval histogram are indeed parallel to $\tau_1 + \tau_2 = 0$. In Fig. 2*a* we see that the scaled interval histograms for this unit tend rapidly toward a narrow and symmetric shape.

The simple Poisson type of model has been used previously in neurophysiology (Kuffler, Fitzhugh, and Barlow, 1957; Grossman and Viernstein, 1961). Unfortunately, it is a purely descriptive model that does not contribute much to our understanding of the detailed mechanisms that may be involved in the firing of a neuron. It is also clear that many units produce spike trains for which the three fundamental predictions based on a Poisson model are grossly incorrect. In this paper we shall study two such examples, one rather briefly, the other in detail.

The Gaussian Model

Consider a train of action potentials which exhibits appreciable periodicity so that there are only relatively small fluctuations in the durations of interspike intervals. Here, the probability density of intervals may "look" very much like that of a sample from a Gaussian population. If this is indeed the case, and if successive intervals τ_1 and τ_2 are independent, then the joint density of successive intervals would be the product

$$I(\tau_1, \tau_2) = (2\pi\sigma^2)^{-1} \exp\{-(2\sigma^2)^{-1}([\tau_1 - E(\tau_1)]^2 + [\tau_2 - E(\tau_2)]^2)\}. \quad (4)$$

Here, $E(\tau_1)$ is the expectation value, or mean duration, of the first interval of the pair and σ^2 is the variance of either interval. According to equation (4), the contours of equal density are circles with a common center at the coordinate point $E(\tau_1), E(\tau_2)$.

The sum of k successive interspike intervals will be a sum of k independent Gaussian variables. According to the well known properties of the Gaussian distribution, this sum will itself be a Gaussian variable with a mean value $kE(\tau)$ and a variance equal to $k\sigma^2$. (Here, $E(\tau)$ and σ^2 refer to the probability density of the basic interspike intervals.)

We may find cases that are in reasonable agreement with these theoretical predictions. As an example, consider the data of Rodieck *et al.* (1962) for Unit *R*-4-10 which show an interval histogram that looks like a Gaussian distribution.[3] In Fig. 1*b* of the present paper we demonstrate that lines of constant probability density in the joint interval histogram are approximately circular. The scaled interval

[2] See Rodieck, Kiang, and Gerstein (1962), Fig. 5*a*.
[3] See Rodieck, Kiang, and Gerstein (1962), Figs. 4*b* and 5*b*.

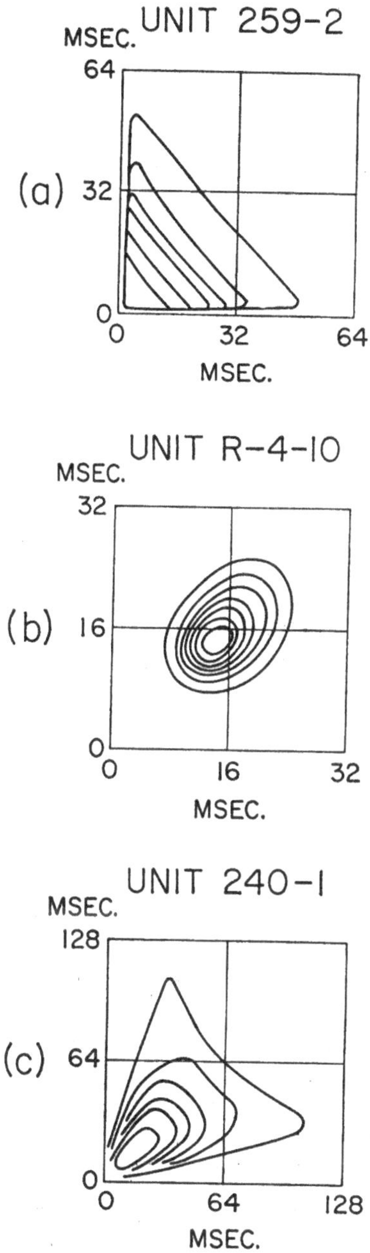

FIGURE 1 Lines of constant density in joint interval histograms for the three units discussed here. (*a*) Unit 259-2; (*b*) Unit R4-10; (*c*) Unit 240-1.

histograms of Fig. 2*b* show that the mean value and variance of the sum of intervals indeed behave as predicted. (Note that the time scale of each successive order of scaled interval is adjusted by a factor of 2, and that this keeps the center (mean) of the successive scaled interval distributions at the same location on the abscissa.)

Long Interspike Intervals

The exponential and the Gaussian distributions share an important property: The asymptotic decrease of the probability density function is quite rapid for large values of τ. In other words, long interspike intervals are extremely rare. However, there are experimental situations in which the density of intervals behaves very differently. In qualitative terms, there may be, in comparison with the number of intermediate length intervals, an excessive number of both fairly short and very long interspike intervals. An example of such data is given by Rodieck *et al.* (1962) for Unit 240-1: their Fig. 5*d* clearly shows an upward curvature of the semilogarithmic plot of the interval histogram. It is interesting to note that similar behavior has been observed by one of the authors (B. B. M.) in several fields of study that are quite remote from neurophysiology (Mandelbrot, 1960, 1963 *a* and *b*; Berger and Mandelbrot, 1963).

Another important characteristic of the data from Unit 240-1 is shown in Fig. 2*c*: with the indicated changes in time scale, the shape of each of the first few successive orders of scaled interval histogram remains approximately the same. This means that a sum of successive interspike intervals has the same probability density—to within a scale factor—as the basic interspike intervals. If we make the simplifying assumption that the lengths of successive intervals are independent (this is not quite true for the short intervals in the data), successive orders of scaled interval histogram are related by $I_2(\tau) = I(\tau) * I(\tau)$. Then the observed invariance of shape under the scaling transformation means that the density of interspike intervals must obey

$$I(\tau)*I(\tau) = (1/2)I(\tau/2). \tag{5}$$

Here, the symbol * denotes the operation of convolution. This extremely restrictive condition may be stated as follows: When the probability density is convolved with itself, the result must have the same functional form, with only a particular change of the time scale. In the conventional terminology of probability theory, the "type" of the interspike interval is invariant under addition.

These properties, as we shall show, suggest a series of models that can describe the spike activity of a single neuron.

THE SIMPLE RANDOM WALK MODEL

Development of the Model

Distribution functions whose densities obey a general relation that is similar to

21

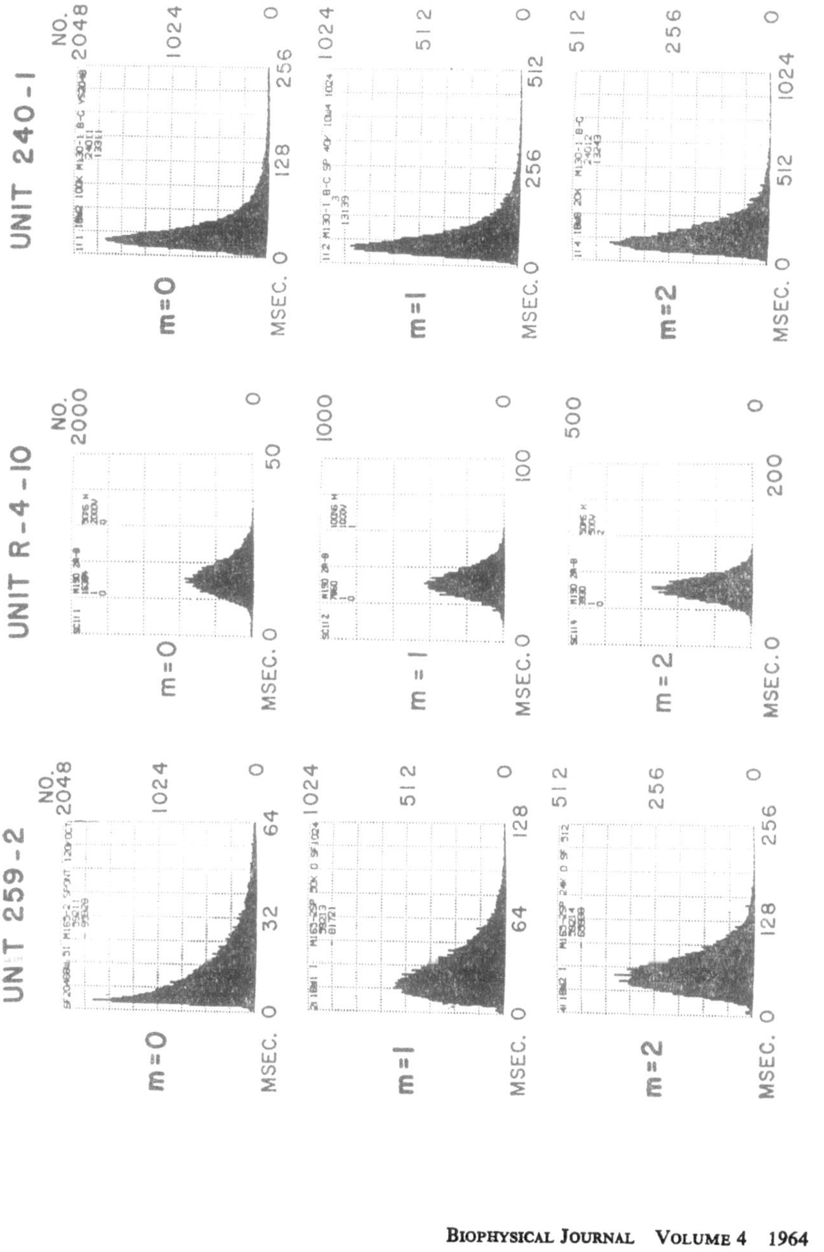

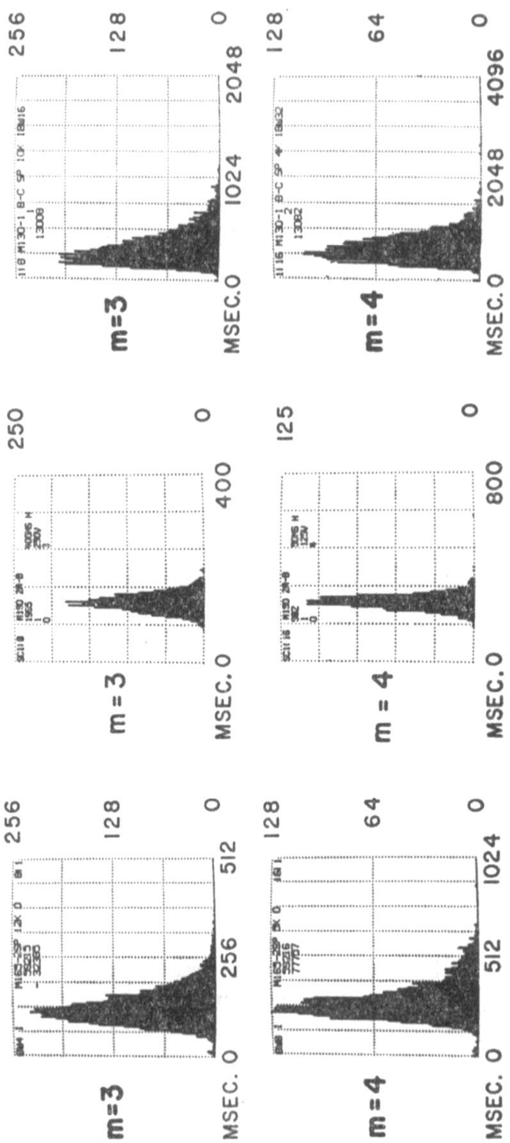

FIGURE 2 Scaled interval histograms for the same units as in Fig. 1. (*a*) Unit 259-2; (*b*) Unit R4-10; (*c*) Unit 240-1.

equation (5) are known as stable distribution functions. The stability property refers to the invariance of the functional form of the distribution function under convolution; in general, there may be a linear change of scale in the variable. These functions have been extensively studied (Lévy, 1940; Gnedenko and Kolmogorov, 1954); we shall postpone more detailed examination of their properties until the section on Stable Distributions. For the present, suffice it to say that only three stable distribution functions are known in closed analytic form: the Gaussian distribution, the Cauchy distribution, and the so called stable distribution of order 1/2.

In attempting to match one of these probability densities to our experimental data, we must keep in mind the following requirements: (*a*) the interspike interval is always a positive quantity, and (*b*) the interspike interval histogram decreases asymptotically more slowly than an exponential function, and is asymmetric about its mode (skewed).

The Gaussian distribution has already been considered; the Cauchy distribution has two long tails, corresponding to τ varying to $-\infty$ and $+\infty$, and thus could not readily be associated with the distribution of an inherently positive quantity. The stable distribution of order 1/2, however, is defined only on the interval $(0, +\infty)$, and also has one of the longest "tails" to be found in probability theory. It is associated with certain solutions in games of coin tossing, or with random walks. In particular, it gives the probability density of first passage times in a one-dimensional random walk that begins some fixed distance from an absorbing barrier.

The random walk problem can be restated in the language of neurophysiology in order to define a model for the spike activity of a single neuron:

1. Let the electrical state of polarization of the somatic and dendritic membrane of the neuron be specified by a single number. As time passes and the electrical state of the membrane varies, the state point will move back and forth along a straight line.

2. Choose a particular point on the line as the resting potential.

3. Choose another point on the line, z_0 units away, as the threshold. (This corresponds to the absorbing barrier in the random walk problem.) If at any time the state point reaches the threshold, the neuron will produce an action potential.

4. Assume that each incoming elemental EPSP (Excitatory Post Synaptic Potential) moves the state point one unit toward the threshold, and that each incoming elemental IPSP (Inhibitory Post Synaptic Potential) moves the state point one unit away from the threshold.

5. Let the average rate of incoming elemental EPSP and elemental IPSP be the same; *i.e.,* at any time there is an equal probability that the state point move either a unit toward or a unit away from the threshold.

6. Immediately after the state point has attained the threshold and caused the

production of an action potential, it returns to the resting potential, only to begin again on its random walk.

Some basically different neuron models which also utilize a state point and a threshold have been previously described. Our random walk model should be compared, for example, with the additive noise models of Viernstein and Grossman (1961) and Verveen (1961). These authors postulate fluctuations of the membrane state (or alternatively of the threshold) which have a Gaussian distribution of amplitudes. They also assume that there is no correlation between successive fluctuations, so that there may be arbitrarily large sudden transitions of the state point. In contrast, transitions in our model are the result of many small steps and may be compared with the lag or lead of the number of heads in the repeated tossing of a coin. Thus the probabilistic characteristics of our model are a reflection of the rate and order of incoming elemental EPSP and IPSP, and do not explicitly require the introduction of a "noisy" membrane or noisy threshold.

Even if the membrane state fluctuations in the additive noise models are band width-limited so as to produce smooth transitions, our random walk model differs by its use of the reset to resting potential. This fundamental distinction causes the two types of model to have quite different densities for reaching threshold. There is, of course, neurophysiological evidence that at least for some neurons reset does take place after the occurrence of an action potential (Eccles, 1957, particularly chapter 2).

It should also be pointed out that the precise rules that govern the motion of the state point in the random walk do not seriously affect the calculation of first passage time. In the model described thus far we have considered a "membrane" with infinite time constant and identical excitatory or inhibitory steps occurring at regular increments of time. Recently work has been done (Fetz and Gerstein, 1963) on models which utilize an exponential time distribution of excitatory and/or inhibitory steps together with finite time constants for the membrane. Even with time constants shorter than typical interspike times, such models can exhibit most of the properties discussed for the random walk model of the present paper. We may conclude that the microscopic rules of a random walk model cannot be uniquely determined from physiological data on spike trains. More detailed intracellular potential measurements than presently available would be needed in order to set such details in the model.

Interval Densities

In order to make theoretical predictions that can be compared with data on the probability density of interspike intervals, we must, in terms of our random walk model, calculate the probability density for first attaining the threshold. If we assume that each elemental movement of the state point in the model is small in comparison with the distance z_0 between the resting potential and the threshold, and if we do not need a description of the detailed motion of the state point, then the problem may be treated with high accuracy as a diffusion process with an absorbing barrier. The normal diffusion current at the absorber will correspond to the probability density of first passage times (interspike intervals) (Chandrasekhar, 1943).

This density has the form

$$I(z_0, \tau) = (4\pi)^{-1/2} z_0 \tau^{-3/2} \exp(-z_0^2/4\tau) \qquad \tau \geq 0$$
$$= 0 \qquad \tau < 0 \tag{6}$$

where τ is the duration of an interspike interval and z_0 has been defined.

Long intervals are so probable in this model that the variable τ has no population moments of any order. Even its mean value is infinite. In other words, a sample mean interval calculated on a finite sample of data would be expected to grow without bound as the sample size increases. These characteristics of the model may be made more intuitively acceptable by considering the random walk: the state point is free to wander arbitrarily far from the threshold, thus allowing an extremely long first passage time. We illustrate this behavior in Fig. 3, which shows a computer simulation of several random walks. With the resting potential 32 steps from the threshold (or absorber), the state point has frequently not reached the firing threshold even after 2560 steps.[4]

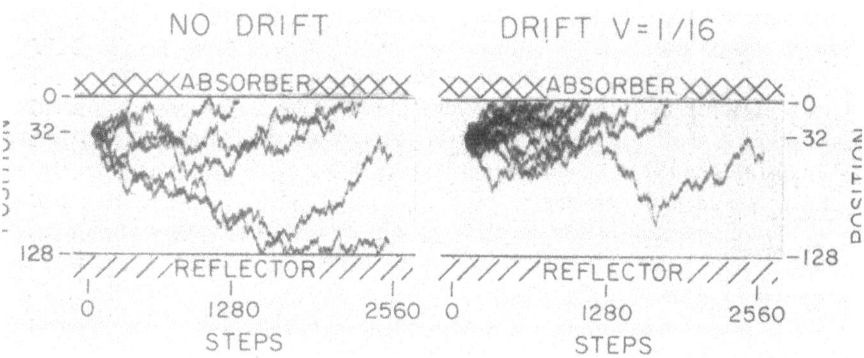

FIGURE 3 Typical random walks in one dimension in a computer simulation of the model.

For the simple random walk model, successive interspike intervals are independent. The joint density of successive intervals, when both τ_1 and τ_2 are large enough to discard the exponential, is of the form

$$I(\tau_1, \tau_2) = C^2 \tau_1^{-3/2} \tau_2^{-3/2} = C^2(\tau_1 \tau_2)^{-3/2}. \tag{7}$$

[4] We have, for convenience, introduced a reflecting barrier into the illustration, which corresponds to some maximum hyperpolarization. This limits the maximum distance that the state point may be from the firing threshold. However, as long as this distance is large compared with z_0, the density of first passage times remains as we have described above.

Clearly, the lines of equal probability are hyperbolae. Their curvature contrasts qualitatively with the straight contour lines of the Poisson (or exponential) model and with the circular contour lines of the Gaussian model.

One may easily verify that, under addition of successive interspike intervals, the density of equation (6) shows the shape invariance required of a stable function. By direct integration, we obtain

$$I(\tau) * I(\tau) = (1/4) I(\tau/4). \tag{8}$$

More generally, for the density of equation (6), the sum of k successive interspike intervals, weighted by k^{-2}, will have the same probability density as the fundamental interspike intervals.

It is not surprising that the simple model that we defined above does not quantitatively fit the data, and that it is necessary to make the model more complex. A number of discrepancies exist between the predictions of the simple model and the data. Perhaps most striking is the fact that the experimentally observed scaling invariance (Fig. 2*c*) implies a weighting factor k^{-1} in contrast with the k^{-2} weighting factor of the simple random walk model.[5] This difference is also made evident by comparing the experimental convolution relation (equation (5)) with the prediction of the simple random walk model in equation (8). Thus, although the model predicts invariance of shape under the scaling transformation, it does so with the wrong change of time scale.

A second discrepancy between data and model is, experimentally, that as k becomes large, the scaling invariance no longer applies (see the last few scaled interval histograms at the bottom of Fig. 2*c*). In other words, the asymptotic decrease of the experimental interval density is much faster than the $\tau^{-3/2}$ decrease predicted by the simple random walk model.

A third discrepancy is that the observed joint density of successive interspike intervals $I(\tau_1, \tau_2)$ does not clearly exhibit the hyperbolic contour pattern that we derived in equation (7). (See Fig. 1*c*.)

We shall now demonstrate that a relatively small modification of the simple random walk model will allow a far better fit of the data.

THE RANDOM WALK WITH DRIFT

Calculation of Interval Density

In the section on The Simple Random Walk Model it was assumed that the average rates of incoming elemental EPSP and elemental IPSP are identical: there was no "bias toward" either input. But it is far more reasonable in a physiological model to assume that these two rates are different, and that there is some excess of either

[5] If $I(\tau)$ were a Gaussian density with zero mean, the required weighting factor would be $k^{-1/2}$.

EPSP or IPSP input. In this case, the probability for the state point to move one unit toward the threshold will be different from the probability for it to move away from the threshold.

In terms of a diffusion process, the difference between these probabilities can be considered a "drift velocity," either toward or away from the threshold. Taking the origin of our coordinate system at the threshold, we must solve

$$\frac{\partial \omega}{\partial t} = \frac{\partial^2 \omega}{\partial z^2} - c\,\frac{\partial \omega}{\partial z}$$

subject to the boundary conditions

$$\omega = \delta(z - z_0) \quad \text{at} \quad t = 0; \quad \omega = 0 \quad \text{at} \quad z = 0 \qquad \text{all} \quad t \geq 0.$$

Here, c represents the drift velocity, and we have taken the diffusion coefficient to be unity. The quantity ω represents the spatial density of the diffusing particles: each such particle represents the state point of our neuron model during the time between some two action potentials. The normal diffusion current at the absorber corresponds to the probability density of first passage times in the model. Thus we find

$$\begin{aligned} I(z_0, c, \tau) &= (4\pi)^{-1/2} z_0 \tau^{-3/2} \exp\left\{-\frac{(z_0 + c\tau)^2}{4\tau}\right\} \qquad \tau \geq 0 \\ &= 0 \qquad\qquad \tau < 0. \end{aligned} \tag{9}$$

We have again used τ to indicate the duration of a first passage time, z_0 to indicate the distance between resting potential and threshold, and c to express the difference between the rate of elemental EPSP and elemental IPSP input to the neuron. Clearly, for $c = 0$ this result reduces to equation (6).

The probability density given in equation (9) is, unfortunately, such that convolution with itself does produce change in the functional form. However, it is easy to show that for a certain range of parameters there is approximate shape invariance under the convolution operation with a 2× expansion of the time scale. This is illustrated in Fig. 4 by scaled interval histograms in a computer simulation of the random walk model with draft.

When drift is included in the random walk model, successive intervals remain independent. The joint density of successive intervals $I(\tau_1, \tau_2)$ is thus given by a product of two terms like equation (9). For reasonable choice of parameters, a numerical evaluation quickly shows that the contours of equal density are less curved than the hyperbolas of equation (7).[6] The contour lines return to near the origin through regions close to the τ_1 and τ_2 axes. The over-all contours are therefore roughly right "triangles" with an inward curved hypotenuse. This curvature be-

[6]Because of the linear term in the exponential of equation (9) for very large values of τ_1 and τ_2 (far beyond values which are experimentally observed), the contour lines become straight lines parallel to $\tau_1 + \tau_2 = 0$.

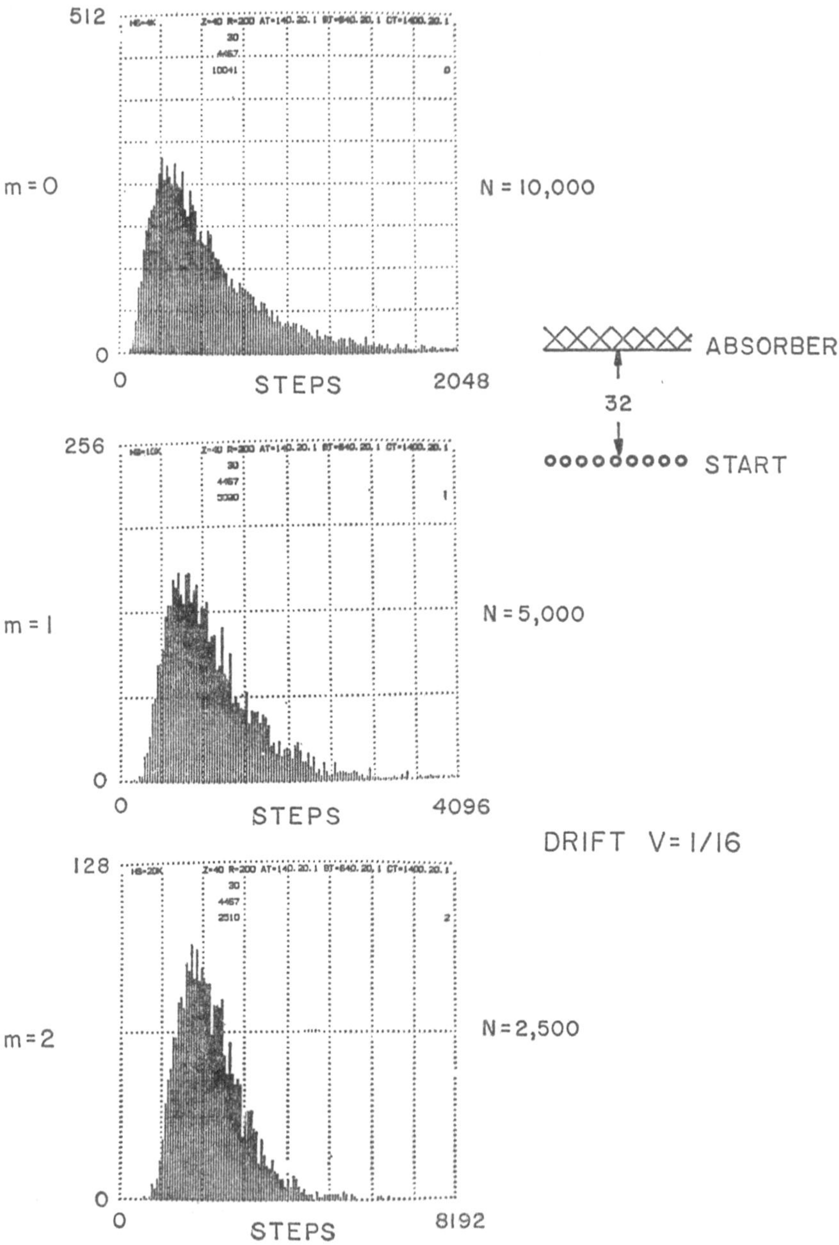

FIGURE 4 Scaled interval histograms for the computer simulation of the model.

comes unnoticeable for the small contour triangles near the origin which correspond to very high probability values.

Comparison of Theory with Data

Let us write the probability density of equation (9) as

$$I(a, b, \tau) = K\tau^{-3/2} \exp\left(-\frac{a}{\tau} - b\tau\right). \tag{10}$$

In this form we have a two-parameter curve to compare with the data; the third parameter is fixed by the normalization condition. Inspection of equations (9) and (10) shows that the parameter a is associated with the square of the resting potential-threshold distance, and that the parameter b is associated with the difference between the rate of elemental EPSP and IPSP which is incident on the neuron membrane.

The fit of equation (10) to the interval histograms of data from three different units is shown in Figs. 5 to 7. Unit 240-1 was located in the cochlear nucleus of a cat under moderate dial-urethane anesthesia; Unit 6-2 was located in the auditory cortex of a completely unanesthetized cat with a chronic implanted electrode. Despite this great disparity of experimental conditions and anatomical location, the interval histograms of the data from both units are well fitted by similar choice of parameter values. The drift parameter in both these causes corresponds to an approximate 4 per cent excess rate of EPSP over IPSP.

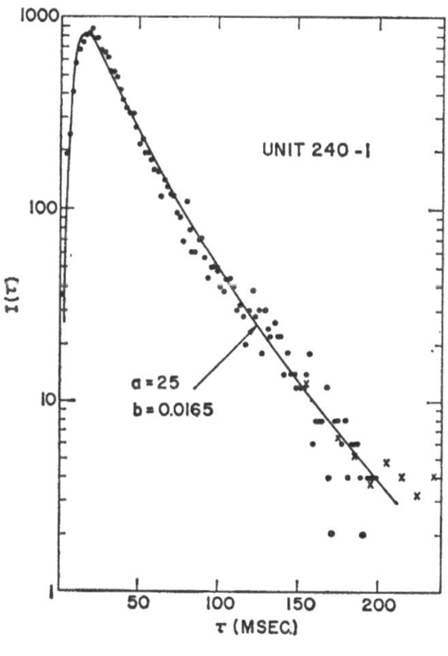

FIGURE 5 Fit of equation 10 to interval histogram of data for Unit 240-1.

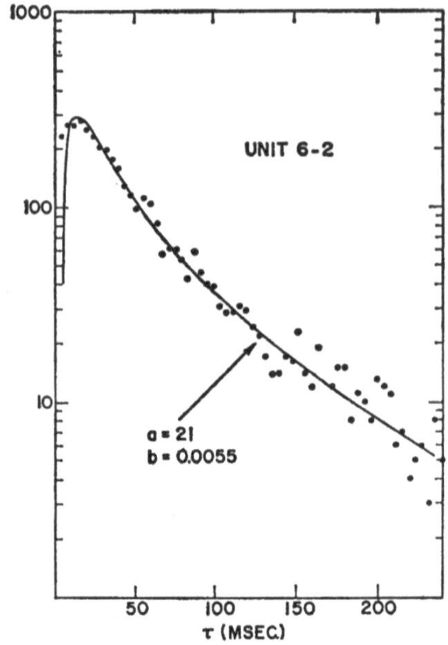

Figure 6 Fit of equation 10 to interval histogram of data for Unit 6-2.

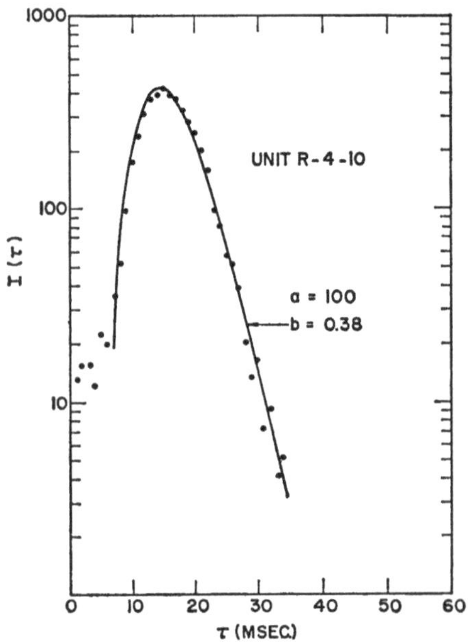

Figure 7 Fit of equation 10 to interval histogram of data for unit R4-10.

Reasonably good agreement between model and experiment is also observed for the joint density of successive interspike intervals from these two units. The contour pattern of the data from Unit 240-1 (Fig. 1*c*) differs from the theoretical prediction (page 52) mainly in that the vertex angle of the contour triangles is less than the predicted 90°. The source of this discrepancy is that in the model successive interspike intervals are completely independent, while in the data there is some correlation between successive intervals when they are short.

In Fig. 7 we show that the Gaussian-like data of Unit R-4-10 can also be fitted by equation (10), but require a very different set of values of a and b: In comparison with the values used in Figs. 5 and 6, we have here increased the a parameter by a factor of 4, and the b parameter by a factor of 23. Physically, this change means that the model for this particular neuron has a higher threshold than in Figs. 5 and 6 and a much larger excess rate of EPSP over IPSP. It is intuitively obvious that under such conditions of high drift velocity the state point of our model will reach the threshold frequently and in an almost periodic way.[7] In order to choose between this fit and the fit by the Gaussian density, we shall have to determine whether, for a large number of units, the parameters of the model exhibit a continuous range of values or tend to cluster in some significant way. Again for this case there is good agreement between model and experiment for the contour lines in the joint density of successive interspike intervals.

A TIME-VARIANT MODEL

In most physiological experiments data are taken not only during spontaneous activity, but in the presence of transient stimuli that are often periodically presented. In recordings from single units one frequently observes correlation between the pattern of action potentials and the instants at which the stimulus is presented (Gerstein and Kiang, 1960). Unless the unit is firing very rapidly in comparison with the rate of stimulus presentation, an interval histogram of the activity should show some evidence of the interval between stimuli. This situation is illustrated in Fig. 8, which shows interval histograms for the activity of a unit from the cochlear nucleus of an anesthetized cat during presentation of clicks at various rates. The mode of the interval histogram of spontaneous activity lies at approximately 10 msec. When clicks are presented at 10/sec., the neuron fires "spontaneously" several times in the course of the 100 msec. between stimuli; thus no evidence of 100 msec. periodicity

[7] Since the distance between the points that represent the resting potential and the threshold is experimentally found to be quite similar for most cells, we should point out that an alternate physical interpretation of the parameters of Fig. 7 is possible. In the foregoing development we have assumed that the variance of EPSP-IPSP rate fluctuations has been taken at some fixed value, and accordingly have set the diffusion constant at unity. If this limitation is removed, we may interpret the change in parameter values in Fig. 7 as follows: variance of the EPSP-IPSP rate fluctuations is smaller, there is a large excess of EPSP rate, and the threshold remains at some "normal" value.

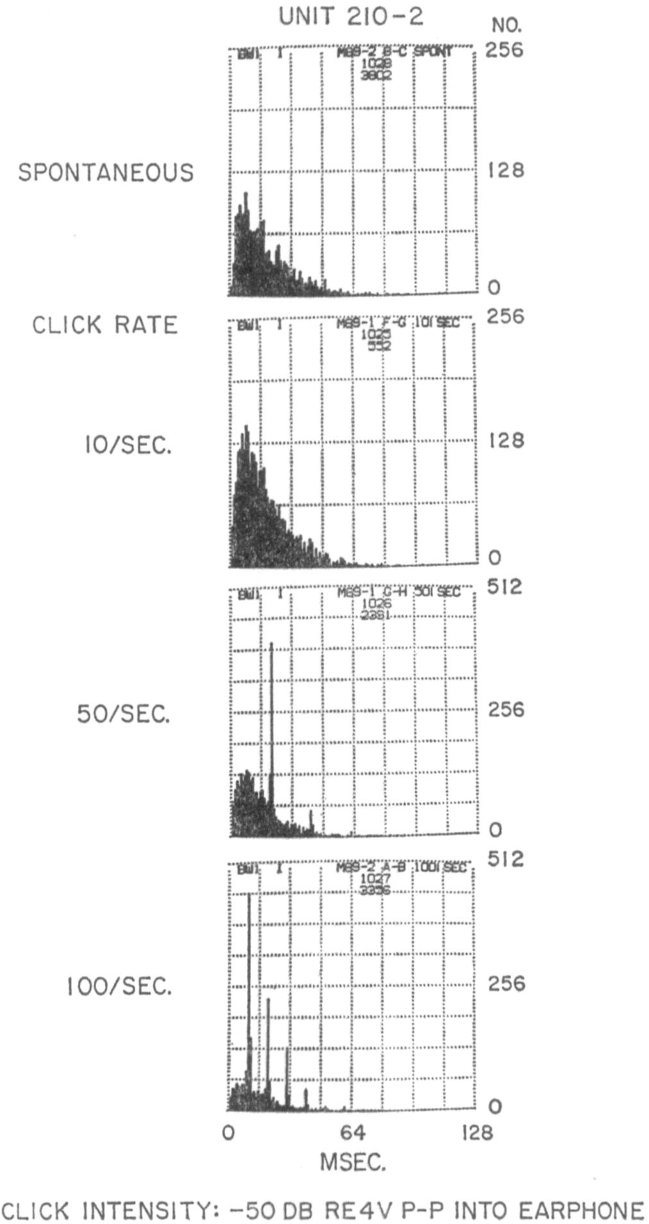

FIGURE 8 Interval histograms for a cochlear nucleus unit during stimulation by clicks at various rates. (courtesy of N.Y-S. Kiang)

G. L. GERSTEIN AND B. MANDELBROT *Spike Activity of Single Neuron* 57 is visible in this histogram. However, at 50 clicks per second, the neuron frequently does not get to fire between stimuli. The 20 msec. periodicity of the stimulus appears as a sharp peak. In fact, the neuron on occasion does not fire for 40 msec., so that a second peak is seen which corresponds to firing on alternate stimuli. The same situation occurs at 100 clicks per second; the many peaks correspond to "skipping" of various numbers of stimuli. It should also be noted that the continuous distribution of intervals which is visible between the peaks decreases as the rate of stimulation increases; an increasing fraction of the firing pattern is time-locked to the presentation of stimuli.

It is interesting to examine the extent to which our random walk model can duplicate such behavior. For this purpose one or more of the parameters in the model might be made to vary with time. In physical terms we might picture the corresponding events at either a variation of the threshold or a variation of the relative amount of incoming EPSP and IPSP. To investigate the latter possibility, we must make the drift parameter c at time-variant quantity. Unfortunately, the diffusion equation with time-variant coefficients is not readily soluble; we have, therefore, examined this problem with the aid of digital computers (TX-0 or TX-2) by means of Monte Carlo methods.

Some recent intracellular measurements in the inferior colliculus (Nelson and Erulkar, 1960; and Erulkar, 1962) have suggested that immediately after the presentation of a discrete auditory stimulus, the membrane potential of many neurons goes through a short period of depolarization, followed by a longer period of hyperpolarization. These measurements can be interpreted in terms of a short excess of incoming EPSP followed by a longer lasting excess of incoming IPSP. To incorporate such data into the model, we let the drift parameter c assume a similar time pattern, as shown in the right-hand side of Fig. 9. One such pattern follows each "stimulus"; at sufficiently high rates of stimulus the pattern will begin to overlay so that there may cease to be a time during which the drift parameter is at the value that corresponds to that of the "spontaneous" condition.

The left-hand half of Fig. 9 shows interval histograms from our random walk model with periodic time variation of the drift parameter. The particular values chosen are not critical; a fairly wide choice of time pattern will produce almost equally good agreement with the physiological data of Fig. 8. As might be expected, it is areas in the drift pattern which are the significant variable (that is, strength of drift × duration).

We have not specifically investigated other possible ways of making our model time-variant. The drift variations that we discussed above seem to be physiologically quite reasonable.

DEFECTS OF THE RANDOM WALK MODEL

Although the model that we have described does seem to fit a wide variety of

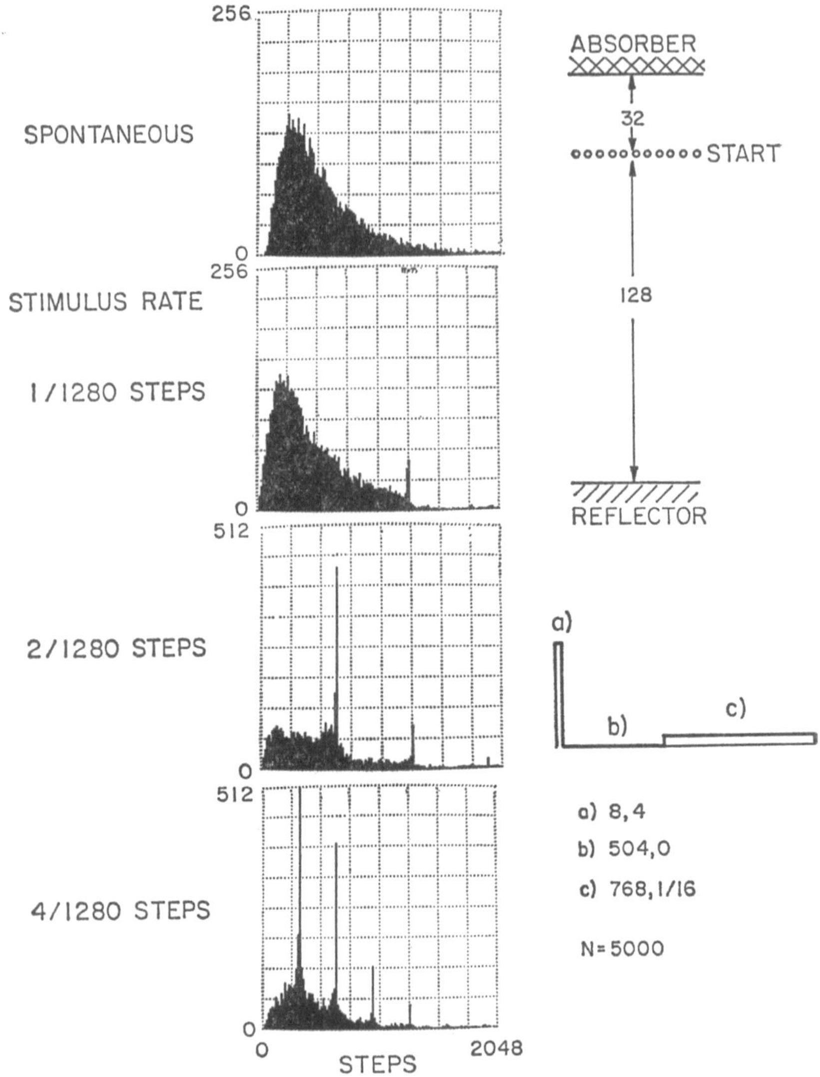

Figure 9 **Interval histograms for the time-variant random walk model; periodic drift changes. Upper inset, the model's boundaries. Lower inset, time course of the drift parameter corresponding to 1 stimulus per 1280 steps; values for each portion are given in the form: duration, drift velocity towards absorber.**

neurophysiological observations, there are a number of obvious defects in its formulation and also in its predictions. By examining these limitations, we shall be able to point out several possible ways in which this type of model may be extended.

(*a*) We have attempted to describe the electrical state of the entire neuron membrane by a single number that locates the state point. This is a gross oversimplification that does not take into account the complex geometry of the neuron membrane and the complicated distribution of synaptic knobs (Rall, 1962).

(*b*) We have assumed a particular set of rules for the random walk in our model: the state point takes some standard size step in response to each incoming elemental EPSP or IPSP, toward or away from the threshold, respectively. As pointed out in the section on The Simple Random Walk Model, a certain narrow range of alternate random walk rules leads essentially to the same results that we have obtained. However, none of these mechanisms takes into account the experimental evidence that the contribution of a single elemental IPSP to the electrical state of the neuron membrane depends very strongly on the particular degree of membrane polarization (Coombs, Eccles, and Fatt, 1955; Eccles, 1957).

(*c*) Each time that the state point in our model attains the threshold, we assume that it returns to the resting potential and the random walk begins again. Thus successive interspike intervals in the model are assumed to be independent. Unfortunately, this is not always the case for a real neuron. For example, some correlation between the durations of successive intervals, particularly when the intervals are short, has been shown by Rodieck *et al.* (1962). (See also Kuffler *et al.* (1957).)

(*d*) No set of parameters in the basic model that we have described will produce an exponential interval histogram.

MORE COMPLEX RANDOM WALK MODELS

The difficulties that we have just described could, in principle, be met by various modifications of the basic random walk model. We shall examine the effects of several different types of structural changes.

A first modification of the basic random walk model is suggested by the known geometric complexity of synapses and of the somatodendritic membrane. Instead of trying to describe the electrical state of these structures by a single number, let us assume that several numbers are needed. Thus we might hope to allow the state to vary over the surface of the neuron.

The random walk will now take place in the same number of dimensions as the number of quantities chosen to describe the location of the state point. We might postulate that such a multidimensional walk be along some set of orthogonal directions, so that the state point can move from point to point in some multidimensional lattice. The simple absorber that we have associated with a threshold now must be taken to be some surface that bounds the region of the lattice in which the state point is allowed to move. As before, whenever the state point reaches the absorber, the model produces an action potential, and the state point is reset to its resting location.

In this guise, it is clear that the particular choice of absorbing surface exerts a profound effect on the behavior of the model. For example, consider the absorbing surface to be some array of the orthogonal coordinate planes. Then, in effect, the model will behave as if several competing but independent one-dimensional random walks are proceeding simultaneously; the one-dimensional walk that first reaches its absorber plane would produce the action potential and reset the state point. In physiological terms this type of model might correspond to a neuron that has two (or more) functionally independent trigger zones or dendritic trees, each of which can, when properly depolarized, initiate the spike discharge.

A concrete example of this type of model is the following: consider a two-dimensional random walk along Cartesian coordinates which is restricted to a single quadrant by choosing the absorbing boundary on the x axis and y axis. If we assume that the resting position of the state point is at $x_0 = y_0$ and that there is no drift, then the density of first passage times takes the form:

$$I(\tau) = \pi^{-1/2} x_0 \tau^{-3/2} \exp(-x_0^2/4\tau)\Phi(x_0/2\tau^{1/2}) \tag{11}$$

where

$$\Phi(y) = \frac{2}{\sqrt{\pi}} \int_0^y e^{-u^2}\, du.$$

The corresponding result obtained by simulating this two-dimensional random walk on a computer is shown in Fig. 10. Qualitatively, at least, this form of the model produces an interval distribution that is similar to that produced by the simple one-dimensional model: the tail of the interval histogram falls off more slowly than would an exponential. Thus, little is gained by increasing the complexity of the model in this way.

A quite different situation is obtained if we choose the absorbing boundary so that it is a more complex function of the coordinates. In such a case the several simultaneous one-dimensional random walks along the orthogonal coordinates will interact. The position of the state point along one coordinate will modify the effective boundary conditions of the random walks along the other coordinates. In physiological terms, this situation might represent a neuron with several interacting trigger zones, or perhaps with several interacting dendritic trees.

We have investigated two random walk models that exhibit this type of interaction: (*a*) a two-dimensional random walk inside a circular absorbing boundary, and (*b*) a three-dimensional random walk inside a spherical absorbing boundary. Each of these problems can, of course, be analytically solved by standard methods. However, the resulting solutions are in the form of series, and are extremely unwieldy for our purposes. Thus, for example the three-dimensional walk spherical boundary yields a probability density of first passage times

$$I(\tau) = \frac{a}{b}\left[\frac{1}{4\pi\tau^3}\right]^{1/2} \sum_{n=-\infty}^{+\infty} [(2n+1)a - b] \exp\left\{-\frac{[(2n+1)a-b]^2}{4\tau}\right\}, \qquad (12)$$

where the resting position of the state point is at radius b and the absorbing boundary at radius a. We assume there is no drift, and set the diffusion constant equal to 1. Clearly, the form of equation (12) is not conducive to a study of the tail of the density, nor of how it behaves in the convolution process of equation (5).

Again we have turned to Monte Carlo methods and have obtained the required probability density of intervals from the TX-2 digital computer. Under these circumstances it was also possible to study the effects of drift, which was arbitrarily chosen along one of the Cartesian axes.

Results from these particular two- and three-dimensional models are shown in Figs. 11 and 12. The tails of these probability densities fall off exponentially (or perhaps even somewhat more rapidly). As in all other cases that we examined, a drift parameter can be used to adjust the general slope of the density's tail. The behavior of both probability densities under the convolution of the scaling process is approximately stable, that is, qualitatively similar to that shown in Fig. 4 for the one-dimensional random walk with drift.

Interesting results can be obtained from a fundamentally different type of two-

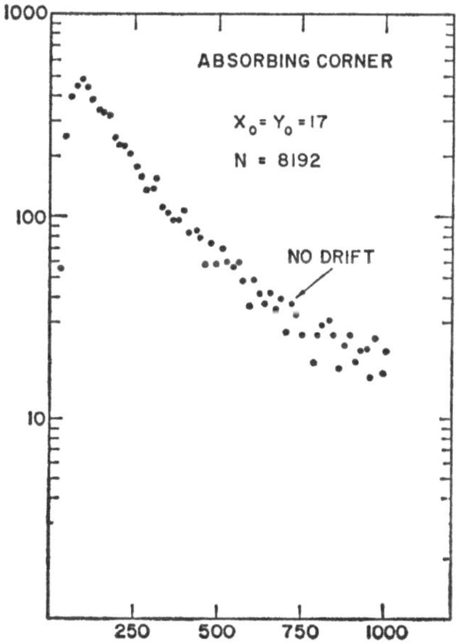

FIGURE 10 Interval histogram of two-dimensional random walk with absorbing 90° corner.

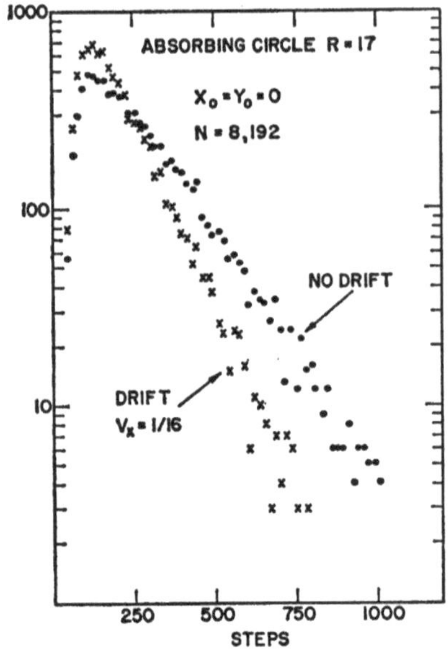

FIGURE 11 Interval histogram of two-dimensional walk with absorbing circle.

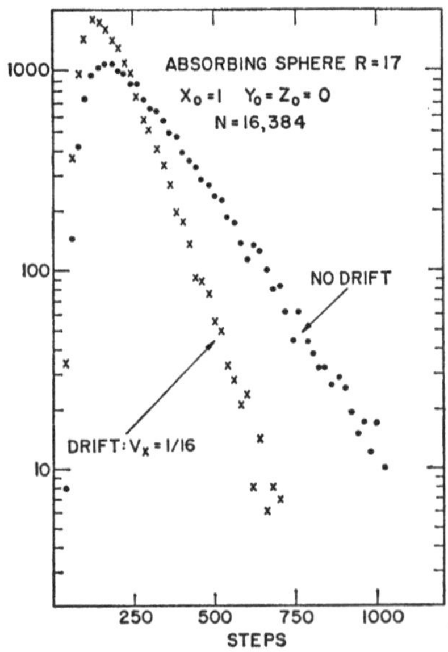

FIGURE 12 Interval histogram of three-dimensional walk with absorbing sphere.

dimensional model that is specified by the following rules: We allow the state point to perform two independent random walks. However, the model will give rise to a firing only after both random walks have reached threshold value; both coordinate numbers of the state point then return to their resting values. (One returns from threshold, the other from some random value after having attained the threshold.) Since the two walks are independent, the probability that the interspike interval is longer than τ will be equal to the product of the probabilities that the two random walks reach the threshold values after a time longer than τ. For large τ

$$Pr\,(\min\cdot(\tau', \tau'') > \tau) \sim [\tau^{-1/2}]^2 = \tau^{-1}.$$

This probability distribution function exhibits the precise invariance property—with a weighting factor k^{-1}—that is required by the experimental data of Unit 240-1. On the other hand, this model is less convincing from a physiological viewpoint and depends critically upon formulation in two dimensions.

In the course of this investigation we tried a number of other modifications of the basic random walk model. In particular, we have examined the one-dimensional case in which the size of the step toward threshold is a constant, while the size of the step away from threshold is a function of the location of the state point. This model results in a density of interspike intervals with an exponential tail, but which differs qualitatively from a Poisson-like distribution. We have also examined a one-dimensional model in which the drift contains a term proportional to the distance of the state point from some point z_1 on the z axis. (Mathematically, this model is reminiscent of the Brownian motion of an elastically bound particle.) Again, this more complicated model yields a density of interspike intervals with an exponential tail.

We have not tried a model with explicit correlation between successive random walks which corresponds to the experimentally observed correlation between successive interspike intervals. This could be done, for example, by letting the threshold value take some physiologically reasonable time course after the production of each spike. We should point out that the concept of absolutely refractory time is built into all the random walk models that we have described, since the state point cannot reach threshold in less than z_0 units of time.

STABLE DISTRIBUTIONS

Properties

In this section we shall examine briefly the general properties and peculiarities of stable distributions. (See Lévy, 1940; and Gnedenko and Kolmogorov, 1954.)

We limit ourselves to the k-fold summation of identically distributed variables (as in scaled interval histograms, for example). Distributions that are invariant to within scale under this type of transformation are called stable and are defined through a relation of the form

$$f(x)*f(x) = f(Ax + B)$$

where A and B are constants.

This process may be iterated to give the desired k-fold summation; in this case A and B can be functions of k only.

We pointed out in the section on The Simple Random Walk Model that for the Gaussian distribution $A(k) = k^{-1/2}$ and for the stable distribution of order 1/2, $A(k) = k^{-2}$. Also, it is known that the Cauchy law yields $A(k) = k^{-1}$.

In general, stable density functions have thus far only been specified in terms of a Fourier characteristic function that involves four parameters. Three of the parameters are measures of "location," "scale," and "skewness" and "peakedness" that replace the usual measures based upon moments. The fourth parameter, usually called α, is limited to $0 < \alpha \leq 2$. The value $\alpha = 2$ corresponds to the Gaussian distribution; $\alpha = 1$, to the Cauchy distribution, and $\alpha = 1/2$, to the stable distribution of order 1/2.

The form of the characteristic function precludes inversion into densities of closed analytic form for any values of α but the three examples given. Numerical methods may, of course, be used to obtain the stable distributions for any other values of α.

An interesting theorem may be proved about the moments of a stable distribution function (Gnedenko and Kolmogoroff, 1954, p. 182). A stable distribution with characteristic exponent α $(0<\alpha<2)$ has finite absolute moments only of order $\delta(0<\delta<\alpha)$. (Note that the first inequality circumvents the Gaussian case.) Thus, while the Gaussian distribution has first, second, (and all higher) moments, the Cauchy distribution and the stable distributions of order 1/2 do not even have a first moment.

These intuitively unpleasant ideas are illustrated in Fig. 13, in which for a random walk process we have plotted sample means as a function of sample length. The non-convergent quality of these measurements is apparent.[8]

A Phenomenological Model

During another investigation, several of the stable densities of non-analytic form were computed numerically (Mandelbrot and Zarnfaller, 1961). In order to avoid

[8] Direct simulation of the random walk model must be carried out with an upper limit for the duration of the walk, because of computer time limitations. Under these conditions, even without drift ($c = 0$), the sample mean and sample variance of the first passage times do converge fairly rapidly. In order to avoid this practical difficulty the values used to produce Fig. 13 were obtained in the following way: Uniformly distributed random numbers between zero and one were chosen, and the corresponding values τ_i were read from the known probability distribution function.

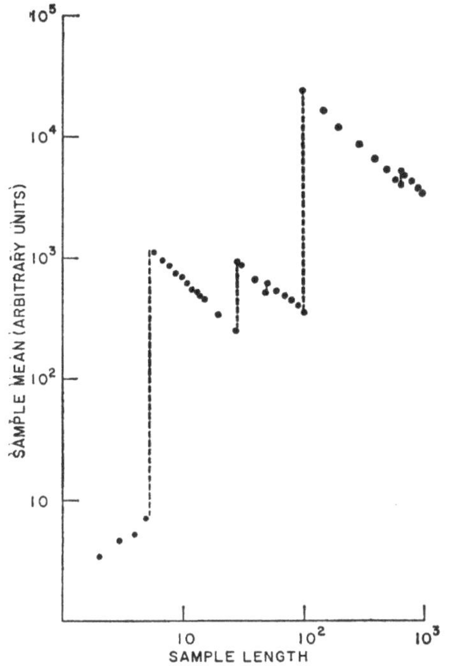

FIGURE 13 Sample mean of the first passage time in the random walk model without drift as a function of sample length. See footnote 8.

a heavy investment in mathematical notation, we shall only mention here that one of these stable densities, a skew form of the Cauchy law, has essentially the same invariance property as that found for the density of interspike intervals of Unit 240-1. The invariance property holds with the desired constants only for small *m*, just as in the data of Unit 240-1; asymptotically the density decreases as τ^{-2} (just as the Cauchy law itself).

In spite of close agreement with certain mathematical properties of the data, this phenomenological model for the spike activity of the neuron is intrinsically unsatisfactory. We know of no mechanism that would generate this stable probability density; hence this "model" can provide, at best, a mathematical shorthand for the data.

CONCLUSION

In this paper we have presented a series of random walk models for the spike activity of a single neuron. More generally, in making use of stable distributions, we have introduced a type of statistics into electrophysiology which offers a new interpretation of the striking "irregularity" of electrical activity in the nervous system. Usually such "irregular" phenomena have been explained in terms of some well behaved process whose properties happen to vary in time; such an explanation requires many more parameters for a description in terms of the usual statistical measures. In contrast, the irregularity is a fundamental property of a variable that is described by a stable distribution; there is no need to invoke properties that vary in time. We have pointed out that all stable variables (except the Gaussian) lack a second moment and many lack even the first moment. We have shown in computing sample mean or sample variances for a stable variable, that the major portion of the calculated quantity can be contributed by a very few data points. Thus, in contradiction to our intuitive feelings, increasing the length of available data for such processes does not reduce the irregularity and does not make the sample mean or sample variance converge.

The general properties of stable distribution functions would seem to make these functions useful statistical tools in neurophysiology. However, it is clear that our measurements and data are not enough to suggest unique models. A more detailed study of the applicability of the stable distributions must await more critical and restrictive types of measurements on neurophysiological data.

We wish to thank the technical staffs of the TX-2 computer, of the TX-0 computer, and of the Research Laboratory of Electronics, Massachusetts Institute of Technology, for their assistance. We are grateful to Professor W. A. Rosenblith for his support and continued interest in this project.

This work was supported in part by the United States Army, the Air Force Office of Scientific Research, and the Office of Naval Research; in part by the National Science Foundation (Grant G-16526); and in part by the National Institutes of Health (Grant MH-04737-02).

Received for publication, March 19, 1963.

REFERENCES

BERGER, J. M., and MANDELBROT, B., A new model for the clustering of errors on telephone lines, *IBM J. Research Develop.*, 1963, **7**, 224.

CHANDRASEKHAR, S., Stochastic problems in physics and astronomy, *Rev. Modern Physics*, 1943, **15**, 1; also *in* Noise and Stochastic Processes, (N. Wax, editor), New York, Dover Publications, 1954.

COOMBS, J. S., ECCLES, J. C., and FATT, P., The electrical properties of the motoneurone membrane, *J. Physiol.*, 1955, **130**, 291.

ECCLES, J. C., The Physiology of Nerve Cells, Baltimore, The Johns Hopkins Press, 1957.

ERULKAR, S. D., 1962, private communication.

FETZ, E., and GERSTEIN, G. L., R-C models for the spike activity of a single neuron, *Quart. Progr. Rep.*, Research Laboratory of Electronics, Massachusetts Institute of Technology, October 15, 1963, No. 71, in press.

GERSTEIN, G. L., and KIANG, N. Y-S., An approach to the quantitative analysis of electrophysiological data from single neurons. *Biophysic. J.*, 1960, **1**, 15.

GNEDENKO, B. V., and KOLMOGOROV, A. N., Limit Distributions for Sums of Independent Random Variables, (K. L. Chung, translator and editor), Cambridge, England, Addison-Wesley Publishing Company, 1954.

GROSSMAN, R. G., and VIERNSTEIN, L. J., Discharge patterns of neurons in cochlear nucleus, *Science*, 1961, **134**, 99.

KUFFLER, S. W., FITZHUGH, R., and BARLOW, H. B., Maintained activity in the cat's retina in light and darkness, *J. Gen. Physiol.*, 1957, **40**, 683.

LÉVY, P., Sur certains processus stochastiques homogènes, *Compositio Math.*, 1940, **7**, 283.

MANDELBROT, B., The Pareto-Lévy law and the distribution of income, *Internat. Econ. Rev.*, 1960, **1**, 79.

MANDELBROT, B., New methods in statistical economics, *J. Political Econ.*, 1963*a*, **71**, 421.

MANDELBROT, B., The variation of certain speculative prices, *J. Business*, 1963*b*, **36**, 392.

MANDELBROT, B., and ZARNFALLER, F., Five place tables of certain stable distributions, *in* IBM Research Report RC-421, 1961.

NELSON, P. G., and ERULKAR, S. D., Intracellular recordings from cat inferior colliculus, *Physiologist*, 1960, **3**, 120.

RALL, W., Electrophysiology of a dendritic neuron model, *Biophysic. J.*, 1962, **2**, No. 2, pt. 2, 145.

RODIECK, R. W., KIANG, N. Y-S., and GERSTEIN, G. L., Some quantitative methods for the study of spontaneous activity of single neurons, *Biophysic. J.*, 1962, **2**, 351.

VERVEEN, A. A., Fluctuation in excitability, research report on signal transmission in nerve fiber, Ph.D. Thesis, Amsterdam, Holland, Netherlands Central Institute for Brain Research, 1961.

VIERNSTEIN, L. J., and GROSSMAN, R. G., Neural discharge patterns in the transmission of sensory information, *in* Fourth London Symposium on Information Theory, (C. Cherry, editor), London, Butterworth and Company, 1961.

Chapter 2
Neuronal Spike Trains and Stochastic Point Processes

Why George Gerstein's Papers are Still Worth Reading

Stefan Rotter

I regret to have no joint publication with George Gerstein. He had just published his first paper in Science when I was born, and several more followed in the same magazine later (Gerstein 1960; Gerstein and Clark 1964; Gerstein and Perkel 1969; Kristan Jr. and Gerstein 1970). While I was in kindergarten and elementary school, he also put out some content-rich publications in the Biophysical Journal (Gerstein and Kiang 1960; Rodieck et al. 1962; Gerstein and Mandelbrot 1964; Perkel et al. 1967a,b; Gerstein and Perkel 1972), which at that time seemed to be the most highly respected journal for ground-breaking work in neuronal data analysis and modeling. Scientifically, George was miles ahead of me. I first read his papers 30 years after they were first published, and even now I am stunned how "modern" they are. My relation to George is literally that of a dwarf standing on the shoulders of a giant.

Among the colleagues of my age, George had the reputation of a neuroscientist, who combined latest recording technology with cutting-edge and highly original computational methods for analysis, pioneering entirely new approaches in brain research with this combination. Needless to stress that in the 1960s amplifiers and computers were still "emerging technologies," and those who made their scientific success depend on them were adventurers in the best sense of the word. My perception is that George stood out among his colleagues by his conceptual strength and his willingness and ability to abstract. Knowingly or unknowingly, he might even have lighted some innovations in mathematics that happened to take place at about the same time.

S. Rotter (✉)
Bernstein Center Freiburg and Faculty of Biology, Univ. of Freiburg, Freiburg, Germany
e-mail: stefan.rotter@biologie.uni-freiburg.de

A. Aertsen et al. (eds.), *Introducing Computation to Neuroscience*, Springer Series in Computational Neuroscience, https://doi.org/10.1007/978-3-030-87447-6_2

I met George for the first time in the early 1990s when he visited the Max-Planck Institute for Biological Cybernetics in Tübingen. At that time, he had some intense collaboration with Ad Aertsen and Valentino Braitenberg on the question how to detect the coordinated activity of neuronal assemblies in multiple single-neuron recordings, and the topic of my own PhD thesis was somewhat related to that. I had frequent encounters with George on scientific meetings, like the conference on Cortical Dynamics organized by Moshe Abeles in Jerusalem in the year 1995. For a young PhD student, it was a great privilege to meet George, but it also meant to be confronted each time with a new formidable intellectual challenge. I was never able to substantially contribute to their solutions right away, so we never published anything together. But my "collaborative distance" to him is anyway small. It just took me a little longer to fully understand and respond to the challenges he pointed out to me.

How can I best express my appreciation of George Gerstein's scientific legacy? I find his early works on the analysis of neuronal spike trains particularly appealing, probably thanks to my own training in math and physics. I have a lot of sympathy for the whole approach of viewing neuronal activity as an abstract type of signal that is to be treated with unbiased detachment and appropriate statistical rigor. The biologist's interpretation in terms of a specific function and meaning is subordinate. It seems that George was among the first to rigorously apply this point of view to his own data analysis, which—and this is yet another important innovation he pursued—was mostly performed with the help of digital computers. Today, of course, this combination is undisputed standard in the field. And biologists interested in the evolution of the nervous system find it now relevant that the same "abstract" type of signal is used to solve a huge variety of problems associated with representation, control, and integration, acting as some sort of "common currency" in the brain.

This embedding of neuronal data analysis into an abstract, technically oriented signal theoretical framework had another practical implication, both in George's own writings and in the papers by others that referred to him: Many technical terms that were before only known to and used by engineers and mathematicians now had their part in biological papers. Prominent examples are the (peristimulus) time histogram and the (interspike) interval histogram (Gerstein and Kiang 1960). These concepts were not commonly used, or even unknown, before George pointed out their usefulness. A few years later, step by step, he promoted more involved mathematical concepts like the Poisson process (Rodieck et al. 1962), random walk models (Gerstein and Mandelbrot 1964), or stochastic point processes (Perkel et al. 1967a,b). He clearly demonstrated their general applicability for the analysis of neuronal spike trains, as their apparent complexity and discouraging variability had now found an almost perfect match in these mathematical models. Needless to emphasize how big the gain was by adding these concepts to the repertoire of neuroscience.

George already early on was aiming at understanding how one neuron influences another one by analyzing simultaneous recordings of two nerve cells. Once more, a concept from stochastic point process theory was found useful to measure the strength of neuronal interaction in terms of statistical dependencies between their respective spike trains: the cross-correlation histogram (Perkel et al. 1967b). A natural generalization of this question was how neurons interact in networks, or more precisely, how the joint activity of all nerve cells in a network reflects the various modes of interaction among its neurons. To illustrate the complex ambiguities that emerge in this situation and to elucidate the power and the pitfalls of new data analysis methods, numerical simulations of small networks of spiking neurons were performed together with Don Perkel (Perkel et al. 1967b; Gerstein and Perkel 1972). This represents yet another methodological innovation that is nowadays firmly established as a complementary technique in neuroscience. I find it hard to imagine that there were times, where all these powerful concepts were essentially unknown and unavailable to neuroscientists.

As a mathematician, I would find it very interesting to know what impact the early research by George and his colleagues has left in the field of mathematics, where new approaches emerge if abstraction has the potential to solve problems in several different fields of application. There is the interesting case of Alain G Hawkes, a mathematician who later became well-known for his analysis of the stochastic behavior of ion channels. His early work, however, dealt with an analytically tractable model of networks of interacting point processes, now known as the "Hawkes process" (Hawkes 1971a,b). It seems safe to claim that networks of spiking neurons in the brain are the most obvious application of Hawkes' model. On top of its obvious relevance for neuroscience, however, the model has experienced a recent renaissance in financial mathematics, with dozens of applications and exciting generalizations. The fact that Hawkes' original model also appeared in a book collection (Hawkes 1972) together with other articles that explicitly discuss pulse coded networks in neuroscience may suggest that the mathematician Hawkes might have known about George's early work as well. A more solid analysis of this claim, however, must be left to historians of science.

How come all these techniques pioneered by George and his collaborators in the 1960s still have this aura of being "modern"? Here is a possible answer: Of course, neuroscience has progressed tremendously in the past decades, mainly in terms of its methodological repertoire. Silicon-based multi-electrode arrays can now be bought off the shelf to effortlessly register the spiking of thousands of cells simultaneously, with very high signal-to-noise ratio and very high temporal resolution. Optical methods complement electrical recordings and provide spatial specificity that makes spike sorting much less of an issue. Data processing and data analysis is run on huge cluster computers, and many tasks can be performed in real time using the latest big data and machine learning technologies. Large-scale numerical simulations of networks are part of the standard repertoire of many experimental labs, complementing the straightforward data analysis in useful ways. Everything is bigger, faster, and more accurate. But everything is run pretty much by the same ideas and concepts that were developed in the 1960s by giants of neuroscience like George Gerstein. This is why I think his papers are still highly worthwhile reading and a great source of inspiration. This is also why, when Sonja Grün and I undertook the effort to gather the repertoire of multiple-electrode data analysis methods into a single textbook (Grün and Rotter 2010), I was extremely happy and, indeed, very proud that George Gerstein, who pioneered the larger part of these methods, was willing to contribute a chapter to our book (Chapter 8 in Grün and Rotter 2010).

References

Gerstein GL (1960) Analysis of firing patterns in single neurons. Science 131(3416):1811–1812. https://doi.org/10.1126/science.131.3416.1811

Gerstein GL, Clark WA (1964) Simultaneous studies of firing patterns in several neurons. Science 143(3612):1325–1327. https://doi.org/10.1126/science.143.3612.1325

Gerstein GL, Kiang NS (1960) An approach to the quantitative analysis of electrophysiological data from single neurons. Biophys J 1(1):15–28. https://doi.org/10.1016/s0006-3495(60)86872-5

Gerstein GL, Mandelbrot B (1964) Random walk models for the spike activity of a single neuron. Biophys J 4(1, Part 1):41–68. https://doi.org/10.1016/S0006-3495(64)86768-0

Gerstein GL, Perkel DH (1969) Simultaneously recorded trains of action potentials: analysis and functional interpretation. Science 164(3881):828. https://doi.org/10.1126/science.164.3881.828

Gerstein GL, Perkel DH (1972) Mutual temporal relationships among neuronal spike trains: Statistical techniques for display and analysis. Biophys J 12(5):453–473. https://doi.org/10.1016/S0006-3495(72)86097-1

Grün S, Rotter S (2010) Analysis of parallel spike trains. Springer series in computational neuroscience, vol 7. Springer US. https://doi.org/10.1007/978-1-4419-5675-0

Hawkes AG (1971a) Spectra of some self-exciting and mutually exciting point processes. Biometrika 58(1):83–90. https://doi.org/10.1093/biomet/58.1.83

Hawkes A (1971b) Point spectra of some mutually exciting point processes. J R Stat Soc 33(3):438–443. https://doi.org/10.1111/j.2517-6161.1971.tb01530.x

Hawkes AG (1972) Spectra of some mutually exciting point processes with associated variables. In: Lewis PAW (ed) Stochastic point processes. Wiley, New York, pp. 261–271

Kristan Jr WB, Gerstein GL (1970) Plasticity of synchronous activity in a small neural net. Science 169(3952):1336–1339. https://doi.org/10.1126/science.169.3952.1336

Perkel DH, Gerstein GL, Moore GP (1967a) Neuronal spike trains and stochastic point processes. I. The single spike train. Biophys J 7(4):391–418. https://doi.org/10.1016/S0006-3495(67)86596-2

Perkel DH, Gerstein GL, Moore GP (1967b) Neuronal spike trains and stochastic point processes. II. Simultaneous spike trains. Biophys J 7(4):419–440. https://doi.org/10.1016/s0006-3495(67)86597-4

Rodieck RW, Kiang NYS, Gerstein GL (1962) Some quantitative methods for the study of spontaneous activity of single neurons. Biophys J 2(4):351–368. https://doi.org/10.1016/S0006-3495(62)86860-X

AN APPROACH TO THE QUANTITATIVE ANALYSIS OF ELECTROPHYSIOLOGICAL DATA FROM SINGLE NEURONS

GEORGE L. GERSTEIN *and* NELSON Y.-S. KIANG
From the Research Laboratory of Electronics, Massachusetts Institute of Technology, Cambridge, and the Eaton Peabody Laboratory of Auditory Physiology, Massachusetts Eye and Ear Infirmary, Boston

ABSTRACT The application of a digital computer to the processing of data from single neurons is described. Examples from experimental data are presented to demonstrate the usefulness of certain types of computations. These methods are placed in a descriptive mathematical framework. Other easily attainable computations are suggested.

INTRODUCTION

Since the days of Ramon y Cajal (1), the neuron has been generally accepted as the basic building block of the nervous system. This view has been important in the interpretation of electrical phenomena recorded in nervous tissue by macroelectrodes. Even the most careful of these recordings poses problems of measurement for which we lack satisfactory theoretical formulations. In the (more or less) intact animal, a macroelectrode records electrical activity from a volume that is large compared to that of a single neuron; the recorded activity is not of necessity functionally homogeneous. Thus there arises the problem of meaningfully relating the electrical activity seen by a macroelectrode to the underlying single neuron activity and, more generally, to a valid indicant of the behavior of the nervous system as a whole.

The introduction of microelectrodes has made it possible to obtain greater spatial resolution in recordings. It has become possible to record electrical events from a single nerve cell or even from certain regions of a given neuron. The temporal characteristics of these events cover the range of "fast" action spikes to DC and "slow"

This work was supported in part by the United States Army (Signal Corps), the United States Navy (Office of Naval Research), the United States Air Force (Office of Scientific Research, Air Research and Development Command), and Research Grant B-1344, National Institute of Neurological Diseases and Blindness, National Institutes of Health, Bethesda.

Dr. Gerstein is Postdoctoral Fellow of the Institute of Neurological Diseases and Blindness, National Institutes of Health, Bethesda.

Received for publication, August 3, 1960.

potential changes. Although any and all of these potentials may have functional significance (2-5), the action potentials or spikes have loomed largest, historically, in the analysis of single unit activity. Spatiotemporal patterns of spike activity have frequently been used as indices of nervous system function.[1] Such patterns have conveniently been studied in sensory systems because these systems are particularly accessible to manipulation.

In stimulus response studies, changes in spike activity must be assessed against a background of "spontaneous activity." Stimulation can lead to subtle changes of the temporal distribution of spikes. When these pattern changes occur with long latency and much variability, it hardly seems wise to define the "response" of a neuron as the casually detectable change in background activity that immediately follows stimulation. As closer and more formal examination of single-unit firing patterns becomes possible, the number of units that fall into the "no response" category diminishes strikingly. Potentially more useful categories can be obtained by more exact description of response pattern changes. For this purpose it appears necessary to utilize probabilistic approaches to the quantification of unitary spike activity.

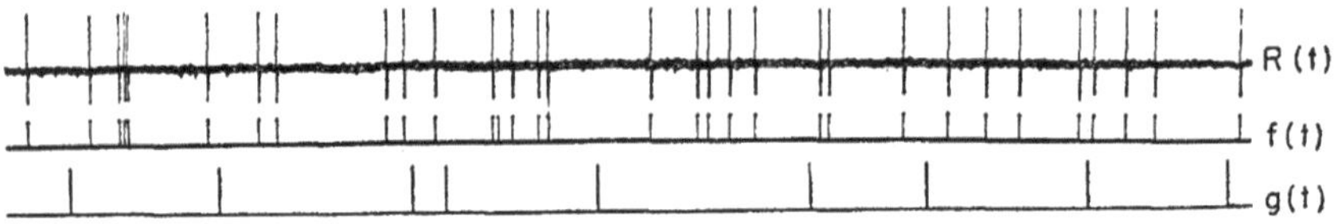

FIGURE 1 Spike and pulse trains discussed in the text. $R(t)$, single-unit activity (spikes) recorded by microelectrodes in cochlear nucleus; $f(t)$, idealization by pulse generator; $g(t)$, externally determined time marker.

It is the purpose of this paper (a) to illustrate, from our experimental work, the use of several methods of data processing which make possible the quantitative description of single neuron firing patterns and (b) to place these methods in a mathematical framework that makes explicit the interpretations that can be drawn from such analyses. We wish particularly to set forth both the power and the limitations of this approach to the description of firing patterns.

ANALYSIS OF SINGLE-UNIT DATA

If the time pattern of action spikes is selected as the subject of interest, other aspects of the electrical activity (such as details in the spike waveform, slow potentials) can be neglected. This selection is illustrated in Fig. 1 where $R(t)$ is a train

[1] Today it seems much less certain that the analysis of discrete events alone will lead to a complete understanding of the behavior of the nervous system. It is likely that an adequate description of nervous system activity can be obtained only through the analysis of *both* the graded electrical activity and the discrete spikes.

of spikes recorded by a microelectrode, and $f(t)$ is the idealization of $R(t)$ produced by letting the spikes trigger an electronic pulse generator. In such a train of identical pulses of unit amplitude all "information" is carried by the times at which individual pulses occur. In this article we shall consider the spike train $R(t)$ as being equivalent to the corresponding pulse train $f(t)$. In order to analyze the time structure of the train it is necessary to choose an origin (or a series of origins) for time measurements. There are two possibilities for this choice: (*a*) events in the spike train itself, or (*b*) externally determined instants (for example, $g(t)$ in Fig. 1).

The simplest application of the first choice leads to a class of analyses that we shall call "interval histograms;" the second choice leads to a class that we shall call "time histograms."

Within the past few years it has become possible to perform such computations on large masses of data with high resolution, accuracy, and speed. In order to obtain maximum flexibility in choosing the particular type of analysis to be used, it is convenient to work with a general-purpose digital computer. A data-processing system of this type which is in use at this laboratory has already been described (6). With this system, computations of the simplest cases of interval and time histograms have yielded useful analyses of single-unit firing patterns in several neurophysiological studies. In particular, we are using:

(*a*) an interval histogram—a histogram of the intervals between *successive* spikes.

(*b*) a poststimulus time (PST) histogram—the distribution of spikes in time relative to the instant of the *most previous* presentation of the stimulus, summed over many repeated stimulus presentations. A peak on an interval histogram shows a preferred interval between firings; a peak on a PST histogram shows a preferred time of firing relative to the stimulus.

(These analyses are, respectively, the second term of equation (3) and the first term of equation (7) below.)

It is not our intention to present the results of completed studies here; rather we wish to indicate certain advantages and disadvantages in the use of these and other methods for the analysis of single-unit data. All of our examples are drawn from work on the auditory system.

In order to formalize most simply the types of computation that we have used, let us assume that the basic spike train results from a process that is statistically stationary in the large. This means, if we are considering an experiment which involves the repeated presentation of the same stimulus, that we assume the statistical properties of all responses to be the same. In other words, we are examining a steady-state process and are not concerned with adaptation effects and other "dynamic" properties that could alter the response to successive identical stimuli. Within the time between presentations of the stimulus, the statistical properties of the process may of course vary. This assumption will avoid difficulties in choosing the segments of data over which calculations are made.

In this framework, interval histograms are selected portions of an estimate of the autocorrelation function for a train of suitably defined spikes. Let the spike train we are investigating be

$$f(t) = \sum_{k=1}^{N} \delta(t - t_k) \tag{1}$$

where $\delta(t)$ is a Dirac δ function (or, alternatively, a unit impulse), and where N is a large number. (For mathematical convenience we are representing the original train of *unit amplitude* spikes by a train of δ functions of unit area, each located at the time of an original spike. To interpret any results we must now consider *areas*.)

An estimate of the autocorrelation of (1) is

$$\varphi_{ff}(\tau) \equiv \int f(t)f(t-\tau)\,dt = \int \sum_{k=1}^{N} \delta(t-t_k) \sum_{l=1}^{N} \delta(t - t_l - \tau)\,dt$$
$$= \sum_{k=1}^{N}\sum_{l=1}^{N} \delta(t_k - t_l - \tau) \tag{2}$$

For convenience, the double sum can be broken up in the following way:

$$\varphi_{ff}(\tau) = N\,\delta(\tau) + \sum_{k=2}^{N} \delta(t_k - t_{k-1} - \tau) + \sum_{k=3}^{N} \delta(t_k - t_{k-2} - \tau) + \cdots . \tag{3}$$

These terms correspond respectively to letting $l=k$, $l=k-1$, $l=k-2$, . . . Since the autocorrelation function is symmetric in τ, we have not written the $-\tau$ terms explicitly.

Thus the estimate of the autocorrelation function of (1) is itself expressed as an array of δ functions. The numerical values of the estimated autocorrelation function would be obtained from the envelope of the δ function *areas*.[2]

For convenience we may measure the continuous values of τ as discrete time values τ_j, $j=1, 2, 3, \ldots$, in such a manner that all τ that satisfy $\tau_j<\tau<\tau_{j+1}$ are measured as τ_j. Thus $\tau_{j+1}-\tau_j \equiv \Delta\tau$ defines a time bin that determines the resolution of the analysis. The choice of bin width is governed by the time scale of the particular phenomena that are being studied. In this case of discrete time values, the *areas* of all δ functions of φ_{ff} which fall between τ_j and τ_{j+1} are added together to give the height of the τ_j—th bar in a histogram.

The first term of equation (3) is a constant contribution to the estimated autocorrelation function at $\tau=0$; its numerical value (area) is N, the total number of spikes in the train. If the time duration of the spike train is known, this term of equation (3) yields the average firing rate in the spike train. The second term is an interval histogram; *i.e.*, a histogram of the intervals between *successive* (k and $k-1$) spikes of the train. The

[2] This usage is common in text-books. See chapter 2, reference 7. The reader who finds our notation obnoxious can perform a similar development in terms of summations only, without using δ functions, integrals, or integral properties, by starting with a spike train

$$f(t_i) = \sum_{k=1}^{N} \Delta(t_i - t_k) \quad \text{where} \quad \Delta(t_i - t_k) = \begin{matrix} 1, & t_i = t_k \\ 0, & t_i \neq t_k \end{matrix}$$. For discrete time shifts

τ_m, equation (2) would be $\varphi_{ff}(\tau_m) = \sum_{k=1}^{N}\sum_{l=1}^{N} \Delta(t_k - t - \tau_m)$. Interpretation of results in terms of areas is then unnecessary. Still other forms of notation are possible. However, the particular resolution of components expressed in (3) and (7) remains valid.

third term is an interval histogram of the intervals between every second (k and $k-2$) spike. All subsequent terms are interval histograms for higher order intervals in the spike train.

A peak on an interval histogram shows a preferred periodicity in the train; *i.e.*, a preferred interval between spikes or groups of spikes. The value of this preferred interval is indicated by the position of the peak on the abscissa. The ordinate measures the number of spike intervals falling into each discrete τ_j time bin.

In a similar manner, the time histograms are selected portions of an estimated cross-correlation function between the spike train under study and an external pulse train that can be determined arbitrarily by the experimenter. Again, let the spike train that we are investigating be

$$f(t) = \sum_{k=1}^{N} \delta(t - t_k) \tag{4}$$

and let the externally determined (not necessarily periodic) pulse train be

$$g(t) = \sum_{l=1}^{N} \delta(t - t_l) \tag{5}$$

Generally, very few, if any, t_k and t_l will have the same values. An estimate of the cross-correlation function of functions (4) and (5) is

$$\begin{aligned} \varphi_{fg}(\tau) &\equiv \int f(t)g(t-\tau)\,dt = \int \sum_{k=1}^{N} \delta(t-t_k) \sum_{l=1}^{N} \delta(t-t_l-\tau)\,dt \\ &= \sum_{k=1}^{N}\sum_{l=1}^{N} \delta(t_k - t_l - \tau) \end{aligned} \tag{6}$$

Although (6) looks formally like (2), the t_k and t_l are *not* the same series in this case, and the double sum cannot be broken down as in (3). Instead we write

$$\varphi_{fg}(\tau) = \sum_{l=1}^{N}\left[\sum_{\substack{k \\ t_l < t_k < t_{l+1}}} \delta(t_k - t_l - \tau) + \sum_{\substack{k \\ t_k > t_{l+1} \\ t_k < t_l}} \delta(t_k - t_l - \tau)\right] \tag{7}$$

Thus the estimate of the cross-correlation function is also expressed as an array of δ functions. Just as for equation (3), the numerical values are obtained from the envelope of δ function *areas*.

As in the discussion of equation (3), the continuous τ can be measured as discrete τ_j, $j=1, 2, \ldots$

The two right hand terms of equation (7) are time histograms; *i.e.*, histograms of the average times of occurrence of the $f(t)$ spikes relative to origins determined by the external $g(t)$ pulses. The first term describes the average times of occurrence of the $f(t)$ spikes relative *only* to the last previous (lth) $g(t)$ pulse as origin. The second term lumps together the average times of occurrence of the $f(t)$ spikes relative to the $l-1$, $l-2$, $\ldots$, $g(t)$ pulses. In other words, a peak on a time histogram shows, on the average, a preferred time of occurrence of the $f(t)$ spikes relative to the $g(t)$ pulses. The value of this preferred time is indicated by the position of the peak on the abscissa. The ordinate measures the number of spikes that fall into each discrete time bin.

This computation serves to sort the spike train under study in a way that detects patterns that are time-locked to the arbitrary external pulse train. Suppose, for example, that the arbitrary external pulse train marks the delivery of stimuli. Then the first term of equation (7) would detect average evoked response patterns that occur during the time *before* the delivery of the next stimulus. Let us call this particular computation a post-stimulus time (PST) histogram. If the perturbations caused by the first stimulus have not died away by the time the next stimulus is delivered, a part of the pattern that is time-locked to the stimulus would be found in the second term of equation (7). (The physical meaning of the two terms of (7) is most easily visualized in the case in which the stimuli are delivered aperiodically but at a mean interval that is shorter than the duration of the time-locked evoked response pattern.)

We cannot overemphasize that both types of histogram that have been discussed involve averaging the data. This means that in performing these computations, to bring out certain aspects of the spike pattern, we have lost other information. For example, we lose the time sequence of the events contributing to a particular histogram. This could be remedied, in principle, by computing separate time and interval histograms for the first, second, ... , spike, or spike pair after the stimulus. Unfortunately, there are so few data in most experiments that such fractional analyses are statistically impractical.

A far more crucial loss of information is hidden in the assumptions involved in averaging: a meaningful average can be obtained only if the events in the pattern that enter into the average have statistical properties which are similar. If the statistical properties of the events vary during the period of the average, then an "average" of the events can still be computed, but there will be a loss of all information about the changes of the statistical properties. In any real neurophysiological experiment such changes of the statistical properties are associated with inevitable physiological variability. Thus a more complete description of the properties of neural firing patterns must eventually necessitate the use of higher order statistical measures.

There is an interesting difference in the interpretation of the two general types of histogram. In the time histogram, we seek significant variations of the distribution as a function of poststimulus time—a distribution that is flat within statistics, *i.e.*, a DC contribution conveys little of interest. Such a flat distribution represents the events in the spike pattern which are not time-locked to the external pulse train, and which are therefore randomly distributed in time relative to the external train. For the interval histogram, however, all parts of the distribution are pertinent because no criteria external to the spike train are used in the computation.

EXPERIMENTAL METHODS

All experiments were performed on cats anesthetized with either dial or nembutal, given either intraperitoneally or intravenously. Recordings were made from one of two places: (*a*) the cochlear nucleus, or (*b*) the auditory cortex. The cochlear nucleus was approached through the intracranial route with removal of cerebellum. Cortical recordings were made with a closed chamber technique similar to that described by Davies (8). Electrodes were either the indium-platinum type described by Dowben and Rose (9) or the tungsten type described by Hubel (10). In all cases the experiments were conducted in sound-proof chambers, with hydraulic remote control of the microelectrode. Clicks were delivered by a PDR-10 earphone connected to the external auditory meatus through a plastic tube. The intensity of the clicks is stated relative to the VDL (visual detection level), of the appropriate evoked response as recorded by a macroelectrode.

Responses were recorded with an FM tape system. For the cortex both a high-pass (200 to 10 kc) and an "unfiltered" (8 to 10 kc) channel were recorded. This allowed investigation of both unit spike potentials and slow potentials.

The processing of single-unit data by the TX-O general-purpose digital computer has already been described. (6) Output from the computer was obtained as a display of a bar graph and titles, and was photographed with a Polaroid camera.

RESULTS AND DISCUSSION

The fundamental objective of our experimental work is the detection of changes that take place in single neuron activity patterns as a result of changes in stimulus conditions. Neurophysiological spike trains can be considered in two categories: *(a)* spontaneous activity, and *(b)* responses to stimuli. Any stimulus condition results in a spike pattern that is a mixture or interaction of both types of activity. The effect of a discrete stimulus on the over-all spike pattern decreases with time after stimulation; the pattern gradually returns to the spontaneous condition. If the stimuli are spaced far apart in time, the over-all contribution of the stimulus-dependent activity (responses) is low relative to the spontaneous activity. If stimuli are delivered at sufficiently high rates, the contribution of spontaneous activity can become relatively small.

The interval histogram emphasizes *internal* timing relations in the spike train, whether the individual spike be spontaneous or in response to a stimulus. The PST histogram emphasizes the time relations of the spikes in the train to the stimulus time marker.

In general, the spike trains with which we deal represent some interaction of spontaneous and responsive activity, so that both the interval and time histograms are useful in describing the patterns of activity.

In practice, for particular samples of data, we have found that one or the other display is more striking in showing differences in activity patterns. This contrast in effectiveness between the two displays can be attributed to the relative contribution of spontaneous and responsive activities.

Figs. 2 and 3 show applications of interval histogram computations. Fig. 2*a* shows single-unit activity recorded from the cochlear nucleus during various rates of click presentation. Note that the unit exhibits spontaneous activity that appears to be suppressed when clicks are delivered at rates around 10 or 20/sec. At 10/sec. the unit appears to fire once and only once to each click. At 20/sec. there appear to be some clicks to which the unit does not respond, although whenever a spike occurs it seems time-locked to the stimulus. Over considerable stretches of time the unit appears to fire to every other click.

At higher rates, from these pictures, it becomes more difficult to describe the pattern of activity with respect to the stimulus. On a faster time base, there appears to be some following with irregular skipping.

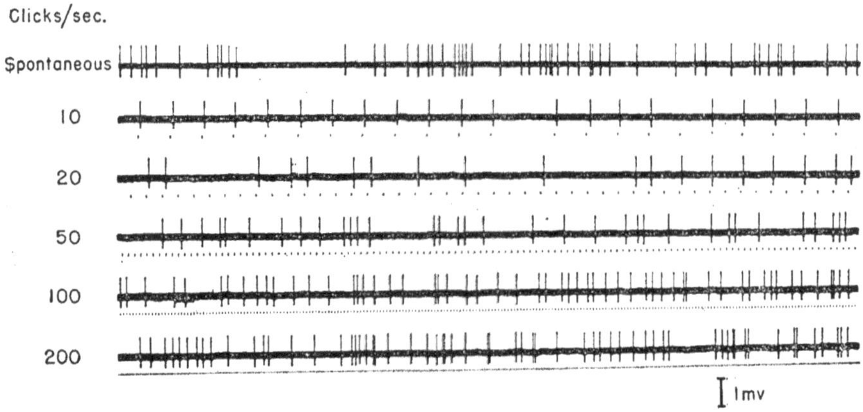

FIGURE 2*a* Activity of a single unit in cochlear nucleus (unit 165-1) during presentation of clicks at various rates. Click intensity 25 db re VDL.

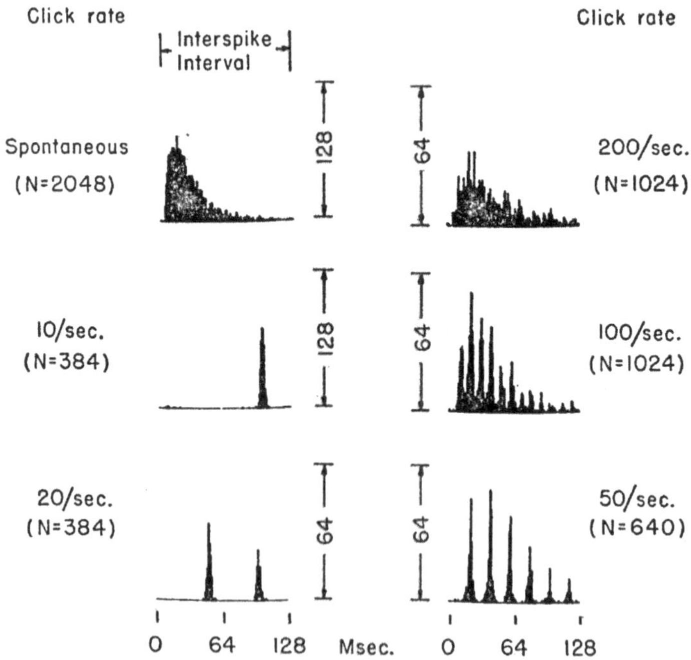

FIGURE 2*b* Interval histograms of the same data from which Fig. 2*a* was obtained. The ordinate on all displays is the number of spikes per unit time interval. *N* is the total number of spikes represented in each histogram.

Fig. 2*b* shows the interval histograms for the same samples of data from which Fig. 2*a* was taken.

The interval histograms show very clearly the marked effect of 10/sec. clicks on the pattern of spontaneous activity. The single sharp distribution at 100 msec. (*i.e.,* time interval between clicks) reflects the one-to-one following seen in Fig. 2*a*. For 20/sec., there is one large peak at 50 msec. (*i.e.,* time interval between clicks) and an additional smaller peak at 100 msec. which corresponds to those intervals when the response skips one stimulus. The relative areas of the two peaks expresses the relative number of times that the unit responds in a one-for-one manner as compared with the number of times that the unit skips one stimulus. Note especially that there is negligible activity with intervals other than at the periods of the stimulus or its subharmonics. At 50/sec. the fundamental period of the stimulus is 20 msec. The first peak in the interval histogram corresponds to this period (*i.e.,* one-for-one responses). Now however, the second peak, representing intervals of 40 msec., is larger than the first peak. This signifies that the unit now fires more often to every other click than to every click. The areas of the peaks at higher interval values indicate the relative occurrence of activity which skips two, three, etc., clicks. At 100/sec., similar observations can be made. At 200/sec., while the individual peaks are no longer clearly separable, the envelope of the interval distribution approaches that of spontaneous activity. It is evident that the interval histogram emphasizes certain "interesting" features of the pattern not easily studied in the original records.

Fig. 3 shows interval histograms for a unit in the auditory cortex (A1). The

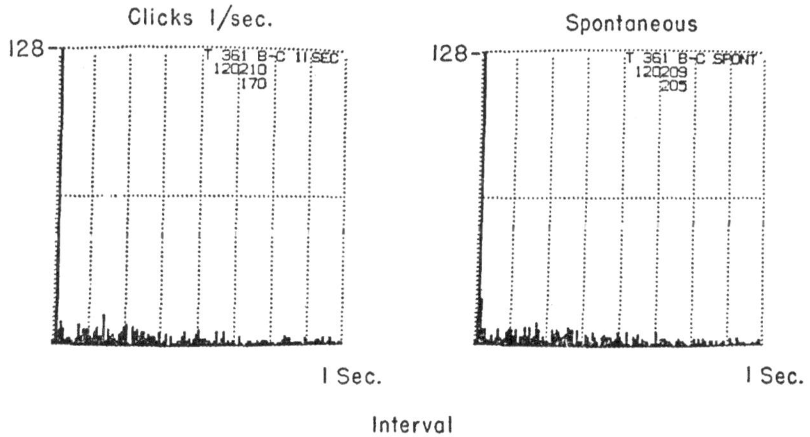

FIGURE 3 Interval histograms for a unit (120-2) in auditory cortex. Left, activity during click presentations. Right, activity during the interleaved spontaneous conditions. Click intensity 20 db re VDL. $N = 576$.

large peak at small interval corresponds to the rapid bursts of 2 or 3 action potentials which were characteristic for this unit. These displays do not show any gross differences between the stimulated and unstimulated conditions. However, observation of the unit's activity during the experiment seemed to indicate that the stimulus had some effect, although there was much spontaneous activity. The similarity of the interval histograms for 1/sec. clicks and the spontaneous condition is probably a reflection of the relatively large "spontaneous" contribution to the over-all activity even during stimulation. This picture is true for most cortical units that we have observed even at higher click presentation rates. The interval histogram does not clearly reflect changes in the pattern of most of our cortical unit responses. Comparison of Figs. 2 and 3 shows that, determined by the particuliar behavior of a unit, the interval histogram may or may not be trivial.

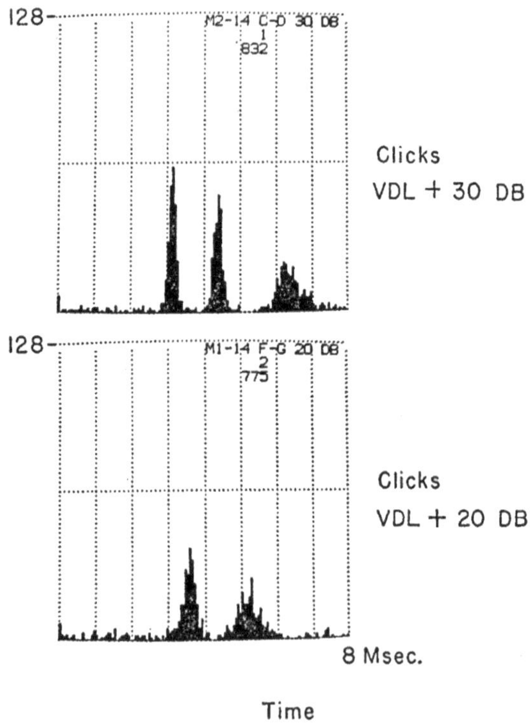

FIGURE 4 PST histograms for a unit in cochlear nucleus (unit K — 173 — 3). Top, activity during presentation of clicks at 30 db re VDL. Bottom, activity during presentation of clicks at 20 db re VDL. $S = 256$, the number of stimulus intervals processed.

Various applications of the time histogram display are shown in Figs. 4 to 6. In each case the time of stimulus presentation or dummy time marker is at the extreme left of the histogram.

Fig. 4 shows the time histogram for a cochlear nucleus unit when two intensities of clicks are used. Under these conditions photographs of the responses show that

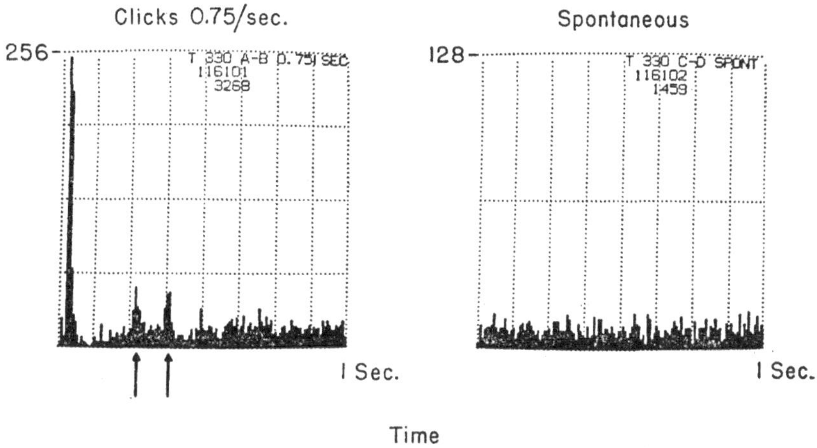

FIGURE 5 PST histograms for a unit in auditory cortex (unit 116-1). Left, activity during clicks presented at 0.75/sec. Right, spontaneous activity. Click intensity 20 db re VDL. $S = 512$.

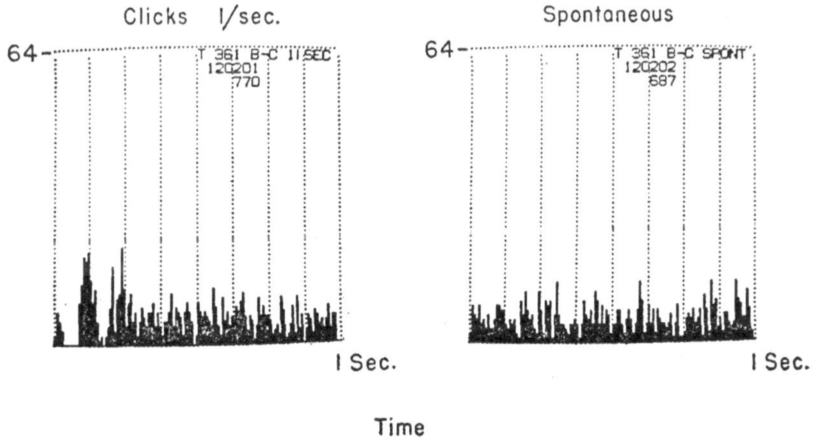

FIGURE 6 PST histograms for the same data as Fig. 3 (unit 120-2 in auditory cortex). Left, activity during clicks presented at 1/sec. Right, interleaved spontaneous activity. $S = 256$.

the unit fires repetitively to each click. The number of repetitives may not be constant. The area under a particular peak in the PST histogram relates to the number of responses that may be identified as that particular repetitive. The increasing width of successive peaks indicates the increasing time jitter of successive repetitives. Not only does the PST histogram show an increase in the number of repetitives with increased stimulus intensity but it also shows that *(a)* the latency of the first response decreases; *(b)* the time between the first response and the first repetitive response decreases; and *(c)* the time distributions become sharper. These features are very difficult to detect in an interval histogram display.

Fig. 5 shows PST histograms for a cortical unit that responds to 0.75/sec. clicks. Here, there is a peak in the distribution around 20 msec. and some evidence for time-locking of later activity in the two peaks indicated by arrows. Immediately after the large first peak there appears to be a relatively silent period of about 150 msec.

Fig. 6 shows PST histograms for a unit whose most apparent reaction to clicks is a silent period followed by a peak in the distribution around 110 msec. This strikingly display of the effect of the stimulus should be contrasted with the corresponding interval histograms of Fig. 3. The samples of data processed for Figs. 3 and 6 are the same. Thus the PST histograms always selects activity time-locked to the stimulus, although it is of little value in studying ongoing spontaneous activity.

We have found the interval and PST histograms useful even in the analysis of response patterns from unanesthetized animals, in which the spontaneous activity is extremely labile.

The type of single-unit data processing that we have used can be advantageously combined with other computational methods in certain problems. For example, Fig. 7 is from a study of the relationship between slow waves and action potentials observed by the same microelectrode. The spikes, for various rates of click presentation, were analyzed by PST histograms. The slow wave activity during the same time was averaged by the ARC computer (11) and is, in each case, plotted above the histogram. The parallel course, in this particular case, of the average PST neuron firing rate and the average of the slow wave is thus made apparent.

There are other computational methods that can be applied to the study of unit activity. As has been shown in the mathematical section, a more complete description of spike trains can be achieved by computing the remaining terms of equations (3) and (7). When these terms in equation (3) (interval histogram) are computed, it is possible to study the distribution of *groups* of spikes. In the case of PST histograms, the additional terms in equation (7) will show the effects of one stimulus that last beyond the instant of presentation of the next and subsequent stimuli. This makes possible the study of responses to patterns of stimuli.

Another extension of our analytical tools is to set more complex criteria for the time markers in the averaging procedures for computing time histograms. This will

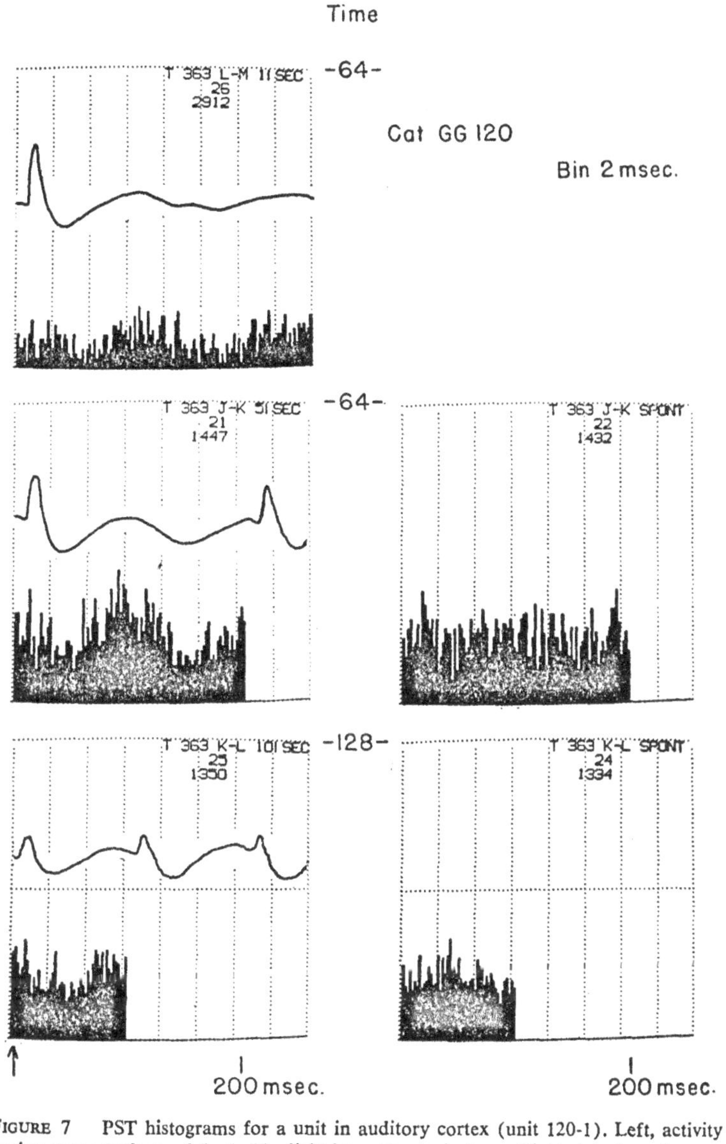

FIGURE 7 PST histograms for a unit in auditory cortex (unit 120-1). Left, activity during presentations of 1, 5, 10 clicks/sec., top to bottom. Right, interleaved spontaneous activity. The curves are the average of the slow activity measured by the same microelectrode during the same time. Click intensity 20 db re VDL. $S = 512$; 1536; 2560 respectively, top to bottom.

enable study of response probabilities which are conditional on certain aspects of the response. In Fig. 4 for instance, one could study the dependence of the latency of the first response on the occurrence of repetitives to the preceding stimulus.

As experimental methods begin to permit the simultaneous recording of many functionally connected neurons, it will become necessary to investigate various conditional statistics between the several activity patterns.

These particular extensions over our present work are relatively easy to achieve with the general-purpose computers that we are using. Complete descriptions of the time properties of a spike train involve variances and other higher order statistics that cannot be discussed here. But the various computation techniques we have outlined begin to permit quantitative tests of specific mathematical models of the properties and interactions of single neurons.

We should like to thank Professor W. A. Rosenblith for his continued advice and interest in these problems, and R. M. Brown for help with problems of instrumentation. We are grateful for the technical assistance of the staffs of the Research Laboratory of Electronics, Massachusetts Institute of Technology, and the Eaton Peabody Laboratory, Massachusetts Eye and Ear Infirmary.

REFERENCES

1. Ramon y Cajal, Recollections of My Life, Philadelphia, American Philosophical Society, 1937.
2. Granit, R., Receptors and Sensory Perception, New Haven, Yale University Press, 1955.
3. Bullock, T. H., Neuron doctrine and electrophysiology, *Science*, 1959, **129**, 997.
4. Rosenblith, W. A., Some quantifiable aspects of the electrical activity of the nervous system, *Rev. Mod. Physics*, 1959, **31**, 532.
5. Adrian, E. D., The Basis of Sensation. The Action of Sense Organs, London, Christophers, 1928.
6. Gerstein, G. L., Analysis of firing patterns in single neurons, *Science*, 1960, **131**, 1811.
7. Mason, S., and Zimmermann, H., Electronic Circuits, Signals, and Systems, New York, John Wiley & Sons, 1960.
8. Davies, P. W., Chamber for microelectrode studies in the cerebral cortex, *Science*, 1956, **124**, 179.
9. Dowben, R. M., and Rose, S. E., A metal filled microelectrode, *Science*, 1953, **118**, 23.
10. Hubel, D., Tungsten microelectrode for recording from single units, *Science*, 1957, **125**, 549.
11. Communications Biophysics Group of the Research Laboratory of Electronics, and W. M. Siebert, Processing Neuroelectric Data, Technology Press Research Monograph and Technical Report 351, Cambridge, Research Laboratory of Electronics, Massachusetts Institute of Technology, 1959.

SOME QUANTITATIVE METHODS FOR THE STUDY OF SPONTANEOUS ACTIVITY OF SINGLE NEURONS

R. W. RODIECK, N. Y.-S. KIANG, *and* G. L. GERSTEIN

From the Center for Communication Sciences, Research Laboratory of Electronics, Massachusetts Institute of Technology, Cambridge, and the Eaton Peabody Laboratory of Auditory Physiology, Massachusetts Eye and Ear Infirmary, Boston. Dr. Rodieck's present address is Department of Physiology, University of Sydney, New South Wales, Australia

ABSTRACT Four different illustrative examples of single unit data from the cochlear nucleus of anesthetized cats are presented. The spontaneous activity of each of these four units can be quantitatively described by the histogram of interspike intervals and by other related measurements. Several descriptive models are suggested by these measurements.

INTRODUCTION

Spontaneous electrical activity, the activity that is observed in the absence of experimentally introduced stimui, has long been of interest to students of the nervous system. Various widely divergent opinions have been expressed concerning the functional significance of such activity (Lowenstein and Sand, 1940), (Bremer, 1953), (Granit, 1955), (Kuffler, FitzHugh, and Barlow, 1957), (Schlag, 1959). These views range all the way from depicting spontaneous activity as merely "noise" in the nervous system, to considering it as a carrier that is modulated to transmit information.

Regardless of one's speculations about the function of spontaneous activity, one may study it in a completely operational way by describing its statistical characteristics. In this paper we shall present some systematic procedures that may be used in the study of spontaneous spike activity of single units. One of the particular measurements, the interval histogram, has been used previously to describe the statistical properties of a sequence of spontaneous nerve impulses recorded from isolated axons in specific chemical environments or from relatively intact preparations. (Brink, Bronk, and Larabee, 1946), (Hagiwara, 1954), (Kuffler, FitzHugh, and Barlow, 1957). The present paper can be regarded as an extension of these earlier attempts to quantify the statistical nature of neural discharge.

We have chosen the cochlear nucleus as the general site for our study, since it is small, delimited, accessible, relatively peripheral, and contains many types of cells. Relatively detailed anatomical descriptions are available (Ramón y Cajal, 1909), (Lorente de Nó, 1933). In addition, some knowledge exists concerning responses of single units in this region to acoustic stimuli (Galambos and Davis, 1943), (Tasaki and Davis, 1955), (Rose, Galambos, and Hughes, 1959), (Grossman and Viernstein, 1961). It should be emphasized, however, that this paper is not intended to be a survey or catalog of unit spontaneous activity in the cochlear nucleus, but to be a description of certain methods that are applicable to the study of unit spontaneous activity in general, with illustrative examples taken from single units in the cochlear nucleus.

METHODS

Experimental Methods. All of our experiments were performed on cats that were anesthetized intraperitoneally with Dial (75 mg/kg). Most of the experiments lasted for 12 to 18 hours; no supplementary anesthesia was given. The cochlear nucleus on one side was exposed by partially removing the cerebellum. Electrodes of the indium-platinum type as described by Dowben and Rose (1953) and modified by Gesteland, Howland, Lettvin, and Pitts (1959) were inserted in the cochlear nucleus. We have been able to locate the site of the recording electrodes only in a gross way. Most of the units reported here were in either the anterior or posterior ventral cochlear nucleus.

All of the recording was done with the cat in a shielded soundproof chamber. In most of the experiments the round-window responses to acoustic stimuli were checked to establish the "normality" of the preparation. Acoustic stimuli were delivered by a TDH-39 (10 ohm) earphone connected to hollow ear bars of the headholder by a plastic tube. Most of the units reported here responded to appropriate sounds.

The spike discharges of single units were recorded on magnetic tape. Units selected for study had to meet the following criteria:

(*a*) The spikes had to be initially negative in polarity with reference to the headholder.

(*b*) The spikes had to remain stable in rate with small advancements of the electrode.

(*c*) The spikes had to be sufficiently well isolated from other activity so that a level detector would be activated only by the spikes of the unit that was under study.

Computational Methods. The spikes selected by the level discriminator were fed directly to the TX-O digital computer, which had been programmed to perform these analyses:

(a) Interval histogram—a histogram of the distribution of intervals between successive spikes. This analysis is illustrated schematically in Fig. 1a.
(b) Joint interval histogram—a histogram of the joint distribution of two successive interspike intervals. This analysis is illustrated in Fig. 1b.
(c) Scaled interval histogram—a histogram of the intervals between every $2^{m\text{th}}$ spike, where m is an integer. (For $m = 0$ this is identical to the interval histogram). This analysis is illustrated in Fig. 1c for three values of m.

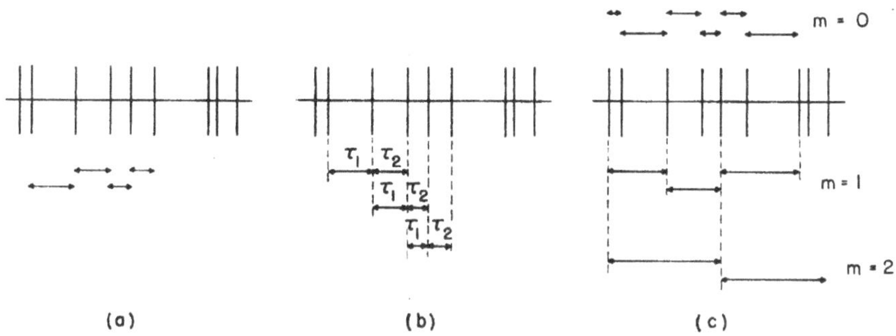

FIGURE 1 Diagram illustrating the computations used in this paper. (a) Interval histogram; (b) Joint interval histogram; (c) Scaled interval histogram of order m.

The results of these analyses appear as displays on the computer-controlled oscilloscope and are photographed. For the ordinary interval and scaled interval histograms the display is in the form of a bar graph (128 bars). The abscissa represents the duration of the time interval; the ordinate represents the number of intervals whose duration falls between τ and $\tau + \Delta\tau$. The $\Delta\tau$ is called the bin width and determines the time resolution of the analysis. For convenience in describing the shape of histograms, we shall frequently use such terms as "rise" and "decay" to describe the envelopes. This terminology in no way implies a time ordering of the events described by the histogram.

Joint interval histograms are displayed as two-dimensional plots: the abscissa represents duration of τ_1, the first of the pair of intervals; the ordinate represents duration of τ_2, the second of the pair of intervals. The number of interval pairs which has occurred with particular (τ_1, τ_2) values is represented by the intensity of the spot at coordinates (τ_1, τ_2).

The analyses described above can be explained more precisely and formally. This has already been done for interval histograms (Gerstein and Kiang, 1960). We shall use similar notation here.

Let the spike train be represented by

$$f(t) = \sum_{k=0}^{N} \delta(t - t_k), \tag{1}$$

where $\delta(t)$ is a Dirac delta function, and N is a large number. Each δ-function (of unit area) is located at the time of a spike in the original train. The interval histogram (of the intervals between successive spikes) may then be written

$$I(\tau) = \sum_{k=1}^{N} \delta(t_k - t_{k-1} - \tau). \tag{2}$$

It is often convenient to measure the continuous values of τ as discrete time values τ_j, $j = 1, 2, 3, \ldots$, in such a way that all τ that satisfy $\tau_j \leq \tau < \tau_{j+1}$ are measures τ_j. Thus $\tau_{j+1} - \tau_j = \tau$ defines a time bin width that determines the resolution of the analysis. For discrete time, the areas of all δ-functions of equation (2) which fall between τ_j and τ_{j+1} are added together to give the height of the τ_jth bar in a histogram.

Although the shape of the interval histogram does provide a method of characterizing some features of spike trains, as shown below, all of the information about the time ordering of particular intervals is necessarily lost.

A particular aspect of such time ordering is the relationship between successive time intervals (Gerstein, 1961). This may be investigated by computing the estimator of the joint probability density for successive time intervals, that is, the joint interval histogram. Formally, we write

$$I(\tau_1, \tau_2) = \sum_{k=2}^{N} \delta(t_{k-1} - t_{k-2} - \tau_1)\, \delta(t_k - t_{k-1} - \tau_2). \tag{3}$$

Here, $I(\tau_1, \tau_2)$ is related to the probability of finding an interval τ_1 followed by an interval τ_2. In a formal sense, the restriction to successive time intervals is arbitrary. We would, however, expect correlation between two intervals only if they are close together in time. A joint interval histogram for two intervals that are relatively far apart in time should show complete independence.

Note that the joint interval histogram can be simply related to the ordinary interval histogram in the following manner:

$$\sum_{\text{all } \tau_2} \frac{1}{N^2} I(\tau_1, \tau_2) = \frac{1}{N} I(\tau_1), \tag{4}$$

in which we have considered the τ to be discrete. This process corresponds to summation of columns in the display (Gerstein, 1961).

Scaled interval histograms (Halliday, 1955) for the intervals between every $2^{m\text{th}}$ spike can be written in a way that resembles ordinary interval histograms:

$$I_2(\tau) = \sum_{k=1}^{N} \delta(t_{2k} - t_{2(k-1)} - \tau) \qquad \text{for} \quad m = 1 \tag{5}$$

$$I_4(\tau) = \sum_{k=1}^{N} \delta(t_{4k} - t_{4(k-1)} - \tau) \qquad \text{for} \quad m = 2 \tag{6}$$

$$\vdots \qquad\qquad \vdots$$

$$I_{2^m}(\tau) = \sum_{k=1}^{N} \delta(t_{2^m k} - t_{2^m(k-1)} - \tau) \qquad \text{for any integer} \quad m \tag{7}$$

The τ may, again, be taken as discrete.

We note that both the ordinary and scaled interval histograms may be considered as terms in a particular expansion of the autocorrelation function of the spike train (Gerstein and Kiang, 1960). There are other, more useful, ways of examining the meaning of scaled interval histograms; these will be shown in a subsequent paper.

The computations that we have discussed can always be obtained for any data but become ambiguous in interpretation unless the data are statistically stationary. We use the term stationary when statistical properties of a short sample of data are independent of the particular choice of sample within a long run. This usage does not exclude short-term variations in statistical properties over time intervals comparable to interstimulus intervals. A simple test of stationarity is to show that the chosen computations do not give widely different results for various samples of data that may be selected from a long run. Fig. 2 shows such a demonstration for interval histograms of one of our units. Since our basic computation is the interval histogram, we have applied this test to each unit that has been studied. The few units that failed to pass this relatively crude test for runs as long as 30 minutes were discarded.

RESULTS AND DISCUSSION

Spike Trains, the Basic Data. Short samples of the data that we shall examine are illustrated in Fig. 3. Each spike train was obtained by isolating a different unit. From these photographs one can observe gross differences in the spike patterns. Thus Unit 261–1 seems to fire in bursts,[1] Unit R–4–10 seems to fire rapidly, almost regularly, with no very long intervals between spikes, etc. It is clear, however, that casual visual inspection will not provide adequate descriptions of the spike trains.

The Interval Histograms. The first computation applied to the spike trains is the interspike interval histogram. Fig. 4 shows such histograms for the four spike trains from which the samples in Fig. 3 were selected. A number of interesting features may be discerned from the shape of these histograms. The histogram for Unit 259–2 appears to be unimodal and asymmetric, with a sharp rise and a slower decay; that of Unit R–4–10 appears to be unimodal, symmetric, and possibly bell-shaped; that of Unit 261–1 appears to be bimodal and asymmetric, while that of Unit 240–1 is unimodal and asymmetric, but on a quite different time scale than that of Unit 259–2.

By matching the interval histograms with the corresponding photographs of spike trains (Fig. 3) one can see that Unit R–4–10 is the unit that fires almost regularly, while Unit 261–1 is the "bursty" unit. The spike trains of Unit 259–2 and Unit 240–1 do not appear to be easily characterizable. The interval histograms of these last two units do appear dissimilar; we shall clarify the differences in the next section.

Replotted Interval Histograms. In order to elucidate the differences between the four interval histograms, the results of the computations may be replotted in special ways. When the histogram of Unit 259–2 is replotted on a semilogarithmic scale, all of the points except those for the lowest intervals fall on a straight line;

[1] Whenever one finds units that discharge in bursts, one must suspect possible motion of the electrode relative to the unit as a result of respiratory movements or arterial pulsation. This may be the case for Unit 261-1, although we have no convincing evidence one way or the other.

the histogram is, therefore, exponential (Fig. 5*a*). The fact that the interval histogram rises rapidly to its mode (at 3 msec.), together with the exponential decay, suggests that the process generating the spike train might be a Poisson process with dead time (Alaoglu and Smith, 1938). To verify this interpretation, it is necessary to establish the independence of successive intervals.

When the histogram of Unit R–4–10 is replotted on a semilogarithmic scale,

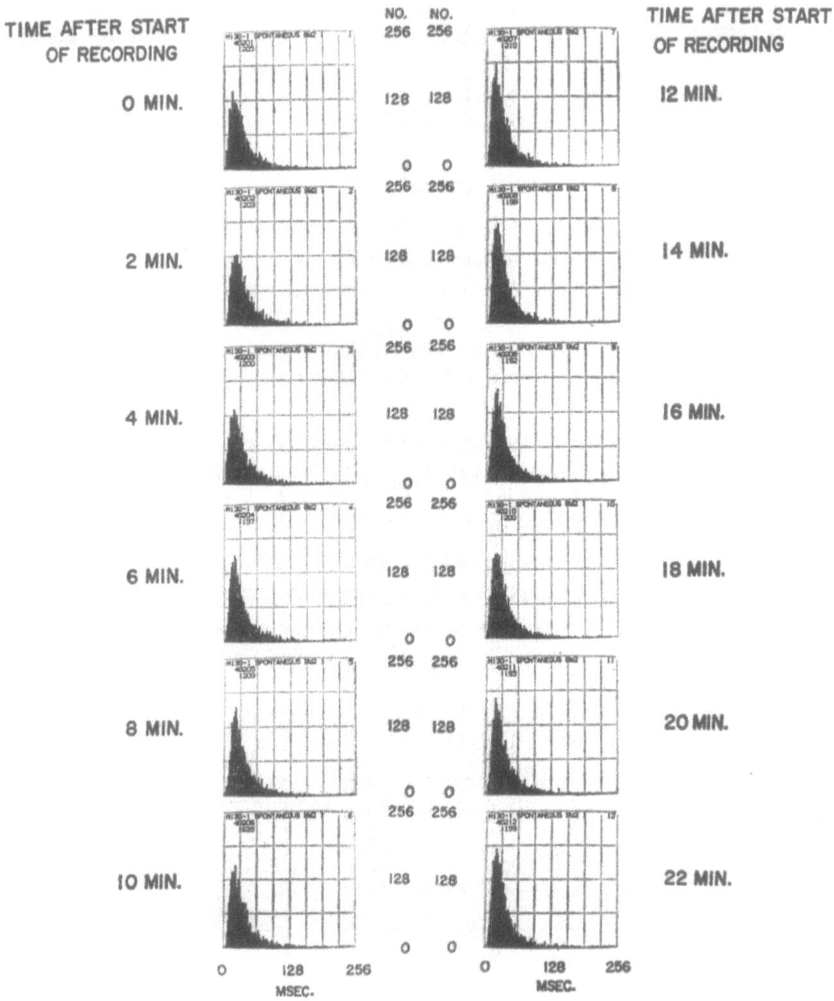

FIGURE 2 Unit 240-1. Interval histograms of successive 2-minute samples from a half-hour length of data.

356 BIOPHYSICAL JOURNAL VOLUME 2 1962

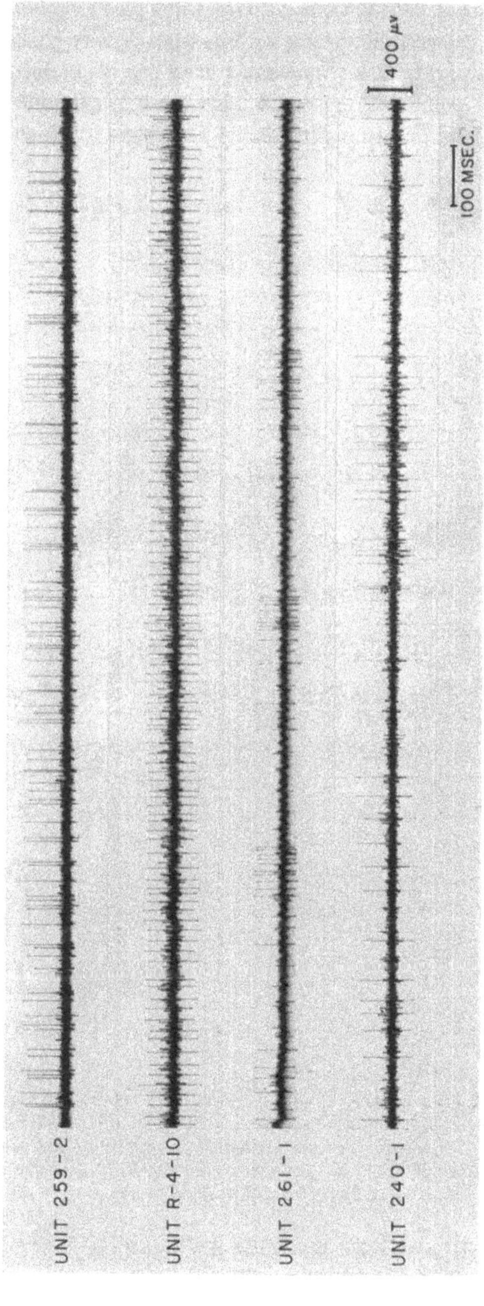

FIGURE 3 Film strips of the spontaneous activity of four selected units.

neither the rise nor the decay shows as a straight-line plot (Fig. 5*b*). The symmetric bell shape, however, suggests replotting on Gaussian density paper. (Onno, 1961). When this is done, as in Fig. 6, a somewhat curved line is obtained. This result suggests that the process generating this spike train is not greatly different from a periodic process with a quasi-Gaussian time jitter.[2] The extent to which successive values

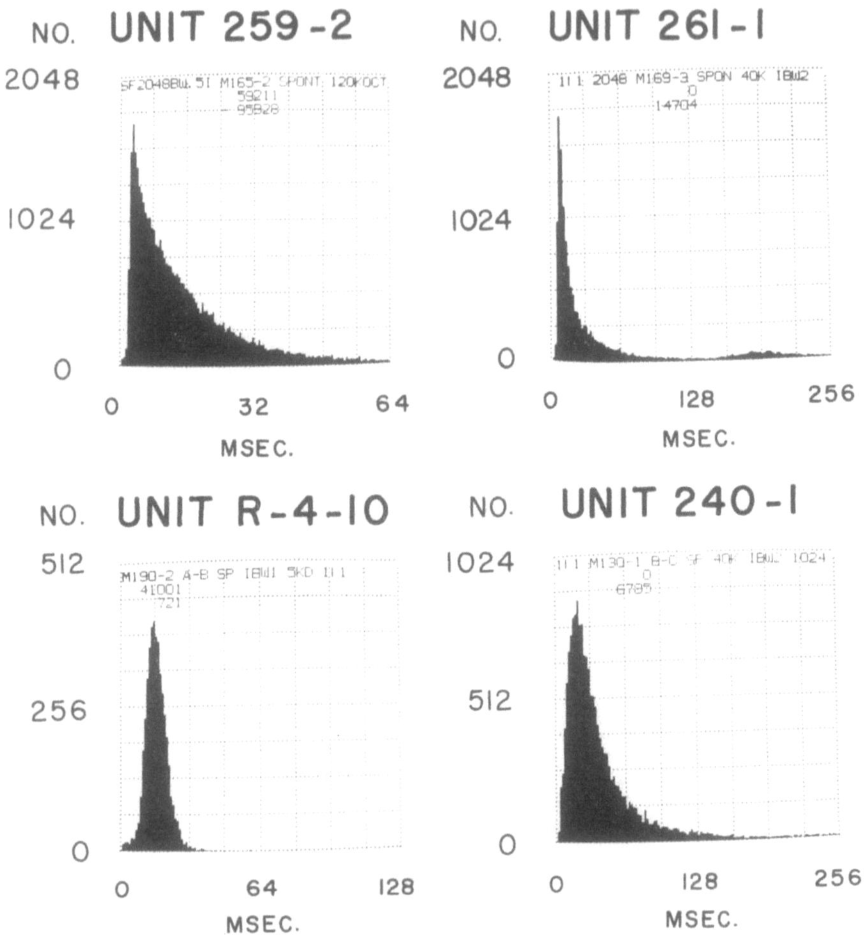

FIGURE 4 Interval histograms for the spontaneous activity of the four units shown in Fig. 3. The number of intervals processed: Unit 259-2, $N = 40{,}960$; Unit R-4-10, $N = 5000$; Unit 261-1, $N = 16{,}384$; Unit 240-1, $N = 16{,}384$.

[2] We shall use "quasi-Gaussian" for a distribution that near its mean is adequately fitted by a Gaussian distribution, but that deviates in the tails.

of the time jitter are statistically independent can be examined by computing the correlation of successive intervals.

Replotting of the histogram of Unit 261–1 on a semilogarithmic scale emphasizes its bimodal aspect (Fig. 5*c*). Such a histogram cannot be the resultant of two steady but different rate processes as, for example, if the activity of more than one unit had been included in the computation. Clearly, the first mode of the histogram at 8 msec. corresponds to the interspike intervals within the bursts, while the second mode at

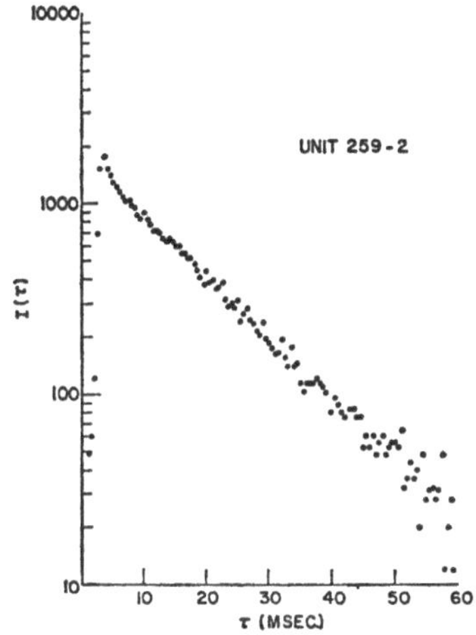

FIGURE 5(*a*)

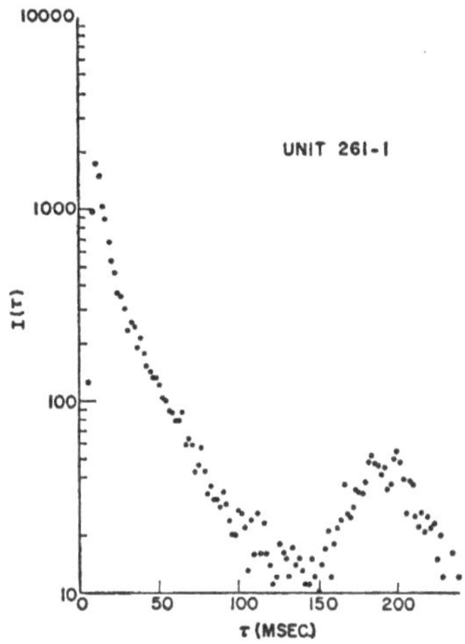

FIGURE 5(*c*)

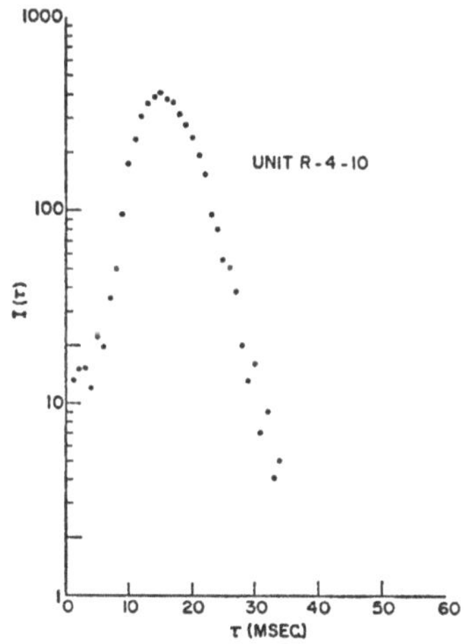

FIGURE 5(*b*)

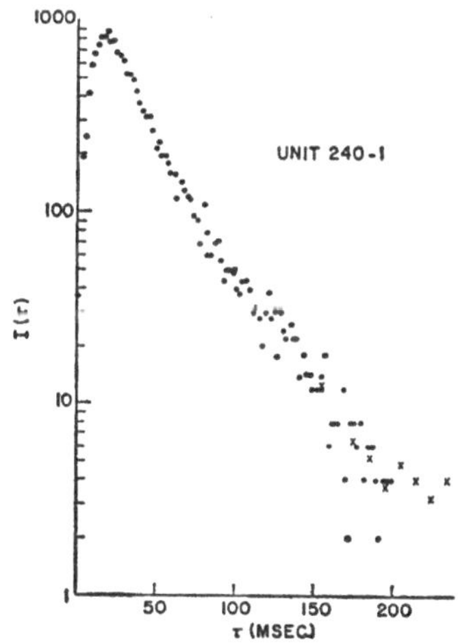

FIGURE 5(*d*)

FIGURE 5 Semilogarithmic replots of the interval histograms of Fig. 4.

approximately 200 msec. corresponds to the interburst intervals. It should be emphasized that the interval histograms cannot be used alone to distinguish between a recurring bursty pattern or a sharp change in the average firing rate of the unit at some time during the data sample. Either the pictures of the spike trains or the stationarity check is also required.

Finally, when the histogram of Unit 240–1 is replotted on a semilogarithmic scale, the decay is clearly seen to be non-exponential (Fig. 5*d*). At this time, we can only characterize the histogram of this unit as unimodal, asymmetric, and non-exponential.

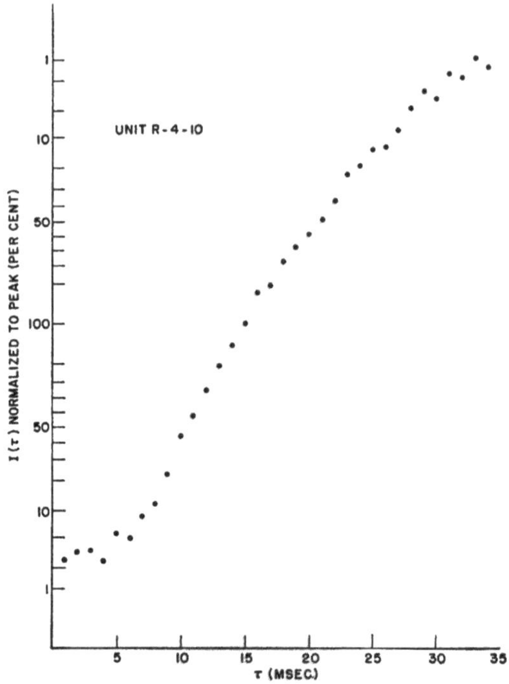

FIGURE 6 The interval histogram of Unit R-4-10 plotted on probability density paper.

Joint Interval Histograms. While the interval histogram gives the range and distribution of time intervals, it does not allow detection of any ordered pattern of intervals.

To study these particular higher-order statistics, we have computed and displayed the joint interval histograms as described in the section on methods. The results of the computations are displayed so that the intensity of each point of the joint histogram represents the number of pairs of intervals which was found with that particular set of values. The joint interval histograms are capable of revealing certain types of recurring time patterns in the spike trains. For example, if there is recurrence of a time interval of short duration followed by one of long duration, the display will show a dark area in the upper left region. If there is recurrence of a long time interval followed by a short time interval, the display will show a dark area in the lower right region. If successive intervals are independent, the joint interval histogram will be symmetric about a 45° line through the origin. This symmetry test is a necessary but not sufficient condition for statistical independence of successive intervals.

The joint interval histogram of the four sample units are shown in Fig. 7. The joint interval histograms of Units 259–2, R–4–10, and 240–1 apparently meet the 45° symmetry test, while the joint interval histogram of Unit 261–1 does not. The joint interval histogram of Unit 261–1 shows dark areas at the upper left and lower right; these show that there is recurrence of a long interval (approximately 200 msec.) preceded by a short interval (8 msec.) and followed by a medium interval (approximately 30 to 60 msec.). There is also a strong tendency for a short interval to be preceded by a short interval and some tendency for a short interval to be preceded by medium or long intervals. These characteristics of the time pattern of the spike train may be easily checked by examining Fig. 3. For a still more systematic study of the spike train, more time-ordering information would be needed. This might be obtained from the order in which a joint interval histogram is filled, or from the joint histogram for non-successive intervals.

For the other three units, a more thorough investigation of statistical independence of sucessive intervals is needed. A definitive proof for statistical independence of successive intervals would be to show that the probability of finding interval 1 followed by interval 2 is just the probability of finding interval 1 multiplied by the probability of finding interval 2, that is,

$$p(1, 2) = p(1)p(2).$$

If, as in most experimental situations, this relationship is not exactly satisfied, it is necessary to estimate the degree of independence. Standard statistical methods can be used (Cramer, 1946) to find the probability that a particular joint distribution be independent. This numerical estimate, however, would not determine the nature of the correlation between successive intervals.

We shall, therefore, substitute another test for independence, which, although less direct, does determine the nature of the correlation. In essence, this test consists of finding the means of each column and each row in the joint interval histogram. Thus we find the mean interval that is preceded (or succeeded) by an interval of a particular given duration. These means are plotted as follows: Column means are plotted *versus* row, and row means are plotted *versus* column. For statistical inde-

pendence, each set of means will fall on a straight line parallel with one of the axes. This test also is necessary, but not sufficient.

This test for the correlation of successive intervals in a joint interval histogram may be developed in the following way. Consider the numbers found in each column (or row) as an ordinary interval histogram. The numbers in a particular (i^{th}) column, properly normalized, represent the probability density of all of the intervals that are preceded by an interval of the particular (i^{th}) duration. Similarly, the normalized numbers in a particular (j^{th}) row represent the probability density of all of the intervals that are succeeded by an interval of the particular (j^{th}) duration.

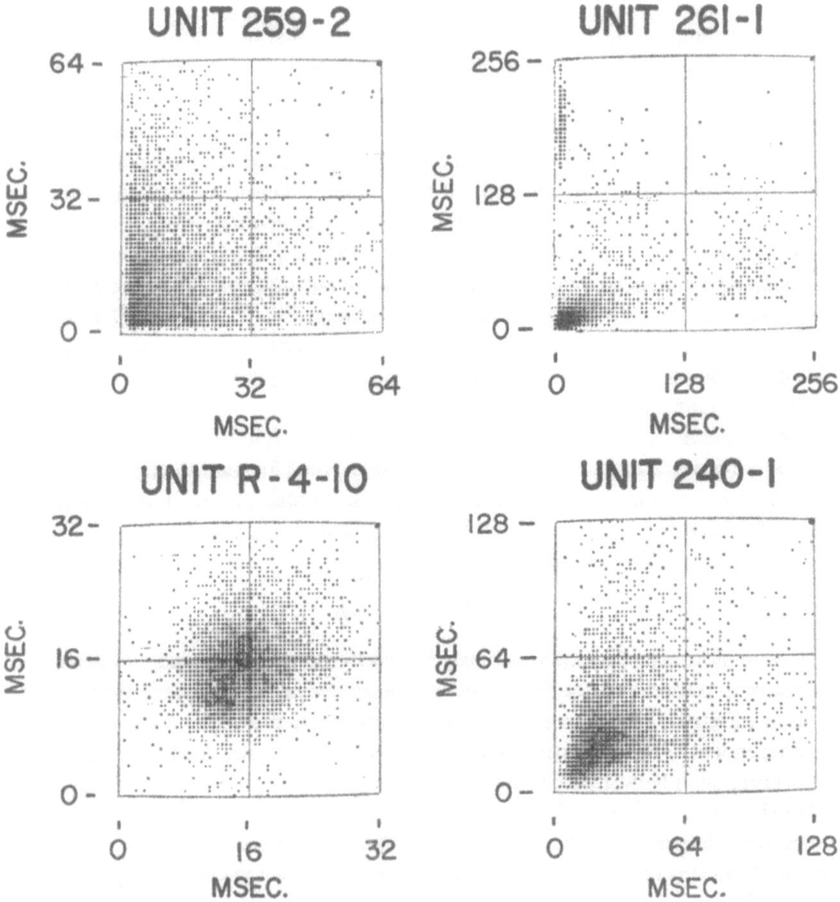

FIGURE 7 Joint interval histograms for the four selected units. Abscissae represent duration of τ_1; ordinates represent duration of τ_2. (See Methods section.) Number of interval pairs processed, $N = 4096$.

In this context, the statistical independence of successive intervals means that the probability density of all of the intervals preceded (or succeeded) by an interval of particular duration does not depend on the length of that particular duration. In other words, with proper normalization, the numbers in every column (or row) should correspond to the same probability density if successive intervals are statistically independent.

Thus our test of independence is reduced to the problem of comparing the probability densities represented by each successive row and column of the joint interval histogram. Two probability densities are the same if they have the same mean, variance, and all higher moments. Thus we may compare the mean, variance, etc. for each successive row and column. Any systematic changes indicate departure from independence and can be directly interpreted.

Constancy of means or variances, etc. represents a necessary condition for independence; sufficiency requires that *all* moments remain constant in this test. In this paper we are discussing applications of this test to only the first moment (mean) of the various probability densities.

Fig. 8 shows the results of such computations for the four joint interval histograms shown in Fig. 7. The row and column means for Unit 259–2 fall on lines parallel to the axes, and thus, support our contention that the discharge pattern of this unit is adequately described by a Poisson process with dead time.

The row and column means for Unit R–4–10 fall on straight lines that are somewhat inclined with respect to the axes. This inclination indicates that there is a weak correlation between sucessive intervals. In the mean short intervals tend to be followed by short intervals, and long intervals by long intervals. Thus the spike train might be generated by a periodic process with quasi-Gaussian time jitter and with some weak correlation between successive intervals.

The row and column means for Unit 261–1 show a distinct difference in their behavior; this is consistent with the asymmetry of the joint interval histogram. Since means of bimodal distributions are extremely poor estimators, we shall not discuss this point in greater detail.

The row and column means for Unit 240–1 fall on a 45° line for intervals less than 30 msec. and on lines almost parallel to the axes for longer intervals. These facts indicate that there is in the mean strong correlation of successive intervals for intervals less than 30 msec., but very little correlation for longer intervals.

This information does not yet permit us to postulate a descriptive model.

Scaled Interval Histograms. With the tests thus far described, a unit such as Unit 240–1 is characterized only by an interval histogram that is unimodal, asymmetric, and non-exponential. Also, Fig. 8 shows that the means of successive intervals are correlated only at small values of interval.

A further test that may have interesting theoretical implications is computation of scaled interval histograms (see Methods). In essence, these are computations of interval histograms for the intervals between every $2^{m\text{th}}$ spike, where m is an integer. In comparing successive orders of scaled interval histograms (successive m), the scales of the ordinates and abscissae are adjusted in the following manner: (a) The vertical scale in N, $N/2$, . . . $N/2^m$. This is necessary because, for a fixed length of total data, each successive order of scaled interval histogram includes only half as many intervals. (*b*) The horizontal scale is T, $2T$, . . . 2^mT for successive orders of scaled interval histograms. These computations for Units 259–2 and 240–1 are shown in Figs. 9 and 10. The behavior of the two units under this set of transformations is qualitatively different. The scaled interval histogram for Unit 259–2 (Poisson-like process) changes in shape considerably, even for the first transformation.

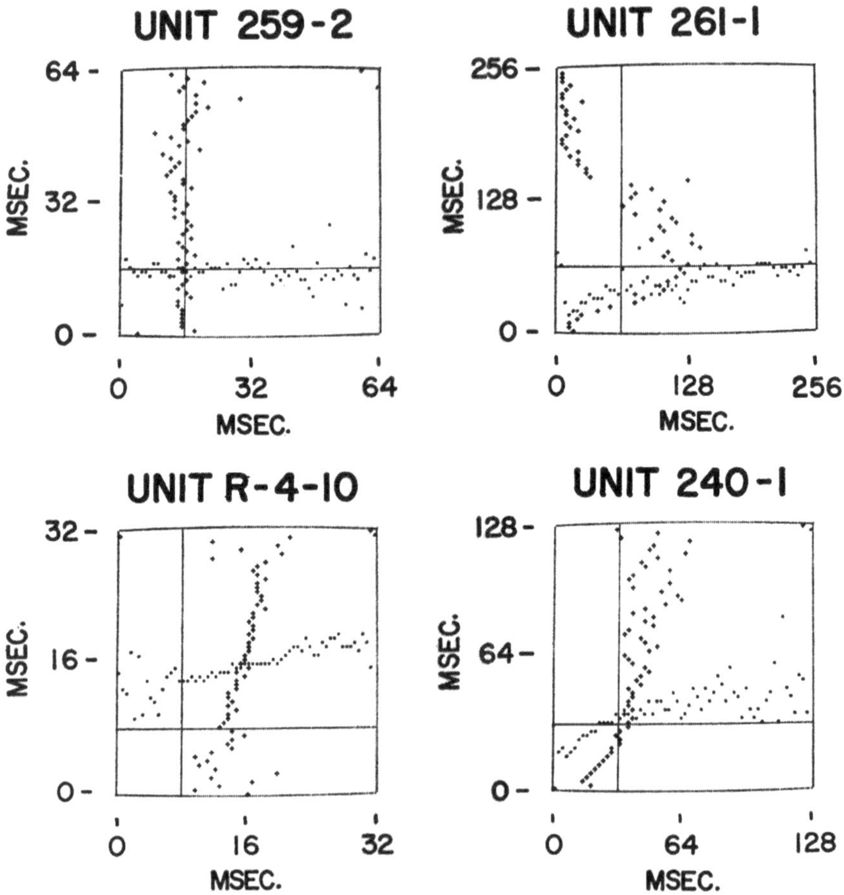

FIGURE 8 Means of each row and column of the joint interval histograms of Fig. 7. Row means (crosses) are plotted against column, column means (dots) are plotted against row.

The scaled interval histogram for Unit 240–1 does *not* change shape appreciably, even for the fourth transformation.

The scaling behavior of Unit 259–2 is consistent with a Poisson-like process (Rainwater and Wu, 1947). The scaling behavior of Unit 240–1 implies a theoretical model that will be described in a subsequent paper.

For Unit R–4–10, scaling is less illuminating, but not inconsistent with the

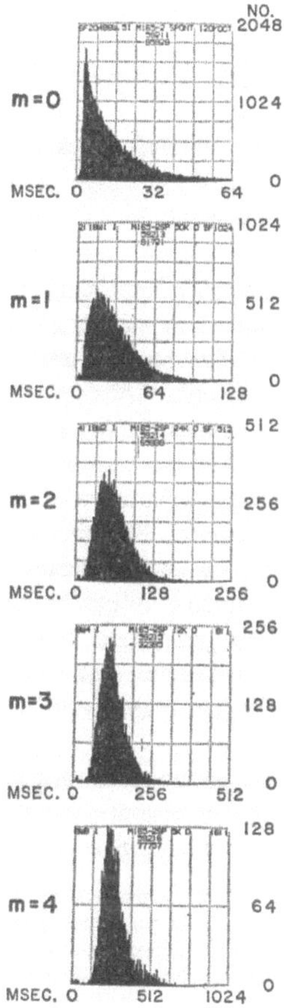

FIGURE 9

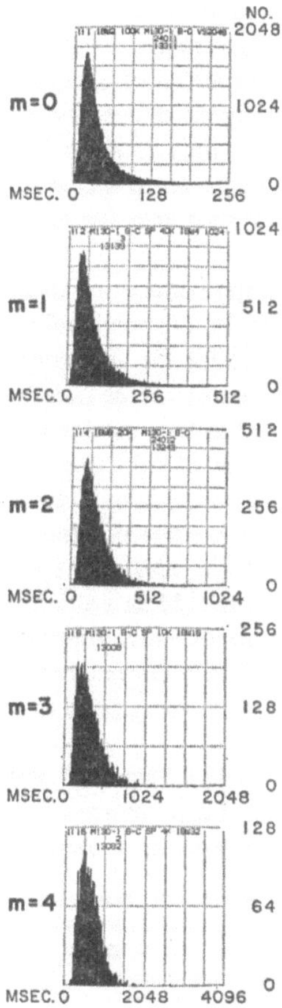

FIGURE 10

FIGURE 9 Scaled interval histograms for Unit 259-2. Note that both vertical and horizontal scales are adjusted by a factor of 2 for each successive order of scaling. $N_0 = 40{,}960$ for $m = 0$; $N_m = N_0/2^m$ for other m.

FIGURE 10 Scaled interval histograms for Unit 240-1. Same scale adjustments as in Fig. 9. Note the approximate invariance of the histogram envelope as m increases.

behavior expected of a periodic process with quasi-Gaussian jitter. As for Unit 261–1, invariance under the scaling transformation is imposible for a bimodal interval histogram.

COMMENTS

We have presented some methods for studying spontaneous discharges of single units. The nature of the data makes any non-statistical description difficult. Several advantageous factors favor statistical study.

(*a*) Large samples of data are easily obtainable.

(*b*) The pattern of spontaneous discharges is usually stationary over the time during which data is obtained.

(*c*) Computers now provide the tools for large-scale statistical studies.

The methods described here seem applicable to the study of single units not only in the auditory system (from which we have drawn our examples) but also in other systems.

Perhaps the patterns of spontaneous activity can provide important clues to the functional roles of individual units. Our examples show how statistical models can be postulated for the spontaneous discharge patterns. Such models may be related to some properties of the functional connections of these cells. Although this paper is not intended to be a catalog of units in the cochlear nucleus, we have attempted to show examples of units that are frequently encountered in this region.[3]

A number of additional aspects of the activity of these units must be investigated before their functional roles can be realistically appraised. The spontaneous discharge patterns must be correlated with responses to systematically presented acoustic stimuli. Most units in the cochlear nucleus appear to exhibit time-locked spike discharges when appropriate acoustic stimuli are presented. Whether or not there are actually discrete categories of spontaneous discharges, and whether or not these categories correspond to different profiles of response patterns to sound is still unknown.

The possible influence of physiological variables on spontaneous discharges must be assessed. More systematic data on the effects of temperature, anesthesia, biochemical manipulation, sleep or wakefulness, "states" of alertness, etc. is needed.

[3] Although a previous paper (Grossman and Viernstein, 1961) reported that 30 out of 31 cells studied in the cochlear nucleus had spontaneous discharges that suggested generation by a Poisson process, our own data, also based on approximately 30 units do not exhibit this overwhelming preponderance. It is interesting that the one exceptional unit in the study by Grossman and Viernstein had a bell-shaped interval histogram. Our own breakdown is approximately: ¼ Poisson, ¼ quasi-Gaussian, several bimodal, several scaling invariant, and ¼ unclassified. In addition many units do not exhibit spontaneous activity and were not included in these numbers. It should be re-emphasized that the meager sampling and inadequate localization of recording site precludes any attempt to deduce the actual numerical distribution of cell types on the basis of these numbers. We have mentioned our proportions only because they differ so markedly from those given by Grossman and Viernstein.

Any or all of these may drastically alter the discharge patterns and, therefore, may need to be carefully controlled, (Brink *et. al.* 1946) (Li, McLennan, and Jasper, 1952) (Li, 1959) (Evarts, 1960).

Finally, the correlation of spontaneous discharge patterns with anatomical knowledge is necessary. It is tempting to inquire whether or not cell types that are distinguishable by histological methods might be correlated with the various types of spontaneous activity.

As the available data on spontaneous activity increase, it is virtually certain that our rather crude analyses will be replaced by more refined techniques. The examples presented here may whet the appetite of theoreticians for more detailed quantitative data.

We wish to thank the technical staffs of the Research Laboratory of Electronics, of the TX-O computer, and of the Eaton Peabody Laboratory of Auditory Physiology, for their generous assistance in this work. We are especially indebted to Professor W. A. Rosenblith for his encouragement and critical interest throughout this study.

This work was supported in part by the United States Army Signal Corps, the Air Force Office of Scientific Research, and the Office of Naval Research; in part by the National Science Foundation (Grant G-16526); in part by the National Institutes of Health (Grant MP-4737), and in part by the National Institute of Neurological Diseases and Blindness (Grant B-1344).

Received for publication, February 26, 1962.

REFERENCES

1. Alaoglu, L., and Smith, N. M., Jr., Statistical theory of a scaling circuit, *Physic. Rev.*, 1938, **53**, 832.
2. Bremer, F., Some Problems in Neurophysiology, London, The Athlone Press, 1953.
3. Brink, F., Bronk, D. W., Larabee, M. G., Chemical excitation of nerve, *Ann. New York Acad. Sc.*, 1946, **47**, 457.
4. Cramer, H., Mathematical Methods of Statistics, Princeton University Press, 1946, 443.
5. Dowben, R. M., and Rose, S. E., A metal filled microelectrode, *Science*, 1953, **118**, 23.
6. Evarts, E. V., Effects of sleep and waking on spontaneous and evoked discharges of single units in visual cortex, *Fed. Proc.*, 1960, **19**, 828.
7. Galambos, R., and Davis, H., The response of single auditory-nerve fibers to acoustic stimulation, *J. Neurophysiol.*, 1943, **6**, 39.
8. Gerstein, G. L., Joint probabilities in firing patterns of a single neuron, Quarterly Progr. Rep. 63, Research Laboratory of Electronics, M.I.T., October 15, 1961, 176.
9. Gerstein, G. L., and Kiang, N. Y.-S., An approach to the quantitative analysis of electrophysiological data from single neurons, *Biophysic. J.*, 1960, **1**, 15.
10. Gesteland, R. C., Howland, B., Lettvin, J. Y., and Pitts, W. H., Comments on microelectrodes, *Proc. IRE*, 1959, **47**, 1856.
11. Granit, R., Receptors and Sensory Perception, New Haven, Yale University Press, 1955.
12. Grossman, R. G., and Viernstein, L. J., Discharge patterns of neurons in cochlear nucleus, *Science*, 1961, **134**, 99.
13. Hagiwara, S., Analysis of interval fluctuation of the sensory nerve impulse, *Japan. J. Physiol.*, 1954, **4**, 234.
14. Halliday, D., Introductory Nuclear Physics, New York, John Wiley and Sons, Inc., 1955, 211.
15. Kuffler, S. W., FitzHugh, R., and Barlow, H. B., Maintained activity in the cat's retina in light and darkness, *J. Gen. Physiol.*, 1957, **40**, 683.

16. LI, C. L., Cortical intracellular potentials and their responses to strychnine, *J. Neurophysiol.*, 1959, **22**, 436.
17. LI, C. L., McLENNAN, H. and JASPER, H., Brain waves and unit discharge in cerebral cortex, *Science*, 1952, **116**, 656.
18. LORENTE DE NÓ, R., Anatomy of the eight nerve. III. General plan of structure of the primary cochlear nuclei, *The Laryngoscope*, 1933, **43**, 327.
19. LOWENSTEIN, O., and SAND, A., The individual and integrated activity of the semicircular canals of the elasmobranch labyrinth, *J. Physiol.*, 1940, **99**, 89.
20. ONNO, PETER, New graph paper for the analysis of Gaussian distributions, *Rev. Scient. Inst.*, 1961, **32**, 1253.
21. RAINWATER, L. J., and WU, C. S., Applications of probability theory to nuclear particle detection, *Nucleonics*, 1947, **1**, 60.
22. RAMÓN Y CAJAL, S., Histologie du système nerveux de l'homme et des vertébrés. I. (1909), Madrid, Instituto Ramón y Cajal, French edition, 1952.
23. ROSE, J. E., GALAMBOS, R., and HUGHES, J. R., Microelectrode studies of cochlear nucleus of the cat, *Bull. Johns Hopkins Hosp.*, 1959, **104**, 211.
24. SCHLAG, J., L'activité spontanée des cellules du système nerveux central, Bruxelles, Editions Arscia S.A., 1959.
25. TASAKI, I., and DAVIS, H., Electric responses of individual nerve elements in cochlear nucleus to sound stimulation, *J. Neurophysiol.*, 1955, **18**, 151.

NEURONAL SPIKE TRAINS AND STOCHASTIC POINT PROCESSES

I. The Single Spike Train

DONALD H. PERKEL, GEORGE L. GERSTEIN, *and* GEORGE P. MOORE

From the Department of Mathematics, The RAND Corporation, Santa Monica, California 90406, the Departments of Biophysics and Physiology, School of Medicine, University of Pennsylvania, Philadelphia, Pennsylvania 19104, and the Departments of Physiology and Engineering and the Brain Research Institute, University of California, Los Angeles, California 90024

ABSTRACT In a growing class of neurophysiological experiments, the train of impulses ("spikes") produced by a nerve cell is subjected to statistical treatment involving the time intervals between spikes. The statistical techniques available for the analysis of single spike trains are described and related to the underlying mathematical theory, that of stochastic point processes, i.e., of stochastic processes whose realizations may be described as series of point events occurring in time, separated by random intervals. For single stationary spike trains, several orders of complexity of statistical treatment are described; the major distinction is that between statistical measures that depend in an essential way on the serial order of interspike intervals and those that are order-independent. The interrelations among the several types of calculations are shown, and an attempt is made to ameliorate the current nomenclatural confusion in this field. Applications, interpretations, and potential difficulties of the statistical techniques are discussed, with special reference to types of spike trains encountered experimentally. Next, the related types of analysis are described for experiments which involve repeated presentations of a brief, isolated stimulus. Finally, the effects of nonstationarity, e.g. long-term changes in firing rate, on the various statistical measures are discussed. Several commonly observed patterns of spike activity are shown to be differentially sensitive to such changes. A companion paper covers the analysis of simultaneously observed spike trains.

INTRODUCTION

Motivation

A neuronal spike train is the sequence of nerve impulses, or action potentials, produced by a neuron, typically observed over a relatively long period of time. The analysis of spike trains has been of increasing interest to neurophysiologists in recent years, stimulated, no doubt, by wide availability of automatic data-processing equipment. Spike-train analysis differs from "classical" electrophysiological methods in that the raw data of interest are not precise voltage measurements, but rather precise measurements of times of occurrence of events. From these essentially temporal data, statistical descriptions are obtained of the output behavior of neurons, from which inferences may be made, in turn, with regard to certain specific types of basic neurophysiological questions. It is not our intention to pursue the questions of interpretation and inference in the present paper (for a recent review, see Moore, Perkel, and Segundo, 1966), but rather to present in systematic fashion the statistical techniques of spike-train analysis, and in particular to point out some of the relevant mathematical assumptions and relationships underlying the computational techniques.

Methodological considerations, however, must not be isolated from inferential and interpretational questions; we have tried to discuss the computational techniques within the larger context of neurophysiological investigation. Our principal working assumptions, as stated in the study mentioned above (Moore, Perkel, and Segundo, 1966), are "(*a*) that there is an enormous wealth of information about the structure and function of the nervous system which can be derived from careful study of the detailed timings of spike events; (*b*) that analysis of these signals can shed light on mechanisms of spike production within the observed cell, on the presynaptic input to the cell, and on the mechanisms by which the latter is transformed into a postsynaptic output; and (*c*) that observation of multiple units can reveal details of interconnections and functional interactions . . . [and] the appreciation that neuronal processes at all levels involve a probabilistic element which must be adequately incorporated if quantitative hypotheses or models of neuronal functions are to be valid. Finally, it is held that only the more detailed analyses of spike timings are appropriate to any quantitative theory of information processing by the nervous system."

Spike-train analysis is applied at several different levels of interpretation, and it is the level of interpretation that largely dictates the choice and depth of statistical techniques for processing of the data. At one level, the statistical measures of the spike train provide a relatively concise characterization of the output of the neuron, which may be used for description, comparison, and classification of nerve cells. At another level, spike-train statistics of a neuron may afford insight into the internal mechanisms underlying spike production; of particular importance here is the comparison of models of spike-production mechanisms, which is typically effected through comparison of the corresponding spike-train statistics. In *inter*neuronal analysis, simultaneously recorded spike trains are compared statistically to reveal information concerning possible connections between neurons, shared sources of activity, responses to stimuli, and synaptic input-output relations. It is in the comparison of neuronal spike trains, however, that we believe the greatest promise of these techniques to lie, despite the greater incursion of mathematical difficulties when compared with single-train analysis.

In summarizing currently used techniques of spike-train analysis, we are further prompted by our feeling that these computational techniques have lacked adequate theoretical underpinning, which has resulted in (*a*) inconsistency of nomenclature and notation in this field, (*b*) difficulties, not always well enough appreciated, in assigning measures of statistical significance to experimental findings, (*c*) presentations of experimental data in several forms that are in fact mathematically derivable from each other, and (*d*) the risk of attributing particular physiological significance to results that illustrate purely mathematical theorems or that are more plausibly attributable to chance effects. With the advent of the high-speed electronic digital computer, it has become feasible to perform on a routine basis the lengthy computations required for spike-train analysis; it is our feeling that in relating these computations to the underlying mathematical, i.e. probabilistic and statistical, theory, the potential usefulness of these techniques will be enhanced.

Basic Terminology and Scope

In every instance in which a detailed examination is made of the timing of neuronal events, we are forced to realize that a certain degree of unpredictability or randomness is present in the underlying process. For some purposes we can afford to overlook this aspect of the record, but for other purposes, and in particular those we focus on here, the question of interest, and indeed the source of greatest information about the process being observed, is the variability and randomness of the spike train.

This very property forces us to describe the spike train in statistical terms and to view the processes underlying it either as inherently probabilistic or as sufficiently complex that we can best and most simply treat them in probabilistic terms. Processes of this type are commonly referred to as *stochastic processes*. Indeed, the transfer and processing of information in nervous systems may be viewed as a repeated alternation, in time and space, of two different types of stochastic processes. The first type is characteristic of the continuous *intraneuronal* fluctuations in significant state variables of each neuron. Typically, a state variable might be the membrane potential as observed at the primary spike-initiating locus, and a model of the underlying process might be described as a complicated type of *random walk*, with continuous time and "displacement" variables. The second type of stochastic process, which is the primary concern of this paper, arises in the study of the times of occurrence of *interneuronally* transmitted action potentials, i.e. the spike train, as commonly observed with either extracellular or intracellular microelectrodes. Because of our "all-or-none" conception of the nerve impulse, each spike is regarded as indistinguishable from the others produced by the same neuron. Furthermore, with each spike can be associated a unique instant of time, e.g. the time of maximum excursion of electrical potential, which can be measured with a high degree of precision. By virtue of the assumed indistinguishability[1] and instantaneity of the

[1] I.e., they are distinguishable only by where they occur in time.

individual spike events, the stochastic process characterizing the spike train can be considered an example of a *stochastic point process*. This process occurs in one dimension, corresponding to the time axis.

In any point process, in which all "events" (spikes, for example) are indistinguishable except for their times of occurrence, it is the elapsed times between events, e.g. the interspike intervals, that exhibit the properties of random variables. These intervals are regarded as being drawn (not necessarily independently) from an underlying probability distribution; if that distribution, together with its parameters, does not vary with time of observation, the stochastic point process is *stationary*.[2] A sample from a stationary point process might be approximated, for example, by a spike train from a spontaneously firing neuron, or from a neuron well adapted to a steady stimulus. A monotonic trend in the firing rate or other parameter is one of the many features of a spike train that may preclude its characterization as stationary.

In most of this paper, spike trains will be considered as realizations of stationary point processes (except for local effects due to stimulation). We cover first the analysis of the spike train produced by a single neuron exhibiting spontaneous or "steady-state" activity. After a presentation of basic concepts and nomenclature, we describe the statistical measures that are independent of the serial order of intervals. We then describe order-dependent statistical measures. Then we discuss some problems of description and interpretation encountered with single spike trains. Next we consider the effects on a spike train of isolated, repeated presentations of a stimulus. Finally, we describe the effects of nonstationarity on the statistical measures described for stationary processes. The presentation is illustrated with examples drawn largely from digital computer simulations (Perkel, 1965).

A companion paper covers the analysis of simultaneously observed spike trains (Perkel, Gerstein, and Moore, 1967).

THE SINGLE SPIKE TRAIN

Stochastic Point Processes: Basic Nomenclature

A stochastic point process, as mentioned above, is a stochastic process "whose realizations consist of a series of point events" (Cox and Miller, 1965). The point events are considered to be instantaneous and indistinguishable (except for position in time); for neuronal spikes, therefore, we consider, for example, the time corresponding to the maximum of the observed action potential to be the time of occurrence, and we ignore all other characteristics of the spike, such as duration, amplitude, undershoot, etc.

[2] Strictly speaking, stationarity is defined in terms of the invariance under translation in time of the joint distribution of numbers of events in fixed intervals of time (Cox and Miller, 1965, pp. 339–340; Cox and Lewis, 1966, p. 59). An equivalent definition in terms of interval distributions is difficult to formulate rigorously, mainly because of the complications introduced by the choice of the starting point for describing the process.

In a *stationary* point process, the underlying probability distributions governing the times of occurrence of the point events do not vary with respect to an arbitrary translation of the time axis.[3] Therefore, accelerations and decelerations in firing rates and effects such as fatigue and adaptation disqualify spike trains from acceptance as realizations of stationary point processes. We consider the detection of nonstationarity and its effects, if present, in the final section of this paper. For a spike train observed in the absence of repetitive stimulation, the assumption of stationarity means, in a practical sense, a neuron that does not display any apparent trend in firing rate and whose "mode" of firing does not exhibit any significant shift from one portion of the record to another.

One important class of stationary point processes, known as *renewal processes*, has the property that the lengths of intervals between events are statistically independent.[4] Neuronal spike trains rarely satisfy this requirement completely; even those spike trains that can adequately be described as stationary often exhibit serial dependence among interspike intervals.

Many results first established in renewal theory have subsequently been generalized to nonrenewal stationary point processes (McFadden, 1962), and in some cases even to nonstationary processes. The terminology of renewal theory, however, has been retained because of its intuitive appeal, and we use it here. Another set of metaphors more appropriate to neuron firings could easily be substituted.

The Poisson, the Erlang, and the Weibull processes are some of the most commonly encountered renewal processes with particularly simple properties. These are discussed amply in the literature (Cox, 1962) and have been applied to the description of certain classes of neuronal spike trains. Most spike trains with independent interval lengths, however, do not fall into any of these mathematically attractive classes.

In the following sections we describe some of the most important and useful statistical measures of spike-train properties.

Order-Independent Statistical Measures

For both renewal and nonrenewal stationary point processes, the (marginal) distribution of intervals between successive events is of paramount importance in characterizing the process. For a renewal process, in fact, the distribution of intervals completely characterizes the process. For finite samples of data, such as an observed neuronal spike train, the interspike-interval histogram serves as an estimator of the "actual" probability density function (pdf).

To construct it, the range of observed interval lengths is customarily divided into

[3] A more explicit definition is given by Cox and Lewis (1966), p. 59.

[4] A renewal process is stationary only if observation begins at a random instant in time, in which case the interval from that instant to the first event has a different distribution from that of subsequent interevent intervals (Cox, 1962). We will assume this condition when discussing renewal processes.

bins of equal width δ; if the ith observed interspike interval T_i satisfies the inequalities

$$(j - 1)\delta < T_i \leq j\delta, \tag{1}$$

then that interval is placed in bin j of the histogram. The bins are thus numbered 1, 2, $\cdots$, J. Letting N_j designate the number of intervals placed in bin j in an observation of N intervals, i.e., $N + 1$ spikes, then the ratios N_j/N are a smoothed estimate of the pdf $f(\tau)$; i.e., they estimate the corresponding integrals

$$\frac{N_j}{N} \approx \int_{(j-1)\delta}^{j\delta} f(\tau)\, d\tau. \tag{2}$$

This quantity is the probability that the duration of a randomly chosen interval lies between $(j - 1)\delta$ and $j\delta$. The estimator for the average value of the pdf within the bin is given by

$$f_j \equiv N_j/(N\delta). \tag{3}$$

Although N interval measurements are used in estimating the usual population parameters such as mean, variance, etc., it is only for a renewal process that the N observations are independent. Measures of precision assigned to these estimates by the standard formulas may be misleading if the process is not a renewal.

For both renewal and nonrenewal processes, there are several functions completely equivalent mathematically to the pdf $f(\tau)$ or its estimator f_j (Cox, 1962, pp. 2–7). These are illustrated in Fig. 1. One of these, the (cumulative) *distribution function*,

$$F(\tau) = \int_0^\tau f(t)\, dt = \text{prob } (T \leq \tau), \tag{4 a}$$

is estimated by

$$F_j = \sum_{k=1}^{j} f_k \tag{4 b}$$

and, in neurophysiological terms, measures the probability that a neuron will have fired by time τ from the last firing.

The *survivor function*, the complement of $F(\tau)$, or

$$\mathfrak{F}(\tau) = 1 - F(\tau) = \text{prob } (T < \tau), \tag{5}$$

is the probability that the neuron will not have fired by time τ.

A third function, the *hazard function*, measures (in the terminology of renewals) the instantaneous risk of failure of a component known to be of age τ. It is given by

$$\varphi(\tau) = f(\tau)/\mathfrak{F}(\tau) = f(\tau)/[1 - F(\tau)]. \tag{6}$$

In the neurophysiological context, the quantity $\varphi(\tau)\Delta\tau$ is the probability that a neuron will fire during the time interval $\Delta\tau$, given that it has not fired prior to the time τ. This function is called the "age-specific failure rate" by Cox (1962). In the neurophysiological literature, it is denoted as c(x) by Poggio and Viernstein (1964) and called by them the "postimpulse probability;" the same function is called "conditional probability" by Goldberg et al. (1964). It is also denoted as $\lambda(x)$ by McGill (1963), who points out some earlier synonyms for the same function: "IRT/OPS," "conditional density function," and "hazard function."

For a Poisson process the hazard function is a constant (Fig. 1 b, bottom). Other processes display positive or negative "ageing" accordingly as $\varphi(\tau)$ increases or decreases with τ. Interspike-interval distributions from pacemaker neurons, for example, characteristically display positive ageing, i.e. an increased "hazard" of firing as a function of time since the last spike (Fig. 1 a, bottom), whereas long-tailed distributions, such as those obtained for some neurons in the auditory system (Gerstein and Mandelbrot, 1964), display negative ageing (Fig. 1 c, bottom). Note that the estimate of the hazard function loses precision for long intervals (Watson and Leadbetter, 1964 *a*, 1964 *b*).

Summaries of the interspike-interval distribution are furnished by various scalar quantities; estimation of these scalars from finite samples of spike-interval data does not differ in any essential respect from estimation using sample data from any other source. Useful quantities are the mean interval μ, the interval variance σ^2, the standard deviation σ, and the coefficient of variation σ/μ. The mean firing rate ρ is defined as the reciprocal of the mean interval. Standard measures of skewness and kurtosis are often useful for describing and classifying interval distributions.

Order-Dependent Statistical Measures

It is of considerable neurophysiological interest to determine whether or not successive interspike intervals are independent in the statistical sense, i.e., whether or not the spike train can be described as a realization of a renewal process. Cox and Lewis (1966, pp. 164–171) discuss two classes of tests for independence of intervals, one based on sample serial correlation coefficients, and the other based on the spectrum of intervals. These authors state that "this is a difficult problem because the null hypothesis is very broad and the alternatives usually not at all clearly specified, and also because the associated distribution problems are hard." Moreover, if the intervals are not independent, such tests offer little information about the type of dependence, and they have so far found little application in spike-train analysis.

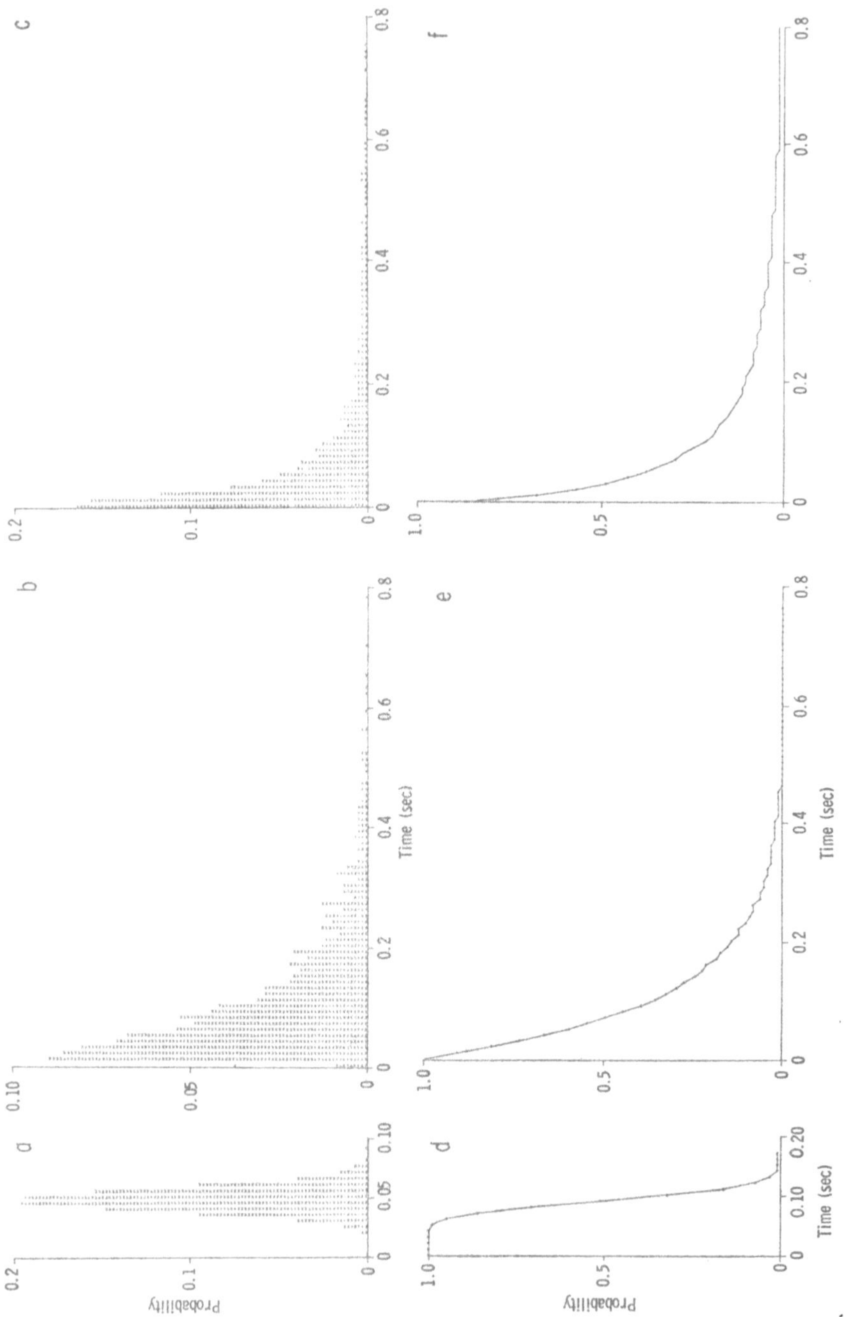

398

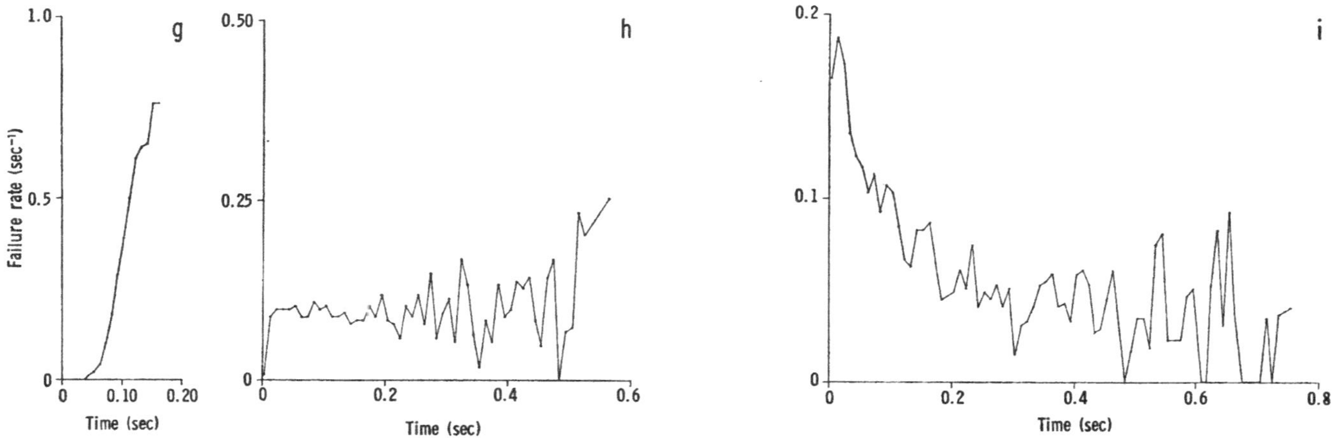

FIGURE 1 Equivalent forms of the interspike-interval distribution. a–c, interspike-interval histogram (estimate of the pdf); d–f, survivor function; g–i, hazard function; a, d, and g drawn from a "pacemaker neuron": independently normally distributed intervals with mean 100 msec and standard deviation 5 msec. b, e, and h from a Poisson process with mean interval 10 msec, conditioned by a normally distributed dead time (refractory period) with mean 10 msec, standard deviation 2 msec. c, f, and i from a log-normal distribution, with pdf $x^{-1}(2\pi\alpha)^{-1/2}\exp\{-[\log(\rho x)]^2/(2\alpha)\}$, with $\rho = 27.2\ \text{sec}^{-1}$ and $\alpha = 2$. Each sample contains approximately 2000 intervals.

Aside from their use in hypothesis testing, certain statistical measures have been used to describe and quantify serial dependence among interspike intervals. We call these measures *order-dependent*, in order to distinguish them from those based solely on the interspike-interval distribution, discussed above. The methods of spectral analysis, recommended by Cox and Lewis (1966; see also Bartlett, 1963), have not been used for analyzing spike-interval data, but their potential utility deserves extensive investigation. Serial correlation coefficients are statistics based on joint distributions of intervals, and another set of measures is based on intervals between nonadjacent spikes. We describe here the latter two classes of techniques.

Use of the *joint interval density* for spike-train analysis was introduced by Rodieck, Kiang and Gerstein (1962). The data are displayed in the form of a scatter diagram, in which the length of an interspike interval is represented by the abscissa, and the length of the next interval in the record by the ordinate. Each point on the diagram then represents a pair of adjacent intervals. An alternate form is the matrix equivalent of Smith and Smith (1965). If successive intervals are independently distributed, then the normalized frequency distribution along the ordinate is the same for each abscissa value, and vice versa. This implies that the corresponding row and column means in the corresponding joint interval histogram have a constant expected value. An observed constancy of row and column means, which is a necessary condition for independence of adjacent intervals, has in practice been used as a sufficient test. Departures from independence are reflected not only in these means but also in the symmetries of the scatter diagram itself. For example, an over-all "upward" trend in the joint-interval scatter diagram (as illustrated in Fig. 7 of Rodieck et al., 1962, units 261-1, R-4-10, and 240-1) indicates positive correlations between successive intervals. This means, loosely, that short intervals tend to be followed by short ones, and long intervals by long ones. A "downward" trend would imply negative serial correlation. The uncorrelated case is illustrated by unit 259-2 in the same figure. The corresponding row and column means are shown in Fig. 8 of the same work.

A quantitative measure of such correlation is furnished by the *serial correlation coefficient* of interval lengths, which is defined as follows: We define the covariance of interval lengths, of lag j, by

$$C_j = E[(T_i - \mu)(T_{i+j} - \mu)](j = \cdots, -1, 0, 1, \cdots), \tag{7}$$

where T_i is the ith interspike interval in an (infinite) stationary train of spikes, with mean interval μ and variance σ^2. Then the serial correlation coefficient ρ_j of order j is the ratio of the corresponding covariance to the interval variance:

$$\rho_j = C_j/\sigma^2. \tag{8}$$

In a finite sample, the mean μ and the variance σ^2 must be estimated from the sample data. To avoid the slight bias introduced by use of sample mean and variance to estimate the corresponding population parameters, more complicated formulas are suggested by Cox and Lewis (1966, pp. 89–92).

The serial correlation coefficient of lag 1 furnishes a single scalar quantity as a summary of the entire diagram of the joint interval distribution of lag 1. Joint interval distributions of larger lags, i.e., for nonadjacent intervals, have not been used in the analysis of spike trains; the corresponding serial correlation coefficients, however, have been used extensively. The set of serial correlation coefficients is usually called the *serial correlogram*; it has sometimes (Hagiwara, 1954) been called the "autocorrelogram" or "autocorrelation," terms that we reserve for a different function (see below).

The expected value of the serial correlation coefficients of all orders (lags) is approximately zero if the intervals arise from a (stationary) renewal process, i.e., if the intervals are drawn independently from a common distribution. For large N, for a renewal process, the quantity $\rho_i/(N-1)^{1/2}$ has a unit normal distribution. However, the distribution of the sample serial correlation coefficient is not known for small samples, and the sample coefficients of various lags are correlated for moderately sized samples (Cox and Lewis, 1966, p. 165). No test of independence of intervals is known which is based jointly on several serial correlation coefficients (P. A. W. Lewis, private communication).

One useful expedient, however, is to subject the sample of interspike intervals to random shuffling, which destroys serial dependence but preserves the order-independent statistics of the sample. Shuffling thus converts the sample to one from the corresponding renewal process. The recomputed serial correlogram for the shuffled train provides a control case, in which the departures from 0 of the serial correlation coefficients are in fact those due to random fluctuations. The net departure may be measured, for example, by the sum of squares of the coefficients. In principle, the shuffling and recomputation can be repeated at length to provide an empirical sampling distribution of the sum of squares under the null hypothesis of serial independence. From this distribution, tests of the independence hypothesis for the unshuffled data can readily be constructed. This kind of procedure is discussed by Cox and Lewis (1966, p. 165) as a permutation test of serial correlation. A refinement discussed there is to replace the observed interval values by ranks or exponential scores. In this way the sampling distribution can be computed once and for all for a given sample size.

The most frequently encountered source of positive contributions to the serial correlation coefficients is a long-term trend in the data; a sufficiently great monotonic increase or decrease in the firing rate over the time of observation will contribute a positive component to each serial correlation coefficient, out to lags of arbitrarily high order. Monotonic trends, of course, are a form of nonstationarity; they are discussed further in a subsequent section.

In a stationary point process that is not a renewal process, the serial correlogram furnishes indications as to the nature of departure from independence among intervals. Local trends in firing rate will introduce positive contributions to the serial correlation coefficients, primarily to the lower lag values (Hagiwara, 1949; Junge and Moore, 1966). Cyclic variations in firing rate, for example, produce a damped oscillation in the serial correlogram, which starts at positive values. Somewhat similar oscillations are produced by fairly regular occurring bursts of spikes, as exemplified by medullary respiratory neurons. Irregular bursts of spikes, such as have been described by Smith and Smith (1965), are characterized by negative serial correlations for low lags, followed by slightly positive and then 0 correlation coefficients. Alternation between long and short intervals, as commonly seen in certain cells in the dorsal column nuclei (Amassian et al., 1962) or in locust wing-muscle motor neurons (Wilson, 1964), gives a strongly negative first serial correlation coefficient with subsequent alternation in sign of the coefficients of higher order. Other types of patterned activity have corresponding signatures in the serial correlogram. Some examples, from computer simulations, are illustrated in Fig. 2.

It must be emphasized that both the joint interval histogram and the serial correlogram must be interpreted with caution and in conjunction with other statistical measures. The positive contribution to low-order correlation coefficients due to local or global trends in firing rate may mask any negative correlation between adjacent intervals that would otherwise be apparent. Segmentation of the data may clarify this situation, but at the expense, of course, of statistical reliability. A gap in the record, which introduces an exceptionally long interval, may grossly affect the interval variance and seriously distort the serial correlogram; this effect is most pronounced in pacemaker neurons (small coefficient of variation of intervals). On the other hand, effects of trends and gaps on the serial correlogram are much less pronounced in data with a large inherent variability (large coefficient of variation), such as a Poisson process with a time-varying rate parameter or neurons with highly irregular firing times. It is important to measure and correct for these distorting effects since significant information about the physiology of the neuron, such as refractory effects, persistence of synaptic effects, etc., may be uncovered through correlational analysis of successive intervals (Firth, 1966; Junge and Moore, 1966; Geisler and Goldberg, 1966).

The use of ranks or exponential scores greatly alleviates the effects of gaps in the record. The use of the estimated spectrum of intervals overcomes some of the effects of trends. The tests for independence of intervals based upon the spectrum of intervals, as described by Cox and Lewis (1966, pp. 67 et seq.), appear to have been neglected by investigators of spike trains. The estimated spectrum gives a single test for independence based on all of the data and thereby overcomes the obstacles mentioned above to the interpretation of the serial correlogram. Computation of the spectrum has been greatly facilitated by a recently devised algorithm (Cooley and Tukey, 1965), which has been incorporated in a set of computer programs by P. A. W. Lewis (1966) for the statistical analysis of series of events.

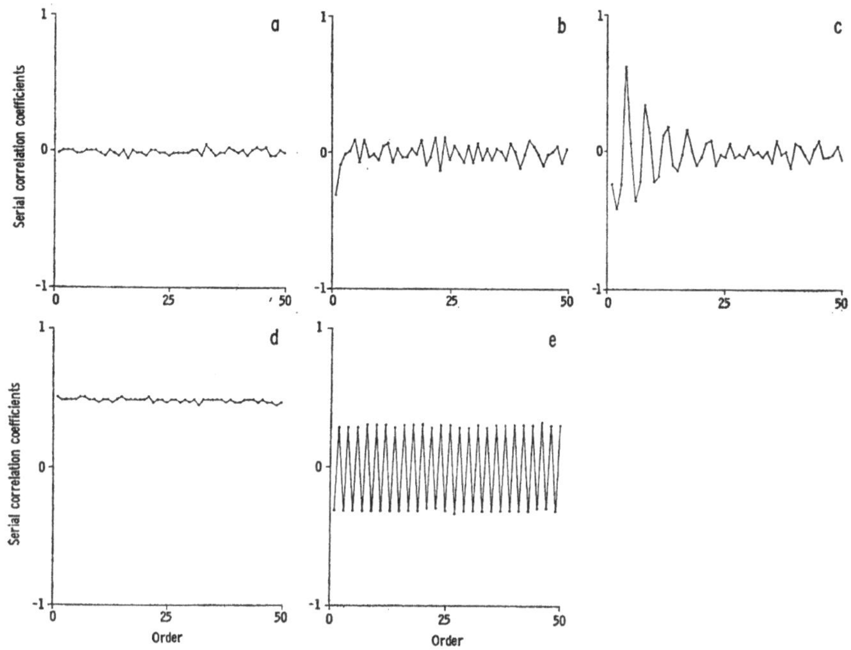

FIGURE 2 Typical serial correlograms. a, independent intervals drawn from a Weibull distribution, with pdf $\alpha\rho(\rho x)^{\alpha-1} \exp -(\rho x)^{\alpha}$; $\rho = 15$ sec^{-1}, $\alpha = 0.6$, 2161 spikes. b, simulated neuron producing irregular bursts. Resting level of membrane potential −70 mv, reset to −100 mv after a spike; exponential recovery with decay constant of 6.93 sec^{-1}. Asymptotic threshold −40 mv, reset to −20 mv after a spike; exponential recovery with decay constant 3.47 sec^{-1}. Two input channels producing 18.5-mv EPSP's: one channel with mean firing rate 2.0/sec, the other with mean firing rate 2.1/sec; standard deviation of intervals 3% of mean. Cell fires with mean interval 2.6 sec, standard deviation 2.9 sec, with sample of 389 spikes. c, simulated neuron producing more regular bursts. Similar to case b, but with standard deviation of intervals in input channels 0.6% of respective means. Cell fires in bursts of four or five spikes; interspike intervals within bursts 0.5–1.0 sec; bursts start at intervals of approximately 10 sec. Sample of 424 spikes has mean interval 2.4 sec, standard deviation 2.8 sec. d, decelerating train. Intervals drawn from a time-dependent normal distribution with $\mu(t) = 0.1 - 0.02e^{-.01t}$ sec, $\sigma = 0.01$ sec. Sample of 2447 spikes. e, alternation between long and short interval lengths. Intervals drawn alternately from distribution with mean 0.11 sec and from distribution with mean 0.09 sec, each with standard deviation 0.015 sec. 5000 spikes.

Joint interval distributions and the corresponding serial correlation coefficients involve time intervals that are defined by two successive spikes. The second class of order-dependent statistical measures that we discuss involves time intervals between nonsuccessive events. Denoting as a first-order interval the elapsed time from an event to the next following event, we may define a second-order interval as the elapsed time between an event and the second following event, etc. An nth-order interval is the sum of n consecutive first-order intervals and is spanned by $(n + 1)$ consecutive spikes (see Fig. 3).

The probability density of the nth-order interval is designated $f_n(\tau)$. The interval density for successive events is thus the first-order density: $f_1(\tau) = f(\tau)$.

In the special case of a renewal process, the higher-order densities may be obtained by successive convolutions of the first-order density because the intervals are independent. Thus, the second-order interval density is given by the convolution integral

$$f_2(\tau) = \int_0^\infty f(t)f(\tau - t)\, dt, \tag{9 a}$$

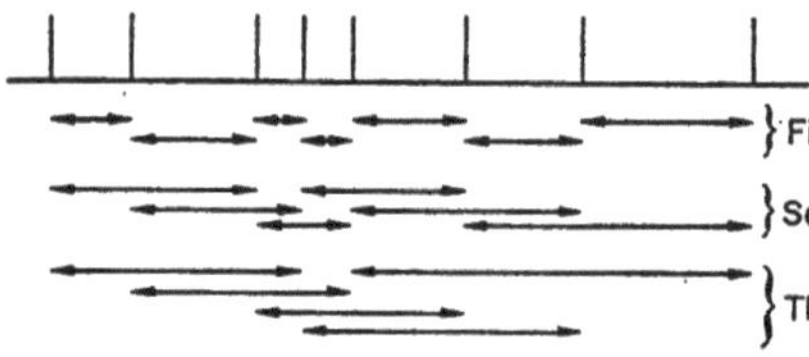

FIGURE 3 Higher-order interspike intervals. A first-order interval is the time difference between adjacent spikes. A second-order interval lies between a spike and the second spike following, etc. Note that an interval of order n spans $n + 1$ spikes. See text.

and in general we have the recursion

$$f_{n+1}(\tau) = \int_0^\infty f_n(t)f(\tau - t)\, dt. \tag{9 b}$$

These equations do not apply when successive interval durations are not independent.

A related function is the *renewal density*, $h(\tau)$, which specifies the probability of encountering any event as a function of time after a given event; i.e.,

$$h(\tau) = \lim_{\Delta\tau \to 0} \text{prob}\,\{\text{an event in } (\tau, \tau + \Delta\tau) \mid \text{an event at } 0\}/\Delta\tau. \tag{10}$$

Since any event encountered must be either the first, second, . . . , etc., event after the event at time 0, it is evident that the renewal density is the sum of the interval densities of all orders:

$$h(\tau) = \sum_{k=1}^{\infty} f_k(\tau). \tag{11}$$

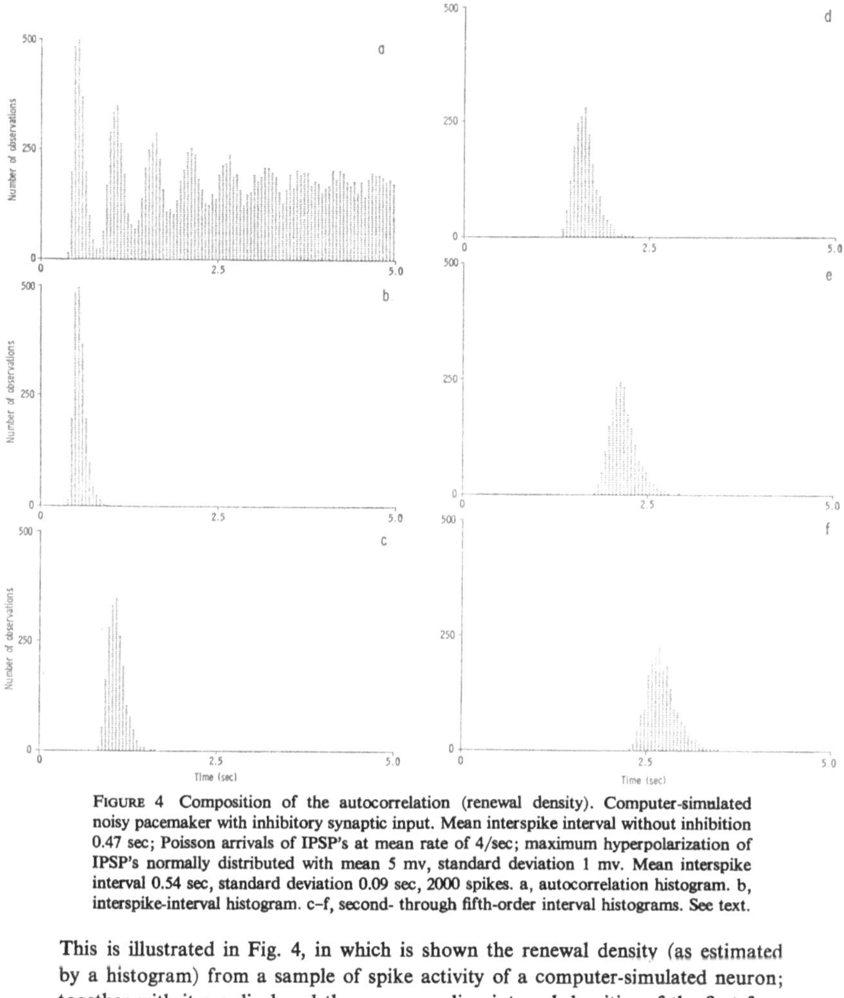

FIGURE 4 Composition of the autocorrelation (renewal density). Computer-simulated noisy pacemaker with inhibitory synaptic input. Mean interspike interval without inhibition 0.47 sec; Poisson arrivals of IPSP's at mean rate of 4/sec; maximum hyperpolarization of IPSP's normally distributed with mean 5 mv, standard deviation 1 mv. Mean interspike interval 0.54 sec, standard deviation 0.09 sec, 2000 spikes. a, autocorrelation histogram. b, interspike-interval histogram. c–f, second- through fifth-order interval histograms. See text.

This is illustrated in Fig. 4, in which is shown the renewal density (as estimated by a histogram) from a sample of spike activity of a computer-simulated neuron; together with it are displayed the corresponding interval densities of the first four orders.

In spike-train analysis, the renewal density is often called the *autocorrelation* (Gerstein and Kiang, 1960), since if the spike train is regarded as a signal of 0 am-

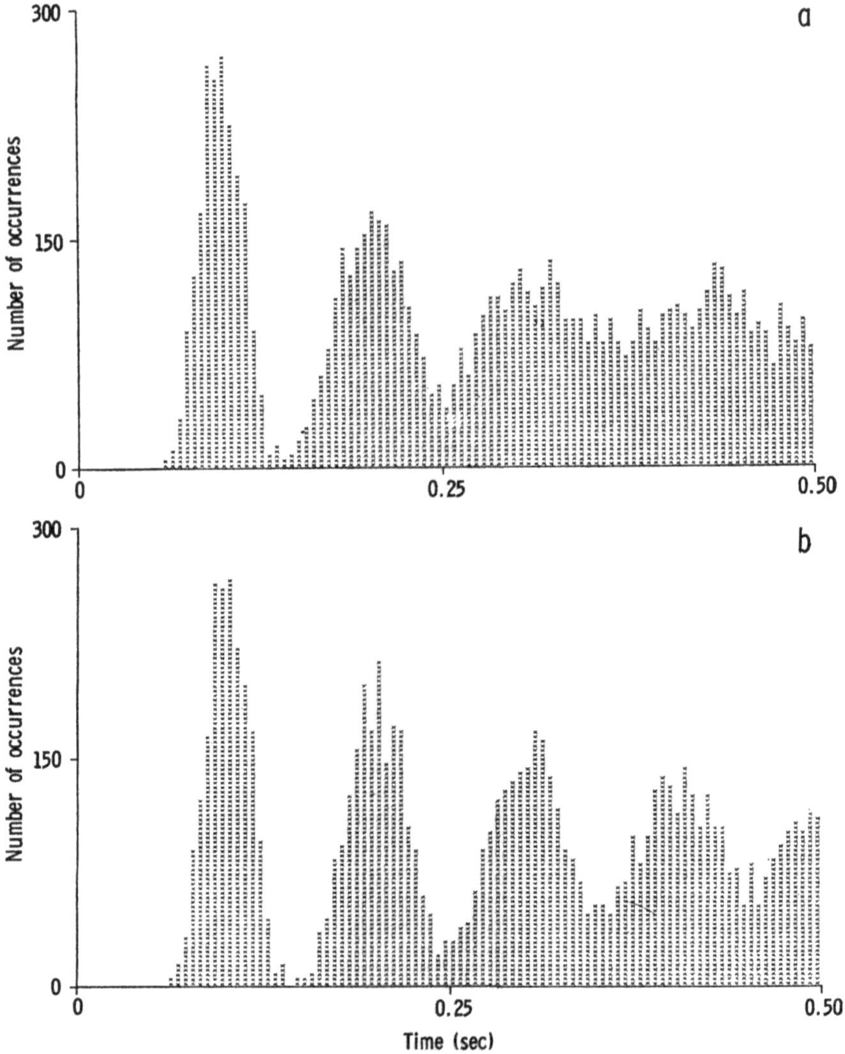

FIGURE 5 Effect of shuffling on peaks in the autocorrelation. Intervals generated by a semi-Markov process with three states with mean intervals as follows: state 1 (short), 0.09 sec;

plitude everywhere except where a spike is present, and if each spike is represented as a Dirac delta function, then the renewal density corresponds to the autocorrelation as ordinarily defined for continuous signals. In the usage of Cox and Lewis (1966), the renewal density is one of a class of "intensity functions." Another synonym that has gained some currency is "expectation density;" it is used by Poggio and Viernstein (1964) and others, following a usage introduced by Huggins (1957). Yet another synonym is "post-firing interval distribution," introduced by Lamarre and Raynauld (1965).

Another property of the autocorrelation is that it "flattens out" to a constant value; i.e.,

$$\lim_{\tau \to \infty} h(\tau) = 1/\mu. \tag{12}$$

In general, this limit is approached slowly for pacemaker neurons (narrow interspike-interval densities), and more rapidly for spike trains with greater variation in interval length. The limiting value is attained for all values of τ for a Poisson process; it can be shown, in fact, that a Poisson process is the renewal process having a constant renewal density.

The representation of higher-order interval densities as corresponding convolutions of the first-order density (Equation 9) holds only for independent intervals. If there is serial dependence (as measured by the serial correlogram), more complicated expressions are required, as given by McFadden (1962). Thus, the sum of the interval densities of various orders (the autocorrelation) for two spike-train sequences having identical interval distributions will be different if in one train the intervals are independent, i.e. a renewal process, and in the other they are dependent. Hence, it is possible to compare the observed autocorrelation with that predicted under the independence hypothesis as a test of that hypothesis. One convenient computer method is that of prolonged random shuffling of the intervals, as discussed above. The autocorrelation of the shuffled train then represents a control case of serial independence. Discrepancies between the unshuffled and shuffled autocorrelation not only furnish a test of serial dependence, but may also indicate the nature of that dependence.

---

state 2 (medium), 0.10 sec; state 3 (long), 0.11 sec. All intervals normally distributed with standard deviation 0.01 sec. Transition matrix

$$\begin{pmatrix} 0.70 & 0.20 & 0.10 \\ 0.45 & 0.10 & 0.45 \\ 0.10 & 0.20 & 0.70 \end{pmatrix}$$

is such that a long interval is most likely to be followed by a long interval, and a short by a short. Mean interval 0.10 sec, standard deviation 0.014 sec, in sample of 2000 spikes; first five serial correlation coefficients: 0.251, 0.184, 0.085, 0.071, 0.040. After shuffling of intervals, first five serial correlation coefficients were −0.018, 0.017, 0.021, −0.041, −0.004. a, autocorrelation histogram, unshuffled data. b, autocorrelation histogram of reconstructed spike train after shuffling of interspike intervals. Note sharpening of peaks. See text.

Shuffling may either enhance or flatten peaks in the autocorrelation. For example, if interspike-interval lengths exhibit negative serial correlation, peaks in the autocorrelation are generally broadened by shuffling. On the other hand, if interval lengths are positively correlated, shuffling of intervals may sharpen peaks in the autocorrelation. An example of the latter effect is shown in Fig. 5, in which the original interval sequence was generated by a three-stage semi-Markov process (Cox and Lewis, 1966, p. 82), according to which relatively long intervals were more likely to be followed by long intervals than by short ones, and vice versa.

To avoid confusion it should be emphasized that the autocorrelation is a function of time; the serial correlogram [Hagiwara's (1954) "autocorrelation"] is a function of the serial position of the interval, an integer. The two functions need not correspond at all. For example, a pacemaker cell that fires at nearly uniform intervals will have a strongly oscillating autocorrelation, whereas the serial correlogram may be positive, negative, oscillatory, or zero.

Long-term trends or slow oscillations in firing rates are ordinarily not obvious in the autocorrelation, but are more typically revealed in the serial correlogram. These and other effects of relatively long-term rate variations are discussed below, in the section dealing with the effects of nonstationarity.

Description, Prediction, and Information

Having presented certain statistical measures that can be applied to individual spike trains, we now introduce briefly some considerations about the adequacy and utility of statistical descriptions of the train. These considerations bear on the use of statistical descriptions both in characterizing and classifying neurons and in comparing observed spike data with those predicted by models.

The simplicity of statistical description of a spike train differs widely from case to case. A single parameter suffices to describe a Poisson process, whereas two are required for an Erlang or Weibull process. If a spike train fits into one of these categories, not only is the characterization of the particular spike train extremely simple, but also the pdf of the interval distribution (and therefore the hazard function, autocorrelation, etc.) may be written as an explicit equation. For less easily described renewal processes, the entire interspike-interval histogram is required to characterize its properties. When successive intervals are not independent, then much more complicated descriptions are needed, except in special cases of highly patterned spike configurations. When higher-order joint interval densities or similarly elaborate statistical measures are required to effect a reasonably complete description, the statistical description itself is unmanageable, and its practical utility is highly questionable. The limit is reached, of course, when the number of parameters in the statistical model equals (or exceeds!) the number of spikes in the sample, at which point statistical analysis loses all justification.

The relationship between the statistical properties of a spike train and the information-handling capability of the neuron is generally complex. Estimates of channel capacity depend strongly upon the particular choice of encoding scheme imputed to the neural structure. Once a choice of encoding scheme has been made, then estimates of channel capacity can be obtained on the basis of interval statistics. For example, if a parameter is thought to be encoded in terms of mean firing rate, then a smaller coefficient of variation of intervals gives rise to a higher channel capacity, etc. Interval statistics, however, cannot of themselves provide a choice of coding scheme. Facile "derivations" of information-handling characteristics of neurons based solely or primarily on spike-interval statistics are usually misleading or worse. For a more extended discussion of these problems, the reader is referred to Moore et al. (1966) and to Segundo et al. (1966).

SINGLE SPIKE TRAINS IN THE PRESENCE OF STIMULATION

In many neurophysiological experiments a controlled series of changes in the physical environment is introduced. We consider here the case of a repeated, relatively short stimulus. In order to detect and evaluate the effect of such a *stimulus train* on the train of spikes, it is now common to compute a post-stimulus-time (PST) histogram (Gerstein and Kiang, 1960). Specifically, the PST histogram shows the probability of firing as a function of time after the stimulus onset. As shown below, this measure is equivalent to a cross correlation between the train of stimulus presentations and the train of spikes. If the stimulus has no effect on the pattern of the spike train, the PST histogram will be flat (subject to the usual statistical fluctuations). On the other hand, if the stimulus does produce a timelocked "evoked response" in the spike-train pattern, the PST histogram will show deviations from flatness. A peak in a PST histogram indicates a higher probability of firing at that particular time after stimulation and can presumably be associated with an excitatory process. Dips in a PST histogram indicate a lower timelocked probability of firing and often are associated with inhibitory or refractory processes.

In interpreting a PST histogram it is obviously necessary to decide on the statistical significance of the observed deviations from flatness. A simple method that has been used (Weiss, 1964) is to compute the mean square deviation of all bins from the mean level of the histogram. Some criterion value for this number can be chosen to distinguish a "flat" PST histogram from one that shows a weak timelocked response pattern.

There is some difficulty in this type of significance test, however, since successive bins in the histogram may not represent independent quantities. For example, since each firing of the neuron is followed by a refractory period, there is always a tendency toward a negative correlation between adjacent bins in the histogram. An empirical control case can be constructed by randomly shuffling the intervals of the spike train and computing the corresponding "flat" PST histogram. The distribution of mean square deviations from mean bin level can be obtained from a set of replications of this procedure, and the criterion value can then be chosen as usual to satisfy a specified error probability. A control case can also be constructed using fictitious times of stimulus presentation in a portion of record where no actual stimulations were presented (Gerstein, 1960; Burns and Smith, 1962).

Because of such correlation effects, it is necessary to verify that features in a PST (or similar) histogram that are suspected of having significance are not simply artifacts of the choice of bin width. Two techniques that may be helpful are (*a*) recomputing the histogram with a different bin width that is not a simple fraction of the original bin width and (*b*) calculating the autocorrelation of the PST histogram. Meaningful features should have a width of several bins. The existence of "wide" features in the PST histogram will be shown by large values near the origin of its autocorrelation function.

Another way of analyzing stimulus effects on a single spike train is to measure the elapsed times or "latencies" between stimulus presentations and the earliest encountered subsequent spikes. Latency and PST studies are discussed in the review paper by Moore et al. (1966).

THE PROBLEM OF NONSTATIONARITY

Basic Concepts

All the statistical measures discussed above, with and without stimulation, carry the implicit assumption that the data are stationary. Specifically, this means that in the absence of stimulation, the spike trains represent realizations of stationary point processes. In the presence of stimulation, it is assumed that the point processes are time dependent, but that the time variation is the same after each stimulus presentation, i.e., that each stimulus presentation represents a new, independent trial. Thus the stimulated cases are "stationary" in a larger sense.

It should be kept in mind that the phrases "stationary data" and "nonstationary data" are, strictly speaking, misnomers. The experimental data are samples, i.e., realizations over finite durations of stochastic point processes; only the (hypothetical) underlying point processes possess the properties of being stationary or nonstationary. In testing data for "stationarity," we are in fact testing whether the assumption of a stationary underlying process is a reasonable one for the body of data in question.

One of the difficulties commonly encountered in neurophysiological investigation is the fact that the behavior of a neuron under study may change significantly during the course of observation, and therefore cannot validly be assumed to arise from a stationary process. Such changes may be exhibited in a gross way, or may be subtle and difficult to detect. The problems of detecting nonstationarity and assessing its effects are vexatious. The very meaning of stationarity depends on the context of the experiment. If, for example, a neuron undergoes a diurnal cycle of activity, a sample of a few minutes' activity may well be accepted as "stationary," whereas a sample of a few hours' activity may show marked trends, and hence be classified as nonstationary. In operational terms, therefore, it may be impossible to distinguish nonstationarity from inadequate sampling.

The most direct, straightforward, and recommended way of dealing with suspected nonstationarity is to segment the data, analyze each segment separately, and apply standard techniques for testing that the several samples were drawn from the same population. In practice, this is often impossible because of an insufficiently long sample.[5] A related technique, which is useful when data processing is accomplished "on line," is to observe the temporal order in which a histogram is built up. For a stationary process, the fractional mean rate of accumulation should be uniform for all portions of the histogram (over a time period that is long with respect to any known periodic variation); any conspicuous systematic departures from this uniformity strongly suggest nonstationarity in the data.

Although the strict definition of stationarity implies that all parameters of the stochastic process are invariant with respect to a random displacement in time, detection and measurement of nonstationarity in spike trains has centered almost exclusively upon one parameter, the firing rate. In the following discussion we restrict ourselves to nonstationarities in firing rates, after pointing out the utility of statistical techniques for investigating nonstationarities in, for example, the variance of intervals, such as Bartlett's test for homogeneity of variance (see Kendall and Stuart, 1961).

There are three principal aspects to the analysis of rate nonstationarities in spike data. The first is the detection of nonstationarity. "We may wish to test the reality of any apparent trends and this is done by testing the hypothesis of no trend" (Cox and Lewis, 1966, p. 37); to this end the reader is referred to a chapter by Cox and Lewis (1966, chapter 3) devoted to the analysis of trends. The second aspect deals with the characterization and measurement of the variations in rate when they have been found to occur. The third aspect refers to the assessment of the effects of rate variations on other statistical measures; we consider below the effects of rate variations on the autocorrelation and the serial correlogram.

Measurement of Rate Variations

In spike trains exhibiting rate variation together with a relatively high degree of variability in interspike intervals, special techniques are necessary to observe the rate variations themselves, unobscured by the "local" fluctuations in interval length. One

[5] Some classes of spike trains, which arise from certain random-walk models of neurons, correspond to renewal processes which do not have finite moments (Gerstein and Mandelbrot, 1964). A renewal process with an infinite mean would correspond to a nerve cell which has a finite probability of remaining silent indefinitely long after a spike. A nonpacemaker cell with wholly inhibitory synaptic input would remain permanently silent; some mixture of inhibitory and excitatory input could result in the cell's firing sporadically, with a long-tailed distribution of interspike intervals, which might not have a finite mean. Finite samples of such a process cannot adequately establish this possibility.

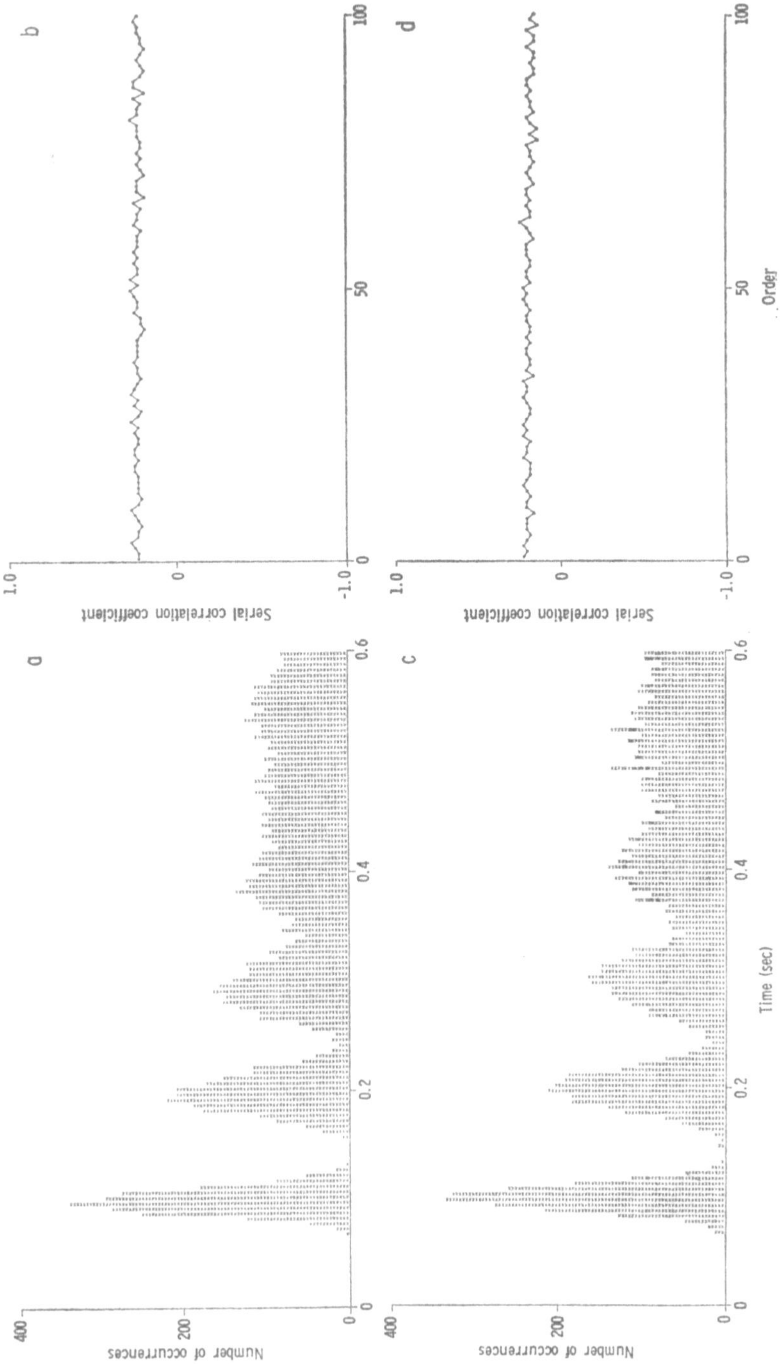
a
b
c
d
Number of occurrences
Serial correlation coefficient
Time (sec)
Order
400
200
0
0.2
0.4
0.6
1.0
-1.0
50
100

412

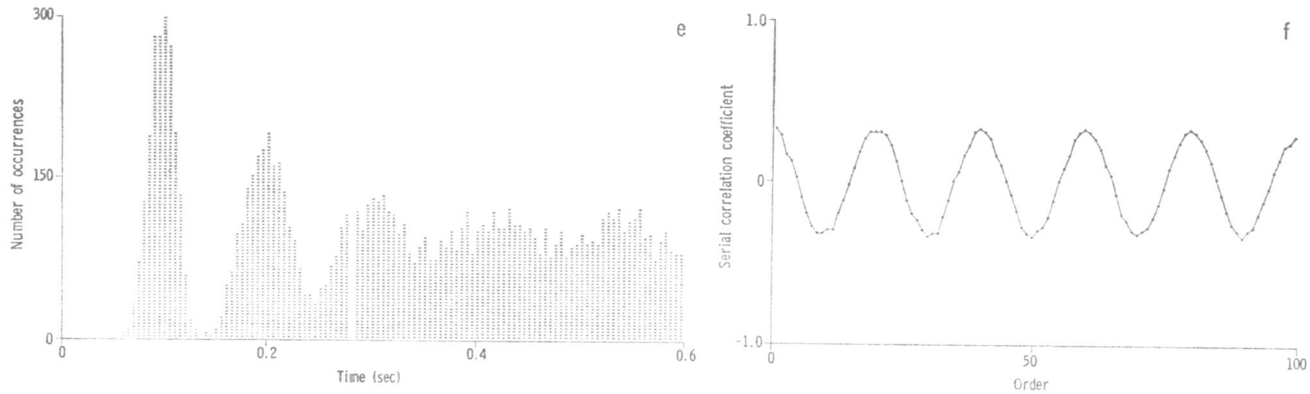

FIGURE 6 Effects of slow rate changes in a pacemaker neuron. On the left, autocorrelation histograms; on the right, serial correlograms of interspike intervals. Intervals are independently normally distributed with time-varying mean, over-all mean interval in sample of 100 msec, and constant standard deviation of 10 msec. All samples are 200 sec long, approximately 2000 spikes each. a–b: *accelerating pacemaker*. Mean interval decreases linearly from 110 msec to 90 msec during observation period. c–d: *decelerating pacemaker*. Mean interval increases linearly from 90 msec to 110 msec during observation period. e–f: *oscillating pacemaker*. Mean interval varies sinusoidally with time, from maximum of 110 msec to minimum of 90 msec, with a 2 sec period.

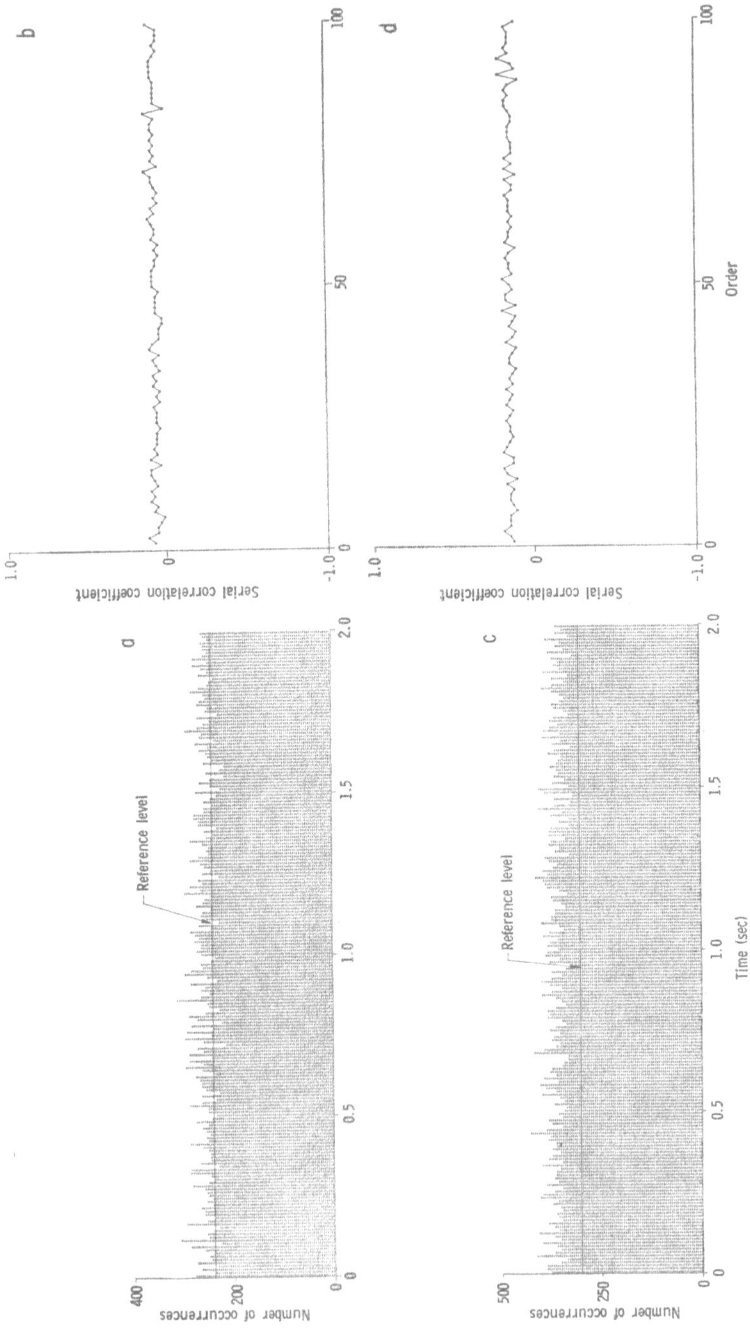
a
b
c
d
Reference level
Reference level
Number of occurrences
Number of occurrences
Serial correlation coefficient
Serial correlation coefficient
Time (sec)
Order

414

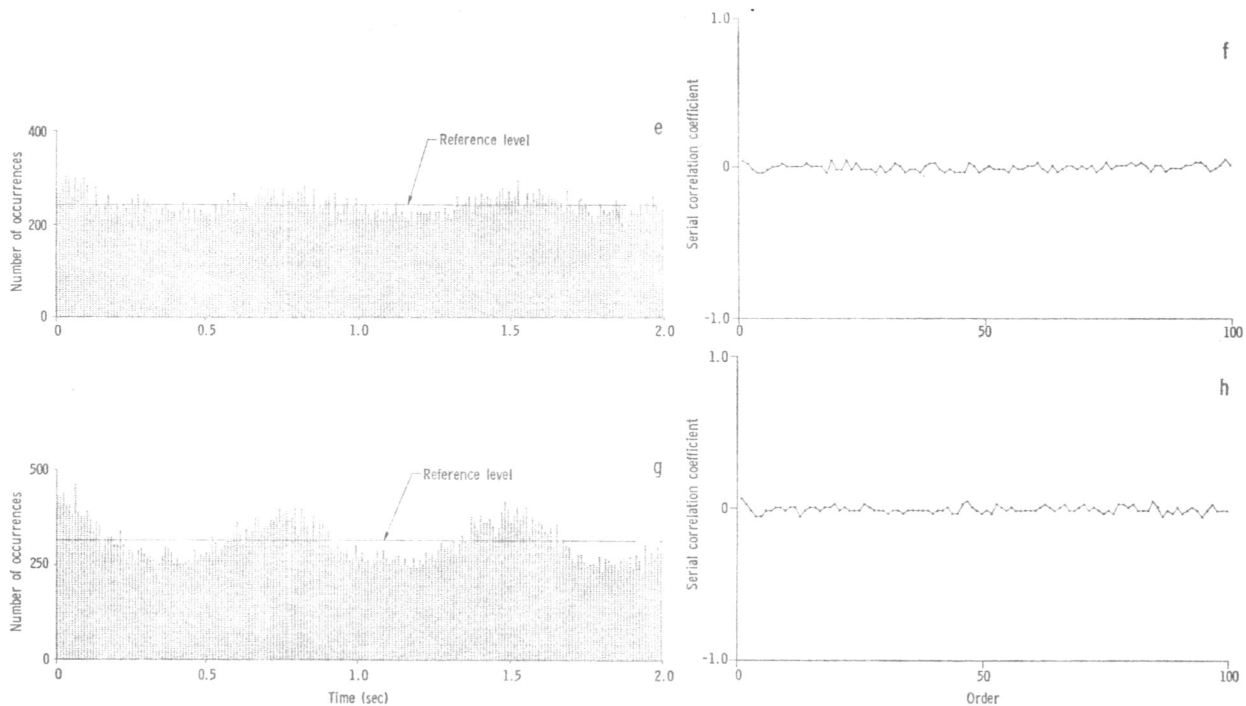

FIGURE 7 Effects of slow rate changes in a nonpacemaker neuron. On the left, autocorrelation histograms; on the right, serial correlograms of interspike intervals. Intervals are generated by a Poisson process with time-varying rate parameter; period of observation is 200 sec, with approximately 2000 spikes each sample. Reference level in autocorrelation is predicted asymptotic value (see text, Eq. 12). a–b: *deceleration*, ±50%. Mean interval increases linearly with time from 50 msec to 150 msec. c–d: *deceleration*, ±70%. Mean interval increases linearly with time from 30 msec to 170 msec. e–f: *oscillation*, ±50%. Mean interval varies sinusoidally from minimum of 50 msec to maximum of 150 msec, with period of 0.75 msec. g–h: *oscillation*, ±70%. Mean interval varies sinusoidally from minimum of 30 msec to maximum of 170 msec, 0.75 sec period.

common technique for examining rate variations in a spike train is to plot each interval length as a function of time (usually taken to be the time of the second spike determining the interval) or of serial number of the interval. If the cell fires fairly regularly, such a plot will clearly reveal the structure of rate variations. If, on the other hand, the interval variance is large, trends are hidden in the "noise."

Moving-average techniques have often been used for smoothing purposes. They are typically based on either a fixed number of intervals or a fixed length of time over which an average rate is computed. A more meaningful type of moving average would be one in which the contribution of each spike to the rate estimate is a decreasing exponential function of the time interval between the occurrence of the spike and the time referred to by the estimate. A "ratemeter" approach of this sort corresponds more closely to the response of an integrating neuron than do the more rigid moving-average techniques. The time constant for the exponential function can be chosen to correspond to the integration period of a neuron known or postulated to receive the output from the observed neuron (Segundo et al., 1966).

If trends are monotonic, estimation methods based on regression analysis are useful; these are described by Cox and Lewis (1966, chapter 3).

Effects of Trends

One important effect of rate changes on a spike train is to increase the variation of the interspike intervals. This effect will be conspicuous, and therefore detectable in the statistics, only if this additional variability of intervals is significantly large as compared with the "intrinsic" variability of the intervals. We illustrate this with two classes of examples: a "noisy pacemaker," with an intrinsic coefficient of variation of 10%, and a Poisson process (intrinsic coefficient of variation 100%), in each of which the mean interval is a function of time.

The pacemaker results are shown in Fig. 6, in which the autocorrelation and serial correlograms are shown for a linearly accelerating train, a linearly decelerating train, and a train with sinusoidally varying mean intervals. In each of these examples, the maximum deviation of mean interval, due to rate changes, was $\pm 10\%$ of the mean interval. The effects on the autocorrelation and on the interval histograms (not shown) are not conspicuous; there is some broadening of the peaks, but this effect is apparent only upon detailed comparison with the exactly corresponding null case of no trend. Given the autocorrelations as observed experimentally, there is no reason for suspecting a trend in the data. The serial correlograms, on the other hand, clearly indicate the rate variations. An elevated serial correlogram, extending more or less uniformly out to high orders, is a specific indicator of monotonic trend. It is to be noted that the effects of monotonic acceleration are indistinguishable from those of monotonic deceleration. The undamped oscillatory nature of the third serial correlogram shown is due to the imposed constancy of period of rate variation; if that period had varied during the observation, the oscillations in the serial correlogram would have exhibited damping. An ingenious method is described by Firth (1966) for separating the effects of trend on the serial correlogram from its "inherent" features. His technique, involving successive differences, is in essence a form of analysis of variance and is applied to cells that fire at extremely regular intervals.

The time-varying Poisson process, on the other hand, requires a rate variation of ±50% before some effects are noticeable, and only at ±70% rate variation do the effects become conspicuous (Fig. 7). The only effect on the autocorrelation of a monotonic trend is to increase its level; the shape remains flat, as in the absence of trend. The predicted asymptotic level for the autocorrelation depends only on the mean observed interval (Equation 12); therefore, the observed autocorrelation histogram, together with this predicted level, can indicate a trend. For the monotonically time-varying Poisson processes illustrated (Fig. 7), this is the only conspicuous effect of the large rate variations, since the corresponding serial correlograms depart only slightly from 0. At these intensities of cyclic rate variation, oscillations are visible in the autocorrelation, but not in the serial correlogram; this is in contrast to the opposite situation observed in the case of a pacemaker with weak oscillations in rate. A monotonically declining autocorrelation histogram, such as the early portions of Figs. 7 c and 7 d, is strongly suggestive of rather severe rate changes in the data, and some detailed statistical features of the spike train may be masked or distorted thereby.

The authors thank Dr. José P. Segundo and Professor Theodore H. Bullock for continued encouragement, discussions, and criticism. We are also very grateful to Dr. P. A. W. Lewis for valuable criticisms and suggestions.
This work was supported by United States Air Force Project RAND, by the National Institutes of Health through grants GM-09608, GM-14633, NB-05606, and NB-06372, and by the National Science Foundation through grant G-21497.

Received for publication 27 December 1966.

REFERENCES

Amassian, V. E., J. Macy, Jr., H. J. Waller, H. S. Leader, and M. Swift. 1962. *Proc. Intern. Union Physiol. Sci., Intern. Congr., 22nd, Leiden.* **3**:235.

Bartlett, M. S. 1963. *J. Roy. Statist. Soc. B.* **25**:264.

Burns, B. D., and G. K. Smith. 1962. *J. Physiol., (London).* **164**:238.

Cooley, J. W., and J. W. Tukey. 1965. *Math. Comp.* **19**:297.

Cox, D. R. 1962. Renewal Theory. John Wiley & Sons, Inc., New York.

Cox, D. R., and P. A. W. Lewis. 1966. The Statistical Analysis of Series of Events. John Wiley & Sons, Inc., New York.

Cox, D. R., and H. D. Miller. 1965. The Theory of Stochastic Processes. John Wiley & Sons, Inc., New York.

Firth, D. R. 1966. *Biophys. J.* **6**:201.

Geisler, C. D., and J. M. Goldberg. 1966. *Biophys. J.* **6**:53.

Gerstein, G. L. 1960. *Science.* **131**:1811.

Gerstein, G. L., and N. Y.-S. Kiang. 1960. *Biophys. J.* **6**:15.

Gerstein, G. L., and B. Mandelbrot. 1964. *Biophys. J.* **4**:41.

GOLDBERG, J. M., H. O. ADRIAN, and F. D. SMITH. 1964. *J. Neurophysiol.* **27**:706.
HAGIWARA, S. 1949. *Rept. Physiograph., Sci. Inst. Tokyo Univ.* **3**:19.
HAGIWARA, S. 1954. *Japan J. Physiol.* **4**:234.
HUGGINS, W. H. 1957. *Proc. IRE.* **45**(Pt. 1):74.
JUNGE, D., AND G. P. MOORE. 1966. *Biophys. J.* **6**:411.
KENDALL, M. G., and A. STUART. 1961. The Advanced Theory of Statistics. Charles Griffin and Co. Ltd., London. 2nd edition. **2**.
LAMARRE, Y., and J. P. RAYNAULD. 1965. *Electroencephalog. Clin. Neurophysiol.* **18**:87.
LEWIS, P. A. W. 1966. *IBM Systems J.* **5**:202.
McFADDEN, J. A. 1962. *J. Roy. Statist. Soc. B.* **24**:364.
McGILL, W. J. 1963. *In* Handbook of Mathematical Psychology. R. D. Luce, R. R. Bush, and E. Galanter, editors. John Wiley & Sons, Inc., New York. **1**:309.
MOORE, G. P., D. H. PERKEL, and J. P. SEGUNDO. 1966. *Ann. Rev. Physiol.* **28**:493.
PERKEL, D. H. 1964. Detection of Functional Interactions Among Neurons: A Technique Using Repetitive Presentations of Stimuli. The RAND Corporation, Santa Monica, Calif. RM-4234-NIH.
PERKEL, D. H. 1965. *In* Biophysics and Cybernetics Systems. M. Maxfield, A. Callahan, and L. J. Fogel, editors. Spartan Books, Inc., Washington, D.C. 37.
PERKEL, D. H., G. L. GERSTEIN, and G. P. MOORE. 1967. *Biophys. J.* **7**: 419.
POGGIO, G. F., and L. J. VIERNSTEIN. 1964. *J. Neurophysiol.* **27**:517.
RODIECK, R. W., N. Y.-S. KIANG, and G. L. GERSTEIN. 1962. *Biophys. J.* **2**:351.
SEGUNDO, J. P., D. H. PERKEL, and G. P. MOORE. 1966. *Kybernetik.* **3**:67.
SMITH, D. R., and G. K. SMITH. 1965. *Biophys. J.* **5**:47.
WATSON, G. S., and M. R. LEADBETTER. 1964 *a. Biometrika.* **51**:175.
WATSON, G. S., and M. R. LEADBETTER. 1964 *b. Sankhya (Indian J. Statist.).* **26**:101.
WEISS, T. F. 1964. A Model for Firing Patterns of Auditory Nerve Fibers. Technical Report No. 418. Research Laboratory of Electronics, Massachusetts Institute of Technology, Cambridge, Mass.
WILSON, D. M. 1964. *J. Exptl. Biol.* **40**:191.

NEURONAL SPIKE TRAINS AND STOCHASTIC POINT PROCESSES

II. Simultaneous Spike Trains

DONALD H. PERKEL, GEORGE L. GERSTEIN, *and* GEORGE P. MOORE

From the Department of Mathematics, The RAND Corporation, Santa Monica, California 90406, the Departments of Biophysics and Physiology, School of Medicine, University of Pennsylvania, Philadelphia, Pennsylvania 19104, and the Departments of Physiology and Engineering and the Brain Research Institute, University of California, Los Angeles, California 90024

ABSTRACT The statistical analysis of two simultaneously observed trains of neuronal spikes is described, using as a conceptual framework the theory of stochastic point processes.

The first statistical question that arises is whether the observed trains are independent; statistical techniques for testing independence are developed around the notion that, under the null hypothesis, the times of spike occurrence in one train represent random instants in time with respect to the other. If the null hypothesis is rejected—if dependence is attributed to the trains—the problem then becomes that of characterizing the nature and source of the observed dependencies. Statistical signs of various classes of dependencies, including direct interaction and shared input, are discussed and illustrated through computer simulations of interacting neurons. The effects of nonstationarities on the statistical measures for simultaneous spike trains are also discussed. For two-train comparisons of irregularly discharging nerve cells, moderate nonstationarities are shown to have little effect on the detection of interactions.

Combining repetitive stimulation and simultaneous recording of spike trains from two (or more) neurons yields additional clues as to possible modes of interaction among the monitored neurons; the theory presented is illustrated by an application to experimentally obtained data from auditory neurons.

A companion paper covers the analysis of single spike trains.

INTRODUCTION

In a companion paper (Perkel, Gerstein, and Moore, 1967), we discuss the statistical analysis of the single spike train. The mathematical framework used is the theory of stochastic point processes, that is, of random processes whose realizations may be described as series of point events occurring in time. In that paper and in a recent review (Moore, Perkel, and Segundo, 1966), it is shown how analysis of spike trains by means of these statistical techniques can shed light on such neurophysiological questions as the mechanisms of spike production, integrative, i.e. input-output, properties of nerve cells, and theories of information processing by the nervous system.

It is, however, in the comparison of two or more simultaneously observed spike trains that we believe lies the greatest potential usefulness for the statistical analysis of the data provided by precise measurement of times of spike events. It has become increasingly feasible to record spike trains simultaneously from several neurons (Gerstein and Clark, 1964; Simon, 1965). These experimental techniques have been motivated by the promise that intercomparison of spike trains will reveal details of synaptic connections and other sources of interaction among the observed neurons. The statistical techniques for analyzing simultaneous spike-train data, however, are not in as satisfactory a state of development as the experimental techniques for obtaining these data. The present paper is offered as a contribution to the statistical theory so that the experimental techniques may be more thoroughly and properly exploited.

There are two fairly conspicuous difficulties associated with two-train analyses. The first has to do with sampling distributions: in the principal statistical measures, e.g. the cross-correlation histogram, the distribution of tallies is not known in general, and except in the special case of two independent Poisson processes, the observations do *not* constitute Bernoulli trials. Hence, confidence limits and tests of hypotheses based on binomial statistics are not valid, and their use, although reported in the neurophysiological literature, may yield erroneous conclusions from the data.

The second difficulty is involved with the interpretation of the statistics when dependencies are found. It is not always clear whether a given cross correlation is better explained by a synaptic connection between the two neurons, mediated perhaps by one or more interneurons, or alternatively by a shared source of input to the two cells. Interpretation of simultaneous spike-train measurements must take cognizance of the several alternative possible explanations for observed statistical signs of dependence and acknowledge the difficulty which arises from the paucity of our quantitative knowledge of the statistical consequences of each of the several postulated alternative possibilities.

In both of the difficulties mentioned and with regard to complications arising from nonstationarities, we have found the use of simulation of interacting neurons on a digital computer to be a powerful and indispensable tool. Many of the examples and illustrations presented, therefore, are drawn from computer simulations (Perkel, 1965).

The interpretation of data taken simultaneously from three sources of spikes is at present considerably more difficult than that of two-train data. One somewhat special case is furnished by experiments in which the third source of point events consists of the times of presentation of a brief stimulus. When these presentations are sufficiently infrequent, the intercomparison of the three trains of events (two spike trains and the stimulus-presentation train) can yield much more information about the underlying functional relationships between the two neurons than can be obtained from analysis of their spontaneous activity alone.

In this paper, then, we discuss first the consequences of the independence hypothesis for two spike sequences. The statistical measures are based on the idea that, for independent trains, spikes in one cell represent random instants in time with respect to the other. We then describe their use in tests for independence and show the effects on the statistical measures of some of the typical types of dependence. We next discuss and illustrate the effects of nonstationarities, in the form of rate trends, on these statistical measures. Finally, we develop a simple theory for dealing with brief, repetitive stimuli in conjunction with the observation of two spike trains; we apply these methods to an example based on experimental data from auditory neurons.

TWO SIMULTANEOUS SPIKE TRAINS

The Problem of Functional Relationship

In the interpretation of spike trains recorded simultaneously from several neurons, the spike trains are ordinarily compared by pairs.[1] The first question to be asked is whether the two trains of spikes are independent. More specifically, we wish to make a statistical test of the null hypothesis that the two trains are drawn from independent point processes. This would imply that the two neurons are functionally unrelated. The test is accomplished, in principle, by computing a histogram, which estimates a suitable function, e.g. either the cross-density or the cross-correlation function described below, and comparing the estimate with the predicted function as based on the assumed independence of the two trains. If the observed and predicted functions differ significantly, the null hypothesis is rejected, and the trains are considered to be dependent.

An observed dependence between two spike trains can arise from one (or both) of two sources: (*a*) functional interaction and (*b*) common input. By functional interaction, we mean any mechanism by which the firing of one neuron influences the probability of firing by the other neuron. Such mechanisms could be synaptic (whether direct or mediated through interneurons), ephaptic, or due to "field effects." By common input we mean any mechanism that simultaneously modulates the firing patterns of both neurons. Such mechanisms could involve synaptic contact from branches of the same axon or field effects from a source other than the two neurons. It must be emphasized that long-term concomitant changes in firing rates, if shared wholly or in part by two neurons, constitute a form of dependence and if

[1] Statistical measures for the concurrent intercomparison of three or more trains are currently under investigation.

sufficiently pronounced, will be detected by statistical test procedures (see discussion below on Types of Dependence).

Independence: Consequences of the Null Hypothesis

According to the null hypothesis, spike trains A and B are independent in the mathematical sense. This means that spikes in train A occur at moments taken at random with respect to train B. In relating the two spike trains we may therefore use some mathematical results about single point processes observed from random moments in time.

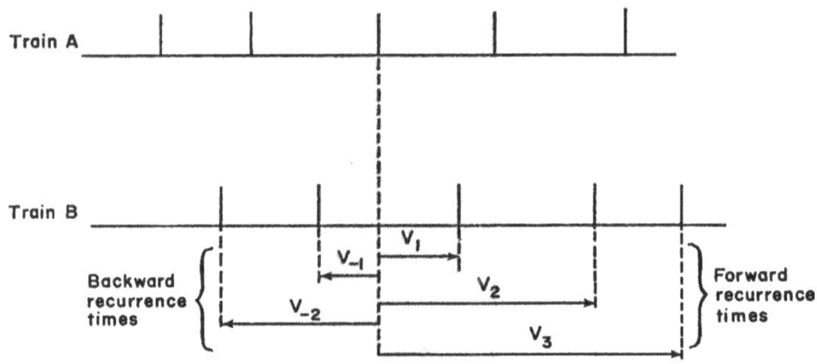

FIGURE 1 Recurrence and waiting times. Train B is a train of spikes. If the events in train A are random instants in time, then V_1 and V_{-1} are the (first-order) forward and backward *recurrence times*, respectively; V_2 and V_{-2} are the corresponding second-order recurrence times, etc. If train A is a spike train, then (in the usage of this study) the corresponding time intervals are called *waiting times*, and are designated W_1, W_{-1}, W_{-2}, etc. See text.

One such result concerns the distribution of so-called *recurrence* times, which are defined as follows. From a random instant in time, we denote by V_1 the time until the next event in a point process and by V_{-1}, the time backward to the most recent event in the point process; V_1 and V_{-1} are known as the *forward* and *backward recurrence times*, respectively (Fig. 1). Each of these times has the same distribution with the pdf

$$g_1(\tau) = [1 - F(\tau)]/\mu = \mathfrak{F}(\tau)/\mu, \tag{1}$$

where μ is the mean interspike interval; $F(\tau)$, as usual, is the cumulative probability distribution of the intervals; and $\mathfrak{F}(\tau)$ is the survivor function. This result is true for both renewal and nonrenewal processes (Cox, 1962; McFadden, 1962).

In terms of two spike trains, this result means that the time from a randomly selected spike in A to the immediately following, or to the immediately preceding, spike in B is distributed with the pdf

$$\eta_1(\tau) = \eta_{-1}(\tau) = [1 - F_B(\tau)]/\mu_B \tag{2}$$

if the two spike trains are independent. A similar relationship holds, of course, for the times between any spike in B and the neighboring spikes in A, with the appropriate change of subscripts.

Generalizing this notion, we may make the corresponding measurements from two simultaneous spike trains, whether or not they are in fact independent. We designate as W_1 the "waiting time" from a spike in train A to the next subsequent spike in train B and as W_{-1}, the time backward to the most recent spike in train B. The distributions of these random variables are specified by their pdf's, $\eta_1(\tau)$ and $\eta_{-1}(\tau)$, which we call the *forward* and *backward cross-interval densities*, respectively. These densities, as usual, are estimated by histograms constructed from the observed spike trains. If the trains are independent, the cross-interval histograms will agree with their predicted forms, according to Equation 2. It is to be noted that this prediction is based only on the interval distribution in train B. If the observed histograms agree with the predictions, however, it can be concluded that the trains are independent only insofar as adjacent spikes are concerned. Effects of one neuron's spikes on the other's spikes may well be delayed to the extent that they will not be revealed in the cross-interval histograms. Departures of the cross-interval histograms from their predicted form may often indicate the type of dependence between the two trains. Furthermore, in cases of indicated dependence, the forward and backward cross-interval histograms will in general be different.

The second mathematical result deals not only with the first-order recurrence times (forward and backward to the adjacent events), but with the sum of these for all orders. Letting V_2, V_3, $\cdots$ refer to the times from a random moment to the second, third, $\cdots$ events encountered in the process B, we may define in general $g_i(\tau)$ to be the pdf of recurrence time V_i. If these pdf's exist (as they do for most physically realizable processes, including spike trains), then it follows from a result of McFadden (1962, Equation 2.21) that the sum of the recurrence densities of all orders is a constant:

$$G_B(\tau) \equiv \sum_{i=1}^{\infty} g_i(\tau) = 1/\mu_B . \tag{3}$$

Here μ_B is the mean interval between events in the process B, which is not necessarily a renewal process. Since the backward recurrence times V_{-i} of all orders have the same distributions as the corresponding forward recurrence times, their sum also has the same distribution as that for the forward times.

We proceed as before to generalize to the case of not necessarily independent spike trains, and we define the *forward* and *backward waiting times of order i*, W_i and W_{-i}, as the time measured from a spike in train A to the *i*th subsequent spike encountered in train B, or backward to the *i*th previous spike in train B, respectively (Fig. 1). The corresponding *cross-interval densities of order i* are denoted by $\eta_i(\tau)$. We further define the *cross-correlation function* $\zeta_{AB}(\tau)$ as the sum of all orders of cross-interval densities:

$$\zeta_{AB}(\tau) = \sum_{\substack{i=-\infty \\ i\neq 0}}^{i=\infty} \eta_i(\tau). \tag{4}$$

Now, reasoning as we did with Equations 10 and 11 in Perkel et al. (1967) for the autocorrelation, we realize that for a given spike in train A, the time, backward or forward, to any spike in train B must be the waiting time of some order i ($i = \cdots, -2, -1, 1, 2, \cdots$). Therefore, the cross correlation represents the probability of encountering any event in train B as a function of time before or after an actual event in train A:

$$\zeta_{AB}(\tau) = \lim_{\Delta t\to 0} \text{prob}\,\{\text{an event in } B \text{ in } (t_0+\tau,\, t_0+\tau+\Delta\tau) | \text{ an event in } A \text{ at } t_0\}/\Delta t. \tag{5}$$

Functions of this kind are called "cross intensity functions" by Cox and Lewis (1966, p. 247).

The same reasoning applies, of course, if train A consists of random instants of time, and so we may write

$$\begin{aligned} G_B(\tau) &= \lim_{\Delta t\to 0} \text{prob}\,\{\text{an event in } B \text{ in } (t_0+\tau,\, t_0+\tau+\Delta t) | \text{ random instant } t_0\}/\Delta t \\ &= 1/\mu_B\,. \end{aligned} \tag{6}$$

There is a parallelism between the function $G_B(\tau)$ and the renewal density (autocorrelation) $h(\tau)$. If we observe a stochastic point process starting at a *random event*, then the probability of encountering spikes at any time thereafter is measured by $h(\tau)$. For large τ, i.e. at long times after the initial observation point, this density becomes a constant, equal to the mean firing rate $1/\mu$ (Equation 12 in Perkel et al., 1967). On the other hand, if we start our observation at a *random instant* in time, rather than at a particular event, the probability density $G_B(\tau)$ of encountering a spike at some later time is already a constant. Heuristically, this means that if we know that a spike has occurred, we can predict (from the statistical properties of the spike train) when subsequent spikes will occur; our prediction becomes worse the farther removed we are from the initially observed spike, until our prediction is based solely on the mean firing rate. But if we start at a random moment, we do not even know the position of one spike, and we can do no better at predicting subsequent spike locations than by using the mean rate. These considerations apply to both renewal and nonrenewal point processes.

The cross-correlation function $\zeta_{AB}(\tau)$ has a useful symmetry relationship. Consider the compound event "a spike in record A is followed at a time X by a spike in record B"; this event may be equivalently described as the event "a spike in record B is preceded at a time X by a spike in record A." The probability of this compound event may be expressed in either of two ways, implying the following symmetry relationship for the cross-correlation:

$$\zeta_{AB}(\tau)/\mu_A = \zeta_{BA}(-\tau)/\mu_B \,. \tag{7}$$

This identity holds even if the two trains are not independent. For this reason, the cross-correlation function is customarily measured in only one direction, from train A to train B, and for both forward and backward waiting times, i.e., for both positive and negative values of τ.

Application to Experimental Data

Thus, in order to test whether two spike trains are independent, we may make the following measurements:

(*a*) The *cross-interval histogram.* We select spikes in train A and construct a histogram of the times to the *nearest* spikes in train B. This histogram can then be compared with the estimated backward and forward recurrence-time densities $\eta_1(\tau)$ and $\eta_{-1}(\tau)$, which in turn are estimated from the interspike-interval histogram of train B, using Equations 4 *b* in Perkel et al. (1967) and Equation 2 above. This procedure must be repeated after interchanging the rôles of the trains.

(*b*) The *cross-correlation histogram.* We select spikes in train A and construct a histogram of the times to *all* spikes in train B, both forward and backward, out to some specified time. This histogram provides an estimate of the cross-correlation function $\zeta_{AB}(\tau)$, which can be compared with the predicted constant value $1/\mu_B$.

If the two spike trains are independent, then the forward and backward cross-interval histograms will be equal to the prediction $[1 - F_B(\tau)]/\mu_B$ and hence equal to each other, within sampling effects. Since these equalities are a necessary but not sufficient condition for independence of the spike trains, the cross-interval histograms have two major uses: (*a*) as a corroboration of independence when indicated by a flat cross correlation, and (*b*) as a means of exploring suspected short-latency interactions, i.e., those occurring with a latency smaller than the mean interspike interval in train B. Cross-interval histograms must be computed separately for train A against train B and vice versa. These measurements are related to the phase histograms of Wyman (1965).

Satisfactory statistical tests of the null hypothesis of independent firings have not yet been completely developed (see Cox and Lewis, 1966, pp. 247–248). There are two sources of difficulty. The first arises from the lack of independence of successive bins in the cross-correlation histogram, as was mentioned for the PST histogram in Perkel et al. (1967). An observation contributing to one bin of the histogram is pro-

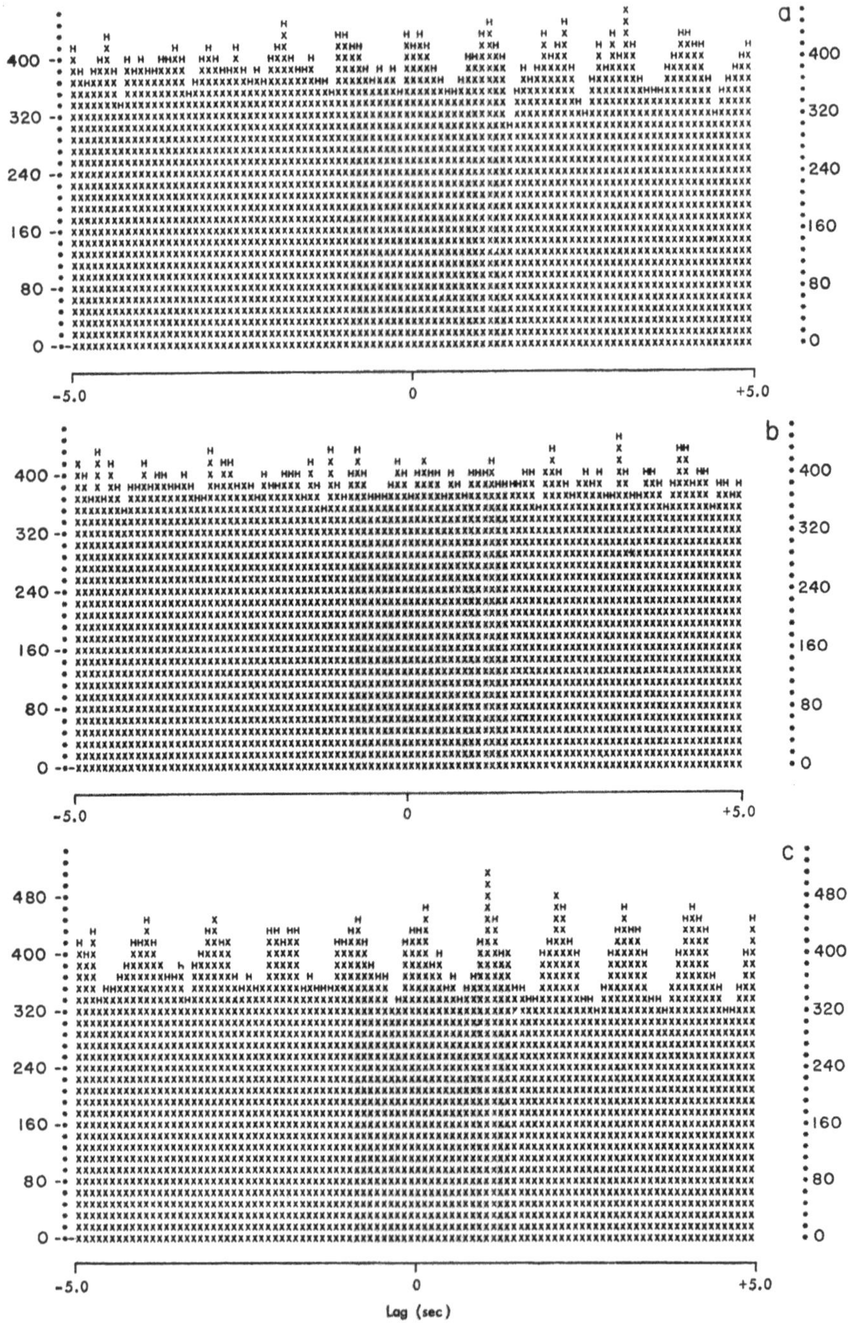
a
b
c
0
80
160
240
320
400
480
-5.0
0
+5.0
Log (sec)

duced by the firing of a neuron; the refractoriness of the neuron will then lower the probability of a contribution to the next bin. The magnitude of this effect depends on the degree and duration of refractoriness, as well as on bin width. If two neurons are in fact independent, consider a bin that has a larger number of spikes than average due to a random fluctuation. The next bin is then more likely to have a smaller number of spikes than average. Thus, the contribution to the sum of squares of deviations from the mean will be larger than if bins represented independent observations. This can lead to false attributions of dependence to cells that are, in fact, firing independently.

A related source of the same type of error arises from the sampling requirements for spike trains of neurons that are pacemakers. Two independently firing pacemakers of nearly the same frequency typically exhibit peaks in the cross-correlation histogram, which oscillate with the period of the pacemakers. The magnitudes and phases of these peaks vary randomly from sample to sample, and they may disappear only with extraordinarily long samples. In Fig. 2, we show examples of cross-correlation histograms taken from two simulated independent pacemakers (normally distributed intervals with standard deviation of interval set at 10% of the mean). Each sample consisted of 4000 spikes in each neuron. The examples show the appearance and occasional disappearance of the peaks, which may be interpreted as a run of phase "locking" arising by chance. It is consequently very difficult to decide from reasonably sized samples whether two pacemakers are independent.

Thus, there are appreciable risks of falsely attributing dependence to independent spike trains, particularly when both neurons are pacemakers. The other possible type of error is false attribution of independence to trains that are in fact dependent. This is much less likely and can arise in either of two ways: (*a*) The dependence may be so weak that its effects are indistinguishable from "noise" in the sample taken. (*b*) A simultaneous combination of positive and negative interactions may coincidentally combine in such a fashion as to cancel each other. If this should happen, however, the two neurons may still be regarded as firing independently since, in an operational sense, the firing of one cannot be used to predict firing times of the other.

Types of Dependence

Dependence may arise from a number of physiologically important processes. We have already distinguished *interaction* and *common input*. The combinations of possibilities are virtually endless.

FIGURE 2 Cross-correlation histograms for two independent pacemakers. Each train has independently drawn normally distributed intervals, with mean 1.0 sec, standard deviation 0.1 sec; each train has 4000 spikes. a, unshuffled data; b, after prolonged random shuffling of intervals in train B; c, after further prolonged random shuffling of intervals in train B. See text.

Two cells may interact through synaptic connections that may be either direct, e.g. an axon collateral of one of the cells forming a synapse with the other, or indirect, i.e., mediated through one or more interneurons. The effects of such interactions on the cross correlation cannot be predicted precisely without a detailed knowledge of the intracellular processes, including postsynaptic potentials, which underlie the production of spikes by each cell. In simple cases, however, the gross effects are intuitively obvious. For example, if cell A makes a single excitatory synaptic connection with cell B, with a mean conduction time w, then we would expect the probability that cell B fires to be enhanced during a period starting at a time w after the occurrence of every spike in cell A. Therefore, we observe a peak in the cross-correlation function $\zeta_{AB}(\tau)$ near the point $\tau = w$. The shape of the peak depends on the details of the synaptic interaction (Fig. 3 a). If the connection is inhibitory, a depression rather than a peak will be observed (Fig. 3 b).

Two cells may receive input from a common source either directly or indirectly. In a simple example, the source might be a cell that makes excitatory synapses with cells A and B at latencies z_A and z_B, respectively. We then expect a peak in the cross-correlation function $\zeta_{AB}(\tau)$ at or near $\tau = z_B - z_A$. The magnitude of the peak will depend on the firing rate of the source cell as well as on the neurophysiological characteristics of the observed cells.

We have examined a number of cases by means of computer simulation to obtain guidelines for the interpretation of dependence observed in physiological experiments. A few representative examples are illustrated in Fig. 3.

When networks are constructed of neurons with basically similar properties and parameters, then the following generalizations may be made on the basis of simulation studies.

(*a*) Common sources of input are more difficult to detect than direct or indirect connections.

(*b*) Indirect connections are more difficult to detect than direct connections.

(*c*) Several different arrangements of functional interaction may lead to the same cross correlation; unique inferential conclusions concerning the anatomy and physiology of the interneuronal dependence may therefore be impossible. Furthermore, moderate degrees of nonstationarity will not appreciably mask the statistical signs of neuronal interaction, nor in general will they lead to false inferences of interaction when none, in fact, is present, as discussed in the next section.

Effects of Nonstationarity

It is shown in the companion paper on single spike-train analysis (Perkel et al., 1967) that the effects of nonstationarities in the form of trends, or rate changes, in a single spike train are most apparent in the serial correlogram, and that they are far less pronounced in an irregularly firing cell than in a pacemaker-like neuron. We confine our discussion of nonstationarities in the comparison of two spike trains to a class of cases in which the nonstationarities show little effect on the single-train statistics, namely that of Poisson processes with time-varying rate parameters. It should be borne in mind that trends exacerbate the problem of interpreting cross correlations between regularly firing cells (Fig. 2).

If rate changes are shared by two otherwise independent neurons, the cross correlation will typically display, more or less symmetrically in the neighborhood of its origin, an elevation above its "null" level (that expected for independent cells). Rather large rate changes must be present (for irregularly firing cells) in order for this elevation to be readily apparent in the histogram. It can be shown, for example, that for two Poisson processes with time-varying parameters, the maximum fractional expected departure of the cross correlation from its "null" level is less than the square of the maximum fractional rate variation. If, for example, the shared rate changes of the two processes vary within 20% of their respective means, then the cross correlation will depart by at most 4% from its predicted value for independent stationary processes based upon observed firing rates (Equation 3). The actual amount of departure depends upon the precise nature of the temporal variation of the rate parameters of the Poisson processes; details will be presented elsewhere.

These results are illustrated in Fig. 4, in which cross-correlation histograms are shown for pairs of Poisson processes subjected to monotonic and oscillatory rate changes. The elevation in level of the cross-correlation histogram near its origin is seen to be much more pronounced under more drastic rate changes. Linear trends result in a uniformly elevated cross correlation, which remains flat. Oscillating trends give rise to an oscillating cross-correlation histogram, if the range of the latter is great enough.

Because the effects of nonstationarity on the cross correlation are of second order, the statistical indications of interaction between two cells (discussed earlier) are not severely affected by rate changes in one or both of the cells, even if they are fairly severe. Inasmuch as stationary conditions are often difficult to maintain in an experiment, this fact is an encouragement to the attempt to elucidate functional interconnections of neurons through statistical comparison of their spike trains.

A special case of nonstationarity, deliberately introduced, is the use of experimentally controlled stimulation of the organism. This provides a new experimental variable, which can markedly enhance the utility of spike correlation measurements, as discussed in the following section.

TWO SPIKE TRAINS IN THE PRESENCE OF STIMULATION

Functional Effects of Stimulation

Interactions between neurons may be altered by presenting a stimulus to the animal; these changes are generally reflected in the cross-correlation histogram. The effects on the cross correlation may arise (*a*) through changed firing rates of one or both cells, (*b*) through direct or indirect synaptic input to both cells from a common

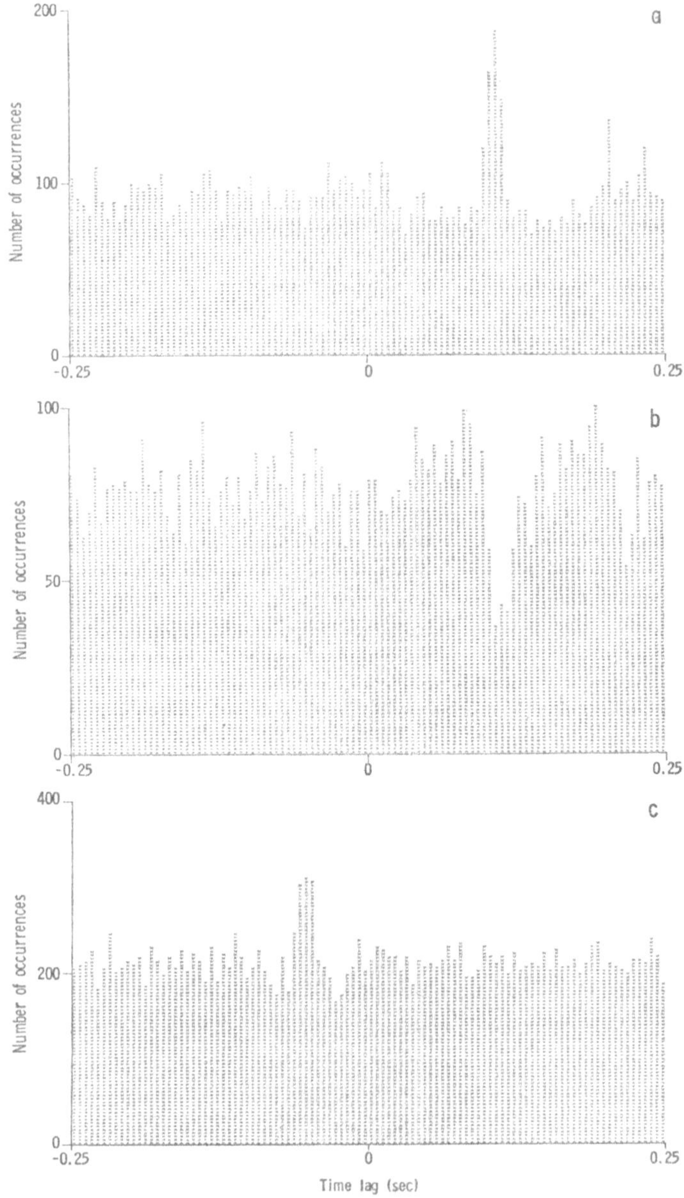
200
100
0
a
Number of occurrences
-0.25
0
0.25
100
50
0
b
Number of occurrences
-0.25
0
0.25
400
200
0
c
Number of occurrences
-0.25
0
0.25
Time lag (sec)

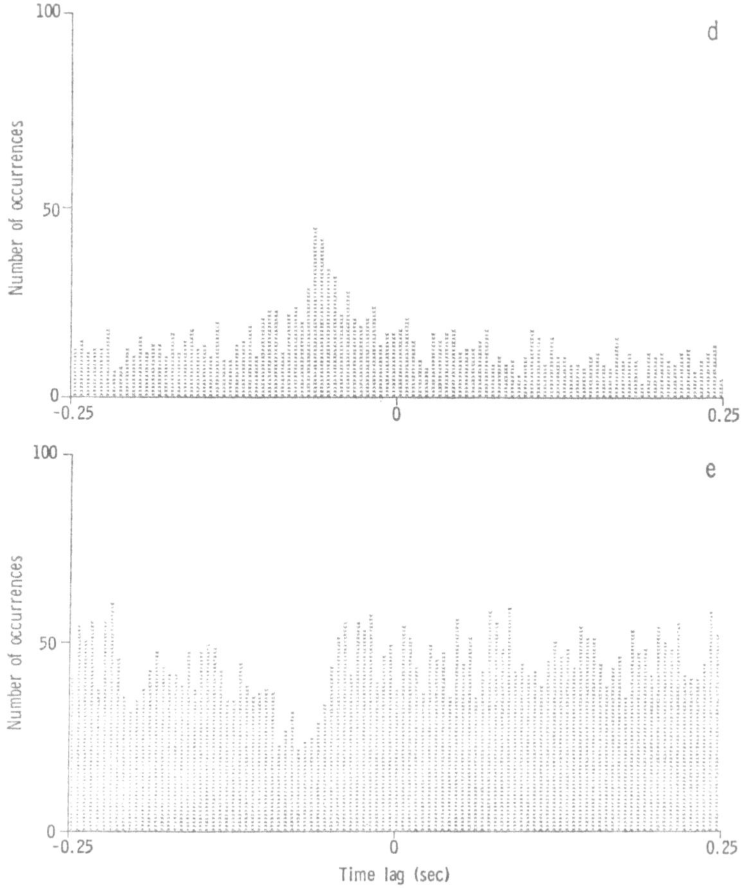

FIGURE 3 Cross-correlation histograms for interacting neurons; data from computer simulations. a, *direct excitation*. Neurons with characteristics similar to those of Fig. 2 b of Perkel et al. (1967). Each neuron excited by an independent Poisson source of EPSP's, amplitude normally distributed with mean 7.5 mv, standard deviation 1.0 mv; mean arrival rate 100/sec. Cell A makes excitatory synapse on cell B with latency of 100 msec; EPSP mean amplitude 10 mv, standard deviation 1 mv. In 250 sec, 2053 spikes produced by cell A, 2286 spikes by cell B. b, *direct inhibition*. Same as Fig. 3 a except A-B synapse is inhibitory (mean amplitude −10 mv, etc.). In 250 sec, 2068 spikes produced by cell A, 1858 spikes by cell B. c, *shared excitation*. Simulated neurons similar to those above, but with no synaptic connections between them. Each independent Poisson source now produces 10-mv EPSP's and is reduced in rate to 25/sec. An additional Poisson source with a mean rate of 150/sec delivers 10-mv EPSP's to both cells, each arriving at cell A 50 msec earlier than at cell B; i.e., each cell receives an average of 175 EPSP's per sec, of which 150 (86%) are shared with the other cell. Each cell produced about 2500 spikes in 150 sec. d, *shared inhibition*. Similar to previous example. Each independent excitatory input produces 10-mv EPSP's at a mean rate of 175/sec, shared channel produces 30-mv IPSP's at a mean rate of 100/sec. Both cells depolarized by DC input to produce spikes. About 550 spikes produced by each cell in 150 sec. e, *shared inhibition and excitation*. Similar to previous cases; cell A is depolarized and receives IPSP's from shared 100/sec channel as in case 3 d; cell B is excited as in example 3 c. Cell A produced 732 spikes and cell B, 2600 spikes in 200 sec.

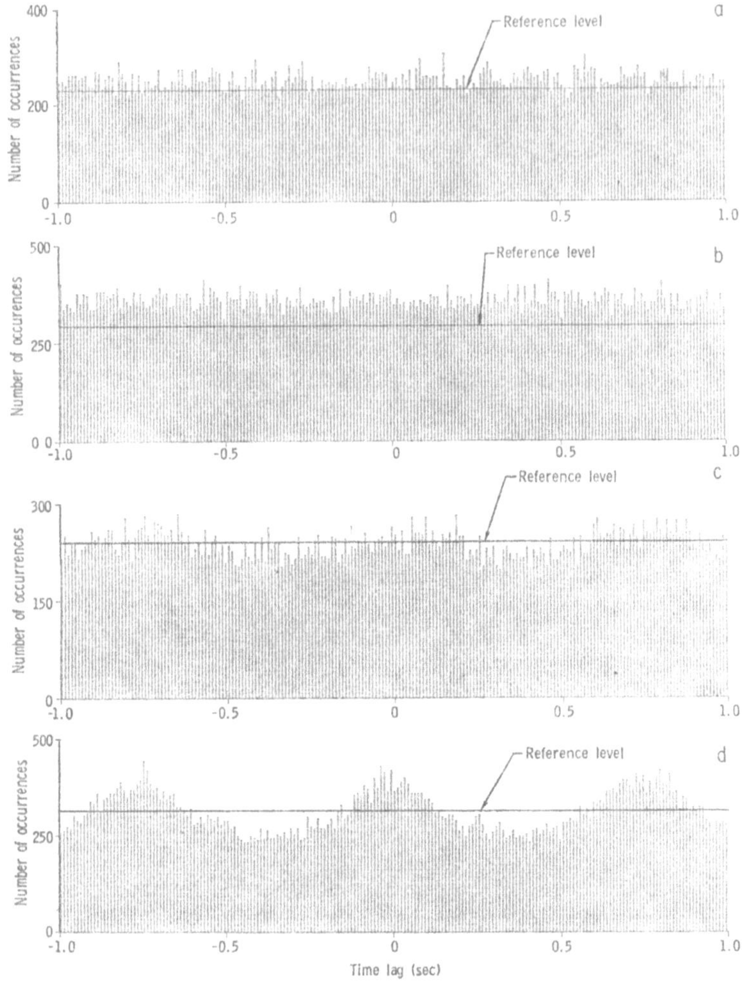

FIGURE 4 Effects of slow rate changes on cross correlations. Cross-correlation histograms between pairs of trains drawn from Poisson processes with time-varying rate parameters. Each member of a pair of trains has identical time variation of mean rate, but each train is generated independently. Reference level in cross correlation is predicted expected level for stationary independent processes, based upon observed mean firing rates (see text, Equation 3). Rate variations same as in Fig. 12 of Perkel et al. (1967). a, deceleration, ±50%. b, deceleration, ±70%. c, oscillation, ±50%. d, oscillation, ±70%.

source (or a set of parallel sources) that responds directly to the stimulus, (*c*) through the effects of the stimulus on interaction pathways between the two observed cells, or (*d*) through any combination of these. It is shown below that the changes in the cross-correlation function can be predicted when the stimulus is repeated periodically, and when mechanisms (*a*) or (*b*), or both, are operative. If the observed changes from unstimulated to stimulated situation agree with those predicted, it may be concluded that mechanism (*c*) is not operative, i.e., that any interaction pathways between the two neurons are not significantly affected by the stimulus.

We now consider the case in which an identical stimulus is presented at regular intervals of duration *P*.[2] These intervals are long with respect to conduction times and the duration of postsynaptic effects; i.e., the observable effects of a given stimulus have essentially died away by the time the next stimulus is presented.

The post-stimulus-time (PST) histogram discussed in Perkel et al. (1967) is an estimate of the firing probability of a neuron as a function of time since the onset of the stimulus. (It is clear that this is a special case of the cross-correlation function, $\zeta_{SA}(\tau)$, where "cell" S always fires at the onset of each stimulus.) It has two additive components: the relatively constant background component, which is due to firings of the observed cell that are not necessarily related to the stimulus, and departures from this level, which are due to the effects of the stimulus, whether directly or indirectly mediated. The background level may be affected by the stimulus, since ρ'_A, the mean firing rate of the cell with (repetitive) stimulus "on," may be different from ρ_A, the mean rate with stimulus "off," accordingly as the stimulus has a net excitatory or inhibitory effect on cell A.

It is clear that the observed cross-correlation function between two cells A and B will, in general, be different under "stimulus-on" and "stimulus-off" conditions. The major classes of effects possible are illustrated schematically in Fig. 5.

In the simplest situation, shown in Case a, cells A and B are independent, but both receive input from the stimulus. (Each arrow represents one or more synaptic connections, with the possible interposition of interneurons; the sketches in Fig. 5 are functional rather than anatomical diagrams.) Cells X and Y represent independent sources of synaptic activity supplying cells A and B, respectively. Cells X and Y may be influenced by the stimulus, as indicated by the broken lines. With the stimulus off, in this situation, the only contribution to the cross correlation $\zeta_{AB}(\tau)$ consists of a different background component (due to rate changes) and a superimposed, nonuniform component due to shared input.

In a more complicated situation (Case b), the two cells may interact. The interaction may be from A to B, from B to A, or in both directions; it may be direct or indirect, single or multiple. There may also be shared input from sources other than the stimulus. All these possibilities are subsumed in the square box *I* in the figure. With the stimulus off, the background effects, interactions, and possible

[2] Irregularly spaced stimulus presentations lead to irrelevant complications in the calculations.

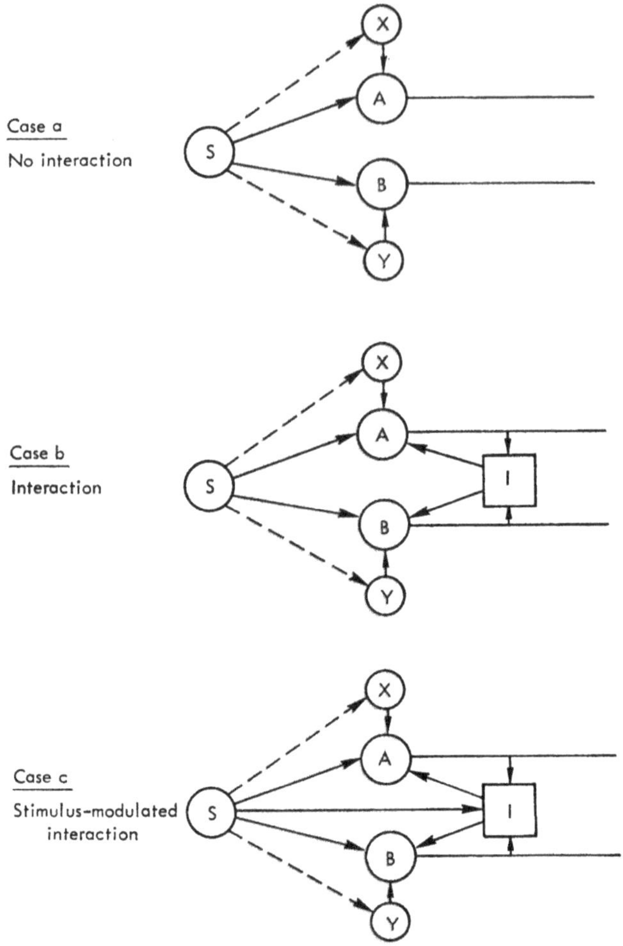

FIGURE 5 Functional relationships between stimulus and observed neurons; schematic. Stimulus S affects neurons A and B either directly or (dotted lines) through intermediate networks X and Y. In Case a, there is no interaction between cells A and B. In Case b, there are interaction pathways I for possibly reciprocal interaction between neurons A and B. In Case c, interaction pathways I are themselves affected by the stimulus S. See text.

shared input contribute to the cross correlation (which need not be flat). When the stimulus is on, this function is affected by (*a*) a changed background level, (*b*) "first-order" shared input, represented by the arrows going directly from the stimulus S to the cells A and B, and (*c*) "second-order" shared input, represented by the pathways from S through A and I to B, and conversely.

A still more complicated situation is shown in Case c, in which the interaction itself is modified by the stimulus. Such modifications, or "modulation," may result from activation of excitatory or inhibitory synaptic connections from S on interneurons in I, causing increased or reduced effectiveness in the transmission of interactions between cells A and B.

Prediction of the Cross-Correlation

In the two simpler situations described above, it is possible to predict the cross correlation in the stimulus-on condition, on the basis of the following measurements: (*a*) the mean firing rates of both cells under both stimulated and unstimulated conditions, (*b*) the observed cross-correlation function with the stimulus off, and (*c*) the observed PST histograms for both cells.

The basic assumption involved in this prediction is that these various modifications of the cross correlation, produced by the stimulus, are sufficiently independent to be additive. This assumption breaks down, for example, when firing rates are high, so that refractory characteristics of the neurons become important.

With stimulus off, the cross-correlation histogram is composed of "background" and "interaction" components. The background component is predicted on the basis of expected random coincidences, as follows.

With a bin width of δ in the cross-correlation histogram, we define a lagged coincidence (with lag time τ) as the occurrence of an event in A at t_o together with the occurrence of an event in B in the interval $t_o + \tau \pm \delta/2$. The lag times τ correspond to midpoints of bins: $\tau_k = \pm k\delta/2$; $k = 1, 2, \cdots$. Then the expected number $N(\tau, \delta)$ of lagged coincidences in the corresponding bin, arising from a segment of A-record of length T, is given by

$$E[N(\tau, \delta)] = \rho_A T \int_{\tau-\delta/2}^{\tau+\delta/2} \zeta_{AB}(x)\, dx, \tag{8}$$

where $\rho_A = 1/\mu_A$ is the mean firing rate in record A. Thus, the estimate of the theoretical density $\zeta_{AB}(\tau)$ is given by

$$\hat{\zeta}_{AB}(\tau) = \frac{N(\tau, \delta)}{n_A \delta}, \tag{9}$$

where n_A is the observed number of spikes in A during T; its expected value is given by $E[n_A] = \rho_A T$, so that we have

$$\zeta_{AB}(\tau) = N(\tau, \delta)/(\rho_A T \delta). \tag{10}$$

If, however, the stationary point processes of the two cells are independent, so that an event in record A occurs without any knowledge of the sequence of events in record B, we have already pointed out that the cross-correlation function is a constant: $\zeta_{AB}(\tau) = \rho_B$. Therefore, for records of finite length, the observed histogram for such independent cells has an expected value of $\rho_A \rho_B T\delta$ in each bin, with "noisy" fluctuations that are reduced in relative magnitude as the sample size increases. This "random," or background, level, produced through randomly occurring lagged coincidences, is the sole component of the cross-correlation function for independent cells and is a contributing component to this function for dependent cell pairs. (See Cox and Lewis, 1966, pp. 246–248).

Rewriting Equation 10, we have

$$N(\tau, \delta) = \rho_A T\delta\zeta_{AB}(\tau) = \rho_A\rho_B T\delta + \rho_A T_\delta \xi_{AB}(\tau), \tag{11}$$

where $\xi_{AB}(\tau)$ represents the departure from background level or the "interaction" component:

$$\xi_{AB}(\tau) \equiv \zeta_{AB}(\tau) - \rho_B . \tag{12}$$

With stimulus on, and with new firing rates ρ_A' and ρ_B', the corresponding contributions to the new lagged coincidences $N'(\tau, \delta)$ are given by two terms:

$$G = \rho_A'\rho_B'T'\delta + \rho_A'T'\delta\xi_{AB}(\tau), \tag{13}$$

taking into account that the new observation time T' may differ from the previous time T.

A third component of the stimulus-on correlation histogram, which is due to shared input from the source, is obtained from the post-stimulus-time histograms. By subtracting out the background components as in Equation 12, we have the net PST density functions

$$\xi_{SA}(\tau) = \zeta_{SA}(\tau) - \rho_A', \tag{14 a}$$

$$\xi_{SB}(\tau) = \zeta_{SB}(\tau) - \rho_B', \tag{14 b}$$

which represent the departures in spike densities of cell A or B from the new mean levels at times τ after the onset of a stimulus.

Now the enhancement (or reduction) of the probability of a lagged coincidence that occurs between cells A and B can come about by an enhancement (or reduction) in delayed coincidences between S and B at a time $t + \tau$, making a contribution $\xi_{SA}(t)\xi_{SB}(t + \tau)$. We must integrate this over the entire period of the stimulus to obtain the enhancement (or reduction) per stimulus presentation of the lagged-coincidence density. Thus, we evaluate the correlation integral

$$[\xi_{SA} \times \xi_{SB}](\tau) = \int_0^P \xi_{SA}(t)\xi_{SB}(t + \tau)\, dt. \tag{15}$$

Because the stimulus is periodic, with period P, we take the arguments modulo P in the correlation integral.[3]

An alternative way of predicting the contribution of shared input from the stimulus to the cross correlation is to isolate those effects that are time-locked to the stimulus. This may be done as follows. One of the records, e.g., record B, is divided into equal segments, each equal to the interval P between stimulus presentations. These segments are thoroughly shuffled, so that their new order is effectively random. Then the cross correlation is recomputed between record A and shuffled record B. The shuffling has destroyed all significant time relationships between the two trains except those related to stimulus presentations. The cross-correlation histogram after shuffling and after subtraction of the background contribution estimates the same quantity as the correlation integral of the PST's:

$$\zeta'_{AB,\mathrm{shuf}}(\tau) - \rho'_B \approx [\xi_{SA} \times \xi_{SB}](\tau). \tag{16}$$

In practice, shuffling need not be performed, but rather the two records are offset in time by an amount sufficient to destroy direct temporal relationships; the amount of offset is, of course, an integral multiple of the interstimulus interval P.

The expected number of lagged coincidences produced by the shared input from the stimulus is given by

$$H = (T'_A/P)\delta[\xi_{SA} \times \xi_{SB}](\tau). \tag{17}$$

The total predicted number of lagged coincidences with stimulus on is given by the sum of terms G and H (Equations 13 and 17):

$$N'_{\mathrm{pred}}(\tau, \delta) = T'\delta\{\rho'_A\rho'_B + \rho'_A\xi_{AB}(\tau) + (1/P)[\xi_{SA} \times \xi_{SB}](\tau)\}. \tag{18}$$

Hence the predicted cross-correlation function is given by

$$\zeta'_{AB,\mathrm{pred}}(\tau) = \xi_{AB}(\tau) + \rho'_B + (\rho'_A P)^{-1}[\xi_{SA} \times \xi_{SB}](\tau). \tag{19}$$

Application to Experimental Data

Comparison of the predicted and observed cross correlations enables us to distinguish among the three situations shown in Fig. 5.

If the cross-correlation function with stimulus off is flat, within statistical limits, we may conclude that there is no interaction (Fig. 5 a). In this case the expected value of $\xi_{AB}(\tau)$ is 0. If the observed cross-correlation function with stimulus on

[3] When the densities ξ are obtained from PST histograms, they represent averages over a bin width. If the cross-correlation histogram has the same bin structure as the PST histograms, there is an imprecision in the approximation of the integral by a sum, as well as a shift in the argument τ by half a bin; the latter may be compensated for by an averaging procedure.

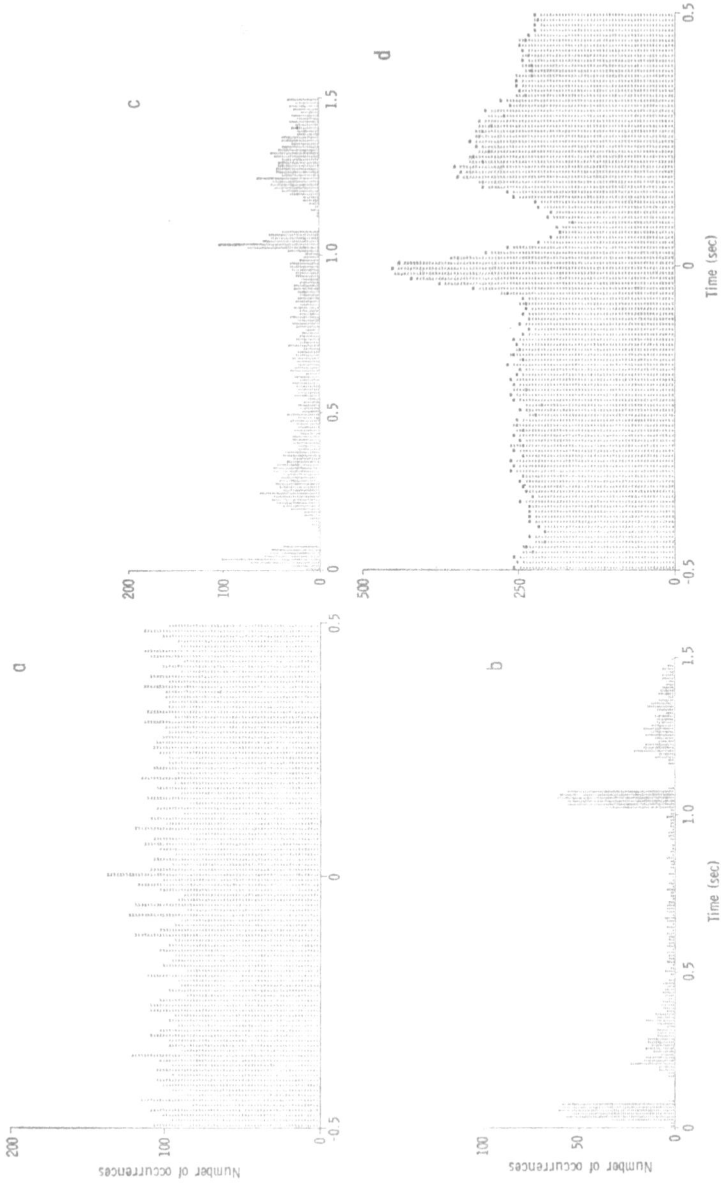

FIGURE 6 Effects of repeated stimulation on the cross correlation. a, cross-correlation histogram between spike trains recorded from two neurons in the cochlear nucleus of the cat; unstimulated. b, post-stimulus-time histogram, cell A. Stimuli consisting of brief tones were presented at 1-sec intervals. c, post-stimulus-time histogram, cell B. d, cross-correlation histogram under repetitive stimulation. Plotted histogram is observed; blackened points correspond to predicted cross correlation (method described in text) based on assumption of absence of potentiating effect on interactions by the stimulus. See text.

does not then agree with that predicted using Equation 19, the appropriate conclusion is that we actually have the situation of Fig. 5 c, where I represents an interneuronal network that is ineffective unless potentiated by the stimulus.

If the cross correlation with stimulus off is not flat, we may conclude that there is interaction, i.e., that I is functioning. If the prediction for the stimulus-on cross correlation agrees with the actual measurement, we may conclude that we have unmodulated interaction (Fig. 5 b). If there is statistically significant disagreement, we may conclude that the interaction is modulated by the stimulus (Fig. 5 c). It may then be possible to distinguish among several different modes of such modulation.

The application of these techniques to the detection of interactions has been investigated through digital computer simulations, and the results have been presented elsewhere (Perkel, 1964).

An example of this technique, drawn from experiments in the cochlear nucleus of the cat, is shown in Fig. 6. The agreement between predicted and observed cross-correlation histograms with the stimulus on indicates that the complex shape of this histogram can be completely explained by stimulus-imposed changes in firing patterns of both units.[4]

In summary, we may distinguish among the three possibilities as follows: (*a*) When there is no interaction, the cross-correlation histogram is flat with stimulus off, and periodic with the period of the stimulus with stimulus on; and the cross correlation can be predicted from mean rates and PST histograms. (*b*) When there is interaction, the cross-correlation histogram with stimulus off is not flat; with the stimulus on, it may not be strictly periodic, but the correlation function can still be predicted. (*c*) When the interaction is modulated by the stimulus, the predicted cross correlation does not agree with observation.

The authors thank Dr. José P. Segundo and Professor Theodore H. Bullock for continued encouragement, discussions, and criticism. We are also very grateful to Dr. P. A. W. Lewis for valuable criticisms and suggestions.
This work was supported by United States Air Force Project RAND, by the National Institutes of Health through grants GM-09608, GM-14633, NB-05606, and NB-06372, and by the National Science Foundation through grant G-21497.

Received for publication 27 December 1966.

REFERENCES

Cox, D. R. 1962. Renewal Theory. John Wiley & Sons, Inc., New York.
Cox, D. R., and P. A. W. Lewis. 1966. The Statistical Analysis of Series of Events. John Wiley & Sons, Inc., New York.
Gerstein, G. L., and W. A. Clark. 1964. *Science.* **143**:1325.
McFadden, J. A. 1962. *J. Roy. Statist. Soc. B.* **24**:364.

[4] The possibility of a small excitatory influence of cell A on cell B, with a latency of about 4 msec, however, cannot be excluded, but this would have a negligible effect on the remainder of the cross-correlation histogram, which extends to ½ sec.

MOORE, G. P., D. H. PERKEL, and J. P. SEGUNDO. 1966. *Ann. Rev. Physiol.* **28**:493.
PERKEL, D. H. 1964. Detection of Functional Interactions Among Neurons: A Technique Using Repetitive Presentations of Stimuli. The RAND Corporation, Santa Monica, Calif. RM-4234-NIH.
PERKEL, D. H. 1965. *In* Biophysics and Cybernetic Systems. M. Maxfield, A. Callahan, and L. J. Fogel, editors. Spartan Books, Inc., Washington, D.C. 37.
PERKEL, D. H., G. L. GERSTEIN, and G. P. MOORE. 1967. *Biophys. J.* **7**: 391.
SIMON, W. 1965. *Electroencephalog. Clin. Neurophysiol.* **18**:192.
WYMAN, R. 1965. *Biophys. J.* **5**:447.

Chapter 3
Visual Functions of Primate Inferior Temporal Cortex

Setting up a Primate Neurophysiology Lab

Charles G. Gross

How I Met George

In 1961, I was a post-doc with H.-L.Teuber at MIT in what was then the Psychology Section of the Department of Economics and Social Science. Under Teuber's leadership, it was to become the Department of Psychology and then the Department of Brain and Cognitive Sciences and arguably the first department of neuroscience in the world.

I had come to MIT to work with Teuber on brain injured patients, but due to some territorial dispute about access to patients I went back to animals. I had done my dissertation on the effects of frontal lobe lesions on behavior in monkeys in Cambridge, England. However, I had come to believe that frontal lobe studies existed in some limbo outside of anatomy, physiology, and psychology. So I decided to move to the temporal lobe, specifically inferior temporal cortex that was known to be involved in visual learning and memory.

I started by studying the effect of lesions of inferior temporal cortex on visual learning. At this time, Hubel and Wiesel had begun their studies of microelectrode recording of neurons in striate cortex so I thought recording from temporal cortex seemed a good way to go. However, I had no clue as how to carry out such a study, so I went to Teuber and told him I wanted to leave his department and go elsewhere to learn how to record from single neurons.

Teuber, who was in the very first stages of building a department, said I should stay and do it at MIT. I told him I had no equipment such as oscilloscopes or

C. G. Gross (✉) (deceased)
Princeton University, Princeton, NJ, USA
Passed away on April 14, 2019.

A. Aertsen et al. (eds.), *Introducing Computation to Neuroscience*, Springer Series in Computational Neuroscience, https://doi.org/10.1007/978-3-030-87447-6_3

electrodes or what not to carry out such a study. He said he would buy whatever I needed. I replied that even if I had the equipment I had no idea, for example, how to even turn on an oscilloscope let alone how to record from isolated neurons in the brain of a monkey. Teuber replied that I should go collaborate with a George Gerstein who knew about such things.

George was a post-doc in the Communications Biophysics Laboratory (headed by Water Rosenblith) of the Research Laboratory of Electronics, MIT, and was recording from the auditory system in cats. When I discovered, he had a Ph.D. in physics from Harvard and had already made major contributions to the quantitative analysis of single-neuron activity I was absolutely delighted with Teuber's suggestion. I introduced myself to George, and he was quite enthusiastic to collaborate on studying the electrical activity of inferior temporal cortex.

What We Did

We bought lots of equipment: oscilloscopes, an 8-track tape recorder, switching circuitry, and all sorts of tools and wires. George built equipment (most of which I used for another 40 years). We made tungsten- and glass-coated electrodes (long before you could buy them). George taught me how to use an oscilloscope, not to step on insulated cables, to measure seven times before cutting and many other useful things that formed the foundations of my subsequent life as a neurophysiologist.

Among George's many mantras about doing science was: "the cerebral cortex is soluble in anesthetics," i.e., record from awake animals engaged in some cognitive task; by contrast recording from animals under anesthesia is likely to be uninteresting and artifactual. So we trained macaques to fixate and discriminate visual stimuli. Our plan was to record from inferior temporal cortex when the monkeys were actively engaged in such behavior.

As a preliminary to single-neuron recording, we recorded slow cortical potentials from interior temporal cortex during performance of a visual discrimination task. We averaged the potentials evoked by the discriminanda and found that there was no effect on them (a) of the meaning of the stimuli, (b) of their physical pattern, (c) of whether the following response was correct, or (d) of the state of learning (Gerstein et al. 1968). That is, we found nothing other than in our hands slow potential was not an obvious key to cortical function, which, perhaps, we should have guessed. (The analysis was done by an early LINC computer that George played a role in adapting to neurophysiology.) So we proceeded to build a laboratory to record from single neurons in temporal cortex.

Before we recorded from our first neuron, George had to leave MIT and take an assistant professorship at the University of Pennsylvania. His departure was due to a dispute George, and some of his colleagues involved in the development of the LINC had with the lab chief Walter Rosenblith over money, status, and power. (How such an unscrupulous operator as Rosenblith who seems to have done little science himself could have subsequently become MIT Institute Professor and provost, and one of the few members of the all three National Academies should be some kind of a lesson.)

George was now a long-distance collaborator, and not around to orchestrate the multiple complexities of recording from single neurons in awake monkeys as they carried out cognitive tasks, a type of experiment that had hardly been done anyplace at this time. So I decided to radically simplify the planned experiment so that I could carry it out without George on site. The simplest experiment I could think of was just to ask whether inferior temporal (IT) neurons would respond to visual stimuli and not to auditory stimuli and, for a control area we recorded from the superior temporal gyrus, believed to be an auditory analogue of IT cortex. Soon another post-doc who had also come to work with Teuber, Peter Schiller, joined George and me. Even for the time, our experiment was beyond simple: it was naïve and simplistic. For example, the standard visual stimuli we used were diffuse light, already known to be rather ineffective for cortical neurons. Moreover, the monkey's eyes were uncorrected and merely covered with a viscous silicone fluid to prevent drying out; the fovea and other retinal landmarks were not located. The animals were immobilized and anesthetized.

By vigorous averaging of the responses, we found IT cells to respond to light in about a quarter of our sample; no IT cells responded to the auditory stimuli. We found the opposite pattern in the superior temporal gyrus. Thus, in spite of the crudeness of our experiment, we manage to show for the first time that cells in inferior temporal cortex were exclusively visual, whereas those in superior temporal cortex were auditory (Gross et al. 1967).

Then with George's support, we returned to his original plan of recording from awake behaving monkeys. We had three intriguing findings. First, most of the inferior temporal neurons responded vigorously when the monkey fixated an attention getting visual stimulus like a small flame. These results suggested that these neurons either were involved in some attentional mechanism, had foveal receptive fields, or both. The second finding was that the receptive field properties of IT units in the awake animal were similar to those we had seen in the anesthetized paralyzed animal. The third finding came from an experiment in which the monkey was fixating and discriminating visual stimuli. We found that the responses of some IT neurons depended on whether the stimulus was presented first or second and if second on whether it followed the same or a different stimulus (Gross et al. 1979).

Postscript

These experiments with George initiated a still expanding and very active field of research in humans and other primates. The basic findings that inferior temporal cortex is a visual area with neurons having foveal receptive fields whose activity can be modulated by attention and experience were eventually repeatedly replicated and extended. Further research indicated that some inferior temporal neurons selectively respond to faces and this developed led to a vigorous industry of studying face processing areas in humans and monkeys and then, more generally, domain specificity in the temporal lobe. George Gerstein played a major role in getting this whole field started (Gross 2005, 2008).

References

Gerstein GL, Gross CG, Weinstein M (1968) Inferotemporal evoked potentials during visual discrimination performance by monkeys. J Comp Physiol Psychol 65(3, Part 1):526–528. https://doi.org/10.1037/h0025813

Gross CG (2005) Processing the facial image: a brief history. Am Psychol 60(8):755–763. https://doi.org/10.1037/0003-066X.60.8.755

Gross CG (2008) Single neuron studies of inferior temporal cortex. Neuropsychologia 46(3):841–852. https://doi.org/10.1016/j.neuropsychologia.2007.11.009

Gross CG, Schiller PH, Wells C, Gerstein GL (1967) Single-unit activity in temporal association cortex of the monkey. J Neurophysiol 30(4):833–843. https://doi.org/10.1152/jn.1967.30.4.833

Gross CG, Bender DB, Gerstein GL (1979) Activity of inferior temporal neurons in behaving monkeys. Neuropsychologia 17(2):215–229. https://doi.org/10.1016/0028-3932(79)90012-5

SINGLE-UNIT ACTIVITY IN TEMPORAL ASSOCIATION CORTEX OF THE MONKEY[1]

CHARLES G. GROSS, PETER H. SCHILLER, CAROLYN WELLS, AND GEORGE L. GERSTEIN

Harvard University and Massachusetts Institute of Technology, Cambridge, Massachusetts; Yale University, New Haven, Connecticut; and University of Pennsylvania, Philadelphia, Pennsylvania

(Received for publication October 17, 1966)

LESIONS IN THE POSTERIOR ASSOCIATION CORTEX of the monkey are known to produce modality-specific deficits on discrimination tasks. Ablation of inferotemporal cortex (area TE) impairs the performance of visual discriminations (4, 5, 14, 15). Ablation of superior temporal association cortex (area TA) interferes with auditory discrimination habits (6, 13, 16). Hence modality-specific "learning," "memory," or "perceptual" functions have been ascribed to these areas.

Inferotemporal cortex receives fibers from prestriate cortex which in turn receives fibers from striate cortex (2, 12). Interruption of this cortico-cortical pathway seems to have the same effects as inferotemporal ablation (15). Furthermore, visual-evoked responses can be recorded from gross electrodes on inferotemporal cortex (7a, 22) and these responses are drastically reduced by ipsilateral striate cortex lesions and ipsilateral optic tract section (22, 22a). Thus it is conceivable that inferotemporal cortex might be a site of further processing of visual information received from prestriate and striate cortex (cf. 10, 11). If this were true, its neurons might have receptive fields and highly complex response properties.

Stimulation of inferotemporal cortex has been reported to affect the visual-evoked response recorded at striate cortex (20). Thus an alternate speculation on the functions of inferotemporal cortex is that it has a centrifugal influence on striate cortex and, conceivably, its units might not respond to visual input at all.

Somewhat less is known about superior temporal cortex. Gross evoked responses to auditory stimuli have been recorded from this area (18) and it has afferent and efferent connections with several portions of the classical auditory system (1, 21, 24).

[1] This investigation began when the first three authors were associated with the Psychology Department, Massachusetts Institute of Technology and the last with the Communications Biophysics Laboratory, Research Laboratory of Electronics, M.I.T. It was supported by grants from the National Institutes of Health (HD01079, HD01907, and NB05606), the National Science Foundation (GB4104), the National Aeronautics and Space Administration (NsG-496), and the U.S. Air Force Office of Scientific Research. Additional facilities were provided by the Psychology Department and the Research Laboratory of Electronics, Massachusetts Institute of Technology.

The present study is an investigation of single-unit activity in areas TA and TE. It concerns the response of units in these areas to auditory and visual stimuli.

METHODS

Surgical procedures. Macaca mulatta or *Macaca speciosa* weighing 3–6 kg were used. Surgery was performed under deep Nembutal anesthesia (40 mg/kg, iv). After insertion of an intravenous cannula, a tracheotomy was performed. The animal's head was fixed by clamping the orbits and roof of the mouth, the temporal skin was cut, a flap turned in the temporal muscle, and a 5/8-inch-diameter hole trephined in the temporal bone near the posterior termination of the zygoma. Removal of the dura mater revealed the superior temporal sulcus and, usually, the vein of Labbé. Thus portions of areas TE (namely, the middle temporal gyrus) and TA were accessible for recording (see Fig. 1). The trephine hole was threaded and a stainless steel well screwed into it and sealed to the surrounding bone with a low-melting-point wax. The incision was then closed with wound clips. The well was filled with mineral oil at 38 C. The monkey was kept on a thermostatically controlled heating pad, and its temperature and heart rate monitored. An intravenous drip of 5% dex-

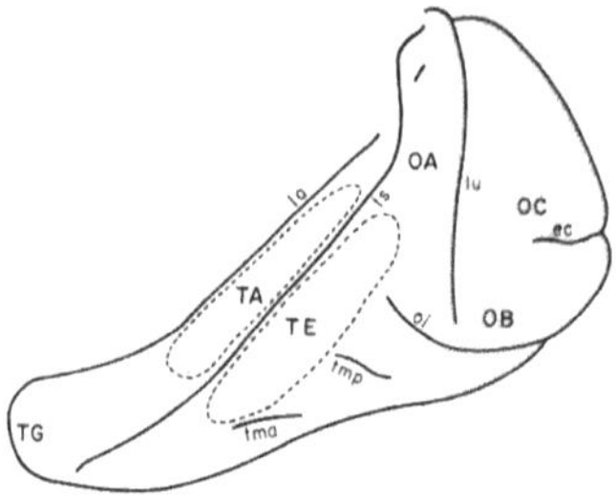

FIG. 1. Areas within which recordings were made are indicated by dashed lines. The capital letters designate cytoarchitectonic areas (23)—ec = ectocalcarine sulcus; la = lateral fissure; lu = lunate sulcus; oi = inferior occipital sulcus; tma = anterior middle temporal sulcus; tmp = posterior middle temporal sulcus; ts = superior temporal sulcus. Derived from Von Bonin and Bailey (23).

trose in Ringer solution was maintained. One or both eyelids were retracted and the pupil dilated with atropine.

In the first series of preparations light Nembutal anesthesia was maintained (ca. 15–30 mg/hr, iv). Usually the animal's eyes were immobilized by intravenous administration of gallamine triethiodide or succinylcholine and the animal maintained by artificial respiration. In these preparations the eyes were covered with a thin film of Corning silicon 200 (viscosity, 60,000 centistokes) or a contact lens. Otherwise the eyes were kept moist with application of Tearosol.

In the second series of preparations Nembutal anesthesia was not maintained after the surgical procedures had been completed. Rather, the head, throat, and leg incisions were packed with Xylocaine (lidocaine hydrochloride) jelly and infused every 4 hr with 2 ml of 0.5% solution of Xylocaine. The animals were immobilized with gallamine triethiodide or succinylcholine when they began to recover from the Nembutal anesthesia. Procedures were otherwise the same as in the first series. Recording was not begun until at least 1 hr after behavioral and electrographic indications that the animal was no longer under Nembutal anesthesia. Several passes were also made under chloralose anesthesia (60 mg/kg, iv).

Recording methods. Insulex-coated tungsten microelectrodes were used for recording. They were inserted in a hydraulic micromanipulator which was fitted into the steel well in the monkey's skull. This closed-chamber device was modified from one designed by Gerstein and Kiang (8). It had two eccentrics which enabled the microelectrodes to be inserted into the brain at any specified point within the inner diameter of the well.

A conventional recording system was used: unity gain emitter follower, Grass P5 amplifier, Tektronix 502 oscilloscope, audio amplifier and loudspeaker. Potentials and stimulus markers were recorded on a Honeywell 8100 FM tape recorder. Isolated units were detected by a Schmitt trigger that provided a digital input for a Mnemotron Computer of Average Transients (CAT) and for a counter. Subsequently the recorded data were reanalyzed in more detail with the CAT or a LINC computer.

Two types of "formal" visual stimuli were used, "flash" and "light." The flash was produced by a Grass PS-2 photostimulator. Its duration was 9μsec. The light was produced by a GE-327 tungsten filament bulb mounted in a 6-inch reflector, covered with a diffusing screen and driven by a Grass S-4 stimulator. A 4.5-v bias was maintained across the bulb. The duration of the light was 1 sec, its intensity 16 ft-L, its rise time 50 μsec, and its fall time 15 μsec. The diffusing screens of both light sources were located 20 inches in front of the monkey's eyes. Stimuli were presented once every 2 sec. In addition, visual stimuli could be projected onto a screen 36 inches in front of the monkey. The background illumination of the screen was .08 ft-L and that of the projected stimuli 5 ft-L. The screen subtended a visual angle of 45°.

Two types of controlled auditory stimuli were used. The first was produced by putting a 200-μsec square pulse into a speaker ("click"). The second was a 1-sec tone burst ("tone"). Its frequency was usually 500 cycles/sec but other frequencies were also used. The intensity of both auditory stimuli was about 70 db re human threshold. They were presented once every 2 sec. Informal stimuli such as jangling keys, handclaps, and monkey calls were also used.

RESULTS

Response to diffuse visual and auditory stimuli: Nembutal anesthesia

Under Nembutal anesthesia inferotemporal units responded only to visual stimuli. Conversely, only auditory stimuli altered the firing of units in superior temporal cortex. In both areas, fewer than half the units studied showed evoked activity (Fig. 2). With one or two possible exceptions, every unit that responded to the brief flash also responded to the 1-sec light and every unit that responded to the click also responded to the 1-sec tone. However, with the particular stimulus parameters employed, the light and click produced stronger responses than the flash or tone.

None of the four formal stimulus conditions produced a clear and consistent change in the mean frequency of firing of any unit. Rather, when the stimulus affected the unit, it did so by altering the poststimulus pattern of firing. Because of the spontaneous activity of the units it was usually impossible to detect this evoked activity by listening to the loudspeaker or observing the oscilloscope. Rather, to detect evoked activity it was usually necessary to compute poststimulus histograms by summing the firing frequencies at different times after the stimulus onset, over successive stimulus presentations. In the case of very weak responses the reality of the response was checked by repeated computations of poststimulus histograms and by comparisons with poststimulus histograms obtained in the absence of stimulus presentation.

Over 90% of the time-locked units responded to the brief flash or click with one or more periods of increased activity. In the remaining cases, all inferotemporal, the response consisted of a period of decreased activity. The pattern of response to the 1-sec light could be classified according to whether the unit *a*) increased or decreased its firing at *b*) the onset or termination of the stimulus. Although some inferotemporal units fell into each of the four classes, over half of the responsive units showed increased activity both at stimulus onset and stimulus termination. Similarly, the great majority of the responsive superior temporal units showed increased activity to both onset and termination of the tone. In both cortical areas the pattern of response of some units that were studied over several hours altered with time. Typical forms of response to light by inferotemporal cells are shown in Fig. 3 and to sound by superior temporal units in Fig. 4.

The peak of the initial change in activity after the onset of the visual stimuli varied from 140 to 340 msec with a mean of about 250 msec. The peak of the initial response to the auditory stimuli occurred at an earlier

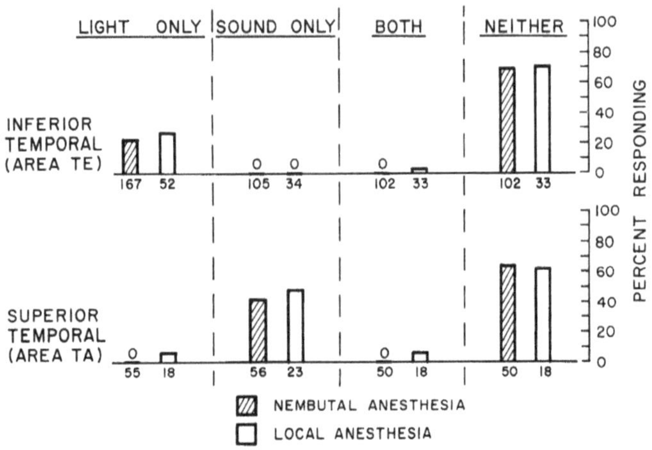

FIG. 2. Proportions of units in inferior and superior temporal association cortex responding to visual and auditory stimuli under Nembutal and local anesthesia. Numbers under each bar are the total number of units from which the illustrated percentages were obtained.

time with a range of 45–370 msec and a mean of about 130 msec. The shortest latency of response to light of any inferotemporal unit was 90 msec and to sound of any superior temporal unit was 30 msec.

In both inferior and superior temporal cortex the evoked activity of some units would decline and disappear after 5–30 min of repeated stimulation. Often, after 5 min of no stimulation or stimulation in the other modality, the evoked activity would reappear (Fig. 5). Such a cycle was not associated with any change of mean frequency of firing.

In a few cases stimulation of the contralateral, ipsilateral, and both eyes were compared. The response to stimulation in each eye alone was similar to but smaller than the response when both eyes were stimulated (Fig. 6).

As shown in Fig. 2, in both superior and inferior temporal cortex, more than half of the units did not respond to any of the stimuli. The proportions shown are subject to at least two qualifications. In the first place, although units varied enormously in their rate of firing, poststimulus time histograms were obtained for constant periods of time (almost always 4 min) rather than for a constant number of spikes. Since the effect of the stimulus was often a

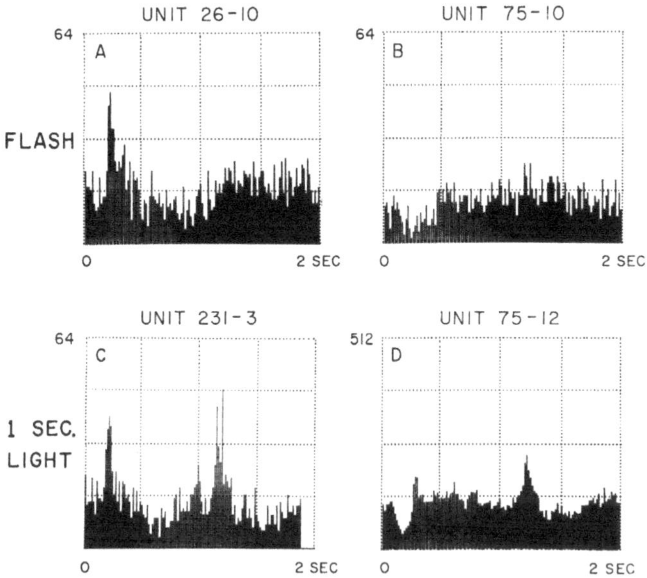

FIG. 3. Poststimulus time histograms from four inferotemporal units during presentation of visual stimuli once every 2 sec. *A*: 9-μsec flash, Nembutal, 75 stimuli. *B*: 9-μsec flash, local anesthesia, 110 stimuli. *C*: 1-sec light, Nembutal, 120 stimuli, *D*: 1-sec light, local anesthesia, 110 stimuli.

weak one, slow units would therefore be much less likely to be classified as time-locked. However, even if units less active than any given rate are eliminated from the sample the proportion of responsive units would not increase above about 75%. In the second place, since the evoked activity of some units disappeared with repeated stimulation and never reappeared after 5–15 min of no stimulation, it is possible that some of the units classified as nontime-locked would have been classified otherwise if they had been studied early in the preparation or after a longer period of no stimulation in the relevant modality. This factor was probably not important since there was no consistent decline in the frequency of finding time-locked units within one preparation or one electrode pass.

Response to diffuse visual and auditory stimuli: no Nembutal anesthesia

The results obtained when the monkeys were paralyzed with gallamine triethiodide or succinylcholine and locally anesthetized with Xylocaine

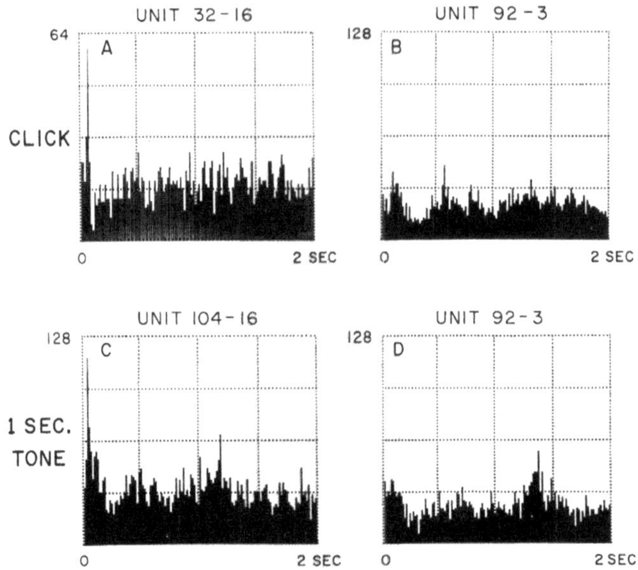

FIG. 4. Poststimulus time histogram from four superior temporal units during presentation of auditory stimuli once every 2 sec. *A*: click, Nembutal, 105 stimuli. *B*: click, Nembutal, 104 stimuli. *C*: 500 cycles/sec tone for 1 sec, Nembutal, 220 stimuli. *D*: 100 cycles/sec tone for 1 sec, Nembutal, 110 stimuli.

were identical in all respects to those obtained under Nembutal (Figs. 4 and 5), except for three units. One unit in inferotemporal cortex responded to both the light and click stimuli. Another unit, in superior temporal cortex, responded to the light stimulus but not to the click. The third unit, also located in superior temporal cortex, responded to both stimuli.

A few cells in inferotemporal cortex were studied under chloralose. Seven responded to the visual stimuli and not to the auditory stimuli and six responded to neither the visual nor the auditory stimuli.

Receptive fields: response to projected stimuli

In the course of examining about 30 inferotemporal units, moving and stationary circles, edges and bars of light were projected onto a screen in

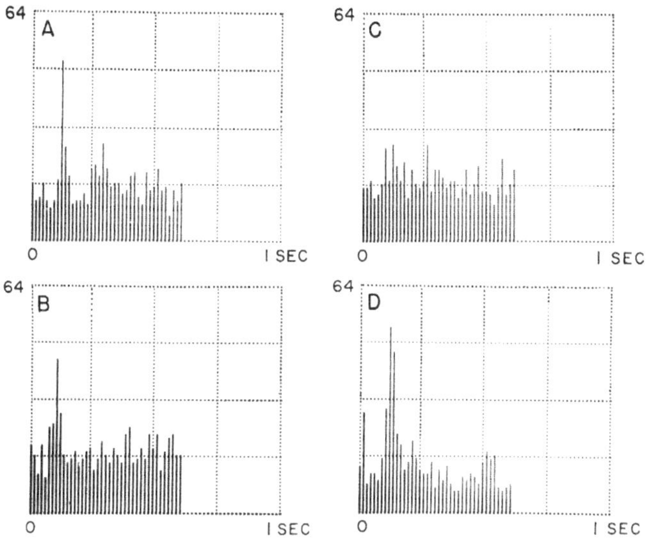

FIG. 5. Poststimulus histograms of an inferotemporal unit showing waning and recovery of response to 9-μsec flash presented once every 2 sec. The light stimulus was presented for 15 min, then no stimulus was presented for 15 min, and finally the light was again presented for 4 min. *A*: 0–4 min. *B*: 5–9 min. *C*: 10–14 min. *D*: 30–34 min. Nembutal anesthesia.

front of the monkey. This did not appear to produce any marked or repeatable changes in pattern or frequency of firing. Since averaging techniques were almost always necessary to detect responses to diffuse illumination, an attempt was made to detect receptive fields by averaging techniques. A bar of light was repeatedly moved across the screen synchronously with the sweep of a Mnemotron Computer of Average Transients. Thus the CAT provided a plot of frequency of firing at each locus of the stimulus in the visual field. By varying the direction of movement and orientation of the

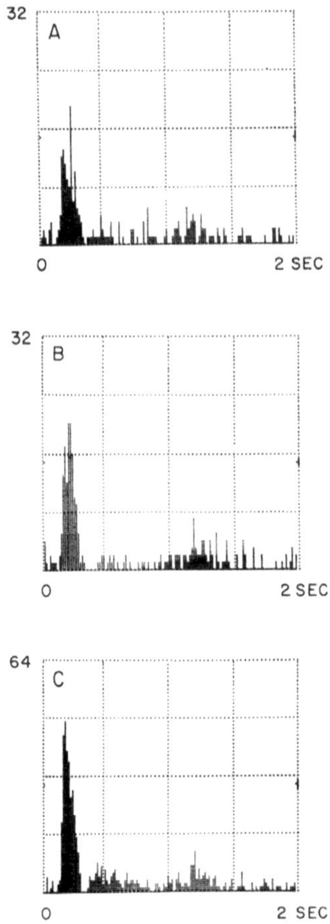

FIG. 6. Comparison of response of an inferotemporal unit to monocular and binocular stimulation by a 1-sec light presented once every 2 sec under Nembutal. *A*: left eye. *B*: right eye. *C*: both eyes.

stimulus, this method had proved successful in plotting receptive fields of units in the striate cortex of the cat and the monkey. However, this method never yielded any clear, consistent and reliable evidence of receptive fields in the about 25 inferotemporal units so studied.

There were only a few units in inferotemporal cortex that seemed to respond to complex moving stimuli. By contrast, relatively "natural" auditory stimuli, such as handclaps and jingling keys, frequently evoked vigorous activity on the part of those superior temporal units that responded to clicks and tone bursts.

Spontaneous activity. All units that were observed were spontaneously active at rates from 50/sec to under 1/sec. The spontaneous rate of firing of a particular unit often varied over a fivefold range within a period of several hours. No relationships were noted between the pattern of spontaneous activity (as measured by an interval histogram) and the response properties or site of the unit. Although there were no significant differences in the mean spontaneous rate as a function of anesthesia, rates of firing in excess of 20/sec were somewhat more common in the preparations that were not under general anesthesia.

Interval histograms were obtained for most units in this study under both spontaneous and stimulated conditions. Since stimuli were presented at a slow rate and since the responses were relatively weak, there was little effect of stimulation on the interval histograms. Approximately a third of the units in our sample fired in rapid bursts with long pauses between bursts. In such cases it was usually not possible to accumulate sufficient data to delineate the interval histogram. The remaining interval histograms were classified according to slope and curvature on a semilogarithmic plot (19). In our sample there were no units with gaussian interval histograms (tail decreasing more rapidly than exponential), and a very small number of units with bimodal interval histograms. There were about twice as many units with long tailed interval histograms (tail decreasing more slowly than exponential) as units with exponential interval histograms (Fig. 7).

Discussion

The finding that inferotemporal units responded to light but not to sound is in striking accord with previous findings that inferotemporal lesions

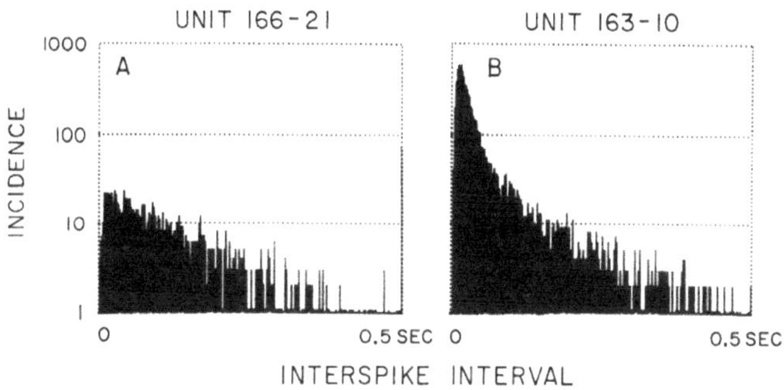

FIG. 7. Interval histograms of two units in inferotemporal cortex recorded under Nembutal anesthesia. Plotted semilogarithmically. In the unit illustrated in *A* the density of the interspike intervals decreases exponentially, whereas in the unit illustrated in *B* it decreases more slowly than exponentially. *A* is based on 1,359 spikes and *B* on 21,745 spikes.

impair visual but not auditory discrimination. Similarly, the response of superior temporal units to sound but not to light fits well with the reports that destruction of this area impairs auditory but not visual discrimination. In fact, the similarities of the evoked unit activity in these two temporal association cortices supports the view that they play analogous modality-specific roles in discrimination learning.

The response latency to light of the inferotemporal units was consistent with a prestriate-striate afferent route. Proof of this possibility will be sought by studying inferotemporal unit response after interruption of this pathway.

The relatively high incidence of units with long tailed histograms in our sample from the monkey should be contrasted with the relative rarity of such units at various locations in the auditory system of the cat (17). Recent modeling work (7, 9) has suggested that long tailed interval histograms might be associated with a rich variety and high rate of excitatory and inhibitory synaptic input. Perhaps this difference in incidence of units with long tailed interval histograms is due to the increased complexity of neurons in the monkey cortex (3, p. 860).

The present findings do not lend specific support to either of the two speculations on inferotemporal cortex mentioned in the introduction. The fact that some inferotemporal units responded to light suggests that the function of this tissue may not be exclusively a centrifugal one. On the other hand, there was no indication of the type of visual processing discovered by Hubel and Wiesel in areas 17, 18, and 19 of the cat (10, 11). This absence of evidence for receptive fields of inferotemporal units may reflect an organization fundamentally different from that found in prestriate and striate cortex. However, in the present experiments the principal stimuli studied in a quantitative fashion, other than diffuse light, were slits, edges, and circles. Although these are adequate stimuli for many striate cortex units, the recent report of Hubel and Wiesel (11) indicates that they are not optimal for units in areas 18 and 19 and hence may not have been appropriate for units in inferotemporal cortex.

Summary

In the monkey, single-unit activity of the superior temporal gyrus (area TA) and the middle temporal gyrus (area TE) was investigated. These areas are known to be necessary, respectively, for normal auditory discrimination learning and for normal visual discrimination learning.

About one-quarter of the area TE units studied responded to diffuse light; virtually none responded to click or tone burst. About half of the area TA units studied responded to the auditory stimuli; virtually none responded to the visual stimuli. Results under Nembutal, chloralose, and local anesthesia were similar.

REFERENCES

1. Bailey, P., von Bonin, G., Garol, H. W., and McCulloch, W. S. Functional organization of temporal lobe of monkey (*Macaca mulatta*) and chimpanzee (*Pan satyrus*). *J. Neurophysiol.* 6: 121–128, 1943.
2. Bailey, P., von Bonin, G., Garol, H. W., and McCulloch, W. S. Long association fibers in cerebral hemispheres of monkey and chimpanzee. *J. Neurophysiol.* 6: 129–134, 1943.
3. Cajal, Ramón y, S. *Histologie du Système Nerveux, de l'Homme et des Vertébrés.* Madrid: Instituto Ramón y Cajal, 1955, vol. II.
4. Chow, K.-L. Effects of temporal neocortical ablation on visual discrimination learning sets in monkeys. *J. Comp. Physiol. Psychol.* 47: 194–198, 1954.
5. Chow, K.-L. Anatomical and electrophysiological analysis of temporal neocortex in relation to visual discrimination learning in monkey. In: *Brain Mechanisms and Learning*, edited by J. R. Delafresnaye. Oxford: Blackwell, 1961, p. 507–525.
6. Dewson, J. H. Complex auditory discrimination and lesions of temporal cortex in the monkey: A progress report. *J. Acoust. Soc. Am.* 39: 1254, 1966.

7. Fetz, E. and Gerstein, G. L. A R.C. model for spontaneous activity of single neurons. Quarterly Progress Report, Research Laboratory of Electronics, Massachusetts Institute of Technology, Oct. 15, 1963, pp. 249–257.
7a. Gerstein, G. L., Gross, C. G., and Weinstein, M. Inferotemporal evoked potentials during visual discrimination performance by monkeys. *J. Comp. Physiol. Psychol.* In press.
8. Gerstein, G. L. and Kiang, N. Y.-S. Responses of single units in the auditory cortex. *Exptl. Neurol.* 10: 1–18, 1964.
9. Gerstein, G. L. and Mandelbrot, B. Random walk models for the spike activity of a single neuron. *Biophys. J.* 4: 41–68, 1964.
10. Hubel, D. H. and Wiesel, T. N. Receptive fields, binocular interaction and functional architecture in the cat's visual cortex. *J. Physiol., London* 160: 106–154, 1962.
11. Hubel, D. H. and Wiesel, T. N. Receptive fields and functional architecture in two nonstriate visual areas (18 and 19) of the cat. *J. Neurophysiol.* 28: 229–289, 1965.
12. Kuypers, H. G. J. M., Szwarcbart, M. K., Mishkin, M., and Rosvold H. E. Occipito-temporal corticocortical connections in the rhesus monkey. *Exptl. Neurol.* 11: 245–262, 1965.
13. Massopust, L. C., Barnes, H. W., and Verdura, J. Auditory frequency discrimination in cortically ablated monkeys. *J. Aud. Res.* 5: 85–93, 1965.
14. Mishkin, M. Visual discrimination performance following partial ablations of the temporal lobe: II. Ventral surface vs. hippocampus. *J. Comp. Physiol. Psychol.* 47: 187–193, 1954.
15. Mishkin, M. Visual mechanisms beyond the striate cortex. In: *Frontiers of Physiological Psychology*, edited by R. Russell. New York: Academic. In press.
16. Neff, W. D. Neural mechanisms of auditory discrimination. In: *Sensory Communication*, edited by W. A. Rosenblith. Cambridge, Mass.: Technology Press, 1961, p. 259–278.
17. Pfeiffer, R. R. and Kiang, N. Y.-S. Spike discharge patterns of spontaneous and continuously stimulated activity in the cochlear nucleus of anesthetized cats. *Biophys. J.* 5: 301–316, 1965.
18. Pribram, K. H., Rosner, B. S., and Rosenblith, W. A. Electrical response to acoustic clicks in monkey: Extent of neocortex activated. *J. Neurophysiol.* 17: 336–344, 1954.
19. Rodieck, R. W., Kiang, N. Y.-S., and Gerstein, G. L. Some quantitative methods for the study of spontaneous activity of single neurons. *Biophys. J.* 2: 351–368, 1962.
20. Spinelli, D. N. and Pribram, K. H. Changes in visual recovery functions produced by temporal lobe stimulation in monkeys. *Electroencephalog. Clin. Neurophysiol.* 20: 44–49, 1966.
21. Sugar, O., French, J. D., and Chusid, J. G. Corticocortical connections of the superior surface of the temporal operculum in the monkey (*Macaca mulatta*). *J. Neurophysiol.* 11: 175–184, 1948.
22. Vaughan, H. G. and Gross, C. G. Observations on visual evoked responses in unanesthetized monkeys. *Electroencephalog. Clin. Neurophysiol.* 21: 405–406, 1966.
22a. Vaughn, H. G. and Gross, C. G. Light evoked responses from the cortex of unanesthetized monkeys and their alteration by lesions of the visual system. In preparation.
23. Von Bonin, G. and Bailey, P. *The Neocortex of Macaca mulatta*. Urbana, Ill.: Univ. Illinois Press, 1947.
24. Whitlock, D. G. and Nauta, W. J. H. Subcortical projections from the temporal neocortex in *Macaca mulatta*. *J. Comp. Neurol.* 106: 183–212, 1956.

Journal of Comparative and Physiological Psychology
1968, Vol. 65, No. 3, 526–528

INFEROTEMPORAL EVOKED POTENTIALS DURING VISUAL DISCRIMINATION PERFORMANCE BY MONKEYS[1]

GEORGE L. GERSTEIN
University of Pennsylvania[2]

CHARLES G. GROSS
Harvard University

AND MEIR WEINSTEIN
University of Pennsylvania[3]

During various visual discrimination tasks, responses from inferotemporal and frontal cortex were evoked by the visual discriminanda themselves. The form of the evoked response from any pair of electrodes remained constant regardless of (a) nature of task, (b) stage of learning, (c) stimulus pattern, (d) stimulus meaning, and (e) S's response.

In the monkey, ablation of neocortex on the inferior convexity of the temporal lobe produces a severe impairment of visual, and only visual, discrimination tasks (Mishkin, 1966). We have previously shown, in the unanesthetized monkey, that visual evoked responses can be recorded from inferotemporal cortex in single unit and slow potential recordings (Gross, Schiller, Wells, & Gerstein, 1967; Vaughan & Gross, 1966). Thus, this cortical area may be involved in the processing, or in the storage of, visual information.

There have been two previous studies of inferotemporal slow potentials during visual discrimination learning. In one, Chow (1961) found a transient decrease in amplitude on inferotemporal EEG and pointed out that this decrease might have simply reflected a change in arousal level. In the other, Stamm and Mahoney (1962) found bursts of 6–8 per second activity associated with the rising portion of the learning curve.

In the present study, averaging techniques were used to examine the responses evoked by visual cues at inferotemporal cortex during visual discrimination learning. An effort was made to keep constant *S*'s state of arousal and attention. The *S* controlled the presentation of the discriminanda and always appeared to fixate them. During recording sessions, which normally lasted 4–6 hr., *S*'s state, as reflected in its response latency, its observed behavior, and its electrocorticogram, remained constant.

METHOD

Subjects and Electrodes

Two immature *M. mulatta* and one immature *M. speciosa* served as *Ss*.

Delgado cortical plate electrodes were used. They consisted of six .5 mm. diameter stainless-steel balls arranged in two rows and mounted in .125 mm. thick polyethylene sheets. Adjacent balls were 5 mm. apart. They were implanted subdurally and unilaterally over (*a*) the middle and inferior temporal gyri and (*b*) sulcus principalis in the frontal lobe. Dura, muscle, fascia, and skin were sutured over the electrodes, and their leads brought out to a miniature plug.

Behavioral Techniques

For each experimental session, *Ss* were placed in a monkey chair in an electrically and acoustically shielded chamber. The discrimination tasks were all go no-go (successive) discriminations with symmetrical reinforcement. The discriminanda were patterns projected into 1 in. diameter buttons by Grason-Stadler stimulus projector units. The .15 cc of water, which was the reinforcement, was all the water *Ss* received on the 6 days each week that they were run; on Day 7 water was available ad lib. The *Ss* faced two vertically-aligned buttons 3 in. apart. At the end of the 7-sec. intertrial interval, a single press of the lower button (observing response) projected one of two stimuli onto the upper button for a maximum of 5 sec. If this projected stimulus was the positive one, and *S* pressed the upper button within 5 sec., it was reinforced. However, if this stimulus was negative, and *S* did not respond, it was reinforced. Conversely, failures to respond to the positive stimulus within

[1] This investigation began in the Psychology Department, Massachusetts Institute of Technology. It was supported by grants from the National Institutes of Health (HD01079, HD01907, and NB05606), the National Science Foundation (GB4104), the National Aeronautics and Space Administration (NsG-496), and the U.S. Air Force Office of Scientific Research.

We would like to thank J. S. Stamm for helpful discussions, and Eugenia N. Gregory for assistance in every phase of this study.

[2] Now at the Johnson Foundation and Department of Physiology, University of Pennsylvania.

[3] Now at the Johnson Foundation, University of Pennsylvania.

5 sec. and responses to the negative stimulus were not reinforced. The S's response or termination of the 5-sec. period, whichever came first, turned off the stimulus. Response to either button during the intertrial interval reset the timer controlling the intertrial interval.

Cerebral potentials were studied in the three behavioral situations. In visual discrimination tasks, Ss were trained until they reached a criterion of 90 correct responses in 100 trials. In the reversal-criterion tasks, each time Ss attained the above criterion, the formerly positive cue became the negative one, and the formerly negative cue became positive. In the reversal-fixed trials tasks, the significance of the stimuli was alternated after every block of 11 or 22 trials.

Prior to implantation, Ss were trained on several pattern discrimination and discrimination reversal tasks. They were also trained on a delayed alternation task, the results of which will not be reported here.

Recording and Data Analysis Techniques

Cortical potentials were amplified by Grass P5 preamplifiers and recorded on a Honeywell F.M. tape recorder and a Schwarzer polygraph. Overall frequency response of the system was 1.5–1,250 cps. At any one time, potentials were recorded from three pairs of inferotemporal points, and, as a control, one pair of frontal points. A fifth channel was used to record stimulus, response, and reinforcement events. The electrical activity in the first 500 msec. after onset of the discriminanda served as basic datum. For each task, a LINC computer calculated the average responses at different stages of learning, and over the entire training period, categorized according to positive or negative stimulus and correct or incorrect response. In addition, on the reversal tasks the evoked responses were averaged in categories according to which stimulus was presented, and whether it had positive or negative meaning.

Results and Discussion

Use of the observing response and long continuous daily experimental sessions appeared to maintain a constant level of arousal and attention. Early trials with a naive S usually gave erratic behavior and large fluctuations in the general level of electrical activity. However, during periods of recording from an acclimatized S, there were never any electrographic or behavioral signs of changes in arousal or motivation.

The chief positive finding of this study is that averaged evoked potentials to small dim visual stimuli can be obtained from bipolar electrodes placed over inferotemporal and frontal cortex. All averaged responses grew with their sample size and reached a constant form after about 20 samples. No such averaged responses appeared in controls where no visual stimulus was pro-

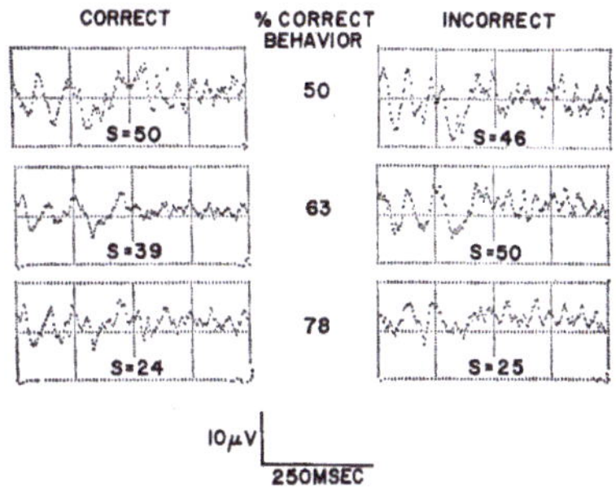

Fig. 1. Evoked responses from S as recorded by bipolar inferotemporal leads during different stages of learning a discrimination between two patterns. (The left and right columns show averages of the evoked responses on trials on which S responded correctly or incorrectly, respectively; at the levels of performance indicated in the center column S = number of evoked responses [trials] in each average.)

jected after the observing response. The average responses from different pairs of electrodes were different, which does suggest a local source for most of the recorded activity. However, from a given pair of electrodes there was no discernible effect on the evoked response (*a*) of the meaning of the stimulus, (*b*) of the physical pattern of the stimulus, (*c*) of the correctness or incorrectness of the response following the stimulus, (*d*) of the stage of learning, or (*e*) of the number of previous trials or problems that day. (See Figures 1 and 2.)

The evoked responses began at least as early as 40 msec. and continued out to about 400 msec. This onset time was roughly comparable to that of responses obtained from inferotemporal, frontal, and striate cortex to a brief intense flash (Vaughan & Gross, 1966). However, single unit response from inferotemporal cortex to both a brief flash and a 1-sec. diffuse light usually had latencies of at least 120 msec. (Gross et al., 1967).

Although the electrical activity in the intertrial interval was not studied in detail, on one discrimination problem there was clear evidence of an association between the occurrence of short 6–8 cps bursts and the middle portion of the learning curve. Stamm and Mahoney (1962) have reported a similar finding.

Since interference with inferotemporal cortex by ablation, cross hatching, circumsection, undercutting, electrical stimulation, and epileptic foci have all been reported to interfere with visual discrimination performance (Mishkin, 1966;

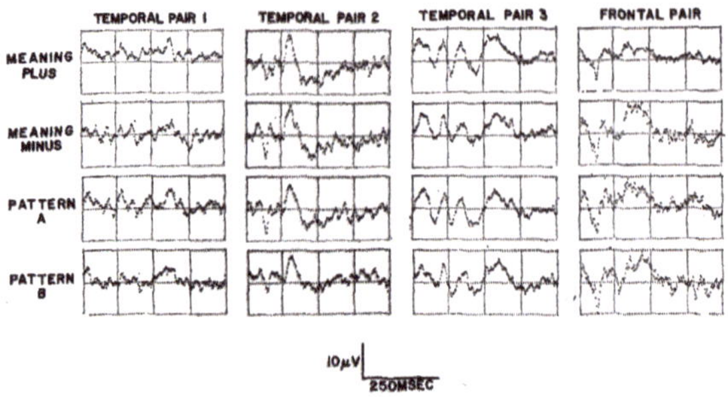

FIG. 2. Evoked responses from *S* as recorded by bipolar inferotemporal and dorsolateral frontal leads during the reversal-fixed trial experiment involving the repeated reversal of a discrimination between horizontal and vertical stripes. (Respectively, the rows show averaged evoked response obtained from trials on which the stimulus presented was the positive stimulus irrespective of pattern, the negative stimulus irrespective of pattern, the horizontal pattern irrespective of meaning, and the vertical pattern irrespective of meaning. These data were obtained from 20 consecutive reversals—a total of 440 trials—and thus each average is of 220 responses.)

Pribram, Blehert, & Spinelli, 1966), it may be inferred that this tissue may be involved in the processing or in the storage of visual information. The failure of either stimulus or experiential variables to affect the evoked response indicates either that this inference is wrong or, much more likely, that gross evoked responses from this tissue are incapable of detecting processing or storage mechanisms.

REFERENCES

CHOW, K. L. Changes of brain electropotentials during visual discrimination learning in monkey. *J. Neurophysiol.*, 1961, **24**, 377–390.

GROSS, C. G., SCHILLER, P. H., WELLS, C., & GERSTEIN, G. L. Single unit activity in temporal association cortex of the monkey. *J. Neurophysiol.*, 1967, **30**, 833–843.

MISHKIN, M. Visual mechanisms beyond the striate cortex. In R. W. Russell (Ed.), *Frontiers in physiological psychology*. New York: Academic Press, 1966. Pp. 93–119.

PRIBRAM, K. H., BLEHERT, S. R., & SPINELLI, D. N. Efferts on visual discrimination of crosshatching and undercutting the inferotemporal cortex of monkeys. *J. comp. physiol. Psychol.*, 1966, **62**, 358–364.

STAMM, J. S., & MAHONEY, W. A. Electrical slow waves from temporal lobe during learning of complex tasks. Paper presented at the *22nd International Congress of Physiological Sciences*, Leiden, 1962. (Abstracted in *Excerpta Medica*, International Congress Series No. 48, 1962.)

VAUGHAN, H. G., & GROSS, C. G. Observations on visual evoked responses in unanesthetized monkeys. *EEG clin. Neurophysiol.*, 1966, **21**, 405–406.

(Received March 20, 1967)

Chapter 4
Plasticity of Neuronal Interactions

Early Days in the Gerstein Lab: Neuronal Plasticity in 1, 2, and 3 Dimensions

William B. Kristan Jr.

George Leonard Gerstein (or GLG, as he signed his notes to us) was a wizard. A kindly one, for sure, but a wizard nonetheless. He could consume and store information faster than any human being I have ever known. And his drive for knowledge was semi-infinite, so he seemed to know everything. That all made him formidable, but what made him a wizard was his ability to fix things. His office was on the fourth floor of the Richards Building and the lab was in the basement. So, when something (the animal, the electronics, the computer) did not work, we would try everything we could imagine to fix it before we would make the long trudge upstairs to ask George to come down to the lab and take a look. Many times, he would come downstairs, sit down at the rig/electronics/computer, twiddle one electrode/knob/switch and the blasted thing would work! Usually, it was an electrode/knob/switch that we had already twiddled several times; it just needed the magical touch of the wizard.

Having arrived at Penn in 1964, George spent a year assembling his lab. He had secured a first-generation lab computer, the LINC (the precursor of the PDP-8), which he had put together himself. And by "put together," I mean that he and a lab tech soldered the components (resistors, capacitors, transistors) onto individual memory cards by hand—all 512 of them! Yes, the total RAM in that computer was 512 bits, each of which could serve either as a program command or as one unit of data storage. (The data could be dumped onto one of two small tape drives, to store

W. B. Kristan Jr. (✉)
Department of Neurobiology, School of Biological Sciences, UC San Diego, La Jolla, CA, USA
e-mail: wkristan@ucsd.edu

A. Aertsen et al. (eds.), *Introducing Computation to Neuroscience*, Springer Series in Computational Neuroscience, https://doi.org/10.1007/978-3-030-87447-6_4

"large" data sets.) I can still picture George trying to track down an intermittent computer problem: he would write a small program that looped though all 512 locations, and then he would open the back of the computer (it was about the size of a home refrigerator) and wiggle the memory cards one at a time until the program hung up. He would then pull that card and fix or replace it. Those were the days when a computer scientist truly knew his machine inside out.

Russ Fernald and I were the first graduate students in George's lab; we both joined the lab in 1965. I wanted to work on the cellular mechanisms of learning and memory. George, who had never worked on any animal other than cats and monkeys, suggested that I work on either mammalian brain slices (there were two people in the world doing slice electrophysiology at that time, and one of them was at Penn) or on an invertebrate nervous system. He had just heard Eric Kandel give a talk on synaptic plasticity in the sea slug, Aplysia californica, and thought it held some promise (how prescient!). I chose Aplysia both because they had huge neuronal cell bodies (20–50 time larger than the largest mammalian neurons in slices) and because Aplysia afforded a better chance to tie any neuronal plasticity I might find to the animal's behavior, a capability that has always been important for me.

For the Aplysia project, I had a rig of my own in my own little room. (People who worked on cats shared the rest of the lab equipment.) Having my own equipment did mean, however, that I had to build everything that I needed, from the Faraday cage to the temperature controller for the saline to the AC and DC amplifiers that I needed to do extracellular and intracellular recordings. Right in the middle of building the electronics from transistors, operational amplifiers came on the market. They were bulky (a single channel was in a box about the size of a deck of cards) and relatively expensive (more than $50 each), but they had two advantages over circuits of transistors: (1) op-amp outputs were linear and (2) successive stages of amplification were independent of one another. The immediate benefit of building my own rig was that whenever anything went wrong, I knew how to fix it. The longer-lasting benefit was that I could readily assemble my own equipment, first as a Grass Fellow at Woods Hole, then as a postdoctoral fellow at Stanford and Berkeley, and ultimately in my own lab at UCSD and in student labs that I taught for many years. I was grateful that George required me to acquire these skills.

During the time that I was in George's lab, he had a very productive collaboration with Don Perkel, who was then at the Rand Corporation in La Jolla, CA. With some collaborators at UCLA (notably George Moore and Jose Segundo), George and Don had developed the use of autocorrelations, cross-correlations, and PST histograms. At that time, PST was short for Post-Stimulus Time. One afternoon over tea in the lab, they puzzled over the fact that in George's cat auditory experiments, they were giving a frequency-modulated signal that repeated at 1 Hz and they were generating histograms over the whole modulated signal. Hence, because the stimulus was on all the time, there was not a true "post-stimulus" response. After a couple of teas and much debate, George came up with the solution: call it a "Peri-Stimulus Time" histogram. This term had the advantage of not needing to change the acronym, its meaning was self-evident, and it would work for responses to punctate data, too. So they introduced this terminology in their next paper and, as far as I know, the terminology caught on—people refer to "Peri-Stimulus Time histograms" with no mention of "Post."

The major idea that George and Don were hatching while I was in George's lab was how to use cross-correlations between two different spike trains to suggest connectivity. They developed a display scheme that they called a "joint PST scatter diagram," which is a way to look at the unfolding of correlations over time. This diagram effectively plots one PST histogram along the X axis and a second PST histogram along the Y axis. In each stimulus cycle, dots are placed at the times of each X and Y spike: if each cell generates only 1 spike in a cycle, one dot is placed at the intersection of the x and y times on the plot; 5 spikes in each cell generate 25 dots, at all the possible (x,y) intersections. Like a histogram, the dots from all the stimulus cycles are plotted on the same diagram. In such a plot, responses to each stimulus form a density parallel to their appropriate axis, and time-locked spikes (i.e., highly correlated spike trains) form a density of dots parallel to the 45-degree line, because they occur at nearly the same times wherever they occur in the stimulus cycle. The advantage of such a plot over simply a cross-correlation between two spike trains is that it can show whether the correlations between the spikes are similar over the whole stimulus cycle.

In addition, George and Don developed a way to visualize the interactions among three simultaneously recorded neurons, generating a "joint PST scatter diagram in 3-D" (Gerstein and Perkel 1969). Each cell's spike train was plotted along one of the three axes, generating dots in a 3-D display: each spike in a given train would produce dots in the 3-D space with each of the spikes in the other trains. Again, stimulus-driven spikes would produce clouds of dots parallel to the appropriate axis, and correlated spikes between any two neurons produce dot clouds parallel to the 45° axis between the two neurons. The big payoff was that correlated spikes among all three neurons produce a dot cloud at 45 degrees to all three axes. To appreciate the orientations of the clouds, the diagrams need to be seen in 3-D. To do so, George figured out how to produce pairs of 2-D images that appear to be 3-D because when the viewer crosses her/his eyes, (s)he sees one image in each eye as it would appear if the image were truly in 3-D. George said the display was "simple trigonometry," but the hard part was writing the algorithm in less than 512 commands with space for the data points. In fact, the memory could not store all the dots in such a diagram, so each dot was displayed singly just for a short time and its location was captured on a Polaroid film. Hence, these images were not seen on the computer display; they appeared only after developing the Polaroid image. Because I was recording intracellularly from three Aplysia neurons at a time for my thesis, my data was an ideal test case for this new scatter plot. These plots are shown in Fig. 2 of our first paper (Kristan Jr. and Gerstein 1970; reprint included). The 3-D images never really caught on, probably because many people have trouble crossing their eyes to perceive a single image. And anyway, the real goal was to generalize the imaging of correlation to large numbers of spike trains. George and Don tried developing multi-axis displays on a 2-D surface (Perkel et al, 1975) (they called such displays "snowflakes"), but these displays proved to have so many ambiguities in interpretation that they were abandoned, although such analysis of correlations among spike trains was a major focus throughout George's remarkable career.

In the past 10 years or so, I have gotten back to doing cross-correlational analysis of multi-unit data (in yet another invertebrate, the European medicinal leech). In writing up our papers, I checked in on the state of the literature on this subject and was pleased to find that the Gerstein and Perkel articles written in the 1960s and 1970s were still the gold standard, more than 40 years later! What a strong testament to these truly visionary intellects.

When I was writing up my thesis research for publication—those innocent, trusting days—the academic tradition in neuroscience was to publish a short paper, as sort of a teaser, followed up by a more complete, more substantive paper later. A further tradition was for the mentor to have his/her name on the first, short paper, and the student would publish the second paper with just his/her own name on it. In some ways, the first paper established your origins and the second paper showed what you could do on your own. George followed this pattern: he wrote the first paper, during a 35-min commute on the old Paoli Local commuter train one morning. He wrote it in ink; he made only about 3 crossed-out corrections on it, and I made none. My contribution was to supply the data figures and the references. About 10 weeks after George's fruitful morning commute, our manuscript was accepted, with no revisions, in Science (Kristan Jr. and Gerstein 1970). I later submitted the second, longer paper to the Journal of Neurophysiology. This one I wrote myself, and it was an abysmal mess. I got it back with a review that said, "The data are interesting, but the text is incomprehensible." After many struggles and rewrites with George and Don Perkel, the paper was finally accepted (Kristan Jr., 1971). My writing has gotten much better over the years (I now get reviews of grant proposals that say how beautifully I write, but reject the proposal anyway!) and my ability to write began with that paper from my thesis.

When I started to think seriously about doing a postdoctoral stint, I was firmly committed to working on "simple" nervous systems, i.e., those of invertebrates. George was in charge of the departmental seminar series that year, and to help me find a good postdoctoral lab, he lined up all the big names in the invertebrate neuroscience field (e.g., Ted Bullock, Eric Kandel, Don Wilson, Don Kennedy, Graham Hoyle) to give seminars in a Medical School department. I was thrilled to meet all of these people, and the Med School folks got their consciousnesses expanded. Based upon these contacts, I was able to arrange a joint postdoc with the Dons (Kennedy and Wilson) of Stanford, which was a tremendous boost to my career.

After 5 years in George's lab, I was proud of all I had accomplished: I learned a lot of electronics, how to use computers to run experiments and analyze data, how to do fine dissection—and I had a PhD to show for it. Then I realized that George, in his 6 years at Penn, had not only gotten tenure, he was a Full Professor of Physiology and Biophysics and had a glowing international reputation as a computational neuroscientist. Not bad for someone who had never taken a physiology course! He picked up all his biological experience as a Postdoctoral Fellow with Nelson Kiang at M.I.T.; the rest he learned on his own. In fact, he told me that he knew at a young age that he wanted to apply mathematical and physical principles to the study of the brain. He decided that he could more readily learn biology on his own, but that he would need help to learn math and physics; that is why he got both his undergraduate and PhD degrees in Physics (at Harvard) and then went into neuroscience. It certainly worked well for him!

The academic legacy of my time in George's lab was that I could take advantage of computational tools as they came along. I never had the mathematical skills to do so on my own, but I understood well enough how they worked to attract outstanding graduate students and postdoctoral fellows who did have these skills, and to find colleagues in Physics, Computer Science, and Engineering at UCSD to collaborate when we needed even more expertise. I did not realize at the time what an unusual combination George was: he had a knowledge of and a respect for both experimental data and data analysis, and he could talk about neuroscience in depth with both experimentalists and theoreticians. The personal legacy was how to treat graduate students as important people who need the time and space to grow. George was a caring and patient mentor and an excellent role model. Times have changed, and the field of neuroscience has become immense, so we need more of the Gerstein-style wizardry today!

References

Gerstein GL, Perkel DH (1969) Simultaneously recorded trains of action potentials: analysis and functional interpretation. Science 164(3881):828. https://doi.org/10.1126/science.164.3881.828

Kristan Jr WB (1971) Plasticity of firing patterns in neurons of Aplysia pleural ganglion. J Neurophysiol 34(3):321–336. https://doi.org/10.1152/jn.1971.34.3.321

Kristan Jr WB, Gerstein GL (1970) Plasticity of synchronous activity in a small neural net. Science 169(3952):1336–1339. https://doi.org/10.1126/science.169.3952.1336

Perkel DH, Gerstein GL, Smith MS, Tatton WG (1975) Nerve-impulse patterns: a quantitative display technique for three neurons. Brain Res 100(2):271–296. https://doi.org/10.1016/0006-8993(75)90483-7

getting, since none of the animals had been offered any ice for 2 weeks.

The rats were made deficient in iron by removal of blood, a process in which many other substances in the body are also reduced. The availability of the iron-poor diet theoretically permitted the restoration of all these substances except iron. Furthermore, an iron-deficient diet given over a long period has been reported to be as effective as blood loss in leading to pagophagia (*1*). Therefore, we attribute the increase in ice-eating in our experiment to the loss of iron.

It is interesting that rats can be trained to eat ice in order to obtain their daily water. Of more interest is the fact that normal rats took a substantial portion of their daily water in the form of ice when water was available (Fig. 1), although this preference appeared to decline over time. The main contribution of the experiment, however, is experimental support for the earlier speculation based upon clinical data that iron deficiency tends to cause ice-eating. Although the experiment suggests no explanation for the phenomenon, other workers have reported that domestic rats suffering from either vitamin or mineral deficiencies show strong preferences for new foods (*3*).

STEPHEN C. WOODS
RICHARD S. WEISINGER
Departments of Physiology and Biophysics and *Psychology, University of Washington, Seattle 98105*

References and Notes

1. P. Lanzkowsky, *Arch. Dis. Child.* **34**, 140 (1959); C. E. Mengel, W. A. Carter, E. S. Horton, *Arch. Intern. Med.* **114**, 470 (1964).
2. R. D. Reynolds, H. J. Binder, M. B. Miller, W. W. Y. Chang, *Ann. Intern. Med.* **69**, 435 (1968).
3. W. Rodgers, *J. Comp. Physiol. Psychol.* **64**, 49 (1967); P. Rozin and W. Rodgers, *ibid.* **63**, 421 (1967).
4. We thank Profs. Walter Makous and Moncrieff Smith for their assistance. Supported by PHS training grant 5T01 GM 00666.

20 July 1970

Plasticity of Synchronous Activity in a Small Neural Net

Abstract. *Electrical activity from three neurons at a time was recorded in a pleural ganglion of* Aplysia. *Synchronous activity could be temporally patterned by electrical stimulation over a single nerve. After a period of pairing with electrical stimuli over another nerve the degree and time course of the synchrony were altered. Such alterations reverted after 5 to 30 minutes.*

An important problem in neurobiology is identification of mechanisms that underlie plasticity of the nervous system. The multiplicity of such mechanisms has been stressed in a recent review (*1*). In particular, it has been shown that a stimulus sequence based upon a classical conditioning paradigm can increase the size of a postsynaptic potential (*2*) or alter the firing patterns of single neurons (*3*). The purpose of the present study was to examine the effects of such stimulus sequences on the activity of a functionally related group of neurons. This report will show a number of ways in which pairing of stimuli over different afferent paths could reversibly modify the probability of nearly synchronous firings. A more complete report is in preparation (*4*).

Experiments were performed with neurons of the mollusk *Aplysia californica*. The circumesophageal ring of ganglia together with maximal lengths of nerves was dissected free and perfused with oxygenated seawater in a controlled-temperature chamber (15° ± 0.5°C). Connective tissue overlying the pleural ganglion was removed, and glass pipette electrodes (0.5*M* K_2SO_4, 5 to 20 megohms) were inserted into three neurons. Intracellular potentials were amplified, displayed on an oscilloscope, and recorded on an FM instrumentation tape recorder. Suction electrodes were used to deliver electrical stimuli to various nerves or connectives. For each experiment two particular nerves were chosen for delivery of a "test" stimulus and a "priming" stimulus. The test stimulus was a 0.1- to 5.0-msec pulse presented every 10 seconds; its strength was adjusted to slightly above the threshold value for production of postsynaptic potentials in the observed neurons. The priming stimulus consisted of a train of five pulses spaced 100 msec apart, presented every 10 seconds. Its strength was arbitrarily chosen as two to three times that of the test stimulus.

A typical experiment consisted of observation of the three impaled neurons during the following stimulus sequence: (i) 5 minutes without stimulus presentation, (ii) 5 to 10 minutes of test stimulus presentation alone, (iii) 4 minutes of paired presentation of test and priming stimuli (the priming stimulus train began 500 msec after each test stimulus), (iv) 5 to 10 minutes of test stimulus presentation alone, (v) 5 minutes without stimulus presentation, and (vi) repetition of conditions (iv) and (v).

We did not vary the timings or order in the presentation of paired stimuli. However, if the condition of the preparation remained stable, we examined nonspecific effects of the priming stimulus that would be analogous to classical sensitization. This was done by presenting again the entire stimulus sequence outlined above except that the priming stimulus alone was given in part (iii). It should be noted that the experimental design always included periods of "rest" between periods of presentations of the test stimulus. This allowed comparison of spontaneous activity before and after stimulus pairing.

Firings of the observed neurons were analyzed by means of the joint peri-stimulus-time scatter diagram (*5*) (Figs. 1 and 2). This statistical display allows detection of direct stimulus effects on the neurons, as well as of synaptic connections between or of common input to the observed neurons (*6*). Such functional relations of the observed neurons can be inferred from bands of altered density which are often observed in the scatter diagrams. These bands result from clustering of points in the scatter diagram. Bands of altered density that are parallel to the coordinate axes represent the direct effects of the stimulus on each of the observed neurons. Bands that are parallel to the 45° diagonal represent either a direct connection between the observed neurons or a common input from a source other than the stimulus. Variations of density *along* a diagonal band suggest that the stimulus is able to modulate the implied pathway (*5, 7*).

The effects of our stimulus sequence on the firings of two pairs of neurons are shown in Fig. 1. The leftmost scatter diagram of Fig. 1a represents the activity of the first pair of neurons during presentation of the test stimulus alone (condition ii). Two dense bars surrounded by low density are visible near and parallel to each coordinate axis. These represent two bursts of firing of each neuron in response to the stimulus. In the region far from each coordinate axis and away from the principal diagonal, there is an approximately uniform density of points. This region represents the non-stimulus-related or spontaneous firings of the two observed neurons. A dense concentration of points is visible along the principal diagonal; the two interruptions of this band correspond to the silent intervals between and after the two stimulus-associated bursts of firing. This band is flanked by two diagonal bands of density lower than the uniform background density; these reflect the refractory properties of the neurons. The dense diagonal band of points shows that there is a high probability for the two observed neurons to fire in close coincidence both during the stimulus-induced activity and during the subsequent spontaneous activity. Such closely spaced firing could be caused either by a synaptic connection between the observed neurons or by a strong input to both neurons from a common source (other than the stimulus). The scatter diagram alone usually does not allow unique choice between these alternatives (*8*). However, since our neurons were observed by means of intracellular electrodes, it was possible to determine that a large common synaptic input was responsible for the nearly coincident firing. Both neurons also fired (presumably independently) at instants when the large common input was not seen.

The second scatter diagram of Fig. 1a represents the activity of the two observed neurons during presentation of the same test stimulus immediately after 4 minutes of paired presentation of test and priming stimuli (condition iv). Stimulus-related firing of both neurons is again visible as two bands of high density near and parallel to the coordinate axes, although the second bands for each neuron have become more diffuse. A dense, interrupted diagonal concentration of points is visible in the period 0 to 2 seconds; this feature of the scatter diagram corresponds to the nearly simultaneous firing of the two neurons during the time that each is influenced by the stimulus and is very similar to that observed in the pre-pairing situation. However, in contrast to the pre-pairing situation, the dense diagonal point concentration is essentially absent during the period 3 to 10 seconds. Thus at times that are far from instants of stimulus presentation, there is now only a much reduced, almost

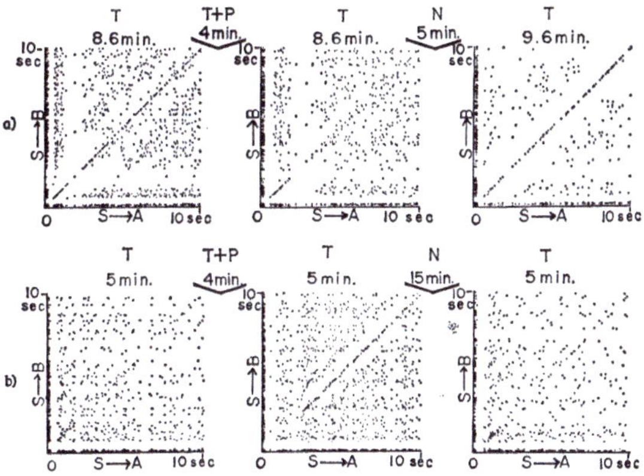

Fig. 1. Joint peri-stimulus-time scatter diagrams for two pairs of neurons. Stimulus conditions and durations of presentation are shown above or between scatter diagrams. T = test stimulus; P = priming stimulus; N = without stimulus. Row (a) shows activity of two neurons for which stimulus pairing produced tonic desynchronization without alteration of phasic synchronization. For each diagram of row (a), the numbers of stimuli, A spikes, and B spikes are: 52, 315, 253; 52, 255, 240; and 58, 234, 203. Row (b) shows activity of two neurons for which stimulus pairing produces tonic synchronization with slight phasic desynchronization. Numbers of stimuli, A spikes, and B spikes: 30, 175, 175; 33, 179, 242; and 26, 144, 145.

chance probability for nearly simultaneous firing of the observed neurons.

The third scatter diagram of Fig. 1a represents the activity of these neurons during later presentation of the same test stimulus after a period of 5 minutes without stimulus presentation (condition vi). Stimulus-related firing of both neurons remains visible as bands of high density near and parallel to the coordinate axes, although the second such band for each neuron has become noticeably diffuse. A dense concentration of points flanked by regions of low density again appears along the diagonal, as in the left scatter diagram. In contrast to the middle scatter diagram, the dense diagonal point concentration is present throughout the period between stimulus instants. Thus, the stimulus sequence applied to the two neurons of Fig. 1a had a desynchronizing effect on their firing only at times remote from stimulus presentation. This effect reverted to the control situation after some time.

The converse situation is depicted for two different neurons in the scatter diagrams of Fig. 1b. All three scatter diagrams show dense bands of points parallel and close to the coordinate axes, corresponding to the direct response of each neuron to the test stimulus. Little, if any, diagonal density is seen in the first scatter diagram, indicating that there is essentially chance probability for nearly simultaneous firing of the two neurons. The second scatter diagram, taken after a period of pairing of the test and priming stimuli, shows a strong diagonal band at times remote from stimulus instants. Thus, after pairing there is a high probability for nearly simultaneous firing except within 2 seconds after the test stimulus presentation. The third scatter diagram corresponds to the activity of these neurons somewhat later after a period without stimulation, and represents a reversion to the properties of the first scatter diagram with essentially chance probability of nearly simultaneous firing.

Similar experiments involving three neurons at a time are shown in Fig. 2. The statistical display was extended by making the scatter diagram in three dimensions (*9*). The origin of each cube is at the center rear, and the three coordinate axes are labeled a, b, c. The edge of the cube represents 10 seconds, the same time scale as used in Fig. 1.

The experiment shown in the left

half of Fig. 2 demonstrates that a stimulus sequence based on a classical conditioning paradigm can reversibly reduce the probability for nearly simultaneous firing of a neuron triplet. The cube pair labeled A represents the activity of three neurons in response to the control presentation of the test stimulus. Bands of increased point density parallel to the three coordinate axes represent the direct effects of the stimulus on the firing of the observed neurons. The lack of symmetry in the scatter diagram shows that the three neurons responded with different temporal patterns. Diagonal bands of increased point density, which are most visible near the ac and ab planes, correspond to the time structure of the probability for nearly simultaneous firing of the corresponding neuron pairs. Finally, a band of increased point density along the principal diagonal of the cube represents the enhanced probability for nearly simultaneous firing of all three neurons.

The cube pair labeled B shows the response of these three neurons to the test stimulus immediately after a period of 20 pairings of the test and priming stimuli. Most of the band structure has faded and the silent period near the origin and near the bc plane has been reduced.

Finally, after 5 minutes without stimulation, the response of the three neurons to the test stimulus is shown in the cube pair C. Most of the features of the point distribution have returned to the pre-pairing situation.

The experiment shown in the right half of Fig. 2 demonstrates conversely that the stimulus sequence can reversibly increase the probability for nearly simultaneous firing of a neuron triplet. Diagonal bands of increased point density occur only in the cubes E, immediately after a period of pairing of priming and test stimuli.

Of the 72 cell pairs in this study that showed alterations in the probability of nearly simultaneous firing during our stimulus sequence, 26 cell pairs were tested for nonspecific effects of the priming stimulus. In half of these cell pairs the priming stimulus alone produced different alterations than had stimulus pairing. Thus, specific effects of stimulus pairing are responsible for at least some of the observations reported here.

The variations of point density along the diagonals of the various scatter diagrams show that the probability of synchronous firing in two or three neurons can itself have a time structure

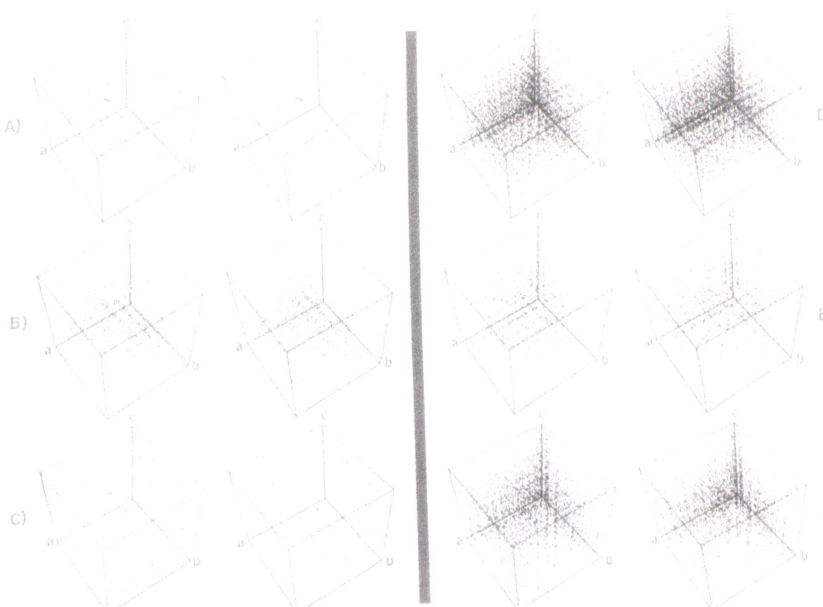

Fig. 2. Joint peri-stimulus-time scatter diagrams for two triplets of neurons. These are stereopairs, and should be merged for optimum viewing. Each scatter diagram is calculated during presentation of test stimulus alone. The test and priming stimuli were paired 20 times between the top and middle scatter diagrams, and a period without any stimulation occurred between the middle and bottom scatter diagrams. The side of each cube represents 10 seconds. Left half represents activity of three neurons that were desynchronized by the stimulus pairing. Numbers of stimuli, A spikes, B spikes, and C spikes: 76, 282, 269, 218; 31, 181, 286, 124; and 41, 168, 165, 148. Right half represents activity of three neurons that were synchronized by the stimulus pairing. Numbers of stimuli, A spikes, B spikes, and C spikes: 32, 329, 310, 273; 39, 278, 253, 192; and 38, 283, 276, 250.

related to the instants of (test) stimulus presentation. We have shown that this time structure can be altered in various ways during presentation of our stimulus sequence. In some cases the alterations of synchronization are not accompanied by alterations of the firing patterns of the individual neurons as measured by the peri-stimulus-time histograms.

It is particularly evident in Fig. 1 that the alterations of the probability of nearly simultaneous firing can be either phasic or tonic with respect to the test stimulus. This suggests that there may be two dissociable sources that produce the nearly simultaneous firings. Although the spike train analysis described so far cannot uniquely determine such sources, we had available intracellular recordings. These showed, in addition to individual synaptic activity, that a large excitatory postsynaptic potential appeared almost simultaneously in all the observed neurons and was responsible for much of the nearly simultaneous firing. Various unsynchronized firings occurred at other times due to unshared excitatory postsynaptic potentials or to intrinsic pacemaker potentials.

Thus, the alterations of probability of nearly coincident firing are caused by alterations in the firing pattern of a source common to all the observed neurons. Under the in vitro conditions of these experiments this yet unidentified interneuron is able to partially synchronize the firings of a population of neurons. Synchronous firing in a neural population and plastic alteration of the time structure of such synchronization are of obvious theoretical interest. However, the detailed connectivity and the behavioral significance of these mechanisms remain to be examined, perhaps by methods used in recent studies of gill withdrawal habituation (*10*).

WILLIAM B. KRISTAN, JR.*
GEORGE L. GERSTEIN
Department of Physiology,
School of Medicine, University of
Pennsylvania, Philadelphia 19104

References and Notes

1. E. R. Kandel and W. A. Spencer, *Physiol. Rev.* **48**, 65 (1968).
2. E. R. Kandel and L. Tauc, *J. Physiol.* **181**, 1 (1965); J. Bruner and L. Tauc, *Symp. Soc. Exp. Biol.* **20**, 457 (1966); R. von Baumgarten and B. Djahanparwar, *Brain Res.* **4**, 295 (1967).
3. J. Bureš and O. Burešová, *J. Neurophysiol.* **30**, 98 (1967); F. Morrell, in *The Neurosciences: A Study Program*, G. C. Quarton, T. Melnechuk and F. O. Schmitt, Eds. (Rockefeller Univ. Press, New York, 1967), p. 452.
4. W. B. Kristan, Jr., in preparation.
5. G. L. Gerstein and D. H. Perkel, *Science* **164**, 828 (1969).
6. For two simultaneously observed neurons (A and B) and a periodic stimulus, the peri-stimulus-time scatter diagram is produced as follows. Let the time of occurrence of the *i*th stimulus be S_i, that of the *j*th spike from neuron A be A_j, and that of the *k*th spike from neuron B be B_k. The ordinate of each point plotted corresponds to the time between a stimulus event and a spike from A (for example $A_j - S_i$), and the abscissa corresponds to the time between the same stimulus event and a spike from neuron B (for example, $B_k - S_i$). For each stimulus event a point is plotted for each combination of S–A and S–B intervals, both of which fall within a specified range. Thus, a spike from neuron A will give rise to as many points in the scatter diagram as there are spikes from neuron B within the specified time range about the stimulus (and vice versa). The total number of points in a scatter diagram is approximately $(N_a \times N_b)/N_s$ where N_a and N_b are the total number of spikes from neurons A and B respectively, and N_s is the number of stimulus events.
7. G. L. Gerstein, in *The Neurosciences: Second Study Program*, F. O. Schmitt, Ed. (Rockefeller Univ. Press, New York, in press).
8. G. L. Gerstein and D. H. Perkel, in preparation.
9. These three-dimensional scatter diagrams are presented as pairs of stereo images. The left image is intended for the left eye. Images may be merged by looking at a distant object and raising the figure into the line of sight. Alternatively, a viewer (model CF-8 Stereoscope, Abrams Instrument Co., Lansing, Mich.) may be used.
10. I. Kupfermann, V. Castellucci, H. Pinsker, E. Kandel, *Science* **167**, 1743 (1970); V. Castellucci, H. Pinsker, I. Kupfermann, E. R. Kandel, *ibid.*, p. 1745.
11. Supported by NIH grants NS-05606 and FR-15.
* Present address: Department of Biological Sciences, Stanford University, Stanford, California 94305.

23 July 1970

Lack of Coincidence between Neural and Behavioral Manifestations of Cortical Spreading Depression

Abstract. *The presence of cortical spreading depression is typically inferred from the presence of hypesthesia. The electrocorticogram and slow-potential change were recorded during cortical spreading depression and it was found that hypesthesia remained long after the cortex recovered from neural depression. Hypesthesia, therefore, is an unreliable indicant of cortical spreading depression; if cortical spreading depression is used as a research tool, neural activity must be monitored. These data offer a special problem for memory transfer studies.*

Cortical spreading depression (CSD) has been used to investigate learning phenomena such as interhemispheric transfer of training (*1*). The assumption in such studies is that the CSD produced by the topical application of potassium chloride is a "reversible lesion" or a "functional decortication." Recently it has been demonstrated that CSD should not be considered to be a functional ablation during the entire depression period, since the cortex gives evidence of at least partial recovery of electrical activity from time to time during the course of the treatment (*2*).

Typically, the behavior criterion for CSD is the development of hypesthesia (diminished sensibility) of the limbs contralateral to the depressed hemisphere (*3*). It is assumed that this unilateral hypesthesia is the behavioral manifestation of an underlying unilateral neural CSD. The purpose of this study is to question the assumption that the presence of hypesthesia is necessarily indicative of the presence of CSD.

Six male hooded rats of the Long-Evans strain weighing 250 to 350 g were used in the first experiment. Animals were surgically prepared for CSD and for chronic electrocorticographic (ECoG) recording. Two pairs of cortical electrodes (stainless steel screws) were placed bilaterally in the skull about 5 mm on either side of the saggital suture: one pair was about 2 mm anterior to the coronal suture and the other pair was about 2 mm anterior to the lambdoid suture. A polyethylene cannula [0.070 inch (0.178 cm) inside diameter, 0.110 inch (0.279 cm) outside diameter] was inserted (*4*) over one hemisphere in the anterolateral parietal bone. In three rats the cannula was over the left cortex and in the other three it was over the right cortex. The cannula was filled with sterile 0.9 percent saline whenever the rat was in its home cage. The electrical activity of both hemispheres was recorded differentially between ipsilateral pairs of electrodes. This made it possible to compare the activity of the depressed hemisphere with the activity of the nondepressed hemisphere for the same animal. An animal was considered to exhibit CSD when the ECoG amplitude reduced by at least one-third of the baseline amplitude. This was ascertained by making a calibrated template of the baseline amplitude and continuously measuring

Brain Research, 100 (1975) 271–296 271

NERVE-IMPULSE PATTERNS: A QUANTITATIVE DISPLAY TECHNIQUE FOR THREE NEURONS

DONALD H. PERKEL, GEORGE L. GERSTEIN, MARK S. SMITH* AND WILLIAM G. TATTON**

Department of Biological Sciences, Stanford University, Stanford, Calif. 94305 and (G.L.G.) the Departments of Physiology and Biophysics, School of Medicine, University of Pennsylvania, Philadelphia, Pa. 19174 (U.S.A.)

(Accepted June 4th, 1975)

SUMMARY

A scatter diagram is described that displays the relative timings of nerve impulses in 3 simultaneously monitored neurons. The technique is a generalization of the cross-correlation histogram for two impulse trains. The time intervals between impulses in different neurons are plotted on triangular coordinates, to yield a Joint Impulse Configuration Scatter Diagram. The resulting 'snowflake' plot shows a pattern of spots and lines, which is interpretable in terms of the functional circuitry among the neurons. Illustrations are given of the snowflakes produced by a variety of three-neuron circuits, which may serve as a preliminary catalog of snowflake types for interpretation of experimental data.

INTRODUCTION

In a number of papers[4,5,10,15,16] we have described display techniques and statistical measures for the timings of nerve impulses. These procedures are intended to allow the experimenter to infer functional interactions that may exist among the neurons whose impulse activity is recorded, using the information inherent in the relative timings of impulses originating in one neuron or in several. Almost all of the techniques previously described are confined in their application to two neurons, with or without periodic stimulation.

* Present address: Yale University School of Medicine, New Haven, Conn. 06510, U.S.A.
** Present address: Department of Medical Physiology, Faculty of Medicine, University of Calgary, Alberta, Canada.

If the impulses in a set of neurons are coordinated, to the extent that their temporal relationships are arranged, at least probabilistically, in characteristic patterns, then the set of neurons constitutes a functional group. Such a group may arise either through synaptic interaction among neurons in the set or through shared input from one or more external sources of activity[2]. A functional group of neurons, then, produces characteristic relative firing patterns. For example, if one neuron excites another, the firing of the presynaptic neuron temporarily increases the firing probability of the postsynaptic neuron within a closely delimited period of time; thus, ordered pairs of firings, within a certain range of latencies, are especially likely. The time course of this altered firing probability is visualized in the cross-correlation histogram for the two impulse trains. Every configuration of neural circuits produces its own such 'signature' in the cross-correlation. If the experimenter is provided with a catalog of cross-correlation histograms for known circuit arrangements, then, when confronted with a set of impulse recordings from neurons known or suspected to be functionally related but for which the detailed arrangement of connections is not known, he may compare his observed histograms with those of the catalog and so eliminate from further consideration all but a few of the possible circuits.

Ideally, such identification should be unique: each histogram should be sufficiently different from those of all other circuits to permit unequivocal selection of the actual pattern of connections. For pairs of neurons, at least, this ideal situation does not exist. For example, a peak in the cross-correlation histogram may be due to an excitatory synaptic connection from one of the neurons to the other; a fairly similar peak may arise, however, from shared synaptic excitation, or even from shared synaptic inhibition from a common source. The hypothesis of a synaptic connection between the two can be tested by examining the longer-range features of the cross-correlation; both pre- and postsynaptic autocorrelations are represented in the cross-correlation histogram, and such features can be examined qualitatively[10] or quantitatively[11]. But not all such ambiguities can be resolved, either in principle or, because of limitations on sample size, in practice.

When we consider larger sets of neurons, the possible number of circuits increases rapidly, as does the information contained in the joint distribution of relative timings between impulses. As a consequence of the principle that joint probability distributions cannot in general be derived from the corresponding marginal distributions, it follows that simultaneous recordings of impulse activity in members of a functional group can be expected to yield more information — in some cases including uniquely obtainable information — than can single-cell recordings or simultaneous recordings of subsects of the group. Some of the examples presented later in this paper corroborate this claim.

Guided by these considerations, we have devised a technique that displays in compact form the relative timings of impulses in 3 neurons monitored simultaneously, and which can often be interpreted, by comparison with known examples, in terms of the functional circuitry of the neurons. An example of this plot — the *joint impulse configuration scatter diagram* — has been presented earlier (ref. 5, Fig. 12). In this paper we describe its construction, characterize many of its properties, and illustrate the kinds of patterns produced by a variety of functional circuits that may exist among 3 neurons. The computer simulation methods used to provide most of the examples have been described elsewhere[12,13,17]. In another paper[3] the technique is applied in detail to a particular class of circuits: a ring of neurons connected by inhibitory synapses. It has more recently been applied to the problem of distinguishing pre- and post-synaptic inhibition[14].

CONSTRUCTION OF THE SCATTER DIAGRAM

Each point on the scatter diagram corresponds to an ordered trio of impulses, one in each of the 3 neurons. Let the times of occurrence of the 3 impulses in trains A, B, and C be designated t_a, t_b and t_c, respectively. The 3 time intervals between the 3 pairs of impulses are related by the equation

$$(t_b - t_a) + (t_c - t_b) + (t_a - t_c) = 0 \tag{1}$$

which expresses the fact that the time between the A-impulse and the C-impulse is the algebraic sum of the times from the A-impulse to the B-impulse and from the B-impulse to the C-impulse. This relation, although trivially obvious, reduces the number of independent variables in the trio (t_a, t_b, t_c) from 3 to 2, and hence allows representation in the plane.

A triangular coordinate system has several advantages for displaying the points. First of all, it is symmetric; no single neuron is the preferred or reference neuron. Secondly, as is familiar to physical chemists who use triangular coordinates for plotting three-phase systems, the sum of the 3 coordinates of a point is a constant. In the system of axes we use, that sum is 0, corresponding to Eqn. 1.

The axes are set at 120° angles with one another, as shown in Fig. 1a. The vertical axis corresponds to the AB interval; the distance corresponding to the interval $t_b - t_a$ is measured upwards if that interval is positive, and downwards if it is negative. Similarly, the BC axis extends in the positive direction from the origin to the lower right, and the CA axis extends positively from the origin to the lower left. To plot a point corresponding to a trio (t_a, t_b, t_c), the AB interval, $t_b - t_a$, is measured along the AB axis, and a perpendicular is drawn. Next, the BC interval, $t_c - t_b$, is measured along the BC axis and a perpendicular is drawn to that coordinate axis. The point is plotted where the two perpendiculars intersect. If now the CA interval, $t_c - t_a$, is measured along the CA axis (with proper attention to algebraic sign), then the perpendicular erected there will pass through the intersection of the other two perpendiculars, *i.e.*, the 3 perpendiculars have a common point of intersection. This result arises from Equation 1 and the choice of coordinate axes; it can be proven by elementary methods.

It is also easily shown that the abscissa and ordinate values of the point in an ordinary Cartesian coordinate system erected with the same linear scale about the same origin are given by $(1/\sqrt{3})\,[(t_c - t_a) - (t_b - t_c)]$ and $t_b - t_a$. Cartesian coordinates are necessary for use of automatic plotting machinery.

For a set of 3 simultaneously recorded trains of impulses, the scatter diagram is made by plotting every possible trio of points whose span does not exceed a limiting value L. The span is defined as the time from the first impulse to the last of the trio. An impulse may be a member of more than one trio, and hence may be represented by more than one point on the scatter diagram. The locus of points representing trios of

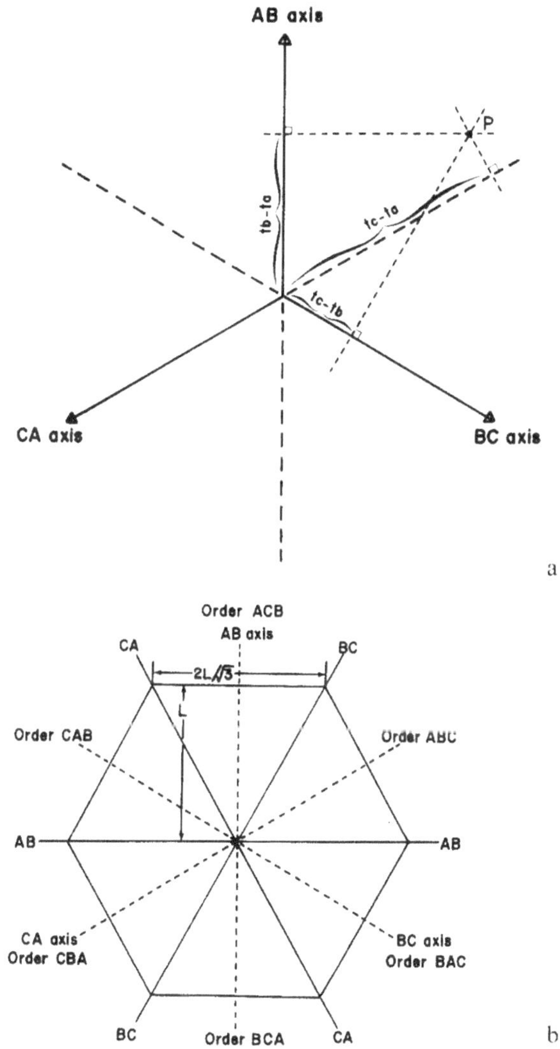

Fig. 1. Triangular coordinate system for joint interval configuration scatter diagram. a: the 3 coordinate axes, set at 120° angles. For every point P, the algebraic sum of its coordinates on this system is zero. See text. b: midway between the coordinate axes (broken lines) lie the coincidence lines (solid), which are the loci for simultaneous impulses in 2 of the 3 trains. If the maximum span of a trio of impulses is L, the points lie within a regular hexagon of side $2L/\sqrt{3}$. Within each of the 6 sectors formed by the coincidence lines, the plotted points correspond to trios of impulses in the serial order indicated.

impulses of span not exceeding L is a regular hexagon of side $2L/\sqrt{3}$; because of the hexagonal outline, the joint impulse configuration scatter diagram is also referred to as a *snowflake*.

INTERPRETATION OF LOCATIONS OF POINTS

The position of a point reflects several properties of the trio of impulse times: the time span, the order of firing of the neurons, and the interval ratio. For example, a point at the origin of coordinates represents simultaneous firing of the 3 neurons. The further a point lies from the origin, the longer the total time span covered by the 3 impulses.

The sequence, or serial order, of the 3 events is represented by the sector of the coordinate system in which the point lies. Considering any two impulse times, *e.g.*, t_a and t_b, it is clear that the upper half of the coordinate plane corresponds to 'A before B', and the lower half of the plane corresponds to 'B before A'. The line of division is the perpendicular to the AB axis passing through the origin; it is the locus of points for which the A and B impulses are simultaneous. For this reason, we refer to it as the *AB coincidence line* (Fig. 1b). Similarly, the BC coincidence line passes from lower left, through the origin, perpendicular to the BC axis, to upper right; above and to the left of this line C precedes B, and below and to the right, B precedes C. Finally, the AC coincidence line passes from lower left to upper right.

The 3 coincidence lines divide the plane into 6 equal sectors; each sector corresponds to a particular order of impulses. For example, any trio of sequence ACB falls in the uppermost sector. The coordinate axes bisect these sectors; in the middle of each sector is the positive or negative branch of one of the coordinate axes. We can therefore designate each sector by the name of coordinate axis passing through it. We make the further convention that, if the negative branch of the axis passes through the sector, we reverse the order of the letters. Thus the sectors, starting at the top and proceeding clockwise, are called AB, AC, BC, BA, CA and CB.

This scheme now furnishes us with a simple mnemonic for determining the sequence of events within each sector. We insert the letter corresponding to the third train between the letters designating the sector, so that the topmost sector becomes ACB. This is now the order of impulses for points plotted within the sector. Similarly, the lowermost sector corresponds to the order BCA, so that, for points in that sector, the inequalities $t_b < t_c$ and $t_c < t_a$ hold. The complete designation of the sectors, shown in Fig. 1b proceeding clockwise from the top, is ACB, ABC, BAC, BCA, CBA, and CAB.

The span of a trio of impulses determines the distance from the origin of the plotted point. Within a given sector, trios of equal span lie on the same line, which is perpendicular to the axis bisecting the sector; for brevity, we call such a line segment a rib (Fig. 2). Thus the locus of all trios of equal span, regardless of order of firing of the trains, is a hexagon composed of 6 ribs. Each adjacent pair of ribs intersects on a coincidence line, and each rib is bisected by a coordinate axis. The boundary of the snowflake is the set of ribs corresponding to the maximum span plotted, of length L.

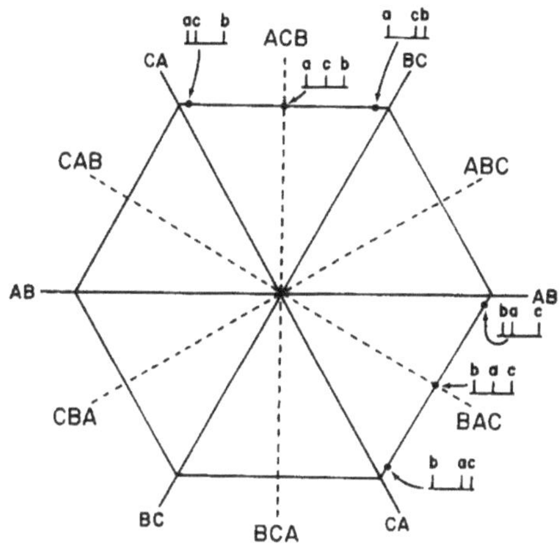

Fig. 2. Position of a point on the snowflake corresponds to the pattern of a trio. On any rib, a point in the center corresponds to equal intervals; a point displaced anticlockwise from the center represents a short–long pattern; a point displaced clockwise corresponds to a long–short pattern.

For a trio with a given sequence and a given span, the position of the point on the rib depends on the pattern of the trio[18], *i.e.*, on the location of the middle impulse relative to the two outer impulses. For example, in the uppermost sector, the order of impulses is ACB. If the C impulse occurs just midway between the A and B impulses (*i.e.*, if $t_c - t_a = t_b - t_c$), the point will appear in the middle of the rib, on the AB axis. If the impulse in C occurs earlier than midway between the A and B impulses ($t_c - t_a < t_b - t_c$), the point will lie to the left; similarly, if the pattern of intervals is long–short ($t_c - t_a > t_b - t_c$), the point will appear to the right. These relationships are illustrated in Fig. 2.

ELEMENTARY NEURONAL CIRCUITS AND THEIR SNOWFLAKES

Null case: 3 independently firing neurons

If the 3 neurons are unconnected and not otherwise coordinated or phase-locked, the points will be distributed over the hexagon with an approximately constant mean areal density, and ordinarily will appear to be scattered at random (Fig. 3a, left). However, as the maximum span increases, each impulse is represented by more and more points. If the firing rates of the 3 neurons differ sufficiently, this multiplicity of representation is made visible by a characteristic 'woven' pattern; the density of the weave in each of the 3 directions depends on the mean firing rates of the neurons (Fig. 3).

Excitatory connections

If an excitatory synaptic connection exists between two neurons, then, for a period after the firing of the presynaptic neuron, there is an increased probability of firing of the postsynaptic neuron, among other observable effects[2]. This leads to an

277

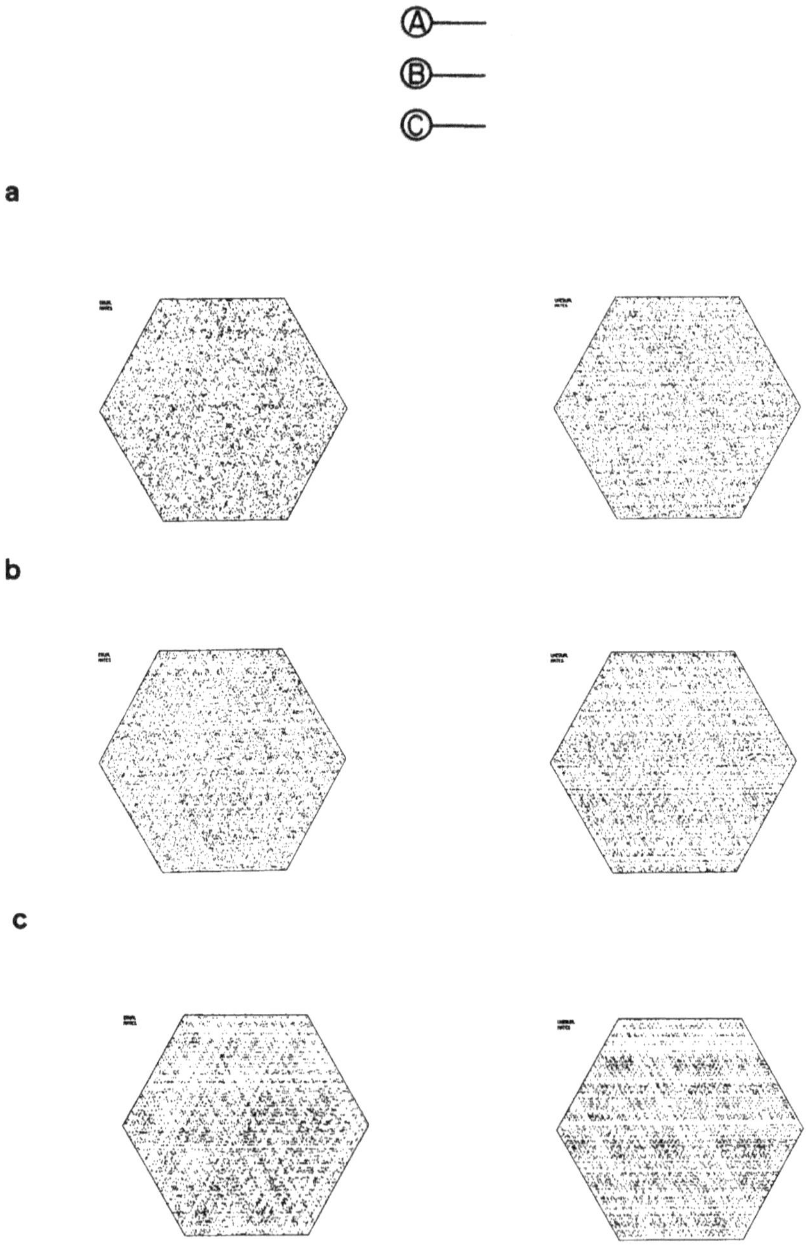

Fig. 3. The null case: 3 independently firing neurons; exponentially distributed intervals. Left, all 3 neurons fire at 5/sec. Right, neuron A fires at 1/sec, B at 5/sec, and C at 10/sec. a: span is 0.5 sec; b: span is 1.0 sec; c: span is 2.0 sec. Note the enhancement of visible texture with increasing span and for large rate differences. Sample lengths have been adjusted so that each scatter diagram contains 5000 points, to facilitate comparison.

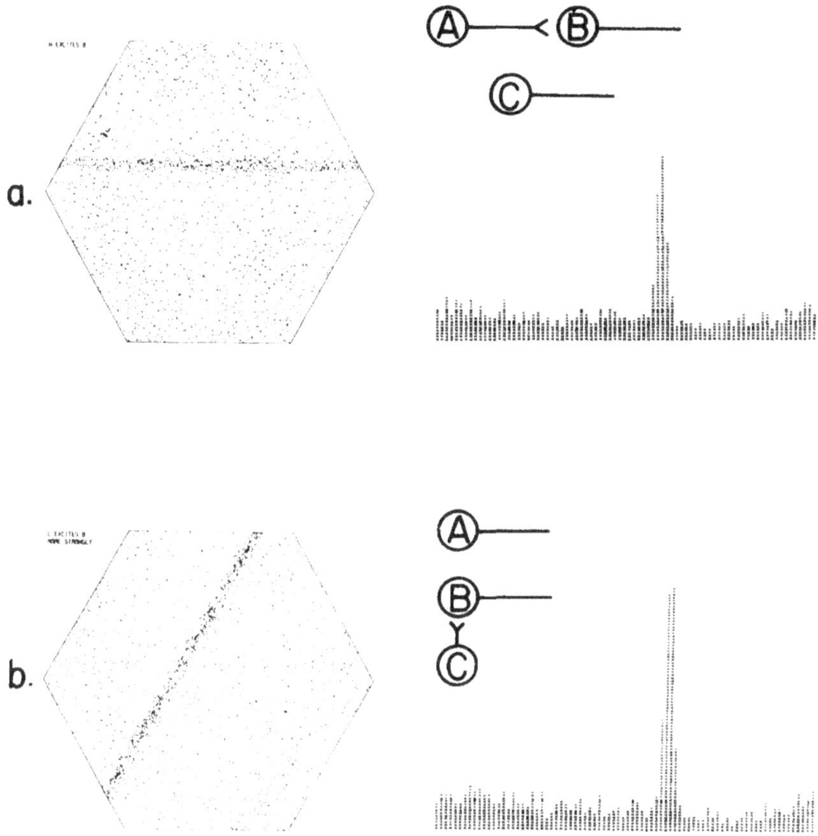

Fig. 4a and b.

increased frequency of occurrence of trios in which the pre- and postsynaptic neurons fire in a fixed order and with a narrowly defined range of interval lengths. For example, if neuron A excites neuron B, and if neuron C is independent of both the others, then, of all possible trios, there will be an excess of instances in which A closely follows B. It follows that there will be a narrow band of increased density of points and, since A impulses precede B impulses, that band will lie in the upper half-plane, in the sectors CAB, ACB and ABC. The band will be located perpendicular to the AC axis, and hence will be horizontal, and displaced upward from the origin by an amount corresponding to the predominant interval between A impulses and B impulses, which in turn reflects impulse propagation time and synaptic delay. In other words, the lagged or delayed coincidence between A and B gives rise to a band of points parallel to the AB coincidence line. It is an extended rib, displaced from the AB coincidence line by a distance proportional to the amount of lag, and in a direction corresponding to the sequence of firing. Finally, because the firings of C are independent of those of A and B, the band will maintain a constant density from left to right (Fig. 4a).

By similar reasoning, an excitatory connection from B to A will yield a dark

279

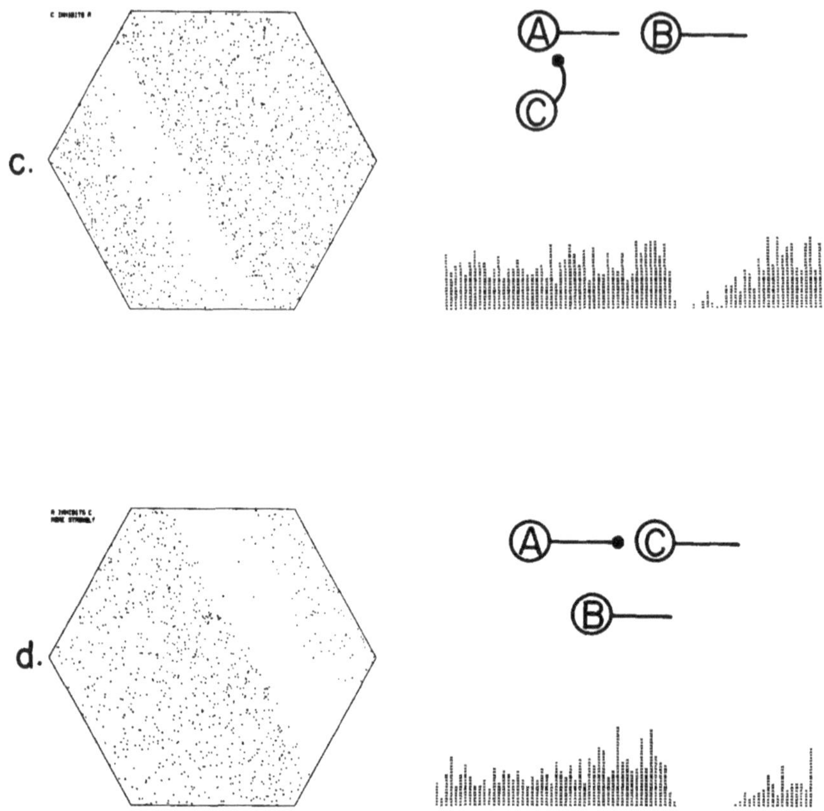

Fig. 4. Single synaptic connections. At left, snowflakes; at right, cross-correlation histograms between pre- and postsynaptic neurons. a: neuron A excites neuron B. b: C excites B, with a larger-amplitude postsynaptic potential than in case a. c: C weakly inhibits A. d: A strongly inhibits C.

horizontal band *below* the origin. An excitatory connection between B and C will give a band parallel to the BC coincidence line, *i.e.*, extending from upper right to lower left. The band will be below and to the right of the origin if B excites C, and above and to the left of the origin if C excites B, etc. (Fig. 4b). If, on the other hand, the snowflake shows a single narrow band of uniform density, parallel to one of the coincidence lines, a reasonable characterization of the circuit is a single excitatory connection between the two appropriate neurons, and independent firing of the third.

The width of the band, the density of its points, and its fine structure reflect the corresponding features of the peak in the cross-correlation histogram between impulse trains of the pre- and postsynaptic neurons, as illustrated in Fig. 4. Thus, the greater the amplitude of the excitatory postsynaptic potential (EPSP), the denser the band and the higher the peak in the cross-correlation function. In the snowflake, however, the persistent (though unrelated) firing of the third neuron is necessary to bring out the relationship between the other two neurons. If the firing rate of the third neuron is very low, the relationship between the other two will not be revealed in the scatter diagram; for this reason, among others, the snowflake is not a replacement for, but rather a supplement to, the cross- and autocorrelation histograms.

Inhibitory connections

In the same way that an excitatory connection gives rise to a band of increased density of points in the scatter diagram, an inhibitory connection produces a band of decreased density. The location, direction, and displacement of the band from the appropriate coincidence line depend as usual on the identity of the pre- and postsynaptic neurons and on the lag between presynaptic impulse and effective inhibitory action.

There are some differences in fine structure between the excitatory and inhibitory bands. The greater the amplitude of an EPSP, the more likely it is to trigger an immediate postsynaptic impulse; the band will then have a sharp, dense leading edge, perhaps followed by a region of decreased density, corresponding to postsynaptic refractoriness. Even a relatively weak inhibitory postsynaptic potential (IPSP) will usually produce a sharply defined leading edge on the band of decreased point density. A strong IPSP will not only lower the density of points in the band to zero, but also will extend the duration of the band. The amplitude of an EPSP is reflected in the sharpness of the leading edge, whereas IPSP amplitude is seen in the broadness of the band. These differences are seen also in the cross-correlation histograms (Fig. 4).

Other influences

The details of the features in a snowflake are modified, just as in a cross-correlation histogram, by a number of characteristics of both pre- and postsynaptic neurons in addition to the synaptic characteristics mentioned above. Among such effects are the degree to which the membrane potential is reset following an action potential, the intensity of noise, the amount of regularity or rhythmicity of discharge, and more generally any departures from flatness in the autocorrelation of each neuron. Most of the influences on the cross-correlation histogram described by Moore *et al.*[10] can be generalized to the snowflake. For the sake of clarity, we have confined our examples here to the simplest cases; most presynaptic neurons (*e.g.*, in Figs. 5 and 8) have discharge patterns approximating a Poisson process. In two examples (Fig. 11e, f) we have illustrated the effect of presynaptic rhythmicity on the snowflake, but in general we have necessarily restricted the scope of this paper to a representative rather than exhaustive catalog. We stress that no such catalog can be applied blindly; the snowflakes supplement rather than supplant such statistical measures as the auto- and cross-correlation histograms and such experimental measurements as intracellular recordings.

KINDS OF CIRCUITS

In recent years there has been an active investigation of relatively small neural circuits, particularly in invertebrate systems[9,21]. As the functional and behavioral correlates of specific neuronal configurations are being determined with increasing precision, it becomes appropriate to examine the possible kinds of neuronal circuits in a systematic manner.

There is a formidable variety of possible circuits that can be formed by synaptic connections among 3 neurons. We present here a systematic classification and enumeration of the basic types, and illustrate particular examples through snowflakes and the corresponding cross-correlation histograms.

Three-neuron circuits in which all of the effective synaptic connections are between neurons in the trio are called *intrinsic* circuits. If one or more neurons other than members of the recorded trio make synaptic contact with 2 or 3 members of the trio, or otherwise act so as to influence the functional relationships among the members of the trio, the trio can be considered to be embedded in a larger functional group, and we designate the circuit as partially or wholly *extrinsic* to our trio. In the remainder of this paper we deal systematically only with intrinsic circuits, and illustrate only certain extrinsic circuits of particular importance.

TABLE I

TYPES OF INTRINSIC CIRCUITS AMONG THREE NEURONS

Circuit	Diagram	Number of generic types	Number of distinguishable circuits
One fiber			
One synapse	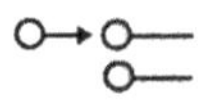	2	12
Two fibers			
Chain	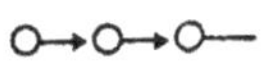	4	24
Divergent	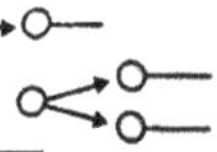	3	12
Convergent	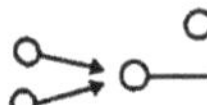	3	12
Reciprocal pair		3	12
		Subtotal	60
Three fibers			
Loop	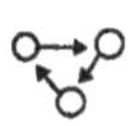	4	16
Divergent-convergent	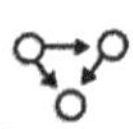	8	48
Reciprocal with input	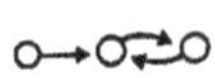	8	48
Reciprocal with output	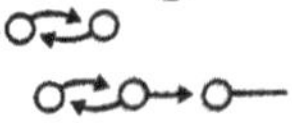	8	48
		Subtotal	160
Four fibers			
Reciprocal plus chain		16	96
Reciprocal plus divergent		10	48
Reciprocal plus convergent		10	48
Chained reciprocal	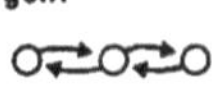	6	48
		Subtotal	240
Five fibers		32	192
Six fibers	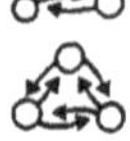	16	128
		Total	728

The most straightforward way of classifying intrinsic circuits is by the number of synaptic connections that exist between neurons in the trio. We have already discussed the one-synapse circuits. There are two *generic types*: one neuron excites or inhibits a second, while the third neuron is independent of the other two. For purposes of enumeration, we assume that all neurons are functionally indistinguishable and that there are only two distinguishable types of connection: excitatory and inhibitory. Recurrent collateral connections, by which impulse activity is fed back to its originating neuron, are excluded from the enumeration.

If we then count all the ways in which the 3 actual neurons A, B, and C can be associated with the different functional roles for neurons in each of the generic types of circuit, we can sum to find the total number of *distinguishable circuits*. For each of the one-synapse circuits, there are 3 functional roles for neurons: presynaptic, postsynaptic and independent; these roles can be filled in 6 different ways by the 3 neurons. Hence, there are 12 distinguishable one-synapse circuits.

In Table I, we illustrate the generic types of intrinsic three-neuron circuits, and give the numbers of generic and distinguishable circuits of each type, for circuits containing up to 6 fibers. Note that in this tabulation we allow at most one type of connection, excitatory or inhibitory, to exist between any ordered pair of neurons; parallel fibers in the same direction are not admitted. This restriction is arbitrary, as is the exclusion of recurrent collaterals; the patient reader is encouraged to extend the enumeration to these cases.

Two-fiber circuits

As illustrated in Table I, there are 4 categories of circuits containing 3 neurons linked by two fibers: chains, divergent circuits, convergent circuits, and the case of a reciprocally connected pair with the third neuron operating independently. We defer discussion of the last circuit to the more general consideration of loops.

At first glance, the snowflakes produced by two-fiber chains, by divergent circuits, and by convergent circuits are essentially similar (Fig. 5); there are two lines of increased or decreased density of points, which intersect at an angle of 60°. Closer analysis reveals characteristic differences which enable rapid and precise interpretation of the snowflake as originating in one of the 48 possible circuits of this group.

In *chain circuits*, the acute (60°) angle encloses the origin; this distinguishes them from divergent and convergent circuits, in which it is the obtuse (120°) angle that encloses the origin. The 6 permutations of neurons comprising the chain can be distinguished by the directions of the two lines, or equivalently by the position of their intersection relative to the origin, as illustrated in the diagram (Fig. 6a). For example, if the lines of increased or decreased density of points both slope downwards, to intersect below the origin, the circuit is of the type B → C → A. For each of the 6 possibilities, either synapse may be excitatory or inhibitory; these 4 possibilities combined with the 6 permutations of neurons yield the 24 distinguishable chain circuits.

For the *divergent circuit* pattern, there are 3 possibilities. If neuron A is presynaptic to B and C, the point of intersection of the two lines in the snowflake is above and to the right of the origin; if B is presynaptic, the intersection is below and to

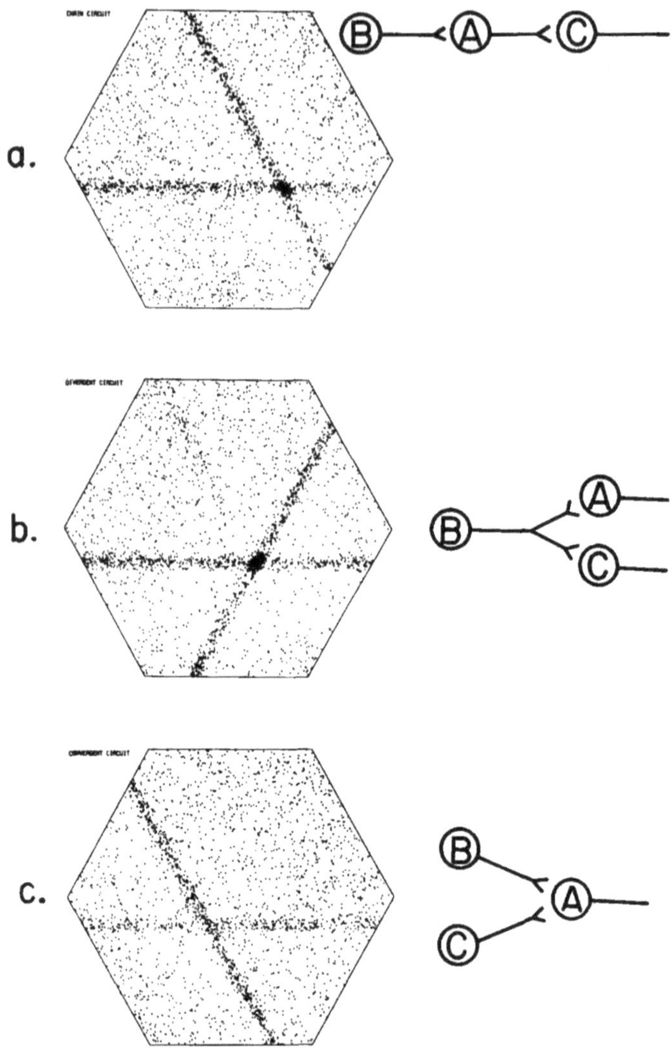

Fig. 5. Snowflakes for two-fiber circuits. a: chain: B excites A, A excites C. b: divergent circuit: B excites A and C. c: convergent circuit: B and C excite A. Note that in all 3 examples, B excites A, giving rise to the horizontal band below the origin.

the right; and if C is presynaptic, the intersection is to the left. These cases are summarized in Fig. 6b.

Convergent circuits follow much the same pattern, but the points of intersection are removed 180° from those of the corresponding divergent cases (Fig. 6c). For example, if neuron C is presynaptic to both the others, the point of intersection is to the left of the origin; if C is postsynaptic to A and B, the point of intersection is to the

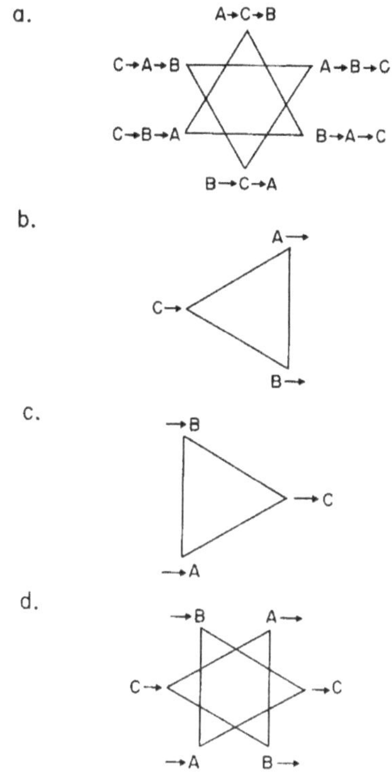

Fig. 6. Positions of intersection of the two bands for various two-fiber circuits. a: chains. The 6 possible chains correspond to the 6 positions indicated, at the points of a magen David (six-pointed star). For example, if the two bands intersect in a 60° angle (implying a chain circuit), forming an X-shaped pattern with the apex below the origin, then B excites C and C excites A. b: divergent circuits. c: convergent circuits. d: divergent and convergent circuits, superimposed. Note that this magen David is rotated 30° from its usual orientation.

right. The 6 possibilities for convergent and divergent circuits are summarized in Fig. 6d.

Once the basic circuits have been determined from the orientations of the lines and the position of their intersection, then the signs of the synapses, the characteristic delays, and other information about the properties of the circuits can be obtained in the usual way by examining the displacement and finer structure of the lines, as discussed previously.

Coincidence detectors. One special case of convergent circuits is of particular interest, for its functional importance as well as for its characteristic snowflake pattern. This is the *coincidence detector*, in which the postsynaptic neuron requires presynaptic impulses in each of its input fibers to arrive within a certain time interval in order to elicit a postsynaptic impulse, and in which more widely separated presynaptic impulses will fail to elicit a postsynaptic firing. A perfect coincidence detector would have a precisely specified resolution time, within which an output impulse would follow with certainty, and beyond which an output impulse would never occur. Using neurons rather than electronic components, it is more appropriate to describe the probability of a postsynaptic impulse as a function of the time interval between the two presynaptic impulses (here we confine our discussion to the purely excitatory case). In any convergent circuit, that probability will be at a maximum at the origin, *i.e.*, for simultaneous arrivals of excitatory impulses. The probability falls off monotonically and more or less gradually, with increasing temporal separation of input impulses (Fig. 7); the shape of this curve depends not only on the integrative properties of the postsynaptic neuron but also on the temporal structure of the input trains (*e.g.*, on the probability that a second impulse will arrive in one presynaptic train before the next impulse arrives in the other) and on the nature of other sources of excitation of the postsynaptic neuron. We may characterize a convergent excitatory neuronal circuit as a coincidence detector if (1) the firing probability is unity at the origin and for some reasonable interval about zero, (2) the firing probability descends from this plateau rather abruptly, and (3) it reaches and maintains a relatively low value thereafter, although this level need not be zero. We may define the resolution of the coincidence circuit as the time at which this function descends to some specified value, *e.g.*, 0.5. Thus, if a coincidence circuit has a resolution of 30 msec, then two presynaptic impulses must arrive with a time interval of 30 msec in order for the probability of a postsynaptic impulse to exceed 0.5 in each instance.

For a neuronal circuit to function well as a coincidence detector, it must fire with high probability within its resolution, and also fire with low probability outside that range (*i.e.*, it should have a low false-alarm rate). The first requirement means that the sum of the amplitudes of the postsynaptic potentials should exceed threshold when they arrive simultaneously; the second requirement implies that neither postsynaptic potential should individually attain threshold. Moreover, the presynaptic impulse

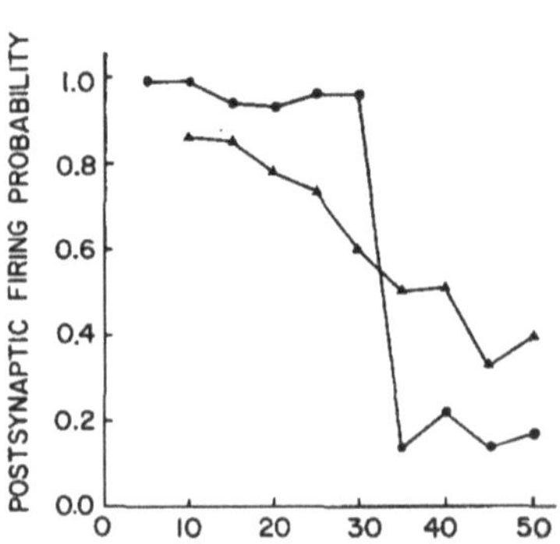

Fig. 7. Coincidence detectors. Probability of firing of postsynaptic neuron as a function of time separating two excitatory input impulses; postsynaptic impulse must occur within 60 msec of first excitatory input to contribute to estimate of probability. Circles correspond to a coincidence detector, having a resolution of about 30 msec; triangles correspond to a convergent circuit that does not function as a coincidence detector. Data taken from computer simulations.

trains must not be too irregular — *i.e.*, the minimum interval between impulses in either train should be greater than the resolution — so that the coincidences detected will not be between impulses in the same train but between those in the two presynaptic trains. Finally, the postsynaptic neuron must not be excited or inhibited by any other neurons than the two presynaptic ones, since other inputs will impair the reliability of the neuron in its role of coincidence detector.

A coincidence circuit of this sort is often referred to as an AND-circuit, since both presynaptic neurons must fire to produce an output. If the postsynaptic neuron has a sufficiently low threshold, then the same arrangement functions as an OR-circuit, in which an impulse in either presynaptic fiber is sufficient to excite a post-synaptic impulse. Actual neurons ordinarily fall in an intermediate position between these cases. A sufficiently high threshold will convert an AND-circuit into one that requires additional excitatory input, from a neuron other than the two presynaptic ones under consideration, in order to produce an output impulse.

The snowflake produced by a nearly perfect coincidence detector is shown in Fig. 8a (top line); the pair of intersecting lines is confined in extent to a chevron. The length of the arms of the chevron correspond to the resolution of the coincidence detector. (Arms of unequal length would arise if the two input channels produced postsynaptic potentials of different amplitudes.) The position of the vertex of the chevron reflects the identities of the pre- and postsynaptic neurons in the circuit (Fig. 6), and its distance from the origin reflects the latency between the time of occurrence of the *second* presynaptic impulse and that of the postsynaptic impulse. The thickness of the chevron reflects the extent of fluctuations in the timing relationships between pre- and postsynaptic impulses, such as those produced by membrane noise.

The coincidence detector provides a clearcut example of a functional relationship among 3 neurons that is made observable through the snowflake diagram, but which is not brought out in the 3 pairwise cross-correlation histograms, as seen by comparing the cross-correlograms for the coincidence detector (Fig. 8a) with those for an ordinary convergent excitatory circuit (Fig. 8b). An additional example is furnished by an anticoincidence detector, in which an interneuron (unobserved) that functions as a coincidence detector itself inhibits the neuron that is directly excited by the other two observed cells (Fig. 8c). Once again, the cross-correlations are indistinguishable from those of a simply convergent circuit or from a coincidence detector.

In general, the snowflakes will reveal more about the underlying neuronal circuit than will the set of cross-correlation histograms whenever the linear features in the snowflakes do not display a uniform density of points along their length, but rather are interrupted or concentrated. Such departures from uniform density are direct consequences of the influence of the third neuron; the situation is entirely analogous to that of diagonal features in the joint peristimulus–time scatter diagram[5], in which variations in point density reflect stimulus-dependent effects on the interaction between the two observed neurons. A similar situation occurs for presynaptic inhibition[14].

This is illustrated in the snowflakes for convergent circuits. In a coincidence (AND) circuit, the density of points in the region of the intersection of the two bands is greater than the density in the bands outside the intersection region. In an ordinary convergent (OR) circuit, the point density is approximately constant, throughout the length of the band as well as in the region of intersection. In an anticoincidence circuit, on the other hand, the point density in the bands is reduced at their intersection; these features may be seen by comparing the snowflakes of Fig. 8.

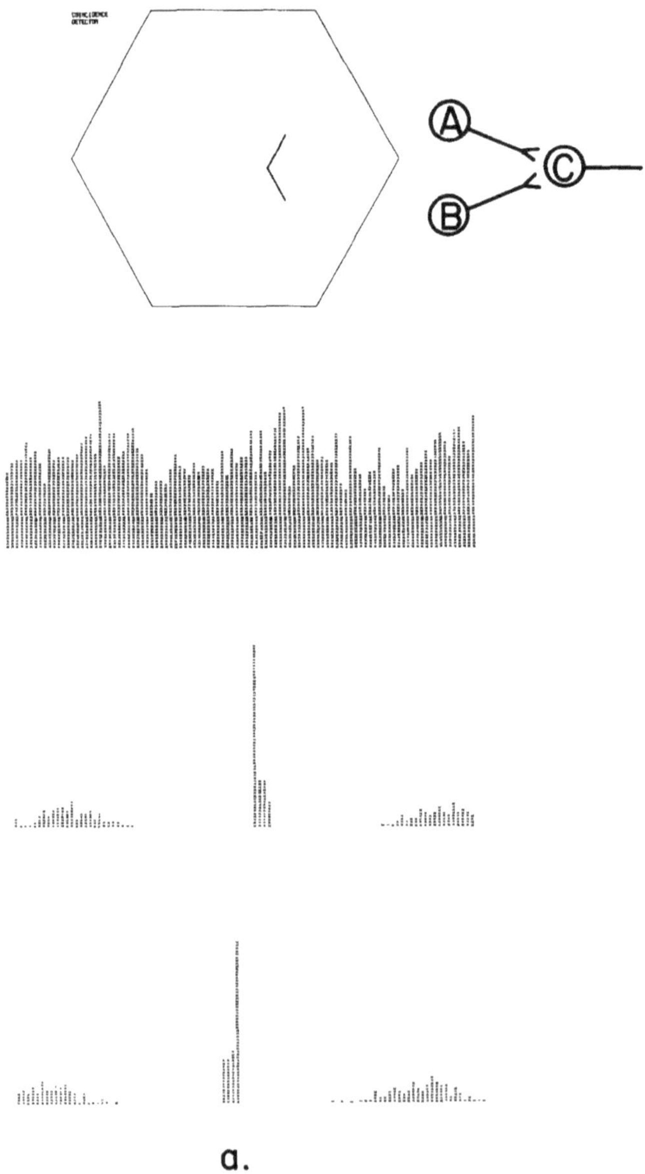

Fig. 8a.

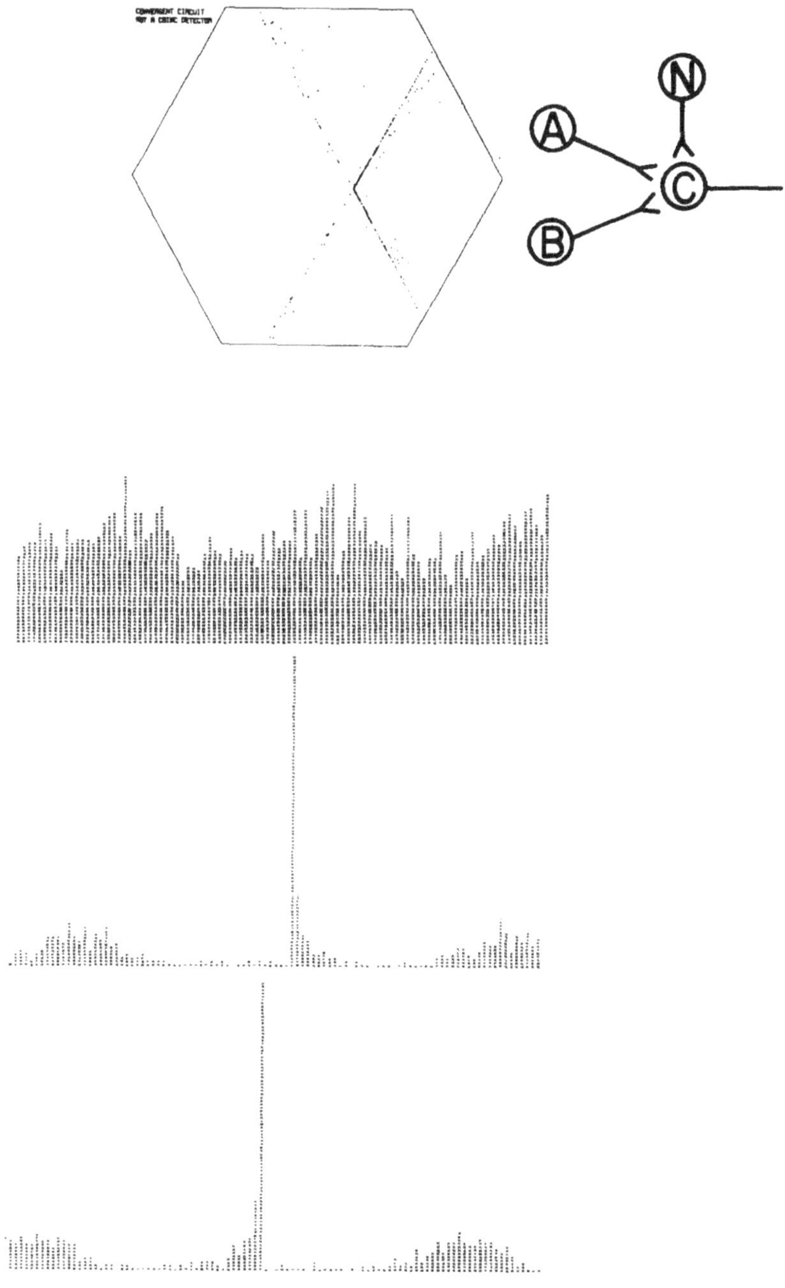

b.

Fig. 8b.

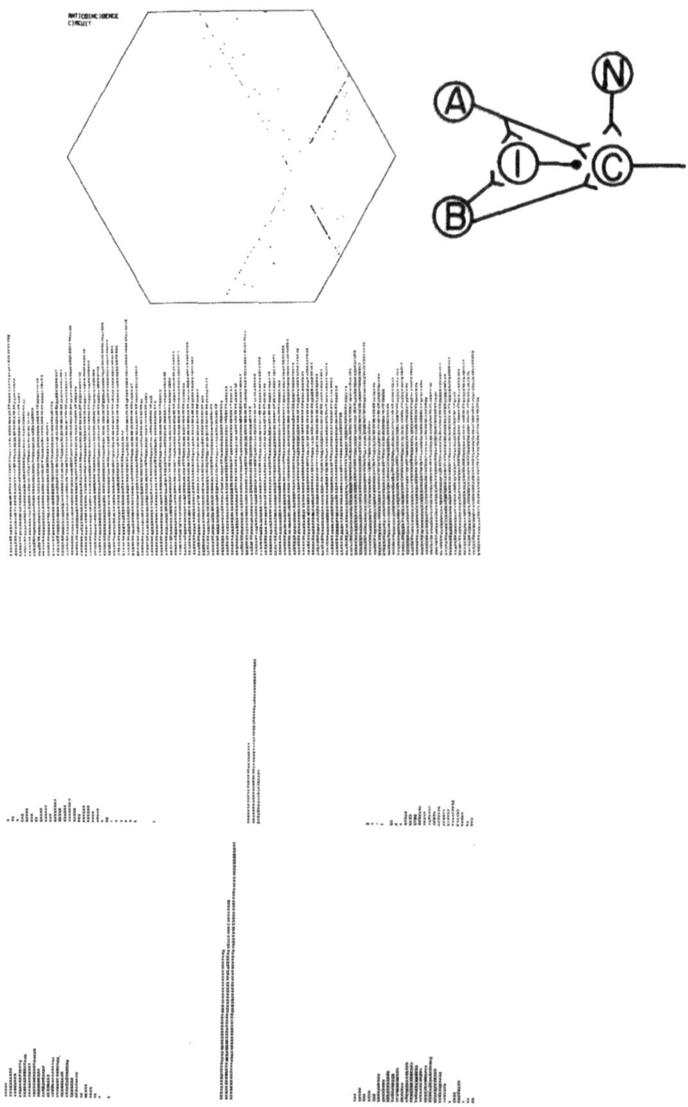

c.

Fig. 8. Convergent circuits. Top line, snowflakes. Lines 2 through 4, cross-correlation histograms between impulse trains A and B, B and C, and C and A. a: coincidence detector. b: convergent circuit, not a coincidence detector. c: like (b), but with a coincidence-detecting neuron, not recorded, inhibiting the postsynaptic neuron. A longer sampling time resulted in a larger number of tallies in the A–B cross-correlogram than for cases a and b. See text. N is an additional neuron exciting C.

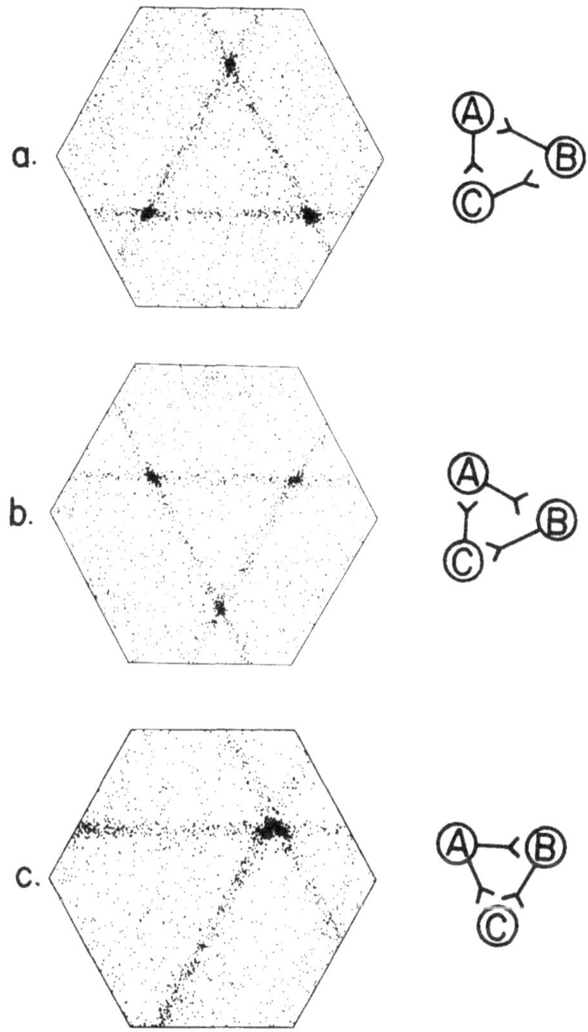

Fig. 9. Snowflakes for three-fiber circuits. a: counterclockwise loop; b: clockwise loop; c: divergent-convergent circuit. See text.

Three- and four-fiber circuits

The snowflake produced by a *loop* incorporates a third band in the scatter diagram for a chain circuit (Fig. 9a, b). Observe that the 'counterclockwise' loop — A excites C, C excites B, B excites A — gives rise to an equiliateral triangle oriented like the upper-case Greek delta, whereas the 'clockwise' loop — A excites B, B excites C, C excites A — produces an inverted delta. When the threshold is high, the sides of the triangle may vanish, leaving only the 3 spots at the vertices. The triangle encloses the origin of coordinates.

A *divergent–convergent circuit* similarly gives rise to an equilateral triangle. It differs from that of a loop in two ways: it is smaller, and it does not enclose the origin (Fig. 9c). In the snowflake for a loop, the sides of the triangle are proportional to the conduction times of impulses between the neurons. For a divergent–convergent circuit, on the other hand, the lengths of the sides are proportional to the pairwise

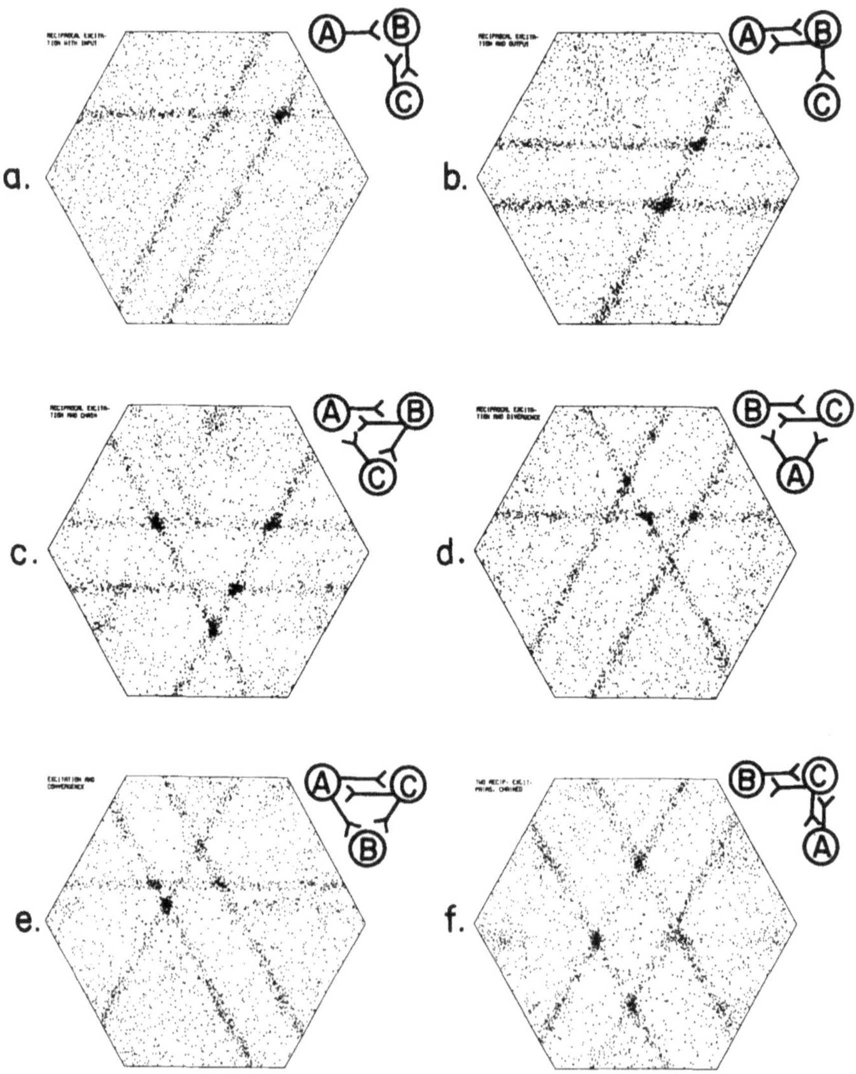

Fig. 10. Snowflakes for circuits incorporating reciprocal excitatory synapses. a: reciprocal pair with input. b: reciprocal pair with output. c–f: four-fiber circuits. c: reciprocal plus chain. d: reciprocal plus divergent circuit. e: reciprocal plus convergent circuit. f: chained reciprocal pairs.

differences in conduction times. If all 3 conduction times are equal, the equilateral triangle degenerates into a point, or a single spot. If the threshold is high, so that the bands disappear, and if the 3 spots at the intersection coalesce into one because of very similar conduction times, the snowflake for a convergent–divergent circuit may be indistinguishable from that of a two-fiber convergent or divergent circuit with high threshold; in favorable cases, the fine structure of the cross-correlation histograms may then serve to distinguish among the possibilities.

Reciprocal connections give rise to a pair of parallel bands that enclose the origin. The two remaining types of three-fiber circuits and the 4 types of four-fiber circuits (Table I) all incorporate reciprocal connections. The corresponding 6 snowflakes, for excitatory connections, are illustrated in Fig. 10. When the bands are clearly defined, the circuits are easily reconstructed systematically, simply by interpreting each band by its angle and position. Such a straightforward procedure works most easily if all the connections are excitatory.

INHIBITORY CONNECTIONS

As already illustrated (Fig. 4c, d), an inhibitory connection gives rise to a band of reduced point density that is generally thicker than the band of increased density from an excitatory connection. Moreover, the stronger the inhibition, the wider the band. It is generally more difficult to detect and measure inhibition than excitation in a snowflake just as in a cross-correlation histogram, and for the same reasons: inhibitory action is displayed only against an ongoing 'background' of excitation or autonomous impulse production, and the effects of inhibition saturate at zero firing probability of the postsynaptic neuron. In order to display the effects of inhibitory connections in a snowflake, a relatively long sample must be used and the range of the snowflake (*i.e.*, maximum span of impulse trios) must be greater than in comparable cases of excitation.

Aside from these methodological considerations, the patterns of snowflakes produced by various neuronal circuit configurations, and the interpretation of the former in terms of the latter, follow the same principles already illustrated for excitation. Fig. 11 illustrates some selected cases, including effects of presynaptic rhythmicity and examples taken from experiment.

If the postsynaptic neurons exhibit postinhibitory rebound — a period of enhanced excitability following synaptic inhibition — then a band of decreased or zero point density will be bordered, on the edge away from the origin, by a band of increased density, just as if the inhibitory connection were accompanied by an excitatory

Fig. 11. Snowflakes of circuits with inhibitory connections. a: A excites B, B inhibits C. b: A excites C, C excites B, B inhibits A. c: A inhibits B, B and A inhibit C. d: A inhibits B and C. e and f: same as cases c and d except that neuron A is a fairly regular pacemaker. Note repetition of bands. g: A inhibits B, with slight postinhibitory rebound; B inhibits C, less strongly. h: A inhibits B and C. Rhythmic firing of A gives rise to the repetition of patterns in the snowflake as in case f. Cases (a–f) from computer simulation; cases (g, h) from visceral-ganglion neurons in *Aplysia californica*, courtesy of Professor J. P. Segundo, Department of Anatomy, University of California, Los Angeles.

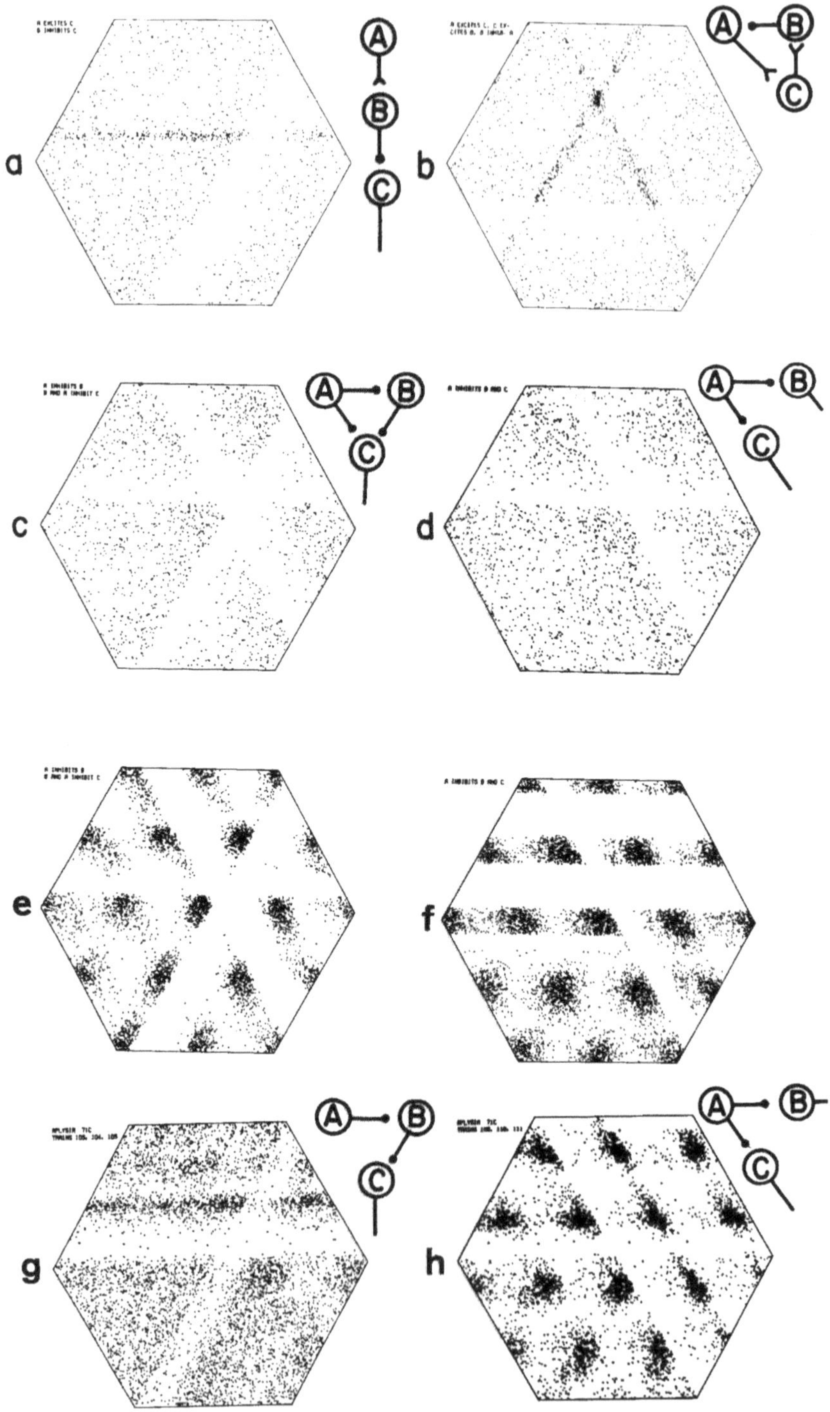

Fig. 11.

connection having a somewhat longer latency. Reciprocally inhibitory circuits, with or without rebound, can give rise to rhythmic patterns of impulse activity; illustration of the corresponding snowflakes is beyond the scope of this introductory paper, but a particular class of ring circuits is investigated in detail by Espinosa[3].

IMPLEMENTATION

No particular programming difficulties arise from the joint impulse configuration scatter diagram; the problems are associated entirely with the graphical display.

At the University of Pennsylvania, a cathode-ray-tube display under the control of a PDP-12 computer (Digital Equipment Corporation) is used to produce the scatter diagrams. The points are not all displayed at one time; rather, the camera shutter is left open while relatively small numbers of points are displayed in sequence.

At Stanford, snowflakes have been made with a line-printer display, with a Versatec electrostatic plotter, and with a CalComp pen-and-ink plotter, all under the control of an IBM 360/67 computer, at the Academic Computing Facility of the Stanford Center for Information Processing.

The line-printer snowflake is rapid and inexpensive, although the resolution is coarse. It has proven highly useful for making surveys of experimentally obtained trios of nerve-impulse trains as well as for experimenting with the range of the snowflake for both experimental and simulated data.

The Versatec plotter operates on line, so that its plots share with the line printer the advantage of being immediately available. Resolution is acceptable, and problems of smearing the image have been overcome, but the low contrast causes difficulties in reproduction. In addition, the costs have been surprisingly high, due to unusually large 'wait' times, attributable to the interface between the computer and the plotter.

All the snowflakes illustrated in this paper were produced on the CalComp plotter, through the use of the Overall Plotting System, a package of Fortran-callable subroutines available at Stanford. The programs produce a magnetic tape, which is subsequently used to control the plotter. Plotting points in temporal sequence is excessively wasteful of travel time for the pen and the drum; in order to plot most economically, the points must be grouped into strips for scanning and sorted according to horizontal position within each strip. Assuming the vertical and horizontal pen velocities with respect to the paper are equal, and that the pen does not move diagonally, one may easily show that the optimal number of horizontal strips is given by $\sqrt{(N/3)}$, where N is the number of points in the plot. If the pen moves directly from point to point (*i.e.*, along a hypotenuse), the optimal number of strips is $\sqrt{(N/\sqrt{3})}$. Similar formulas can be derived for unequal vertical and horizontal pen velocities. Sorting within the strip was performed using the partition-exchange algorithm ('quick sort') described by Knuth[6]. Resolution is acceptable with this technique and contrast and reproducibility were excellent, although this method is the most costly of the three.

APPLICATION

Computer simulation[12,13,17] has been used to prepare selective catalogs of the statistical signs of functional interaction between two neurons, as visualized in the cross-correlation histogram[10,16,19] and in peristimulus-time scatter diagrams[4,5]. These predictions have subsequently been tested in a neuronal system in which simultaneous penetration of two neurons was possible[1]. These methods have been used for working out neuronal circuitry in systems in which dual penetrations are not obtainable, and which therefore have had to rely solely on extracellularly derived records[2,7,8,20–24].

The 'snowflake', which extends these techniques to trios of neurons, is proposed as a supplementary method, furnishing additional power and sensitivity, for making physiologically useful inferences on the basis of nerve-impulse timings[3,24]. Such techniques have become more accessible to the neurophysiologist through the increased availability of laboratory-sized digital computers.

ACKNOWLEDGEMENTS

This work was supported through National Institutes of Health Research Grants NS 09744 (D.H.P.) and NS 05606 (G.L.G.), and through a Medical Research Council of Canada Centennial Fellowship (W.G.T.).

Paul B. Goldstein wrote an earlier version of the Versatec programs; Stephen K. Riffenburgh wrote the current versions of programs for the line printer, Versatec, and CalComp plotters. We are most grateful for their assistance. We also thank José P. Segundo for his valuable suggestions regarding the manuscript and for permission to use the data in Fig. 11g and h. The authors are also indebted to Urs R. Wyss for pointing out errors in the original version of Table I, and to Richard P. DeFrancisci for carrying out an independent enumeration of circuit types for 3 neurons.

REFERENCES

1 Bryant, H. L., Jr., Ruiz Marcos, A., and Segundo, J. P., Correlations of neuronal spike discharges produced by monosynaptic connections and by common inputs, *J. Neurophysiol.*, 36 (1973) 205–225.
2 Dickson, J. W., and Gerstein, G. L., Interactions between neurons in auditory cortex of the cat, *J. Neurophysiol.*, 37 (1974) 1239–1261.
3 Espinosa, E., J. I., *Simulation Studies in Neural Oscillators*, M. S. Thesis, The Moore School of Electrical Engineering, Philadelphia, Pa., 1974, 112 pp.
4 Gerstein, G. L., and Perkel, D. H., Simultaneously recorded trains of action potentials: analysis and functional interpretation, *Science*, 164 (1969) 828–830.
5 Gerstein, G. L., and Perkel, D. H., Mutual temporal relationships among neuronal spike trains: techniques for display and analysis, *Biophys. J.*, 12 (1972) 453–473.
6 Knuth, D. E., *The Art of Computer Programming, Vol. 3, Sorting and Searching*, Addison-Wesley, New York, 1973, 722 pp; see pp. 116 *et seq.*
7 Kristan, W. B., Jr., Plasticity of firing patterns in neurons of *Aplysia* pleural ganglion, *J. Neurophysiol.*, 34 (1971) 321–336.
8 Kristan, W. B., Jr., Characterization of connectivity among invertebrate motor neurons by cross correlation of spike trains. In F. O. Schmitt and F. G. Worden (Eds.), *The Neurosciences: Third Study Program*, MIT Press, Cambridge, Mass., 1974, pp. 371–377.

9 MAYNARD, D. M., Simpler networks, *Ann. N.Y. Acad. Sci.*, 193 (1972) 59–72.

10 MOORE, G. P., SEGUNDO, J. P., PERKEL, D. H., AND LEVITAN, H., Statistical signs of synaptic interaction in neurons, *Biophys. J.*, 10 (1970) 876–900.

11 PERKEL, D. H., Spike trains as carriers of information. In F. O. SCHMITT (Ed.), *The Neurosciences: Second Study Program*, Rockefeller Univ. Press, New York, 1970, pp. 587–596.

12 PERKEL, D. H., A computer program for simulating a network of interacting neurons. I. Organization and physiological assumptions, *Comput. Biomed. Res.*, in press.

13 PERKEL, D. H., A computer program for simulating a network of interacting neurons. III. Applications, *Comput. Biomed. Res.*, in press.

14 PERKEL, D. H., Presynaptic inhibition: detection through statistical analysis of impulse trains, *Brain Research*, 96 (1975) 330–336.

15 PERKEL, D. H., GERSTEIN, G. L., AND MOORE, G. P., Neuronal spike trains and stochastic point processes. I. The single spike train, *Biophys. J.*, 7 (1967) 391–418.

16 PERKEL, D. H., GERSTEIN, G. L., AND MOORE, G. P., Neuronal spike trains and stochastic point processes. II. Simultaneous spike trains, *Biophys. J.*, (1967) 419–440.

17 PERKEL, D. H., AND SMITH, M. S., A computer program for simulating a network of interacting neurons. II. Programming aspects, *Comput. Biomed. Res.*, in press.

18 SEGUNDO, J. P., PERKEL, D. H., AND MOORE, G. P., Spike probability in neurons: influence of temporal structure in the train of synaptic events, *Kybernetik*, 3 (1966) 67–82.

19 SEGUNDO, J. P., PERKEL, D. H., WYMAN, G., HEGSTAD, H., AND MOORE, G. P., Input-output relations in computer-simulated nerve cells. Influence of the statistical properties, strength, number, and interdependence of excitatory presynaptic terminals, *Kybernetik*, 4 (1968) 157–171.

20 SMITH, D. O., Central nervous control of presynaptic inhibition in the crayfish claw, *J. Neurophysiol.*, 35 (1972) 333–343.

21 SOKOLOVE, P. G., AND TATTON, W. G., Analysis of postural motoneuron activity in the crayfish abdomen. I. Coordination by premotoneuron connections, *J. Neurophysiol.*, 38 (1975) 313–331.

22 TATTON, W. G., AND SOKOLOVE, P. G., Analysis of postural motoneuron activity in the crayfish abdomen. II. Coordination by excitatory and inhibitory connections between motoneurons, *J. Neurophysiol.*, 38 (1975) 332–346.

23 TATTON, W. G., AND SOKOLOVE, P. G., Intersegmental gradient of activity in an invertebrate postural system, *Brain Research*, 85 (1975) 86–91.

24 WIENS, T. J., AND GERSTEIN, G. L., Cross connections among crayfish claw efferents, *J. Neurophysiol.*, 38 (1975) 909–921.

25 ZUCKER, R. S., KENNEDY, D., AND SELVERSTON, A. I., Neuronal circuit mediating escape responses in crayfish, *Science*, 173 (1971) 645–650.

Chapter 5
Repeating Patterns in Spike Trains and Functional Groups of Neurons

Designing Methods for Analyzing Single- and Multi-Neuron Recordings

Judith E. Dayhoff

It all started in August of 1976. At the age of 21, I had finished my bachelor's degree at Duke University and, in addition, an entire year of courses in the Biophysics PhD program at the graduate level at the University of Pennsylvania. It was clear that my interests encompassed intense mathematics and computing as well as the biological sciences. I was given the task of accomplishing three preceptorships in biophysics and to decide among them who should be my PhD advisor. The first two were completed and had resulted in a non-match to my interests as well as a list that I compiled of possible matching professors—all were visited and eliminated as possible thesis advisors. The last person on my list was Professor George Gerstein. Everything seemed to be at stake when I made an appointment with him and walked into his laboratory.

After explaining that I was interested in mathematics and computing as well as biophysics, I asked whether he had a thesis project that might be appropriate. He responded immediately that he had a topic available. It was to examine nerve spike train data and to ask the question as to whether there might be repeating patterns in the data, or any temporal patterns, that could possibly reflect the workings of the nervous system. The project, he thought, would result in graduation regardless of the outcome, because it would be a new result to be able to test—at all—for the first time—whether there were temporal patterns in the spike trains, i.e., non-random

J. E. Dayhoff (✉)
Silver Spring, MD, USA
e-mail: jdayhoff12@gmail.com

A. Aertsen et al. (eds.), *Introducing Computation to Neuroscience*, Springer Series in Computational Neuroscience, https://doi.org/10.1007/978-3-030-87447-6_5

spike timing modulations in neural recordings. Regardless of whether the answer was "yes" or "no," the project would be considered to be completed.

I accepted the project and began the work immediately with Professor Gerstein. I was fascinated by the possibilities of how the brain might work and how signaling or any form of computation could arise within its confines. George wanted to know what signaling could be taking place from a single neuron as sent to another and saw this question as being part of the larger question of determining what groups of neurons could be doing in the nervous system. At that time and for another few decades, it was customary for neuroscientists to perform research on only one neuron at a time and to measure the responses of only single neurons to stimuli. The prevailing instrumentation could take only averages of the firing rates of the recorded neurons and typically measured only one channel, which was only one neuron or a few closely placed neurons. Although this limited experimental setup was used by many neuroscientists to creatively explore and query the responses of single neurons, it could not determine what non-averaged signals were produced (such as temporal patterns) or what groups of neurons (more than one neuron) were doing together.

George paraded the idea of the "functional group" of the nervous system. He wanted to know what size group was truly the size that played a functional role, with key significance as to perception or pattern production or cognitive processing and thinking. He had previously published a study with Professor Donald Perkel on finding groups of neurons that fired together among simultaneously recorded spike trains (Gerstein et al. 1978). However, he was deeply concerned that he could not discern, with that method, those groups that fired with n simultaneous spikes significantly without picking up the subgroups with (n-1) or fewer first. He had the idea that perhaps the groups of neurons, size n for example, could appear without much premonition from smaller groups. Surely, he did not want to miss anything! That key size of group—the proverbial functional group that must be sought—should not be missed.

Although I listened to George's predicament about finding the basic "functional groups" of the nervous system with deep concern, we could not yet think of a new approach to solving the synchronous group identification problem, and we set aside that work until later. I continued on my search for repeating patterns in single spike trains. Several ideas were tossed around, one of which would involve a time shuffling or time-randomized version of the spike train to compare against the recorded train and then to determine, via the comparison, whether the original data contained more repeating patterns than its randomized version. After several other approaches appeared to be not as productive, I started to shuffle spike trains, first encoding them as lists of interspike intervals. Immediately, I could see that there were stark differences between the randomized versions and the original data trains. As a result, I quickly began the task of deciding how to develop a compendium of information about each train (shuffled or not) to be used in making a comparison. Although we were looking for repeating patterns, similar to sparse beads on a string, there would be many extra patterns between them; how would we know which to ignore and which to make note of? Additionally, where could we start on such a project when there were so many patterns in the first place? In addition to all of the patterns in a spike train, there were all of the patterns that did not even appear in that spike train! Plus an infinitude of different length interspike intervals on which to base or describe those patterns! One could drown in this project in 1976!

Something had to be condensed to be able to accomplish our goals, and it had to be condensed simultaneously with developing the spike train shuffling and measurements. As a result, we developed several approaches to this new methodology, which were all used at the same time: (1) Represent the spike train pattern intervals as quantized versions of the original interval lengths, which resulted in a list of integers (or "letters"), with patterns being represented discretely as "words" (Dayhoff and Gerstein (1983a), Fig. 2); (2) Consider only those patterns that appeared in the data spike trains or their shuffles, without looking at any of the other patterns; (3) Consider all of the patterns with the same length at the same time, and when done, consider all of the patterns of the next length; (4) Count the number of patterns of each length, q, that occur at least n times for the range of values of n. This step is important because it avoids having to consider the count for every single pattern separately, and in addition, it utilizes the "at least" counts (which are preferable) rather than the "exactly" counts; and (5) Produce statistical testing and evaluations at the end. By making these refinements, I could still address the overall properties of the spike train, now a sequence of discrete integers ("letters"), and still not lose sight of the goal of searching for the important favored patterns, or "beads on a string," in which the "beads" could be sparsely placed. Additionally, many types of non-random structure would be detectable in the results and would provide more information on the spike trains. In the end, my result was a discrete, applicable, and verifiable method of identifying favored patterns in sequences of discrete data (letter sequences or integer sequences) that was cogent for use on nerve spike trains and even other sequence data.

We applied the new method, which we called the quantized Monte Carlo (QMC) method, to two datasets that had been collected in George's lab, one from crayfish claw (Lindsey and Gerstein 1979) and the other from cat visual cortex (Gerstein and Michalski 1981). We found non-random and "favored" patterns in both. After applying the QMC method to each spike train, we subsequently swept the identified patterns across the spike train in a non-quantized form to identify additional matches and pattern repetitions with more variation, which we found.

The next task was to interpret the findings. We had indeed found favored, repeating patterns, or other non-random structure, in single spike trains. These could be interpreted in a number of ways, the first of which was to imply specific meaning or specific biological connections. However, in our interpretations, we had to allow for whether fluctuations other than the insertion of patterns could result in repeating patterns; whether the brain might deliberately produce such non-randomness for a specific use or not; whether such patterns or fluctuations could be used in some way; and whether the brain required statistically significant repetitions or whether it simply "knew" what it was looking for in the first place, without excess repetition. Carefully, we wrote about all of these considerations, hoping to document all possibilities that might arise for the sake of future researchers (Dayhoff and Gerstein 1983a,b).

The papers were a success upon publication and triggered a worldwide response of hundreds of "reprint request" cards that arrived in the mail, in which scientists asked for a copy to be mailed to them. This clever technique allowed us to connect with our readers even before the era of electronic publication and communication. In the 1980s, I gave many presentations of my work and, in 1984, made a lecture tour of Europe to further communicate my results.

The synchronous group problem was finally tackled during a postdoctoral assignment that I did in the laboratory of Professor Donald Perkel at Stanford University. Before I left George's lab, we discovered a paper on a clustering approach called gravitational clustering, and we wondered whether that approach would be useful for tackling the problem of finding synchronously firing groups without being influenced by the presence of smaller groups. The original gravitational clustering method (Wright 1997) was a static method that could be used on a set of vectors to determine the clusters. However, it could be expanded in a natural way to neuronal spike trains in which the clusters fluctuated over time. This new approach appeared to address the possible limitations of accretional methods, in principle, because it treated all of the neurons the same from the outset and did not rely on building from subsets (size $< n$) to attain the final group of "n" neurons that would be identified as a group firing in synchrony. An additional asset to this method was that real neurons might signal one another and gather into dynamically operating "functional" groups that changed over time, with the same neuron participating in one group for a while and then in another group during a different time interval. The "gravity method" could calculate a display of the changing nature of the synchronous groups and the migration of a neuron from one group to another as time passed.

I began the first coding of the gravitational clustering method in Don Perkel's laboratory at Stanford in the winter of 1982. That initial effort was quickly expanded into a complete study on the gravitational method and its verification with simulated data, with George visiting often to move the results forward. Again, we simulated data to test the methodology, this time to demonstrate simultaneous (simulated) spike trains that did—or did not—display functional (synchronous) group behavior. We validated the method and published the work in The Journal of Neuroscience, the Society for Neuroscience's premier research journal (Gerstein et al. 1985). Subsequent to that publication, years of research that utilized "gravity" was conducted in both Gerstein's laboratory and the Lindsey/Morris Laboratory at the University of South Florida Tampa, in which "gravity" was used to detect synchronously firing groups and their changes and expanded to perform the detection of underlying circuits; in addition, group-wise spike patterns were detected along with the synchronies. (See also Chap. 5 this Volume.) The results were both plentiful and prolific (Lindsey and Gerstein 2006; Gerstein et al. 1989; Morris et al. 2001).

I have been asked to include how my subsequent research was enlightened by the work that I did with George Gerstein. After obtaining my PhD, I developed, worked with, tested, and characterized a variety of analysis tools, using novel approaches, for the study of biological systems as well as medical applications.

Many of my research projects involved new computational methods for a spectrum of applications and the development of models that furthered the understanding of timing relationships in neural computation. In 1990, I authored a book on "artificial" neural networks, entitled Neural Networks Architectures: An Introduction (Dayhoff 1990), which summarized the basic available models for building analysis tools inspired by neural computing in the form of networked processing elements, simple models of neurons. This book was well received across a large community of readers and appeared at a turning point of technology, at which a "paradigm shift" was taking place between the use of a single complex computing processor and the possible use of many simpler processors ("neurons") networked together. Next, I lectured this material at the graduate level at the University of Maryland and at the National Institutes of Health. In the 1990s, I performed research on the temporal structure contained in several different neural computing models. I worked with Dr. Dalton Lin on the subject of adaptive time-delay neural networks, characterizing their inherent capabilities and generality and demonstrating applications in the areas of trajectory production, spatiotemporal pattern recognition, and time series approximations. I also explored attractor networks with dynamic attractors (complex oscillations), as opposed to the fixed point simpler attractors in the Hopfield network, and I looked at the integrative properties of pulse-coupled networks as well as improving the speed of the gravity calculation (Dayhoff, 1994). In the 2000s, I considered an approach to medical applications that utilized a neural network followed by a ROC plot (Dayhoff and DeLeo 2001). Then, in 2007, I published a paper on a synchronous group network, a configuration that could be embedded inside a synfire chain and still accomplish what appears to be general function approximation (Dayhoff 2007). Overall, the focus and directionality of George Gerstein's work was an anchor point for me in these many relevant research projects.

I remember George as a brilliant colleague, a friendly co-worker, a wise boss and advisor, and an individual with outstanding personal character. He raised a family simultaneous with some of his best work and frequently invited guests from around the world to visit. Of course, he will be greatly missed.

References

Dayhoff JE (1990) Neural network architectures: an introduction. Van Nostrand Reinhold Co., New York

Dayhoff JE (1994) Synchrony detection in neural assemblies. Biol Cybern 71(3):263–270. https://doi.org/10.1007/BF00202765

Dayhoff JE (2007) Computational properties of networks of synchronous groups of spiking neurons. Neural Comput 19(9):2433–2467. https://doi.org/10.1162/neco.2007.19.9.2433

Dayhoff JE, DeLeo JM (2001) Artificial neural networks: opening the black box. Cancer 91(S8):1615–1635. https://doi.org/10.1002/1097-0142(20010415)91:8+<1615::AID-CNCR1175>3.0.CO;2-L

Dayhoff JE, Gerstein GL (1983a) Favored patterns of spike trains. I. Detection. J Neurophysiol 49(6):1334–1348. https://doi.org/10.1152/jn.1983.49.6.1334

Dayhoff JE, Gerstein GL (1983b) Favored patterns of spike trains. II. Application. J Neurophysiol 49(6):1349–1362. https://doi.org/10.1152/jn.1983.49.6.1349

Gerstein GL, Michalski A (1981) Firing synchrony in a neural group: putative sensory code. In: Székely G, L'abos E, Damjanovich S (eds) Advances in physiological sciences, vol 30. Neural communication and control, Pergamon Press, Akademiai Kiado, Debrecen/Budapest, pp 93–102. https://doi.org/10.1016/b978-0-08-027351-8.50012-3

Gerstein GL, Perkel DH, Subramanian KN (1978) Identification of functionally related neural assemblies. Brain Res 140(1):43–62. https://doi.org/10.1016/0006-8993(78)90237-8

Gerstein GL, Perkel H, Dayhoff E (1985) Cooperative firing activity in simultaneously recorded populations of neurons: detection and measurement. J Neurosci 5(4):881–889. https://doi.org/10.1523/JNEUROSCI.05-04-00881.1985

Gerstein GL, Bedenbaugh P, Aertsen AMHJ (1989) Neuronal assemblies. IEEE Trans Biomed Eng 36(1):4–14. https://doi.org/10.1109/10.16444

Lindsey BG, Gerstein GL (1979) Interactions among an ensemble of chordotonal organ receptors and motor neurons of the crayfish claw. J Neurophysiol 42(2):383–399. https://doi.org/10.1152/jn.1979.42.2.383

Lindsey BG, Gerstein GL (2006) Two enhancements of the gravity algorithm for multiple spike train analysis. J Neurosci Methods 150(1):116–127. https://doi.org/10.1016/j.jneumeth.2005.06.019

Morris KF, Shannon R, Lindsey BG (2001) Changes in cat medullary neurone firing rates and synchrony following induction of respiratory long-term facilitation. J Physiol 532(2):483–497. https://doi.org/10.1111/j.1469-7793.2001.0483f.x

Wright WE (1977) Gravitational clustering. Pattern Recogn 9:151–166

JOURNAL OF NEUROPHYSIOLOGY
Vol. 49, No. 6, June 1983. *Printed in U.S.A.*

Favored Patterns in Spike Trains. I. Detection

JUDITH E. DAYHOFF AND GEORGE L. GERSTEIN

Departments of Biophysics and Physiology, University of Pennsylvania, Philadelphia, Pennsylvania 19104

SUMMARY AND CONCLUSIONS

1. Traditional spike-train analysis methods cannot identify patterns of firing that occur frequently but at arbitrary times. It is appropriate to search for recurring patterns because such patterns could be used for information transfer. In this paper, we present two methods for identifying "favored patterns"—patterns that occur more often than is reasonably expected at random.

2. The quantized Monte Carlo method identifies and establishes significance for favored patterns whose detailed timing may vary but that do not have extra or missing spikes.

3. The template method identifies favored patterns whose occurrences may have extra or missing spikes. This method is useful when employed after the results of the first method are known.

4. Studies with simulated spike trains containing known interpolated patterns are used to establish the sensitivity and accuracy of the quantized Monte Carlo method. Certain trends with regard to parameters of the detected patterns and of the analysis methods are described.

5. Application of these methods to neurophysiological data has shown that a large proportion of spike trains have favored patterns. These findings are described in the accompanying paper (3).

INTRODUCTION

It is generally accepted that spike trains are carriers of information, especially when dealing with long-axon neurons. (We use "information" in the general, nontechnical sense, thus avoiding Shannon's (13) definition.) Under such circumstances, spikes are assumed to be all of the same amplitude and duration, so that information must be carried by the duration and sequences of interspike intervals.

Some aspects of spike activity are easily associated with events that can be observed or imposed in the laboratory (i.e., stimuli, movements, etc.). However, even when such events are not taking place in the laboratory, many parts of the nervous system continue to function in what is known as ongoing, or spontaneous, activity. Ongoing activity is usually observed in the intervals between stimuli, as long as these are presented at sufficiently slow rates. It is possible that in the course of ongoing activity, the nervous system is transmitting information about events that are not directly observed (or observable) in the laboratory. If such internal events repeat, we might expect, under appropriate circumstances, to find that the ongoing firing contains corresponding recurring firing patterns. The objective of the present work has been to find whether favored firing patterns (patterns that recur excessively) appear in single spike trains during ongoing activity.

The traditional armamentarium of spike-train analysis methods (4, 10–12) does not offer tools to detect recurring patterns in general. The poststimulus time (PST) histogram is limited to detection of stimulus time-locked responses and yields response probability averaged over all stimulus presentations. This averaged probability does not necessarily reflect repeating patterns. Favored patterns, on the other hand, are not average phenomena over the entire spike train; they are individual repetitions of relatively short sequences in the spike train and are not necessarily related to known stimulus times.

 0022-3077/83/0000-0000$01.50

In the present paper we outline two detection methods and appropriate significance tests to identify favored patterns (if they exist) in a single spike train. When these methods are applied to experimental data, it turns out that an appreciable fraction of spike trains do indeed contain favored patterns, as described in the accompanying paper (3).

Although we have detected favored patterns in a number of experimental situations, this does not prove that these patterns represent coding of certain as yet undefined events. Detection of favored patterns does, however, contribute to the growing evidence for patterns as a candidate neural code (3, 9). It is also quite likely that the information in one spike train must be considered in conjunction with other spike trains in some sort of ensemble coding scheme. The work in this paper is a necessary first step in examining such more-complex possibilities.

What constitutes a pattern?

Suppose a nerve cell is generating a particular firing pattern as part of a neural coding scheme. The pattern may not be exactly the same each time it is produced either because of biological noise or because the variability itself may be part of the code. These two sources of variability force different coding-decoding schemes in the nervous system. The particular biological origin of the variability is not relevant to the process of favored pattern detection and analysis. However, the actual type of variability may affect the choice of mathematical techniques to be used.

There are a number of different ways in which the successive occurrences of a pattern could vary. *1*) Variation in temporal spacing. The same number of spikes comprise each occurrence of the pattern, but the interspike intervals vary in length. If several occurrences of the pattern are aligned, all of the spikes appear to be approximately in the same place. *2*) Extra or missing spikes. If such pattern occurrences are aligned, only some of the spikes appear to be in approximately the same place at each occurrence. *3*) Time-scaled patterns. These pattern occurrences have the same relative spacing between successive spikes but may appear stretched or contracted in time. It is plausible that time-scaled patterns might encode information in such a way that stretched or contracted patterns stand for the same or similar information. (An analogy would be changing drive speeds on playback of a tape recording; information is preserved although the time-scale is changed.)

Another situation that complicates recognition of a pattern is temporal overlap. Suppose that a certain stimulus causes a particular nerve cell to fire a particular pattern (as in Fig. 1). It is possible that the stimulus could occur twice in a row, the occurrences so closely spaced that the nerve cell cannot complete its pattern response to the first stimulus before the second stimulus arrives. The nerve cell could do many things, including the following four possibilities: *1*) Complete the pattern response to the first stimulus and ignore the second stimulus. *2*) Complete the pattern response to the first stimulus. The cell then gives a delayed response to the second stimulus so that a second occurrence of P is immediately after the first. *3*) Complete the pattern response to the first stimulus. The cell then gives only a terminal portion of the pattern P. The initial section of the second stimulus response has been suppressed by the

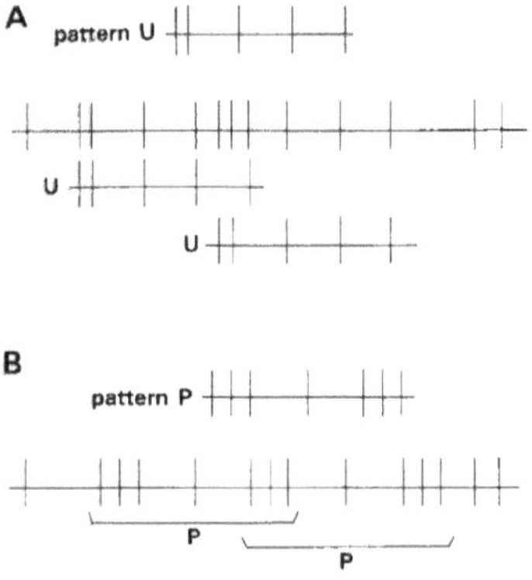

FIG. 1. *A*: the general case of pattern overlap. Occurrences of pattern U are redrawn under the spike train. Note that the portion of spike train associated with each occurrence of the pattern contains at least one extra spike. *B*: a special case. The beginning and end of the pattern P have the same temporal structure. Two occurrences of P are shown to overlap so that neither contains extra spikes.

first stimulus response. *4*) The cell responds to both stimuli, superimposing the second pattern response over the first one. In this case, the second pattern begins before the generation of the first pattern has been completed, forming overlapping pattern occurrences. Examples are given in Fig. 1.

Preliminaries to analysis

Quantization is implicit in the use of time measurement for numerical computation. However, the optimal description of patterns may require a coarser time resolution than determined by the experimental measurement precision.

Quantization is achieved by the process illustrated in Fig. 2. First, the spike train is represented as a list of successive interspike interval values. Then a quantization resolution x is chosen. By means of this quantization resolution, the spike train is translated into a list of integers by making any interval between 0 and x ms a 1, between x and $2x$ ms a 2 and, in general, between $(i-1)x$ and ix ms an i. The spike train then becomes a list of integers. Note that the numerical value of a quantized integer j is not intended to approximate the actual interval value; it is merely a label that reflects the range of values that fall into the jth bin. It would make no difference to the analysis if quantized intervals were labeled with letters or other nonnumeric characters.

A candidate for a pattern word can also be quantized. It becomes a short list of integers. Suppose we are interested in finding occurrences of a pattern P, which, when quantized, becomes 1 4 2 5. P occurs whenever the word 1 4 2 5 appears in a contiguous block in the quantized spike train. These concepts are shown in Fig. 2.

Note that quantization does not allow identification of pattern occurrences with extra or missing spikes. When quantized, pattern occurrences with extra or missing spikes become a different list of integers than the original quantized pattern. The template method described below shows how to identify pattern occurrences that can sometimes have extra or missing spikes. This method does not require quantization.

Even if a pattern does not have extra or missing spikes, quantization has certain consequences. On the positive side, the quantization allows the occurrence of the pattern to have interval lengths that vary within the quantization bin. Thus we automatically accommodate many types of variability. On the

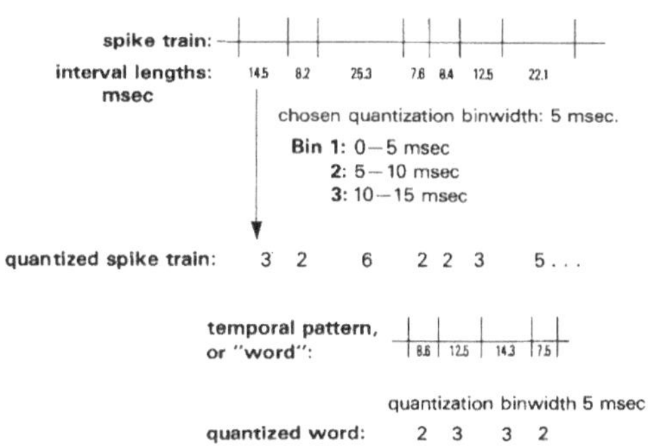

FIG. 2. Spike-train quantization. Successive interspike interval lengths are translated into a list of integers by using a chosen quantization bin width. In this example, a quantization bin width of 5 ms is used. An interval between 0 and 5 ms becomes the integer 1; an interval between 5 and 10 ms becomes the integer 2, etc. The resulting list of integers is the quantized spike train in which each integer reflects an approximate interval length. A temporal pattern word (below) that is quantized becomes a short list of integers. Note that a different quantization bin width may yield a different list of integers.

negative side, the quantization causes loss of some pattern identifications because an interval that is near a quantization bin boundary will fall into different bins when there is variability. These considerations suggest that several different bin widths should be used routinely in the analysis process on each spike train.

Defining excess occurrences of a pattern

If experimental data show recurrence of patterns, these can be considered favored only if an appropriate significance test demonstrates that they recur unusually often. In this work we have based such a significance test on comparison between the original and shuffled versions of the experimental spike train.

In the shuffled versions, the position of each interspike interval does not depend on the values of other intervals. Shuffling the spike train is achieved by a randomized reordering of exactly the same set of interspike intervals that comprise the original data spike train.

One advantage of comparing the original and shuffled spike trains is that they have the same interval histogram. The shuffled spike train differs from the data spike train solely in that the shuffling process destroys the ordering of the intervals, not the actual values of the intervals.

The effect of preserving the interval histogram in the random model is demonstrated by the following example. Suppose the spike train has a large number of occurrences of a certain length interspike interval, say 5 ms. Suppose also that the pattern word 5.0, 5.0, 5.0 occurs 20 times in the spike-train data. We know that at random we can expect to find many occurrences of pattern words containing 5-ms intervals simply because there are many 5-ms intervals. Are the 20 occurrences of the pattern 5.0, 5.0, 5.0 due solely to the fact that there are so many 5-ms intervals, or is 20 occurrences more than one would expect at random? Shuffling the spike train gives us the appropriate random model to answer this question because the shuffled, random model contains the same number of 5.0-ms intervals as the data spike train but produces sequences of three such intervals solely by chance.

METHODS

In this paper we report two analysis methods, only the first of which involves quantization. The second method (template method) can deal with pattern occurrences that have extra or missing spikes. A third method of considerable mathematical interest has been developed but is less practical for long spike trains; it is reported elsewhere (2).

Quantized Monte Carlo method

The data spike train is represented as a list of interspike intervals, and the interval values are quantized at a chosen quantization resolution.

Next, the computer program does the following counts: For each length t ($t = 2, 3, \ldots, 10$) and each value k ($k = 2, 3, \cdots$), the number of t-letter words occurring at least k times is counted. The program starts with two-letter words, counting the number of two-letter words occurring at least twice, at least three times, etc. Eventually a value k is reached such that the number of two-letter words occurring at least k times is 0. The counting of two-letter words can then stop. (Example counts are shown in Figs. 3 and 4.)

The computer program then deals with three-letter words, counting the number of three-letter words occurring at least once, at least twice, at least three times, etc. The same counting process is done for words of longer length. In our analysis, we count words of length 2 through 10; in a typical spike train of 1,500 intervals, there generally are no words of 10 or more letters that occur more than once.

The spike train is shuffled 99 times, and for each shuffled version the same counts are made. The purpose of the shuffled counts is to find whether there are any values of t and k for which the number of t-interval words occurring $\geq k$ times is larger in the data spike train than the number found in all 99 shuffled versions of the same spike train. Any such count is statistically significant at the 1% significance level.

The statistical test used distinguishes between the following two hypotheses. Null hypothesis (H_0): the data spike train is apparently generated by a purely random process equivalent to shuffling. Alternate hypothesis (H_1): the spike train is apparently the result of a process that causes certain patterns to recur systematically.

A statistical test at the 1% significance level is achieved by the following. Let c = number of t-letter words occurring at least k times in the unshuffled spike train. Let n_i = number of t-letter words occurring at least k times in the ith shuffled version of the spike train. Let $m = \max(n_1, n_2, \cdots, n_{99})$. If $c \leq m$, then accept H_0. If $c > m$, then reject H_0 at the 1% significance level. Conclude

length (spikes)	k	number of words of given length occurring at least k times: unshuffled count	minimum and maximum counts over 99 shuffles: min	max
3	2	25	21	30
3	3	14	10	17
3	4	8	5	10
3	5	5	0	8
3	6	3	0	4
3	7	0	0	1
4	2	16	10	19
4	3	12	8	17
4	4	4	2	6
4	5	2	0	4
4	6	0	0	0
5	2	8	0	12
5	3	3	0	4
5	4	2	0	0
5	5	0	0	0

FIG. 3. An example where the excessively recurring patterns can be identified uniquely. Two 5-spike words occur at least 4 times in the unshuffled spike train. No 5-spike words occurring at least 4 times were found in 99 shuffles of the spike train. Thus, both the 5-spike words occurring at least 4 times in the unshuffled spike train occur unusually often; these two words can be identified by a computer program.

that c, the number of t-letter words occurring at least k times in the unshuffled spike train data, is unusually high.

The significance level is 1% because in this test we assume initially that the null hypothesis H_0 is true. If H_0 is true, then the count c is included in the estimate of the probability density for $f(y)$, the probability that there are y t-letter words occurring at least k times in a shuffle of the spike train. The histogram of the values $(n_1, n_2, \ldots, n_{99}, c)$ forms an estimate of the probability density function $f(y)$. If at least 99% of the probability density in the histogram occurs at values less than c, then H_0 is rejected at the 1% significance level, and one can conclude that the count c is unusually large. At least 99% of the probability density in the histogram is at values less than c whenever $c > m$.

In summary, for each length t (t from 2 to 10) and each value k ($k = 2, 3, \cdots$): if c, the number of t-interval words occurring $\geq k$ times in the unshuffled spike train is higher than the corresponding counts for all 99 shuffled versions of the spike train, then it is "unusually high" by the statistical test at the 1% significance level. One can conclude that among those t-letter words occurring at least k times in the unshuffled spike train, some or all of them were generated by a nonrandom process. Figures 3 and 4 show examples of counts from original and shuffled spike trains.

Note that with the quantized Monte Carlo method there are two distinct situations in which a count can be significantly high. In the first situation, all the corresponding shuffled counts (for the same values of t and k) are zero. This means that all patterns contributing to the unshuffled count are favored patterns. In the second situation, some of the corresponding shuffled counts are greater than 0 but all are less than the un-

length (spikes)	k	number of words of given length occurring at least k times: unshuffled count	minimum and maximum counts over 99 shuffles: min	max
3	2	54	50	61
3	3	36	27	42
3	4	21	11	33
3	5	14	3	17
3	6	5	0	7
3	7	0	0	1
4	2	35	20	45
4	3	20	15	25
4	4	15	2	17
4	5	10	0	4
4	6	1	0	1
4	7	0	0	0

FIG. 4. An example where the quantized Monte Carlo method identifies a group of patterns, some (but not necessarily all) of which occur excessively. The number of 4-spike words occurring at least 5 times is found to be 10 in the unshuffled spike train; this is statistically significant at the 1% level. Only 0–4 such words were found in each of 99 shuffles of the spike train. Hence, it seems that some (but not necessarily all) of the 10 4-spike words occurring at least 5 times recur systematically as a result of a nonrandom process.

shuffled count. This means that only some of the patterns contributing to the unshuffled count are favored patterns. The quantized Monte Carlo method does not identify which of these patterns are so favored.

Figure 3 is an example where the excessively recurring patterns can be identified uniquely. Two four-letter words occur at least 4 times in the unshuffled spike train. In all 99 shuffles of the spike train, no four-letter words are found to occur 4 times. Thus, both of the four-letter words occurring at least 4 times in the unshuffled spike train occur unusually often (according to the statistical test described above). A simple computer program can now identify the particular two four-letter words that occurred at least four times.

Figure 4 shows an example in which the quantized Monte Carlo method identifies a group of patterns, some (but not necessarily all) of which occur excessively. The number of three-letter words occurring at least 5 times is found to be 10 in the unshuffled spike train. The corresponding counts in the shuffled spike trains are between 0 and 4. Note that the three-interval words occurring at least 5 times in the shuffled trains do not have to be a subset of the 10 three-letter words occurring at least 5 times in the unshuffled train.

Of the 10 three-letter words occurring at least 5 times in the unshuffled train, it seems that 0–4 of these recur as a result of a random process (since 0–4 such words appeared at random in the shuffled spike trains). Apparently, the remaining words contributing to the unshuffled count of 10 were actually favored as a result of a nonrandom process in the nervous system. We do not know, however, which of the 10 three-interval words recur 5 times as a result of being generated by a nonrandom process. We can, however, use the 10 words as template patterns for the template method (below); The template method might then show if there are additional "noisy" occurrences and which of the patterns selected by the Monte Carlo method are most likely to be significant.

Template method

This method is based on a sliding comparison of the chosen template with the entire spike train.

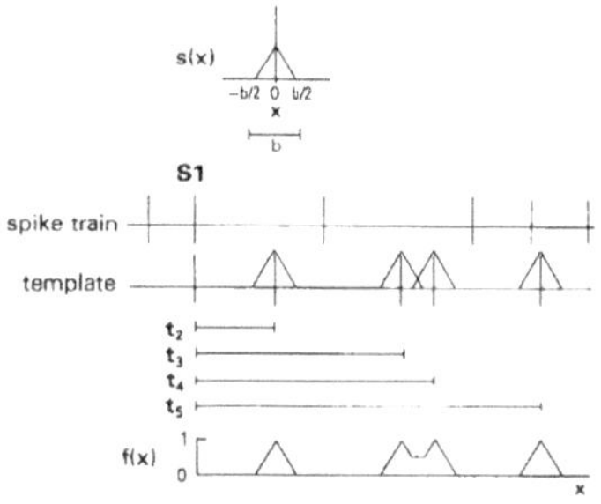

FIG. 5. Comparing the template to the spike train. A triangle of base b is drawn centered on each spike except for the first in the template. The function $f(x)$ is the sum of the triangle heights at each point in the template. This function is used in computing the match value M.

A weighting function based on the template allows detection of imperfect matches.

Figure 5 illustrates the way the template is compared to each section of the spike train. We first compute a template function $f(t)$ by the following process. Choose a value b. Then define a function $s(x)$

$$s(x) = \begin{cases} 1 - 2x/b & \text{if} \quad 0 < x < b/2 \\ 1 + 2x/b & \text{if} \quad 0 > x > -b/2 \\ 0 & \text{otherwise} \end{cases}$$

When $s(x)$ is plotted, it is a triangle of width b and height 1, centered at $x = 0$. It is appropriate to choose b as a reasonable fraction of the average interspike interval.

Let n = the number of spikes in the template. Let t_k be the time between the first spike and the kth spike of the template ($k = 2, 3, \ldots, n$).

Define

$$f(x) = \sum_{k=2}^{n} s(x - t_k)$$

Conceptually, the construction of $f(x)$ can be thought of in the following way. On the template, draw triangles centered about each spike except the first. These triangles have width b and height 1 (see Fig. 5). If two triangles overlap, sum their height in the overlapping areas. The resulting function is $f(x)$.

We use $f(x)$ to compare the template to a section of the spike train. First choose a spike S1 in the spike train with which to initiate the comparison. Let h_k be the time between spike S1 and the kth spike after S1 in the spike train. Then

$$M = \sum_{k=1}^{\infty} f(h_k)$$

It is clear that comparisons yielding higher values of M reflect a better match between the template and the spike train section to which it is compared. In our work we have chosen threshold values for M above which the template is said to "match" the spike train. Note that extra spikes in either the spike train or the template do not result in a decrement of M, and may in fact increase M. Hence, a section of the spike train that approximately matches the template pattern but that may contain a few extra or missing spikes can still be considered a match, providing the value of M for the comparison is above threshold.

Sometimes the neuron will fire considerably more rapidly than in the template. When this happens, the match value M can be very high simply because so many spikes occur in the spike train that some are bound to be near the template spikes and hence result in increments to M. Figure 6 shows an example where M is above threshold as a result of bursting activity in the spike train and not as a result of the neuron generating a match to the template pattern. Such a burst necessarily produces a large match value for just about any template with relatively low firing frequency and reasonable triangle widths. Since we wish to avoid identifying areas of bursting as matches to the template, we can allow for only 1 or 2 extra spikes to occur in a match to the template. If a section of the spike train contains more extra spikes, then that section is not considered a match regardless of the value of M. This is an arbitrary but useful restriction.

Note that a number of variations of the template method have potential utility. Alternative match functions can be proposed. In our match function, all spikes have equal weight except for the first, which is aligned. It is possible to weigh the spikes unevenly or to align a spike other than the first. The triangle function $s(x)$ shown above is linear in x; one could use functions that are neither triangular nor linear. We have not evaluated such variation.

Template statistical test

A formal statistical test is needed to determine whether a template pattern occurs excessively (with extra or missing spikes) in a data spike train.

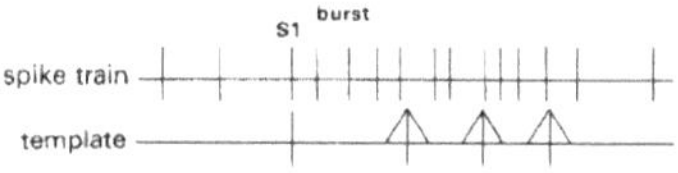

FIG. 6. Comparing the template to a burst in the spike train. Many spikes in the train are near spikes in the template; consequently the matching value M will probably be above the preset threshold.

The test distinguishes between the following hypotheses: null hypothesis (H_0), the data spike train is apparently generated by a purely random process equivalent to shuffling; and alternate hypothesis (H_1), the spike train is apparently the result of a process that causes the template pattern to recur systematically.

First, the template pattern is compared to every position in the unshuffled data spike train, and c, the number of matches, is determined. Next, the spike train is shuffled 99 times, and the template is compared to every position in every shuffled version of the spike train. Here we use shuffled versions that retain one occurrence of the template. The computer counts v_i = the number of times the template matches in the ith shuffled version of the spike train.

A statistical test at the 1% significance level is achieved by the following. Let $m = \max(v_1, v_2, \ldots, v_{99})$. If $c > m$, then reject H_0. The number of occurrences of the template pattern is unusually large.

The statistical test is at the 1% significance level because of the following reasoning. Initially we assume H_0 is true. Then a histogram of (v_1, v_2, . . . , v_{99}, c) approximates the probability distribution for $f(x|\geq 1)$ = the probability that the template occurs x times in a randomly shuffled version of the spike train (given that it occurs at least once.) Using this approximation, one can see that if $c > m$, then the probability of getting at least c occurrences of the template at random is less than 0.01.

Since the template selection methods only select templates that occur at least once, it is best to use a random model in which one occurrence of the template is retained and the rest of the spike train intervals are shuffled. This will, in essence, result in testing whether c is an unusually high number of occurrences of the template given (since the template is taken from the spike train) that there is at least one occurrence of the template. If we did not retain a single occurrence of the template, we would sometimes conclude that a single occurrence is unusually frequent, a patently ridiculous statement.

Choosing templates

Ideally, one would like to test every pattern that occurs in the spike train as a template. Unfortunately, testing every pattern in the unshuffled spike train against shuffled spike trains takes so much computer time that it is impractical. It is necessary to choose a subset of all patterns occurring in the spike train and test only those for significance. In this paper we choose as templates the potentially favored patterns identified by the quantized Monte Carlo method. This allows search for noisy versions of these patterns as well as the possible identification of the subset of Monte Carlo counts that is favored. Other methods of template choice are possible.

Suppose that the quantized Monte Carlo method finds that the number of t-interval words occurring $\geq k$ times is unusually high (for some particular values t and k). As explained above, there are two cases. Case 1. All these pattern words occur excessively if none of the 99 shuffles of the quantized spike train have any t-letter words occurring $\geq k$ times. Case 2. Only a subset of these pattern words occur excessively if some of the 99 shuffles have at least one t-letter word occurring $\geq k$ times.

If the conditions of case 1 are met, all the identified pattern words are statistically significant. We use these identified patterns as templates, comparing each pattern to every position in the spike train. Those sections of the spike train where the match value M is above threshold will be occurrences of a particular template that may have extra or missing spikes. We may examine whether these occurrences have systematic omissions of the same spike or whether an extra spike tends to occur in the same place.

If the conditions of case 2 are met, the quantized Monte Carlo method indicates only that an unknown subset of certain patterns occurs excessively. In this situation the template method may be able to identify some of the subset. This would result if some of the patterns turn out to be favored by the template method. Note, in general, that only some of the favored patterns would be so identified when they recur with extra or missing spikes. Other patterns of the original group indicated by the quantized Monte Carlo method in this case either may not be favored or may not repeat with extra or missing spikes.

The two methods we have given for finding favored patterns are complementary. Favored patterns found with the quantized Monte Carlo method are not necessarily identified as favored by the template method because variability of the pattern occurrences are treated differently. Thus, for a given pattern word that occurs multiply in the data spike train, the shuffled spike trains used for significance testing could contain few or no counts by the quantized Monte Carlo method while still containing a large number of approximate matches by the template method. In this case, the quantized Monte Carlo method could imply a favored pattern (or set of patterns), while the template method implies that the pattern is not favored in template form.

If a pattern is favored by both methods, then we can infer that a nonrandom process has generated both the quantized occurrences (without extra or missing spikes) and the template occurrences (with extra or missing spikes). If a pattern is favored in quantized form but not in template form, then we can infer that a nonrandom process has generated only occurrences without extra or missing spikes.

False positive errors in two methods

Even in a spike train known to be random, certain patterns may be found to occur in excess by either statistical test described above due to false positive errors. In the case of the quantized Monte Carlo method, false positives take the form of excess counts. In the case of the template method, the false positive refers directly to a particular pattern. In this paper we have done statistical testing at the 1% significance level; therefore, we must assume a false positive rate of 1%.

When examining real neurophysiological data with the quantized Monte Carlo method, spike trains with favored patterns yield many counts that are greater than expected at random. In contrast, the false positive errors generally show as occasional isolated counts that are only slightly greater than the counts expected at random. This difference in count behavior suggests that effective use of the quantized Monte Carlo method should examine many pattern lengths and, hence, many different counts. When using the template method, only a small number of templates are usually chosen from the quantized Monte Carlo method. As explained above, there is no a priori reason to expect that a pattern favored in precise form (and identified by the quantized Monte Carlo method) should necessarily be excessive in template form. Consequently, we expect no bias from the method of selection and the 1% expected false positive rate remains valid.

Shuffles and correlation between intervals

In both methods described above we have used shuffles of the data spike train to provide a significance test for favored patterns. Such shuffling preserves the original distribution of intervals in the data spike train but of course destroys the original order. If there was correlation between successive intervals, this is lost. Thus we might expect, if such correlation is indeed present and if repeated patterns occur in the data spike train, that the shuffled versions will provide a deceptively low count and, hence, will give a falsely high indication of favored pattern significance.

It turns out that this is not a significant problem. Many real spike trains do indeed demonstrate some degree of serial dependence in their interval structure. However this dependence rarely goes further than the adjacent interval. There is almost no serial correlation of order two or greater. In contrast, when our two methods are applied to real spike trains (see the companion paper (3)), we find that the favored patterns, if any, usually have four or more intervals. Hence, real data tend to have patterns that are much longer than the reach of serial correlations that are on occasion present. Hence we feel that the simple shuffles used in these papers are reasonable for the significance tests.

Note that it is a relatively simple matter to generate randomized spike trains that retain a pairwise serial correlation but are otherwise similar to a shuffled spike train. Such Monte Carlo randomizations could be used in place of shuffles for cases where pairwise serial correlations are strongly present. In this case, the only difference to the quantized Monte Carlo and template methods described here would be in the program module that generates the randomized spike trains.

An associated problem in using shuffled spike trains is the occurrence of long-term nonstationarities of mean spike rates. We address this question in the companion paper (3).

Implementation of calculations

The programs for pattern identification and for significance testing were set up in Fortran IV on a DEC LSI-11/2 with floating point chip, 56 kilobytes of memory and a 5 megabyte (RK05 type) hard disk. The resulting system could analyze a 5- to 10-min recording of about 1,500 spikes in 6–10 (and sometimes more) hours of computing time. Such analysis would include the quantized Monte Carlo counts and only a limited set of candidate patterns for the template method.

This rather unimpressive performance can be greatly improved by reprogramming and by using more recent technology. In the original programs no effort was made to optimize loop structure, overlay segments, hashing algorithms, threaded versus in-line code, or any of the other possible sources of inefficiency because we were interested in getting a working tool for the initial physiological investigation. Now that we know favored patterns occur quite frequently (see the companion paper (3)) it is worth reprogramming.

Note that the original implementation was on the slowest and smallest LSI-11. We now have available an LSI-11/23 with a 256 kilobyte memory. This should allow a considerable restructuring of memory usage with reduction of disk access. In addition, there is an intrinsic speed improvement in the central processor. We have not as yet rebuilt this analysis system but believe that a time reduction of 5–10 would come relatively simply in the LSI-11/23, and that a faster processor (such as the recently announced 32-bit Hewlett-Packard scientific microprocessor) with additional memory would improve matters by at least another factor of 10. Thus, although we have not yet done so, it would be possible to attain a quasi real-time analysis system, i.e., a 5-min recording would take 5 min to analyze.

A somewhat different strategy might be effective for interactive use during an experiment. The objective in this situation is to find laboratory manipulations (stimuli?) that are associated with particular patterns. We would suggest dispensing temporarily with the significance testing and presenting a running account to the experimenter of all patterns of suitable length and precision that have happened more than say 2 or 3 times. There is probably much to be learned about the "threshold" conditions in such an interactive situation. Presumably the significance testing of candidate-favored patterns could come later.

PATTERN DETECTION SENSITIVITY ON SIMULATED DATA

Before applying the pattern detection methods to real spike trains, it is essential to assess their performance with respect to sensitivity and reliability by analyzing simulated data with known repeating patterns. By this means it is possible to set lower limits to the number of pattern repetitions required for detection. Physiologically significant recurrences might of course take place below such a detection threshold.

The pattern detection sensitivity can be expressed in terms of three different types of thresholds. These three concepts are valid only if we know explicitly what the added pattern was and, hence, the concepts are appropriate only in reference to analyzing simulated, rather than real, data. *1*) Pattern detection threshold for the complete pattern, the minimum number of occurrences of a chosen pattern that are necessary for the entire pattern (rather than fragments of the pattern) to be detected. *2*) Fragment-reconstruction threshold, the minimum number of occurrences of a chosen pattern that are necessary to detect enough fragments of the pattern that the entire pattern can be reconstructed. *3*) Nonrandomness threshold, the minimum number of occurrences of a chosen pattern that are necessary for the rejection of the hypothesis that the spike train was generated by a random process equivalent to shuffling.

The above thresholds can depend on the pattern length, the particular temporal spacing of spikes in the pattern, the amount of variability in the pattern at its occurrences, whether the pattern occurrences have occasional extra or missing spikes, and the quantization bin width. The three thresholds given above sometimes have different values for the same pattern (2).

Generating data with known patterns

Simulated spike trains containing known favored patterns were generated in the following way. First a spike train of 800 spikes was generated by a computer program using a Poisson process (requiring a random number generator). The interspike intervals thus had an exponential distribution, and the value of each interval was independent of the lengths of previous and subsequent intervals.

Next a chosen pattern was interpolated a number of times into the spike train. Sometimes the pattern was exactly the same each time it was interpolated; sometimes the pattern varied. Each time the program was run, it generated a simulated spike train having a recurring pattern, which varied a chosen amount from occurrence to occurrence. (Each interval was normally distributed with chosen mean and standard deviation.) The program user could choose the pattern, the number of times the pattern was interpolated into the random spike train, and the amount of variation between pattern occurrences.

Parameters that affect sensitivity

Figures 7–10 show some results obtained from applying the quantized Monte Carlo method to simulated spike trains with interpolated patterns. Although the interpolated patterns had varying interval lengths, none contained extra or missing spikes.

From our studies on pattern detection threshold, we can make four major generalizations about the dependence of detection threshold for the complete pattern on pattern length, pattern variation, and quantization bin width.

1) For a given quantization bin width, the pattern detection threshold for the complete pattern increases as the amount of variation between occurrences of the favored pattern increases. This trend is directly illustrated in Fig. 8; it is also evident in Fig. 7. The trend was observed over all lengths of pattern tested in the range three to nine intervals. Note that pattern detection thresholds for a given length

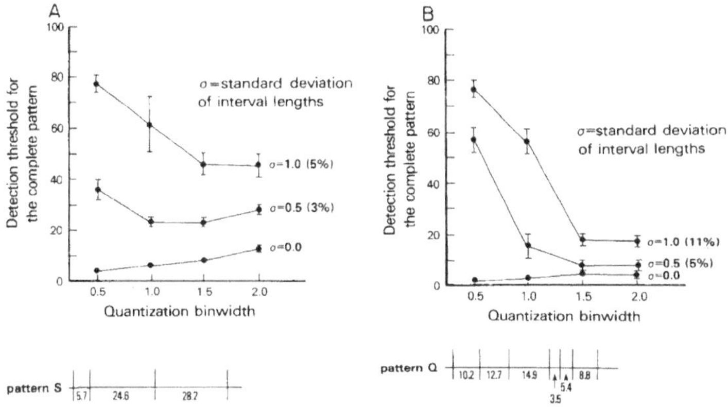

FIG. 7. Detection threshold for the complete pattern as a function of quantization bin width. Pattern detection results (quantized Monte Carlo method) on example patterns of three (*A*) and six (*B*) intervals. Three different degrees of variation for each pattern are shown. The value of σ in each case is the standard deviation of all intervals. In each simulated spike train, the standard deviation of all intervals of the interpolated pattern was the same.

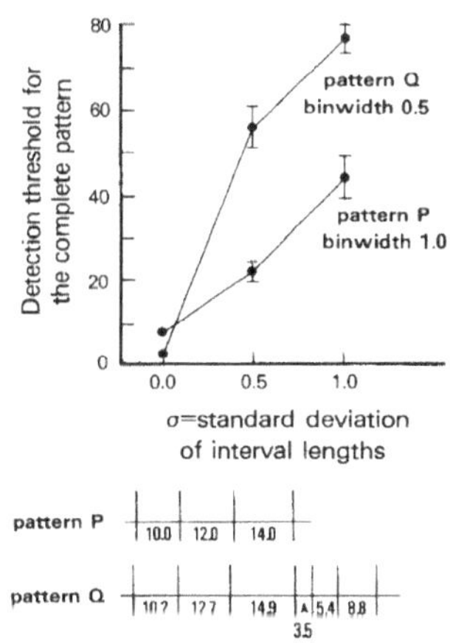

FIG. 8. For a given quantization bin width, the pattern detection threshold increases as the amount of variation between occurrences of the favored pattern increases.

pattern are similar regardless of the detailed pattern.

2) The bin width at which a given pattern is detected best tends to increase as the amount of variation in the pattern increases. This trend is illustrated in Fig. 9 and also holds over all patterns tested regardless of length. Note that the bin width at which a pattern is detected best is the bin width that requires the fewest occurrences of the pattern for detection. At higher standard deviations, patterns tend to be detected best at larger bin widths. When the standard deviations are lower, the patterns tend to be detected best at smaller bin widths. The conclusions that can be drawn from this trend are discussed later.

3) The minimum detection threshold for the complete pattern over all bin widths tested increases as the amount of variation in the pattern increases. This trend is illustrated in Fig. 10. Note that at a given amount of variation, the pattern will have different detection thresholds at different quantization bin widths. Figure 10 considers the minimum value of threshold over all bin widths tested.

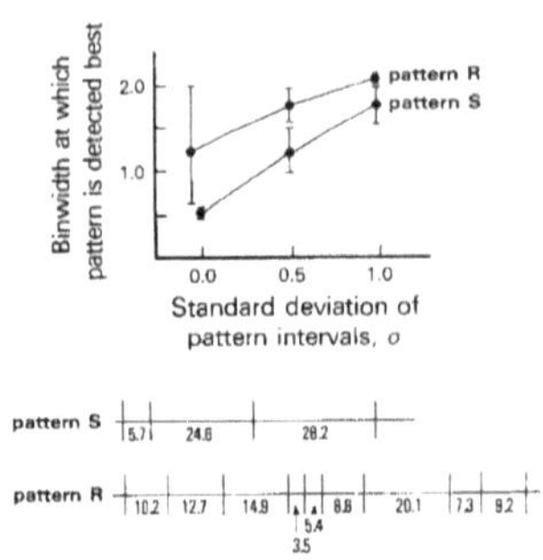

FIG. 9. The bin width at which a given pattern is detected best as a function of the amount of variation in the pattern. The bin width at which a given pattern is detected best is the bin width tested that requires the fewest interpolated occurrences for detection of the complete pattern. Patterns with higher standard deviations in interval lengths tend to be detected best at larger bin widths. Patterns with lower standard deviations in interval lengths tend to be detected best at smaller bin widths.

4) At the optimum bin width, patterns of six and nine intervals of a given variability seem to be more easily detected than patterns of three intervals with comparable variability. This is shown in Fig. 7.

Bin widths and pattern variation

When analyzing spike-train data by the quantized Monte Carlo method, it is essential to use several different bin widths on each spike train. Each bin width yields a different quantized version of the spike train. If a favored pattern is detected best at a smaller quantized bin width, we conclude that the pattern has a relatively small degree of variation from occurrence to occurrence. If a favored pattern is detected best at a larger quantized bin width, we conclude that the pattern has a relatively large degree of variation from occurrence to occurrence. These conclusions are based on the observed trend that the bin width at which a given pattern is detected best tends to increase as the amount of variation in the pattern increases.

What underlies this observed trend?

As an example, suppose a nerve cell is generating an excessive number of occurrences of a pattern that consists of two successive intervals. The first interval varies from 11 to 14 ms, and the second from 22 to 26 ms. At a quantization bin width of 10 ms, this pattern would always be 2 3. At the quantization bin width of 1 ms, the pattern could be any of the following

11 22	12 22	13 22	14 22
11 23	12 23	13 23	14 23
11 24	12 24	13 24	14 24
11 25	12 25	13 25	14 25
11 26	12 26	13 26	14 26

If, for example, the pattern is generated only 10 times in the observed length of spike train, it is unlikely that one would detect the pattern at the 1-ms quantization bin width because each occurrence of the pattern may quantize to a different entry in the above list. It is likely that one would detect the pattern at the 10-ms quantization bin width because all repetitions would quantize to 2 3.

Fragments

We have observed that the quantized Monte Carlo method identifies fragments of relatively long interpolated patterns that have substantial amounts of variation. As frag-

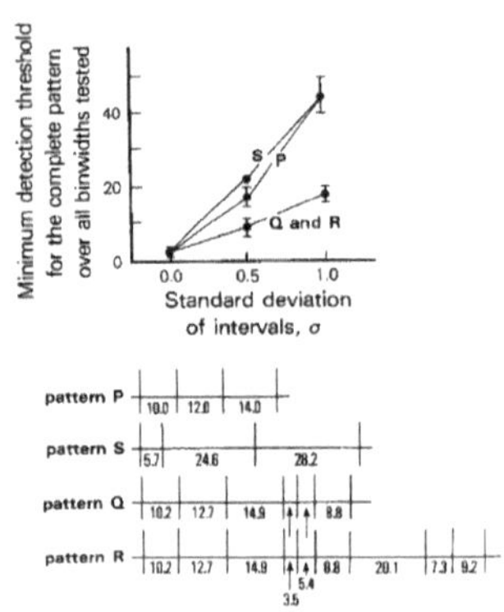

FIG. 10. Minimum detection threshold (for the complete pattern) over all bin widths tested as a function of the amount of variation of the pattern. In other words, the ordinate is the number of interpolated occurrences of each pattern required for detection with at least one of the bin widths used in these studies. This minimum threshold increases with increased pattern variation.

ments are identified, the detection of the whole pattern may or may not be taking place. The detected fragments of the patterns tested did not have less variation than the rest of the interpolated pattern.

These detected fragments can be used in several ways to reconstruct the longer interpolated pattern. Sometimes the fragments have matching sections that, if aligned properly, define the larger interpolated pattern. If such an alignment is not possible, then the environment of each fragment occurrence can be examined to find similarities that could be used to construct a potential template for the larger pattern.

A threshold can be defined for the above reconstruction processes as the number of interpolated long patterns required for their detection by reconstruction with fragments. This threshold is often lower than the detection threshold for the complete pattern.

Note that in some cases the spike train is clearly nonrandom but the original pattern cannot be reconstructed from detected fragments. For this situation we can define a "nonrandomness threshold" that is often lower than the fragment reconstruction threshold.

Sometimes the quantized Monte Carlo method indicates excessive numbers of words that include the interpolated pattern but are longer. This "superword" problem is considered in detail in Ref. 2.

Up to this point we have discussed simulated data with interpolated occurrences of a single favored pattern. The quantized Monte Carlo method can sometimes detect a group of patterns, each of which occurs in slight excess; the excess in this situation may be so slight that a single such pattern alone would not be detected. The group would be detected because the corresponding quantized Monte Carlo count is augmented by the number of patterns in the group.

Sensitivity summary

We have demonstrated that the quantized Monte Carlo method can be used to detect excessively recurring patterns in spike trains provided the patterns occur sufficiently often and do not vary too much from occurrence to occurrence.

With pattern variability, the number of occurrences detected by the quantized Monte Carlo method is (unfortunately) usually considerably less than the known number of interpolated pattern occurrences. This is because the variability of the pattern causes the interpolated pattern occurrences to quantize to different lists of values. If enough of the interpolated pattern occurrences quantize to the same list of values, the pattern is assessed as significant. Note that it may be possible to identify most of the remaining occurrences by the template method.

Patterns that are exactly the same each time they occur are easiest to detect. In the simulated spike trains, detection of patterns of three to nine intervals with no variance always required under five occurrences in a typical spike train of 800–1,000 spikes.

If there is a substantial amount of variation between pattern occurrences, then the patterns become harder to detect. In our examples, patterns with variation were sometimes not detected unless there were 10–30 occurrences. For a given degree of variability, longer patterns required fewer occurrences for detection. A three-interval pattern with 3% variability and a six-interval pattern with 11% variability each required 20 recurrences for detection. Fortunately, if there are fewer occurrences of the pattern than it takes to detect the complete pattern, one is likely to detect enough fragments of the pattern to demonstrate that the spike train was not generated at random but rather has some patterning phenomena. In some cases, it may be possible to reconstruct the original interpolated pattern from its detected fragments.

The quantized Monte Carlo method can also be used to study spike trains that contain groups of patterns, each of which occurs in slight excess. It may be that none of these patterns occur excessively enough to be identified individually as a favored pattern. A spike train with such a group of patterns can be shown to differ significantly from the random model provided that enough patterns occur sufficiently often.

DISCUSSION

We have assumed in this paper that favored patterns, those that repeat an excessive number of times according to appropriate statistical tests, can be related to events of information transfer. We have presented two practical methods which, under various circumstances, can be used to detect and quantify favored patterns if they exist. The sensitivities of our methods have been examined with the aid of artificially constructed spike trains before these methods were applied to real neuronal data.

In addition, an exact method for seeking favored patterns has been presented elsewhere (2). This method does not use the Monte Carlo approach but rather bases its statistical test on a theoretically derived probability formula. We do not recommend this method for the analysis of spike trains longer than 500 spikes because it requires evaluation of formulas that are both difficult and lengthy to calculate.

We have pointed out before that the traditional methods of spike-train analysis are ineffective for the detection of favored patterns. It was therefore necessary to develop the new methods presented here, which are not limited to pattern phenomena that are time-locked to laboratory events.

It is particularly interesting when favored patterns are not associated with known laboratory controlled events but occur during ongoing or spontaneous portions of the neuronal activity. In this case we have no a priori knowledge of when these favored events would occur. If we assume that these sequences represent unknown repeated events within the organism, we should examine the times at which these sequences occur and seek whether any new observations in the laboratory can be correlated with these particular moments. We are implicitly assuming that the occurrences of a favored pattern mark an internal event in the organism, and thus make the event available for study.

Once favored patterns are identified, we usually align them according to the first spike for display and comparison. At this stage we can examine both variance and average of these patterns. Such a comparison could not have been made, of course, before the identification process. Until the identification, any averaging would involve the entire spike train. Since the favored sequences in fact represent a very small fraction of the overall activity of the neuron, any process involving averaging over the entire spike train (as in traditional spike train analysis) cannot give general access to favored patterns. Note that the above process for comparing identified pattern occurrences is analogous to alignment of spike responses to stimuli according to stimulus times.

The general mathematical problem of finding nonrandom structure in an interval list (the idealization of a spike train) has been examined in the statistical literature (5). An example application is in testing random number generators. Generally, these methods are less satisfactory than the ones we have developed because either they tend to break up some short sequences that may exist or they produce a single number that expresses the degree of overall nonrandomness but does not identify explicitly the sequences occurring in excess. The question of extra or missing spikes in pattern occurrences has never been treated either mathematically or physiologically.

To our knowledge there has been only one other type of approach to the favored pattern problem in spike trains (1, 6–8, 14). In these papers, a new sequence is derived from the spike train by marking a + when an interval is longer than the previous interval, and a − when an interval is shorter than the previous interval. The resulting series of +'s and −'s is then tested to see if it reflects a nonrandomness of sequential structure. This nonrandomness is reflected by a single number whose significance is expressed by a χ^2 statistic. The method does not provide a statistical means for finding which +/− patterns cause the nonrandomness. One can merely infer a list of candidate patterns that are in short supply or in excess.

Unfortunately, conversion of the spike train to a +/− sequence loses a very large amount of information. For example, intervals 1 2 3, or 10 20 30, or 5 25 26 are all translated into ++. It seems unlikely that all such patterns have the identical biological significance and hence merit recognition as a single pattern. Note that although the first two of these example patterns are time-scaled versions of each other, the third is not. Although it seems possible that the two time-scaled patterns could have similar biological significance (perhaps not over such a wide range) and could be lumped as one pattern class, it seems unlikely that unrelated patterns such as 5 25 26 should be put into the same class. The inverse transformation of identified +/− patterns back into interval sequences has not been given in these papers.

We would like to point out that, at the present state of knowledge, any spike train is a potential candidate for examination by the new pattern-detection methods we have described. There are at present no criteria for selecting spike trains that are more or less likely to have favored patterns. In contrast, when studying stimulus-response relationships in the nervous system, we make an a priori choice of data by finding the stimulus that drives the neuronal response or by finding a neuron that responds to a predetermined stimulus. Application of the favored pattern methods to arbitrarily selected available spike train data has shown that the phenomenon of favored patterns is quite common (see the accompanying paper (3)).

The present paper has been confined to the search for favored patterns in single spike trains. Since presumably neurons tend to work together in assemblies, it seems likely that there are favored patterns in a more general sense than has been considered here. Such favored patterns would consist of simultaneous patterns (not necessarily the same) along several axons. It is possible that the set of neurons involved in a particular pattern would impinge on a single target neuron and that each type of multiunit pattern could produce different postsynaptic consequences. Experimental approaches to this problem are made difficult by the need to record from an appropriate (and unknown) set of neurons. At this point we have no way of knowing a priori which if any set of neurons is appropriate. Analytic solutions of this problem require generalizations or extensions of methods presented in this paper or, alternatively, some fundamentally new approaches.

ACKNOWLEDGMENTS

We thank Dr. Herbert Wilf for his helpful mathematical advice throughout this work.

This work was supported in various parts by National Institutes of Health Grants NS 05606, 5-T32-07229, RR-05415, and RR-07083.

Present address of J. E. Dayhoff: Fairchild Advanced Research and Development Laboratory, MS 30-590, 4001 Miranda Ave., Palo Alto, California 94304.

Received 10 February 1982; accepted in final form 4 January 1983.

REFERENCES

1. Brudno, S. and Marczynski, T. J. Temporal patterns, their distribution and redundancy in trains of spontaneous neuronal spike intervals of the feline hippocampus studied with a non-parametric technique. *Brain Res.* 125: 65–69, 1977.
2. Dayhoff, J. E. *Detection of Favored Patterns in Nerve Spike Trains* (Ph.D. Thesis). Philadelphia: University of Pennsylvania, 1980.
3. Dayhoff, J. E. and Gerstein, G. L. Favored patterns in spike trains. II. Application. *J. Neurophysiol.* 49: 1349–1363, 1983.
4. Gerstein, G. L. Functional association of neurons: detection and interpretation. In: *The Neurosciences: Second Study Section Program*, edited by F. O. Schmitt. New York: The Rockefeller University Press, 1970, p. 648–661.
5. Knuth, D. E. *The Art of Computer Programming, Seminumerical Algorithms.* Reading, MA: Addison-Wesley, 1968, vol. 2.
6. Marczynski, T. J. Pattern recognition in sequences of neuronal spike intervals; a non-parametric approach and its implications. *Materia Med Pol.* 8: 97–107, 1976.
7. Marczynski, T. J., Burns, L. L., and Marczynski, G. T. Neuronal firing patterns in the feline hippocampus during sleep and wakefulness. *Brain Res.* 185: 139–160, 1980.
8. Marczynski, T. J. and Sherry, T. J. A new analysis of trains of increasing and decreasing neuronal spike intervals treated as self-adjusting sets of ratios. *Brain Res.* 35: 533–538, 1971.
9. Perkel, D. H. and Bullock, T. H. Neural coding. *Neurosci. Res. Program Bull.*, Vol. 6, No. 3, 1968.
10. Perkel, D. H., Gerstein, G. L., and Moore, G. P. Neuronal spike trains and stochastic point processes. I. The single spike train. *Biophys. J.* 7: 391–418, 1967.
11. Perkel, D. H., Gerstein, G. L., and Moore, G. P. Neuronal spike trains and stochastic point processes. II. Simultaneous spike trains. *Biophys. J.* 7: 419–440, 1967.
12. Perkel, D. H., Gerstein, G. L., Smith, M. S., and Tatton, W. G. Nerve-impulse patterns: a quantitative display technique for three neurons. *Brain Res.* 100: 271–296, 1975.
13. Shannon, C. E. and Weaver, W. *The Mathematical Theory of Communication.* Urbana: University of Illinois Press, 1949.
14. Sherry, C. J. and Marczynski, T. J. A new analysis of neuronal spike intervals based on inequality testing. *Int. J. Neurosci.* 3: 259–270, 1972.

0270-6474/85/0504-0881$02.00/0

Printed in U.S.A.

The Journal of Neuroscience
Vol. 5, No. 4, pp. 881–889
April 1985

Cooperative Firing Activity in Simultaneously Recorded Populations of Neurons: Detection and Measurement[1]

GEORGE L. GERSTEIN,*,[2] DONALD H. PERKEL,‡ AND JUDITH E. DAYHOFF§

* *Departments of Physiology and of Biophysics, School of Medicine, University of Pennsylvania, Philadelphia, Pennsylvania 19104,* ‡ *Department of Psychobiology, University of California, Irvine, California 92717, and* § *Judith Dayhoff and Associates, P. O. Box 4029, Mountain View, California 94040*

Abstract

Recent advances in techniques for chronic recording from multiple extracellular microelectrodes allow simultaneous observation of firings of substantial populations of neurons. We describe a new conceptual representation of cooperative behavior within the observed neuronal population. This representation leads to a new technique for detecting and studying functional neuronal assemblies that are characterized by temporally related firing patterns. The representation may be applied to both dynamic and long-term aspects of cooperativity. The basic idea is to map activity of neurons into motions of particles in a multidimensional Euclidean space. Each neuron is represented by a point particle located in this space. In the simplest version of the mapping, each nerve impulse results in an increment in a "charge" associated with that particle; between firings the charges decay. The force exerted by any such particle on any other is, by analogy with some physical forces, proportional to the product of their charges and may depend on the Euclidean distance separating them. The force on a particle directly affects its velocity rather than its acceleration, as with actual particles in a viscous medium. These forces result in aggregation of those particles that correspond to neurons tending to fire together; separate clusters represent independent cooperative groups. Modification of the charges and forces permits inclusion of inhibitory interactions. Identification, measurement, and display of the resulting clusters can be performed with any of a number of algorithms. We illustrate the application of this approach to populations of computer-simulated neurons having both direct and indirect excitatory coupling.

A popular experimental paradigm in the study of brain function, used extensively for the last two decades, is based on electrophysiological recording of single neuron activity. Such work examines neural firing patterns as a function of stimulus, motor, or behavioral conditions and generally is used to measure a receptive (or motor) field or a tuning curve for the neuron. Many such single neuron measurements, carried out sequentially, together with various modes of anatomical localization, are then used to *infer* the properties of an assembly of neurons in the system under study. Although such measurements allow examination of firing patterns down to a time scale of milliseconds, they provide no direct evidence about the functional connectivity among the sequentially observed neurons. This must be done indirectly by inference based on shared or hierarchical response properties of the several neurons. Similarly, the role of the neuronal assembly in the information handling (stimulus analysis or behavioral) process can only be viewed in terms of the properties of the individual neurons. A striking result of such work by Mountcastle (1957), Hubel and Wiesel (1962), and others has been the discovery of widespread modularity in many portions of the nervous system; the modularity is based on the spatial clustering of neurons with similar stimulus response properties (for example, cortical columns or slabs). However, the internal operations and dynamic aspects of connectivity within these modular assemblies remain largely unknown because, so far, only sequential measurement has been employed.

Direct investigation of neuronal assemblies, of their possibly time varying organization and extent, and of their dependence on stimulus or behavioral circumstances, can only be based on simultaneous but separable observation of many individual neurons. Such work involves a number of additional technical and conceptual difficulties (reviewed in Gerstein et al., 1983), but is becoming increasingly practical. Thus, it is now becoming possible to implant chronic multiple microelectrodes or to use analogous optical techniques (Grinvald et al., 1981) to record the spatiotemporal patterns of firing in a substantial population of neurons. These powerful experimental techniques, together with the appropriate analyses, should allow detailed study of cooperativity among subpopulations of neurons as well as the relation of such cooperativity to experimental and behavioral parameters.

However, the very profusion of data produced in such experiments creates serious problems of analysis and interpretation; these problems are conceptual as well as practical. The principal conceptual problems are (*1*) defining cooperativity or functional grouping among neurons and (*2*) formulating quantitative criteria for recognizing and characterizing such cooperativity. The chief practical problem is screening the massive data base in order to detect and identify putative neuronal groups as defined through measures like temporally related firing. The goal is to measure the extent and nature of the cooperativity within each group, the degree of independence from other such groups, the relationship to known modularity and individual neuron response characteristics, and the relationship of group properties to experimental circumstances.

One hopes ultimately to relate such measures to the underlying physiological mechanisms that select and modulate dynamic group properties. Although such mechanisms are, undoubtedly, ultimately based on synaptic modulation and its biophysical substrates, one may speculate that mechanisms affecting groups of neurons will themselves involve activity of groups of neurons. Thus, we probably

Received March 16, 1984; Revised September 24, 1984;
Accepted October 3, 1984

[1] This work was supported by National Institutes of Health Grants NS 05606 (G. L. G.), NS 16451 (D. H. P.), and NS 17754 (D. H. P.), and by System Development Foundation Grant SD00013 (G. L. G.). We thank Dr. A. Aertsen for helpful discussions.

[2] To whom correspondence should be addressed.

need to understand groups of neurons interacting with other groups of neurons.

It should be stressed that the principal tools available to date for the examination of related firing among neurons involve cross-correlation of the firing of two or at most three neurons at a time (reviewed in Gerstein et al., 1983). These tools allow inference of effective connectivity and have been used by many investigators in many parts of the nervous system. If, however, we record from populations of 10 or more neurons, such pairwise analysis becomes exceedingly tedious simply because of combinatorial arithmetic. In addition, neuronal group properties must then be *inferred* from combinations of neuronal pair properties, a process that is subjectively difficult. We need theoretical means to treat a neuronal group as a single entity.

We have previously addressed certain problems of neural groups (Gerstein et al., 1978). Conceptually, the new approach (which is related to the gravitational clustering technique of Wright, 1977) is clearer than the older method and entails minimal reliance on subjective analysis. It provides greater flexibility in the criteria for related firing of two or more neurons. We emphasize that the new approach deals with a neuronal assembly as an entity rather than as a set of neural pairs; it differs fundamentally from analysis by pair cross-correlation methods (Perkel et al., 1967). Quasi-on-line implementation is possible, so that experimental strategy can be modified while the neural assembly is still under observation.

After describing the basic approach together with some modifications that enhance performance of the calculations, we examine some simple neuronal circuits whose firing behavior was simulated on a computer. Finally, we indicate some generalizations of the method that make accessible other aspects of neuronal cooperativity.

The Basic Method

Representation of neurons as point particles. We begin by representing each of N neurons by a fictitious particle, located at a point in N-dimensional Euclidean space. At any time, then, the vector position of the particle corresponding to neuron i is written as

$$\mathbf{x}_i = (x_{i1}, x_{i2}, x_{i3}, \ldots, x_{iN}) \qquad (1)$$

where x_{ij} is the j'th coordinate value for particle i.

Initially, all of the particles are set to be equidistant. Specifically, each particle is located at a vertex of an N-dimensional hypercube or, most simply,

$$\begin{aligned} \mathbf{x}_1 &= (\sqrt{2}/2, 0, 0, \ldots) \\ \mathbf{x}_2 &= (0, \sqrt{2}/2, 0, \ldots) \\ \mathbf{x}_3 &= (0, 0, \sqrt{2}/2, \ldots) \qquad (2) \\ &\ldots \\ \mathbf{x}_N &= (0, 0, 0, \ldots, 0, \sqrt{2}/2) \end{aligned}$$

The initial distance between any two particles is then equal to 1.

For the calculations shown in this paper, we have scaled the initial distance between any two particles to be 100.

Dynamics of charges. When a neuron fires, its particle acquires a finite increment of a fictitious "charge," analogous to an electrical or gravitational charge. Between firing times the charge on particle i obeys the differential equation

$$dq_i/dt = -q_i/\tau \qquad (3)$$

for which the solution is

$$q_i(t) = q_i(0)\, e^{-t/\tau} \qquad (4)$$

in which t is measured from just after the last augmentation. The time constant τ is a parameter chosen by the investigator. The transformation from firing history to charge is illustrated in Figure 1.

In the simplest version of this scheme, the particles attract each other with a force proportional to the product of their charges. Note

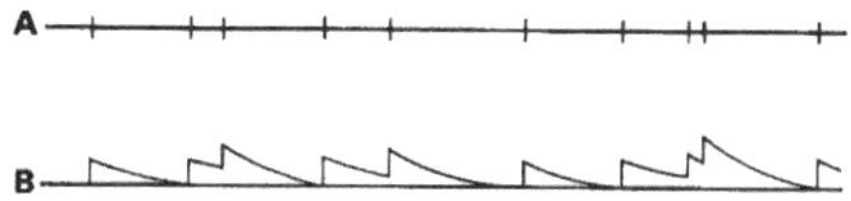

Figure 1. Conversion of spike train (*A*) to charge (*B*) used in the gravitational calculation. At the time of each neural impulse, the existing charge in the representation of that neuron is incremented by a fixed amount. The magnitude of this charge increment is different for each neuron and is inversely related to its average neuronal firing rate. Between neural impulses the charge decays with a fixed time constant (same for all neurons).

that in contrast with electricity and gravitation, the charges fluctuate; they decay continually and are augmented at the instants the corresponding neurons fire. The force may depend on distance between particles. The individual forces exerted on a particle add vectorially, and the particles move under the influence of the changing forces according to a particular dynamic law; the most convenient one here is comparable to motion through a viscous medium.

As a result of the fluctuating charges and attracting forces, the particles representing any group of neurons that tend to fire simultaneously will exert mutually attractive forces and will be drawn closer together. After a sufficient time these sets of synchronously firing neurons will be represented by clusters of particles in N-space. The clusters can then be identified, distinguished, and characterized by standard algorithms of cluster analysis.

Propulsive fields. The propulsive field produced by particle j at the location currently occupied by particle i has the normalized unit direction in N-space given by

$$\mathbf{r}_i(j) = (\mathbf{x}_j - \mathbf{x}_i)/s_{ij} \qquad (5)$$

where s_{ij} is the Euclidean distance between particles i and j:

$$s_{ij}^2 = \| \mathbf{x}_j - \mathbf{x}_i \|^2 = \sum_{k=1}^{N} (x_{jk} - x_{ik})^2 \qquad (6)$$

The amplitude of the propulsive field is proportional to the charge on j and may fall off with distance by a dimensionless attenuation function $A(s_{ij})$. (The most familiar such function would be an inverse-square law, $(s_0/s_{ij})^2$).

In general, then, the propulsive vector field at point i exerted by all other particles is given by

$$\mathbf{f}_i = \sum_{j \neq i} q_j\, \mathbf{r}_i(j)\, A(s_{ij}) \qquad (7)$$

We have found that elimination of the attenuation factor is convenient and desirable, and thus we have set $A(s_{ij}) = 1$.

Motion under the propulsive field. The force vector acting on a particle is proportional to the product of its charge and the propulsive field, i.e., to $q_i\mathbf{f}_i$. Although it would be possible to imagine the particles as being accelerated by these forces, like point masses in a gravitational field, this would involve second-order differential equations. Since we are not concerned with a realistic prediction of trajectories of fictitious particles in a hypothetical N-dimensional space, but rather with ensuring that attractive forces will induce particles to approach one another, the simplest dynamic rule is to imagine the particles as moving in a viscous fluid. We also assume that they always travel at terminal velocity, which is proportional to the force exerted. With these assumptions, the dynamic equation for particle i is simply

$$d\mathbf{x}_i/dt = \sigma\, q_i(t)\, \mathbf{f}_i(t) \qquad (8)$$

in which the single scalar parameter σ subsumes the amplitude of the force, the reciprocal of the "fluid" viscosity, etc. The dimension of σ is velocity per squared unit of charge.

It may be noted that the points do not move in an arbitrary fashion throughout the N-space. In fact, they are initially placed on a hyperplane of dimension N-1, and will henceforth remain on that

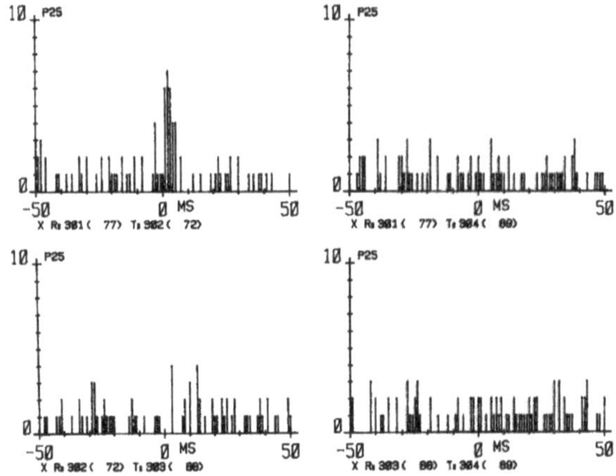

Figure 2. Cross-correlograms for a group of spike trains that were produced by computer simulation. These and similar spike trains are analyzed in Figures 3 and 4. Each group of data contains 10 spike trains, each Poisson like, with a mean rate of 10(±2) spikes/sec. Neuron 1 has an excitatory synaptic connection to neuron 2 (replication probability 0.25); the other eight neurons are completely independent. Total data length is 8.5 sec, with 70 to 100 spikes in each train. The cross-correlogram in the *upper left* is for the synaptically coupled neuron pair; the other three correlograms are typical for independent neuron pairs. (There are 41 other similar histograms between the other independent pairs).

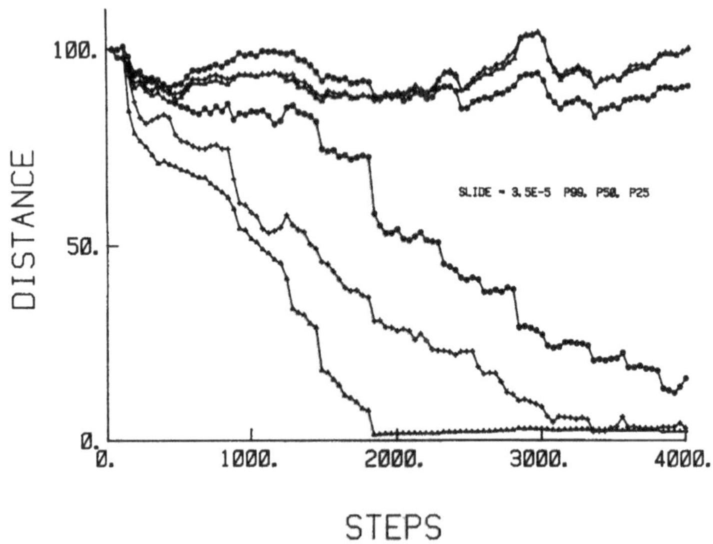

Figure 3. The time to gravitational aggregation decreases as the strength of synaptic connection increases. Distance between selected pairs of points is shown as a function of step number (time) in the gravitational calculation. All pairs start at the distance of 100 arbitrary units; each step in the calculation corresponds to 2 msec of laboratory time. Time constant for charge decay was set at 10 msec, and the distance moved per unit force was 3.5×10^{-5} units. The three descending *curves* show distances between points that represent neurons connected with (*left* to *right*) decreasing synaptic strength (replication probabilities 0.99, 0.50, 0.25, respectively). Distances between typical points representing independent neurons are at the *top* of the figure and remain essentially unchanged as the computation progresses.

hyperplane. In the discussions and calculations of this paper we have not made use of this slight reduction of dimensionality, and have described the movements in the full *N* space. We believe this convention simplifies the initial visualization of our problem as well as its initial mathematical formulation. We may anticipate some small improvement in computation time if the calculation is expressed in the *N*-1 dimensional subspace.

Having set up the explicit set of ordinary differential equations (*8*),

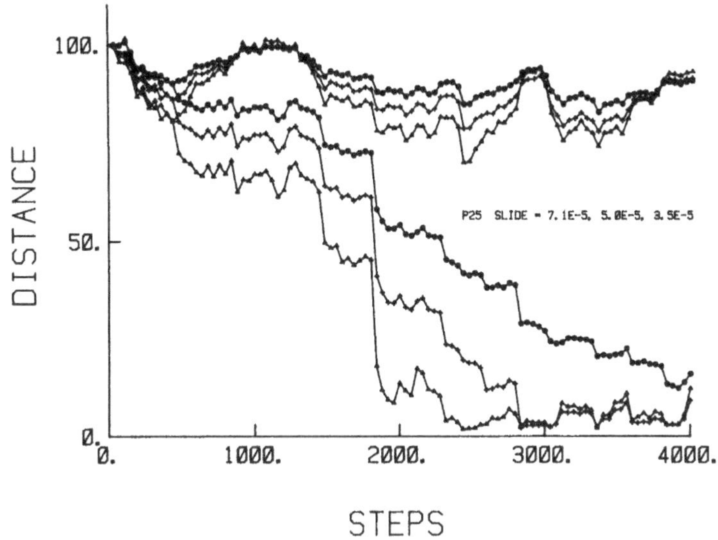

Figure 4. The time to gravitational aggregation decreases as the distance moved per unit force increases. Axes are as in Figure 3. The three descending curves represent recomputations of the data with weakest synaptic connection (replication probability 0.25) shown in Figure 3, with three different values for the constant σ. From left to right, $\sigma = 7.1 \times 10^{-5}$, 5.0×10^{-5}, and 3.5×10^{-5}, respectively. Distances between points representing a typical pair of independent neurons are shown at the *top* for the same three calculation parameters. Note the increase in variance as the value of σ increases.

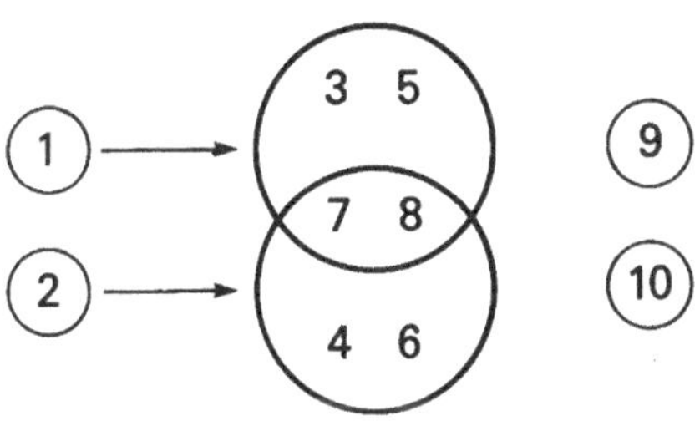

Figure 5. A Venn diagram of more complicated neuronal connectivity that was simulated in a set of 10 spike trains. These data are analyzed in Figures 6, 7, 8, and 9. Neurons *1* and *2* are drivers. Neurons *7* and *8* receive input from both drivers, whereas neurons *3,5* and *4,6* receive input only from one driver. Neurons *9* and *10* are independent.

we can use any standard numerical integration scheme to follow the positions of the particles in response to the recorded history of firing patterns of the corresponding neurons. We emphasize again that we are not concerned with dynamic realism but are using these dynamics only as a means to allow the appropriate particles to aggregate; thus, the simplest algorithm suffices. Using Euler integration with a time step h, we have the simple formula

$$\mathbf{x}_i(t + h) = \mathbf{x}_i(t) + h\sigma q_i(t)\mathbf{f}_i(t) \qquad (9)$$

We note that in this simple version there are two essential parameters that may be manipulated by the investigator. The time constant for decay of charge, τ, governs the strictness of the degree of synchrony required to produce aggregation. The other parameter, σ, may be thought of as a measure of mobility of particles and can be varied to modulate the rate of coalescence.

Modifications to the Algorithm

Independence of rate. In the representation described, in which every particle receives the same increment of charge when its associated neuron fires, the propulsive influence of that particle is proportional to the neuron's mean firing rate. Such a rate dependence may not be desirable and can be eliminated by a simple convention. We set the increment of charge added at the time of each action potential to be the mean interval (the reciprocal of ρ, the mean firing rate) for the neuron represented by that particle. It is then easy to show that the mean value of charge is the same for all particles, and, for exponential decay of charge, is in fact numerically equal to τ, the time constant for decay of charge. Other types of rate normalization are possible and will be contrasted and evaluated elsewhere.

Repulsion and attraction. Another undesirable property of the original formulation is that all particles attract and will eventually aggregate. In other contexts, this phenomenon has been used directly; Wright (1977) has used time until aggregation to define clusters. For our application, it is preferable that the particles corresponding to independently firing neurons neither attract nor repel one another on the average. This can be accomplished by replacing the charge q in equations 7 to 9 by an effective charge $(q - \theta)$, where θ is a reference level of charge, so that the force between particles i and j will be proportional to $(q_i - \theta_i)(q_j - \theta_j)$. If both neurons have recently fired, so that both charges are above their reference levels, the particles will attract. If one neuron has not fired

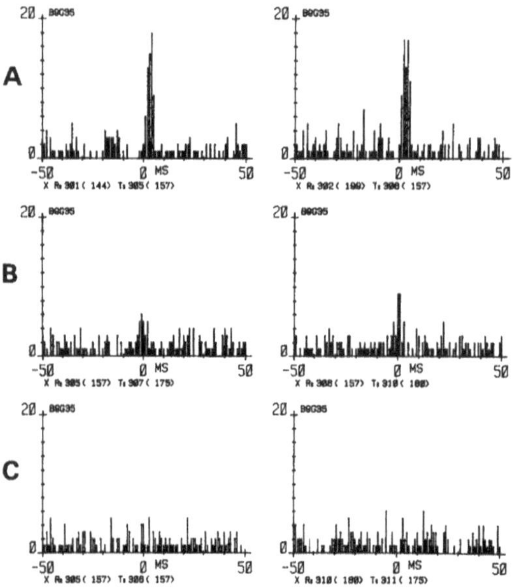

Figure 6. Typical cross-correlograms of spike trains from computer simulation of the neural assembly shown in Figure 5. All 10 spike trains were Poisson like (mean rates, 10 ± 2 spikes/sec; duration, 17 sec). Synaptic connections were simulated by replication probability of 0.35. *A*, Two neuron pairs, each consisting of driver and driven. *B*, Two neuron pairs each consisting of two neurons driven from the same driver. *C*, Two pairs of independent neurons.

recently and the other has, the corresponding particles will repel. Finally, if neither neurons has fired recently, so that both charges are below their reference levels, the particles will attract weakly. These are precisely the properties needed to ensure aggregation of those particles whose neurons tend to fire synchronously and, more weakly, of those particles whose neurons tend to be silent synchronously.

The appropriate value for the reference charge θ is the mean value of charge for that neuron. If, as suggested above, the increment of charge has been set equal to the mean firing interval of the corresponding neuron, then the mean charge is the same for all particles, and, if charge decays exponentially, is numerically equal to τ. The net effect of these modifications is that q is replaced by $(q - \tau)$ in equations 7 to 9:

$$\mathbf{f}_i = \sum_{j \neq i} (q_j - \tau)\, \mathbf{r}_i(j) \qquad (10)$$

$$\mathbf{x}_i(t + h) = \mathbf{x}_i(t) + h\, \sigma\, [q_i(t) - \tau]\, \mathbf{f}_i(t) \qquad (11)$$

Close approaches. A numerical complication of this model arises when the distance between any two particles becomes small. Euler integration and similar techniques give rise to unwanted oscillations of closely neighboring particles. This is especially severe when using a spatial attenuation function having a singularity at the origin, such as the inverse square law. A simple remedy is to introduce a minimal distance between particles within which the net force is zero. Note that, once two particles are within this minimal distance and cease to interact, they continue to be influenced by more distant particles through the firing of their corresponding neurons. A continuous spatial attenuation function, which gradually goes to zero at the origin, may also be used.

Results

Factors affecting the aggregation process. The speed and variability in the aggregation of particles corresponding to interacting neurons depend on the strength and nature of the interaction and on the choice of parameters used in the computations. We illustrate these effects with computer simulation of a simple neuronal circuit consisting of a pair of synaptically connected neurons together with a set of independently firing control neurons.

In the simulations all independent impulse trains were Poisson processes. The excitatory synaptic interaction was approximated by choosing impulses at random in the presynaptic train, with a specified probability, and copying each selected impulse to the postsynaptic train, imposing a delay chosen from a suitable probability distribution. In addition, the nearest following original impulse to the copied impulse in the postsynaptic train was deleted in order to preserve mean firing rate. Thus, the postsynaptic impulse train contained a partial, noisy, delayed replica of the presynaptic train. This class of spike train simulator is, or course, only intended to accurately model spike trains with temporal interrelationships similar to those found in actual recordings; it is not intended to model the underlying integrative dendritic processes.

Mean firing rates were between 8 and 12 impulses/sec; the sample was 8.5 sec long. The delay function was a rectangular probability distribution from 1 to 5 msec. Cross-correlation histograms are shown in Figure 2 for a replication probability of 0.25.

In Figure 3, we illustrate the effect of strength of synaptic interaction on the aggregation process. The strength was varied by modulating the replication probability, using values of 0.25, 0.50, and 0.99. The strongest interaction corresponds to a complete but temporally noisy replication of the presynaptic impulse train. The figure shows, for each case, the time course of the distance between the points corresponding to the pre- and post-synaptic neurons as well as those of an unrelated control pair. It is clear that, for unrelated neurons, the points execute a random walk while remaining at an approximately constant distance. The points for an interacting pair, on the other hand, approach each other consistently although somewhat irregularly. The average approach velocities increase with

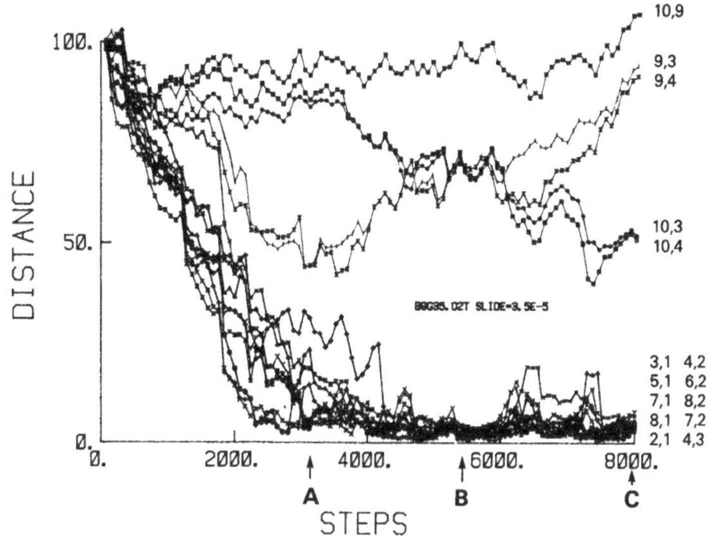

Figure 7. Gravitational computation for the 10-spike train data set described in Figures 5 and 6. *Ordinate,* Distance between points representing selected neuron pairs; *abscissa,* steps in the computation (2 msec/step). Three step (or time) values labeled *A, B,* and *C* are selected on the *abscissa* for reference to Figure 8. See the text.

increasing strength of synaptic interaction; therefore, the two points coalesce most rapidly for the strongest interaction. In the case of strongest interaction, coalescence was achieved after about 40 impulses (4 sec).

In addition to the intrinsic strength of the neuronal interaction, the rate of coalescence is influenced by the parameters of the computation. In particular, the velocity parameter strongly affects the rate of coalescence for a given pair of interacting neurons as well as the degree of random fluctuation in distance between each pair of points. These effects are shown in Figure 4, in which the impulse trains incorporating the weakest interactin (0.25 probability of replication) were analyzed using three different values of the velocity parameter.

Another important factor influencing the aggregation process is the time constant for decay of charge. If the time constant is chosen to be small, aggregation will result only for interactions having correspondingly short latencies and time jitter. As the time constant increases, interactions having longer latencies and/or greater jitter will be included in the aggregation process. If, however, the time constant becomes comparable to average intervals between impulses, sensitivity and selectivity to temporal interactions are effectively lost. If nothing is known about the nature of interactions among the neurons under study, therefore, it is highly advisable to analyze the data several times, using different values of the time constant as appropriate.

The two other parameters that must be chosen for the computation are the time step h and the approach distance below which the attractive force is set to zero. Clearly, the time step must be smaller than the time constant for decay of charge. The critical approach distance should be greater than the maximum increment in distance in a single time step of the computation, which is given approximately by

$$\Delta x_{max} \simeq h\sigma/\rho^2_{min} \qquad (12)$$

where ρ_{min} represents the minimum firing rate over the ensemble of neurons. Choice of this critical distance may be used as a criterion for significant interaction; judicious use of such a criterion at a value larger than that given by equation 12 can save computation time.

Neuronal population containing several types of interaction. A more complex and extensive example is provided by applying the algorithm to a set of neurons more diversely connected than in the previous example. Here we use a computer simulation of 10 neurons, consisting of (*a*) two independently firing presynaptic neurons, (*b*) a set of three postsynaptic neurons for each presynaptic one, one neuron in each of these sets being driven by both presynaptic neurons, and (*c*) two independently firing control neurons. This example provides direct synaptic interaction as well as two degrees of shared input, as seen in the schematic circuit (Fig. 5). The impulse trains were generated by the same rules as in the previous example; the sample was made 17 sec long having up to 200 impulses for each neuron. Typical cross-correlation histograms are shown in Figure 6.

Once again, we illustrate the aggregation process through plots of pairwise distances as a function of time. The identities (see Fig. 5) of the several pairwise distances are indicated at the *right* of Figure 7 and fall into four groups. At the top, the distance between points *9* and *10* (which represent the control neurons) remains essentially unchanged throughout the calculation. Next from the top are the distances for *9,3* and *9,4* (representing in each pair one control neuron and one postsynaptic neuron); these distances initially decrease, but later increase. The third group from the top are the distances for *10,3* and *10,4* (the other control neuron and one postsynaptic neuron); these distances slowly decrease through the calculation. Finally, at the bottom are the pairwise distances for various transynaptic pairs, as well as for the two drivers (*1,2*) and

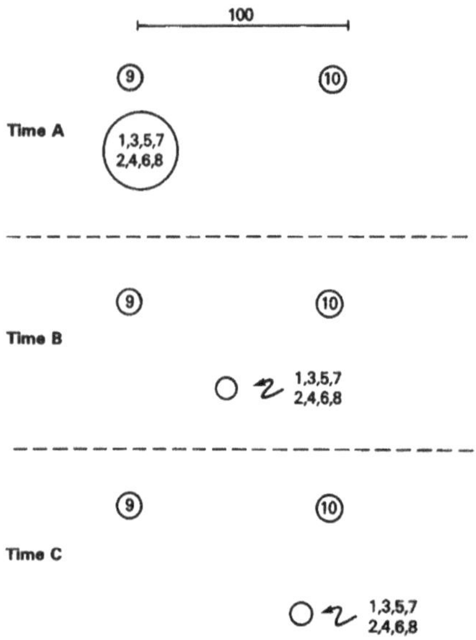

Figure 8. Approximate projection of the 10 points (representing the 10 neurons) from the 10-space onto an appropriate plane. The projection is shown at the three times (*A*, *B*, and *C*) identified in Figure 7. The *scale* showing a distance of 100 is at the *top*. At *Time A* the group of points representing interacting neurons has not yet fully aggregated; however, the loose group is closer to *point 9* than to *point 10* (these represent the independent neurons). At *Time B* the aggregation is complete, and equidistant from *points 9* and *10*. At *Time C* the tight aggregation has drifted closer to *point 10*.

for an independent postsynaptic pair (*4,3*); all of these distances demonstrate similar rapid coalescence.

The reason there is no distinction in the behavior of any of these pairs is that all synaptic interactions have been made the same strength. Thus, each driver attracts equally (and is attracted by) each of its postsynaptic partners. This strong interaction will indirectly bring all of the postsynaptic partners of the same driver close to each other, even though their own direct interaction is weaker. Since neurons 7 and 8 receive input from both drivers, they will serve as an attractant for the groups of points aggregating around each driver, thus accounting for the eventual proximity of 1,2 (independent drivers) and 3,4 (independent postsynaptic pair). Thus, pairwise distances can identify a subset of interacting neurons but are not in themselves sufficient to infer the circuit or its relative synaptic strengths.

The pairwise distances shown in Figure 7 can be interpreted in terms of clustering of the points moving in the representational space of the calculation. This is most readily done by projecting the positions of our 10 points as they move in the 10 space onto some appropriate plane. (Considerations and methods of choosing such a projection plane will be discussed elsewhere). Note at the *bottom* of Figure 7 that we have marked three times with letters *A*, *B*, and *C*. At time *A* (Fig. 8, *top*) we see that the projected points representing interacting neurons have loosely coalesced, and that this rather large cluster is nearer to point *9* than to point *10* (these two points represent the independent control neurons). At time *B* (Fig. 8, *middle*), the points representing the interacting neurons have coalesced into a tight cluster which is about halfway between the two points which represent the control neurons. Finally, at time *C*, the tight cluster has drifted closer to point *10* than to point *9*.

The results shown in Figures 7 and 8 were obtained using the impulse trains from the two presynaptic neurons. Clearly, a typical physiological recording may fail to include signals from the neurons presynaptic to the recorded neurons. Figure 9 shows the results of analyzing the same data as in Figures 7 and 8, but excluding the two presynaptic impulse trains. Although the rate of coalescence is somewhat slower because neuron pairs are now only related by shared input rather than by direct synaptic contact, the interacting neurons can still be clearly distinguished from the control cases. It is striking that omission of the driving neurons from the analysis has so little effect on the selectivity and sensitivity; the same set of interacting neurons is detected in the two examples.

Implementation. The gravity representation was programmed in Fortran and initially run on a Digital LSI 11/23 (RT-11) computer. Under these conditions run times are very unsatisfactory: with typical parameters, analysis of 16 sec of 10-spike trains (as shown in this paper) took almost 3 hr. (Note that the calculation time behaves as the third power of the number of neurons). Presumably some small improvements in program efficiency are possible. However, when the *same* program and data were implemented on a Digital VAX 780 (VMS), run times were 5 min, while on a Data General MV 10000 (AOS/VS), run times fell to 3 min. This is not yet real time but would allow this form of analysis with a short enough time lag to be useful *during* an experiment. Additional computational speed can probably be obtained with array processor hardware, since the gravity calculation can be reduced to matrix manipulations. (This possibility is worth investigating for laboratory microcomputers as well as the cited minicomputers). At the present stage of development, access to a powerful and unloaded minicomputer seems essential for this problem; the microcomputers inhabiting many neurophysiological laboratories are too slow for practical results.

Discussion

We have described a method for detecting tendencies toward synchronous firing in groups of neurons. As shown by the examples, the approximate firing synchrony can be produced by a variety of mechanisms, such as direct synaptic input or by shared input. The latter is typically much harder to detect (Perkel et al., 1967).

Although in the examples illustrated we have made use of distances between pairs of points to demonstrate the aggregation process, we must consider the underlying migration of points in the full *N*-space of this representation. Total information about the behavior of the neuronal population is contained in the time history of the position coordinates of the *N* points. For $N > 3$, this process cannot be directly visualized, and we must resort to a suitable projection or clustering technique.

An intrinsic strength of this representation is that weak neuronal interactions are not masked by strong interactions. The strong direct synaptic interactions reveal themselves at an early stage of the calculations, whereas the more diffuse effects of shared input (assuming the shared driver is off stage) become evident only after a longer period of observation. The ultimate aggregation of particles in this representation is hierarchically organized, so that strong neuronal interactions rapidly produce tight clusters. These clusters themselves become aggregated more slowly in response to the weaker neuronal interactions. The original tight clusters remain and do not interfere with the aggregation process resulting from the weaker neuronal interactions. Thus, the dynamic evolution of the aggregation can be used to illuminate attributes of the neuronal system. The advantages of observing the dynamic evolution of the aggregations are analogous to the procedures recommended by Wright (1977).

The shapes of the clusters formed in *N*-space can be important clues to the nature of the underlying neuronal interactions. For

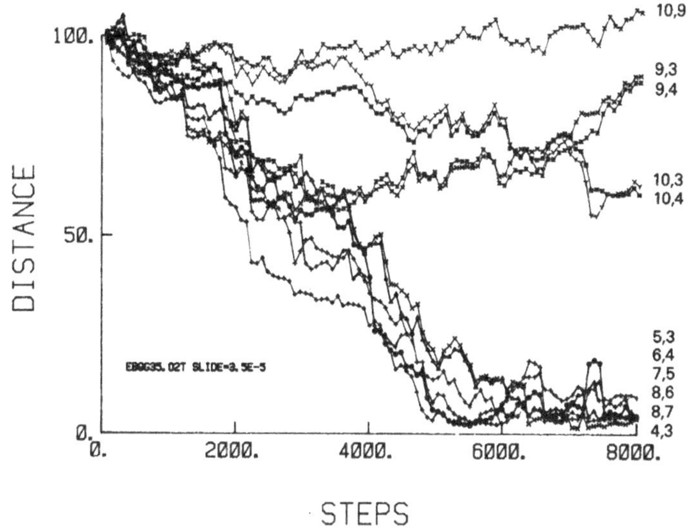

Figure 9. Distances in the gravitational computation for the same data as in Figure 7, but without spike trains 1 and 2 (the drivers). Axes and labels are as in Figure 7. Note that aggregation is slower than in Figure 7, although other features remain the same.

example, if a set of neurons is driven by a single, shared presynaptic excitatory neuron, the corresponding points will form a single cluster. A different set of neurons, similarly excited by a different driver, will form a similar but independent cluster. If there are neurons that receive excitatory input from both drivers, the two clusters will be joined by a bridge, in the fashion of a dumbbell. Points in the neck of the dumbbell correspond to the neurons that receive input from both drivers. This configuration of points resembles a Venn diagram in the *N*-space.

A different example is provided by a systematic pattern of sequential firing by a set of neurons within the population. Such a sequence might be the result of a chain of synaptically coupled neurons or, alternatively, of an appropriate pattern of delays from a common source of excitation. In the *N*-space, the resulting configuration of points will resemble a snake or sausage, since there will be aggregation only of pairs of points corresponding to adjacent neurons in the chain. It will be an elongated figure but is unlikely to be a straight line; therefore, detection of such a configuation will require more extensive analysis than needed for more compact or symmetrical clusters.

A useful assessment of significance in the aggregation process can be provided by including one or more impulse trains in the data base that are known to be causally unrelated to any of the other trains. These could be derived from another experiment, or produced artificially on a computer. The corresponding points should remain at approximately a constant distance from all other points, neither aggregating nor separating significantly, as illustrated in our examples.

Up to this point we have assumed implicitly that the pattern of neuronal interactions remains unchanged throughout the period of recording. If in fact the interaction pattern is itself changing, then the consequences of such changes may become apparent in the evolution of the aggregation process. For example, if a strong neuronal interaction ceases to operate, then the corresponding cluster of particles will begin to disperse. If the new interaction pattern persists for a long enough time, this will become apparent in a new pattern of aggregation. The transformation from one pattern of aggregation to another can be followed. If the interactions are too labile, there will be insufficient time for reaggregation to be observed, although our examples show rapid migration. In some such cases a useful strategy might be to segment the data and to restart the analysis process at those times when disaggregation is first detected.

Consider an experiment in which some number of stimuli (or behavioral conditions) recur cyclically. The data can be segmented on the stimulus boundaries, and a run of the analysis process can be carried out separately for *each* type of stimulus condition. The "pasting together" of the segments representing the same stimulus condition does not cause any boundary problem to our analysis process. In this guise we can address the question of whether different patterns of aggregation (representing particular values of neuronal interaction) are associated with different stimulus conditions. Other approaches to the problem of detecting time-varying coupling for two neurons subjected to repetitive stimulation are described by Gerstein and Perkel (1972) and by Schneider et al. (1983).

Generalization of charge and of particle dynamics. For conceptual and computational simplicity we have assumed a single descending exponential function to describe the dynamics of "charge." We may generalize the definition and dynamics of charge to produce more appropriately shaped functions to produce performance more sensitive to particular types of neuronal interaction. For example, a low-order gamma function may replace the single exponential. It exhibits a single smooth maximum with a delay after spike time that may be chosen appropriately. Such functions would mimic known temporal properties of postsynaptic potentials. The modified charge function, z_i, would replace q_i in equation 8. However, in equation 7, which defines the propulsive vector field, we would continue to use the q_i as originally defined. The effect of this generalization is to enhance sensitivity to synaptic interactions having time courses similar to the new charge function. The reason for retaining the original q_i in equation 7 is to associate synaptic effects upon other neurons with the firings of neuron j. Note that this modification means that both old and new charge functions must be calculated and maintained for all particles.

None of the above methodology explicitly addresses inhibitory influences among the observed neurons. Our basic technique may readily be generalized for this purpose. The idea is to redefine the second type of charge so as to reflect silence rather than firing. Such a charge function may be designed with the following properties: (*1*) it is small for short intervals; (*2*) it is small for long intervals; (*3*) it smoothly rises to a maximum for a selected intermediate interval length; (*4*) duration and shape within the interval are selected appropriately. This new charge again replaces q_i in equation 8 rather than q_j in equation 7 because we are selecting for the silent period of neuron i, which is putatively postsynaptically inhibited.

In the variations discussed so far, we cannot infer causal relations among the neurons represented by an aggregation of points. For certain classes of neuronal circuitry, this goal can be achieved through a further extension of these modifications. Specifically, we associate two points rather than one with each neuron. Each point bears one class of charge only; we refer to them as q-points and z-points. In order to emphasize inhibitory interactions, we define the z-charge in terms of silences, as described above. We now specify two variations on the particle kinematics. In the first version, the q-points remain fixed at their original positions (at which they are mutually equidistant), while the z-points move as described by equations 7 and 8, as modified by inclusion of the z-charge in equation 8. With these rules, we can interpret an aggregation of z-points around a particular (distinct) q-point as direct evidence for divergent inhibition: the neuron corresponding to the q-point directly inhibits the neurons represented by the z-points clustering about it.

The second variation is the dual of the first. Here, q is replaced by z in equation 7 and not in equation 8. Also, the z-points are constrained to their original position, and the q-points are allowed to migrate. We may interpret a cluster of q-points about a single z-point as a convergent inhibitory circuit: the z-point corresponds to a neuron that receives inhibitory synaptic input from each of the neurons represented by the q-points clustered about it.

These variations are useful only after an initial survey of the neuronal population has been performed using the basic representation, i.e., using the original definition of charge and a single set of points. This initial screening can serve to identify appropriate subsets of the original neuronal population to which it will be useful to apply the types of variations described.

We emphasize that even these variations, which may use subsets of the original population, are performed in a higher-dimensional space, in which particle dynamics are governed by all interactions simultaneously. None of these approaches is equivalent to a set of pairwise comparisons, such as cross-correlation histograms (Perkel et al., 1967).

State of neuronal population. We have discussed our representation of neurons as migrating points in N-space in the context of questions about functional circuitry in the neuronal population and about dynamic changes in that circuitry. However, this representation can also lead to a quantitative notion of the state of the neuronal population. By analogy with the phase-space representation used, for example, in statistical mechanics, we can abstract a still higher-level representation than the one we have described. Specifically, we can incorporate the instantaneous positions and migration velocities of all N points into a single vector, of dimension $2N^2$. In an abstract space of this dimension, this vector is represented by a moving point. The trajectory of the point represents the temporal evolution of the state of the system, which in turn embodies the changing set of functional interactions among the neurons as manifested by their impulse activity.

For example, approximately closed or repeating trajectories of this state point would correspond to cyclic alterations in the organization of the neuronal population. Such changes in organization could arise from actual changes in effective connectivity as well as from alterations in thresholds, membrane excitability, and the like, or from changes in input external to the observed population of neurons.

We have mentioned these variations and extensions of our basic representation to emphasize that it is not merely a data-processing technique. On the contrary, it provides a unifying conceptual framework for understanding and depicting the activity of an entire neuronal population as an entity rather than as a collection of individual or pair properties. In particular, use of this type of representation can make concrete and measurable the notion of cell assembly that goes back to Hebb (1949). Further conceptual and computational development of this representation should prove fruitful.

References

Gerstein, G. L., and D. H. Perkel (1972) Mutual temporal relationships among neuronal spike trains. Biophys. J. *12:* 453–473.

Gerstein, G. L., D. H. Perkel, and K. N. Subramanian (1978) Identification of functionally related neural assemblies. Brain Res. *140:* 43–62.

Gerstein, G. L., M. J. Bloom, I. E. Espinosa, S. Evanczuk, and M. R. Turner (1983) Design of a laboratory for multineuron studies. IEEE Trans. Systems, Man, and Cybernetics *SMC-13:* 668–676.

Grinvald, A., L. B. Cohen, S. Lesher, and M. B. Boyle (1981) Simultaneous optical monitoring of activity of many neurons in invertebrate ganglia using a 124 element photodiode array. J. Neurophysiol. *45:* 829–840.

Hebb, D. O. (1949) *The Organization of Behavior*, John Wiley & Sons, New York.

Hubel, D. H., and T. N. Wiesel (1962) Receptive fields, binocular interaction and functional architecture in the cat's visual cortex. J. Physiol. (Lond.) *160:* 106–154.

Mountcastle, V. B. (1957) Modality and topographic properties of single neurons of cat's somatic sensory cortex. J. Neurophysiol. *20:* 408–434.

Perkel, D. H., G. L. Gerstein, and G. M. Moore (1967) Neuronal spike trains and stochastic point processes. II. Simultaneous spike trains. Biophys. J. *7:* 419–440.

Schneider, J., R. Eckhorn, and H. Reitboeck (1983) Evaluation of neuronal coupling dynamics. Biol. Cybernet. *46:* 129–134.

Wright, W. E. (1977) Gravitational clustering. Pat. Recogn. *9:* 151–166.

Chapter 6
Application of the Gravitational Clustering Method

Neuronal Assemblies in the Respiratory System

Bruce Lindsey

In 1975, I was a first year post-doc in George's lab, learning how to make neurophysiological recordings and the methods for spike train analysis that George and his collaborators, particularly Donald Perkel, had developed. George gave his trainees considerable latitude in choosing and working on projects, and I had taken on the question of "ensemble neural coding" and the generation of motor neuron proprioceptive fields in the crayfish claw control system. This involved measuring synaptic potentials generated in claw motor neurons by an array of parallel sensory channels and using cross-correlation and other tools to assess their influence. Specific motor neurons could be unambiguously identified from axonal recordings at the periphery near the claw muscles, and I was probing the neuropil of the first thoracic ganglion with a sharp pipette electrode to get good records of synaptic potentials in the motor neurons.

One morning George came into the lab and cheered, "I have a little gift for you." He had coded and loaded a program into the lab's PDP-12 computer. This "app"—as it would now be called—delayed the intracellular voltage signal by a user-defined time, so that signals from the motor neurons in the ganglia appeared after the action potentials recorded at the periphery. By using the peripheral spikes to trigger the sweep of an oscilloscope, I could now search for time-locked "bumps" in the signal from the electrode to help find the relevant neuronal processes for recording. The program was a useful tool and demonstrated to me in an immediate and tangible way

B. Lindsey (✉)
Department of Molecular Pharmacology & Physiology, Morsani College of Medicine, University of South Florida, Tampa, FL, USA
e-mail: blindsey@health.usf.edu

A. Aertsen et al. (eds.), *Introducing Computation to Neuroscience*, Springer Series in Computational Neuroscience, https://doi.org/10.1007/978-3-030-87447-6_6

Fig. 6.1 Meeting George in Newcastle for the Spike Train Workshop in 2006 (private photo by B. Lindsey)

that one can "play" with time inside a computer—and manipulate events recorded in the nervous system. I could not then imagine that the work on the crayfish would be the first in a series of projects with George that would span four decades (Fig. 6.1).

After joining the Department of Physiology at the University of South Florida, Tampa, I continued to work on the crayfish as a model system for studying the role of spike synchrony as a mechanism for gain control in multi-path parallel circuits. In the earlier work, we had represented the synchronous firing of two motor neurons as logical "AND" events to evaluate their coordinated discharge as a function of claw joint angle and time after opening step movements. I extended this approach and coded an algorithm for "logical" cross-correlation analysis, where trigger events could be either logical "AND" or "NOT" events, representing, respectively, coincident or nearly coincident spikes across multiple channels, or the occurrence of events at times when other monitored neurons did not discharge. I visited George in 1980 to discuss my results, and he suggested additional types of "surrogate" spike trains that would be useful as null hypothesis "controls." Around that time, Roger Shannon and I had started a collaboration to study the brain stem network for breathing. With Lauren Segers, a new graduate student in the department, we were recording at multiple sites in the pons and ventral respiratory column, increasing the numbers of electrodes, amplifiers, and tape recorders as resources became available.

So it was a special day in 1985 when I opened an issue of the Journal of Neuroscience and read the first paper on gravitational clustering by George, Donald Perkel, and Judy Dayhoff (Gerstein et al. 1985). (This paper is just a few pages away—

I hope you enjoy it!) Here was a conceptual approach that could screen multiple spike trains for evidence of functional associations—and allow visualization of the dynamics of their interactions through time. The method represents each neuron as a particle in *N* space. Each particle carries a time-varying charge that is a filtered version of the corresponding spike train. The charged particles exert forces on each other and move in the *N* space. The resulting trajectories reflect neuronal timing relationships. I was very excited by this work and by another paper by George and Ad Aertsen (Gerstein and Aertsen 1985) that soon followed (see also Chap. 7 this Volume).

The following year George gave us the source code for the gravity algorithms and visited our lab to discuss a collaboration to evaluate the method using spike trains from the respiratory brain stem. We watched 8mm movies showing the trajectories of particles in projections from gravity *N* space. After an initial study with George evaluating the efficacy of the gravity method for detecting functional interactions among the neurons we were studying (Lindsey et al. 1989), we began a series of experiments with graduate student Kendall Morris and two visiting scientists, Yvonne Hernandez and Akiko Arata. David Millhorn and colleagues had recently described an enhanced drive to breathe that persisted well after being evoked by repeated intermittent stimulation of the carotid chemoreceptors. The induction and expression of this "respiratory memory," termed long-term facilitation, likely involved raphe circuits and motivated our work using gravitational clustering and other methods of spike train analysis to study raphe assemblies and their dynamics.

We first needed a method to define particle aggregation (or "repulsion" for inhibition) caused by non-random timing relationships in the spike trains. Taking a cue from the "shift" control we had used earlier, we decided to generate sets of surrogates by shifting and rotating spike trains 1 to n-1 by 1 to n-1 time blocks (or respiratory cycles), respectively. This procedure obliterated the consequences of any synchrony present on a short (msec)-time scale. A plot of the minimum and maximum distance between each pair of particles at each time step in all the surrogate gravity calculations formed a "cone" that provided a Monte Carlo estimate of particle aggregation attributable to factors other than excess short-time scale synchrony. We first tested the approach using spike trains from simulated networks with defined connections. (George had previously encouraged me to always have network simulation tools on hand for evaluating the circuits that we inferred from correlation analysis. He recommended that I contact Ronald MacGregor. George had collaborated with him on a study of memory storage capacity in neural networks. Professor MacGregor promptly replied to my inquiry and sent along several pages of FORTRAN code, which formed the core of our network simulator.)

This method was described in the first of two papers published with George (Lindsey et al., 1992a,b) (Fig. 6.2). The main result was the identification of correlational assemblies in midline regions of the brain stem and evidence for excitatory and inhibitory connections appropriate for roles in the generation of spike synchrony. The second paper documented the reconfiguration of these assemblies with time and the phases of breathing. George had suggested an approach for sepa-

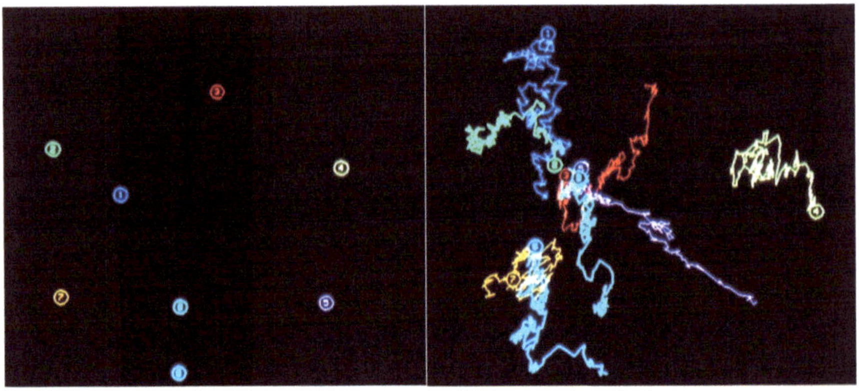

Fig. 6.2 Screen images for a gravity result in the late 1980s. Later they appeared in the Journal of Neurophysiology 1992 (Fig. 16 of Lindsey et al. (1992a))

rating spike trains into inspiratory and expiratory "chunks" and then concatenating them with an interval of at least 4 times the gravity charge time constant between each chunk so that spikes in adjacent chunks would not influence the result. We found that neurons with no apparent respiratory modulation of their individual firing rates collectively exhibited respiratory phase-dependent impulse synchrony, an emergent property with many implications.

While using the gravity method, we frequently noted intervals when the distances between gravity particles decreased in abrupt steps or jumps—moments of high-velocity condensation indicative of impulse synchrony. The jumps were reminiscent of the logical "AND" events, and we conjectured that subtle patterns of transient synchrony were embedded in what would otherwise be considered only tonic or "resting" activity subject to "up or down" firing rate modulation. To screen for repeated patterns of spike synchrony in identified correlational assemblies, we constructed two-dimensional arrays. Each row contained the aggregation velocity of a particle pair during successive intervals of the gravity calculation. Each column had values for all the pairs at each time interval and served as a template that was compared with each subsequent column for a "match." Sets of columns with similar velocity values that occurred more frequently than in any of the surrogate trials were displayed compactly as a set of vectors, each with a common origin. Vector direction indicated the pair of neurons represented, vector length the aggregation velocity of the corresponding particle pair. Because of the transient appearance of the vectors in animations of successive planes through time, we termed these graphs "spark plots." We had numerous communications with George during this time, particularly about surrogates. This work, published as Lindsey et al. (1997), aided the detection of recurring transient assembly configurations in the respiratory network during baroreceptor stimulation and following carotid chemoreceptor-evoked long-term facilitation, as described in papers by Arata et al. (2000) and Morris et al (2001).

Our final work together began with a visit to Penn during which we discussed a modification of the rules for assigning charge increment times to the gravity particles. The short-time constants that we used to restrict aggregation to highly synchronized neurons reduced the sensitivity for delayed correlations. Longer-time constants that could reach long-lagged relationships would reduce selectivity. I had been working on an approach that offsets the charge increment times for each pair to deal with the mixtures of short- and long-lag correlations that commonly occurred in our parallel spike train recordings. We decided to call this "tuned" gravity. George had developed a new approach to visualize particle movements in the N space. The idea was to keep track of not only the distance between particles, but also the distance that each particle of the pair has moved from its starting point. (Each particle in the N space is initially made equidistant from the others.) With three numbers now associated with each particle pair at each step of the calculation, the trajectory of these numbers as they evolve can be displayed in a three-dimensional coordinate system. By using particular charge rules introduced earlier by George and Ad Aertsen, the new 3-D display permitted parsing of the different possible relationships between the observed neurons. The results are directly interpretable in terms of the types of effective connectivity. These two additions to the gravity tool kit and their application to data from simulated networks and in vivo recordings were reported in Lindsey and Gerstein (2006).

The different views of non-random timing relationships afforded by gravitational clustering and the other spike train analysis methods pioneered by George, supplemented by newer approaches for surrogate generation, have remained central to our work on network operations that generate and tune the motor pattern for breathing. During a visit with George just a few years ago, he reiterated that there remains work to do in the gravity N space.

References

Arata A, Hernandez YM, Lindsey BG, Morris K, Shannon R (2000) Transient configurations of baroresponsive respiratory-related brainstem neuronal assemblies in the cat. Journal of Physiology 525(2):509–530. https://doi.org/10.1111/j.1469-7793.2000.t01-1-00509.x

Gerstein GL, Perkel H, Dayhoff E (1985) Cooperative firing activity in simultaneously recorded populations of neurons: detection and measurement. J Neurosci 5(4):881–889. https://doi.org/10.1523/JNEUROSCI.05-04-00881.1985

Gerstein GL, Aertsen AM (1985) Representation of cooperative firing among simultaneously recorded neurons. J Neurophysiol 54(6):1513–1528. https://doi.org/10.1152/jn.1985.54.6.1513

Lindsey BG, Gerstein GL (2006) Two enhancements of the gravity algorithm for multiple spike train analysis. Journal of Neuroscience Methods 150(1):116–127. https://doi.org/10.1016/j.jneumeth.2005.06.019

Lindsey BG, Shannon R, Gerstein GL (1989) Gravitational representation of simultaneously recorded brainstem respiratory neuron spike trains. Brain Research 483(2):373–378. https://doi.org/10.1016/0006-8993(89)90183-2

Lindsey BG, Hernandez YM, Morris KF, Shannon R, Gerstein GL (1992a) Respiratory-related neural assemblies in the brain stem midline. J Neurophysiol 67(4):905–922. https://doi.org/10.1152/jn.1992.67.4.905

Lindsey BG, Hernandez YM, Morris KF, Shannon R, Gerstein GL (1992b) Dynamic reconfiguration of brain stem neural assemblies: Respiratory phase-dependent synchrony versus modulation of firing rates. Journal of Neurophysiology 67(4):923–930. https://doi.org/10.1152/jn.1992.67.4.923

Lindsey BG, Morris KF, Shannon R, Gerstein GL (1997) Repeated patterns of distributed synchrony in neuronal assemblies. Journal of Neurophysiology 78(3):1714–1719. https://doi.org/10.1152/jn.1997.78.3.1714

Morris KF, Shannon R, Lindsey BG (2001) Changes in cat medullary neurone firing rates and synchrony following induction of respiratory long-term facilitation. Journal of Physiology 532(2):483–497. https://doi.org/10.1111/j.1469-7793.2001.0483f.x

RAPID COMMUNICATION

Repeated Patterns of Distributed Synchrony in Neuronal Assemblies

B. G. LINDSEY, K. F. MORRIS, R. SHANNON, AND G. L. GERSTEIN
Department of Physiology and Biophysics and Neuroscience Program, University of South Florida Health Sciences Center, Tampa, Florida 33612; and Department of Neuroscience, School of Medicine, University of Pennsylvania, Philadelphia, Pennsylvania 19104

Lindsey, B. G., K. F. Morris, R. Shannon, and G. L. Gerstein. Repeated patterns of distributed synchrony in neuronal assemblies. *J. Neurophysiol.* 78: 1714–1719, 1997. Models of brain function predict that the recurrence of a process or state will be reflected in repeated patterns of correlated activity. Previous work on medullary raphe assembly dynamics revealed transient changes in impulse synchrony. This study tested the hypothesis that these variations in synchrony include distributed nonrandom patterns of association. Spike trains were recorded simultaneously in the ventrolateral medulla, n. raphe obscurus, and n. raphe magnus of four anesthetized (Dial), vagotomized, paralyzed, and artificially ventilated adult cats. The "gravitational" representation of spike trains was used to detect moments of impulse synchrony in neuronal assemblies visualized as variations in the aggregation velocities of particles corresponding to each neuron. Template matching algorithms were developed to identify excessively repeating patterns of particle condensation rates. Repeating patterns were detected in each animal. The reiterated patterns represented an emergent property not apparent in either corresponding firing rate histograms or conventional gravity representations. Overlapping subsets of neurons represented in different patterns were unmasked when the template resolution was changed. The results demonstrate repeated transient network configurations defined by the tightness and duration of synchrony in different combinations of neurons and suggest that multiple information streams are conveyed concurrently by fluctuations in the synchrony of on-going activity.

INTRODUCTION

Theories of sensory processing, motor control, and memory retrieval propose that systems of neurons engaged in such tasks encode information in transient distributed synchrony (Arbib 1995). Patterns of coordinated activity may recur if similar operations are repeated (Abeles and Gerstein 1988; Dayhoff and Gerstein 1983; Frostig et al. 1990; Mainen and Sejnowski 1995). In a previous study on medullary raphe assembly dynamics, we noted transient changes in impulse synchrony apparently unrelated to monitored physiological variables (Lindsey et al. 1992b). We conjectured that these fluctuations in synchrony included nonrandom patterns of associations predicted under the hypothesis that raphe assemblies operate as an equilibrium seeking system in the regulation of cardiorespiratory function and have roles in the induction and expression of respiratory memory (Lindsey et al. 1992a; Millhorn 1982; Morris et al. 1996a,b). Preliminary accounts have been presented (Lindsey and Gerstein 1996; Lindsey et al. 1996).

METHODS

Data acquisition

Materials and methods have been described in detail (Lindsey et al. 1994; Morris et al. 1996a). All experiments were performed under protocols approved by the University of South Florida's Animal Care and Use Committee. Four adult cats of either sex initially were anesthetized with thiopental sodium (22.0 mg/kg iv). Anesthesia was maintained with Dial-urethane (allobarbital; Ciba; 60.0 mg/kg; urethane, 240 mg/kg). Blood pressure and respiration were monitored continuously. Animals were given additional Dial-urethane if there was an increase in blood pressure or respiration in response to periodic noxious stimuli (toe pinch). Animals received dexamethasone (2.0 mg/kg) and atropine (0.5 mg/kg). Arterial blood pressure, pH, PO_2, PCO_2, and $[HCO_3^-]$ and end-tidal CO_2 were monitored and maintained within normal limits. Core body temperature was maintained at 38.0 ± 0.5°C. Animals were ventilated, vagotomized, and paralyzed to reduce brain stem movements with a bolus of gallamine triethiodide (2.2 mg/kg) followed by constant infusion (0.4 mg · kg^{-1} · h^{-1}). Multiple single neuron spike trains were recorded extracellularly in the medullary raphe nuclei and ventrolateral medulla with planar arrays of tungsten microelectrodes. Some data samples were recorded during a change in blood pressure produced by inflation of an embolectomy catheter in the descending aorta. At the end of the experiments, cats were overdosed with pentobarbital sodium and perfused intracardially with 0.9% NaCl, followed by 10% neutral-buffered formalin solution. Brain stem sections were prepared and examined.

Data analysis

Correlational neuronal assemblies were first identified with the gravity method, which provides a conceptual framework for the analysis and representation of groups of simultaneously monitored neurons and their dynamic associations (Gerstein and Aertsen 1985; Gerstein et al. 1985; Lindsey et al. 1989, 1992a,b; Strangman 1997). Each of N neurons is represented as a particle in N space. Each particle is initially equidistant from all other particles and has a time varying charge that is a filtered version of its spike activity. The charged particles exert forces on each other and move as though in a viscous fluid. The trajectories of particle aggregation may indicate interesting temporal modulation of neuronal timing relationships that would be lost in a summary or averaging procedure, such as ordinary cross-correlation analysis.

Each particle travels at terminal velocity, which is proportional to the product of its charge, the propulsive force generated by all other particles (a vector quantity), and a user-defined coefficient that represents the reciprocal of viscosity, i.e., particle mobility. (See equation 8 in Gerstein et al. 1985 and equation 5 in Gerstein and Aertsen 1985.) The sign of the propulsive force was set to cause attraction if the particles simultaneously had a high charge (i.e., if the neurons they represented were firing in close temporal contiguity), as would be the case with a shared input or excitatory interactions. Aggregation indicated nonrandom spike train timing relationships in a time range set by the time constant of the charge decay parameter used in the computation.

Two procedures ensured that transient increases in short-time

0022-3077/97 $5.00

scale synchrony, not changes in firing rate, were responsible for the significant particle aggregation patterns. First, the charge decay constants used were short compared with interspike intervals; second, the charge corresponding to a sliding window of local firing rate was subtracted from each raw charge function, thus creating an effective charge function with approximately zero mean charge.

The variable that determined particle mobility was adjusted to prevent close approaches. Gravity particle condensation profiles were generated to detect recurring transient assembly configurations. Successive elements in each row of a two-dimensional array contained the aggregation velocity of each pair of particles during an interval represented by the column. The velocities in each column served, in turn, as a template that was compared with all other columns for a match. Velocity values were nulled if the particle pair never exhibited significant aggregation or if they were less than a nulling threshold of either 3.25 or 10% of the maximum velocity detected in the data set. These values were selected to filter out arbitrarily small fluctuations in aggregation velocity. Significant patterns were detected in all animals with each threshold value.

Each data set was analyzed at three different temporal resolutions defined by the number of consecutive plotted time steps used to calculate each particle aggregation velocity vector. Thus each temporal resolution was associated with a different number of template columns. Seven different match criteria based on velocity thresholds, ranging from a noisy match to a nearly perfect duplication were used in each search and tested for significance. *1*) More than 80% of the nonnulled velocities in the template column matched in the target column; a 90% cut-off was used in most pattern searches. *2*) Criterion 1 met and >75% of the matched pairs had velocities within a range of ±25% of the corresponding template velocities. *3*) Criterion 1 met and the percentage of extra nonnulled velocities in the target column was <5% of the total pairs in the group evaluated. *4*) Criterion 2 and 3 met. *5*) More than 99% of the nonnulled velocities in the template column matched in the target column and the percentage of extra nonnulled velocities in the target column was <5% of the total pairs in the group evaluated. *6*) More than 99% of the nonnulled velocities in the template column matched in the target column and there were no extra nonnulled velocities in the target column. *7*) More than 99% of the nonnulled velocities in the template column matched in the target column, there were no extra nonnulled velocities in the target column, and >75% of the matched pairs had velocities within a range of ±25% of the corresponding template velocities. A column had to have at least two velocities to be used as a template. A template could match under more than one criterion.

The null hypothesis was that the number of repeating patterns found when a particular template column was compared with the other columns was not greater than expected by chance. We constructed 100 pair-wise shuffled condensation profile sets and searched them for the maximum number of matching columns. This provided a Monte Carlo type estimate of the number of recurrences of a given pattern expected by chance ($P < 0.01$) in identical distributions of condensation rates associated at random. The null hypothesis was rejected if the maximum number of matches for a particular column was greater than the Monte Carlo estimate.

RESULTS

Correlational assemblies were first identified in four animals by screening simultaneously monitored neurons with the gravity method. Data from one sample, displayed in eight firing rate histograms (Fig. 1*A*), were collected during a perturbation of blood pressure. Several particles in the gravitational representation aggregated as illustrated in the animated projections of particle trajectories (Fig. 1*B*). Labeled circles in the last frame of the movie show final particle positions; trails can be followed back to initial positions in the projected space. Because of information loss in such projections from the *N* space, the distance between each pair of particles always was plotted as a function of time and evaluated for significance (Fig. 1*C*). The heavy black line (⇗) in the particle distance as a function of time (PDFT) graph defines the aggregation of *particles 1* and *4*. The light black lines indicate the minimum, maximum and mean distances between *particles 1* and *4* at each time step in 100 different data sets. The original correlations were obliterated in each data set by shifting and rotating spike times by different amounts. These minimum and maximum lines provide a Monte Carlo estimate of significance for aggregation caused by short-term correlation, as compared with aggregation attributable to all other factors (e.g., "random movements" due to incomplete correction for zero mean charge) (see Lindsey et al. 1992a). The null hypothesis was that significant aggregation did not occur; it was rejected for all periods in the data when the interparticle distance trajectory fell below the minimum line. Each horizontal bar in Fig. 1*D* is a graphical summary of the times when the corresponding pair of particles was significantly close. The gray row (⇗) corresponds to the labeled line plotted in Fig. 1*C*. Screening for aggregation velocities and pattern properties was limited to particles that met the aggregation significance test and therefore represented members of putative neuronal assemblies.

The graph of all particle pair distances revealed transient moments of synchrony among several neurons; these appeared as coincident steps or jumps in corresponding plots (e.g., ⇒ in Fig. 2*A*). Data for *neurons 4* and *1* are plotted in black (⇙) for comparison with Fig. 1*C*. A template matching algorithm was developed to determine whether combinations of neurons exhibited such recurring moments of impulse synchrony or near synchrony more frequently than would be expected by chance. We initially chose a time bin equal to the minimum plotted interval in the PDFT graph, or multiples thereof, over which to calculate an average aggregation velocity for the particle pairs. The results are shown in the particle condensation profile (Fig. 2*B*) where each row shows the aggregation velocities of a particular pair during consecutive time intervals. Each column allows comparison of aggregation velocities of all particle pairs at a particular time. The values in each column served in turn as a template, illustrated in a tutorial figure (Fig. 2*C*), that was compared with other columns for a match. The value for a pair was set to zero in the absence of net movement of pair elements toward each other during an interval or if the velocity was below a nulling threshold. One of two such threshold values (METHODS) was used in each pattern search to remove small velocities from further consideration. The seven different match criteria based on velocity thresholds were tested concurrently. A significant pattern was recorded if the number of matches detected for some template column in the original data set exceeded the number found for that same template in 100 data sets, each of which had different randomly shuffled rows of pair condensation velocities.

Assemblies from each animal exhibited transient configurations of constituent neurons. To detect patterns of distrib-

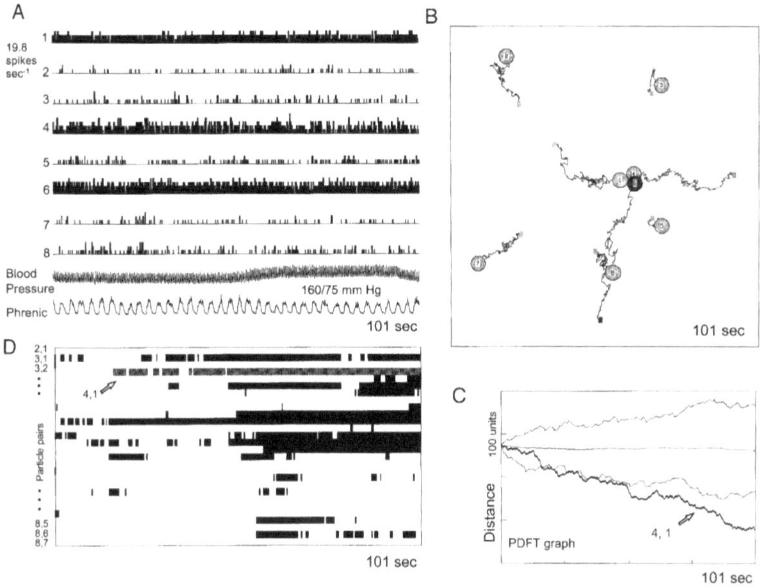

FIG. 1. *A*: firing rate histograms for 8 simultaneously recorded spike trains with blood pressure and integrated efferent phrenic nerve activity to indicate phases of respiratory cycle. Rate calibration corresponds to highest bins. *Neurons 1–3* and *5–8* were monitored in nucleus raphe magnus; *cell 4* was recorded in ventrolateral medulla. *B*: final frame of an animated projection of the particle trajectories from the *N* space to a plane for the spike data in *A*. Labeled circles show final particle positions; trails can be followed to initial positions of particles shown as colored rectangles. Calculation parameters: acceptor and effector charge decays forward; decay time constant 10.0 ms. Numbers of spikes in sample are 1: 686; 2: 85; 3: 135; 4: 674; 5: 170; 6: 851; 7: 100; and 8: 186. *C*: —— (⌀) is particle distance as a function of time (PDFT) plot for *particles 1* and *4*. Initially each particle was equidistant (100 arbitrary units) from all other particles. *Top* and *bottom* —— define maximum and minimum distances between particles in 100 shifted control data sets at each time step. Thus probability that aggregation of *particles 1* and *4* was due to random coincident spikes was <0.01. *Middle* —— indicates mean particle distance at each time step for the 100 control calculations. *D*: ■ in each row indicate if and when distance between each pair of particles was less than expected by chance ($P < 0.01$). Row corresponding to the PDFT plot in *C* is labeled (⌀).

uted synchrony with different durations, all data sets were searched with the temporal resolution set to different multiples of the smallest plotted gravity step time. One to three consecutive plotted time steps were used to calculate each particle pair aggregation velocity. Single-step durations ranged from 119 to 546 ms (average: 289 ± 127 ms; mean ± SD). Two-step durations ranged from 239 to 1,129 ms (average: 578 ± 254 ms). Three-step durations ranged from 358 to 1,693 ms (average: 867 ± 380 ms). This procedure also reduced the possibility that repeated patterns would remain undetected because of the sampling frequency. Significant numbers of repeated patterns were detected in samples of seven or eight simultaneously recorded neurons from each animal for each time resolution used (Table 1). In three of four animals, patterns were detected with each of the seven criteria. For the numbers of neurons analyzed in this work, criteria 4 and 7 were similarly stringent and required the closest matches. No significant matches were found in one animal with criterion 7. The reiterated patterns represented an emergent property that was not apparent in either corresponding firing rate histograms or the conventional gravity representations.

Particle aggregation patterns that recurred with a frequency greater than that of any of the shifted data sets were displayed as sets of vectors, each with a common origin and at the time of occurrence as illustrated in Fig. 2*D*. Vector length indicated the aggregation velocity; vector direction served solely to identify the neuron pair represented. The sets of vectors are termed "spark" patterns because of their transient appearance in animations of successive planes through time. Repeated sparks indicative of recurring transient assembly configurations are illustrated in Fig. 3. These examples were derived from the data illustrated in the previous figures. The two sparks in Fig. 3*A* were detected when velocities were calculated during 1.218-s intervals (3 plotted time steps) and matched according to criterion 7. In this

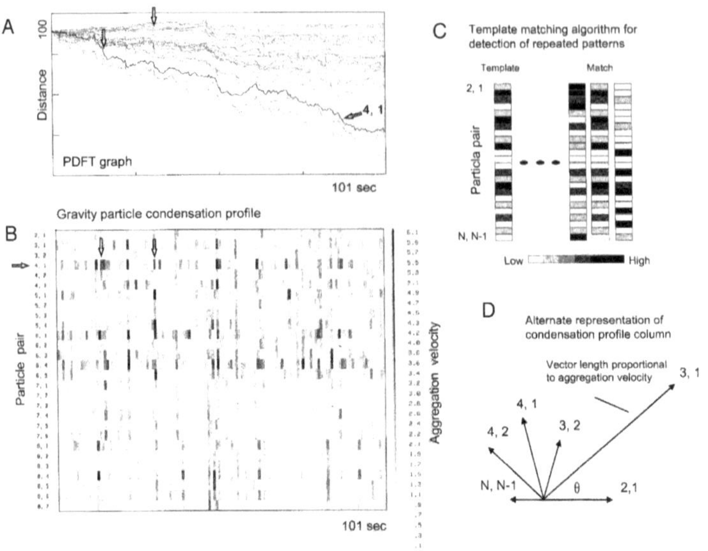

FIG. 2. *A*: PDFT graph of aggregation of all constituent pairs in group represented in Fig. 1*B*. Condensation of pair 4,1 is plotted in black (⇙); other pairs exhibited concurrent moments of high velocity aggregation (e.g., ⇒). *B*: gravity particle condensation profile derived from PDFT data in *A*. Successive elements in each row of this 2-dimensional array are shaded to represent the relative aggregation velocities of each pair of particles during an interval (812 ms) represented by the particular column. ⇒, correspond to times similarly labeled in *A*. *C*: cartoon to illustrate the template matching method used to detect repeated assembly configurations: the set of velocity values in each column served in turn as a template that was compared to all other columns for a match. A pattern was recorded if the number of matches detected for each template column in original data set was greater than the maximum number of matches found between the same template and results derived from 100 pair-wise shuffled condensation profiles. *D*: illustration of how an individual condensation profile column may be represented in compact form as a set of vectors with a common origin. The direction of each vector in a plane identifies the neuron pair represented; vector length indicates the aggregation velocity of corresponding particles.

sample, no other patterns repeated significantly under criteria 4–7 at this temporal resolution. The blood pressure trace and integrated efferent phrenic nerve activity in Fig. 1*A* were plotted again on the side panel. Time is represented along the *z* axis in these spark plots. The rear panel of the plot shows a phase plane of the smoothed integrated phrenic signal (*x* axis) against its first derivative (*y* axis). This particular spark pattern was confined to an interval with control

TABLE 1. *Summary of frequency of matched templates and numbers of repetitions as a function of template interval duration and match criteria*

Temporal resolution	Patterns	Criteria 1	2	3	4	5	6	7
1 step	Matched templates	209 ± 15	6 ± 4	35 ± 12	2 ± 2	30 ± 12	25 ± 14	2 ± 2
	Repetitions	23.0 ± 19.7	2.1 ± 1.9	3.7 ± 3.4	1.2 ± 0.6	3.2 ± 2.5	3.4 ± 2.6	1.2 ± 0.6
2 steps	Matched templates	98 ± 10	1 ± 1	11 ± 5	1 ± 1	8 ± 5	7 ± 5	1 ± 1
	Repetitions	12.8 ± 3.6	1.9 ± 1.9	3.0 ± 2.8	1.0 ± 0.0	2.4 ± 2.1	2.6 ± 2.2	1.0 ± 0.0
3 steps	Matched templates	60 ± 12	<1	5 ± 3	<1	5 ± 3	4 ± 3	<1
	Repetitions	9.7 ± 6.8	1.4 ± 1.1	3.6 ± 2.8	1.0 ± 0.0	2.3 ± 1.2	2.4 ± 1.2	1.0 ± 0.0

The first, third, and fifth rows give the number (mean ± SD) of template column patterns with significant numbers of matches found in a total of 17 subsets of spike data from four cats for three different temporal resolutions. The second, fourth, and sixth rows show the mean ± SD number of repetitions of each significant template. A column could match more than one template. Nulling thresholds were set to 0.03. The threshold for the variable that determined criterion 1 was set to 0.9 rather than 0.8. Row and column labels indicate the temporal resolutions and match criteria, respectively. See RESULTS for further details.

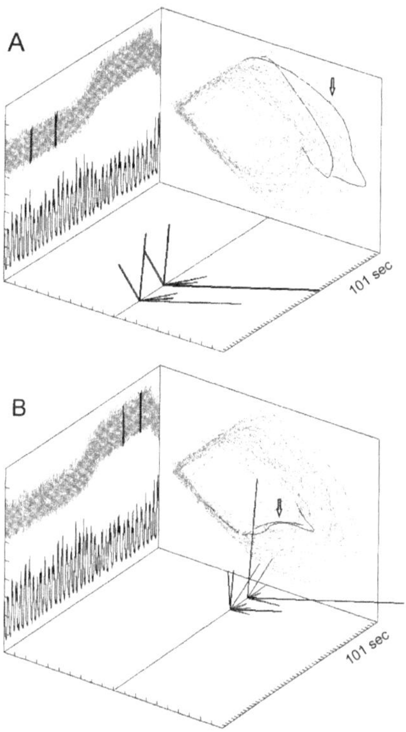

FIG. 3. *A*: spark plots display repeated particle condensation profiles indicative of transient neuronal assembly configurations. Data were derived from the results illustrated in Figs. 1 and 2. The sparks matched according to criterion 7. Blood pressure and integrated efferent phrenic nerve signals in Fig. 1*A* are plotted on side panel. Time is represented along *z* axis. The phase plane on the rear panel is a plot of the smoothed integrated phrenic signal (*x* axis) against its first derivative (*y* axis). The early expiratory phases of 2 respiratory cycles during which the spark patterns occurred are indicated by black segments in the phase plane (⇓). The following particle-pair aggregation velocities, calculated during a 1.218-s interval, were represented in the first spark: 4, 1: 1.47; 4, 2: 0.52; 4, 3: 0.61; 6, 4: 1.18; 7, 4: 0.58. Corresponding values in the second spark were: 2.25, 0.51, 0.70, 1.07, 0.55. *B*: another spark pattern repeated only during the interval with elevated blood pressure; it was confined to the inspiratory phases of 2 respiratory cycles (arrow in phase plane). These patterns matched by criterion 3. Pair aggregation velocities in the first spark, calculated during 0.812-s intervals, were 3, 1: 0.31; 4, 1: 0.89; 4, 2: 0.36; 5, 1: 0.34; 6, 1: 1.19; 6, 4: 0.78; 6, 5: 0.73. Corresponding values in the second spark were: 3, 1: 0.52; 4, 1: 2.65; 4, 2: 0.14; 5, 1: 0.29; 6, 1: 0.98; 6, 4: 2.36; 6, 5: absent.

blood pressure (side panel) and to the early expiratory phase of the respiratory cycle (⇓, rear panel).

A second spark pattern repeated (criterion 3) exclusively during the period of altered baroreceptor stimulation (Fig. 3*B*). This second pattern was detected when velocities were calculated during 0.812-s intervals (2 plotted time steps); it was confined to the inspiratory phase of the respiratory cycle (⇓, rear panel). No other patterns repeated significantly under criteria 3–7 at this temporal resolution. The spike rates of all eight neurons during respiratory cycles concurrent with the increase in blood pressure were not significantly different from control cycles (*t*-test, $n1 = n2 = 15$, $P > 0.05$).

DISCUSSION

The gravity method previously demonstrated that medullary raphe neurons with no respiratory modulation of their individual firing rates collectively exhibited respiratory phase-dependent synchrony (Lindsey et al. 1992b). Those results suggested the hypothesis that rate and synchrony codes are transmitted in the impulse traffic of raphe assemblies. In this work, different spark patterns that repeated more frequently than expected by chance were unmasked in the same time series when the interval duration used to calculate each velocity was changed. These data demonstrate repeated transient network configurations defined by the tightness and duration of synchrony in different combinations of neurons and suggest that multiple information streams are conveyed concurrently by fluctuations in the synchrony of on-going activity. The metabolic efficiency of such synchrony coding, as compared with firing rate modulation, remains to be determined.

Searches for recurring patterns of coordinated activity were always confined to elements of correlational assemblies. Spark patterns detected without such prior screening may be of interest in their own right. The less stringent match criteria allow for the possibility that velocities of different particles representing independent neurons may increase at particular moments because of spikes coincident with the concurrent activity of neurons in the identified assemblies.

This work extends earlier efforts to assess the functional consequences of coincident and near-coincident spikes in concurrently active parallel channels (Gerstein 1970; Stevens and Gerstein 1976; Lindsey 1982; Murthy and Fetz 1994). The constellations of gravity particles defined by repeating spark patterns may represent moments when a (partially represented) set of neurons reads, stores, or transmits information encoded by distributed synchrony. In some contexts, it may be useful to consider these moments discrete events, bearing in mind that each spark is an abstraction of many parallel processes. Approaches such as spark triggered averaging of physiological measurements can link transient assembly configurations to changes in behavior or state.

The spark plot inherits the advantages of the gravity method and offers, in compact form, the information needed to reconstruct the net movements of particles in the gravity *N* space. The method is scaleable with respect to the number of neurons represented and the temporal resolution of each spark interval. Computation time for the initial gravity calculations scales as the cube of the number of neurons represented. The core pattern detection algorithm scales as the number of template columns squared times the number of neuron pairs. Time for the various Monte-Carlo significance tests increases in direct proportion to the number of iterations required. With current data acquisition technology and typical workstation performance, analysis of 100 simultaneously recorded neurons can be achieved readily. Real-time visualization is currently not practical in most laboratories. However, multiprocessor computers and other optimizations should permit at least initial screening for patterns during experiments.

Other data sets, such as alternate representations of the gravity *N* space or synchronized video, may be projected on to the back panel of the spark animation sequence. The algorithms can incorporate readily other match criteria and can be extended to identify recurring patterns of patterns or hyperpatterns of spatiotemporal activity. Spark sequences then would serve as the template for "visualization" of sequential fluctuations of spatiotemporal patterns of synchrony as implied, for example, in holographic memory models (Pribram 1971).

The authors thank J. Gilliland, C. Orsini, T. Krepel, R. McGowan, and X. Li for technical assistance.

This work was supported by National Institute of Neurological Disorders and Stroke Grant NS-19814.

Address for reprint requests: B. G. Lindsey, Dept. of Physiology and Biophysics, University of South Florida Health Sciences Center, 12901 Bruce B. Downs Blvd., Tampa, FL 33612-4799.

Received 17 April 1997; accepted in final form 4 June 1997.

REFERENCES

ABELES, M. AND GERSTEIN, G. L. Detecting spatiotemporal firing patterns among simultaneously recorded single neurons. *J. Neurophysiol.* 60: 909–924, 1988.

ARBIB, M. A. (Editor). *The Handbook of Brain Theory and Neural Networks.* Cambridge, MA: MIT Press, 1995.

DAYHOFF, J. E. AND GERSTEIN, G. L. Favored patterns in spike trains. I. Detection. *J. Neurophysiol*. 49: 1334–1348, 1983.

FROSTIG, R. D., FROSTIG, Z., AND HARPER, R. M. Recurring discharge patterns in multiple spike trains. *Biol. Cybern.* 62: 487–493, 1990.

GERSTEIN, G. L. Functional association of neurons: detection and interpretation. In: *The Neurosciences: Second Study Program,* edited by F. O. Schmitt. New York, NY: The Rockefeller University Press, 1970, p. 648–661.

GERSTEIN, G. L. AND AERTSEN, A.M.H.J. Representation of cooperative firing activity among simultaneously recorded neurons. *J. Neurophysiol.* 54: 1513–1528, 1985.

GERSTEIN, G. L., PERKEL, D. H., AND DAYHOFF, J. E. Cooperative firing activity in simultaneously recorded populations of neurons: detection and measurement. *J. Neurosci.* 5: 881–889, 1985.

LINDSEY, B. G. Measurement and dissociation of joint influence of action potentials in concurrently active parallel channels on motor neuron activity in crayfish. *J. Neurophysiol.* 47: 1060–1173, 1982.

LINDSEY, B. G. AND GERSTEIN, G. L. Detection of recurring patterns of impulse synchrony in neuronal assemblies (Abstract). *FASEB J.* 10: A410, 1996.

LINDSEY, B. G., HERNANDEZ, Y. M., MORRIS, K. F., SHANNON, R., AND GERSTEIN, G. L. Respiratory related neural assemblies in the brain stem midline. *J. Neurophysiol.* 67: 905–922, 1992a.

LINDSEY, B. G., HERNANDEZ, Y. M., MORRIS, K. F., SHANNON, R., AND GERSTEIN, G. L. Dynamic reconfiguration of brain stem neural assemblies: respiratory phase-dependent synchrony versus modulation of firing rates. *J. Neurophysiol*. 67: 923–930, 1992b.

LINDSEY, B. G., MORRIS, K. F., SHANNON, R., AND GERSTEIN, G. L. Transient configurations of brainstem cardio-respiratory neuronal assemblies: detection and display of repeated motifs. *Soc. Neurosci Abstr.* 22: 1642, 1996.

LINDSEY, B. G., SHANNON, R., AND GERSTEIN, G. L. Gravitational representation of simultaneously recorded brainstem respiratory neuron spike trains. *Brain Res.* 483: 373–378, 1989.

LINDSEY, B. G., SEGERS, L. S., MORRIS, K. F. , HERNANDEZ, Y. M., SAPORTA, S., AND SHANNON, R. Distributed actions and dynamic associations in respiratory-related neuronal assemblies of the ventrolateral medulla and brainstem midline: evidence from spike train analysis. *J. Neurophysiol.* 72: 1830–1851, 1994.

MAINEN, Z. F. AND SEJNOWSKI, T. J. Reliability of spike timing in neocortical neurons. *Science* 268: 1503–1506, 1995.

MILLHORN, D. E. Stimulation of raphe (obscurus) nucleus causes long-term potentiation of phrenic nerve activity in cat. *J. Physiol. Lond.* 381: 169–179, 1982.

MORRIS, K. F., ARATA, A., SHANNON, R., AND LINDSEY, B. G. Long-term facilitation of phrenic nerve activity in the cat: responses and short-time scale correlations of medullary neurones. *J. Physiol. Lond.* 490: 463–480, 1996.

MORRIS, K. F., SHANNON, R., AND LINDSEY, B. G. Evidence for change of respiratory phase variance and changes in synchrony of raphe neuron assemblies during long term facilitation (Abstract). *Physiologist* 39: 186, 1996.

MURTHY, V. N. AND FETZ, E. E. Effects of input synchrony on the firing rate of a three-conductance cortical neuron model. *Neural Comp.* 6: 1111–1126, 1994.

PRIBRAM, K. H. *Languages of the Brain.* Englewood Cliffs, NJ: Prentice Hall, 1971.

STEVENS, J. K. AND GERSTEIN, G. L. Interactions between cat lateral geniculate neurons. *J. Neurophysiol.* 39: 239–255, 1976.

STRANGMAN, G. Detecting synchronous cell assemblies with limited data and overlapping assemblies. *Neural Comp.* 9: 51–76, 1997.

ELSEVIER

Journal of Neuroscience Methods 150 (2006) 116–127

JOURNAL OF NEUROSCIENCE METHODS

www.elsevier.com/locate/jneumeth

Two enhancements of the gravity algorithm for multiple spike train analysis

Bruce G. Lindsey[a,*], George L. Gerstein[b]

[a] Department of Physiology and Biophysics and Neuroscience Program, University of South Florida Health Sciences Center, 12901 Bruce B. Downs Blvd., Tampa, FL 33612-4799, USA

[b] Department of Neuroscience, School of Medicine, University of Pennsylvania, Philadelphia, PA 19104, USA

Received 12 October 2004; received in revised form 3 June 2005; accepted 15 June 2005

Abstract

The gravity method for neuronal assembly analysis represents each neuron as a particle in *N*-space with a time varying charge that is a filtered version of the corresponding spike train, with appropriate rules for forces between and movements of the charged particles. Resulting trajectories reflect neuronal timing relationships. The usual short time constants in the filter restrict aggregation to highly synchronized neurons and reduce the sensitivity for delayed correlations; long time constants in the filter reduce selectivity. Here we describe an enhancement that modifies rules for assigning charge increment times to allow mixtures of short and long lag correlations. Charge increments for each pair are offset from the actual spike times by time lags defined by features in corresponding cross-correlograms; no such charge offsets are invoked if the correlogram is flat. Tuning increases charge products and aggregation of long lag correlated pairs. A second enhancement uses a new three-dimensional display of particle pair trajectories to parse the type of neuronal relationship. For each pair, we record and display the interparticle distance and the distance each particle moves from its original location in the *N*-space. The resulting trajectories cluster according to the type of interaction between the represented neurons. Results from simulated networks and in vivo multi-site recordings show that these modifications detect assembly properties not identified by the standard methods.

Keywords: Neuronal assembly analysis; Spike train; Correlations; Multi-site recordings

1. Introduction

Temporal relationships among spike trains are increasingly the subject of neurophysiological study as advances in large electrode array and imaging technologies allow in vivo and in vitro multi-site recordings of impulses in many single neurons. The detection of non-random timing relationships with correlation analysis has provided important insights into the parallel and serial organization of the brain's neuronal circuits as they operate, changing effective connectivity relationships with time, task, and experience, and information transfer in the nervous system (Eggermont, 1990; Nicolelis, 1999; Lindsey et al., 2000; Gerstein and Kirkland, 2001). Numerous challenges remain in the acquisition, "spike sorting", and analysis of parallel recordings from large arrays (Buzsáki, 2004). For example, it is extremely important for such studies to have truly single neuron recordings. Poor sorting of waveforms leads to erroneous inferences of functional relationships among the recorded neurons (Gerstein, 2000).

Several approaches are now available for the analysis of multiple single neuron spike trains (Brown et al., 2004). First described in 1985, the gravity transformation has remained one of the few direct visualization methods for examination of dynamic interactions among a group of simultaneously monitored neurons (Gerstein et al., 1985; Gerstein and Aertsen, 1985). It translates direct analysis of multiple spike train activity into a clustering problem in an *N*-space. Each of the N neurons is represented as a particle, initially all located equidistant in an *N*-space. Each particle is given a time varying charge (Q_i) by convolving the corresponding spike train with an appropriate kernel. In the original versions of the gravity calculation, this kernel was taken to be an exponential with a short time constant like 10 ms.

* Corresponding author. Tel.: +1 813 974 2590; fax: +1 813 974 3079.
E-mail address: blindsey@hsc.usf.edu (B.G. Lindsey).

0165-0270/$ – see front matter © 2005 Elsevier B.V. All rights reserved.
doi:10.1016/j.jneumeth.2005.06.019

B.G. Lindsey, G.L. Gerstein / Journal of Neuroscience Methods 150 (2006) 116–127 117

Forces between particles are made independent of distance and defined through appropriate rules and normalization on charge products ($F_{ij} = Q_i Q_j$), so that the total force on each particle depends on an instantaneous vector sum of pair forces between it and all the other particles. Subsequent motion of each particle is defined by assuming a viscous medium and using an equation of motion where particle velocity is proportional to the total force on it (these rules are of course a purely mathematical construct and do not correspond to any physically realizable system, but are nevertheless very useful). After the gravitational transformation has worked on a suitable length of recorded data, particles representing those neurons that have non-random timing relationships in their firings will aggregate in the *N*-space.

The gravity method has revealed emergent properties of assemblies (Lindsey et al., 2000; Gerstein and Kirkland, 2001), demonstrating, for example, respiratory phase-dependent impulse synchrony among neurons with no respiratory modulation of their individual firing rates (Lindsey et al., 1992b). That early result from gravity analysis supported the view that information may be represented by spike timing relationships (Gerstein, 1970) in addition to firing rate changes that traditionally have been measured by neurophysiologists, since Adrian (1932). When used in conjunction with pattern detection enhancements, the gravity tool has also aided detection of recurring transient configurations associated with sensory processing, motor control, and the expression of memories (Lindsey et al., 1997; Arata et al., 2000; Morris et al., 2001, 2003).

Various modifications of the rules used in the original gravity version have been explored and can improve performance in various ways. These include: (a) introducing two distinct (and time reversed) kernels to define the particle charge when it is being moved (acceptor) in contrast to when it is contributing to movement of other particles (effector); this makes the particle representing a presynaptic neuron stand still while the particle representing a postsynaptic neuron moves towards it (or the reverse); (Gerstein and Aertsen, 1985; Gerstein, 1999; Keating and Gerstein, 2002, submitted 2004); (b) improving sensitivity by defining force as Q_{ij}^m, where m is a small odd number, or by taking account of the charge on a third particle (Baker and Gerstein, 2000); (c) making the force between two particles dependent on their distance (Dayhoff, 1994); (d) measuring particle aggregation velocities rather than distances (Lindsey et al., 1997; Gerstein, 1999); (e) measuring distance that a particle has moved from its original position (Keating and Gerstein, 2002, submitted 2004).

Different signal propagation times and synaptic properties within a network can result in sets of correlations with mixtures of short and long time lags (e.g., Li et al., 1999). However, the usual short charge kernel decay time constants used in gravity analysis restrict aggregation to highly synchronized neurons and reduce the sensitivity for delayed correlation; use of long time constants reduces selectivity. None of the gravity variations described above addresses this problem directly. Here we describe modification of the rules for assigning charge increment times to deal with mixtures of short and long-lag correlations; this will be called "tuned gravity".

A second enhancement described here defines a new three-dimensional display method that allows parsing the different possible relationships between the observed neurons. Position (and clustering) of trajectories in this display is directly interpretable in terms of the type of "effective connectivity", i.e. terms like shared input, pre-post, independent-pre, independent-post, etc.

We also document the usefulness of these modifications with spike train data from network simulations and in vivo multi-site recordings. Preliminary accounts of some of this work have been published (Lindsey and Gerstein, 2000, 2002; Keating and Gerstein, 2002).

2. Materials and methods

Spike train data from circuits with defined connectivity simulated with the program SYSTM11 (MacGregor, 1987) were used to compare standard gravity with tuned gravity enhancements. Neurons received random excitatory and inhibitory inputs from outside the network in addition to inputs from other specified single neurons. If the membrane potential of any cell, calculated according to a standard equivalent circuit model, exceeded threshold, an action potential was recorded in that cell, the potassium conductance was activated, and the targets of the cell received a synaptic input current. Simulated networks included connections that generated correlations between some spike trains with short time lags (0–2.0 ms). To generate long time lagged correlations, selected spike trains were shifted forward in time by either 10.0 or 20.0 ms relative to the other spike trains. This procedure retained the same spike sequences as the original simulator output. This approach allowed exact validation tests of tuning algorithms: a backward time shift converted the simulated long-time lag correlations back to their original temporal relationships.

Spike trains for demonstration of the second enhancement of gravity used a simple simulator. Initially each neuron in the desired circuit was provided with a Poisson or Gamma spike train with a range of realistic average firing rates. In a second step delayed and noisy probabilistic copies of spikes were made from pre-synaptic to post-synaptic trains. The copy probabilities were low and set to mimic typical variabilities and cross-correlations observed in real data. This is a completely non-physiological way of generating the desired spike trains, but produces all the cross-correlograms between trains that are appropriate for the circuit diagram being simulated.

Materials and experimental methods used to acquire the in vivo data have been described previously together with other results from the spike train data sets evaluated herein (Lindsey et al., 1997; Arata et al., 2000). Experiments were performed under protocols approved by USF's Animal Care and Use Committee. Briefly, adult cats of either sex were initially anesthetized with sodium thiopental (22.0 mg kg^{-1}, IV). Anesthesia was maintained with Dial-urethane (allobarbital: Ciba, 60.0 mg kg^{-1}; urethane, 240 mg kg^{-1}). Blood pressure and respiration were monitored continuously. Animals were given additional Dial-urethane if there was an increase in blood pressure or respiration in response to periodic noxious stimuli (toe pinch). Animals received dexamethasone (2.0 mg/kg) and atropine (0.5 mg kg^{-1}). Arterial blood pressure, pH, PO_2, PCO_2, and $[HCO_3^-]$ and end-tidal CO_2 were monitored and maintained within normal limits. Core body temperature was maintained at 38.0 ± 0.5 °C. Animals were ventilated, vagotomized and neuromuscular blocked to reduce brainstem movements with a bolus of gallamine triethiodide (2.2 mg kg^{-1}) followed by constant infusion (0.4 mg kg^{-1} h^{-1}). Multiple single neuron spike trains were recorded extracellularly in the medullary raphe nuclei and ventrolateral medulla with planar arrays of tungsten microelectrodes. At the end of the experiments, cats were overdosed with sodium pentobarbital and perfused intracardially with 0.9% NaCl, followed by 10% neutral-buffered formalin solution. Brainstem sections were prepared and examined.

2.1. *Tuned gravity*

Spike train data from simulations and in vivo experiments were evaluated with standard cross-correlation analysis (Perkel et al., 1967) and with the gravity method (Gerstein et al., 1985; Gerstein and Aertsen, 1985), a tool for the analysis and visualization of groups of correlated neurons as reviewed briefly in the Introduction. In the *N*-space used for the analysis, there will be aggregation of particles representing neurons that tend to fire in near synchrony. Significance of particle aggregation was evaluated using confidence limits defined with an empirical Monte-Carlo method using appropriate shuffling (Lindsey et al., 1992a).

Charge kernels with short decay time constants are normally used to restrict aggregation to highly synchronized neurons. However, these kernels reduce the sensitivity for any possible delayed synchrony. The 'tuned' gravity option is a modification that aids analysis of data sets with mixed short and long-time lag correlations, as would be generated, for example, by the "ball and stick" model shown in Fig. 1A, designated *circuit 1*. In all gravity calculations, the movement of a particular particle in the *N*-space is calculated by evaluating a vector sum of pair forces with all other particles. For 'tuned' gravity, the charge kernels used for each such pair of particles (for convolution with the corresponding neuron's action potential times) are offset by an amount determined by time lags to peaks or troughs in corresponding cross-correlograms (e.g., Fig. 1B). This generally means we must maintain N-1 separate charge functions for each particle, one for use in each possible pairing during the calculations.

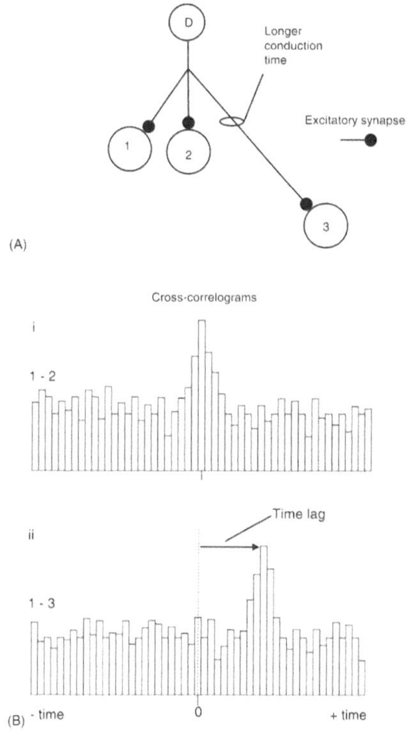

Fig. 1. (A) "Ball and stick" representation of *circuit 1* used to generate *data set 1* with mixed lag correlations. Three neurons receive excitatory inputs from a shared driver (D) neuron; the lengths of the lines connecting the neurons indicate relative conduction times. Each target neuron is assumed to receive other independent inputs (not shown). (B) The simulated cross-correlogram in (B(i)) has a central peak indicative of the illustrated shared input relationship of neuron pair 1–2; the offset peak for the pair 1–3 in (B(ii)) reflects inputs from the shared driver with short and long conduction times.

We used two slightly different algorithmic approaches for shifting charge functions and achieved identical results with both. The first approach stored the entire standard charge history for each particle in memory and used a table of previously calculated lag times to read a correspondingly offset value of the appropriate charge during calculation of the charge products. The second approach pre-computed the (offset) charge functions or kernels needed for each spike in each pair of neurons. This algorithm does not require calculation and storage of complete charge histories and is, therefore, more computationally efficient. Most computation time is consumed

B.G. Lindsey, G.L. Gerstein / Journal of Neuroscience Methods 150 (2006) 116–127 119

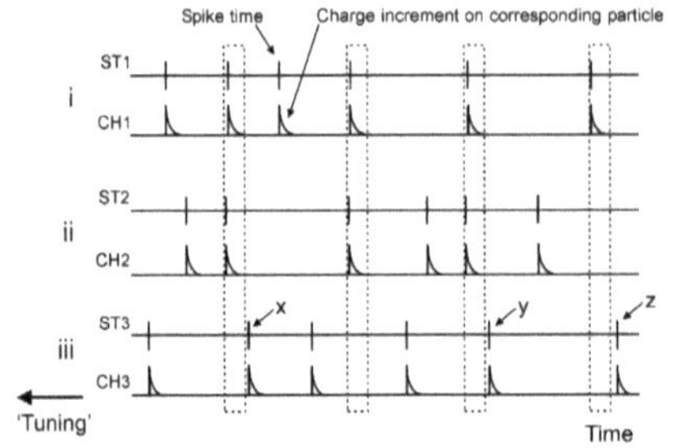

Fig. 2. Gravity 'tuning' by time-shifting particle charge increments an amount equal to the time lags to features in corresponding cross-correlograms (see text).

by calculating the particle movements; specifying particular charge rules is a short operation.

This 'tuning' procedure is illustrated in Fig. 2. For simulated *data set 1* we show three sets of traces (i–iii) with spike times (ST1–ST3) for the three neurons represented in Fig. 1A and their corresponding gravity particle charges (CH1–CH3). The long conduction time between the common driver neuron and neuron 3 results in correlated spikes (spikes x, y, and z in trace ST3) that fall outside the interval when charges for neuron pair 1–2 overlap (dashed boxes). Tuning shifts the charges for neuron 3 (arrow near trace CH3) by an interval equivalent to the lag-to-peak in correlogram Fig. 1B(ii). The tuning procedure aligns the charges of the three neurons, increasing the values of charge products, and thus aggregation of particle pairs representing neurons with temporally offset correlations (Fig. 3).

In this study, no charge offsets were used for pairs of neurons with flat cross-correlograms; charge kernels equivalent to standard gravity were applied to the corresponding gravity particles. We note here and consider further in Section 4 that a flat cross-correlogram, as a time average, may obscure transient correlations, and that a complete analysis of mixed time lag correlations requires iterative calculations of the tuned gravity function and a systematic shifting of charge kernel offsets (Keating and Gerstein, 2002, submitted 2004).

2.2. Three-dimensional display

Gravity runs of the data were taken with both symmetric and anti-symmetric kernels to examine the pair distance functions versus time. We then chose the kernels and run time parameters that produced the clearest bundling of pair distance trajectories. The particle position data from such runs were then used to make a "conventional" projection display and finally to make the new 3D display. Trajectory bundles in this display were identified in relation to the known circuit diagram that had been used to produce the simulated spike train data.

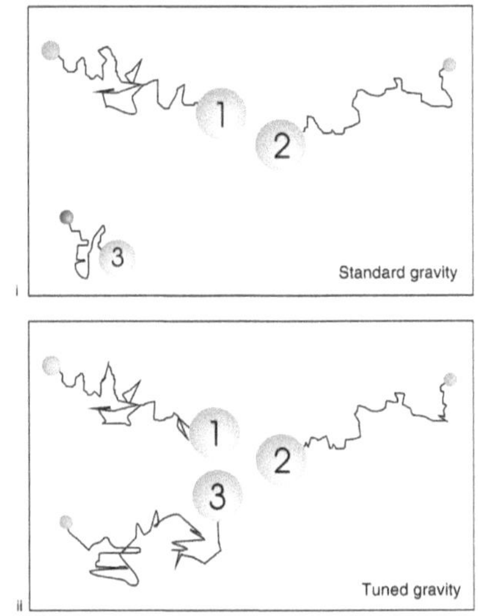

Fig. 3. (i) For *data set 1* particle projections illustrate that the "delayed" temporal association was not detected using the standard gravity method. (ii) Shifting the charges for neuron 3 relative to 1 and 2 by the time lag-to-peak in the corresponding cross-correlogram (Fig. 1B(ii)) results in aggregation of particle 3 with particles 1 and 2.

3. Results

The results are presented in two parts. The first section compares standard and tuned gravity analyses of data from simulated networks with known connectivity and from spike trains recorded in vivo. The second section presents the steps leading up to the 3D results using simulated spike trains from a relatively simple network.

3.1. Tuned gravity

Simulated networks with defined connections and mixed conduction times were used to generate spike trains for testing the tuning modification to the gravity method. Model *circuit 2* used to generate the spike data designated *data set 2*, is represented in Fig. 4(i). Each circle corresponds to a single neuron influenced by independent populations of 30 excitatory and 30 inhibitory "fiber" inputs (not shown). Additional excitatory fiber inputs (circles labeled "D") projected to the indicated subsets of target neurons; the fiber population with divergent connections to cells labeled 8 and 9 was intermittently on between the 120th and 140th seconds of the trial, which lasted 180 s. The indicated "long" conduction times were generated by post-simulation 10.0 ms shifts of spike occurrence times.

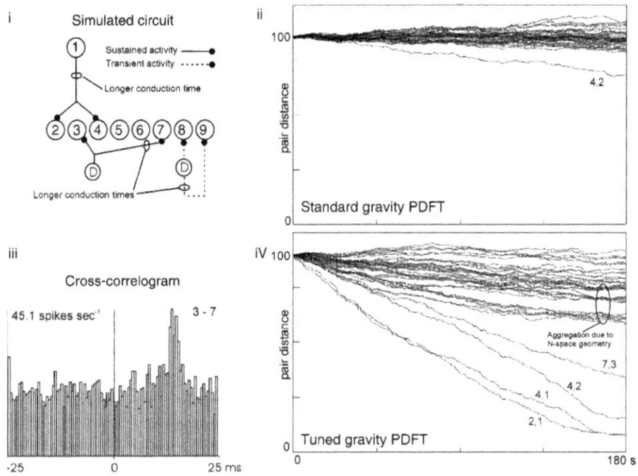

Fig. 4. Tuned gravity enhancements improve sensitivity and selectivity of the gravity algorithm in the detection of "known" mixed time-lag neuronal correlations. (i) Diagram of simulated circuit 2 with different conduction times used to generate spike train *data set 2* for evaluation of the 'tuning' algorithm. Each single neuron represented by a circle (1–9) received 30 independent excitatory and 30 independent inhibitory fiber inputs (not shown). Each excitatory fiber input had four synaptic terminals, all with an equilibrium potential of 115 mv, strengths of 0.0099, time constants of 1.5 ms, and a firing probability of 0.03 at each half ms time step in the simulation. Each inhibitory fiber input had five synaptic terminals, all with an equilibrium potential of −25 mv, synaptic strength of 0.0099, a time constant of 1.5 ms, and a firing probability of 0.05 at each half ms time step in the simulation. Other simulated connections included excitatory connections between neurons 1 and its two target cells, 2 and 4 (each with four terminals: equilibrium potential 115 mv; synaptic strength of 0.045, time constant, 1.5 ms). The circles labeled D represent additional populations of inputs to the indicated subsets of target neurons: Neuron pairs 3 and 7 and 8 and 9 each shared two additional excitatory fiber inputs (two terminals per fiber: equilibrium potential 115 mv; synaptic strength of 0.045, time constant, 1.5 ms and a firing probability of 0.03 at each half ms time step in the simulation). This additional input to neurons 8 and 9 was transiently active during an interval between 120 and 140 s into the simulation run. (ii) Particle distance as a function of time (PDFT) plot from standard gravity analysis of *data set 2* from *circuit 2* represented in panel (i). Each trace plots the interparticle distance (y-axis) for a particular pair as a function of time. Initially each particle was equidistant (100 arbitrary units) from all other particles. Standard gravity analysis revealed significant aggregation of particles 2 and 4, indicative of non-random spike timing relationships between the spike trains of the corresponding neurons. No other particles aggregated more than expected by chance; gravity parameters: acceptor and effector charge decays forward; charge decay time constant 5.0 ms; no rate normalization. Numbers of spikes in sample: (1) 1817; (2) 2203; (3) 3687; (4) 2243; (5) 1799; (6) 1826; (7) 3715; (8) 2004; (9) 2004. (iii) Example of cross-correlation histogram *for data set 2*. Firing rates (spikes s^{-1}) refer to the largest bin in the plot. Correlogram for pair 3–7 with offset peak; lag 13.5 ms; DI, 7.8; 3684 reference spikes, 3712 target spikes. (iv) Significant aggregation of additional particles was observed when charge kernels were tuned using time lags to corresponding correlogram peaks. All pairs of neurons linked with sustained connectivity in the simulated circuit were identified (labeled plots); the distances between some particle pairs representing uncorrelated (and unconnected) neurons also aggregated because of N-space geometry (plots grouped within the superimposed lasso). See Fig. 5 and text.

The particle distance as a function of time (PDFT) plot from standard gravity analysis of *data set 2* shown in Fig. 4(ii) indicated that only particles 2 and 4 aggregated significantly during the gravity run; lagged correlations were not detected. *Data set 2* was then evaluated after tuning the charge kernels for each pair, using the time lag to the pair's correlogram peak to define the tuning shift. Pre-screening with standard cross-correlation analysis detected significant features indicative of non-random timing relationships that reflected circuit connections. The time lags to peaks in correlograms matched the temporal offsets of the target neuron spike trains. For example, in Fig. 4(iii) the cross-correlogram peak for follower cell pair 3–7 shows a peak offset from the origin that reflects the different conduction times from the driver source to each of the follower cells.

Fig. 4(iv) shows that the tuning enhancement detected lagged correlations not identified by the standard gravity method. Here we used the same charge kernel size and time constants as in Fig. 4(ii). (*Note*: For all pairs without a significant correlogram feature, the standard placement of kernels at the times of spikes was used.) The "driver–follower" pairs 2–1 and 4–1 aggregated together, along with pair 4–2 representing the two follower cells. Particles representing the follower cell pair 3–7 also condensed. Significant aggregation of particles representing the transiently linked pair 8–9 was not observed.

The distances between some particle pairs representing uncorrelated (and unconnected) neurons in *data set 2* were also reduced, as indicated in the plots "bundled" by the superimposed "lasso" in Fig. 4(iv). Because of N-space geometry,

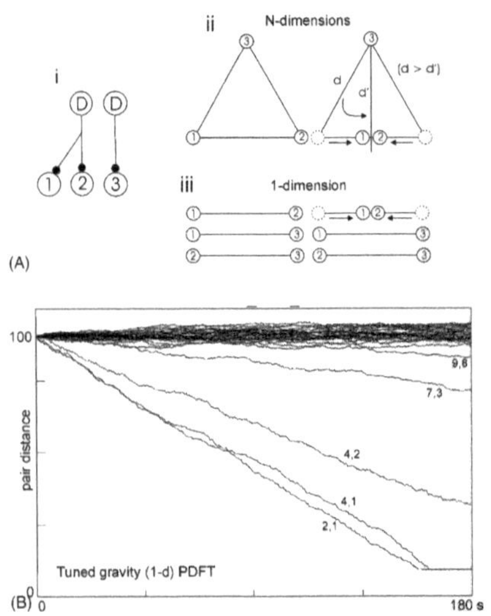

Fig. 5. (A) In the standard gravitational representation, aggregation due to correlation can also reduce distances between particles representing uncorrelated neurons because of N-space geometry. (i) Circuit 3 and corresponding *data set 3* represent two neurons with shared 'driver' (D) inputs and an independent third cell. This "dragging" effect is eliminated when each pair-wise correlation is evaluated in one-dimensional space; see text for details. (B) All network connections in *data set 2* (Fig. 4) including the transient (20 s) functional linkage between neurons 8 and 9, were detected when the tuning option was combined with the rule that limited particle movement to one-dimensional pair-wise charge products.

aggregation of some particles due to correlation may also reduce distances between particles representing uncorrelated neurons. This property is illustrated in tutorial Fig. 5A. Model *circuit 3* (Fig. 5A(i)) represents neurons 1 and 2 with shared inputs and an independent cell *3*. Fig. 5A(ii) shows tutorial *data set 3*: as corresponding particles 1 and 2 cluster in the N-space, their respective distances from particle 3 decrease slightly. This aggregation due to geometry is eliminated when each pair-wise correlation in *tutorial data set 3* is evaluated in one-dimensional space, using only each pairs' charge product to define the force acting on that pair, as shown in Fig. 5A(iii). By removing the effects of (random) charges on other particles and reducing the analytical space for each pair to one dimension, this option can suppress "noise" that limits selectivity and sensitivity of analyses, albeit at a cost of reduced information about overall assembly behavior.

Tuned gravity analysis of *data set 2* was run again, this time using the option that limited charge forces and movement for each pair to one dimension. Fig. 5B shows that all network connections, including the transient (20 s) functional linkage between neurons 8 and 9, were detected without false positives.

Similar results (not shown) were obtained with other data sets from simulated networks with defined connectivity and mixed conduction times. This work with simulated circuits was not intended to be an exhaustive exploration of the limits of sensitivity and selectivity of the tuning option. We note that the method did not detect transient lagged correlations similar to, but weaker than, that generated for pair 8–9 in *data set 2*. In those cases, the brief interval of correlated activity was "washed out" because of the time averaging inherent in the cross-correlation method. Hence, no charge kernel offset was applied for that pair. The results of the simulated circuit analysis did establish that tuning, particularly when used in conjunction with rules that use only single pair charges to define aggregation forces, can identify transient and mixed time-lag timing relationships missed by gravity parameters that would normally be used to detect spike synchrony with the gravity tool.

3.2. Tuned gravity analysis of in vivo data

With the results from the simulated circuits having established the utility of the tuning enhancement to the gravity method, we next applied it to in vivo multi-site spike train recordings. The aim was to determine if additional information on network properties could be extracted with the modifications. Spike train data from one animal, designated *data set 4*, consisted of the impulse times of eight simultaneously recorded brainstem; firing rate histograms are shown in Fig. 6A. *Data set 4* was first screened for lagged correlation features. Time lags to features detected with cross-correlation analysis were used to define the charge kernel time shift in subsequent tuned gravity calculations. A representative cross-correlogram for neurons (Fig. 6B) shows an offset peak with a negative 21 ms time lag. The correlated neuron pair 2–1 from *data set 4* is also represented in the PDFT plot shown in Fig. 6C. The correlation was not detected with standard gravity, but was identified using the tuned option with the same charge parameters.

Results for all pairs in *data set 4* from both standard and tuned gravity are presented in Fig. 7A(i) and (ii), respectively. Black regions in each row of the graphs indicate when the distance between the represented pair of particles was less than expected by chance ($p < 0.01$). The row indicated by the arrow on the left indicates the data from neuron pair 2–1 represented by the PDFT plots in Fig. 6C.

When tuned gravity forces were constrained to charge products for individual pairs and particle movements were limited to one dimension, the number of significantly aggregated pairs was reduced (compare Fig. 7A(ii) and (iii). These differences presumably reflect the loss of higher dimensional relationships and, potentially, a reduction in false positives associated with the N-space geometry.

All three plots in Fig. 7A indicate when the particles pairs had aggregated more than expected by chance. This con-

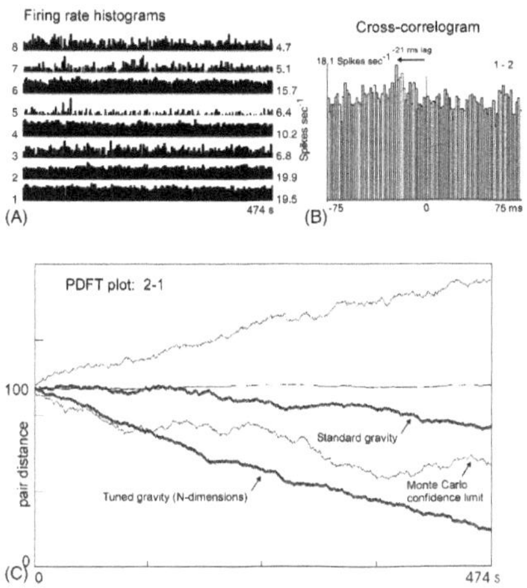

Fig. 6. The tuned gravity method identified functional associations not identified with the standard method among neurons recorded in vivo. (A) Firing rate histograms from eight simultaneously recorded, "spontaneously active" (no experimental stimulation), medullary raphe neurons designated *data set 4*. (B) Cross-correlogram for neurons 1 and 2 in this data set showing an offset peak with a 21 ms negative lag relative to the origin. (C) PDFT plots (labeled heavy lines) for spike trains 1 and 2 represented in panel B. The particles corresponding to neurons 1 and 2 did not aggregate significantly with the standard method; significant aggregation was detected after tuning. Significance of particle aggregation was evaluated using confidence limits defined with an empirical Monte-Carlo method using appropriate shuffling (Lindsey et al., 1992a). Here, the interspike intervals in each spike train were randomly shuffled to remove temporal correlations and the gravity calculation was repeated. This process was iterated 100 times. The top and bottom thin lines define maximum and minimum distances between particles in the 100 shuffled "control" data sets at each time step. Thus, the probability that the aggregation of particles 1 and 2 was due to random coincident spikes was less than 0.01. The middle thin line indicates the mean particle distance at each time step for the 100 control data sets. Gravity parameters: acceptor and effector charge decays forward; charge decay time constant 5.0 ms, no rate normalization. Numbers of spikes in sample: 1: 6937; 2: 6384; 3: 1589; 4: 3524; 5: 402; 6: 5378; 7: 727; 8: 1064.

densed "state" of the particles reflects their charge *histories*. However, at any moment, the particles' *aggregation velocities* are more directly related to the product of their charges and the represented spike synchrony of the corresponding neurons. Steps or jumps in the distances between particles in the gravitational representation mark transient moments of impulse synchrony in spike trains. Thus, dynamic associations among particular combinations of neurons are reflected in the fluctuations of aggregation velocities.

The plane surface shown in Fig. 7B(i) represents a 'snapshot' of gravity particle aggregation velocities for each pair of neurons from *data set 4* represented in Fig. 7A(ii) for tuned gravity analysis in N-dimensions. Velocities, calculated for each interval, are indicated by the relative heights (z-axis) of the respective peaks. The position of each peak along the x-axis indicates the time shift used to tune the charge kernels for the corresponding particles.

"Compressing" the plane in the y-axis, by summation of velocity peaks with the same lag, provides an abstracted "summary histogram" of the associations among the represented neurons during the corresponding time step in the analysis (here, step 667 of 1000 plotted steps). These summary histograms, when plotted sequentially and "stacked" in the y-axis as in Fig. 7B(ii), provided a view of the fluctuations in synchrony at different time lags for the neurons over the course of the experimental observation.

An increase in the number of detected mixed lag correlations and their dynamics was also observed when the tuning algorithm was applied to a second set of 10 simultaneously recorded medullary neuron spike trains (not shown).

3.3. Three-dimensional display

A fundamental problem in using the gravity transformation is how to best visualize the movements and positions of the particles in the N-space. Initial proposals used particle pair distances as a function of time or quasi-geometric projections from the N-space to a plane. A more recent paper has made use of parallel coordinates (Stuart et al., 2002). All these approaches identify aggregating particles (i.e. interacting neurons), but require additional work to identify neuronal groups larger than pairs. More important, these displays are not easily used to directly identify the nature of the non-random timing relationship in firing. We present here another method of display for the motions in the N-space that addresses some of these problems.

To demonstrate the new display we generated spike train data, designated *data set 5*, with a simulator set up to mimic model *circuit 4* shown in Fig. 8A. There are three groups, each consisting of three synaptically coupled neurons, and one independent neuron. In each of the three groups, one neuron drives the other two so that there are direct synaptic interactions as well as shared inputs. Firing rates and synaptic weights were set to values similar to those encountered in real data. No "external" perturbations as from a stimulus were used.

Fig. 8B(i) shows evolution of particle pair distances (PDFT) when identical effector and acceptor kernels are used. During this condition both members of any interacting particle pair move symmetrically towards each other in the N-space. If we examine pair distances at 50 s (arrow) there are three and possibly four groups of trajectories (lassos w–z). The most rapidly aggregating group (with greatest slope) corresponds to the six synaptically connected neuron pairs in circuit 5 (Fig. 8A). The next slower group corresponds to the three neuron pairs that are related by shared input. Separating these two groups is not that certain, and sorting the upper two trajectory groups according to functional relationships is impossible.

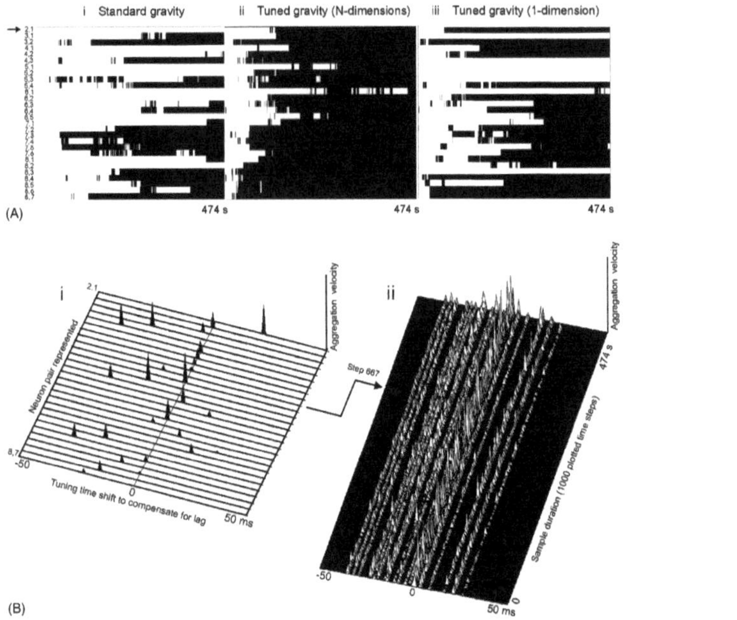

Fig. 7. (A) The three plots (i–iii) show further results of gravity analysis of *data set 4*. Black regions in each row of the graphs indicate when the distance between each pair of particles (from pair 2–1in the top row to pair 8–7 in the bottom row) was less than expected by chance ($p<0.01$) as described in the legend of Fig. 6. The three panels represent results for: (i) standard gravity analysis; (ii) tuned gravity analysis in N-dimensions; and (iii) tuned gravity analysis in one-dimension. The top row indicated by the arrow at left indicates data for the pair represented in Fig. 6B. (B(i)) The plane surface represents a 'snapshot' of gravity particle aggregation velocities for each pair of neurons represented in panel A(ii). Particle aggregation velocities, calculated for a particular interval, are indicated by the relative heights (z-axis) of the respective peaks. The position of each peak along the x-axis indicates the time shift used to tune the charge kernels for the corresponding particles. (B(ii)) Plane shows fluctuations in particle-pair aggregation velocities (z-axis) – and thus firing synchrony – as a function both of tuning time shifts (x-axis) and time in the recording (y-axis). Each peak in a y-axis experimental time "slice" (e.g. at the end of plotted time step 667) represents the sum of aggregation velocities for all pairs with the same kernel-tuning time shift in the corresponding pair-wise plot such as shown in the left plane.

Fig. 8B(ii) shows evolution of particle pair distances when time reversed effector and acceptor kernels are used. During the chosen condition, a particle representing a postsynaptic neuron moves toward the particle representing its presynaptic driver in the N-space; the driver particle stays essentially stationary. Relative movements of particles representing neurons that are not interacting move only because of geometric effects. The most rapidly aggregating trajectory group (lasso w) now corresponds to both the directly connected and shared input neuron pairs and does not separate them. The other trajectory groups are very difficult to sort out, particularly as they seem to separate and rejoin several times.

The advantage of the time reversed effector and acceptor kernels is only seen when we make the quasi-geometric projection from the N-space to a plane as shown in Fig. 9A. The geometric projection is not to a fixed plane in the N-space but to a plane that is dynamically defined by an appropriate function of the current position of all particles; hence apparently all particles move, not only the ones representing the postsynaptic neurons. We have not been able to devise a projection function of this type for more than ten neurons, so that such projections are useful only with recordings of small groups of neurons. The arrangement of initial projected particle positions is indicated in panel 9A(i). The final projected positions in panel 9A(ii) replicate accurately the functional groupings in the circuit diagram of Fig. 8A, but do not indicate the type of relationships within each group.

Optimal performance in terms of speed of aggregation for a given length data set is somewhat dependent on the neuron identities assigned to the starting positions in N-space.

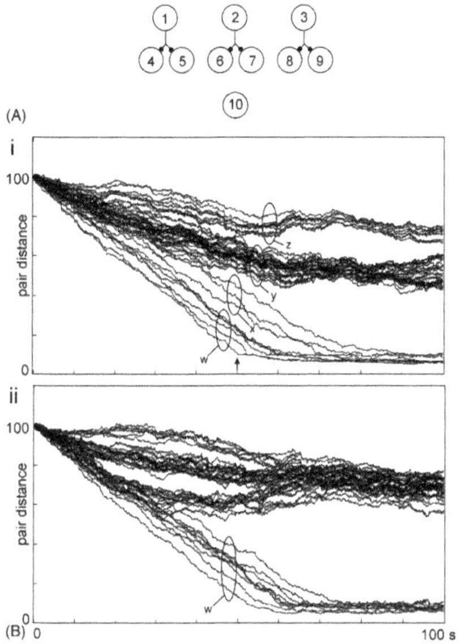

Fig. 8. (A) Schematic of spike simulator circuit used to generate *data set 5*. (B) (i) Particle pair distance trajectories. Symmetric kernels. *Both* particles move when spikes are in near coincidence irrespective of their temporal order. (ii) Pair distance trajectories. Time reversed kernels. Only particles with later spike (i.e. representing a *postsynaptic* neuron) are subjected to non-zero average force and move in the underlying 10-space (see text).

It is advisable to try several such assignments with unknown data. We note that there are other, non-geometric projection methods in the literature. Some of such methods, like projection pursuit, are computationally very intensive and their results depend strongly on the initial positions (Sammon, 1969; Jones and Gibson, 1987). A recently reported approach uses principal components analysis to reduce the particle trajectories in the N-space to a 2-space and also explores the use of parallel coordinates to show particle positions in the N-space (Stuart et al., 2002).

A completely different approach to visualizing the particle movements in the N-space is to keep track of not only the distance between particles but also the distance that each particle of the pair has moved from its starting point (the N particles in the N-space are initially made equidistant from each other). This means we have three numbers associated with each particle pair at each time point of the calculation; graphical display of the trajectories as these three numbers evolve can be made in a three-dimensional coordinate system as shown for *data set 5* in Fig. 9B. Again we have used the time reversed effector and acceptor kernels so that in the N-space any particle representing a post-synaptic neuron moves towards an almost stationary particle representing the corresponding pre-synaptic neuron.

The z-coordinate (vertical) measures the distance between particles of each ij pair. The x-coordinate is the distance the ith particle has moved from its starting point; the y-coordinate represents the distance the jth particle has moved from its starting point. As evident in Fig. 6B the trajectories fall into quite distinct bundles, final positions indicated by the color blobs. Identities in terms of the relationships in the circuit diagram of Fig. 8A are indicated, along with the number of such pair trajectories in parentheses.

The three trajectories that come down the principal diagonal represent the three neuron pairs with shared input (example 5,4); here the distance between particles of each pair has become small, and each particle has moved far from its starting point. Trajectories for the synaptically coupled pairs (example 4,1) also ended with small inter-particle distances but moved down along the left back wall of the cube since one particle did not move from its original position. The two sub-bundles here are an accidental result from the probabilistic generator program used by the simulator. Depending on the numbering of the neurons some or all of these trajectories could come down the right back wall. The point is that although the distance between the particles has become small, one particle stays close to its original position; this condition is satisfied by a descending trajectory on either back wall.

Inter-particle distances for all other pairs remained large, but the trajectories formed into bundles that could be related to the relationship (or independence) of the two neurons being represented. Thus, trajectories for pairs representing pre-synaptic and *other* post-synaptic neurons (example 6,1) stayed on the left back wall, one particle moving little. Trajectories for pairs representing post-synaptic and *other* post-synaptic neurons (example 6,4) came towards the top front corner since both particles moved from their starting positions. Trajectories for pairs representing independent and post-synaptic neurons (example 10,4) stayed on the right back wall, again one particle showing little movement. Finally trajectories for pre-synaptic and other pre-synaptic (example 2,1) or independent and pre-synaptic (example 10,1) stayed near the upper back corner, i.e. neither particle moved much, and the inter-particle distance remained large.

Maximum utility of the new 3D display is realized, as above, for mirror kernel rules, so that with a synaptic connection between two represented neurons, only one particle will move. In analysis of unknown data it is appropriate to first define any trajectory bundles in terms of the implied particle movements, i.e. movement of each particle from its starting point and the distance between the two particles. Then, using information from distance plots and projection plots which identify neuron pairs that have excess near synchrony, it is possible to start assigning functional relationships as done in the above example. There will be ambiguity about any trajectories that stick to a back wall, since which of the two particles

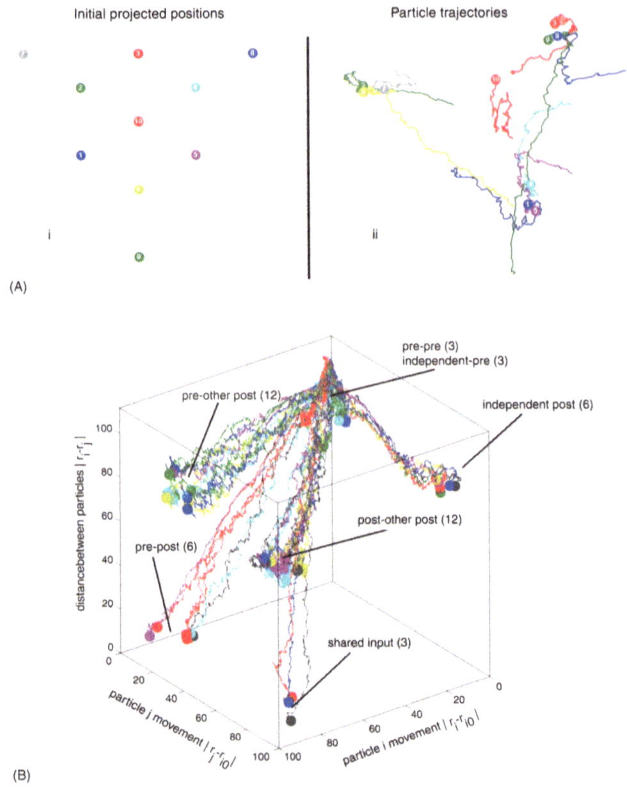

Fig. 9. (A)(i) Initial projected positions of particles for gravity analysis of *data set 5*. (ii) Projected particle trajectories. Only particles with later spike (i.e. representing a *postsynaptic* neuron) are subjected to non-zero average force and move in the underlying 10-space. All particles appear to move in the projection plane because the plane is dynamically defined by the evolving current position of all particles (see text). (B) Three-dimensional representation of pair trajectories. z-coordinate: distance between the particles in pair, x-coordinate: distance one particle has moved from its origin position, y-coordinate: distance the other particle has moved from its origin position. Only the particles representing *postsynaptic* neurons are subjected to a non-zero average force and show significant movement (see text).

does *not* move depends on the numbering scheme. Thus, in the above example the trajectory bundles for independent-post and pre-other post could be mixed on either back wall, but there would be no other functional interpretations. Similarly, the pre-post bundle could have appeared on either or partially on both back walls, although this also would not affect the functional interpretation. With the mirror kernel rules, functional assignment of the remaining three trajectory bundles is unambiguous, and would not depend on the numbering scheme.

4. Discussion

4.1. Tuned gravity

The gravity tuning procedure allowed the use of short charge decay time constants in the gravity calculations, thereby maintaining high selectivity for non-random temporal relationships without diminishing the sensitivity for temporally delayed interactions. This enhancement revealed mixed time lag correlations in brainstem neuronal assemblies, confirming and extending previous observations realized with the gravitational representation (Lindsey et al., 1992b, 1997); such correlations may be associated with dynamic gain control and memory processes (Arata et al., 2000; Morris et al., 2001, 2003). The "reduced dimension" option, which constrained charge products to individual particle pairs and limited movements to one dimension, removed two sources of noise and reduced the likelihood of both type one (false positive) and type two (false negative) errors. Aggregation solely due to the properties of N-space geometry was blocked.

The tuning method improves the ability to detect and evaluate mixed time-lag fluctuations in neuronal interactions. Because the "default" for a neuron pair is no tuning, this enhancement does not decrease the possibility of detecting neurons that are weakly or transiently correlated; the advantages of standard gravity still apply. We were cognizant of the possibility that brief correlations would not be detected during the initial screening with standard cross-correlation, having previously unmasked transient relationships not identified by cross-correlation when the gravity method was applied to the same data (Lindsey et al., 1992b). A complete initial screen requires a systematic "brute force" series of gravity analyses that step through all physiologically relevant time lags (e.g., Keating and Gerstein, 2002).

Tuning should not be limited to pairs with offset cross-correlogram peaks indicative of shared influences or excitatory interactions; it is also appropriate when offset correlogram troughs or dips suggestive of inhibitory process are detected. Indeed, because of the relative insensitivity of spike train analysis methods in general for detecting inhibition, tuning should be advantageous.

4.2. Three-dimensional display

In using the gravity transformation the N-dimensional space in which particles move precludes direct visualization and interpretation of trajectories. The evolution of particle pair distances with various rules for the gravity transformation (as in Fig. 8B) can be used to detect neuron pairs that are firing with near synchrony, and more particularly to detect changes in neuronal synchrony during the recording. However, it is hard to sort out different functional groupings if more than one is present, and it is difficult to identify the types of interaction between neurons.

A more direct approach to the possibility of multiple groupings is given by the projection of particle positions from the N-space to a suitably defined 2-space (a plane). This is obviously not a unique process and depends critically on the method of projection and the choice of final plane. The particular choices shown in Fig. 9A worked well, but would not be expected to be optimal for all situations. However, although projections of this sort may deal with multiple groupings, they cannot easily resolve the nature of the interactions. Fig. 9A shows the functional neuronal groupings correctly but does not directly indicate the synaptic relationships within each group.

By retaining the distance between particles in the N-space as well as the distance each has moved from its original position, and by displaying the history of these three variables as trajectories in a cube (Fig. 9B) we can interpret the trajectories in terms of the type of temporal interactions among the neurons. If there are changes in the interactions during the recorded data the trajectories will indicate that. But in this display we lose the direct explicit demonstration of multiple functional groupings. It remains important to make use of all available display methods.

In summary, there are by now a considerable number of rule modifications of the original gravity analysis that serve to improve various aspects of selectivity or sensitivity under various circumstances. The additional modifications presented in this paper complement those previously described, and should serve to increase the flexibility of the gravity tool approach to analysis of neural network dynamics. Interpretation of the gravity transformation continues to require several different partial views of what transpires in the N-space as exemplified by the several variations in this paper. We note that the gravity transformation in any of its various forms, through movements and interactions of particles with spike train related charges, defines a time varying N-dimensional metric on temporal correlation of all simultaneously recorded spike trains. Future development of multi-variate or graphical methods that directly analyze this metric and the evolution of particle clustering behavior in the full N-space are essential.

Acknowledgements

We thank two anonymous reviewers for their helpful comments, and R. Shannon, K.F. Morris, and P. Barnhill for their assistance in the acquisition and management of the in vivo data. This work was supported by Javits Neuroscience Investigator Award NS19814, grant NS46062 as part of the NSF/NIH Collaborative Research in Computational Neuroscience Program, NIH MERIT grant MH46428 and grant DC01249.

References

Adrian ED. The mechanism of nervous action. In: Electrical studies of the neurone. London: Oxford University Press; 1932. p. 103.

Arata A, Hernandez YM, Lindsey BG, Morris KF, Shannon R. Transient configurations of baroresponsive respiratory-related brainstem neuronal assemblies in the cat. J Physiol (London) 2000;525:509–30.

Baker SN, Gerstein GL. Improvements to the sensitivity of gravitational clustering for multiple neuron recordings. Neural Comput 2000;12:2597–620.

Brown EN, Kass RE, Mitra PP. Multiple neural spike train data analysis: state-of-the-art and future challenges. Nat Neurosci 2004;7:456–61.

Buzsáki G. Large-scale recording of neuronal ensembles. Nat Neurosci 2004;7:351–446.

Dayhoff JE. Synchrony detection in neural assemblies. Biol Cybern 1994;71:263–70.

Eggermont JJ. The correlative brain: theory and experiment in neural interaction. Berlin: Springer-Verlag; 1990. p. 307.
Gerstein GL. Functional association of neurons: detection and interpretation. In: Schmitt FO, editor. The neurosciences: second study program. New York: The Rockefeller University Press; 1970. p. 648–61.
Gerstein GL, Aertsen AMHJ. Representation of cooperative firing activity among simultaneously recorded neurons. J Neurophysiol 1985;54:1513–28.
Gerstein GL, Kirkland KL. Neural assemblies: technical issues, analysis, and modeling. Neural Netw 2001;14:589–98.
Gerstein GL, Perkel DH, Dayhoff JE. Cooperative firing activity in simultaneously recorded populations of neurons: detection and measurement. J Neurosci 1985;5:881–9.
Gerstein GL. Correlation based analysis methods for neural ensemble data. In: Nicolelis MAL, editor. Methods for neural ensemble recordings. Boca Raton: CRC Press; 1999. p. 157–78.
Gerstein GL. Cross correlation measures of unresolved multi-neuron recordings. J Neurosci Methods 2000;100:41–51.
Jones M, Gibson R. What is projection pursuit (with discussions). J R Stat Soc 1987;150:1–36.
Keating JG, Gerstein GL. Fingerprinting network activity through the use of the gravitational transformation of spike trains. Program No. 610.4. 2002 Abstract Viewer/Itinerary Planner. Washington, DC: Society for Neuroscience, 2002 [Online].
Keating JG, Gerstein GL. Network activity in rat somatosensory cortex during reaching, Submitted for publication.
Li Z, Morris KF, Baekey DM, Shannon R, Lindsey BG. Multimodal medullary neurons and correlational linkages of the respiratory network. J Neurophysiol 1999;82:188–201.
Lindsey BG, Gerstein GL. Dynamics of mixed time lag correlations: detection using modifications of the gravity method. FASEB Exp Biol 2002. Abst. Program No. 127.1.
Lindsey BG, Gerstein GL. Tuned gravity method for analysis of neuronal assemblies. Soc Neurosci Abst 2000;26. Program No. 748.7.
Lindsey BG, Hernandez YM, Morris KF, Shannon R, Gerstein GL. Respiratory related neural assemblies in the brain stem midline. J Neurophysiol 1992a;67:905–22.
Lindsey BG, Hernandez YM, Morris KF, Shannon R, Gerstein GL. Dynamic reconfiguration of brain stem neural assemblies: respiratory phase-dependent synchrony versus modulation of firing rates. J Neurophysiol 1992b;67:923–30.
Lindsey BG, Morris KF, Segers LS, Shannon R. Respiratory neuronal assemblies. Respir Physiol 2000;122:183–96.
Lindsey BG, Morris KF, Shannon R, Gerstein GL. Repeated patterns of distributed synchrony in neuronal assemblies. J Neurophysiol 1997;78:1714–9.
MacGregor RJ. Neural and brain modeling. New York: Academic Press; 1987. p. 643.
Morris KF, Baekey DM, Nuding SC, Segers LS, Shannon R, Lindsey BG. Invited review: neural network plasticity in respiratory control. J Appl Physiol 2003;94:1242–52.
Morris KF, Shannon R, Lindsey BG. Changes in medullary neurone firing rates and synchrony following induction of respiratory long-term facilitation in the cat. J Physiol (London) 2001;532:483–97.
Nicolelis MAL. Methods for neural ensemble recording. Boca Raton: CRC Press; 1999. p. 257.
Perkel DH, Gerstein GL, Moore GP. Neuronal spike trains and stochastic point processes. II. Simultaneous spike trains. Biophys J 1967;7:419–40.
Sammon JW. A nonlinear mapping for data structure analysis. IEEE Trans Comput 1969;C-18:401–9.
Stuart L, Walter M, Borisyuk R. Visualization of synchronous firing in multi-dimensional spike trains. BioSystems 2002;67:265–79.

Chapter 7
Dynamics of Neuronal Interactions

Interactions Between Neurons and with George Gerstein

Ad Aertsen

I first met George Gerstein in the early summer of 1981 when George visited Peter Johannesma, my PhD supervisor at the University of Nijmegen. George and Peter were old friends from the times they both worked on random walk models of single-neuron spiking activities. I used the occasion to proudly present George a copy of my PhD thesis that I had just submitted to the Physics Department a week earlier. Also, I collected all my courage to ask him if he could imagine me doing a postdoc in his lab at UPenn. George's immediate response to my question was typically George, as I would learn later on. Yes, he would be willing to give this a thought, but not before I had not visited his lab plus at least three other labs and tested the waters at each of these.

Thus, later that year, after the defense of my thesis and a vacation with my family following it, I organized my first trip to the USA, to visit George's lab in Philadelphia, plus several other labs at the East and West Coasts. Needless to say that also after these several lab visits, I stuck to my original plan to do my postdoc with George. Since George was kind enough to live with that prospect, we sat down and drafted a plan for my project, both regarding its science and for obtaining funding for it. As it turned out, the first was the easiest, the second quite a bit harder.

The reason I wanted to do my postdoc with George and learn from his expertise with multiple-neuron recordings and their analysis resided in a disappointing finding in my PhD thesis work. There I had studied the stimulus-response relations of neurons in the auditory systems of cats and frogs. Using cross-correlation analysis

A. Aertsen (✉)
Bernstein Center Freiburg and University of Freiburg, Freiburg, Germany
e-mail: aertsen@biologie.uni-freiburg.de

A. Aertsen et al. (eds.), *Introducing Computation to Neuroscience*, Springer Series in Computational Neuroscience, https://doi.org/10.1007/978-3-030-87447-6_7

along the Wiener–Volterra framework on nonlinear systems theory, we measured and compared the spectro-temporal receptive fields (STRF) of neurons for different types of acoustic stimuli, including Gaussian white noise, randomized sequences of tones, and a variety of natural sounds (Aertsen and Johannesma, 1981a). Much to our chagrin, however, and in spite of applying the most sophisticated corrections for nonlinearities we could think of at the time, we found that many of the neurons' receptive fields, especially those of neurons in more central areas, were not stimulus-invariant. That is, receptive fields of neurons determined with one stimulus type, say tones, did not allow us to make meaningful predictions regarding these neurons' receptive fields determined with another stimulus type, say natural sounds (Aertsen and Johannesma 1981b). After thinking about this negative result for a long time, I had come to the conclusion that we had reached the boundaries of meaningful single-neuron studies. Instead, we needed to study multiple neurons at a time and try to incorporate the presumably nonlinear, stimulus-dependent interactions between these neurons to understand and interpret their receptive field properties under varying stimulus conditions. Having read George's papers on neuronal interactions and multiple-neuron studies, it was obvious to me that I should turn to his expertise, if I ever wanted to make relevant progress on this ambitious research goal—one that would, in fact, occupy me for the rest of my scientific life.

Having settled the science part, it was time to find funding for enabling me to do this postdoctoral project, that is, to raise enough funding to pay for my salary to help me sustain my small family (my partner Mieke and our daughter Woosje) during this expedition. I soon had to find out, though, that Dutch funding agencies at the time had a peculiar condition for funding a postdoctoral project abroad: funding was only provided to candidates who could provide convincing evidence of a position in Dutch academia after returning from the postdoc. As I could not deliver any such evidence and EU funding did not exist yet at the time, I had to turn to George for help. And, very kindly, George helped me by paying my salary from a newly acquired research grant and by paying for my travel to the USA by inviting me as a speaker to a conference he organized, together with Don Perkel, at Stanford University. In retrospect, I suspect that the unfriendly conditions of the Dutch funding agencies contributed to the fact that, after my postdoc, I never considered returning to the Netherlands.

Having science and funding assured, we now only needed to get my family and me to the USA. Here, a new interesting problem presented itself. Since Mieke and I were not married at the time, my postdoc J1-visum did not allow her and our daughter to come and live with me in the USA. Once again, George helped out: via a befriended Professor of Cultural Anthropology at Haverford College, Mieke magically obtained a J1-visum for an Anthropology research project there. This was one of my first encounters with "fake news" from the USA, but one that helped us tremendously—all in the spirit of George's motto: "Up the opposition!" When I then inquired, whether we would need any documents to bring our daughter's cat, George replied that it would be helpful if we could document that our cat did not have any obvious communist inclinations.

Once I arrived in Philadelphia in May 1983, I was given a desk in one of George's lab offices in the Richards Building at Hamilton Walk. I learned that this building was a National Historic Landmark in American architecture, designed by the famous architect Louis Kahn. To be honest, I never perceived it as a landmark. To me, it felt more like a run-down, over-crowded, derelict research building, close to its collapse. But here, I also learned from my fellow students an interesting US technique to control the office temperature: by carefully adjusting and upregulating both the heating and the air conditioning, any desirable temperature could be reached and maintained, irrespective of outdoor conditions—and irrespective of energy demands. Luckily enough, I did not spend too much time in this office and, instead, spent most of my time in George's lab, down in the basement. Here, George had his experimental setup and also his computers. In fact, when I arrived, George had just bought a new big computer from his latest research grant. The first thing he did was to open it up, complain about its terrible design, and start to rewire it from scratch. This took him several weeks, until he was satisfied with its performance. The computer, afterward, never looked the same and remained in a somewhat "opened" condition—but it worked.

During my postdoc with George, I mainly worked on two projects with him. The first was a reasonably straightforward study on the differential sensitivity of cross-correlation analysis of two-neuron spike train recordings for excitatory vs. inhibitory interactions among the recorded neurons in the low firing rate regime, typical for the neocortex (Aertsen and Gerstein 1985). Here, we learned that from analyzing such recordings, excitatory interactions should be much easier to detect than inhibitory ones, determined by the low firing rate regime (in a high firing rate regime, it should be the other way around)—a finding that was supported and also later corroborated by many physiological experiments. In the second project, we followed up on an earlier study by George, together with Judy Dayhoff and Don Perkel, in which they had developed a new method of analyzing spike time correlations between any number N of simultaneously recorded neurons (Gerstein et al. 1985a). In this method, called "gravitational clustering," they had mapped the problem of finding temporal correlations among the spike times of N neurons into the spatial problem of finding clustering among hypothetical charged particles in an N space, with the time course of the particles' charges being determined by the corresponding neurons' spike time recordings. My contribution to this second project consisted of two parts (Gerstein and Aertsen 1985). First, I played around with different temporal rules of charge assignment, trying to figure out, if such rules could help teasing apart the signatures of different topologies of interactions among the neurons, such as direct connections, shared inputs, and other types of network motifs. Second, in an attempt to go beyond pair-distance analyses, I tried visualizing the particles' trajectories in the N space by studying different (nonlinear) projections onto an appropriately chosen (possibly moving) plane, such that visual inspection of the resulting movies (Fig. 7.1) could possibly help us discerning different patterns of interactions among the neurons. For a more in-depth discussion of the gravitational clustering method, I refer to the chapters in this book by Stuart Baker (Chap. 13, this volume), who later improved the gravitational clustering method and made it

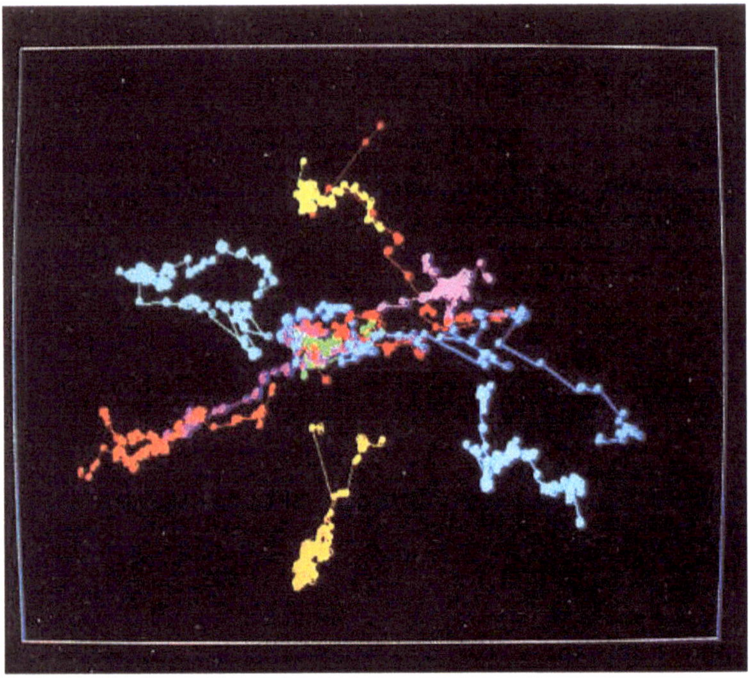

Fig. 7.1 2D projection of the trajectories of 10 particles, representing the spiking activities of 10 simultaneously recorded neurons. The closer the particles at the end of "gravitational clustering," the more correlated the spiking activities of the corresponding neurons

computationally more efficient, and by Bruce Lindsey (Chap. 5, this volume), who applied the method to his recordings from respiratory networks in the cat brain stem.

After finishing my postdoc with George in August 1984, I returned to Europe, to start my own group in the lab of Valentino Braitenberg at the Max Planck Institute in Tübingen. Here, among other things, I continued to work on the gravitational clustering approach, trying to apply it to multiple-unit recordings from monkey visual cortex, made by Jürgen Krüger in Freiburg and, together with Tobias Bonhoeffer, my first PhD student in Germany, studying the clustering dynamics by trying to normalize the correlation dynamics for stimulus-induced spike rate variations of the neurons recorded from. In the course of this study, we found that the math of the gravitational clustering rules could be relatively simply mapped onto the evolution of a neuronal network connectivity matrix, subject to a Hebbian plasticity rule. The results of this study, both the theoretical findings and the results from analyzing the monkey recordings, we published in the Proceedings Book of a beautiful conference in Naples and Amalfi (Aertsen et al. 1987), organized by Eduardo Caianiello (according to Valentino, his co-organizer, financed by whitewashed Mafia funds). We also discovered that publishing our results this way helped to ensure that nobody ever noticed our insightful findings.

Meanwhile, the confounding effects of stimulus- and behavior-induced nonstationarities of neuronal firing rates on correlation-based measures of their interactions kept bothering me. It was obvious that these interactions were dynamic, but how to separate the interaction dynamics from the individual neurons' firing rate dynamics? I tried several things but did not really make much progress. Then, in the autumn of 1986, George Gerstein joined us in Tübingen, for a sabbatical year in Valentino's lab—and we decided to make this a main issue. Building on his earlier work with Don Perkel, on drawing two-dimensional raster displays of the joint spiking of pairs of neurons (Gerstein and Perkel 1969, 1972), we developed a technique to quantify such joint spiking diagrams and, more importantly, to normalize the observed "raw" correlations for any underlying stimulus- and/or behavior-induced modulations of the neurons' firing rates (Aertsen et al. 1989). The result we called the Joint-PST (Fig. 7.2), a 2D histogram, also in reference to the PST (the Peri-Stimulus Time histogram), developed many years earlier by George and Nelson Kiang to capture the single-neuron spiking dynamics in response to a sensory stimulus (Gerstein and Kiang 1960). To evaluate the significance of features in the JPST histogram, we used a measure called "surprise." This was originally introduced into neuroscience by Charles Legéndy and further developed by my Tübingen colleague Günther Palm.

Fig. 7.2 George Gerstein (l.) and Ad Aertsen, inspecting early results of Joint-PST analysis of spike trains from two simultaneously recorded neurons. Picture taken early 1987, during the sabbatical year of George at the MPI for Biological Cybernetics in Tübingen

In fact, during George's sabbatical in Tübingen, the three of us together wrote a paper on how to measure the statistical significance of spike correlations (Palm et al. (1988); for further details I refer to Günther's Chapter in this book). Finally, to make sure that George also had some hardware to play with in Tübingen, Andreas Kreiter set out to build a real-time single-board spike sorter, in his Diploma thesis project under George's and my supervision (Kreiter et al. 1989).

We applied the Joint-PST in a number of studies, first on recordings from simulated networks to calibrate its performance, and then on various sets of experimental data from in vivo spike train recordings. However, it took me until 1991, during a sabbatical year in the labs of Eilon Vaadia and Moshe Abeles at the Hebrew University in Jerusalem, to make a convincing case for truly dynamical, stimulus- and behavior-dependent neuronal interactions in the brain of an awake, behaving monkey (Aertsen et al. 1991)—a result we were very excited about, but which took us several more years to get it decently published (Vaadia et al. 1995).

Also after this sabbatical year in Tübingen, George and I remained in close contact. Thus, he spent several extended research visits in my lab, first at the Weizmann Institute in Israel (a visit he combined with a longer stay in Moshe Abeles' lab at the Hebrew University, where they continued their search for repeating spike patterns, see Moshe's Chapter in this book), and later also at the University of Freiburg, after I moved there in 1996. Over the years, I was fortunate enough to be able to write several more papers and book chapters with George (Aertsen et al. 1986; Aertsen and Gerstein 1991; Gerstein et al. 1985b, 1989). I remember each of them with fondness and as a privilege to be able to interact with an incredibly creative and rigorous mind. I am grateful for this—and I miss him.

References

Aertsen AMHJ, Gerstein GL (1985) Evaluation of neuronal connectivity: Sensitivity of cross-correlation. Brain Res 340(2):341–354. https://doi.org/10.1016/0006-8993(85)90931-X

Aertsen AMHJ, Gerstein GL (1991) Dynamic aspects of neuronal cooperativity: Fast stimulus-locked modulations of 'effective connectivity'. In: Krüger J (ed) Neuronal cooperativity. Springer series in synergetics, vol 49, pp 52–67. https://doi.org/10.1007/978-3-642-84301-3_4

Aertsen AMHJ, Johannesma PIM (1981a) The spectro-temporal receptive field: A functional characteristic of auditory neurons. Biol Cybern 42(2):133–143. https://doi.org/10.1007/BF00336731

Aertsen AMHJ, Johannesma PIM (1981b) A comparison of the spectro-temporal sensitivity of auditory neurons to tonal and natural stimuli. Biol Cybern 42(2):145–156. https://doi.org/10.1007/BF00336732

Aertsen A, Gerstein GL, Johannesma P (1986) From neuron to assembly: Neuronal organization and stimulus representation. In: Palm G, Aertsen A (eds) Brain theory. Springer, Berlin, Heidelberg. https://doi.org/10.1007/978-3-642-70911-1_3

Aertsen A, Bonhoeffer T, Krüger J (1987) Coherent activity in neuronal populations: Analysis and interpretation. In: Caianiello ER (ed) Physics of cognitive processes, vol 42. World Scientific Press, Singapore, pp 1–34

Aertsen AM, Gerstein GL, Habib MK, Palm G (1989) Dynamics of neuronal firing correlation: modulation of "effective connectivity". J Neurophysiol 61(5):900–917. https://doi.org/10.1152/jn.1989.61.5.900

Aertsen A, Vaadia E, Abeles M, Ahissar E, Bergman H, Karmon B, Lavner Y, Margalit E, Nelken I, Rotter S (1991) Neural interactions in the frontal cortex of a behaving monkey: Signs of dependence on stimulus context and behavioral state. J Hirnforsch 32(6):735–743

Gerstein GL, Aertsen AM (1985) Representation of cooperative firing among simultaneously recorded neurons. J Neurophysiol 54(6):1513–1528. https://doi.org/10.1152/jn.1985.54.6.1513

Gerstein GL, Kiang NS (1960) An approach to the quantitative analysis of electrophysiological data from single neurons. Biophys J 1(1):15–28. https://doi.org/10.1016/s0006-3495(60)86872-5

Gerstein GL, Perkel DH (1969) Simultaneously recorded trains of action potentials: Analysis and functional interpretation. Science 164(3881):828. https://doi.org/10.1126/science.164.3881.828

Gerstein GL, Perkel DH (1972) Mutual temporal relationships among neuronal spike trains: Statistical techniques for display and analysis. Biophys J 12(5):453–473. https://doi.org/10.1016/S0006-3495(72)86097-1

Gerstein GL, Perkel H, Dayhoff E (1985a) Cooperative firing activity in simultaneously recorded populations of neurons: Detection and measurement. J Neurosci 5(4):881–889. https://doi.org/10.1523/JNEUROSCI.05-04-00881.1985

Gerstein GL, Aertsen A, Bloom M, Espinosa I, Evanczuk S, Turner M (1985b) Multi-neuron experiments: Observation of state in neural nets. In: Haken H (ed) Complex systems - Operational approaches. Springer, Berlin, Heidelberg, pp 58–70

Gerstein GL, Bedenbaugh P, Aertsen AMHJ (1989) Neuronal assemblies. IEEE Trans Biomed Eng 36(1):4–14. https://doi.org/10.1109/10.16444

Kreiter AK, Aertsen AMHJ, Gerstein GL (1989) A low-cost single-board solution for real-time, unsupervised waveform classification of multineuron recordings. J Neurosci Methods 30(1):59–69. https://doi.org/10.1016/0165-0270(89)90075-7

Palm G, Aertsen AMHJ, Gerstein GL (1988) On the significance of correlations among neuronal spike trains. Biol Cybern 59(1):1–11. https://doi.org/10.1007/BF00336885

Vaadia E, Haalman I, Abeles M, Bergman H, Prut Y, Slovin H, Aertsen A (1995) Dynamics of neuronal interactions in monkey cortex in relation to behavioural events. Nature 373:515–518. https://doi.org/10.1038/373515a0

it is possible to predict from this study that single-kernel differences would be heritable and that breeding programs, such as mass selection and single recurrent selection which make use of high heritability, would probably be effective for modification of corn oil quality. Consequently, a high-resolution NMR selection technique for unsaturation may prove as advantageous as the wide-line NMR single-kernel technique in speeding the development of new corn strains.

In 1962 Johnson and Shoolery (*8*) described a high-resolution NMR method for determining the average molecular weight and iodine value of natural fats dissolved in carbon tetrachloride. With this method the NMR integral signal from C(1) and C(3) glyceride protons provides an internal standard from which the olefinic and total number of hydrogen atoms are measured. From these, an average molecular weight and iodine value are calculated. Agreement between NMR and Wijs iodine values is remarkably good.

To demonstrate viability under the conditions of our method, we dried whole kernels of 25 commercial corn hybrids to 4 percent moisture and subjected them to 95°C for periods up to 90 minutes. Others were soaked in Freon-113 for periods up to 6 days. Subsequent germination tests showed a high degree of tolerance to low moisture heating (94 percent survival) and to the solvent treatment (98 percent survival).

For NMR experiments proposed earlier (*9*), individual corn kernels or excised germ were mounted on a Teflon-glass pedestal, transferred to 12-mm tubes and covered with Freon-113 to minimize the large discontinuities in magnetic susceptibilities within the sample cell. Samples were examined with a Varian model HA-100-15 high-resolution spectrometer equipped with a wide gap magnet. Typical spectra from kernels of commercial-dried (15 percent corn moisture), laboratory-dried (4 percent corn moisture), and laboratory-dried corn soaked in Freon-113 for 6 days are shown in Fig. 1 with NMR chemical shift assignments. The spectrum from the commercial-dried kernel (Fig. 1A) was not usable because of the broad absorption signal. The laboratory-dried kernel (Fig. 1B) exhibited a poor NMR spectrum; resolution and signal-to-noise ratio were less than satisfactory. Surprisingly, heating the dry sample up to 95°C to alter the mobility of the fat did not improve resolution. By contrast, solvent treatment of dried seeds altered the mobility of the oil sufficiently to provide very good signal-to-noise ratios and adequate resolution to separate the olefinic and glyceride methylene protons (Fig. 1C). Proper integration of spectra from single seeds permits determination of iodine value and average molecular weight by means of the Johnson-Shoolery (*8*) calculation. These data indicate that sample drying and solvent treatment are necessary to obtain usable spectra. Variations in NMR spectra resolution have been observed between seeds which are attributed to differences in magnetic susceptibilities between the oil and the corn germ matrix as well as between the seed and surrounding solvent. Minimizing these should improve spectra resolution.

The spectra shown in Fig. 1 were obtained with 250-second sweep time. Through use of NMR radio-frequency pulse techniques and Fourier transformation of the data (*10*), it is possible to greatly reduce instrument time. Studies with the Varian FT-100 Fourier transform accessory (Fig. 2) show a 10-second examination of an excised corn germ which was fitted into a 5-mm tube and surrounded by Freon-113. These results suggest that unsaturation in the whole kernel can be measured in times as short as 10 seconds.

T. F. CONWAY
George M. Moffett Research Laboratories, Corn Products Company, Argo, Illinois

L. F. JOHNSON
Instrument Division, Varian Associates, Palo Alto, California

References and Notes

1. T. F. Conway and R. J. Smith, in *Developments in Applied Spectroscopy*, J. R. Ferraro and J. S. Ziomek, Eds. (Plenum Press, New York, 1963), p. 115; D. E. Alexander, L. Silvela S., F. I. Collins, R. C. Rogers, *J. Amer. Oil Chem. Soc.* **44**, 555 (1967).
2. I. Zupančič *et al.*, *Acta Chem. Scand.* **21**, 1664 (1967).
3. L. F. Bauman, T. F. Conway, S. A. Watson, *Science* **139**, 498 (1963); *Crop Science* **5**, 137 (1965).
4. J. Dumanović and V. Trifunović, *Contemp. Agr.* (Novi Sad) **7-8**, 521 (1964).
5. L. Silvela S., "Improvement in recurrent selection methods for oil in maize through nuclear magnetic resonance analysis," thesis, University of Illinois, Urbana (1965).
6. B. Brimhall and G. F. Sprague, *Cereal Chem.* **28**, 225 (1951).
7. C. G. Poneleit, "Diallel analysis of fatty acids in corn (*Zea mays* L.) oil," thesis, Purdue University, Lafayette, Ind. (1968).
8. L. F. Johnson and J. N. Shoolery, *Anal. Chem.* **34**, 1136 (1962).
9. A. L. Elder, D. M. Rathmann, T. F. Conway, *Qual. Plant. Mat. Veg.* **15**, 67 (1964).
10. R. R. Ernst, in *Advances in Magnetic Resonance*, J. S. Waugh, Ed. (Academic Press, New York, 1966), vol. 2, p. 108.

9 January 1969

Simultaneously Recorded Trains of Action Potentials: Analysis and Functional Interpretation

Abstract. *A new kind of statistical display, the joint peri-stimulus-time scatter diagram, facilitates the analysis and interpretation of two or more simultaneously recorded trains of action potentials. The display is a generalization of the cross correlation and the peri-stimulus-time histogram, and it reflects specific underlying neuronal interactions. The technique yields quantitative measures of interaction in terms of effectiveness of synaptic connections.*

Much of the known picture of the nervous system has been obtained from single neurons individually studied for an extended period of time. On the other hand, anatomical, physiological, and behavioral evidence overwhelmingly suggest that integrative functions in the nervous system are performed by simultaneous activity in groups of neurons. It has become possible to record several trains of action potentials (spike trains) simultaneously and to analyze their temporal relationships. The results of these analyses may, in turn, be interpreted in terms of connections which underlie the integrative process in the group of neurons under observation (*1*, *2*). We describe here a method of performing such analyses for experiments in which a repeated stimulus is presented to a system of neurons and in which two or more spike trains are simultaneously monitored. This method is more powerful and sensitive than that suggested by Perkel *et al.* (*2*, p. 429) and permits a relatively direct functional interpretation.

Our basic data consist of the times of occurrence of three sets of events: a train of periodic (identical) stimuli, and two trains of action potentials (A and B). Let the time of occurrence of the ith stimulus be S_i; that of the jth spike of train A be A_j; and that of the kth spike of train B be B_k. We construct a scatter diagram of the joint occurrences of spikes in trains A and B relative to the times of stimulation. Thus, the ordinate of each point plotted corresponds to the time between a stimulus event and a spike in A (for example, $A_j - S_i$), and the abscissa corresponds to the time between the same stimulus event and a spike in B (for example, $B_k - S_i$). For each stimulus event, a point is plotted for each combination of S-A and S-B intervals, both of which fall within a specified range. Thus, in general, a spike in train A will give rise to as many points in the scatter diagram as there are spikes in train B within the specified time range about the stimulus (and vice versa). This computation and plot have been done with a Linc computer as well as with considerably larger machines.

We call this plot the joint peri-stimulus-time (PST) scatter diagram (*3*) (Figs. 1 and 2). In fact, the scatter diagram represents an estimate of an underlying joint PST density; the two PST histograms estimate a related set of marginal densities (*4*).

The scatter diagram represents a generalization of the PST histograms of each of the two spike trains, and of the cross-correlation histogram. If we impose a square grid of convenient spacing over the scatter diagram, then the set of tallies by columns is related to the PST histogram of train B, and the corresponding set of tallies by rows is related to the PST histogram of train A. Furthermore, tallies along the direction parallel to the principal diagonal are proportional to the cross-correlation histogram, with a suitable normalization of the time axis. The scatter diagram contains information beyond that provided by the two PST and the cross-correlation histograms and cannot be reconstructed from these histograms alone, no matter how fine the resolution.

Each of the several possible types of functional connection to and between the observed neurons produces a characteristic signature in the scatter diagram. For example, two independently firing neurons which are totally unaffected by the stimulus will produce a uniform scatter of points. If, in addition, the stimulus produces an increase in firing probability at fixed latency in neuron A only (such as would be produced by an excitatory synaptic pathway), then a horizontal band of higher point density will be seen in the scatter diagram. The location of the band is a measure of the latency of the excitation; the width and structure of the band depend on the details of the excitatory mechanism. If the stimulus produces a fixed latency decrease in firing probability of neuron A (for example, due to synaptic inhibition) there will be a horizontal band of lower point density than the general background density in the scatter diagram. If the stimulus affects neuron B and not neuron A, then only a corresponding set of vertical bands will be produced.

A direct synaptic pathway from neuron A to neuron B will produce a 45° band of different density lying below the principal diagonal at a distance proportional to the latency of the interaction. A synaptic pathway from neuron B to neuron A will produce a similar band lying above the principal diagonal.

If the synaptic pathways between A and B can be affected by the stimulus (as by stimulus influence on an interneuron), then the diagonal band will show changes in density along its length. For example, if the stimulus inhibits an excitatory interneuron between A and B, the diagonal band will be reduced in density over a segment corresponding to the latency and duration of the inhibition (Fig. 1).

A final principal class of interaction is represented by a shared source of input to A and B which is independent of the stimulus. This gives rise to a diagonal band which is difficult to distinguish qualitatively from the diagonal band produced by a direct excitatory pathway from A to B or from B to A. However, for a given strength of synapse, shared excitatory input produces a much less dense and a wider band than a direct synaptic connection between the two neurons. The density difference and other structural details of the band usually enable a choice to be made between the alternatives (*2*, p. 431). If such a source of shared input is itself modified in a time-locked way by the stimulus, these modifications will be reflected in density changes along the diagonal band.

If combinations of these elementary influences occur, each of the corresponding bands will ordinarily appear in the scatter diagram (Figs. 1 and 2). The scatter diagram can yield quantitative estimates of synaptic effectiveness in the inferred connections between neurons if the points within particular bands and band intersections are counted.

We may estimate the contributions

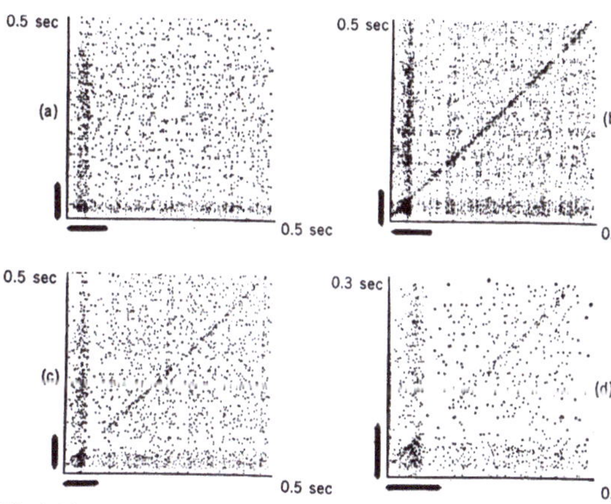

Fig. 1. Joint PST scatter diagrams in computer-simulated neural networks. (a) Neurons 1 and 2 with common excitatory followed by inhibitory input from a periodic stimulus. Duration of the stimulus is indicated by dark bars. Note vertical and horizontal bands of higher density. (b) Same situation as (a) but with added excitatory interaction from neuron 1 to neuron 2. Note diagonal band of higher density. (c) Same situation as (b) except that the excitatory interaction from neuron 1 to neuron 2 is itself inhibited by the stimulus. In contrast to (b) the diagonal band is interrupted during the latter portions of stimulus. (d) Enlargement of lower left region of (c). Details of each band structure allow inference about various synaptic input sequences.

of only the direct stimulus effects to the scatter diagram by constructing the cross product of the marginal probability densities (obtained from the two PST histograms). The resulting bivariate density, which contains no diagonal bands, serves as a control. It represents the hypothetical case in which the two neurons fire independently except for shared input caused by the stimulus. Comparison of this control with the observed joint density facilitates the detection of direct interactions and shared input from sources other than the stimulus.

The method described can be extended as follows: When more than two neurons are monitored together with a stimulus train, each spike train can be assigned its own Cartesian coordinate. The resulting *N*-dimensional scatter diagram, although presenting some difficulties in display, can be interpreted in a completely analogous manner to that of the two-dimensional diagrams. For example, a concentration of points about a line parallel to the principal diagonal of the coordinate system corresponds to a set of interactions, or shared input unrelated to the stimulus, involving all of the monitored neurons.

The scatter diagram described here is a supplement to rather than a substitute for the PST and cross-correlation histograms. Weak interactions may be more readily detected in the appropriate histogram. Examination of the histograms can serve as a guide to the detailed interpretation of particular features in the scatter diagram. This combined approach should be particularly applicable, for example, to experiments which seek to associate plasticity of behavior with changes in neuronal interactions.

George L. Gerstein
Departments of Biophysics and Physiology, University of Pennsylvania, Philadelphia 19104

Donald H. Perkel
Department of Mathematics, RAND Corporation, Santa Monica, California 90406

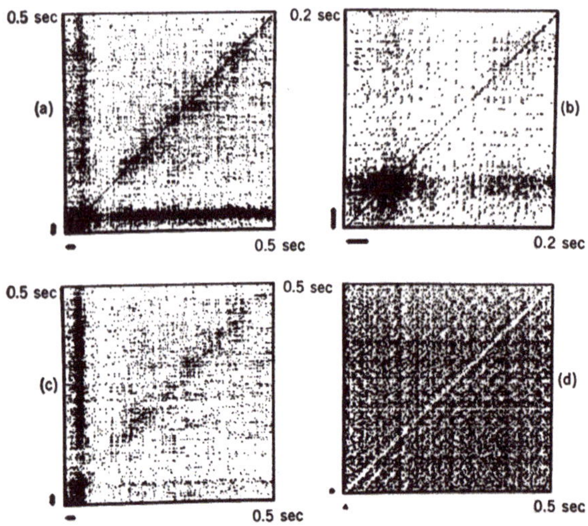

Fig. 2. Joint PST scatter diagrams. Units in auditory cortex of unanesthetized cat. Stimuli were 20-msec noise bursts, two per second, and are indicated by the dark bars. (a) Units 1 and 3: strong vertical and horizontal bands represent the stimulus-caused activity of each unit individually. The different latencies, time courses, and fine structures of the two responses are apparent. The narrow diagonal band represents a direct excitatory interaction between the two neurons with a latency of 3 msec. The wide diagonal band of increased density in central and upper right portion is most plausibly attributed to a weak shared excitatory input. The absence of this wide diagonal band in the lower left portion of scatter diagram indicates inhibition of the inferred common source for some time after the stimulus. (b) Enlargement of lower left portion of (a); differences in vertical and horizontal structure of region of high density and variation of density along the thin diagonal band, presumably reflecting stimulus-related reduction of firing, are shown. (c) Units 2 and 3, same preparation. The wide diagonal band again indicates that the stimulus inhibited shared input from nonstimulus sources. The shared input reaches unit 3 after unit 2, in that the wide diagonal band lies above the principal diagonal. There is no evidence for a direct synaptic connection. (d) Two Purkinje cells in the vermis of anesthetized cat cerebellum during an auditory stimulation by clicks (one click per second). There are no significant vertical or horizontal features, indicating that these units did not respond to the stimulus. Multiple diagonal bands are visible and presumably represent shared input and time delays associated with parallel-fiber input (2).

References and Notes

1. G. P. Moore, D. H. Perkel, J. P. Segundo, *Ann. Rev. Physiol.* **28**, 493 (1966).
2. D. H. Perkel, G. L. Gerstein, G. P. Moore, *Biophys. J.* **7**, 419 (1967).
3. We propose "peri-stimulus-time" instead of the older restrictive term "post-stimulus-time." The initials PST are thereby preserved.
4. The joint PST density may be defined by the equation

$$\xi_{ab}(t, u) = \lim_{\substack{\Delta t \to 0 \\ \Delta u \to 0}} \frac{1}{\Delta t \, \Delta u} P$$

where P equals probability [spike in A in $(v + t, v + t + \Delta t)$ and spike in B in $(v + u, v + u + \Delta u)$, given the stimulus event at v]
5. C. C. Bell and R. J. Grimm, personal communication.
6. Supported by PHS grants NB 05606, NB 07325, FR-15, and by United States Air Force Project RAND.

6 February 1969

Scanning Electron Microscopy of Cells

Abstract. *The scanning electron energy-analyzing microscope uses characteristic energy losses to provide picture contrast. At different levels of energy loss particular structures are distinguished with high contrast in an unstained section of a cell.*

Electron microscopy of ultrathin sections of biological tissue generally requires electron-opaque staining to obtain suitable contrast for detailed analysis. A heavy atom that selectively attaches to certain cell components (such as nuclei, mitochondria, and lysosomes) is typically used to stain the specimen. The contrast provided by such staining for examination with a conventional electron microscope depends primarily on specimen density. An ideal mechanism for obtaining contrast would be one in which different structures could be shown suc-

MUTUAL TEMPORAL RELATIONSHIPS AMONG NEURONAL SPIKE TRAINS

Statistical Techniques for Display and Analysis

GEORGE L. GERSTEIN *and* DONALD H. PERKEL

From the Departments of Physiology and Biophysics, School of Medicine, University of Pennsylvania, Philadelphia, Pennsylvania 19104, and the Department of Biological Sciences, Stanford University, Stanford, California 94305

ABSTRACT We describe a statistical technique, the joint peristimulus time (PST) scatter diagram, for the analysis of data from simultaneously recorded neurons subjected to repeated stimulation. Distinguishable features in the scatter diagram are related to effects of the stimulus on the observed neurons and to functional relations among the neurons. Properties of this measure and its variants are described and practical aspects of its application to experimental data are discussed.

INTRODUCTION

In recent years, increasing numbers of studies have been performed using simultaneous recordings of the activity of several neurons, thus opening a new level of neural integration to direct experimental investigation (Bullock and Horridge, 1965, p. 257). The interactive role of groups of neurons in carrying out integrative functions was stated by Sherrington: "It may be objected that it is not nerve impulses which are likely to tell us what we want to know of the brain. . . . Activity of the brain involves great numbers, not to say, vast numbers, of nerve cells cooperating. Yet the means of securing that cooperation is by impulses via the nerve fibers connecting cells" (Sherrington, 1951, p. 219). Although these ideas may be traced back even beyond Descartes, there continues to be only indirect evidence suggesting that the nervous system does indeed operate by means of relatively small functional groups of neurons.

The basic problems requiring investigation are those of identifying such functional groups and of characterizing their activity in terms of information processing or production of behavior. A higher order problem is that of plasticity of such functional neural groups.

Because functional grouping implies some degree of temporal organization in the activities of the constituent neurons, it follows that experimental evidence for the existence of such groups may be furnished by quantitative characterizations of the mutual temporal relationships in the electrical activity of the involved neurons. The further understanding of the operating principles of neuronal groups, once identified, is crucially dependent on detailed knowledge of the underlying circuitry.

Detailed description of functional neuronal connectivity is only partially furnished by classical neuroanatomical techniques, or by a combination of electrophysiological signal tracing and such newer methods as fluorescent dye injection and scanning electron microscopy. The approach described in this paper is intended as an adjunct to these more familiar methods; it can yield functional "circuit diagrams" of the connections among the observed neurons. Such a circuit diagram may, of course, not be unique, but rather may represent one member of a class of equivalent circuits.

These techniques, statistical in nature, must be imbedded in an appropriate experimental design in order to allow inferences of biological significance to be drawn. Application of these techniques to isolated fragments of data is likely to be a futile exercise.

The measures we describe involve the simultaneous recording of spike trains from several neurons, typically in conjunction with repetitive stimulation. As such, we treat the nerve impulses as point events (Perkel et al., 1967 *a*, *b*). The trains of events may be derived from extracellular recordings, or through intracellular recordings in which classes of postsynaptic events are resolved. It is important to realize that data obtained simultaneously from several neurons inherently contain kinds of information not available from an equivalent body of observations obtained serially from the same neurons.

In this paper we first introduce a class of statistical measures for multiple-neuron impulse data. We illustrate the effects of various assumed connections, describe control calculations, discuss various quantitative aspects, and extend these ideas to larger numbers of neurons. Examples are drawn both from computer simulation and from appropriate animal experiments; the latter are identified in the legends to the figures.

JOINT PST SCATTER DIAGRAM

The simplest statistical measure of the type under discussion is the joint PST scatter diagram (Gerstein and Perkel, 1969). This is a two-dimensional representation, in the plane, of the temporal relations among three simultaneously recorded sets of events: a train of periodic, identical stimuli and two trains of action potentials (*A* and *B*). We represent these data as a scatter diagram of the joint occurrences of spikes in trains *A* and *B* relative to the times of stimulation.

The construction of this scatter diagram from the data is illustrated in Fig. 1. The ordinate of each point plotted corresponds to the time between a stimulus event and a spike in *A*; the abscissa corresponds to the time between the same stimulus event and a spike in *B*. These coordinate values are indicated by the time intervals x_i and y_j in the figure. For the interstimulus interval shown we plot the six

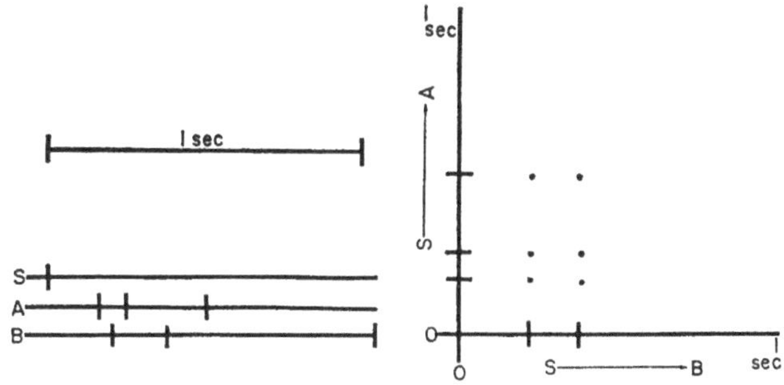

FIGURE 1 Method for constructing the joint peristimulus scatter diagram. The times of occurrence of impulses from neurons A and B during one stimulus interval are shown schematically at left; the corresponding contributions to the scatter diagram are shown at right. See text.

points (x_1, y_1), (x_1, y_2), (x_1, y_3), (x_2, y_1), (x_2, y_2), and (x_2, y_3); these are indicated by dots. This process is continued for all stimulus intervals in the record.

Clearly each spike gives rise, in general, to more than one point in the scatter diagram. The exact number depends on the number of spikes in the other train that occur during the same stimulus interval. In particular, if during one stimulus interval one of the neurons should fail to fire, no points will be added to the scatter diagram.

The resulting scatter diagram represents an estimate of an underlying joint PST density, defined as follows:

$$\xi_{AB}(t, u) = \lim_{\substack{\Delta t \to 0 \\ \Delta u \to 0}} [1/(\Delta t \Delta u)] \text{ prob } \{\text{spike in } A \text{ in } (t, t + \Delta t)$$

$$\text{and spike in } B \text{ in } (u, u + \Delta u) \mid \text{stimulus at zero}\}, \quad (1)$$

where the two time variables t and u are measured from a stimulus event. These variables may take negative values corresponding to the prestimulus period.

Tallies of points along various directions in the scatter diagram estimate certain well-known functions that are used to characterize unit activity. For example, the set of point tallies made in columns parallel to the principal diagonal (45° line), taking proper account of boundary effects, corresponds to the cross-correlation histogram between neurons A and B. Because of obvious geometrical relations, the time scale of this cross-correlation histogram is in the ratio $1:\sqrt{2}$ to that of the PST scatter diagram.

The sets of point tallies made in vertical and horizontal columns do not in general correspond to the PST histograms for the individual neurons. This may be shown by writing the joint density as the product of a conditional density and a marginal density:

$$\xi_{AB}(t, u) = \xi_{A|B}(t, u)\xi_B(u), \quad (2)$$

$$= \xi_{B|A}(u, t)\xi_A(t), \quad (3)$$

where the marginal densities $\xi_A(t)$ and $\xi_B(u)$ are the ordinary PST densities for the neurons indicated by the subscripts. The conditional density is defined as:

$$\xi_{A|B}(t, u) = \lim_{\Delta t \to 0} (1/\Delta t) \text{ prob } \{\text{spike in } A \text{ in } (t, t + \Delta t) \mid \text{spike in } B \text{ at } u \text{ and stimulus at zero}\}. \quad (4)$$

This function and the analogously defined $\xi_{B|A}(u, t)$ are functions of both time variables. Only when one neuron's firing probabilities are completely independent of the firings of the other neuron does its conditional density revert to the corresponding marginal density:

$$\xi_{A|B}(t, u) = \xi_A(t) \qquad \text{(under independence)}. \quad (5)$$

Comparison of equation 5 with equations 2 and 3 shows that independence implies mutual independence. In the case of independence, the joint PST density is the product of the corresponding marginal (ordinary PST) densities. This forms the basis for a control calculation described below. Moreover, only with such independence are the vertical and horizontal tallies proportional to the individual PST histograms.

In the ensuing examples of PST scatter diagrams, we conjoin the corresponding cross-correlation and PST histograms, computed from the original data.

SIGNS OF CONNECTIVITY IN THE SCATTER DIAGRAM

Each of the several possible types of functional connection to and between the observed neurons produces a characteristic signature in the joint PST scatter diagram. We illustrate these relationships with a series of examples of increasing complexity, drawn in these instances from computer simulations.

The null case of two neurons each of which fires entirely independently of the other and of the stimulus is shown in Fig. 2 *a*. Although the underlying density function is uniform over the plane, the scatter diagram, derived from a finite sample, frequently shows a clothlike texture. This effect is particularly evident when the number of spikes per stimulus interval is relatively large, in which case each spike gives rise to a set of points strung out along a line parallel to one of the coordinate axes. In the null case, both PST histograms and the cross-correlation histogram are flat to within the usual sampling fluctuations.

If the two neurons continue to fire independently, but if one is affected by the stimulus, we obtain the scatter diagram shown in Fig. 2 *b*. The vertical band of greater point density corresponds to an excitatory influence of the stimulus on neuron *B*. The latency of this effect is measured by the displacement of the vertical band to the right of the ordinate; the duration of the excitatory effect is measured by the width of the band. These stimulus effects appear in the corresponding PST histogram as the position and width of a peak. The uniform density of the band along its length is a consequence of the lack of any influence of the stimulus on neuron *A*, as shown by the flatness of the corresponding PST histogram.

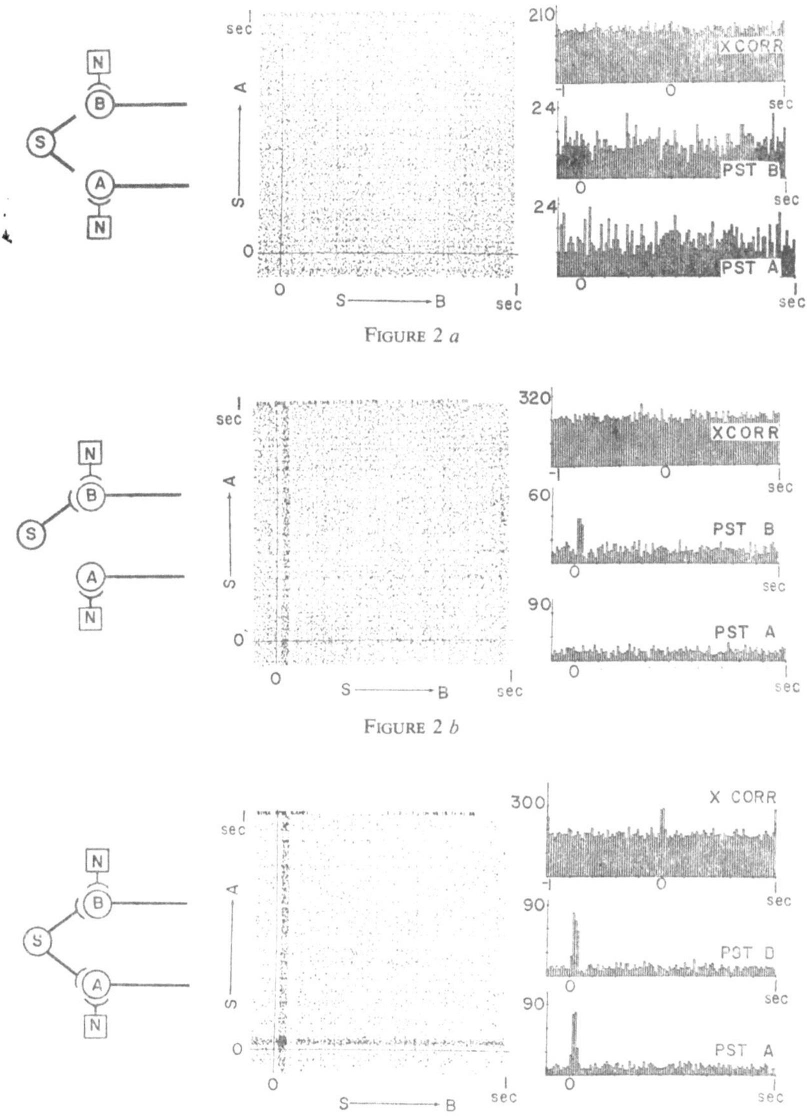

FIGURE 2 *a*

FIGURE 2 *b*

FIGURE 2 *c*

FIGURE 2 Stimulus effects. For each connectivity diagram, the corresponding joint peri-stimulus scatter diagram, cross-correlation histogram, and the two PST histograms are shown. The observed neurons are *A* and *B*. (*a*) Control case; stimulus has no effect and neurons are independent. (*b*) Stimulus excites neuron *B* only. (*c*) Stimulus excites both neurons.

If the stimulus excites both neurons, which otherwise fire independently, the scatter diagram of Fig. 2 *c* results. The latencies and durations of the stimulus influences are indicated separately for the two neurons by the displacements and widths of the horizontal and vertical bands. The peaks in the two PST histograms have corresponding interpretations. Because the stimulus affects both neurons, each band exhibits density variations along its length. The peak in the center of the cross-correlation histogram is a consequence of the shared input to the two neurons from the stimulus; its width corresponds to that of the area of intersection of the two bands. This peak is repeated at the stimulus interval, as seen at the extreme left and right of the histogram.

We now consider the case in which the stimulus has no effect on either neuron, but in which neuron *B* excites neuron *A* through a synaptic connection (Fig. 3 *a*). There are no horizontal or vertical bands because there are no stimulus effects; this is also seen in the flatness of the PST histograms. The narrow diagonal band, parallel to and slightly to the left of the principal diagonal, is the signature of the synaptic interaction. The latency and duration of the effect of *B* on *A* are reflected in the displacement and width of the diagonal band. The uniformity of its density along its length is another consequence of the lack of stimulus effects. The cross-correlation histogram shows a corresponding peak which is narrower and more pronounced than that arising in the previous case from shared stimulus excitation (Fig. 2 *c*).

The case of shared synaptic input from a source other than the stimulus, in the absence of stimulus effects and of direct interaction, is shown in Fig. 3 *b*. Again, there are no horizontal or vertical bands, and the PST histograms are flat. The diagonal band is seen to be somewhat broader and more diffuse than in the previous case of direct interaction. It is also uniform along its length. Its position reflects the effective difference in conduction times from the common source to the two neurons. Its width depends on the durations of the excitatory effects of the shared input on the two neurons, as well as other factors such as the prevailing level of background excitation. The cross-correlation histogram shows a somewhat wider peak than in the case of direct synaptic interaction between *B* and *A*; however, this peak is indistinguishable from that due to the shared excitatory stimulus effects of Fig. 2 *c*, whereas the two scatter diagrams are conspicuously different.

If, in addition to direct synaptic interaction between the two neurons, we also allow the stimulus to affect both neurons, all features described previously are visible in the scatter diagram and associated histograms (Fig. 3 *c*). The horizontal and vertical features, representing stimulus effects, are clearly distinguishable from

FIGURE 3 Interaction effects. (*a*) No stimulus effect; neuron *B* excites neuron *A*. (*b*) No stimulus effect; neuron *C* (not directly observed) excites neurons *A* and *B*. (*c*) Stimulus excites both neurons; neuron *B* excites neuron *A*. (*d*) Stimulus excites *A* and *B*; neuron *B* excites neuron *A* via interneuron *C*; stimulus inhibits neuron *C*.

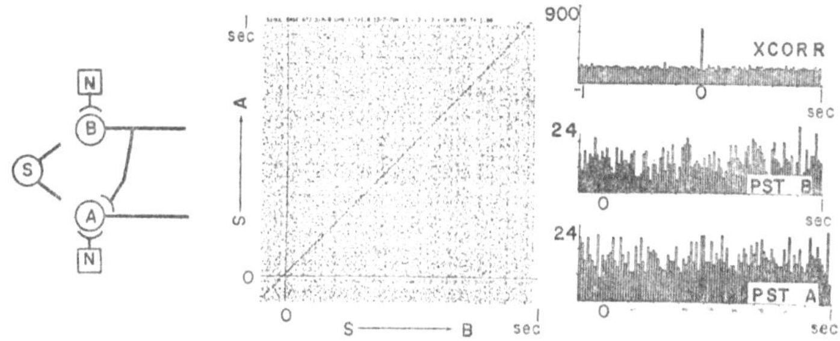

FIGURE 3 *a*

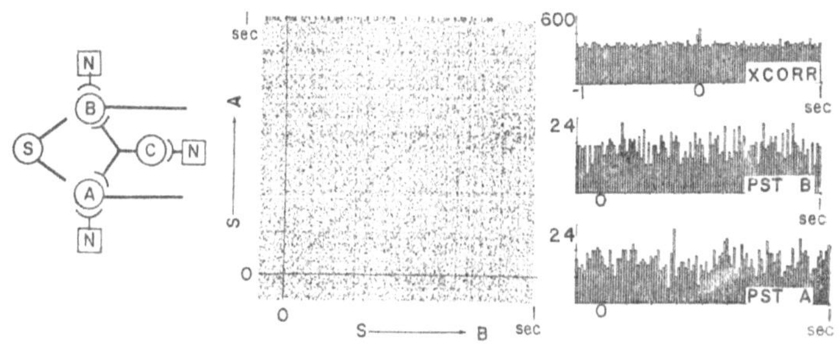

FIGURE 3 *b*

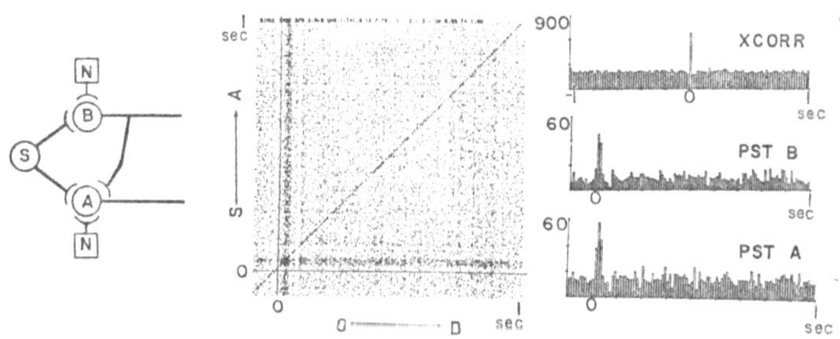

FIGURE 3 *c*

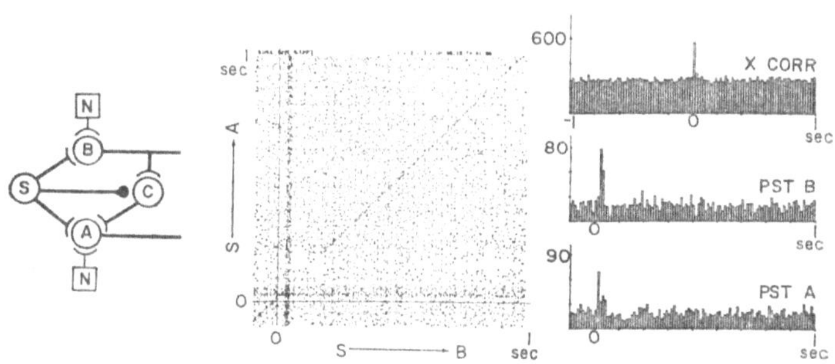

FIGURE 3 *d*

the diagonal features, which result from nearly simultaneous firing. The cross-correlation histogram shows a single peak in which the two types of effect cannot clearly be distinguished. Each band exhibits modifications of point density in regions of intersection with other bands.

Suppose now that the stimulus, which excites neurons *A* and *B*, also has an inhibitory effect on an excitatory interneuron which connects neurons *B* and *A*; thus the stimulus serves to diminish the interaction for a brief time (Fig. 3 *d*). Clearly distinguishable in the scatter diagram are the horizontal and vertical bands of increased density that result from the direct stimulus effects. The diagonal band, representing the interaction between *B* and *A* is interrupted because of the modulation of the interaction pathway.

As we have seen in the above example, an inhibitory influence decreases the point density in the scatter diagram. This is true in general: inhibitory stimulus effects give rise to vertical and horizontal bands of *decreased* point density; inhibitory interaction effects give rise to a diagonal band of decreased point density. Shared inhibitory effects, however, give rise to a diagonal band of somewhat *increased* point density in the scatter diagram, just as they give rise to a peak in the cross-correlation histogram (Perkel et al., 1967 *b*). Since decreased point density is relative to some prevailing background point density, it is visible in the scatter diagram only when the spontaneous activity is sufficiently great to provide the necessary contrast.

This series of possible connections among neurons can be extended to more complex situations. We now turn to some special situations which produce markedly different classes of features.

SOME SPECIAL CASES

In the examples considered above, we have assumed that both neurons exhibit spontaneous activity, and that stimulus effects and interaction effects account for only a certain fraction of the total activity. Moreover, we have assumed that the effects of the stimulus on each neuron are independent of those acting upon the other. Relaxation of either of these assumptions may lead to strong modifications in the scatter diagram.

In the absence of spontaneous activity, impulses will be produced by the neurons shown in Fig. 4 only when elicited by the stimulus. Although each stimulus presentation need not have a high probability of evoking spikes in either neuron, essentially all points in the scatter diagram will fall in the area corresponding to the intersection of the horizontal and vertical bands discussed previously. Neither the PST nor the cross-correlation histograms noticeably show background values outside the single peak. The addition of a direct excitatory connection from *B* to *A* would give rise to points lying entirely within the small rectangular area shown and hence would be generally undetectable.

If only one of the neurons, say *B*, exhibits spontaneous activity, then a horizontal band will be produced, while the vertical band remains in its "collapsed" state. Only if both neurons fire spontaneously will there appear the typical distribution of points over the plane.

When the responses of the two neurons to the stimulus are not independent,

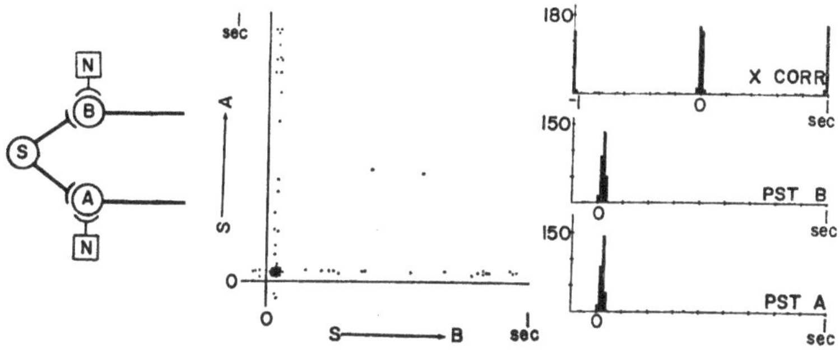

FIGURE 4 Stimulus excites both neurons in the absence of nearly all spontaneous activity.

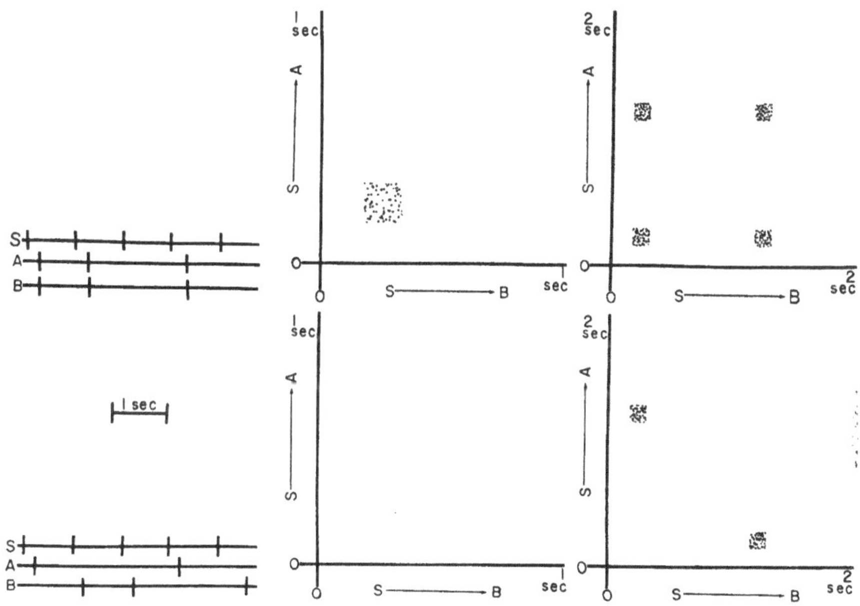

FIGURE 5 Correlated responses to the stimulus (semischematic). Upper row: in response to stimulus both neurons either fire at constant latencies or else both remain silent. The right-hand scatter diagram covers twice the time span of the left-hand diagram. Lower row: in response to the stimulus, one or the other neuron fires at its characteristic latency; both neurons never fire during the same stimulus interval.

the scatter diagram is modified in a different fashion. We illustrate this with two extreme cases. In the first situation, suppose that each stimulus presentation either causes both neurons to fire, or neither neuron to fire. As seen in the upper set of schematic records of Fig. 5, the firings of neurons *A* and *B* are maximally correlated. The corresponding scatter diagram shows a dense concentration of points at a location corresponding to the peaks in the two PST histograms and in the cross-correlation histogram, as also seen in Fig. 4. If, on the other hand, each stimulus presentation causes only one or the other neuron to fire, then no points whatever appear in the scatter diagram (Fig. 5, bottom). The two PST histograms, however, would be indistinguishable from those of the previous situation; the cross-correlation histogram would be empty. This somewhat surprising result may be clarified by extending the range of the scatter diagram to include two (or more) stimulus intervals, as illustrated at the right of Fig. 5.

Diagonal features in the 135° direction are also possible, under circumstances in which the poststimulus latency of one neuron is negatively correlated with that of the other. Suppose, for example, that two neurons fire fairly regularly and out of phase. If the stimulus is presented at roughly the same frequency, a small latency in one neuron is then associated with a large latency in the other neuron, and vice versa. This produces a band of increased point density oriented at 135°. Direct stimulus effects on these neurons will complicate this picture.

Complications of another sort can arise through serial correlation of stimulus effects. These topics are covered in some detail in the section on control calculations for the joint PST scatter diagram.

DIFFERENTIATION BETWEEN INTERACTION AND COMMON INPUT

A region of altered diagonal density in a PST scatter diagram, like the corresponding peak or valley in the cross-correlation histogram, represents coordination in the firing of the two neurons. Two major physiological mechanisms can produce such coordination: direct interaction between the neurons, typically synaptic (Fig. 3 *a*), and shared input from a common source (Fig. 3 *b*). On the basis of the PST scatter diagrams and the cross-correlation histogram it may be difficult to distinguish between the two mechanisms unless additional information is available, such as that provided by intracellular electrodes.

As previously illustrated (Perkel et al., 1967 *b*), the peak in a cross-correlation histogram is broader and lower when due to shared input of a given strength than when due to direct synaptic interaction of the same strength. A similar difference is seen in the diagonal bands of the PST scatter diagram; Fig. 6, drawn from computer simulation, illustrates these differences. In the network, one neuron provides excitatory synaptic input to two neurons each of which has additional independent excitatory input. For clarity, the intensity of the periodic stimulus was reduced to negligible values. In the left-hand diagram, measurements were taken from the

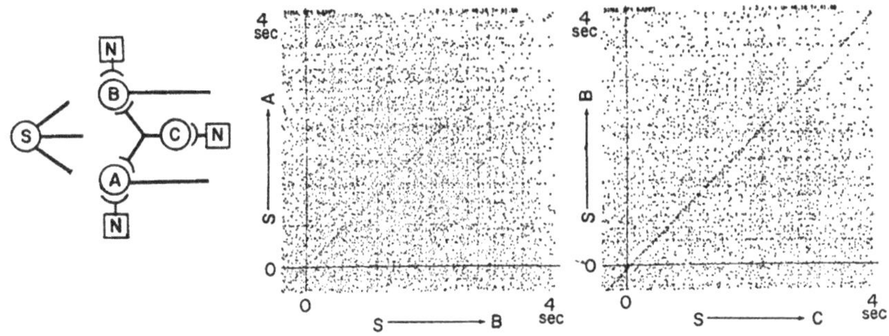

FIGURE 6 Comparison of effects of shared input with those of direct interaction. The left-hand scatter diagram is for neurons *A* and *B*, which share input from neuron *C*; the right-hand scatter diagram is for neurons *B* and *C*, which are directly connected.

two postsynaptic neurons. The diagonal band corresponding to the shared input is relatively broad and diffuse. In the right-hand diagram, taken from a pre- and a postsynaptic neuron, the diagonal band is narrower and more intense. Diagonal bands of both kinds have been observed in physiological experiments, and sometimes even in the same scatter diagram (Gerstein and Perkel, 1969).

In one particular situation the ambiguity is completely resolved. A narrow diagonal band that is *precisely* centered on and symmetric about the principal diagonal can only be the result of common input over conduction paths with identical delays. If, however, this band does not have a symmetric profile, some admixture of direct interaction is indicated.

The ambiguity between the effects of shared input and direct interaction are exacerbated by the introduction of interneurons in the direct pathway. Such interneurons will, in general, broaden and weaken a diagonal band in the scatter diagram, thus making it resemble even more closely the diagram for shared input. A sufficiently long chain of interneurons will wash out the observable features of a scatter diagram.

CONTROLS

In order to draw more quantitative conclusions from the features observed in the joint PST scatter diagram, we introduce control calculations. These controls exhibit just those features of the scatter diagram that arise from independent stimulus effects on the two neurons, i.e. those effects that are predictable in principle solely from the two PST histograms, which estimate the marginal PST densities. The difference between the joint PST scatter diagram and its control therefore represents all other factors influencing the joint firing of the two neurons, including direct interaction and shared input.

In principle, the appropriate control could be constructed simply by cross-multiplying the two PST histograms; however, because of their finite bin structure, the resulting two-dimensional histogram is difficult to compare directly with the original scatter diagram.

An equivalent control statistic, which consists of a scatter diagram rather than a two-dimensional histogram, is constructed as follows. One of the two spike trains is segmented at the instants of stimulus presentation, and these segments are subjected to a random shuffle. This shuffled train is used together with the unaltered train (of the other neuron) to construct a joint PST scatter diagram, in the usual manner. It is an estimate of the product $\xi_A(t)\xi_B(u)$. As seen in the example (Fig. 7), the control diagram retains the vertical and horizontal features of the original diagram with virtually no alteration. The diagonal features, however, are eliminated as such; their constituent points are rearranged and may take the form of vertically or horizontally oriented configurations.

In some situations the effects of the stimulus on one neuron may depend on the response of the other neurons. If this is the case, the control scatter diagram will also alter vertical and horizontal features of the original. One example is shown in the bottom row of Fig. 5.

In the majority of experimental situations, a simpler method suffices for constructing the control. Rather than segmenting and shuffling one spike train, we merely shift it along the time axis by one or more complete interstimulus intervals, after which we construct a scatter diagram as usual. Examples of the "shift control" are given by the off-diagonal squares in the extended scatter diagrams of Fig. 11. The shift control differs significantly from the shuffle control only if there is significant serial dependence in the effects of successive stimulus presentations. The most likely source of such serial dependence encountered experimentally is a long-term trend in the firing rates of both neurons. Thus, at the beginning of such an experiment, a stimulus may elicit more spikes in each train than at the end. In the presence of such trends, care should be exercised in the choice and interpretation

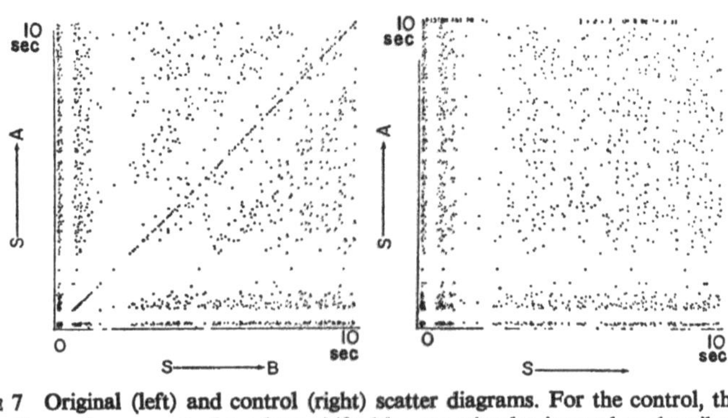

FIGURE 7 Original (left) and control (right) scatter diagrams. For the control, the nerve impulse times of one neuron have been shifted by one stimulus interval as described in the text. Note the disappearance of all diagonal features. Data from isolated pleural ganglion of *Aplysia californica* (Kristan, 1971).

of the control. Various gating mechanisms, unlikely to be encountered in the laboratory, can give rise to more pronounced differences between the two types of control.

To recapitulate, control scatter diagrams wash out all diagonal features of the original. If at a given stimulus presentation the response of one neuron depends on the response of the other, then, in addition, vertical and horizontal features are altered in either type of control. If stimulus effects are serially dependent, the two types of control will differ. All these effects may occur in a given experiment.

A particularly simple statistic for the comparison of actual and control joint PST scatter diagrams is the ratio of the actual number of points plotted to that predicted on the basis of the independence hypothesis. The predicted number is $N_A N_B / N_S$ (Kristan and Gerstein, 1970), where N_A and N_B are the numbers of spikes recorded from the two neurons, and N_S is the number of stimuli. Approximate significance levels may be derived from Poisson counting statistics. Refractory and other physiological phenomena undermine some of the assumptions underlying Poisson statistics, but this is rarely a practical problem unless stimulus presentations are too frequent.

A somewhat more complicated numerical comparison between actual and control joint PST scatter diagrams can be used to make inferences about neurons not under direct observation and even to estimate synaptic effectiveness. For example, suppose that an unobserved neuron is suspected to be interposed between the stimulus and the two observed neurons (Fig. 8). All three synapses are excitatory, act independently, and have the spike-eliciting probabilities P_i, P_a, P_b, as indicated. We observe both a horizontal and a vertical band in the scatter diagram and its control together with corresponding peaks in the PST histograms. In the control scatter diagram, the expected number of points within the horizontal band is equal to the product $N_S P_i P_a$; the expected number in the vertical band is $N_S P_i P_b$. In the actual scatter diagram, the number of points in the rectangular area of band intersection is given by the product $N_S P_i P_a P_b$. The three indicated point counts determine estimates of all three probabilities. If, for example, $P_i = 1$, then the postulated interneuron probably does not exist. If $P_i < 1$, the indicated diagram of neuronal connections is reasonable. If $P_i > 1$, the indicated diagram is not correct, and probably should include a stimulus-modulated interaction be-

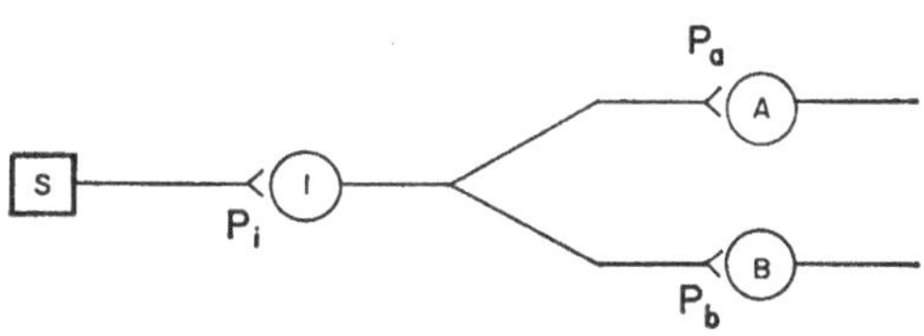

FIGURE 8 Stimulus excitation via an unobserved interneuron. Impulse-eliciting probabilities are indicated by P_i, P_a, P_b for the corresponding synapses. See text.

tween the two observed neurons. Spontaneous activity calls for a simple modification of this procedure.

TIME-COURSE OF SYNCHRONOUS FIRING

As we have seen, the joint PST scatter diagram depicts the direct effects of the stimulus as vertical and horizontal bands, whereas all other effects of interaction

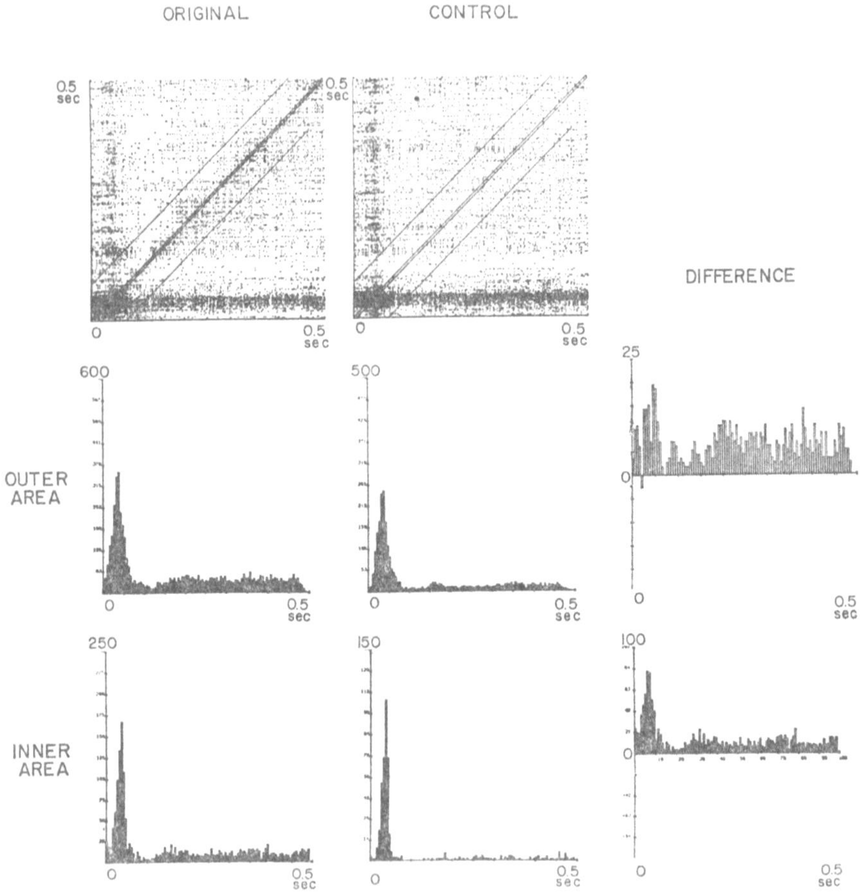

FIGURE 9 PST coincidence histograms. Upper row: original and control scatter diagrams. Superimposed lines delineate outer and inner areas about the principal diagonal. Middle row: PST coincidence histograms derived from the outer areas of the corresponding scatter diagrams; the difference histogram is shown at the right. Lower row: same as middle row, but for the inner areas of the scatter diagrams. Auditory cortex of unanesthetized cat; stimuli were periodic noise bursts.

or shared input appear as diagonally disposed structures. It is useful to examine the time-course of the latter effects relative to the stimulus.

The statistical measure we use corresponds to the marginal density that is estimated by the set of tallies along the direction perpendicular to the principal diagonal. The range with respect to the principal diagonal over which tallies are accumulated may be chosen so as best to study a particular interaction. The resulting histogram is called the PST coincidence histogram, because the events that contribute to it are not spikes, but rather are near coincidences of spikes from the two neurons.

As an example, we turn to experimental data from the auditory cortex of an awake cat (Fig. 9). We interpret the joint PST scatter diagram as implying both a direct interaction between the two neurons (narrow diagonal line) and a shared input to the two neurons (broad, less dense diagonal band), in addition to the direct stimulus effects (vertical and horizontal bands). The PST coincidence histograms shown are calculated using different regions about the principal diagonal. One histogram is taken from the outer area indicated in the scatter diagram; it excludes the inner region. In the other histogram, the region encompasses only the narrow diagonal feature attributed to direct interaction.

These histograms suggest that the phasic effects of the stimulus may be different for the broad and narrow bands; however, a more quantitative statement depends on the comparison with a control coincidence histogram, taken from the corresponding region of the control PST scatter diagram. Such control PST coincidence histograms are also shown in Fig. 9, together with histograms of the algebraic difference between corresponding originals and controls. From these, it is evident that the time-course of the process responsible for the narrow diagonal band is significantly different from that responsible for the broad band. This conclusion cannot be reached by mere inspection of the original joint PST scatter diagram.

The PST coincidence histogram measures only one aspect of the train of near coincidences. In general, this train may be treated by statistical methods appropriate to any train of point events (Perkel et al., 1967 *a*); for example, the interval histogram and autocorrelogram of this train are usually essential for detailed interpretations.

THREE OR MORE NEURONS

The joint PST scatter diagram can also be constructed for three neurons. A three-dimensional cartesian coordinate system is set up in which each axis corresponds to the time from stimulus presentation to the firings of one of the neurons. Positions of points are defined in a completely analogous way to that of the two-dimensional case; the resulting cloud of points is most conveniently visualized by plotting a stereoscopic pair of perspective views (Fig. 10), produced by a standard computer program "package" for graphic display, using well-known results from analytic geometry to provide the slightly discrepant projections.

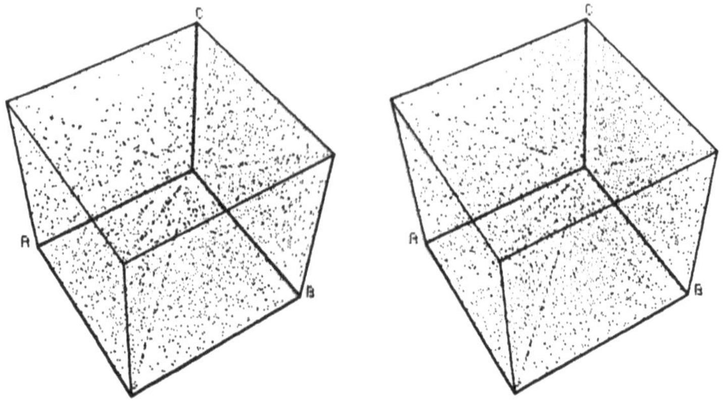

FIGURE 10 Joint peristimulus scatter diagram for three neurons; stereoscopic pair. Origin is the most distant vertex from the viewer. Note point concentrations along diagonals of three faces of the cube as well as along the principal diagonal of the cube. The left-hand cube should be viewed by the left eye. Data from isolated pleural ganglion of *Aplysia californica* (courtesy of W. B. Kristan, Jr.).

If the stimulus has no effect on the firing of any of the neurons, the cube corresponding to one stimulus cycle is uniformly filled with points. If one or more of the neurons fire several times during each stimulus cycle, a clothlike texture results as in the two-dimensional case.

Each neuron that responds with fixed latency to the stimulus gives rise to a plane of altered density. For example, if neuron *B* responds, the corresponding plane of points is parallel to the *AC* plane. If two neurons respond, two such planes will be generated, and their line of intersection will be particularly visible. If neurons *A* and *B* respond to the stimulus, this line of points will be parallel to the *C* axis. If all three neurons respond, the three resulting planes will intersect pairwise in three mutually perpendicular lines, and a local concentration of points will be seen around their locus of common intersection. Where the effects of stimulus are inhibitory rather than excitatory, regions of decreased point density will appear instead of regions of increased point density.

Coordinated firings of the neurons give rise to various diagonal features in a systematic manner. If neurons *B* and *C* tend to fire synchronously, the resulting plane of increased point density passes through the *A* axis and through the bisectrix (45° line) of the *BC* plane. If all three neurons tend to fire synchronously, the three corresponding planes will intersect on the line which forms the principal diagonal of the cube. Fixed latencies of coordinated firing give rise to a similar system of planes which lie parallel to those of the strictly synchronous case.

In typical physiological cases, all the above features may be present simultaneously, and become apparent in the stereoscopic view (Fig. 10). In a manner analogous to the two-neuron case, modulating effects of the stimulus are reflected by systematic changes of point density as a function of the distance from the origin.

If four or more neurons are included, direct visualization of the corresponding joint PST scatter diagram is unfortunately impossible. Examination of the N-dimensional space may be made in a number of indirect ways with the help of a computer. Without recourse to visualization, the computer program can search for altered point densities in the relevant hyperplanes, planes, lines, and points in N-space. Nearly synchronous firings of two or more neurons may themselves be treated as a single series of events (as in the calculation of the PST coincidence histogram above). That series can be used with two other spike trains to generate three-dimensional scatter diagrams of the usual sort. Estimation of probabilities through tallies and construction of controls may then be done in a manner completely analogous to the two-dimensional case.

NONPERIODIC STIMULUS

In most of the above examples, the domain of scatter diagrams and PST histograms has been exactly one stimulus interval. If we extend the domain to encompass more than one stimulus interval, as previously mentioned, a checkerboard-like pattern will emerge. Ignoring the minor contributions of end effects, the original square is reproduced in all its detail at successive squares that straddle the principal diagonal of the plane. The new squares that appear at off-diagonal locations are the control scatter diagrams that have been described above as being produced by shifting one spike train an integral number of stimulus intervals with respect to the other. Each such square is itself replicated along the direction parallel to the principal diagonal. Thus the entire diagram is invariant with respect to stepwise translation along the principal diagonal.

This invariance depends critically on the strict regularity of stimulus presentations. If there is minor departure from strict regularity, there will be a gradual blurring of stimulus-related features with increasing distance from the origin. In the case of highly irregular stimuli, clarity is preserved only in that portion of the original scatter diagram which lies within the time of the minimum interstimulus interval. These properties are illustrated in Fig. 11. For this reason, as well as for computational convenience, strictly regular stimulation should be used whenever possible in studying mutual temporal relations among neurons.

The blurring which we have described can be characterized mathematically, assuming that each stimulus presentation has independent effects. In the strictly periodic case we may interpret the replication of the pattern as the convolution of the primary square with the autocorrelation of the stimulus train, which in this case is a sequence of delta functions. In this context, the primary square delineates the "basic" response to the stimulus. With relaxation of strict regularity, we may continue to describe the resulting pattern in terms of convolution of the basic response with the autocorrelation of the stimulus train. The autocorrelation now consists of a series of increasingly broad peaks (except for the single delta function

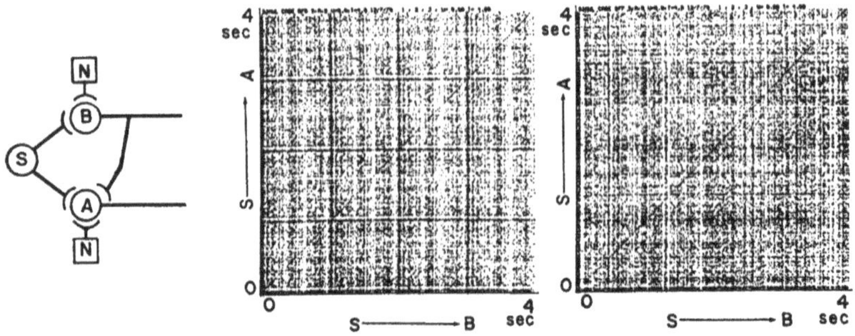

FIGURE 11 Effects of a nonperiodic stimulus. The left-hand scatter diagram extends over four stimulus cycles under conditions of strictly regular stimulus presentation. In the right-hand scatter diagram, which extends over the same time span, the stimulus was presented at random intervals with a mean of 1 sec and a standard deviation of 0.1 sec.

at the origin), which is responsible for the observed blurring. Convolution equations of this type are described by Perkel (1970).

In the extended "checkerboard" diagram, coordinated firings will produce their characteristic diagonal features only in the primary square and its replications along the principal diagonal of the plane. If stimuli are irregular, the consequent blurring will affect the representation of coordinated firing only with regard to stimulus modulation effects. In such a case, squares on the diagonal far removed from the primary square will show no variation of density along the direction of the principal diagonal. Variations of density along the orthogonal direction (135° line) continue to reflect the features of the cross-correlation histogram between the two spike trains.

The irregular stimulus train is conceptually not different from the spike train of a third neuron. Thus an experiment in which three spontaneously active neurons are monitored (ignoring any stimulation) may also be analyzed with joint PST scatter diagrams. The choice as to which spike train plays the role of the stimulus train is arbitrary; the three possible scatter diagrams contain equivalent information.

The asymmetry inherent in using one neuronal spike train in lieu of the stimulus for plotting a scatter diagram can easily be removed by the use of triangular coordinates. The axes of such a system, which intersect at 120° angles, represent the time differences $T_A - T_B$, $T_B - T_C$, and $T_C - T_A$; each point corresponds to such a triplet, and stimulus times (if any) are ignored. This variant of the joint PST scatter diagram can be understood equivalently as the projection of the PST cube (stimulus plus three spike trains) onto a plane perpendicular to the principal diagonal, taking proper account of boundary effects. In such a diagram (Fig. 12), synchronized firings between pairs of neurons give rise to radial concentrations of points midway between coordinate axes, while stimulus effects produce concen-

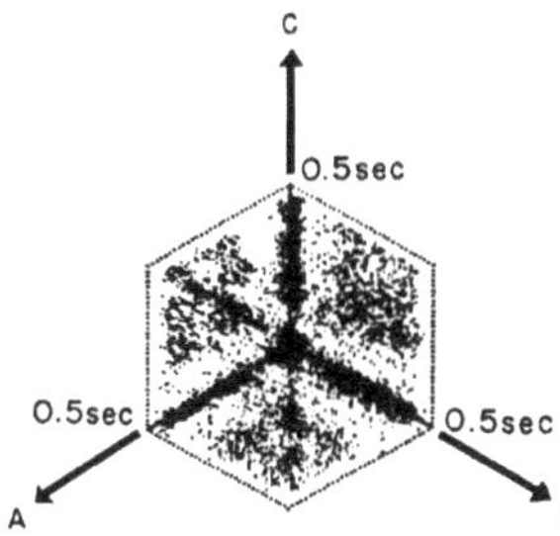

FIGURE 12 Triangular coordinate scatter diagram of the joint impulse activity of three neurons without reference to stimulus. This is also equivalent to the projection on a plane perpendicular to the principal diagonal of the three-dimensional joint PST scatter diagram. The three axes shown are those of the joint PST scatter diagram; the corresponding axes of the triangular coordinate system defined in the text are at right angles to the former. Cat auditory cortex. See text.

trations along lines parallel to the coordinate axes. Synchronous firings of all three neurons give rise to concentrations of points about the origin; systematic latency patterns displace the region of concentration. This hexagram is the natural generalization of the cross-correlation histogram to the case of three simultaneously observed spike trains. The generalization to the case of four spike trains can be plotted with tetrahedrally disposed coordinate axes. Such generalizations of the cross-correlation histogram directly indicate multiple interactions among the observed neurons. These interactions cannot be measured by combinations of pairwise comparisons alone.

DISCUSSION

As we have pointed out above, in order to infer detailed connectivity by analysis of spike timings, it is essential to make use of data recorded simultaneously from two or more neurons. Serial recordings from the same neurons provide only that information which is equivalent to the control calculations we have described; these contain only effects of the stimulus itself on each neuron individually. On the other hand, the full power of the joint PST scatter diagram for investigating interactions is only realized when it is used in conjunction with cross-correlation histograms and PST histograms, together with the appropriate controls. Such additional information as provided by autocorrelation histograms, interval histograms, serial correlograms, and the like is a desirable adjunct. Moreover, intracellular recordings may furnish a complementary set of basic data. These may serve as an independent corroboration of some aspects of the inferred neuronal connectivity, and also may help resolve ambiguities between direct interaction and shared input.

In fact, given sufficient stability of recording conditions, it is possible to generalize the joint PST description of spikes to the case of two or more continuous recordings (e.g., obtained through intracellular or gross electrodes). Interpretation of vertical, horizontal, and diagonal features is analogous to that for the scatter diagrams (C. E. Molnar, private communication). The time relations between a spike train and a simultaneously recorded continuous process (from the same or a different source) may also be analyzed through the same basic technique. Measures of this kind, in the absence of repetitive stimulation, have been used by Frost and Elazar (1968). Such measures delineate the time structure of correlation, with respect to the stimulus, and at the same time allow quantitative comparison with appropriate controls.

The principal strength of the joint PST scatter diagram over and above the cross correlation is the ease with which stimulus effects are abstracted. Isolation of stimulus effects through cross correlograms and PST histograms alone is cumbersome (Perkel et al., 1967 *b*); however, since the cross correlation is a marginal density of the joint PST distribution, it follows that the cross-correlation histogram is typically a more sensitive indicator of weak interactions than is the scatter diagram. Interpretation of features of the cross-correlation histogram has recently been reviewed (Moore et al., 1970). The ambiguity in the statistical signs of direct connection vis-à-vis shared input is common to the cross-correlation histogram and the joint PST scatter diagram. Although in particular situations this ambiguity can be readily resolved (Moore et al., 1970), a generally applicable technique has not yet been devised.

On the theoretical side, there are two principal directions in which further development is needed to improve the quantitative application of these measures. The first of these concerns sampling statistics, both for the original scatter diagram and for its various controls. There are at present no rigorous ways of evaluating the statistical significance of departures from the control diagram. The difficulties involve the choice of sampling time and the assignment of elemental areas in which to compare tallies.

The second direction for further development anticipates the availability of substantially larger numbers of simultaneously recorded spike trains. Although the methods we have described can in principle be extended indefinitely, the practicality of the associated computations decreases rather rapidly as the number of spike trains increases. A radically new approach is needed in order to define and measure a time-dependent "state function" characterizing the activity of a large group of neurons. Interesting attempts along these lines have appeared (Caianiello et al., 1968; Cowan, 1968). It would seem that phenomenological descriptions of neural activity are most usefully formulated within the context of the informational and behavioral function of the neural population under observation.

Despite these limitations, the joint PST scatter diagram and its associated measures are being applied in a number of current investigations of sensory coding, of motor pattern generation (Smith, 1970; Zucker, 1971), and of neural plasticity (Kristan and Gerstein, 1970; Kristan, 1971). These measures are particularly applicable to studies of plasticity arising through modifications of neural connectivity, not only because they provide quantitative measures of effective connectivity, but also because the necessary data can be provided through relatively nondestructive extracellular recording methods.

Note Added in Proof. An essentially identical scatter diagram has been described by Lyamin (1971).

This work was supported through National Institutes of Health grants NS 05606, FR 15, and NS 09744.

Received for publication 26 July 1971.

REFERENCES

BULLOCK, T. H., and G. A. HORRIDGE. 1965. Structure and Function in the Nervous System of Invertebrates. W. H. Freeman and Co., Publishers, San Francisco, Calif.

CAIANIELLO, E. R., A. DE LUCA, and L. M. RICCIARDI. 1968. *In* Neural Networks. E. R. Caianiello, editor. Springer-Verlag KG., Berlin, W. Germany.

COWAN, J. D. 1968. *In* Neural Networks. E. R. Caianiello, editor. Springer-Verlag KG., Berlin, W. Germany.

FROST, J. D., JR., and Z. ELAZAR. 1968. *Electroencephalogr. Clin. Neurophysiol.* **25**:499.

GERSTEIN, G. L., and D. H. PERKEL. 1969. *Science (Washington).* **164**:828.

KRISTAN, W. B., JR. 1971. *J. Neurophysiol.* **34**:321.

KRISTAN, W. B., JR., and G. L. GERSTEIN. 1970. *Science (Washington).* **169**:1336.

LYAMIN, E. A. 1971. *In* Mathematical Models of Biological Systems. G. M. Frank, editor. Izdatelstvo "Nauka," Moscow.

MOORE, G. P., J. P. SEGUNDO, D. H. PERKEL, and H. LEVITAN. 1970. *Biophys. J.* **10**:876.

PERKEL, D. H. 1970. *In* The Neurosciences: Second Study Program. F. O. Schmitt, editor-in-chief. The Rockefeller University Press, New York. 587.

PERKEL, D. H., G. L. GERSTEIN, and G. P. MOORE. 1967 *a*. *Biophys. J.* **7**:391.

PERKEL, D. H., G. L. GERSTEIN, and G. P. MOORE. 1967 *b*. *Biophys. J.* **7**:419.

SHERRINGTON, C. S. 1951. Man on His Nature. Cambridge University Press, London, England. 2nd edition.

SMITH, D. O. 1970. Motor control of the crayfish claw. Ph.D. Dissertation. Stanford University, Stanford, Calif.

ZUCKER, R. S. 1971. A neuronal circuit mediating escape behavior in the crayfish. Ph.D. Dissertation. Stanford University, Stanford, Calif.

JOURNAL OF NEUROPHYSIOLOGY
Vol. 61, No. 5, May 1989. *Printed in U.S.A.*

Dynamics of Neuronal Firing Correlation: Modulation of "Effective Connectivity"

A. M. H. J. AERTSEN, G. L. GERSTEIN, M. K. HABIB, AND G. PALM
(With the Collaboration of P. Gochin and J. Krüger)
Max-Planck-Institute for Biological Cybernetics, D-7400 Tübingen, Federal Republic of Germany; Department of Physiology, University of Pennsylvania, Philadelphia, Pennsylvania 19104-6085; Department of Bio-Statistics, University of North Carolina, Chapel Hill, North Carolina 27514; Vogt-Institute for Brain Research, University of Düsseldorf, D-4000 Düsseldorf, Federal Republic of Germany; Neurlogical Clinic, University of Freiburg, D-7800 Freiburg, Federal Republic of Germany.

SUMMARY AND CONCLUSIONS

1. We reexamine the possibilities for analyzing and interpreting the time course of correlation in spike trains simultaneously and separably recorded from two neurons.

2. We develop procedures to quantify and properly normalize the classical joint peristimulus time scatter diagram. These allow separation of the "raw" correlation into components caused by direct stimulus modulations of the single-neuron firing rates and those caused by various types of interaction between the two neurons.

3. A newly developed significance test ("surprise") is applied to evaluate such inferences.

4. Application of the new procedures to simulated spike trains allowed the recovery of the known circuitry. In particular, it proved possible to recover fast stimulus-locked modulations of "effective connectivity," even if they were masked by strong direct stimulus modulations of individual firing rates. These procedures thus present a clearly superior alternative to the commonly used "shift predictor."

5. Adopting a model-based approach, we generalize the classical measures for quantifying a direct interneuronal connection ("efficacy" and "contribution") to include possible stimulus-locked time variations.

6. Application of the new procedures to real spike trains from several different preparations showed that fast stimulus-locked modulations of "effective connectivity" also occur for real neurons.

INTRODUCTION

When examining the relation between activities of two neurons in a multi-neuron recording, a most useful tool has been the cross-correlation of the two spike trains (29; for reviews, see Refs. 12 and 20). This calculation measures the probability of firing of the "target" neuron at various times relative to the firing of the "reference" neuron, in which the probability is determined by averaging over many occurrences of the two spikes.

This procedure detects the (delayed) coincidence of neural firing among different neurons, which can be regarded in two ways. The first way is to regard such coincidences as a sign of a possible neural code being used by the working brain: a code based on the cooperative action of two or more impulse-carrying pathways (30). A second way is to use the near-coincidence to infer functional connectivity among the underlying elements. Such knowledge can be extracted by the experimenter, but it is not obvious how that should be available to the brain under investigation. Results of this sort of analysis are traditionally interpreted as indicative of "effective connectivity" among the observed neurons. Appropriate control calculations are made to allow the separation of direct stimulus effects from other contributions to the "raw" correlation. The usual strategy has been to subtract the direct stimulus effects [estimated by the so-called "shift predictor" (10, 11, 29)] from the "raw" cross-correlogram, yielding a time-averaged "residual" correlation due to neural origin. Sources of the latter are further subdivided into "direct connections" and "shared input"; this subdivision is based on inspection of the correlogram shape. A review of topics related to cross-correlation of spike trains together with many references to the original literature can be found in Glaser and Ruchkin (16).

Many of the concepts and measurements used with spike trains have been developed and calibrated with the use of pulse trains from simulated neuronal circuits. In that case, the measurements for a given circuit can be predicted: this is the "forward problem." We are fully aware that there is no unique solution to the "inverse problem," i.e., it is impossible to determine uniquely the underlying circuit by spike-train analysis (or any other method that avoids exhaustive enumeration of all states of all elements). Our goal is to define the minimum, simple neuronal model that would replicate the experimentally observed features of measurements made on simultaneously recorded spike trains. Thus, when we make the jump from observed, coincident spike events to a statement of "*effective connectivity*" between two neurons, this should be taken as an abbreviated description of an equivalent class of neuronal circuits.

Most applications of cross-correlation have assumed that the system is stationary at all time scales. This is, of course, rarely the case for real neurons. Variations may occur at three rough time scales: short (<1 s), medium (between 1 and 1,000 s), and longer. The variations on short and medium time scales may be tightly related to stimulus conditions; the long-term variations may be related to phenomena like learning. Tools more complex than ordinary cross-correlation, which integrates over time, are appropriate to deal with each of these nonstationarity time scales. For example, the gravitational clustering algorithm (3, 13,

14) and modifications of it (4) can demonstrate variations on the medium time scale.

In this paper we will examine the possibilities for dealing with short time scale, stimulus-locked variations of near-coincident firing. Such variations of near-coincident firing are to be expected in most cases of effective stimulation, if only because the individual neuron spike rates will be modulated. After correction for such individual rate modulations, any residual variation of the near-coincident firing can be interpreted in terms of corresponding variation of the interactions between the observed neurons. We will demonstrate the detection, quantification, and interpretation of such residual stimulus-time-locked variations in near-coincident firing.

CRITICAL REVIEW OF PREVIOUS APPROACHES TO DYNAMIC CORRELATION

Most electrophysiological measurements on the nervous system involve some kind of time averaging. The reason for this is the apparent stochastic nature of the observed activity. Presumably, the stochastic behavior of the neurons consists partly of statistical fluctuations ("noise"), but may also include genuine (rapid) modulations of activity and/or connectivity. The latter would imply that correlation of firing is a time-varying function; obviously, the usual cross-correlogram would collapse such time variations and present the average values. What is really wanted, but presently unattained, is a new class of measurement that would be essentially instantaneous so as to deal with the modulations of activity, but yet that would be able to discard the statistical fluctuations

The joint peristimulus time scatter diagram

A partial solution to the above dilemma can be developed from the idea of the *peristimulus time (PST) histogram*: let us consider a time-dependent measure of correlation, averaged over many repetitions of the same stimulus and time-locked to the instant of stimulus presentation. This would allow detection of any time structure in the correlation that is related to the instant of stimulus presentation, and yet, being a special form of average, would cope with the statistical fluctuations. Obviously, any modulations (or long-term changes) that are not time-locked to the stimulus would not be detected by such an approach.

An early version of the required calculation was named the *joint peristimulus time (PST) scatter diagram* (10, 11). The basic idea is to create a two-dimensional scatter diagram of the firings of the two neurons relative to each stimulus onset. Technically, this diagram is related to the time-dependent cross-correlation function of the two spike trains. Each dot in the scatter diagram represents a (delayed) coincidence of the two spike trains during one stimulus period, as indicated in Fig. 1*A*, together with the underlying spike trains and their contributions to what will ultimately become PST histograms.

As this process is carried out for many repetitions of the same stimulus, we build up dot densities in the scatter diagram and along the PST axes (Fig. 1*B* shows a scatter diagram for data from a neuronal simulator circuit that will be described in more detail below; note that in such a scatter diagram we do not recognize possible superposition of dots). Regions of high or low density lying in bands parallel to the scatter diagram axes represent stimulus-locked activation or suppression of either neuron; the variation of dot density along such a band represents the stimulus-induced modulation of firing rate in the corresponding neuron. Regions of high or low density lying parallel to the principal diagonal represent excess or deficient amounts of delayed coincident firing, in which the time delay equals the distance from the high- (or low-) density band to the diagonal; the variation of dot density along such a band represents the stimulus-induced modulation of the near-coincident firing at that particular delay. These various features are illustrated for various simulations of simple neuronal circuits in the catalog of joint-PST scatter diagrams, PST histograms, and cross-correlograms given by Gerstein and Perkel (11).

For ease of explanation, we assumed that the time axes of the Joint-PST scatter diagram start at stimulus onset and extend over an interval equal to the stimulus duration. Clearly, one may deviate from this simple scheme. By selecting an appropriate time window, possibly offset with respect to the stimulus trigger, one may "zoom in" on some arbitrary fraction of the stimulus period. One may also go beyond the time duration of the stimulus by incorporating suitably chosen pre- and/or poststimulus intervals, thus enabling explicit comparison of stimulus and nonstimulus conditions within a single scatter diagram. In the latter case, one should, however, make sure that in each of the trials the extrastimulus interval still represents the same experimental conditions; this is especially important when triggering on repeated occurrences of a particular stimulus, temporally embedded in a (pseudo-) randomly structured stimulus ensemble with possibly varying "neighboring" stimuli and/or interstimulus pauses.

We can estimate the density in a scatter diagram with various histograms. A Cartesian grid to define bins over which tallies are made (schematically indicated in Fig. 1*C*) leads to the *joint-PST histogram* (JPSTH) and the two ordinary *PST histograms* (Fig. 1*F*). (To allow easy timing comparisons in subsequent modifications of this figure, the PSTHs along the *axes* are preserved without change in all versions.) Tallies over the diagonally arranged set of bins indicated in Fig. 1*D* lead to a histogram of the time-locked, near-coincident firing shown in Fig. 1*G*. This measure shows the stimulus-locked modulation of near-coincident firing in the same way as the PSTH shows the stimulus-locked modulation of single-neuron firing. Where appropriate, we will call this diagonal histogram the *PST coincidence histogram*. We found it generally convenient to smooth the PST coincidence histogram to reduce statistical variation, especially when we choose a narrow band in which to make the tallies. This would be the case when we study the time distribution of the contributions to a narrow peak (or trough) in the cross-correlogram The smoothing procedure is equivalent to the assumption that variation along the PST coincidence histogram is slow compared to the binwidth. As smoothing function, we use a gaussian with sigma of four (sometimes two) bins; in each case, this

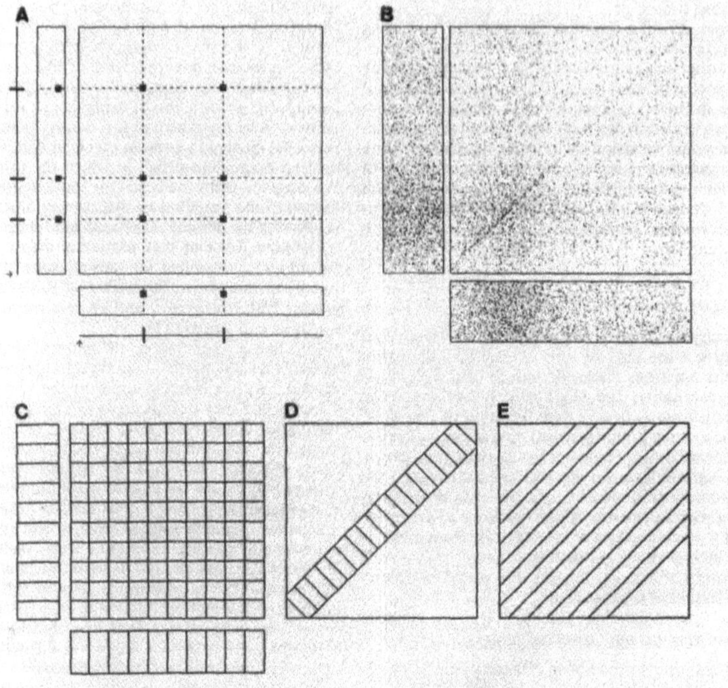
A
B
C
D
E

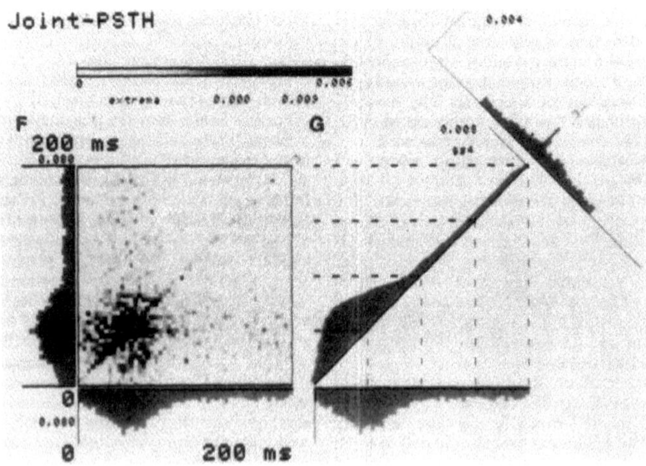
Joint-PSTH
F
G
200 ms
0
0
200 ms

particular value as well as the location and width of the selected diagonal band are indicated in the figure (cf. Figs. 1, *F* and *G*).

Finally, tallies over the set of para-diagonal bins indicated in Fig. 1*E*, suitably normalized for the different length of each bin, estimate the ordinary *cross-correlation histogram* (Fig. 1*G*). Note that this operation is equivalent to an additional time averaging that inexorably washes out any possible stimulus-induced modulation. We will call this time-averaged measure the (ordinary) cross-correlogram.

Clearly, the statistical variance of counts in the cross-correlogram bins will increase with the distance from the center bin, due to the decreasing length of the corresponding para-diagonal JPSTH bins; for this reason, we only show the central part of the cross-correlogram. An alternative would be to use para-diagonal bins of identical size, which inevitably would be the smallest of all values used here. This, however, would amount to an unnecessary sacrifice of statistical reliability in the center region of the cross-correlogram; in fact, that would be most unfortunate, because precisely this region is the more interesting part of the histogram in view of the latencies usually associated with (direct) interneuronal connections. If the spike train data would originate from an experiment that allows extending the time base of the JPSTH to encompass more than the duration of the stimulus itself, the para-diagonal JPSTH bins could be arranged such that they all have the extent of the stimulus duration, simply by going beyond the stimulus boundaries for off-diagonal bins. This would be equivalent to cutting a square (or rectangular) section from a temporally extended joint-PST histogram, the section being oriented parallel to the main diagonal and extending equally wide below and above it. The next logical step would be to take this section and rotate it clockwise over 45°; the result is an alternative temporal arrangement of the joint-PST histogram, with the x-axis representing running time and the y-axis corresponding to "difference time." [A more formal description of such operation has been given in the context of nonlinear signal and systems theory (1).] As noted earlier, however, such extension of the time window beyond the stimulus duration is only allowed if experimental conditions in the extrastimulus interval are the same in all trials. Because at this stage we do not want to impose any restrictions on the temporal experiment design, we adopted the above described procedure of compensating for variation in the extent of para-diagonal bins.

The actual display arrangement of the PST coincidence histogram and the cross-correlogram (along and perpendicular to the diagonal in Fig. 1*G*) was chosen to emphasize their logical relationship to the ordinary PST histograms and the joint-PSTH (Fig. 1*F*). The additional grid of horizontal and vertical broken lines in Fig. 1*G* should accommodate comparisons between the time course of the PST histograms and the diagonal histogram.

Prediction of direct stimulus effects

Just as in ordinary cross-correlation, the joint-PST histogram (JPSTH) will in general contain contributions from several sources of (delayed) coincident firing. Those contributions that come purely from stimulus-related modulations of single-neuron firing rates can be predicted. Comparison of the "raw" JPSTH with this predictor defines a "residual" that represents the intrinsic neuronal dependencies alone. Note that the predictor reflects the null hypothesis that spike events in the two neurons are statistically independent, only the firing probabilities of the two neurons are related to the stimulus.

In the past, as a practical matter, the prediction of purely stimulus-related effects in both joint-PST scatter diagrams and in ordinary cross-correlations has usually been calculated by the so-called "shift (or shuffle) predictor" (10, 11, 29). This predictor is based on the idea that most neuronal interactions occur on a much shorter time scale than the time elapsing between two successive stimuli. Thus (for the case of periodic stimuli) if we shift one of the two spike trains by one (or more) stimulus periods, the spike trains will still contain all direct stimulus-induced modulations at the single-neuron level; neural interaction effects will, however, be destroyed, because the time shift is much larger than the delays usually involved in neural interactions. Similar effects can be achieved by cutting one spike train

FIG. 1. Stimulus-locked dynamic correlation of neuronal firing. *A*: principle of generation of a 2-dimensional scatter diagram of the firings of 2 neurons relative to each stimulus onset (indicated by arrow on horizontal and vertical axis): each dot in the scatter diagram represents a (delayed) coincidence of the 2 spike trains during 1 stimulus period. The spike trains and their contributions to the single-unit dot displays are indicated along the x- and y-axes. *B*: as this process is carried out for many repetitions of the same stimulus, we build up the *joint peristimulus time (PST) scatter diagram* and, along the *axes*, the single-unit dot displays. Spike trains used for Fig. 1 were obtained from a neuronal simulator; the simulated neuronal circuit is depicted in Fig. 2A, stimulus and constant connectivity in Fig. 2, *A* and *B*. *C*–*E*: different binning schemes for estimation of dot densities in the scatter diagram and the single-unit dot displays. *C*: Cartesian grid of bins over which tallies are made results in the joint-PST histogram (JPSTH) and the 2 ordinary PST histograms. *D*: tallies over a (para-)diagonal arrangement of bins lead to a histogram of the time-locked near-coincident firings: the PST coincidence histogram. *E*: tallies over the set of (long) para-diagonal bins, suitably normalized for the different length of each bin, estimate the time average of near-coincident firing as a function of relative delay: the ordinary cross-correlation histogram. *F*: *joint-PST histogram* (JPSTH matrix) and the 2 ordinary *PST histograms* along its x- and y-axis (binwidth: 4 ms). Values in the JPSTH matrix are displayed by using gray levels: the higher the value, the darker the gray, as indicated in the gray wedge with the associated values. The tic mark above the gray wedge corresponds to the value zero. All counts were divided by the number of stimulus presentations. *G*: *PST coincidence histogram* (running from *lower left* to *upper right*) and the ordinary *cross-correlation histogram* (running from *upper left* to *lower right*). The PST coincidence histogram was smoothed using a gaussian with σ of 4 bins; this particular value (gs4), as well as the location and width of the selected diagonal band are indicated in *F* and *G*. The position of true coincidence (zero delay) in the cross-correlogram coincides with the intersection point of the PST coincidence histogram and the cross-correlogram; it is indicated by a tic mark near the diagonal band marker above the correlogram.

on stimulus markers, shuffling the sections, and concatenating them. If we now recompute either the joint-PST scatter diagram or the ordinary cross-correlogram for such manipulated spike trains, we have a prediction for the purely stimulus-related effects. Note that a prediction calculated in this way will have exactly the same kind of statistical variation as the original joint-PST scatter diagram or cross-correlogram, so that comparison must be made between two equally "noisy" functions. Such comparisons, for the scatter diagram, are necessarily visual rather than quantitative.

In the original papers on this subject (10, 11, 29), alternate predictors based on the PST histograms were in fact proposed. The PST histogram is appropriate for this purpose, because it is an estimate of the purely stimulus-induced modulation of the single-neuron firing. Such predictors are intrinsically "smooth" in comparison with the "raw" joint calculations, because the PST histograms are themselves already averages over many trials. Because such predictors are binned, it is mandatory to also transform the joint-PST scatter diagram into a histogram (see Fig. 1, *C* and *F*) to allow for *quantitative* rather than pictorial comparison. (The binning issue did not arise for the usual cross-correlogram, because historically it has always been presented as a binned histogram rather than as a plot of dots along a line.)

In the original papers, the preference for the shift predictor relative to the PST histogram-based predictor was explained by noting that "a two-dimensional histogram is difficult to compare with the original scatter diagram" (Ref. 11, p. 464). From this, in itself correct observation, the authors, however, chose to generate a predictor that was itself a scatter diagram rather than using a binned predictor and transforming the original scatter diagram into a histogram.

This choice implies a comparison between two scatter diagrams, which is, at best, a qualitative process. Humans perform the task by "squinting" at the two scatter plots, in effect smoothing them with a low-pass spatial filter. Obviously, the smoothing could be carried out quantitatively (at some cost in computer time) by passing an appropriate two-dimensional gaussian over the scatter diagram, thus attaining any desired degree of "spatial" resolution (31). A much more convenient and rapid way to attain similar results is by binning. This has the additional advantage that appropriate choice of bins transforms a spike train into the theoretically more attractive representation of a (0, 1)-process: per stimulus presentation at most one count is added to any histogram bin.

Once we have an original measurement that is a histogram, it is appropriate to choose a predictor that is also a histogram for comparison. This leads to the alternate, and superior, choice of predictors, i.e., those based on the PST histograms; a binned histogram of a shift predictor would be far noisier. For the ordinary cross-correlogram, the preferred predictor is the cross-correlogram of the two PST histograms. For the JPSTH, the predictor is the cross-product of the two PST histograms, taken bin by bin. All histograms should be normalized for the number of trials (stimulus presentations).

It turns out that the shift predictor and the PST-based predictor are tightly related, and that the above discussion of relative merits of predictors holds only for the simple shift predictor (i.e., involving any one order of shift), which is, in fact, the one commonly used. It is possible to work with a binned compound shift predictor, i.e., an average over shifts of different order so as to arrive at a statistically smoother predictor. (For such a procedure in the context of ordinary cross-correlograms, see Ref. 8.) It is easy to show that an average over the set of all possible shift predictors (i.e., ALL different orders of shift, including 0) is entirely equivalent to the PST-based predictor for both ordinary and joint cross-correlation (27).

All the above, however, does not directly address a more fundamental problem, associated with the application of the ordinary cross-correlogram to stimulus-driven spike trains. In principle, the cross-correlation of two stochastic processes is a function of two time arguments, t_1 and t_2, namely, the different observation times for each of the two processes. Only under special conditions concerning stationarity does this reduce to a function of a single time argument, i.e., the time shift $t_1 - t_2$, such as is done in the ordinary cross-correlogram. Processes that are called "jointly stationary in the wide sense" do fulfill these conditions (6, 28). Such processes are required, amongst others, to have an expected value that is independent of time. Because the very purpose of presenting "adequate" stimuli is to influence the neuron's firing behavior to a maximum extent (exemplified by large excursions in the PST histogram), the requirement of a constant firing rate throughout the entire stimulus presentation seems to impose a strong, if not contradictory, condition on the experimental data. Consequently, it is not a priori clear whether the current practice of applying ordinary cross-correlation to stimulus-driven spike trains is allowed at all, nor how the results should be evaluated. The issue of stimulus normalization, which only arises for stimulus-driven data to begin with, merely underlines this problematic aspect of the ordinary cross-correlogram of nonstationary spike trains.

COMPARISON OF "RAW" AND PREDICTED JOINT-PST HISTOGRAMS

For ease of explanation in what follows, we will assume that the stimulus and consequent single-neuron rate modulations are periodic. It is trivial to lift this restriction once the arguments and calculations are understood. We assume that all spike trains are "periodically stationary," i.e., that individual trials are statistically indistinguishable. This excludes long-term trends. Finally, we assume that binwidth is chosen such that in each trial we collect at most one spike per bin. We call these numbers of spikes per bin per sweep $n_i^{(k)}(u)$, where u is discrete time (bin number) relative to the kth, i.e., most recent occurrence of the stimulus marker, and i indicates the neuron. Our restriction on binwidth means that n can only take values of zero or one; these define the fundamental events with which we will be working. Note that at this stage we make no assumptions whatsoever about the way in which direct stimulus effects and neural effects combine.

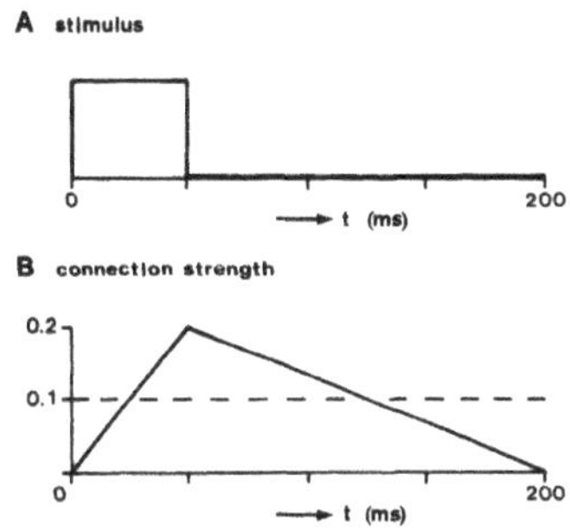

FIG. 2. Time course of stimulus (*A*) and connectivity (*B*) used in simulations of neuronal circuits in Figs. 3*A* and 4, *A* and *B*. The broken line in *B* indicates the constant excitatory connection strength of 0.1 for the circuit in Fig. 3*A*; the continuous line corresponds to the stimulus-modulated connectivity (time average, 0.1) used for the circuits in Fig. 4, *A* and *B*.

To illustrate the various possibilities for comparison of "raw" and predicted joint-PST histograms (JPSTH), we will make use of spike trains produced by a neuronal simulator program. The design of this simple simulator has been described (2), the current version allows for stimulus modulation of both individual firing rates and of the strength of interneuronal connectivity. Let us for the moment continue to work with the data whose "raw" JPSTH is shown in Fig. 1*F* and repeated in Fig. 3*C*. The simulated circuit consisted of two neurons, both spontaneously active (some 4 spikes/s), both excited by a stimulus (schematically depicted in Fig. 2*A*), and with neuron 1 exciting neuron 2 with constant efficacy (i.e., probability of inducing a spike from the "postsynaptic" neuron) of 0.1 as indicated in Figs. 2*B* and 3*A*. Average spike rates were set to 7 spikes/s to mimic typical cortical recordings.

With the above given assumptions, we can develop some convenient notation. Let the stimulus-locked event density be $E_{\text{stim}}[n_i(u)]$; the k index is omitted because we are considering an expectation. The estimator of this quantity is the ordinary PST histogram

$$\langle n_i(u)\rangle = \frac{1}{K}\sum_{k=1}^{K} n_i^{(k)}(u) \qquad (1)$$

Note that because of the normalization for stimulus number, values in the PST histogram are restricted to the range zero to one. With the same notation, the contribution to the joint-PST histogram by the events during the kth stimulus period is given by

$$n_{ij}^{(k)}(u, v) = n_i^{(k)}(u)n_j^{(k)}(v) \qquad (2)$$

These joint events are also zeroes or ones. In analogy to the single-neuron situation, we can now introduce the stimulus-locked joint event density $E_{\text{stim}}[n_{ij}(u, v)]$ and its estimator, the JPSTH

$$\langle n_{ij}(u, v)\rangle = \frac{1}{K}\sum_{k=1}^{K} n_{ij}^{(k)}(u, v) \qquad (3)$$

Note that because of the normalization for stimulus number, also the JPSTH values are restricted to the range zero to one.

Under the null hypothesis of no interaction between the neurons, it can be formally shown that the cross-product matrix of the individual PST histograms (i.e., bin by bin) is an appropriate estimator of the expected JPSTH (27). This predictor is given by

$$\tilde{n}_{ij}(u, v) = \langle n_i(u)\rangle\langle n_j(v)\rangle \qquad (4)$$

and is shown in Fig. 3*B*. Its values also lie between zero and one. As promised, this is a very smooth function in comparison with the "raw" JPSTH; it shows clear horizontal and vertical features, but no diagonal features. Nevertheless, its diagonal histogram, the predicted PST coincidence histogram, shows a clear, stimulus-locked modulation of near-coincidence firing. This is the amount and time course of near-coincident firing that would be expected purely on the basis of the single-neuron firing rates and their modulations by the stimulus.

A number of ways can be envisioned to compare the raw and predicted JPSTH. Because we want to test the null hypothesis of no interaction, we are particularly interested in the dissimilarity between the two. The simplest measure for this dissimilarity is the algebraic difference

$$\begin{aligned} D_{ij}(u, v) &= \langle n_{ij}(u, v)\rangle - \tilde{n}_{ij}(u, v) \\ &= \langle n_{ij}(u, v)\rangle - \langle n_i(u)\rangle\langle n_j(v)\rangle \qquad (5) \end{aligned}$$

The values of this function lie between minus and plus one. *Equation 5* is equivalent to the usual definition of the cross-covariance function (6, 28); we shall refer to it accordingly. This function is shown in Fig. 3*D*. We note that the bands of high counts parallel to the x- and y-axes were present both in the "raw" and predicted JPSTH; the operation of taking an algebraic difference in the cross-covariance removes these features of net positive values. The diagonal feature was present only in the "raw" JPSTH; therefore, it is not surprising that it persists in the cross-covariance. There is still a modulation along the diagonal; the time course of this modulation roughly follows that in the two ordinary PST histograms shown along the axes. We will return to this modulation later on. Note that the average of the background in both the cross-covariance and, consequently, in the differential correlogram (at *upper right*) has been shifted to zero by the subtraction operation. The variance of the background may not be uniform, however, and will depend on the detailed firing rates as shown by the PST histograms. This means that different portions of the cross-covariance may show different "texture" (notice the horizontal and vertical bands of texture around the zero value).

The range of values comprising both the statistical variance and the "real" features in the algebraic difference of the cross-covariance will clearly depend on the details of the data. To compare the importance of various features in different pieces of data (e.g., corresponding to the presentation of different stimuli to the same two neurons), we need an appropriate normalization. In other words, we need a data-dependent measuring stick to arrive at a data-*in*dependent measure of departure from the null hypothesis. To give qualitative meaning to the extent of such departure, let

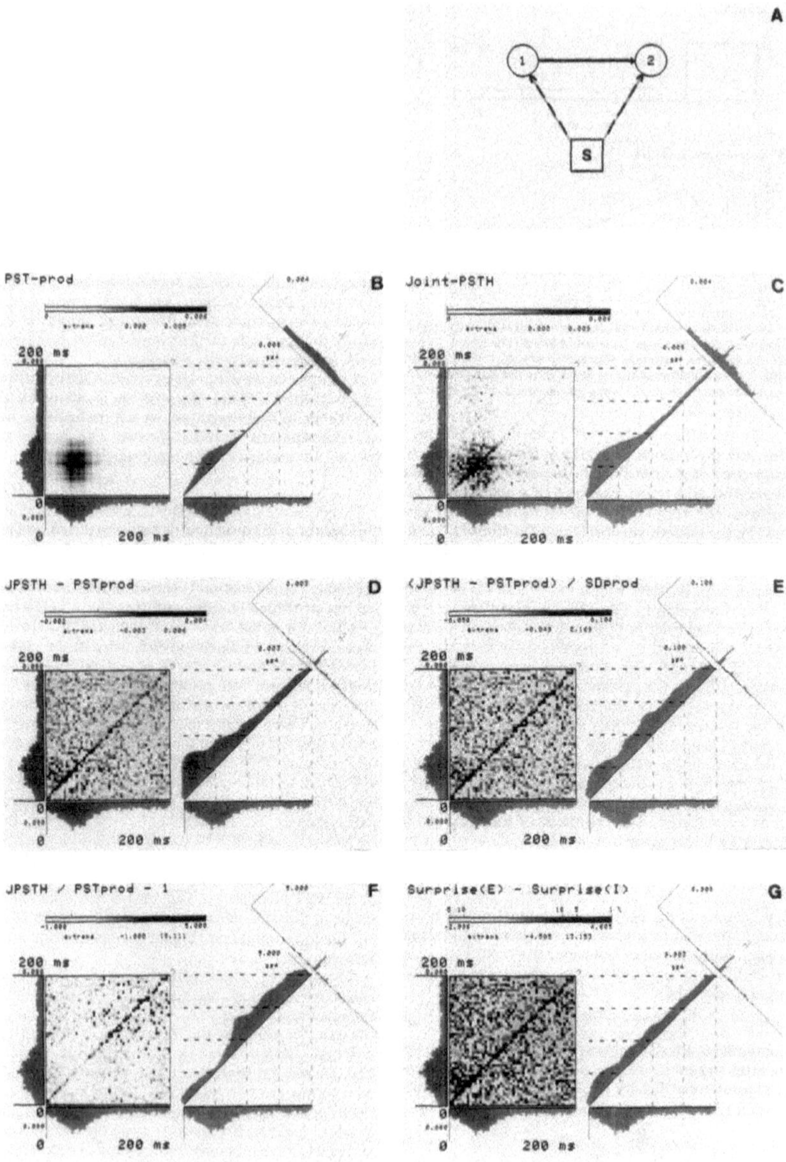
A
1
2
S
PST-prod
B
Joint-PSTH
C
JPSTH - PSTprod
D
(JPSTH - PSTprod) / SDprod
E
JPSTH / PSTprod - 1
F
Surprise(E) - Surprise(I)
G
200 ms
0

us in addition require from an appropriate normalization procedure that the resulting values in diagonal features bear a monotonic relation to the "effective connectivity" in the underlying neuronal network: the stronger the connectivity, the larger the departure, and vice versa. This additional and relatively mild requirement allows us to interpret temporal variations of diagonal features in a properly normalized JPSTH in terms of modulations of "effective connectivity."

One attempt to achieve this normalization goal is to measure the departure from predicted value relative to that predicted value. This means we want to divide the cross-covariance by the cross-product of the PST histograms

$$R_{ij}(u, v) = \frac{D_{ij}(u, v)}{\tilde{n}_{ij}(u, v)} = \frac{\langle n_{ij}(u, v)\rangle}{\langle n_i(u)\rangle\langle n_j(v)\rangle} - 1 \qquad (6)$$

Note that the values of this function are not a priori limited to any particular range. This quantity has been called "scaled cross-correlation density" by Kuznetsov and Stratonovich (21; see also Ref. 7). We show the results of this calculation in Fig. 3*F*; it is clearly unsatisfactory. It has reintroduced vertical and horizontal features in the off-diagonal regions: a clear modulation of variance (note the "blur" in Fig. 3*F*, coinciding with the original horizontal and vertical features from Fig. 3*D*) where there should be only uniformity. In addition, the diagonal histogram is not uniform along its length, but rather has clearly lower values in the early part of its time course, i.e., where the PST histograms have high values. Such variation along the diagonal is hard to reconcile with the known constant strength of connection in the simulated neuronal circuit. We have also applied this attempted normalization to other simulated spike train data (with known underlying circuitry), and in all cases the results of this procedure were similarly inadequate. In summary, this attempted stimulus normalization procedure seems to be decompressing too strongly in regions corresponding to low firing rates of either or both neurons and, therefore, will not be shown again.

A second possible normalization of the cross-covariance is inspired by standard statistical procedures. Here we divide the cross-covariance by the standard deviation of the predictor. Under the null hypothesis this standard deviation is simply the cross-product of the standard deviations of the PST histograms

$$\tilde{s}_{ij}(u, v) = s_i(u)s_j(v) = \left\{\frac{1}{K}\sum_{k=1}^{K}(n_i^{(k)}(u) - \langle n_i(u)\rangle)^2 \cdot \frac{1}{K}\sum_{k=1}^{K}(n_j^{(k)}(v) - \langle n_j(v)\rangle)^2\right\}^{1/2} \qquad (7a)$$

As can be observed from comparison of *Eqs. 2–5* and *7a*, an equivalent formulation of this quantity is the square root of the product of the two single-neuron autocovariances, each matrix considered only along its diagonal

$$\tilde{s}_{ij}(u, v) = \{D_{ii}(u, u)D_{jj}(v, v)\}^{1/2} \qquad (7b)$$

It should be noted that there is a simple relation between the predictor in *Eq. 4* and the standard deviation in *Eq. 7a*. With the assumptions stated at the outset, the expectations $E[n_i(u)]$ and $E[n_i^2(u)]$ are equal for any u because we are dealing with a "zero-one" process. From this it is easily shown that the standard deviation matrix is

$$\tilde{s}_{ij} = s_i s_j = \{\langle n_i\rangle(1 - \langle n_i\rangle)\cdot\langle n_j\rangle(1 - \langle n_j\rangle)\}^{1/2} \qquad (8)$$

where we omit the time bin reference (u, v) for brevity. For $\langle n_i\rangle$ and $\langle n_j\rangle$ small compared to one (the usual case), say both equal to ϵ, it follows that the predictor (*Eq. 4*) behaves as ϵ^2, whereas the standard deviation (*Eq. 8*) behaves as ϵ. Thus effectively the standard deviation is the square root of the predictor. Furthermore, the ratio of maximum to minimum values of expression (*Eq. 8*) will be the *square root* of the corresponding ratio in the predictor of expression (*Eq. 4*). The standard deviation thus has a smaller relative dynamic range than the predictor; it is a very smooth function, and we omit explicitly showing it here. Such smoothness is an obvious requirement for any division process in order not to increase the "noise" level.

The normalization of the cross-covariance using the standard deviation of the predictor as scaling matrix is, therefore, given by

$$C_{ij}(u, v) = \frac{D_{ij}(u, v)}{\tilde{s}_{ij}(u, v)} = \frac{D_{ij}(u, v)}{\{D_{ii}(u, u)D_{jj}(v, v)\}^{1/2}} \qquad (9)$$

Note that this function, unlike the relative scaling operation in *Eq. 6*, is a true *normalization*, i.e., it delivers numbers between plus and minus one. *Expression 9* is referred to in the literature as the "normalized cross-covariance function" or "correlation function coefficient" (4, 6); it has also been called "cross-correlation surface" (18). Finding the above terminology unsatisfactory for various reasons,

FIG. 3. Various possibilities for comparison of "raw" and predicted joint-PST histograms (JPSTH). *A*: the simulated neuronal circuit, giving rise to the spike train data used in Figs. 1 and 3, consisted of 2 neurons, both excited by a stimulus, and with neuron 1 exciting neuron 2 with constant efficacy (see also Fig. 2, *A* and *B*). Average spike rates were set to 7/s to mimic typical cortical recordings. Total numbers of events: 2,774 (neuron 1, *x*-axis), 2,951 (neuron 2, *y*-axis), 2,000 (stimulus triggers). *B*: estimator of expected JPSTH under the null hypothesis of noninteracting neurons: the cross-product matrix of the individual PST histograms (*Eq. 4*). Display format of this and all following figures as in Fig. 1, *F* and *G*. *C*: "raw" JPSTH (*Eq. 3*) as shown in Fig. 1, *F* and *G*, and repeated here for ease of reference. *D*: cross-covariance function (*Eq. 5*): the algebraic difference of "raw" (*C*) and predicted JPSTH (*B*). *E*: normalized JPSTH (*Eq. 9*): normalization of the cross-covariance (*D*) using the standard deviation of the predictor (cross-product of the standard deviations of the PST histograms) as scaling matrix. *F*: scaled cross-correlation density (*Eq. 6*): ratio of cross-covariance (*D*) and predicted JPSTH (*B*). *G*: significance of correlation: algebraic difference between the surprise for "excitation" (i.e., for an excess joint count compared with the individual counts), and the surprise for "inhibition" (i.e., for a deficit of joint counts). The surprise of a value is defined as the negative natural logarithm of the probability for finding that value or a more deviant one (*Eqs. A1.1* and *A1.2*). Surprise matrices were scaled for display between −2.996 (corresponding to a significance level of 5% for "inhibition") and +4.605 (similarly of 1% for "excitation") as indicated along the gray wedge.

we propose to call the result of *Eq. 9* the *normalized* JPSTH. We show the result of the calculation in Fig. 3*E*. The result is clearly better than that of Fig. 3*F*. No obvious horizontal or vertical features reappear· the off-diagonal "texture" is completely uniform. Furthermore, the diagonal feature is approximately constant along its length, as it should be for this neuronal circuit of constant connectivity. This is also shown by the normalized PST coincidence histogram in Fig. 3*E*.

Integration along the diagonal (tallies in the para-diagonal bins as in Fig. 1*E*) results in a histogram that we call the *normalized cross-correlogram*. It should be noted that this normalization in general can *not* be achieved by any operation on the usual cross-correlogram: the proper procedure fundamentally involves the temporal decomposition of near-coincident firing in the JPSTH, its subsequent dynamic correction (subtraction and scaling), and reintegration along the diagonal. The order of these various operations can *not* be interchanged The usual procedure for correcting the ordinary cross-correlogram for stimulus effects has been to subtract the shift predictor correlogram or PST predictor correlogram (with or without additional scaling for numbers of events). It is clear that the normalized cross-correlogram, i.e., a calculation from the normalized JPSTH, will generally give a different and more correct result.

SIGNIFICANCE TESTING: SURPRISE

In the normalized JPSTH (*Eq. 9*), we have a calculation that is appropriate for the intercomparison of different pieces of data with the null hypothesis However, we now need a measure of significance to interpret the deviations from the null hypothesis. The usual approach to significance testing consists of comparing the actual outcome of an experiment with the postulated distribution of values under the null hypothesis. In our case, we need to identify this postulated distribution for the JPSTH value in a given bin.

The point of departure is to calculate the distribution of possible JPSTH values for a particular bin (u, v), given *1*) the null hypothesis of independence, *2*) the values at the corresponding bins in the two PST histograms, and *3*) the number of trials (stimulus presentations) over which these PST histograms were gathered It is possible to give this distribution in closed form, expressed in terms of only these algebraic quantities (for details, see Ref. 27). By substituting the experimental values for these quantities, we obtain an explicit distribution appropriate for this particular piece of data; it generally turns out to be different from Gaussian, especially for realistic, relatively low numbers of stimulus presentations. We can now test the significance of our actual experimental JPSTH value against this theoretically derived, but experimentally particularized, distribution: the significance of a value is defined as the probability for finding that value or a more deviant one. One can view this significance test in two different ways: *1*) we test the significance of deviations of the "raw" JPSTH from the null hypothesis (i.e., the predictor), or equivalently *2*) we test the significance of deviations of the normalized JPSTH from the (predicted) value of zero.

An informative display of the results of the significance test (27) can be made by using the information-theoretical concept of "surprise" (22, 26). The surprise of a value in this context is the negative natural logarithm of the probability for finding that value or a more deviant one (formal expressions are given in Appendix 1). It is, therefore, directly related to the usual statistical notion of "level of significance." For instance, a 5% level of significance corresponds to a surprise of 2.996 ($= -\ln 0.05$); a 1% level corresponds to a surprise of 4.605. The logarithmic transformation serves to expand the scale in the interesting region in which the probability density has low values; this is somewhat comparable to the use of a decibel scale. Moreover, introduction of the logarithmic scale allows a more sensible comparison of different values of significance; the same measure is assigned to equal ratios of probabilities rather than to equal algebraic differences.

We can define a surprise measure for "excitation," $S(E)$, i.e., for an excess joint count compared to the individual counts, and comparably, a surprise for "inhibition," $S(I)$, i.e., for a deficit of joint counts. We could show these measures separately. However, if for a particular bin one of the measures is high, the other is necessarily close to zero. Thus, in case of a significant "excitation" (or "inhibition"), the algebraic difference $S(E) - S(I)$ practically equals $S(E)$ (or $-S(I)$, respectively). For reasons of display economy we have combined the two measures accordingly and show the algebraic difference $S(E) - S(I)$ for all (u, v)-bins in the form of a single matrix (Fig. 3*G*). Note that "excitation" leads to positive values of this combined measure, whereas "inhibition" leads to negative values; the larger the (absolute) value in the surprise matrix, the more statistically significant the normalized JPSTH in the corresponding bin. (In Appendix 1 we give explicit relations between the surprise values for significant "excitation" or "inhibition" and the corresponding values of the normalized JPSTH.)

We will scale all surprise matrices so that the positive range is set to 4.605 (corresponding to a significance level of 1% for "excitation") and the negative range to −2.996 (similarly, 5% for "inhibition"). The reason for this asymmetry is the difficulty in detecting inhibition as compared with excitation for the usual small counts (2, 25, 27).

All normalizations introduced above, as well as the surprise measure for significance are bin-by-bin measures. Yet in the above exposition, we have regularly referred to "features" in the matrix. In other words, we are implicitly searching for regions of approximate "spatial" coherence to assign meaning; isolated bins that reach extreme values are considered "noise." This implicit additional requirement for "spatial" coherence should somehow be incorporated into the formal bin-by-bin measure of significance. Just as in the normalizations, we have introduced this regional aspect into the surprise significance measure by using a moving average (gaussian) along the corresponding diagonal histogram of surprise values. The underlying hypothesis for this smoothing process is that the values in individual bins are independent[1]; recall that the surprise

[1] This (null) hypothesis is based on the assumption that the time interval between the bins considered is large enough such that the single neuron's spike-generating mechanism will not produce appreciable correlations

calculation is bin by bin. In such a situation, the joint probability for having any particular combination of bin values is simply the product of the individual probabilities. Here we encounter another advantage of the logarithmic surprise measure of significance: the joint surprise for any particular bin constellation is found simply by adding the corresponding individual surprise values. In our specific case, we require a large proportion of high values over some limited region of the surprise matrix; the addition of surprise values in such a region translates into the smoothing operation. Note that the added surprise values, however, cannot directly be interpreted as a measure for statistical significance. This is quite obvious, because the addition would lead to amazingly low significance probabilities. The correct statistical procedure for significance analysis of combined surprise values is the subject of current investigation.

We can now interpret the surprise measure in Fig. 3*G*. The surprise histogram shows no coherent structure except along the diagonal, where many points surpass the 1% "excitation" criterion. The corresponding diagonal histogram is relatively flat along its length. We take this result to indicate that the diagonal feature seen in the normalized JPSTH (Fig. 3*E*) is indeed a significant departure from the null hypothesis; the two neurons are not to be considered independent, and their interaction (which is "excitatory") is not modulated by the stimulus. This means that our calculations are capable of recovering the detailed structure that had been built into the simulated circuit. Although there are occasional single bins in the remainder of the surprise histogram that reach extreme values (note extrema indicated in Fig. 3*G*), we do not consider these to be significant according to our extended criterion which incorporates the additional requirement of "spatial" coherence in the surprise matrix.

CALIBRATION BY SIMULATED NEURONAL CIRCUITS

As we continue to calibrate the sensitivity of the normalized JPSTH and the surprise significance, we turn to several additional simulated neuronal circuits, as shown in Fig. 4, *A* and *B*. Both circuits have stimulus-locked modulation of the connectivity between the neurons. The objective is to investigate whether the normalized JPSTH allows detection of such stimulus-modulated connectivity, even when this modulation is masked by direct stimulus influences on each of the two neurons.

The simpler case is shown in the *left* column of Fig. 4. (The layout of each column in Fig. 4 matches the *right-hand* column of Fig. 3, as well as the columns of Fig. 5.) For this simulation, the average firing rates are ~4/s; this is a lower rate than in Fig. 3, because there are no direct stimulus effects on the neurons. The connectivity was a triangular function of time as shown in Fig. 2*B*, with the peak after the first quarter of the stimulus cycle. The numbers were specifically chosen to give an average connectivity of 0.1, the same value as the constant connectivity in Fig. 3. Fig. 4, *C*, *E*, and *G*, in that order, shows the "raw" JPSTH, the normalized JPSTH, and the surprise measure. The PST histogram along the *x*-axis, corresponding to neuron 1 (the "driver"), is flat, as expected from Fig. 4*A*. The connection between neurons 1 and 2 is sufficiently weak (although physiologically realistic) to make the PST histogram along the *y*-axis (neuron 2, the "driven") also appear flat. The only feature in the "raw" JPSTH (Fig. 4*C*) is diagonal, and its intensity, shown in the PST coincidence histogram, matches what we know of the modulation of connectivity in the circuit. Given that the PST histograms are both flat, the normalized JPSTH (Fig. 4*E*) simply reflects a changed scale and base line. Finally, the surprise measure shows a diagonal feature that has many bins exceeding the 1% criterion level. The corresponding diagonal histogram also follows the known connectivity function. To interpret these smoothed values, we also calculated a comparable histogram for the diagonal stripe that is two bins laterally displaced. That histogram does not rise above one-tenth of the peak values here. From this we conclude that the diagonal feature in the normalized JPSTH is highly significant; we have recovered the structure of the simulated circuit.

The *right-hand* column of Fig. 4 presents a combination of direct stimulus modulation of the two firing rates and the stimulus-modulated connectivity that we have just analyzed. The circuit is shown in Fig. 4*B*. The "raw" JPSTH of Fig. 4*D* shows the expected vertical, horizontal, and diagonal features. The PST histograms show the direct stimulus effects, but again there is no obvious signature of the modulated connectivity in the driven PST histogram (along the *y*-axis). The corresponding PST coincidence histogram shows a clear stimulus modulation of the near-coincident events; presumably this represents both the modulation of the firing rates and of the connectivity.

The direct stimulus effects are removed in the normalized JPSTH of Fig. 4*F*; only the diagonal feature persists. Note that Fig. 4*F* looks generally like Fig. 4*E*, its counterpart for the circuit that only has stimulus-modulated connectivity: in both cases, the normalized PST coincidence histogram closely follows the known modulation profile of the connectivity. Finally, the significance measure in Fig. 4*H* indicates that the diagonal feature in the normalized JPSTH represents a highly significant departure from the null hypothesis of no interaction. Therefore, even in this case of mixed direct and indirect stimulus effects, we have consistently recovered the simplest network that is sufficient to explain these data; this network appears to be identical to the (known) simulated circuit.

QUANTITATIVE ASSESSMENT OF "EFFECTIVE CONNECTIVITY"

It should be emphasized that in all of the above development of normalization, no assumptions whatsoever were

among these bins. A possible source for such correlations would be the neuron's refractory mechanism: the firing in one bin would be negatively correlated with the firing in adjacent bins (especially because we assume our bins small enough to have at most one spike per bin per trial). Such negative correlations would manifest themselves as para-diagonal features of decreased coincidence counts in the auto-JPSTH, which is obtained by taking the same single-neuron spike train along both the *x*- and *y*-axis of the JPSTH. What we are, in fact, assuming here is "sparse firing": firing rates are so low that, even in stimulus-driven activity, the intervals between adjacent spikes are (considerably) larger than the neuron's refractory period. With this assumption, which does not appear implausible for cortical neurons with typical, low firing rates, we can extend the null hypothesis to the effect that also the spikes in different bins for the same neuron are independent.

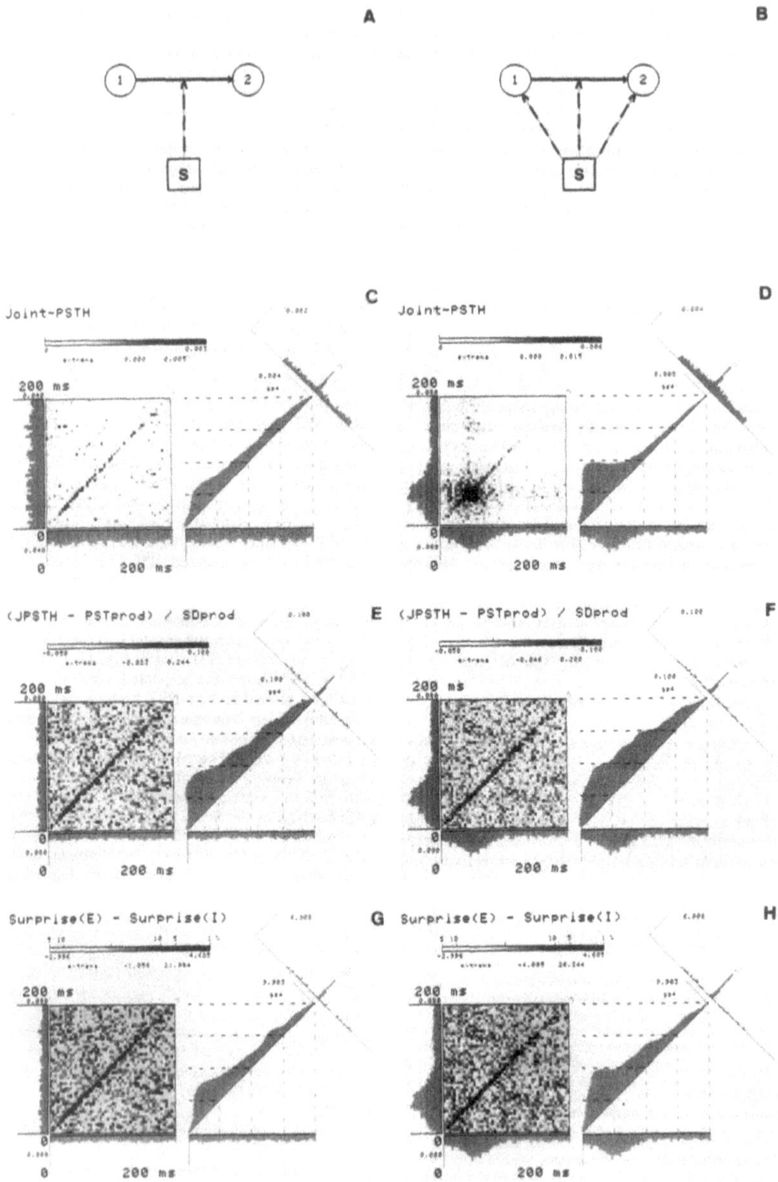
A
B
1
2
S
C
D
Joint-PSTH
200 ms
0
200 ms
E
F
(JPSTH - PSTprod) / SDprod
G
H
Surprise(E) - Surprise(I)

made regarding the structure of the underlying neuronal network: the approach is essentially *model free*. Under these conditions, the proper strategy is to test the data against the null hypothesis of no interaction; the possible outcome of this test is basically "yes" or "no" to some significance level.

Once a significant departure from the null hypothesis of independence has been established, we can move to a second, model-related stage of interpretation. So far, we have only implicitly and qualitatively assigned meaning to the magnitude of the departure from the null hypothesis: the larger the departure, the stronger the "connectivity." This notion can be made more explicit and quantitative by making specific assumptions about the underlying network structure. This means a *model-based* approach and aims at evaluation of the model's parameters. From an excess (or deficit) of correlated events, we can make the conceptual transition to a statement of "effective connectivity." The latter is expressed as a minimal neuronal circuit model that would reproduce such correlation measurements.

An important example of such a model would be the simple circuit in which neuron *a* (*x*-coordinate in the JPSTH) drives neuron *b* (*y*-coordinate) with a specifically defined (linear or nonlinear) synaptic transfer function. The normalized JPSTH in this case will show an above-diagonal feature that is displaced from the principal diagonal by an amount equal to the latency of the connection. In fact, we used just this model for the simulations that produced the data analyzed in Figs. 1, 3, and 4.

To quantitatively assess the strength of direct interaction among neurons, Levick et al. (23) introduced the notions of "efficacy" and "contribution." Efficacy was defined as the fraction of spikes from the driver that are time related to spikes from the driven neuron, contribution as the fraction of spikes from the driven neuron that are time related to spikes from the driver. These two quantities remain the principal quantitative measures for "connectivity" developed so far; consequently, they are widely used in the experimental multi-unit literature (references can be found in Refs. 12 and 20). Adopting a model-based approach, we may generalize the quantities of efficacy and contribution to include possible stimulus-locked time variations. Such manipulations allow the assignment of specific numerical values for this type of dynamic "effective connectivity." A discussion along these lines is given in Appendix 2; here we will only list the final results.

The dynamic generalization of efficacy and contribution can be expressed in terms of the experimentally observable JPSTH and the single-neuron PST histograms

$$e(u, v) = \frac{\langle n_{ab}(u, v)\rangle - \langle n_a(u)\rangle\langle n_b(v)\rangle}{\langle n_a(u)\rangle(1 - \langle n_a(u)\rangle)} \qquad (A2.11)$$

$$c(u, v) = \frac{\langle n_{ab}(u, v)\rangle - \langle n_a(u)\rangle\langle n_b(v)\rangle}{\langle n_b(v)\rangle(1 - \langle n_b(v)\rangle)} \qquad (A2.12)$$

These expressions for the dynamic efficacy and contribution can be related in an interesting way to the normalization of the JPSTH. When we compare *Eqs. A2.11* and *A2.12* with *Eqs. 8* and *9*, we observe that the normalized JPSTH takes an "intermediate" position between the complementary pair of efficacy and contribution. In contrast to the efficacy and contribution, the expression for the normalized JPSTH is symmetric with respect to both neurons, as one should expect for a model free descriptor. The "intermediate" position of the normalized JPSTH between efficacy and contribution takes the form of a geometrical average

$$(\text{normalized JPSTH})^2 = \text{Efficacy} \times \text{Contribution} \qquad (10)$$

MULTI-NEURON RECORDINGS FROM REAL NEURONS

In the above, we have described a new, extended quantification of the old joint-PST Scatter Diagram (10, 11) in terms of the "raw" JPSTH, the normalized JPSTH, and the surprise measure of significance. These quantities were calibrated by the use of spike trains derived from simulations of particular neuronal circuits. Analysis of these simulated spike trains showed that stimulus-time-locked variations of connectivity can indeed be untangled from the direct stimulus modulations of individual firing rates. In this section, we present illustrative examples of the "inverse problem": we analyze the spike trains from real neurons to infer an as yet unknown "effective connectivity."

We have taken data from two preparations: cochlear nucleus of the rat and visual cortex of the cat, both lightly anesthetized. These data were available on tape, and we did not perform any exhaustive or systematic searches. The only criterion for selection of data to be examined was an "adequate" number of firings over an "adequate" number of stimulus presentations (for obvious statistical reasons). For more detail of the experimental arrangements, original intent, and results on other issues than dealt with here, see Refs. 17, 19, and 20, respectively. Cochlear nucleus experiments were done with a repeated long sequence of 50-ms tone pips with different frequencies and amplitudes. Data corresponding to the repetitions of a particular tone pip near best frequency, to which the two observed neurons responded with an on-burst and an off-burst (see PST his-

FIG. 4. Evaluation of dynamic correlation and its significance for spike trains from 2 additional simulated neuronal circuits. *A* and *B*: schematic representation of the 2 simulated circuits. In both circuits, neuron 1 excites neuron 2 with identical stimulus-locked modulation of interneuronal connectivity (cf. Fig. 2*B*). In addition, the circuit in *B* has direct stimulus modulation of the single-unit firing rates (cf. Fig. 2*A*); this circuit thus presents a combination of the 2 circuits in Figs. 3*A* and 4*A*. The layout of each column in Fig. 4 matches the *right-hand* column of Fig. 3. Furthermore, to enable comparison across results for different circuits, scaling within the lower 2 horizontal rows of Fig. 4 was set to identical values (scaling between rows is different though). Binwidth in all cases was 4 ms. Total numbers of events were as follows: *left*: 1,578 (neuron 1, *x*-axis), 1,911 (neuron 2, *y*-axis), 2,000 (stimulus triggers); *right*: 2,658 (neuron 1, *x*-axis), 3,221 (neuron 2, *y*-axis), 2,000 (stimulus triggers). *C, E*, and *G* (*left*): the "raw" JPSTH, the normalized JPSTH, and the significance of correlation for the spike trains generated by the circuit in *A*. *D, F, H* (*right*): similarly for the spike trains generated by the circuit in *B*.

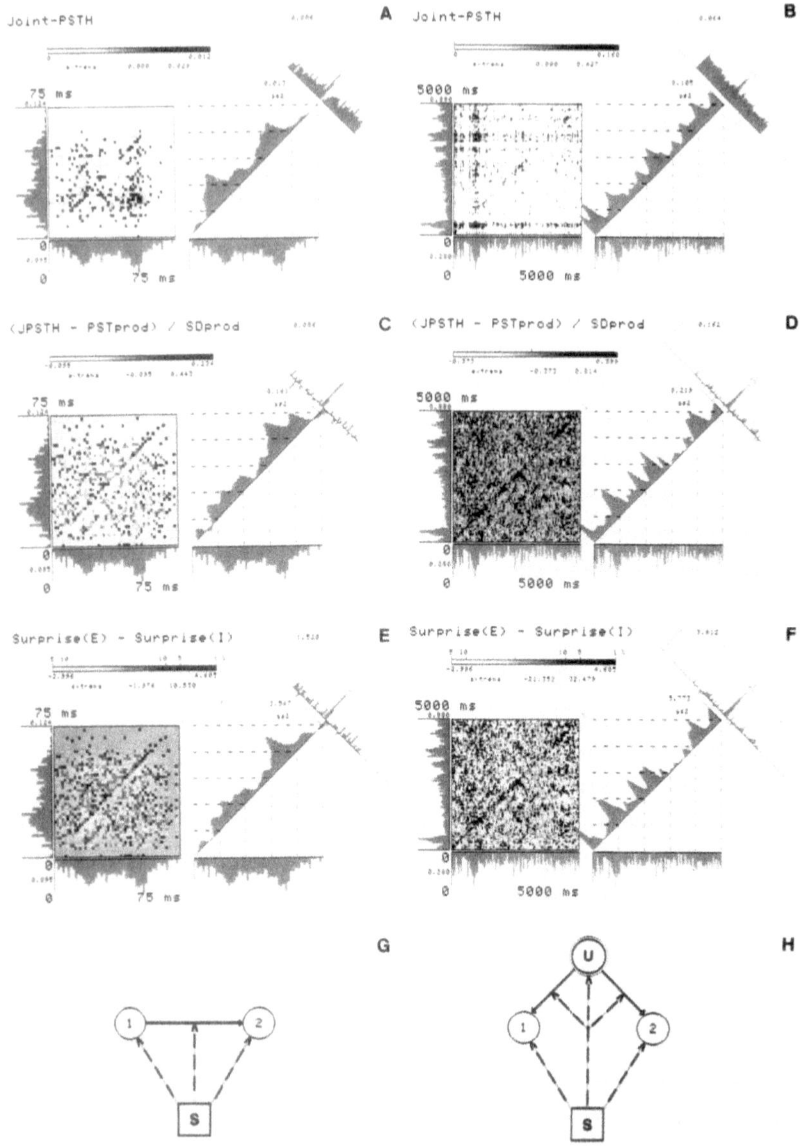
Joint-PSTH
A
Joint-PSTH
B
75 ms
0
0
75 ms
5000 ms
0
0
5000 ms
(JPSTH - PSTprod) / SDprod
C
(JPSTH - PSTprod) / SDprod
D
75 ms
0
0
75 ms
5000 ms
0
0
5000 ms
Surprise(E) - Surprise(I)
E
Surprise(E) - Surprise(I)
F
75 ms
0
0
75 ms
5000 ms
0
0
5000 ms
G
H
1
2
S
U
1
2
S

tograms in *left-hand* column of Fig. 5), was sorted out for analysis by the JPSTH. Cortical data involved repetitions of long sequences of moving bar stimuli (different orientations; one cycle of moving back and forth lasted 5 s), and was similarly sorted to isolate responses to the repetitions of a particularly effective stimulus (in this case optimal bar orientation).

We show results of the JPSTH analysis in Fig. 5. The *left* column is for a pair of rat cochlear nucleus neurons, the *right* column for a pair of cat visual cortex neurons. The layout is similar to that in the preceding figures, with the "raw" JPSTH on top, followed by the normalized JPSTH and the surprise measure for significance.

Both these sets of data show in the "raw" JPSTH all features that we have encountered above for the simulated spike trains: features parallel to the axis, corresponding to direct stimulus effects on the single-neuron firing rates, and diagonal features corresponding to the correlation of firing. The normalized JPSTHs in both cases remove most of the horizontal and vertical features, and leave only the diagonal feature. In both cases, when we examine the diagonal feature with the normalized PST coincidence histogram (tallied over the band indicated in the figure), we observe that the normalized correlation is clearly time varying along the diagonal. The significance of these effects is shown in the surprise matrices, which clearly confirm significance of the diagonal features.

To "explain" the diagonal features, we make the conceptual transition to the notion of "effective connectivity": the minimal neuronal circuit model that will reproduce these features. In such a minimal model there is a monotonic relation between diagonal feature strength and connectivity. Thus time variations along the diagonal features of the normalized JPSTHs in Fig. 5, *C* and *D*, are interpreted as indicators of time modulated "effective connectivity."

The notion of "effective connectivity" usually can be split into two types of circuitry: direct interaction and shared input. This difference is reflected in characteristics of the diagonal feature such as width and (a)symmetry around the principal diagonal. The inferred "equivalent" circuits based on these criteria are shown in Fig. 5, *G* and *H, bottom*. The narrow, slightly above-diagonal feature in the normalized JPSTH for the cochlear nucleus pair (Fig. 5*C*) is interpreted as the signature of a direct, excitatory synaptic connection from neuron 1 to neuron 2, its time course is taken to reflect stimulus modulation of the connection; we thus arrive at the model circuit in Fig. 5*G*, similar to our simulated circuit in Fig. 4*B*. Such a direct connection is not visible in the cortical data (Fig. 5, *D* and *F*): the diagonal feature is symmetrical around time shift 0, suggesting shared neuronal input. The strong time variations along the diagonal are taken to reflect stimulus modulation of the (unobserved) source(s) of shared input and/or of the connections from source(s) to neurons 1 and/or 2 (cf. Fig. 5*H*).

In these two bodies of data, without systematic or exhaustive search, we have encountered other neuron pairs with similar stimulus-modulated "effective connectivity." We have also encountered neuron pairs in which the normalized PST coincidence histogram is relatively flat, indicating that the "effective connectivity" is not modulated in a stimulus-locked fashion. Obviously, we have also encountered neuron pairs that showed no diagonal features at all; these presumably were noninteracting neuron pairs.

DISCUSSION

In this paper we have developed a quantification procedure for the study of stimulus-locked, time-dependent correlation of firing between two neurons. This procedure starts from the ancient joint-PST scatter diagram and makes it possible to unravel and quantitatively describe direct and indirect stimulus effects on correlated firing of two neurons. The direct stimulus effects are shown by the predictor for the JPSTH; this predictor is the cross-product of the two single-neuron PST histograms and represents the null hypothesis of independent firing. The indirect stimulus effects (residual correlation) are described by the normalized JPSTH (*Eq. 9*); this is the algebraic difference between the "raw" JPSTH and the predictor, with subsequent dynamic scaling by the cross-product of the standard deviations in the single-neuron PST histograms. Finally, we have demonstrated the use of a new measure for significance of features in the normalized JPSTH: surprise.

We have investigated several other possible normalization procedures (one of which is explicitly considered in this paper; see also Ref. 27). In comparison, these alternatives proved to be unsatisfactory in performance. The normalization of the JPSTH that we have selected is basically a test statistic for departure from the null hypothesis. It starts

FIG. 5. Evaluation of dynamic correlation and its significance for physiological spike trains recorded from 2 neurons simultaneously. The *left* column is for a pair of rat cochlear nucleus neurons, the *right* column for a pair of cat visual cortex (area 17) neurons. In each of the experiments, spike trains were recorded during repeated presentation of an effective stimulus. Total numbers of events were as follows: *left*: 511 (neuron 1, *x*-axis), 481 (neuron 2, *y*-axis), 242 (stimulus triggers); *right*: 756 (neuron 1, *x*-axis), 2,131 (neuron 2 *y*-axis), 75 (stimulus triggers). The layout is similar to that in Figs. 3 and 4: the "raw" JPSTH on top (*A* and *B*), followed by the normalized JPSTH (*C* and *D*) and the significance of correlation (*E* and *F*). At the *bottom*, we show the inferred "effective" circuitry (*G* and *H*). The U symbol in *H* denotes unobserved other neuron(s). Scaling is done independently for each of the figure elements. Note the different time resolutions for these 2 sets of data: binwidth was 1.5 ms (*left*) and 50 ms (*right*), sweep duration 75 ms (*left*) and 5,000 ms (*right*). The lack of counts in the band along the main diagonal of the "raw" JPSTH for the cochlear nucleus neuron pair (*A*) and, hence, the negative values in the cross-correlograms of *C* and *E*, are an artifact: the multi-unit spike trains in this case were recorded on a single microelectrode and subsequently sorted on the basis of spike waveforms. The spike sorter used in these experiments has a dead time that discards temporally overlapping spikes. This dead time, admittedly, is a most unfortunate artifact of all spike sorter algorithms developed so far. This holds in particular, because it affects the central features in the cross-correlogram, which presumably are the signs of direct interactions. The alternative of using data from different microelectrodes, with consequently larger interneuron distance, however, has the disadvantage of only rarely showing direct interaction among recorded neurons (such as shown here).

from the deviation of actual stimulus-locked correlation from its expectation, but weighs such deviation, bin by bin, for the a priori expected statistical spread as expressed by the single-neuron standard deviations. This normalization is intrinsically model free; accordingly, it is symmetric with respect to the two neurons.

Assumptions

Throughout the development of the normalization procedure, we have made two major assumptions. First, we have assumed that spike trains during successive stimulus presentations are "periodically stationary," i.e., that individual trials are statistically indistinguishable. This assumption in itself is not particularly new: any analysis of (single- or multi-unit) spike trains involving averaging over stimulus sweeps (as in, e.g., the common PST histogram or any measure derived from it) is based on the same assumption. The second assumption involves the choice of binwidth in the analysis window: bins should be taken small enough such that per trial we collect at most one spike per bin. This assumption is essentially connected to the probabilistic nature of our approach: it allows the interpretation of trial averages of various counts as probabilities, i.e., with values between zero and one. This implies that also the second assumption is not restricted to the present issue: any analysis procedure of spike trains that treats a trial average (e.g., a PST histogram) as an event density in the probability sense must necessarily make the same assumption to guarantee interpretable values between zero and one.

The assumption of "periodic stationarity" excludes long-term trends or other sources of variability on medium to large time scales. Clearly, if such sources should be present, our predictors for the JPSTH (*Eq. 4*) and its statistical variation (*Eq. 7*) would be incorrect and, consequently, the normalized JPSTH would yield erroneous results. The way in which the predictors and, consequently, the normalized JPSTH are affected is highly dependent on the nature and form of the nonstationarities. A simple example may serve to illustrate this point. In the case of a trend in overall firing rate of only one of the two neurons, while the other one remains constant, the PST-based predictor (*Eq. 4*) still is a correct estimator of the expected coincidence counts under the null hypothesis [the variance estimator (*Eq. 7*), however, in general is no longer correct]. For a linear trend in *both* neuron firing rates, however, the PST-predictor (*Eq. 4*) can either be an underestimate or an overestimate, depending on whether the trends in the two neurons go in the same or in obverse directions. The result thus is an incompletely normalized JPSTH: remaining horizontal/vertical features and/or possibly delusive suggestions of modulation of "effective connectivity" (as we have sometimes encountered). Careful screening of the spike train data for such nonstationarities, therefore, is mandatory (such screening was performed for all results presented here). Especially in data from the cortex, where medium- to long-term neuronal variability, not under the control of the experimenter and highly correlated within groups of neurons, has been demonstrated (5), and in spike data from awake, behaving animals, the possible effects of nonstationarities across stimulus presentations on any kind of average measure should seriously be considered. If necessary (and possible), one may use the procedure of "slicing" the spike trains into different sections, each one periodically stationary, and compare the results across sections. A more general analysis of the effects of nonstationarities and the development of appropriate control calculations is currently in progress.

Implementation

Software realizations of the various calculations we have described were implemented in Fortran 77; a wide range of graphic display devices is trivially accomodated. Computation time on any current computer is negligible: reading the spike data and calculation (without specific attempts at optimization) of the "raw" JPSTH plus its normalized version, the surprise matrix and the ordinary PST histograms, as well as the diagonal histograms and cross-correlograms for each of the three matrices for spike trains involving some 3,000 spikes over 2,000 sweeps (Fig. 4, *D, F,* and *G*) on the Institute's Vax-750 took ~35 CPU-seconds; comparable timing should be accomplished on any modern microcomputer. Both display and publication printing, however, are more difficult. We have used a Vectrix-384 color display device (672 × 480 pixels, 512-color palette) attached serially to our computers. This type of device with its serial connection requires a considerable time to build the kind of pictures used in this paper. Currently available display devices that are directly attached to the computer bus eliminate this time problem. We have found the use of color in these displays quite useful in the laboratory and for making slides. Publication costs of such color material, unfortunately, remain prohibitive, so that we had to use gray scales here. The results on the screen are quite satisfactory; however, we have found the technical process of getting from the gray screen to the printed page to be quite difficult and disproportionately time consuming.

"Effective connectivity"

The normalized JPSTH erases all vertical and horizontal features (signatures of direct stimulus effects) and leaves only the diagonal features as it should. These diagonal features represent correlation of firing that goes beyond stimulus-induced modulations of single-neuron firing rates. To "explain" such residual correlation, we invoked the notion of "effective connectivity": the minimal neuronal circuit model that will reproduce the diagonal features. Note that "effective connectivity" should not be understood as a unique statement of the actual anatomic connectivity, because more than one arrangement (involving, e.g., extra interneurons) could provide the same overall behavior.

The usual approach when interpreting neuronal correlations in terms of "effective connectivity" is to separate two types of circuitry: direct interaction and shared input. In the present context, we can base this separation on the characteristics of the diagonal features. In Appendix 2 we have elaborated some of the quantitative measures that are appropriate for the case of direct neuronal interaction. We generalized the commonly used measures of "efficacy" and "contribution" to incorporate stimulus-time-locked variations. It turns out that the normalized JPSTH is the geometrical mean of these two measures. We note that the variance in the denominator of the expression for the normalized JPSTH (*Eq. 9*) was originally introduced for statistical reasons. It is interesting (and reassuring) that the variance now also appears as the result of probabilistic reasoning underlying dynamic generalization of efficacy and contribution.

One of the benefits of the normalized JPSTH procedure is an improved way to normalize the ordinary cross-correlogram to remove the direct effects of stimulus. We have shown that tallies over long bins parallel to the diagonal of the normalized JPSTH provide a better estimate of such a normalized cross-correlogram than possible with any procedure applied directly to the "raw" cross-correlogram. Of course, this procedure loses all stimulus-locked time structure of the correlation—it is a time average. The effects of this were clearly demonstrated in Figs. 3, 4, and 5.

Modulation of "effective connectivity"

Application of the normalized JPSTH to simulated spike trains in which we knew the underlying circuit allowed recovery of the known "effective connectivity." In particular, we showed that the normalized JPSTH has the sensitivity to detect stimulus-locked modulation of "effective connectivity," even when strongly masked by direct stimulus rate modulations. We then applied this procedure to several sets of spike trains from physiological recordings in different preparations. Without exhaustive search, we found examples of both rapidly modulated and constant "effective connectivity"; these findings occurred both for direct interaction and for shared input. We observed time constants for these stimulus-locked modulations as low as tens of milliseconds for cochlear nucleus; in cortex the time constants seem to be somewhat longer. Similar observations have recently been made in multi-neuron recordings from a number of different laboratories and preparations; the phenomenon seems to be robust across animals, brain areas, stimulus modalities, and detailed recording procedures (15 and Gerstein and Aertsen, in preparation).

We suggest that it is appropriate to incorporate rapid modulation of "effective connectivity" into brain theory and modeling. Note that there are various physiologically reasonable mechanisms for achieving such rapid modulation of "effective connectivity," in terms of both direct connections and shared input. One possible type of mechanism would involve rapidly modulated synaptic strength (24), but there are many alternative possibilities that will work with constant synaptic strength (e.g., Ref. 9). One possibility is that the observed time variations in "effective connectivity" among two neurons are due to shared input from a large network of neurons in which the stimulus induces a reverberation. The underlying system-theoretical question in choosing the proper model to represent residual correlation is whether one regards a description in terms of a small network with varying connectivity or in terms of a large network with constant connectivity as the simpler one. Both types of description are always logically possible for data such as presented in this paper. Further theoretical and experimental work is needed to select among the various possibilities.

APPENDIX 1

Surprise

The surprise for "excitation" $S(E)$, given a count z of m coincidences in a particular bin is defined as (27)

$$S(E) = -\ln(\text{prob}\,[z \geq m]) \qquad (A1.1)$$

Similarly, for "inhibition" it is defined as

$$S(E) = -\ln(\text{prob}\,[z \leq m]) \qquad (A1.2)$$

Under the assumption that the coincidence counts z are approximately normally distributed, we can derive an explicit relation between the surprise values for significant "excitation" (or "inhibition") and the corresponding large positive (or negative) values of the normalized JPSTH. The assumption of normality is more realistic, the larger the number of stimulus presentations; see Ref. 27 for a discussion on this issue. In that case, we get for significant "excitation"

$$\text{prob}\,[z \geq m] = 1 - \text{erf}\,(m') \qquad (A1.3)$$

where erf denotes the error function, and m' is the standardized coincidence count, i.e., after subtraction of the expected value and scaling by standard deviation. For large m' this can be approximated by

$$\text{prob}\,[z \geq m] \approx a \exp(-bm'^2) \qquad (A1.4)$$

for some appropriate constants a and b. Thus we obtain for the surprise

$$S(E) \approx bm'^2 + c \qquad (A1.5)$$

for some appropriate c. Analogous reasoning for the case of significant "inhibition" leads to the same result.

Therefore, the significant values in the surprise matrix should be approximately proportional to the square of the normalized JPSTH. This relationship amounts to a "sharpening" of peaks (or troughs) in the surprise matrix, as compared with the corresponding peaks (or troughs) in the normalized JPSTH matrix, as indeed can be observed in our Figs. 4, *E* and *G*, *F* and *H*, and 5, *C* and *E*.

APPENDIX 2

Quantitative assessment of connectivity: "efficacy" and "contribution"

In this appendix, we will generalize the concepts of *efficacy* and *contribution*, which were first introduced by Levick et al. (23). These concepts are meaningful only in a context in which one neuron drives another. We will assume that neuron a (x-axis of the JPSTH) drives neuron b (y-axis). The signature of this situation in the normalized JPSTH is an above-diagonal feature, with displacement from the principal diagonal corresponding to the connection's latency.

It will be convenient to adopt a notation in terms of probabilities rather than correlation densities. This is trivially realized, because after appropriately binning the spike trains, we obtain a (0, 1)-process, for which expectation values (correlations) equal probabilities

$$E[n_a(u)] = p\,(a = 1 \text{ in } u, u + \Delta) = p_u(a)$$

$$E[n_b(v)] = p\,(b = 1 \text{ in } v, v + \Delta) = p_v(b)$$

$$E[n_{ab}(u, v)] = p\,(a = 1 \text{ in } u, u + \Delta \text{ and } b = 1 \text{ in } v, v + \Delta) = p_{uv}(a, b) \qquad (A2.1)$$

As in the rest of this text, the time indices (u, v) refer to the time since the most recent occurrence of the stimulus marker.

The classical quantities to describe a direct connection are efficacy (effectiveness) and contribution (23). *Efficacy* was defined as the fraction of driver spikes that are related to the driven spikes by virtue of a precise time delay (number of near-coincidences divided by total number of driver events). *Contribution* was defined as the fraction of driven spikes that are time related to the driver spikes (number of near-coincidences divided by total number of driven events). Note that the quantities, as defined by Levick et al. (23), are numbers and not functions of time; the implicit assumption is that they are never modulated by the stimulus.

We now propose a time-dependent generalization of these notions. Because in our model we have assigned neuron *a* to be the driver and neuron *b* to be the driven, our first approximation to a time-dependent generalization is

$$e = \frac{p(a, b)}{p(a)} = p(b|a) \qquad (A2.2)$$

and

$$c = \frac{p(a, b)}{p(b)} = p(a|b) \qquad (A2.3)$$

For brevity, we have omitted the time arguments (u, v); it is understood, however, that *Eqs. A2.2* and *A2.3* are restricted to (u, v)-values corresponding to the above-diagonal feature in the normalized JPSTH; this is where the model applies initially. We will lift this restriction later.

On closer inspection of *Eqs. A2.2* and *A2.3*, it becomes clear that these expressions actually *over*estimate the magnitude of efficacy and contribution. This is most easily realized when we gradually decrease the strength of the connection; in the limit of completely independent neurons a and b, we would have

$$e = p(b|a) = p(b) \qquad (A2.4)$$

and

$$c = p(a|b) = p(a) \qquad (A2.5)$$

which, in the case of spontaneous activity and/or other unobserved drivers, would not yield the values of zero that are intuitively appropriate for this case of "unconnected" neurons. Clearly, in the original definitions (*Eqs. A2.2* and *A2.3*), we are also counting "accidental" coincidental events: those *b*-events that would have occurred anyway near *a*-events in the absence of the (*a*, *b*)-connection. Evidently, we need a "correction" term to compensate for these "accidental" coincidences.

Let us for the moment concentrate on the efficacy; the contribution measure can be treated in completely analogous fashion, and we will give the corresponding results later. A possible estimate of the portion of *Eq. A2.2* that represents accidental events could be $p(b)$. The probability of an *a*-driven event would then be obtained by subtracting $p(b)$ from the conditional probability $p(b|a)$. This would resolve the paradoxical result of *Eq. A2.4* for independent neurons; the resulting efficacy in that case would be zero. Actually, this correction is precisely what is accomplished by the subtraction of a PST-based predictor. This has been common usage in the context of usual cross-correlograms (29; for reviews see Refs. 11 and 19): to measure the truly driven time-related events, one takes the area of what is left of a peak in the cross-correlogram after subtraction of the predictor correlogram.

Once again, however, this procedure makes an overestimate, this time of the number of "accidental" coincidences; hence, we are *under*estimating the connection's efficacy. This becomes especially clear when we examine another limiting case: full dependence of *b* on *a*, i.e., all *b*-events are due to *a*-events, and there are no "accidental" *b*-events. In this case, taking $p(b)$ to represent the "accidental" coincident events would in fact assign all (driven) *b*-events to be "accidental" and, hence, would measure the efficacy of the connection to be an erroneous zero rather than the correct positive value. In the general case, $p(b)$ is an overestimate of having an "accidental" coincidence, because besides these "accidental" coincidences, it also incorporates the *a*-driven events.

What we actually need is an estimate for the probability $p(b')$ of a *b*-event as it would be in the hypothetical situation that the (*a*, *b*)-connection were not present. Let us sort the total (*a*, *b*)-coincidences into two classes: those that are driven, and those that are "accidental." This corresponds to the assumption that the processes responsible for these events are additive. The probability of truly driven coincidences can then be written as $p(a, b) - p(a, b')$. The second term, by definition, represents independent events and can be written as a product: $p(a)p(b')$. To calculate the hypothetical $p(b')$ we take the total $p(b)$ and subtract from it the above given probability of truly driven coincidences. This leads to the following recursive relation

$$p(b') = p(b) - (p(a, b) - p(a)p(b')) \qquad (A2.6)$$

which can be explicitly solved for $p(b')$

$$p(b') = \frac{p(b) - p(a, b)}{1 - p(a)} = \frac{p(\bar{a}, b)}{p(\bar{a})} = p(b|\bar{a}) \qquad (A2.7)$$

Thus, under the given assumptions, the hypothetical probability $p(b')$ is equal to the measurable probability of having a *b*-event conditional on *not* having an *a*-event (i.e., $\bar{a}$). For the efficacy we then obtain

$$e = p(b|a) - p(b|\bar{a}) \qquad (A2.8)$$

For the earlier discussed case of independent neurons, the conditional probability $p(b|a)$ simply equals the unconditional $p(b)$; therefore, *Eq. A2.8* in this case results in an efficacy of zero, as it should. For strongly connected neurons, the second limiting case discussed earlier, $p(b|\bar{a})$ is zero, so that the efficacy is correctly $p(b|a)$. Hence, unlike the two earlier attempts to formally define efficacy, *Eq. A2.8* takes proper care of both limiting cases of very weak and very strong connections.

For the more usual intermediate case of partial connectivity, i.e., some *b*-events are directly due to *a*, and some are not, *Eq. A2.8* can be further elaborated

$$e = \frac{p(a, b)}{p(a)} - \frac{p(\bar{a}, b)}{p(\bar{a})}$$
$$= \frac{p(a, b)}{p(a)} - \frac{p(b) - p(a, b)}{1 - p(a)}$$

Collecting terms, we obtain for the efficacy

$$e = \frac{p(a, b) - p(a)p(b)}{p(a)(1 - p(a))} \qquad (A2.9)$$

Similarly for the contribution measure we obtain

$$c = \frac{p(a, b) - p(a)p(b)}{p(b)(1 - p(b))} \qquad (A2.10)$$

Both of these measures can be evaluated directly from the observed spike data by using the JPSTH and the single-neuron PST histograms by using *Eq. A2.1*

$$e(u, v) = \frac{\langle n_{ab}(u, v)\rangle - \langle n_a(u)\rangle\langle n_b(v)\rangle}{\langle n_a(u)\rangle(1 - \langle n_a(u)\rangle)} \qquad (A2.11)$$

$$c(u, v) = \frac{\langle n_{ab}(u, v)\rangle - \langle n_a(u)\rangle\langle n_b(v)\rangle}{\langle n_b(v)\rangle(1 - \langle n_b(v)\rangle)} \qquad (A2.12)$$

These new measures, as all others defined in this paper, have been derived and should be evaluated bin by bin. Originally, in this section, we restricted the (u, v) time values to only cover the range of the off-diagonal feature in the normalized JPSTH. We started this appendix with the model assumption that neuron a drives neuron b with a definite timing relationship, as demonstrated by the single off-diagonal feature. The implicit assumption is that there is no other indication of driving in the normalized JPSTH: efficacy and contribution outside the feature region are understood to be zero. Because *Eqs. A2.11* and *A2.12* yield exactly this result, there is no reason to restrict the (u, v)-values to any particular range.

Similarly, we restricted this discussion to the case of an off-diagonal feature in the normalized JPSTH. Obviously, the extended efficacy and contribution notions (*Eqs. A2.11* and *A2.12*) could also be applied blindly to *any* normalized JPSTH matrix, including those having a feature that straddles the principal diagonal. Such a feature is the signature of shared input to the two neurons. However, in such a case, blind application of efficacy and contribution will assign meaningless parameters to an inappropriate model. Meaningful interpretation of efficacy and contribution are restricted to the case of off-diagonal features.

With *Eqs. A2.11* and *A2.12*, we have achieved the goal of a generalized time-dependent quantitative descriptor of "effective connectivity": the evaluation across the (u, v)-plane provides a quantitative characterization of the connection, both along the diagonal (time throughout the stimulus period) and at right angles to the diagonal (different delays in the near-coincidence). From these generalized, dynamic measures of effective connectivity, we can return to time-independent measures similar to those originally proposed by Levick et al. (23). This can be done by the appropriate integrations along the diagonal in a broad enough stripe, much the same as was done in going from the normalized JPSTH to the normalized cross-correlogram (cf. Fig. 1, *E* and *G*). The result of this procedure reduces each of our dynamic quantities to a single number: its time average over the relevant range of running time u integrated across the time difference $v - u$.

This work was started independently in several laboratories and brought together during 1986–87 during a Sabbatical stay by G. L. Gerstein at the Max-Planck-Institute for Biological Cybernetics.

We are grateful to Prof. Valentino Braitenberg for this opportunity. T. Bonhoeffer was involved in early phases of this project. We thank V. Staiger for his skillful photography of endless series of display screens and V. Braitenberg, A. Kreiter, and A. Schüz for critical reading of the manuscript.

Portions of the work were supported by System Development Foundation Grant SDF-00013 to G. L. Gerstein) and by Office of Naval Research Grant ONR-N-00014-83-K-387 to M. Habib and G. L. Gerstein.

Address for reprint requests: A. M. H. J. Aertsen, Max-Planck-Institute for Biological Cybernetics, Spemannstrasse 38, D-7400 Tübingen, West Germany.

Received 17 February 1988; accepted in final form 9 January 1989.

REFERENCES

1. AERTSEN, A. M. H. J., JOHANNESMA, P. I. M., BRUIJNS, J., AND KOOPMAN, P. W. M. Second order representation of signals. In: *Localization and Orientation in Biology and Engineering*, edited by D. Varjú and H.-U. Schnitzler. Berlin: Springer-Verlag, 1984, p. 38–41.
2. AERTSEN, A. M. H. J. AND GERSTEIN, G. L. Evaluation of neuronal connectivity: sensitivity of crosscorrelation. *Brain Res.* 340: 341–354, 1985.
3. AERTSEN, A., GERSTEIN, G., AND JOHANNESMA, P. From neuron to assembly: neuronal organization and stimulus representation. In: *Brain Theory*, edited by G. Palm and A. Aertsen. Berlin: Springer-Verlag, 1986, p. 7–24.
4. AERTSEN, A., BONHOEFFER, T., AND KRÜGER, J. Coherent activity in neuronal populations: analysis and interpretation. In: *Physics of Cognitive Processes*, edited by E. R. Caianiello. Singapore: World Scientific Publishing, 1987, p. 1–34.
5. BACH, M. AND KRÜGER, J. Correlated neuronal variability in monkey visual cortex revealed by a multi-electrode. *Exp. Brain Res.* 61: 451–456, 1986.
6. BENDAT, J. S. AND PIERSOL, A. G. *Random Data: Analysis and Measurement Procedures.* New York: Wiley, 1971.
7. VAN DEN BOOGAARD, H., HESSELMANS, G., AND JOHANNESMA, P. System identification based on point processes and correlation densities. I. The nonrefractory neuron model. *Math. Biosc.* 80: 143–171, 1986.
8. EPPING, W. J. M. AND EGGERMONT, J. J. Coherent neural activity in the auditory midbrain of the grassfrog. *J. Neurophysiol.* 57: 1464–1483, 1987.
9. ERB, M., PALM, G., AERTSEN, A., AND BONHOEFFER, T. Functional versus structural connectivity in neuronal nets. In: *Strukturbildung und Musteranalyse.* Proc. Neunter Kybernetik-Kongress (DGK), March 19–21, 1986. Goettingen, FRG (abstract), 1986, p. 23.
10. GERSTEIN, G. L. AND PERKEL, D. H. Simultaneously recorded trains of action potentials: analysis and functional interpretation. *Science Wash. DC* 164: 828–830, 1969.
11. GERSTEIN, G. L. AND PERKEL, D. H. Mutual temporal relationships among neuronal spike trains. *Biophys. J.* 12: 453–473, 1972.
12. GERSTEIN, G., BLOOM, M., ESPINOSA, I., EVANCZUK, S., AND TURNER, M. Design of a laboratory for multi-neuron studies. *IEEE Trans. Systems, Man and Cybernetics* SMC-13: 668–676, 1983.
13. GERSTEIN, G. L., PERKEL, D. H., AND DAYHOFF, J. E. Cooperative firing activity in simultaneously recorded populations of neurons: detection and measurement. *J. Neurosci.* 5: 881–889, 1985.
14. GERSTEIN, G. L. AND AERTSEN, A. M. H. J. Representation of cooperative firing activity among simultaneously recorded neurons. *J. Neurophysiol.* 54: 1513–1528, 1985.
15. GERSTEIN, G. Information flow and state in cortical neural networks: interpreting multi-neuron experiments. In: *Organization of Neural Networks: Structures and Models*, edited by W. von Seelen, U. Leinhos, and G. Shaw. Weinheim, FRG: VCH Verlagsgesellschaft, 1987, p. 53–75.
16. GLASER, E. M. AND RUCHKIN, D. S. *Principles of Neurobiological Signal Analysis.* New York: Academic, 1976.
17. GOCHIN, P., KALTENBACH, J., AND GERSTEIN, G. Stimulus linked temporal changes in interactions of neurons in rat dorsal cochlear nucleus. *Soc. Neurosci. Abstr.* 12: 1269, 1986.
18. HABIB, M. K. AND SEN, P. K. Non-stationary stochastic point-process models in neurophysiology with applications to learning. In: *Biostatistics: Statistics in Biomedical, Public Health and Environmental Sciences*, edited by P. K. Sen. Amsterdam: Elsevier/North-Holland, 1985, p. 481–509.
19. KRÜGER, J. A 12-fold microelectrode for recording from vertically aligned cortical neurones. *J. Neurosci. Meth.* 6: 347–350, 1982.
20. KRÜGER, J. Simultaneous individual recordings from many cerebral neurons: techniques and results. *Rev. Physiol. Biochem. Pharmacol.* 98: 177–233, 1983.
21. KUZNETSOV, P. I. AND STRATONOVICH, R. L. A note on the mathematical theory of correlated random points. *Izv. Akad. Nauk SSSR Ser. Math.* 20, 167–178 (1956). Also in: *Nonlinear Transformations of Random Processes.* New York: Pergamon, 1965, p. 101–115.
22. LÉGENDY, C. R. Three principles of brain function and structure. *Int. J. Neurosci.* 6: 237–254, 1975.
23. LEVICK, W. R., CLELAND, B. G., AND DUBIN, M. W. Lateral geniculate neurons of cat: retinal inputs and physiology. *Invest. Ophthalmol.* 11: 302–311, 1972.
24. VON DER MALSBURG, C. Nervous structures with dynamical links. *Ber. Bunsen-ges. Phys. Chem.* 89: 703–710, 1985.
25. MELSSEN, W. J. AND EPPING, W. J. M. Detection and estimation of neural connectivity based on crosscorrelation analysis. *Biol. Cybern.* 57: 403–414, 1987.
26. PALM, G. Evidence, information, and surprise. *Biol. Cybern.* 42: 57–68 1981.
27. PALM, G., AERTSEN, A. M. H. J., AND GERSTEIN, G. L. On the significance of correlations among neuronal spike trains. *Biol. Cybern.* 59: 1–11, 1988.
28. PAPOULIS, A. *Probability, Random Variables, and Stochastic Processes.* Tokyo: McGraw-Hill Kogakusha, 1965.
29. PERKEL, D. H., GERSTEIN, G. L., AND MOORE, G. P. Neuronal spike trains and stochastic point processes. II. Simultaneous spike trains. *Biophys. J.* 7: 419–440, 1967.
30. PERKEL, D. H. AND BULLOCK, T. H. Neural coding. *Neurosci. Res. Program Bull.* 6: 1968.
31. SCHNEIDER, J., ECKHORN, R., AND REITBÖCK, H. Evaluation of neuronal coupling dynamics. *Biol. Cybern.* 46: 129–134, 1983.

4 IEEE TRANSACTIONS ON BIOMEDICAL ENGINEERING, VOL. 36, NO. 1, JANUARY 1989

Neuronal Assemblies

GEORGE L. GERSTEIN, PURVIS BEDENBAUGH, AND AD M. H. J. AERTSEN

(Invited Paper)

***Abstract*—This paper examines the concept of neuronal assembly as it has appeared in selected portions of the literature. The context is experimental access to real neuronal assemblies in working brains, as made possible by recent technological progress. One current measure of assembly organization is based on correlation of firing among neurons; recent observations show that such correlations can vary rapidly. In this paper, we demonstrate that dynamic firing correlation can be caused either by dynamic changes in neuronal connection strengths or, alternatively, by the effects of an unobserved (large) pool of other neurons. The static connectivity within the pool appears to be important in determining these effects.**

Introduction

It is probably fair to say that the bulk of our knowledge about the nervous system is on the level of single neurons and their molecular and membrane mechanisms. Over the last 35 years much has been learned about firing patterns of single neurons in various parts of the nervous system under a variety of stimulus and behavioral conditions. More recently, a large body of work on receptors, channels, and modulators has served to explain and to emphasize the importance of endogenous properties of individual neurons. In addition, anatomical studies with modern tracer methods have provided a great deal of information about gross connectivity between different nuclei and brain regions. In vertebrate nervous systems, however, detailed local anatomical connectivity has only been worked out in a few structures like cerebellum.

Long before emergence of all the current complexities of single neuron anatomy and physiology, it was recognized that the computational processes constantly carried out by the nervous system could not be explained by the properties of single neurons alone. Sherrington, in the 1930's, verbalized the idea that neurons must cooperate in fulfilling a complex task. This concept has been completely assimilated into neurobiological thought and explanation, but only in the most general way. Experimental access to the properties of real neuronal assemblies has largely been impossible until quite recently. Thus, the development of the assembly concept has been somewhat unfettered. It is the purpose of this paper to examine and compare the various suggested concepts of neuronal assembly, and to address the problems of experimental verification.

Manuscript received June 7, 1988; revised July 25, 1988. This work was supported by Grant SDF 0013 from the System Development Foundation and Grant N00014-87-K-0766 from the Office of Naval Research.

G. L. Gerstein and P. Bedenbaugh are with the Department of Physiology, University of Pennsylvania, Philadelphia, PA 19104.

A. M. H. J. Aertsen is with the Max-Planck-Institute for Biological Cybernetics, Tuebingen, West Germany.

IEEE Log Number 8824412.

Conceptual Models of Assemblies

A. Motor Pools

Sherrington [10] suggested that the motor neurons serving a particular muscle form a pool of potentially active cells, and that there would be appropriate recruitment of neurons within this pool as the excitatory input (presumably from various sources) increased. The detailed distribution of such input, or of its effects, was assumed to differ across the motor pool, so as to "fractionate" the motor neuron population into a "discharge zone" of firing cells and a "subliminal fringe" of cells that were not sufficiently excited to fire. Inputs which excited (spatially) overlapping populations of neurons would be able to interfere or add nonlinearly. Thus, "occlusion" of discharge would result from two inputs that caused overlapping discharge zones, producing a combined output that is smaller than the sum of the separate outputs. Similarly, "facilitation" of discharge would result from two inputs which caused overlapping subliminal fringes, producing a combined output that is larger than the sum of the separate outputs.

The accuracy of this conceptual model was tested by Lloyd [36], [37] with experiments in which two dissected dorsal root twigs were stimulated separately or simultaneously at various strengths. The resulting discharge was observed as a slightly delayed potential on the corresponding ventral root. Under the assumption that amplitude of this ventral root potential was proportional to the number of neurons in the motor pool that were brought to discharge, Lloyd showed all the phenomena that Sherrington had proposed.

For our present purposes, we note that the neurons forming or having membership in a motor pool are defined exclusively through commonality of output projection (i.e., to a particular muscle). There is also an implied nonuniform distribution of the input effects. There is no requirement for connections or interactions *among* the member neurons in the pool, although these are of course present [38]. For an illuminative example of assembly coding in the context of the motor system, we refer the reader to the recent literature on coding of saccadic eye movements in the superior colliculus (e.g., model oriented: Gisbergen *et al.* [22]; experiment oriented: Lee *et al.* [35]).

0018-9294/89/0100-0004$01.00

B. Hebbian Assemblies

In 1949, in his seminal book *The Organization of Behavior*, Hebb [28] introduced the concept of the cell assembly and postulated mechanisms of both perception and learning: "It is proposed first that a repeated stimulation of specific receptors will lead slowly to the formation of an 'assembly' of association-area cells which can act briefly as a closed system after stimulation has ceased; this prolongs the time during which the structural changes of learning can occur and constitutes the simplest instance of a representative process (image or idea)" [28, ch. 4]. The changes mentioned here refer to "the structural change at the synapse which has been assumed to take place in learning. . .The general idea is an old one, that any two cells or systems of cells that are repeatedly active at the same time will tend to become 'associated,' so that activity in one facilitates activity in the other." This "neurophysiological postulate" later came to be known as "Hebb's rule"; similar formulations, however, can already be found in McCulloch and Pitts [40].

Regarding the formation of the cell assembly, Hebb stresses the statistical nature of both the timing of impulses and the connections to be enhanced, and how "in a larger system a statistical constancy might be quite predictable"; on anatomical considerations Braitenberg [5], [7] observes that the cerebral cortex contains all the necessary equipment and therefore would seem a good place for cell assemblies.

An important property of the Hebbian cell assembly in view of what is known about human perception, pattern recognition, and the like, is the capacity for "pattern completion": because of the strengthened connectivity, excitation of only part of the assembly will result in an "ignition" of the entire assembly [5]. The cell assembly, as a postulated cooperative group of neurons, formed by learning and sustained by later use, thus presents an alternative to the hierarchically organized scheme of the "cardinal" neuron [4]. In order to describe the temporal sequences of assembly activity involved in perception, Hebb introduces the "phase sequence"; this point was further developed by Braitenberg [5] in the form of a threshold control mechanism, which in its dynamic effects might be called a "pump of thoughts."

The notions of cell assemblies and modifiable synapses were later formalized [9], [8], [44], [39] and recently led to an outburst of theoretical work on neuronal networks with learning properties. In this context, it is interesting to note that many contemporary workers make a distinction between the Hebb rule and a "teacher" system [46]. However, in rereading the original books [28], [29] we noted that these two concepts were very much intermixed. In fact, every illustration that Hebb gives involves a convergence of synapses (so as to make the postsynaptic cell fire), although he did not explicitly separate the pupil and teacher synapse.

It should be stressed that both cell assembly and the modifiable synapse were introduced on theoretical grounds. Over the years, indirect evidence had accumulated regarding the actual presence of Hebbian synapses in the brain. Only recently, however, has there been a direct demonstration of a Hebbian-type modifiable synapse in hippocampus slices [27] and (Bonhoeffer *et al.* in preparation).

C. Cortical Columns

The notion of the "cortical column" as a basic functional module of the neocortex was introduced by Mountcastle [42] on the basis of physiological experiments in the cat's somatic sensory cortex. He observed that the (static) receptive field properties (RF location on body surface and modality, i.e., nature of adequate driving stimulus) of single neurons were distributed over the cortex in a most characteristic fashion: the cortex appeared to be organized in vertical columns, running through all six cortical layers, such that neurons in a column share the same receptive field properties. Each local column is thus specified by RF location and modality; rows of columns (or slabs) with the same RF location and those with the same modality specification are arranged in more or less orthogonal fashion. Comparable observations were made for the visual cortex ([31]; reviewed [32]): in area 17 there is an orderly map of visual space onto the cortical surface; within this map there appears to be a segregation into a columnar system, with the columns characterized by ocular dominance and by the orientation of the preferred bar stimulus. Within a single ocular dominance column (width approximately 400 μm), there is a gradual progression of preferred bar orientation, with 1 mm displacement along the cortex surface covering the full range of 180° orientation shift ("orientation hypercolumn"). There is evidence that auditory cortex is organized into columns associated with sound frequency and localization of the sound source in space [41], [33].

In a review of the columnar organization of the neocortex, Mountcastle [43] discusses a number of general characteristics, which for our current purposes we summarize as follows: 1) the basic functional module for input–output processing in cortex is the cortical column: a vertically arranged group of cells, heavily interconnected along the vertical axis, sparsely in horizontal directions (however, see [21]); 2) the columnar system allows for an orderly representation of several variables simultaneously in a 2-D map—the depth dimension is occupied with the task of processing and distribution of inputs and outputs; 3) the order parameters of the columnar organization vary with the particular cortex area, and are determined, on the one hand by the specific thalamo–cortical input (e.g., ocularity), on the other hand (mainly) by intracortical processing (e.g., orientation tuning); 4) ordered sets of columns in different cortical areas (as well as modules in subcortical structures) are interconnected specifically, thus allowing for preservation of topological relations; 5) the columnar functional organization allows for a partially-shifted overlap across a topographical representation-dynamic isolation of the active elements of a column is achieved by lateral, pericolumnar inhibition.

Mountcastle [43] both further subdivides the column into smaller subunits, the "minicolumns," and groups columns for different order parameters into larger entities, the "macrocolumns." These levels of organization and the above-mentioned horizontal interactions among columns with the same cortical area are not undisputed, however [30], [6], [20]. Note that the functional columnar organization scheme as sketched here is not trivially connected to the modular architectonics of the neocortex as described by Szentagothai [50], [51]. That module is a honeycomb-like cortical structure with diameter around 300 μm, anatomically defined by the arborization of cortico-cortical afferents rather than by the termination of specific thalamo-cortical afferents.

D. Correlation of Firing

Suppose that we are able to observe simultaneously the electrical activity of some number of neurons which are embedded in a functioning nervous system—the multineuron experiment. We may now define membership in an assembly on the basis of correlated firing, i.e., whether there are any preferred timing relationships between and among the firings of the observed neurons. The simplest such comparison we can make is of the relationship between two spike trains. The appropriate tool is cross correlation, which is a measure of delayed coincidences between spikes in the two trains. If the correlogram is flat, then no values of delayed coincidence are favored. This uniform level of correlation corresponds to the null hypothesis of independent neuron firing; its value can be simply calculated from the two firing rates, the observation interval and the bin width used in the correlogram. When peaks or valleys appear in the correlogram, there are favored values of delayed coincidence, and we may infer some relationship between the two neurons. Note that the cross-correlation measurement is an average statement over the entire piece of data being analyzed. Such averaging is necessitated by the stochastic and variable nature of neuronal firings.

Preferred values of delayed coincidence are potentially important to the operations of the brain. If two neurons converge synaptically upon a third, the probability that the postsynaptic neuron will fire will be much higher for near coincident, rather than individual, input spikes. Temporal summation in a dendrite is the likely underlying mechanism, i.e., an endogenous property of the single neuron. Thus, many neurons are effective detectors of (slightly) delayed coincidence of input, making available a powerful mechanism for computation and coding.

For the experimenter, however, the measurement of correlation among spike trains can be used to abstract the logical relationships between the observed neurons. By appropriate manipulations and control calculations we may parse the observed "raw" correlogram into contributions caused by direct stimulus effects on the two neurons and contributions caused by neural interactions. The latter can, on the basis of peak width and timing, usually be parsed into contributions from reasonably direct connection between two neurons (possibly mono- and/or polysynaptic) and contributions from shared input that originates in an unobserved, but not stimulus related source. The final result of all this massage is a highly simplified block diagram—the "effective connectivity." We stress that this is an abstraction: it is the simplest neuron-like circuit that would produce an equivalent correlogram. In particular, the "effective connectivity" may be only a subset of the actual anatomical connectivity since it deals only with connections and relations that are active during the time of measurement. It is useful to describe the "effective connectivity" with a connectivity matrix W of "effective" synaptic weights. Matrix element Wij would represent the "effective" influence exerted by neuron i on neuron j. Note that for real brains, in contrast to most physically oriented models (e.g., spinglasses) the connectivity matrix W is in general not symmetric. In subsequent portions of this paper we will no longer use the double quote around effective connectivity; the meaning, however, is intended to be explicitly as above.

An excellent review of the mathematics and procedures surrounding such uses of cross correlation can be found in the book by Glaser and Ruchkin [23], along with many references to the original literature.

In developing and using such procedures over the years, our laboratory has shown that correlated firing of simultaneously recorded neurons can indeed be clearly demonstrated in a number of tissues, and that various interesting effective connectivities may be inferred [49], [11], [12]. A number of other laboratories have made similar observations; reviews can be found in Abeles [1], Gerstein *et al.* [17], and Kruger [34].

Thus, cooperativity of neuron firing is a readily demonstrated phenomenon, and can be used as a defining indicator of neural assemblies and assembly processes.

E. Other Definitions of Assembly

We obviously do not intend to give an exhaustive listing here. However, two additional approaches are important.

1) Palm [44] has written a book concerned with some of the mathematics of neuronal assemblies. His definition is largely in accord with that of Hebb, except that it is applied most particularly to circuit arrangements for associative memory. Such assemblies are characterized by heavy feedback upon themselves, and are able to learn, store, and differentiate a large number of input stimuli in a particularly efficient manner.

2) Shaw [47], [48] proposed to examine neuronal assemblies using the idea of ergodicity. Consider the firing pattern of a single neuron in response to some repeated short stimulus. Even though the stimulus is constant for each repetition, the response of the neuron will vary widely, presumably because of the stochastic properties of neuron firing. The usual solution to this variability has been to compute the PST (peri-stimulus-time) histogram, an average measure of the firing probability relative to the

instant(s) of stimulus presentation. In order to attain a PST histogram that is reasonably smooth, it may often be necessary to average over some 30–50 (or more) repeated stimulus presentations. Now, it is perfectly obvious that the organism owning the neurons does not require 50 repetitions of the stimulus in order to detect and perhaps react in some appropriate way; a single presentation is usually enough. Since the requisite information is not present in the firing of a single neuron during the single stimulus presentation, we must assume that ergodicity holds: there should be a neural assembly whose summed firing for a single stimulus carries the same information as does the PST histogram of a single neuron after many stimulus presentations.

Using this idea, Shaw considers the PST histogram of a single neuron's response firing to be the indirect measure of an assembly response. As a further step, he then makes cross correlations between different (more or less simultaneously) measured PST histograms (Shaw *et al.*, in preparation). Features in such correlograms are interpreted as interassembly processes.

Comparisons

The several views of neuronal assemblies sketched out above have originated over a period of 50 years. Yet they share a number of properties.

Both neuron pools and cortical columns emerged directly from experimental observation. In both, membership in the assembly is based on commonality of individual neuron properties. In the case of the motor pool it is a shared output purpose: all the neurons that can contribute to the activity of a single muscle. In the case of the cortical column it is a shared response property: all the neurons that respond to a particular stimulus parameter specification. Neither concept of assembly directly addresses possible interactions among the members of the assembly; rather, the neurons involved are treated as strictly parallel in their functions.

Both the Hebbian and correlational concepts originated theoretically, and were eventually validated by experiment. In both, the primary defining property is interaction of individual neurons. Coactivation of neurons is the direct observable in both assembly concepts, and this, of course, includes coactivation from shared stimulus response properties as in cortical columns. But alternate sources of coactivation (i.e., nonstimulus shared input or direct (synaptic) interaction) do exist, so that these concepts of assembly go far beyond cortical columns. Experimental access to assembly properties through measurement of neuronal coactivation or firing correlation necessarily requires parallel rather than serial observation of the neuron population. This has only become possible in the relatively recent past after a great deal of technological development [17], and is carried on in relatively few laboratories.

The ergodic model of assembly makes no direct assumptions and allows no inference about interactions among the member neurons. The PST histogram of one neuron is set equal to a population response and the assembly is considered to move between active and inactive stable states [48]. There is no specification of how this population response is produced or how it is assesssed in the nervous system, however.

It should be obvious that the several conceptual models of neuronal assemblies overlap only partially in what they address. Considerable further exploration is needed to allow conversion between the models.

Concepts of Assembly Dynamics

In all the above, we have considered basically static or only slowly varying assemblies. A considerable body of work exists for slow modification of cortical columns by developmental manipulations like deprivation. Spatial extent (relative size) of columns can be profoundly affected, as well as some of the response properties of the individual neurons. On a similar time scale, the Hebbian assembly first comes into being through its training period, the initial stimulations or experiences that mold the internal connectivity. In terms of the correlation model of assembly, we would therefore expect to see only constant effective connectivity in a mature assembly. The matrix of effective synaptic weights *Wij* should be constant on time scales of (approximately) minutes to hours.

Such a static picture of assembly organization is intuitively unsatisfactory, given the rapidity with which the nervous system deals with its constantly varying environment. How could we determine experimentally whether assembly organization is really so static, or whether there are rapid changes in reaction to varying context or purpose?

The correlation concept of assembly, together with the implicit requirement for simultaneous observation of many neurons can be used to examine dynamics of assembly organization. However, new, additional analytic tools are required. Since we remain faced by the stochastic or noisy characteristics of neuron firing, average measures remain inevitable, but it is possible to divide the data periods over which the averages are taken in order to get at dynamics that are temporally related to stimulus presentation. Two relevant tools that we have recently developed are described briefly here.

Tools for Detection of Assembly Dynamics

Tool 1: The gravity transformation. The basic difficulty in using cross-correlation measures to examine assembly properties among some number of observed neurons is purely combinatorial. Since the basic measurement is for a pair of spike trains, there is an enormous proliferation of correlograms to be examined should even a modest number like ten neurons be simultaneously recorded [the number grows as $0.5N(N - 1)$]. The gravity tool is a transformation that allows the entire group of neurons under study to be evaluated for correlation of firing in a single computation, rather than in a (large) set of pair computations.

In the gravity transformation, each of the N neurons is

represented by a particle in an N space; all particles are initially equidistant. Each particle carries a charge which is incremented by every firing of the corresponding neuron, and then subsequently decays with an appropriate time constant. The charge history of a particle thus looks like a low-pass filtered version of the corresponding neuron's spike train. We now set operating rules that make particles attract each other with a force proportional to the product of their charges, and specify a viscous medium. Then particles corresponding to neurons with appreciable correlation of firing will move towards each other with a velocity proportional to the degree of firing correlation. Should there be more than one group of neurons showing (separate) firing correlations, the particle system will condense into a corresponding set of clusters. Particles corresponding to independently firing neurons will continue to wander around the N space in a way reminiscent of Brownian motion. As in the simple cross-correlation calculation, part of the result, i.e., of the clustering, will be caused by direct modulation of the individual neuron firing rates by the stimulus. Appropriate correction is possible, so that we can study the firing coordination that is not directly caused by stimulus.

Thus, the original problem of temporal correlation structure among N spike trains has been transformed into a spatial clustering problem which is easily followed and interpreted. Considerably more detailed descriptions of the gravity transformation [19], [16] and modifications of it [2] have been published. The computation is in use at several laboratories, and its particularly effective in screening blocks of data to identify those spike trains which should be studied in more detail.

The gravity transformation is effective for examining dynamics of the correlation structure at several time scales. Within a particular condensation, which may cover a period where the stimulus was presented many times, there may be changes in aggregation speed of some clusters. To the extent that this can be made significant compared to statistical fluctuations, such changes of aggregation speed correspond to changes in the underlying neuronal interaction strengths.

At a somewhat different time scale, we may compare condensations for two different stimulus conditions. The best experimental design for this purpose would be to present the two stimuli alternately and repeatedly or in a pseudorandom order. The two stimulus conditions are thus interleaved in a way which eliminates slow drift (of anything) as a confounding factor in the analysis. The recorded data must be edited by "cutting and pasting" so as to produce two continuous strips, each involving presentation of only one stimulus condition. The aggregations and the speed with which they are attained in the two condensations provide average measures that allow comparison of the assembly organizations in the two stimulus conditions. Although these are measures averaged over many repeated presentations of stimulus, the interleaved design means that any organizational changes that are detected occur on a time scale of the interstimulus time.

Tool 2: The joint PST histogram. We return our attention to a single pair of spike trains. Suppose these two neurons correspond to clustering particles in the gravity representation or that there is a significant peak in the correlogram, i.e., evidence for one or more of the several types of neuronal interaction that can be represented by effective connectivity. Are the delayed coincidence events that contributed to the correlation peak distributed randomly in time? Or, in other words, is the effective connectivity constant or varying? One approach to this class of problem is to treat these particular delayed coincidence events as just another spike train, and then to use the usual tools of spike train analysis to examine the temporal structure of this higher order spike train [13], [18].

A related approach to the problem, particularly useful to examine possible stimulus effects on the effective connectivity is construction of the joint PST histogram. This is a recently improved version of the old joint PST scatter diagram [18], [3]. The basic construction is shown in Fig. 1. We cut the data at the beginning of each stimulus presentation, and then plot each of the two spike trains along a cartesian axis, placing the time of stimulus at the origin. On the plane we plot points at the locations corresponding to the logical AND of the two spike train segments. This operation is repeated for each stimulus presentation.

The points in the plane are now binned at some appropriate resolution. Bins falling along (near) the diagonal correspond to near coincidental events in the two spike trains, i.e., to events that are signatures of the several forms of neuronal interaction. Bin counts for the diagonal region [see Fig. 1(b)] represent the time locked average for near coincidences in the same sense that the PST histogram represents the time locked average of the single firings. We label these counts the PST Coincidence histogram, and present them as in the right portion of Fig. 4 along the diagonal. It can also be shown that summation over each stripe of paradiagonal bins in the joint PST histogram, normalized for the number of bins in each sum, gives the cross correlogram of the two spike trains. This is plotted at the upper right of the right portion of Fig. 4. In addition, for time reference, the ordinary PST histograms for each spike train are presented at the left edge and twice across the bottom of the display; all the relevant time relationships are simultaneously open for inspection and comparison.

As in ordinary cross correlation, the "raw" joint PST histogram includes the effects of direct stimulus modulation of the individual firing rates of the two neurons. We have shown that an optimal correction for this factor is to subtract the bin by bin product of the two PST histograms, and then to divide by the bin by bin product of the standard deviations of the two PST histograms [3], [45]. After this correction the residual correlation effects (direct connection and shared nonstimulus input) are uncovered and we can examine their stimulus-locked time course.

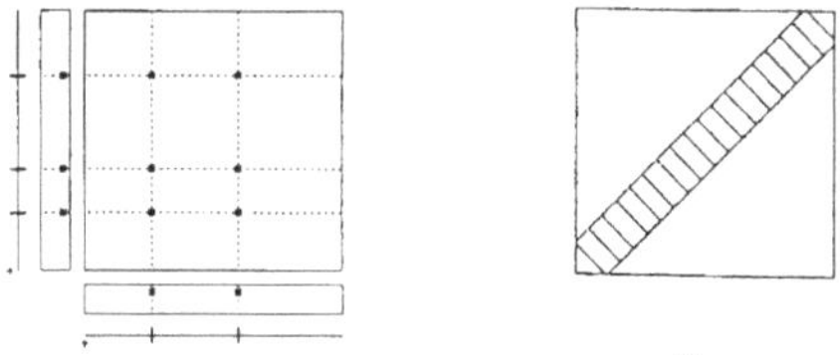

(a) (b)

Fig. 1. Joint peri-stimulus-time histogram. (a) Method of construction. Time of occurrence of impulses from the two units relative to a stimulus presentation are shown schematically on the bottom and left axes. The contribution of each spike to its cell's PST histogram is shown in the rectangular boxes, and their joint contribution to the JPST histogram is shown in the square box. A JPST histogram like that shown in Fig. 4 gradually emerges. (b) Diagonal region of JPST. Joint events that fall into the boxed region occur at about the same time and are the "nearly coincident" events of the sort that would lead to a gravity condensation. When the firing of the horizontal axis cell significantly precedes the firing of the vertical axis cell, the event falls into the upper left triangle. Similarly, when the firing of the vertical axis cell significantly precedes the firing of the horizontal axis cell the event falls into the lower right triangle (modified from [3]).

Such a corrected JPST histogram corresponding to Fig. 4 is shown in Fig. 5.

Measured Assembly Dynamics

Using the above tools, we have examined various appropriate simulations as well as a number of recordings from real multineuron experiments. The simulations involved several model neurons coupled by synapses that either were constant in time or alternatively (in other cases) were modulated in strength by the stimulus. In both of these situations, the joint PST histogram procedure was correctly able to recover the known effective connectivity and its stimulus-time locked structure [3].

Real recordings, when examined with both the gravity tool and with the joint PST histogram in some cases show constant effective connectivity, while in other cases there were dynamic changes. When they appear, the modulations of effective connectivity seem to occur on two different time scales. The gravity tool as described above gives a short-term moving average measure of the correlation structure among the observed neurons over many repeated presentations of the same stimulus; we interpret this as a short-term moving average value of the effective connectivity for the particular stimulus condition. The gravity condensations for the same neurons under two different stimulus conditions are often different, suggesting different effective connectivities. However, since the experiments are run with alternate repeated presentation of the two stimulus conditions, the assembly organization must switch back and forth approximately every half second, i.e., the stimulus presentation interval [14], [15].

Stimulus effects on the effective connectivity measured by the joint PST histogram, when they occur, are on a much shorter time scale, around tens of milliseconds [15], [3], [24].

Interpretations

Given that real neurons often show dynamic changes in correlation structure in association with stimulus, we have the problem of interpreting such changes in terms of assembly organization. In the course of developing the gravity and joint PST histogram tools, we used simulation to verify that they could detect and quantify modulation of connections if these had been included in the simulation. Conversely, when connections were constant in those simple simulations, the gravity and joint PST histogram measured constant correlation and hence constant effective connectivity. Thus, one possible interpretation of stimulus dependent correlation structure among real neurons would be to suggest that the connections among the neurons are changing their effective value: the W_{ij} are dynamic. Although the Hebbian scheme for synaptic change is too slow to explain such observations, mechanisms like presynaptic inhibition have an adequately fast time scale.

A different possible level of interpretation of the correlation dynamics is in the following analogy [15]. Assume that each stimulus sets up a disturbance in the neuronal mass that spreads in the manner of ripples in a pond. Different stimuli would correspond to ripples starting from different locations. As these disturbances travel through, they will cause correlated activity. After the corrections for direct stimulus modulation of firing rates we would still expect to find different residual correlation for different stimuli because the disturbances have traveled somewhat different paths, so that particular connections are differentially effective. It is possible that optical detection methods [25], [26] could be used to further examine such possibilities among large masses of neurons.

Recently, we have found another class of possible interpretations of dynamic residual correlations, i.e., a new set of mechanisms which can cause correlation dynamics without explicitly invoking changes in synaptic effectiveness.

We demonstrate this result by examining spike trains from a simulation of 100 model neurons in a square array, but with anisotropic connections so as to allow different propagation paths for stimuli applied at different locations. Specifically, the simulation was organized to mimic "synfire chains" [1] arranged to carry excitation from the top toward the bottom of the neuron array. Each horizontal row of neurons received input only from the row above, and transmitted only to the row below. By means of additional independent input, each neuron was set to a spontaneous firing rate of about five spikes per second. Stimuli were applied either to the top row of neurons, or to the leftmost column of neurons in the square array. The stimuli were 100 ms duration independent impulse trains drawn from identical exponential interval distributions. A total of 2499 stimuli were presented one every 400 ms. During the stimulus the cells in the network fired at about 15 spikes/second.

The parameter of interest for the present purposes is the probability of interrow connections, i.e., how many neu-

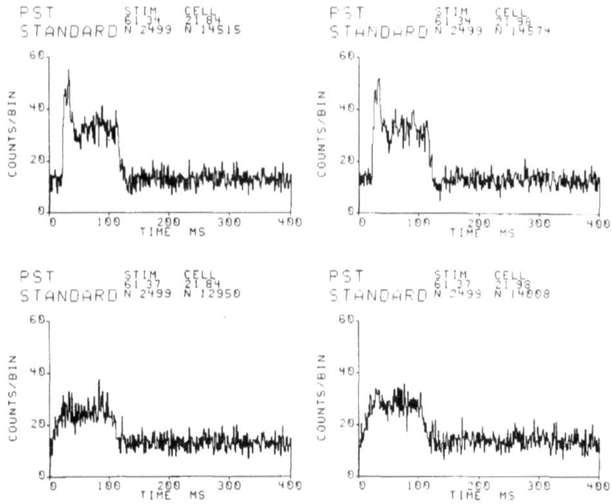

Fig. 2. PST histograms of pre- and postsynaptic cells under two different stimulus conditions. Top, stimulus presented to the top of the network; bottom, stimulus presented to the side of the network. Left is presynaptic and right is postsynaptic. See text for organization of the network.

rons in the row above feed a particular neuron. This quantity can be termed the convergence.

The figures shown here are drawn from a simulation with a high degree of convergence. Typical PST histograms for two neurons near the bottom of the net are shown in Fig. 2. The neurons have similar, but not identical PST histograms for a particular stimulus, and there is a qualitative difference for the two stimuli: stimulation at the top of the array produces a rapid rise with overshoot in the time-locked firing probability, while stimulation at the left edge of the array produces a much more gradual rise and fall in the firing probability.

Corrected cross correlograms for the same two neurons are shown in Fig. 3, including the two stimulus conditions defined above as well as spontaneous activity. It is clear that the three cross-correlogram peaks are quantitatively different, the largest corresponding to the top stimulus, the smallest to spontaneous activity. Thus, the residual correlation is a function of the stimulus condition; we would classify this situation as stimulus dependent effective connectivity, although the underlying synaptic connectivity is known to be quite constant.

The raw joint PST histogram for the same data is shown in Fig. 4. As expected, the raw PST coincidence histogram (diagonal in the right part) shows a stimulus-locked modulation, some of which can presumably be explained by the individual PST histograms. The corresponding normalized joint PST histogram is shown in Fig. 5. Note that the normalized PST coincidence histogram still shows a fairly strong stimulus-locked modulation. Again, we would classify such findings as showing stimulus dependent effective connectivity with constant synaptic connectivity.

If, however, the convergence parameter in the simulation is taken to be smaller, so that fewer neurons converge onto a given neuron in the row below, the same group of calculations show little effects of stimulus condition. We do not show these results explicitly here, but summarize them as follows. The PST histograms are very similar to those shown in Fig. 2. The corrected cross correlograms (unlike Fig. 3) are almost the same for all stimulus conditions. The raw joint PST histogram looks very similar to Fig. 4, while the normalized joint PST histogram shows a much flatter PST coincidence histogram than does Fig. 5. More detailed simulations as well as analytic calculations (Boven *et al.*, in preparation) confirm this effect of convergence and the activity carried by it on the modulation of effective connectivity.

We have therefore demonstrated at least one situation in which a network with known constant connections among the neurons will mimic the effects of dynamic effective connectivity. The relevant conditions are a high degree of convergence from a population of neurons that are all activated by the stimuli. On the other hand, when the degree of convergence is lower, we would correctly recover the actual constant synaptic connections.

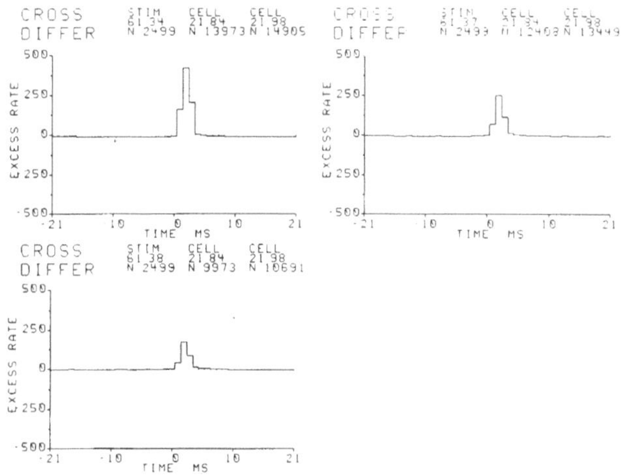

Fig. 3. Cross correlograms under two different stimulus conditions and spontaneous activity. The correlograms were corrected for stimulus modulation of the cells' firing rates by subtracting the cross correlation of the PST histograms. These differences were then normalized (divided) by the product of the number of presynaptic spikes and the bin width. This process yields the excess firing rate (in spikes per second) of the post synaptic cell per presynaptic spike. The 99 percent significance levels for these data at the scale used are invisibly small; i.e., these peaks are extremely significant. The corrected correlogram is largest for stimulation with the grain of the connections (top stimulus—top left), smaller for stimulation orthogonal to the grain of the connections (side stimulus—top right), and smallest for no stimulus (spontaneous activity—bottom left). The cross correlograms shown are for the two cells whose PST histograms are shown in Fig. 2. See text for organization of the network.

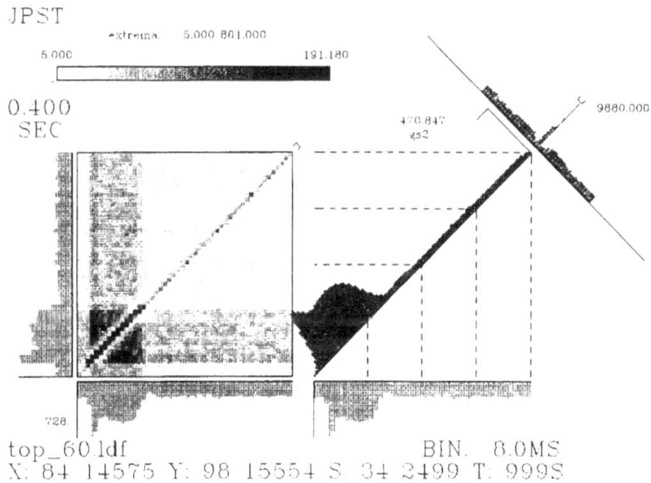

Fig. 4. Uncorrected "raw" joint peri-stimulus-time histogram. The brightness of small squares in the box at left code the number of delayed coincident spike events occurring at various poststimulus times [see Fig. 1(a)]. To the left and below the box are the PST histograms for each of the two spike trains. At right center diagonal is the smoothed PST histogram for the higher order train of near coincident spikes as produced by summation in the bins shown in Fig. 1(b). At the upper right is the cross correlogram constructed by summation within paradiagonal stripes of the JPST in the box at left and normalizing for the number of small squares along that diagonal.

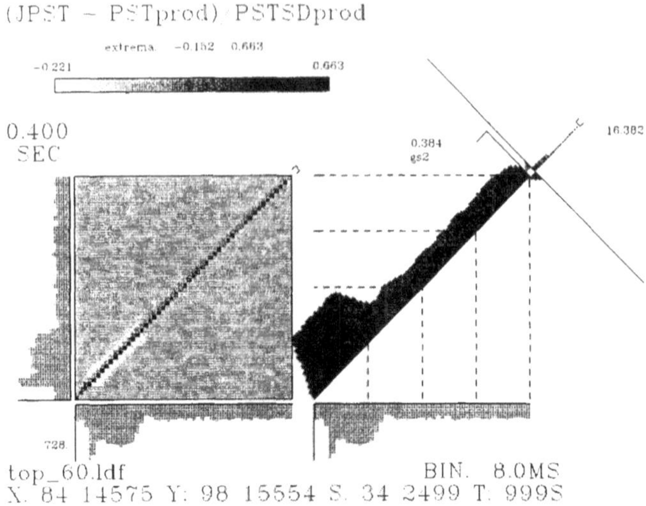

Fig. 5. Normalized joint peri-stimulus-time histogram. This is a redisplay of the information in Fig. 4 after correcting for the effects of stimulus modulation of the firing rates of the two neurons. The binwise product of the two individual PST histograms was substracted from each small square of the JPST of Fig. 4. The value of each small square was then divided by the appropriate bin by bin product of the standard deviations of the PST histograms. Note that even after this correction the apparent connection strength shown along the diagonal at right center is modulated in response to the stimulus, whereas all traces of the single unit firing rate modulations (broad horizontal and vertical bands in Fig. 4) have disappeared.

Discussion and Conclusion

In this paper, we have examined the concept of neuronal assembly in a number of different incarnations. It turns out that each of these versions of the concept is sufficiently different that it is hard to make logical bridges.

Neither the motor pool concept or the well-known cortical column concepts of assembly depend on or imply anything about the internal structure of information flow. They are defined entirely through properties related to external input or output. In contrast, both the Hebbian and correlation models are defined mainly through interactions among the member neurons. Both also depend on input, the Hebbian model for its original development, the correlation model in relation to parsing the interactions.

All four assembly concepts allow overlaps of subassemblies. In each case, a single neuron can be a member of several different assemblies, and can participate in each of their activities. As in the previous paragraph, this multiple allegiance is defined in terms of internal circuitry or relations only in the Hebbian and correlation models.

Dynamics, stimulus related or otherwise, are implicit in all four assembly models. However, they can be detected explicitly only in the Hebbian and correlation models. Time scales for the Hebbian model are long, while for the correlation model they are unspecified, i.e., long or short.

In the last part of this paper we have examined the extent to which a system with constant connections can appear to be dynamically organized. We have here demonstrated that at least one class of connections, i.e., highly convergent systems, can give all the measured signatures of dynamic organization. This result can be viewed as a failure of the measurements to be sufficiently sensitive, and hence as a call for further development of tools that could perhaps do the job better. On the other hand, we have shown that dynamics in the relations of two neurons can be the reflection of group activity that is elsewhere in the net, and hence not under direct observation. This, then, is a potential tool for measurement of emergent, large population properties.

Finally, none of the four assembly models explicitly deals with a way to define "state." This concept has proven vital for the successful description, understanding and prediction of physical systems, and there is no reason to believe it any less crucial for neuronal systems. Unfortunately, the types of state descriptors that are used on physical systems are not directly applicable to the neuronal situation. Additional work on both real and artificial neuronal assemblies is essential.

References

[1] M. Abeles, *Local Cortical Circuits—Studies of Brain Function, Vol. 6.* New York: Springer, 1982.

[2] A. Aertsen, T. Bonhoeffer, and J. Kruger, "Coherent activity in neuronal populations: Analysis and interpretation," in *Physics of Cognitive Processes*, E. R. Caianiello, Ed. Singapore: World Scientific Publishing, 1987, pp. 1-34.

[3] A. M. H. J. Aertsen, G. L. Gerstein, M. Habib, G. Palm, P. Gochin, and J. Kruger, "Dynamics of neuronal firing correlation: Modulation of 'effective connectivity'," *J. Neurophysiol*, to be published.

[4] H. B. Barlow, "Single units and sensation: A doctrine for perceptual psychology?," *Perception*, vol. 1, pp. 371-394, 1972.

[5] V. Braitenberg, "Cell assemblies in the cerebral cortex," in *Theoretical Approaches to Complex Systems—Lecture Notes in Biomathematics, Vol. 21*, R. Heim and G. Palm, Eds. New York: Springer, 1978, pp. 171-188.

[6] —, "Charting the visual cortex," in *Cerebral Cortex, Vol. 3*, A. Peters, and E. G. Jones, Eds. New York: Plenum, 1985, pp. 379-414.

[7] —, "Two views of the cerebral cortex," in *Brain Theory*, G. Palm and A. Aertsen, Eds. New York: Springer, 1986, pp. 81-96.

[8] G. S. Brindley, "The classification of modifiable synapses and their use in models for conditioning," *Proc. Roy. Soc. London.*, vol. B 168, pp. 361-376, 1967.

[9] E. R. Caianiello, "Outline of a theory of thought processes and thinking machines," *J. Theor. Biol.*, vol. 2, pp. 204-235, 1961.

[10] R. S. Creed, D. E. Denny-Brown, J. C. Eccles, E. G. T. Liddell, and C. S. Sherrington, *Reflex Activity of the Spinal Cord.* London: Oxford Univ., 1932.

[11] J. W. Dickson and G. L. Gerstein, "Interactions between neurons in the auditory cortex of the cat," *J. Neurophysiol.*, vol. 37, pp. 1239-1261, 1974.

[12] I. Espinosa and G. L. Gerstein, "Cortical auditory neuron interactions during presentation of 3-tone sequences: Effective connectivity," *Brain Res.* vol. 450, pp. 39-50, 1988.

[13] R. D. Frostig, Y. Gottlieb, E. Vaadia, and M. Abeles, "The effects of stimuli on the activity and functional connectivity of local neuronal groups in the cat auditory cortex," *Brain Res.*, vol. 272, pp. 211-221, 1983.

[14] G. L. Gerstein, "Interactions within neuronal assemblies: Theory and experiment," in *Brain Organization and Memory: Cells, Systems and Circuits*, J. L. McGaugh, N. M. Weinberger, and G. Lynch, Eds. London: Oxford Univ., 1988.

[15] —, "Information flow and state in cortical neural networks: Interpreting multi-neuron experiments," in *Organization of Neural Networks*, W. v.Seelen, G. Shaw, and R. Leinhos, Eds. Weinheim: VCH Verlagsgesellschaft, 1987, pp. 51--75.

[16] G. L. Gerstein and A. M. H. J. Aertsen, "Representation of cooperative firing activity among simultaneously recorded neurons," *J. Neurophysiol.*, vol. 54, pp. 1513-1528, 1985.

[17] G. L. Gerstein, M. J. Bloom, I. E. Espinosa, S. Evanczuk, and M. R. Turner, "Design of a laboratory for multineuron studies," *IEEE Trans. Syst. Man, Cybern.*, vol. SMC-13, pp. 668-676, 1983.

[18] G. L. Gerstein and D. H. Perkel, "Mutual temporal relationships among neuronal spike trains," *Biophysical J.*, vol. 12, pp. 453-473, 1972.

[19] G. L. Gerstein, D. H. Perkel, and J. E. Dayhoff, "Cooperative firing activity in simultaneously recorded populations of neurons: Detection and measurement," *J. Neurosci.*, vol. 5, pp. 881-889, 1985.

[20] C. D. Gilbert, "Horizontal integration in the neocortex," *TINS*, vol. 8, pp. 160-165, 1985.

[21] C. D. Gilbert and T. N. Wiesel, "Clustered intrinsic connections in cat visual cortex," *J. Neurosci.*, vol. 3, pp. 1116-1133, 1983.

[22] J. A. M. van Gisbergen, A. J. van Opstal, and A. A. Tax, "Collicular ensemble coding of saccades based on vector summation," *Neurosci.*, vol. 21, pp. 541-555, 1987.

[23] E. M. Glaser and D. S. Ruchkin, *Principles of Neurobiological Signal Analysis.* New York: Academic, 1976.

[24] P. M. Gochin, G. L. Gerstein, and J. A. Kaltenbach, "Dynamic temporal properties of synaptic interaction in rat dorsal cochlear nucleus" to be published.

[25] A. Grinvald, "Real-time optical mapping of neuronal activity: From single growth cones to the intact mammalian brain," *Ann. Rev. Neurosci.*, vol. 8, pp. 263-305, 1985.

[26] A. Grinvald, B. M. Salzberg, V. Lev-Ram, and R. Hildesheim, "Optical recording of synaptic potentials from processes of single neurons using intracellular potentiometric dyes," *Biophys. J.*, vol. 51, pp. 643-651, 1987.

[27] B. Gustaffson, H. Wigstrom, W. C. Abraham, and Y.-Y. Huang, "Long-term potentiation in the hippocampus using depolarizing current pulses as the conditioning stimulus to single volley synaptic potentials," *J. Neurosci.*, vol. 7, pp. 774-780, 1987.

[28] D. O. Hebb, *The Organization of Behavior—A Neuropsychological Theory.* New York: Wiley, 1949.

[29] —, *A Textbook of Psychology.* 3rd ed. Philadelphia: Saunders, 1972.

[30] D. H. Hubel, "Columns and their function in the primate visual cortex," in *Theoretical Approaches in Neurobiology*, W. E. Reichardt, T. Poggio, Eds. Cambridge, MA: M.I.T., 1981, pp. 109-115.

[31] D. H. Hubel and T. N. Wiesel, "Receptive fields and functional architecture of monkey striate cortex," *J. Physiol.*, vol. 195, pp. 215-243, 1968.

[32] D. H. Hubel and T. N. Wiesel, "Functional architecture of macaque monkey visual cortex" (Ferrier lecture), *Proc. Roy. Soc. London*, vol. B. 198, pp. 1-59, 1977.

[33] T. J. Imig and H. O. Adrian, "Binaural columns in the primary field (A1) of cat auditory cortex," *Brain Res.*, vol. 138, pp. 241-257, 1977.

[34] J. Krüger, "Simultaneous individual recordings from many cerebral neurons: techniques and results," *Rev. Physiol. Biochem. Pharm.*, vol. 98, pp. 177-233, 1983.

[35] C. Lee, W. H. Rohrer, and D. L. Sparks, "Population coding of saccadic eye movements by neurons in the superior colliculus," *Nature*, vol. 332, pp. 357-360, 1988.

[36] D. P. C. Lloyd, "Reflex action in relation to pattern and peripheral source of afferent stimulation," *J. Neurophysiol.*, vol. 6, pp. 111-119, 1943.

[37] —, "On the relation between discharge zone and subliminal fringe in a motoneuron pool supplied by a homogeneous presynaptic pathway," *Yale J. Biol. Med.*, vol. 18, pp. 117-121, 1945.

[38] —, "Facilitation and inhibition of spinal motoneurons," *J. Neurophysiol.*, vol. 9, pp. 421-438, 1946.

[39] C. von der Malsburg, "Nervous structures with dynamical links," *Ber. Bunsenges. Phys. Chem.*, vol. 89, pp. 703-710, 1985.

[40] W. S. McCulloch and W. Pitts, "A logical calculus of the ideas immanent in nervous activity," *Bull. Math. Biophys.*, vol. 5, pp. 115-133, 1943.

[41] J. Middlebrooks, J. C. Dykes, and M. M. Merzenich, "Binaural response-specific bands in primary auditory cortex (A1) of the cat: topographical organization orthogonal to isofrequency contours," *Brain Res.*, vol. 181, pp. 31-48, 1980.

[42] V. B. Mountcastle, "Modality and topographic properties of single neurons of cat's somatic sensory cortex," *J. Neurophysiol.*, vol. 20, pp. 408-434, 1957.

[43] —, "An organizing principle for cerebral function: the unit module and the distributed system," in *The Mindful Brain*, G. M. Edelman and V. B. Mountcastle, Eds. Cambridge, MA: M.I.T., 1977, pp. 7-50.

[44] G. Palm, *Neural Assemblies—Studies of Brain Functions, Vol. 7.* New York: Springer, 1982.

[45] G. Palm, A. M. H. J. Aertsen, and G. L. Gerstein, "On the significance of Correlations among neuronal spike trains," *Biol. Cybern.*, vol. 59, pp. 1-11, 1988.

[46] D. E. Rumelhart, G. E. Hinton, and R. J. Williams, "Learning internal representations by error propagation," in *Parallel Distributed Processing. Explorations in the Microstructure of Cognition, Vol. 1: Foundations*, D. E. Rumelhart and J. L. McClelland, Eds. Cambridge, MA: M.I.T., 1986, pp. 318-362.

[47] G. L. Shaw and D. J. Silverman, "Simulations of the trion model and the search for the code of higher cortical processing," in *Computer Simulation in Brain Science*, R. Cotterill, Ed. Cambridge, UK: Cambridge Univ. Press, 1988.

[48] G. L. Shaw, E. Harth, and A. B. Scheibel, "Cooperativity in brain

function: Assemblies of approximately 30 neurons," *Exp. Neurol.*, vol. 77, pp. 324-358, 1982.
[49] J. K. Stevens and G. L. Gerstein, "Interactions between cat lateral geniculate neurons," *J. Neurophysiol.*, vol. 39, pp. 309-326, 1976.
[50] J. Szentagothai, "The neuron network of the cerebral cortex: a functional interpretation" (Ferrier lecture), *Proc. Roy. Soc. London*, vol. B. 201, pp. 219-248, 1978.
[51] J. Szentagothai, "The modular architectonic principle of neural centers," *Rev. Physiol. Biochem. Pharm.*, vol. 98, pp. 11-61, 1983.
[52] D. H. Perkel, G. L. Gerstein, and G. P. Moore, "Neuronal spike trains and stochastic point processes: II—Simultaneous spike trains," *Biophys. J.*, vol. 7, pp. 419-440, 1967.

Purvis Bedenbaugh was born on July 11, 1961. He received the B.S.E. degree in biomedical engineering from Duke University, Durham, NC in 1983 and the M.S. degree in bioengineering from Clemson University, Clemson, SC in 1984. He is currently in the Ph.D. program in bioengineering at the University of Pennsylvania, Philadelphia, PA

He became interested in neuroscience while working in the laboratory of B.S. Nashold, Jr., of the Division of Neurosurgery, Duke University School of Medicine. His current interests are in quantitative study of neural assemblies and integrative activity in the nervous system.

George L. Gerstein was born in Berlin, Germany, in 1933 and emigrated to the United States in 1938. He received the B.A., M.A., and Ph.D. degrees in physics from Harvard University, Cambridge, MA, in 1952, 1954, and 1958, respectively.

He switched interests to neurobiology during a postdoctoral stay at the Massachusetts Institute of Technology, Cambridge, MA. Since 1964 he has been a Professor in the Physiology Department at the University of Pennsylvania, Philadelphia, PA. He has worked on the neurobiology of the auditory and visual systems, with particular emphasis on the development of spike-train analysis as a way to study coding and information transfer. During the past decade he has been involved in experimental and theoretical investigation of neuronal assemblies as elements of computational processes in the brain.

Ad M. H. J. Aertsen was born in Oosterhout, The Netherlands, in 1948. He received the M.Sc. degree in physics from the Utrecht State University, The Netherlands, in 1973, and the Ph.D. degree in physics from the Nijmegen University, The Netherlands, in 1981.

From 1983 to 1984 he spent a postdoctoral year with George Gerstein at the Department of Physiology, University of Pennsylvania School of Medicine, Philadelphia. Since 1984 he has been a Research Associate in Valentino Braitenberg's Department at the Max-Planck-Institute for Biological Cybernetics, Tuebingen, Federal Republic of Germany. His current research interests focus on the analysis of neurobiological signals (single- and multineuron spike trains, optical recordings), primarily from the auditory and visual system, in relation to experimental and theoretical aspects of neural information processing.

Chapter 8
Significance of Spike Train Correlations

George's Stay in Tübingen: Interactions Between Neurons and with Some Untypical Germans

Günther Palm

My encounter with George Gerstein was relatively short, only about a year, when he was staying in Tübingen to cooperate with us, i.e., the group of Valentino Braitenberg with Almut Schüz, Joseph Rauschecker, Ad Aertsen, and myself at that time. Still I was deeply impressed by his character and his attitude to neuroscience.

Since I had joined the Max Planck Institute in Tübingen and was drawn into neuroscience, my strongest interaction had been with neuroanatomy. I did have the chance to visit some physiology labs and was impressed but perhaps also confused by the many things they had to worry about and take care of. In contrast to these experiences, when I met George I was impressed by the clarity and simplicity of his ideas. Contrary to most neurophysiologists, he seemed to be truly interested in how the brain tissue he studied really works, rather than how to tell a nice and convincing story about it.

When George came to Tübingen, he was interested in measuring interactions between neurons, possibly by synaptic connections, but essentially expressed in spike correlations. I vividly remember discussions about the three basic "scenarios" for creating a correlation: A influences B, B influences A, or A and B are both influenced by a common cause C. This was accompanied by simple drawings of 3 or 4 neuron networks and the resulting correlations. At that time, we were also discussing George's "gravity" method, so we also tried to imagine how the neurons would move in "gravity space." During these imaginative discussions, it

G. Palm (✉)
Institute of Neural Information Processing, Ulm University, Ulm, Germany
e-mail: guenther.palm@uni-ulm.de

A. Aertsen et al. (eds.), *Introducing Computation to Neuroscience*, Springer Series in Computational Neuroscience, https://doi.org/10.1007/978-3-030-87447-6_8

was always clear (in the back of our minds) that we were vastly oversimplifying the real situation, where the experimenter has recorded spike trains from a few neurons among a nervous tissue containing many thousands of unobserved neurons. Of course a single spike in one of the recorded neurons can hardly generate a spike in another recorded neuron on its own, without the help of other neurons. And of course a spike correlation from A to B (with a positive delay) can also be an indirect effect via one or several intermediate neurons (or groups of neurons). And it is not even unlikely that two or perhaps all three of the above scenarios for correlation co-occur.

George's modesty, admitting the vast amount of our ignorance while still trying to find methods of measuring and quantifying the interactions between neurons, was very impressive and convincing, so this is probably what I learned from him. It also confirmed my own view of statistics that in order to show a significant interaction (or some other kind of mechanism supposed to be at work), one does not need to and perhaps should not employ any (more or less) detailed model of it, one should rather invent an appropriate test statistics, which implies a set of predictions that are very surprising under the hypothesis that there is no such mechanism (Palm 2012). This hypothesis can usually be modeled much more easily. This line of reasoning led to our papers on the significance of correlations or co-occurring spikes in the JPSTH (Palm et al. 1988; Aertsen et al. 1989), and later also to investigating higher order interactions (Martignon et al. 1993, 1995), and to simulations showing that such correlations, sometimes called "functional connectivity," have no simple direct relation to real anatomical connectivity (Aertsen et al. 1994).

References

Aertsen AM, Gerstein GL, Habib MK, Palm G (1989) Dynamics of neuronal firing correlation: modulation of "effective connectivity". J Neurophysiol 61(5):900–917. https://doi.org/10.1152/jn.1989.61.5.900

Aertsen A, Erb M, Palm G (1994) Dynamics of functional coupling in the cerebral cortex: an attempt at a model-based interpretation. Physica D Nonlinear Phenomena 75(1–3):103–128. https://doi.org/10.1016/0167-2789(94)90278-X

Martignon L, von Hasseln H, Palm G (1993) Modelling stochastic networks: data to the connectivity structure. In: Herzog O, Christaller T, Schütt D (eds) Grundlagen und Anwendungen der Künstlichen Intelligenz—17. Fachtagung für Künstliche Intelligenz, Informatik aktuell, Subreihe Künstliche Intelligenz. Springer, Berlin, pp 50–58. https://doi.org/10.1007/978-3-642-78545-0

Martignon L, von Hassein H, Grün S, Aertsen A, Palm G (1995) Detecting higher-order interactions among the spiking events in a group of neurons. Biol Cybern 73(1):69–81. https://doi.org/10.1007/BF00199057

Palm G (2012) Novelty, information and surprise. Springer, Berlin. https://doi.org/10.1007/978-3-642-29075-6

Palm G, Aertsen AMHJ, Gerstein GL (1988) On the significance of correlations among neuronal spike trains. Biol Cybern 59(1):1–11. https://doi.org/10.1007/BF00336885

Biol. Cybern. 59, 1–11 (1988)

Biological
Cybernetics

On the Significance of Correlations Among Neuronal Spike Trains

G. Palm [1], A. M. H. J. Aertsen [1], and G. L. Gerstein [2]

[1] Max-Planck-Institut für Biologische Kybernetik, Spemannstrasse 38, D-7400 Tübingen, Federal Republic of Germany
[2] Department of Physiology, University of Pennsylvania, Philadelphia, PA 19104-6085, USA

Abstract. We consider several measures for the correlation of firing activity among different neurons, based on coincidence counts obtained from simultaneously recorded spike trains. We obtain explicit formulae for the probability distributions of these measures. This allows an exact, quantitative assessment of significance levels, and thus a comparison of data obtained in different experimental paradigms. In particular it is possible to compare stimulus-locked, and therefore time dependent correlations for different stimuli and also for different times relative to stimulus onset. This allows to separate purely stimulus-induced correlation from intrinsic interneuronal correlation. It further allows investigation of the dynamic characteristics of the interneuronal correlation. For the display of significance levels or the corresponding probabilities we propose a logarithmic measure, called "surprise".

Introduction

In the last years there has been an increasing interest in multi-unit recordings of neuronal activity, i.e. simultaneous and separable observation of up to some 30 spike trains. Several groups have developed multi-electrode set-ups for this purpose, using various recording technologies (Gerstein et al. 1983; Krüger 1983; Grinvald 1985).

The main advantage of multi-unit recordings in comparison to single-unit recordings is the opportunity to study the interactions among several neurons as they react to a stimulus, rather than just the isolated reaction of each single neuron to the stimulus. The most important quantitative measure in extracellular recordings for the strength of interaction between neurons is the correlation of firing activity. Neuronal spikes are regarded as stereotypic events; correlation thus simplifies to counting near-coincident events. The results are displayed as correlograms.

Coincidence of firing and correlation analysis of multi-unit data has been discussed since the 1960s (Perkel et al. 1967; Gerstein and Perkel 1969, 1972; Perkel et al. 1975; Glaser and Ruchkin 1976), and there are well-established procedures for interpretation of such calculations (Abeles 1982; Gerstein et al. 1983; Krüger 1983). When the results of several different experiments are to be compared, it is essential to have procedures for correction and normalization of the calculations. In particular it is necessary to compensate for variations in individual neuron firing rates, since these directly influence the observed near-coincidence rates. These variations of individual firing rates are conveniently measured with the Peri-Stimulus-Time (PST) histogram. For ordinary cross-correlation the usual procedure to compensate for direct stimulus effects on firing rates is to subtract a so-called "shift predictor" (or "PST histogram based predictor") from the raw correlogram. Residuals are then interpreted as indication of "effective connectivity". More complex normalizations are appropriate for the Joint PST histogram, a particular stimulus-locked temporal decomposition of the correlogram (Gerstein and Perkel 1969, 1972): besides subtracting a PST histogram based predictor, the matrix is scaled, bin by bin, by the product of the variances of the two PST histograms (Aertsen et al. 1988).

The kind of normalization procedures just described offers a measure of deviation from a null hypothesis; the size of such deviations is a measure of their significance. In our case, where we want to demonstrate neuronal interaction, the appropriate null hypothesis is *(1)* two completely *in*dependent neurons that *(2)* respond to the stimulus exactly as shown by the two experimentally observed PST histograms. More specifically: significance for positive

2

deviation from the null hypothesis is determined from the probability of having a near-coincidence count that is at least as high as the observed count, and significance for negative deviation is determined from the probability of having a near-coincidence count that is at most as high as the observed count.

In this paper we will show that these probabilities for the Joint PST histogram (JPSTH) can be expressed in closed form. These results also allow explicit calculation of the distributions of correlation values after application of any of the various types of normalization that have been proposed. These calculations therefore provide a sound basis for the evaluation of significance in correlation measurements and, hence, for the significance of their interpretation. Procedures for the analysis of firing correlations in multi-neuron recordings have benefited from these developments. Recent work along these lines is producing evidence for dynamic cooperativity as an emergent property of neuronal assembly organization in the brain (Gerstein 1987; Aertsen et al. 1988; Gerstein et al. 1988).

1 Calculations of the Probability Densities of Coincidence Counts

The experimental design is usually the following: A certain stimulus is repeatedly presented for a number of times, $i=1,\ldots,n$. During each presentation of the stimulus, spike trains are recorded from two neurons.

Our calculations are based on the following set of assumptions:

1. Spikes are considered as identical *events* for which only the time of occurrence is relevant.

2. We impose discrete *time-bins*, such that each spike train is transformed into a 0–1 *process*. This requires that the time-bin be so small that there will be *at most one spike* in each bin. We will let the time marker associated with each stimulus presentation fall directly onto one such bin.

3. Firing activity during different (repeated) stimulus presentations is statistically indistinguishable. This means there are no long term nonstationarities in the data.

The recording time T per stimulus presentation is divided into J time-bins $B_1,\ldots,B_J$. Thus the total recording time $n\cdot T$ is divided into $n\cdot J$ bins, and B^i_j denotes the j-th bin during the i-th stimulus presentation (see Fig. 1). The activity of each neuron, averaged over the many stimulus presentations, is given by the corresponding PST histogram; counts are generally different from bin to bin and reflect the stimulus-locked firing rate modulation for that particular neuron.

Assumption 2 has the following mathematical consequence: the numbers of spikes X^i_j observed in the bins B^i_j are random variables with values in $\{0,1\}$. Assumption 3 means that the conditions at a particular bin (lower index j) are the same for all successive stimulus presentations (upper index i): the variables $X^1_j,\ldots,X^n_j$ are independent and identically distributed (i.i.d.).

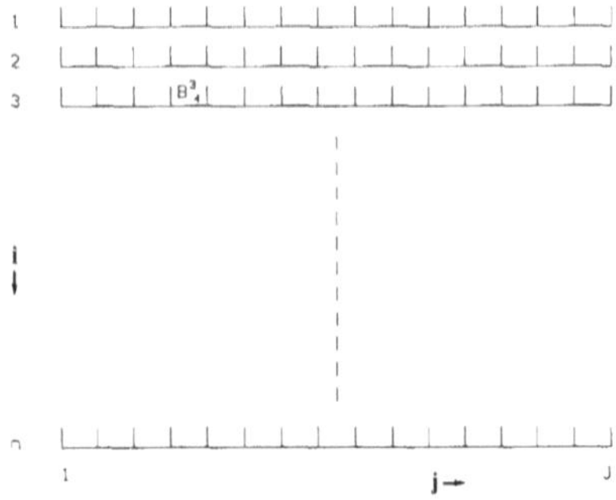

Fig. 1. Scheme of time binning, applied to spike trains, recorded during successive stimulus presentations. Further explanation in text

The mathematical problem can now be formulated as follows: We want to analyze (delayed) coincidence counts between two such variables (which usually would be from two different, simultaneously recorded neurons). We will use shorthand notation X and Y to indicate the two variables at the selected bins; from here on we will implicitly refer to this particular bin selection. Over the sequence of stimulus presentations, we then have two sequences of random variables $X^1,\ldots,X^n$ and $Y^1,\ldots,Y^n$, both independent and identically distributed in $\{0,1\}$.

The null hypothesis for the following calculations is that X^i and Y^i are independent.

1.1 Calculation in Terms of Single Unit Firing Probabilities

The distribution of X^i is determined by the firing probability $p=\mathrm{pr}[X^i=1]$; similarly the distribution of Y^i is given by the probability $q=\mathrm{pr}[Y^i=1]$. In this subsection we assume that these two probabilities are given.

Note that normally these probabilities are not given, and the best one can do is to assume that they are identical to the average spike frequencies, evaluated in the appropriate bins, of the two neurons under the stimulation used in the experiment. This case is treated more appropriately in the next subsection (1.2).

Given p and q it is straightforward to calculate both the probability density and the (cumulative) distribution of the coincidence count. First we note that for the specific selected bins, $X = \sum_{i=1}^{n} X^i$ is the spike count for the X^i-sequence, $Y = \sum_{i=1}^{n} Y^i$ is the spike count for the Y^i-sequence, and $Z = \sum_{i=1}^{n} X^i \cdot Y^i$ equals the coincidence count. For the probability density of the coincidence count we may write:

$$\mathrm{pr}[Z=m] = \sum_{k=1}^{n} \mathrm{pr}[Z=m \mid X=k] \cdot \mathrm{pr}[X=k]. \tag{1}$$

These conditional probabilities can be determined as follows: if $X=k$ we may as well assume that $X^1, \ldots, X^k = 1$ and $X^{k+1}, \ldots, X^n = 0$, and therefore that the coincidence count Z equals $\sum_{i=1}^{k} Y^i$ and is binomially distributed.

Thus we may write:

$$\mathrm{pr}[Z=m \mid X=k] = b(m; k, q) = \binom{k}{m} \cdot q^m (1-q)^{k-m} \tag{2}$$

and

$$\mathrm{pr}[Z=m] = \sum_{k=1}^{n} b(m; k, q) \cdot b(k; n, p). \tag{3}$$

We can rewrite the probability density (3) into the corresponding cumulative distribution, simply by appropriate summation.

Equation (3) can be easily evaluated; some examples of the cumulative distribution for different values of n, p, and q are shown in Fig. 2a. Clearly every value of Z between 0 and n occurs with non-zero probability, but in fact the probability that Z is larger than $n \cdot p$ or $n \cdot q$, is very low and therefore in Fig. 2a we only displayed the curves for this lower range of Z-values.

Next we calculate the expectation of Z and Z^2, and the variance of Z:

$$\begin{aligned} E[Z] &= \sum_{k=1}^{n} E(Z \mid X=k) \cdot \mathrm{pr}[X=k] \\ &= \sum_{k=1}^{n} q \cdot k \cdot \mathrm{pr}[X=k] \\ &= q \cdot E[X] = n \cdot p \cdot q, \end{aligned} \tag{4}$$

$$\begin{aligned} E[Z^2] &= \sum_{k=1}^{n} E(Z^2 \mid X=k) \cdot \mathrm{pr}[X=k] \\ &= \sum_{k=1}^{n} (kq + k(k-1)q^2) \cdot \mathrm{pr}[X=k] \\ &= (q-q^2)E[X] + q^2 E[X^2] \\ &= q(1-q) \cdot n \cdot p + q^2(np + n(n-1)p^2) \\ &= n^2 p^2 q^2 + n(pq - p^2 q^2), \end{aligned} \tag{5}$$

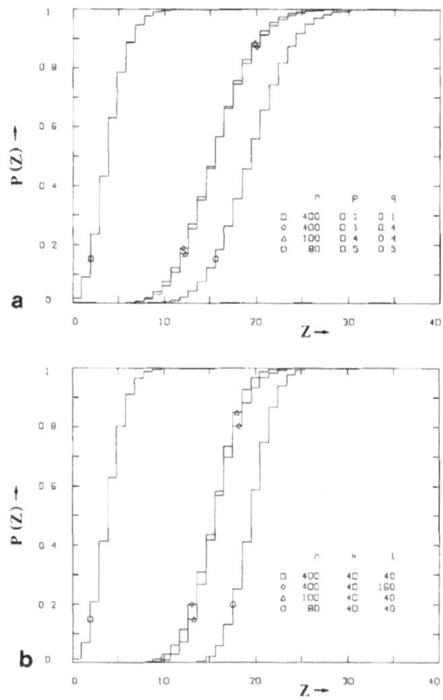

Fig. 2a and b. Examples of cumulative distribution function of coincidence count Z for different numbers of stimulus presentations n and neuronal firing rates. **a** Calculation in terms of single unit firing probabilities p and q. **b** Calculation in terms of observed single unit spike counts k and l

and

$$\mathrm{Var}(Z) = n(pq - p^2 q^2). \tag{6}$$

Actually, the probability given in (3) is not really relevant in the typical experimental situation, because there one does not know the underlying probabilities p and q. One usually knows only the actual spike counts in the selected bins. If additional information were available, it could be used to make reasonable estimates for p and q for use in the above equations. For example, if one may assume that the firing probability is about the same in neighbouring bins, it is reasonable to average over these bins to determine p and q. It may sometimes be of interest to use this kind of neighbourhood sampling, especially for the variability correction that is mentioned in the introduction. Another example of additional information might be the availability of a model for the firing probabilities.

4

1.2 Calculation from the Observed Single Unit Spike Counts

In the usual experimental case, we do not have additional information; it is only possible to make the estimates in terms of the counts themselves, i.e. by using as estimates $p=k/n$ and $q=l/n$. In this case one might as well work directly in terms of the counts, rather than indirectly in terms of functions of these same counts.

Thus the appropriate mathematical description of the usual situation is that the experimenter just has four counts available: $X=k$, $Y=l$, and $Z=m$ and the number of stimulus presentations n. From these four numbers k, l, m, n we want to know the significance of the coincidence count m, given the firing counts of the two neurons are k and l during the n stimulus presentations. In this case one has to calculate the distribution of Z under the condition that $X=k$ and $Y=l$.

Without loss of generality we can assume that $k \leq l$ and, additionally, that $Y^1, \ldots, Y^l = 1$ and $Y^{l+1}, \ldots, Y^n = 0$.

The probability $\mathrm{pr}[Z=m \mid X=k \text{ and } Y=l]$ is the probability that m out of the k X^i-s which are 1, occur before l, while the remaining $k-m$ of these k X^i-s occur after l, i.e.

$$\mathrm{pr}[Z=m \mid X=k \text{ and } Y=l] = \frac{\binom{l}{m} \cdot \binom{n-l}{k-m}}{\binom{n}{k}}. \tag{7}$$

In this expression it may happen that $n-l<k-m$, i.e. $n+m<k+l$; in that case we define

$$\binom{n-l}{k-m} = 0.$$

Expression (7) can be easily evaluated (see Fig. 2b for some examples of the cumulative distribution for different values of n, k, and l; observe the similarity to Fig. 2a) and we can again calculate the mean and the variance of the corresponding distribution:

$$E(Z \mid X=k,\ Y=l) = \frac{k \cdot l}{n}, \tag{8}$$

$$E(Z^2 \mid X=k,\ Y=l) = \frac{k \cdot l(k-1)(l-1)}{n \cdot (n-1)} + \frac{k \cdot l}{n}, \tag{9}$$

and

$$\mathrm{Var}(Z \mid X=k,\ Y=l) = 1/(n-1) \cdot k(1-k/n) \cdot l(1-l/n). \tag{10}$$

If we express this in terms of rates, $\kappa = k/n$ and $\lambda = l/n$, we may write

$$E(Z \mid X=k,\ y=l) = n \cdot \kappa \cdot \lambda \tag{11}$$

and

$$\mathrm{Var}(Z \mid X=k,\ y=l) = n^2/(n-1) \cdot \kappa(1-\kappa) \cdot \lambda(1-\lambda). \tag{12}$$

2 Normalized Correlations

Given the counts $X=k$, $Y=l$, $Z=m$ and n as in the preceding section, several formulae have been used in the literature to normalize the coincidence counts (e.g. Kuznetsov and Stratonovich 1956; Gerstein and Perkel 1969, 1972; Habib and Sen 1985; Boogaard et al. 1986; Aertsen et al. 1987, 1988). The aim of such normalization is to allow interpretation within an experiment and comparison across experiments.

Among these normalization procedures the earliest idea was the so-called "shift or (shuffle) predictor" (Perkel et al. 1967; Gerstein and Perkel 1969, 1972): take any permutation of stimulus presentations $\Pi: \{1, \ldots, n\} \to \{1, \ldots, n\}$ and form the shuffled correlation $Z'_\Pi = \sum X^i Y^{\Pi(i)}$. Now one can consider the difference $D'_\Pi = Z - Z'_\Pi$ as a correlation that is corrected for pure stimulus effects. Since this procedure depends on the choice of the permutation Π, an obvious idea is to average this over all possible permutations Π. This leads to the expression

$$D' = \frac{1}{n!} \cdot \sum_{\Pi} D'_\Pi = Z - \frac{1}{n!} \cdot \sum_{\Pi} \sum_{i=1}^{n} X^i Y^{\Pi(i)}$$

$$= Z - \sum_{i=1}^{n} X^i Y = Z - (X \cdot Y)/n. \tag{13}$$

In other words: the average over the set of all possible "shift (or shuffle) predictors" is equivalent to a predictor based on the cross product of the PST histograms.

In the following we simply list the various expressions which have been suggested and which will be considered here, also including one new suggestion (*e*).

(*a*) Subtraction of the expected coincidence [cf. (13)]:

$$D = Z - (X \cdot Y)/n. \tag{14}$$

(*b*) Division by the expected coincidence:

$$Q = Z \cdot n/(X \cdot Y). \tag{15}$$

(*c*) Subtraction of the expected coincidence, followed by division by the expected coincidence:

$$R = D \cdot n/(X \cdot Y). \tag{16}$$

(*d*) Subtraction of the expected coincidence, followed by division by the product of the estimated standard deviations of X and Y. This leads to a correlation coefficient:

$$C = D/\{X(1-X/n) \cdot Y(1-Y/n)\}^{1/2}. \tag{17}$$

In the quantity C, the difference D has been normalised to values in the range $[-1, 1]$.

(*e*) Subtraction of the expected coincidence, followed by division by the conditional standard deviation of Z (given $X=k$ and $Y=l$) as determined in (10):

$$S=\sqrt{(n-1)}\cdot D/\{X(1-X/n)\cdot Y(1-Y/n)\}^{1/2}$$
$$=\sqrt{(n-1)}\cdot C. \tag{18}$$

S is essentially the standardization of Z.

Note that, like the various counts X, Y, and Z, these normalized correlations are stochastic variables. In the following we will consider their distribution and first and second moments.

2.1 Calculation in Terms of Single Unit Firing Probabilities

Given the firing probabilities p and q as in Sect. 1.1, it turns out to be a hard mathematical problem to determine the distribution or moments of these statistics. In fact, only for D were we able to determine the distribution in closed form.

Analogously to (1) we have

$$\mathrm{pr}[D=d]=\sum_{k=1}^{n}\mathrm{pr}[D=d\,|\,X=k]\cdot\mathrm{pr}[X=k]. \tag{19}$$

If $X=k$, we may again assume that $X^1,\ldots,X^k=1$ and $X^{k+1},\ldots,X^n=0$. Thus for $X=k$ we have

$$D=Z-X\cdot Y/n=\sum_{i=1}^{k}Y^i-(k/n)\cdot\sum_{i=1}^{n}Y^i$$
$$=(1-k/n)\cdot\sum_{i=1}^{k}Y^i-(k/n)\cdot\sum_{i=k+1}^{n}Y^i$$
$$=(1-k/n)\cdot S_1-(k/n)\cdot S_2. \tag{20}$$

Under the condition that $X=k$, the variable D is apparently a linear combination of two independent, binomially distributed random variables S_1 and S_2, where the distribution of S_1 is $b(s_1;k,q)$ and that of S_2 is $b(s_2;n-k,q)$. From these distributions one can compute $\mathrm{pr}[D=d|X=k]$, and from (19) we then obtain $\mathrm{pr}[D=d]$. Some examples of the cumulative distribution of D are shown in Fig. 3, which will be discussed in Sect. 3.

2.2 Calculation from the Observed Single Unit Spike Counts

In the case where we work with the conditional probabilities given that $X=k$ and $Y=l$ as in Sect. 1.2, we can easily determine the probability density of any of the above correlation measures. Let us consider, for example, the correlation coefficient C. Under the given conditions for the individual spike counts $X=k$ and $Y=l$, the only remaining stochastic variable in the value of C is the coincidence count $Z=m$. Hence, we may simply write:

$$\mathrm{pr}\left[C=\frac{m-kl/n}{\{k(1-k/n)\cdot l(1-l/n)\}^{1/2}}\,\middle|\,X=k\text{ and }Y=l\right]$$
$$=\mathrm{pr}[Z=m\,|\,X=k\text{ and }Y=l]. \tag{21}$$

Similar reasoning can be used to obtain probability densities for the other correlation measures.

Equations (7)–(12) also lead us to the expectation and variance of each of the above measures under the condition that $X=k$ and $Y=l$. In the following equations we will no longer explicitly indicate this condition.

(*a*) $E(D)=0$; $\mathrm{Var}(D)=\mathrm{Var}(Z)$, (22)

(*b*) $E(Q)=1$; $\mathrm{Var}(Q)=(1-\kappa)(1-\lambda)/(\kappa\lambda(n-1))$, (23)

(*c*) $E(R)=0$; $\mathrm{Var}(R)=\mathrm{Var}(Q)$, (24)

(*d*) $E(C)=0$; $\mathrm{Var}(C)=1/(n-1)$, (25)

(*e*) $E(S)=0$; $\mathrm{Var}(S)=1$. (26)

For large n, the measures C, Q, and R have decreasing variance with increasing n. The correlation coefficient C has the smallest variance as long as κ and λ are below 1/2 (which is the usual case for physiological data). The standardized coincidence count S has to have variance 1 by definition: it therefore must have values outside the range $[-1,1]$. Z itself and of course D are the only coincidence measures whose variance increases with n.

2.3 Significance of Experimentally Observed Correlations

Having derived explicit expressions for the distributions of the various normalized correlation measures, we can now address the significance of finding a particular value for any of these measures. Here we can use the ordinary approach to this problem. The significance of downward deviation (negative correlation measure) is the probability of finding that value or a lower one; this can simply be read off the lower end of the plotted corresponding cumulative distribution function. The significance of an upward deviation is the probability of finding that value or a higher one; this can simply be read off the upper end of the plotted cumulative distribution function.

3 Comparison to the Gaussian Distribution

In this and the following sections we will concentrate the discussion on the expressions in terms of observed single unit spike counts (as in Sects. 1.2 and 2.2), since only in this case can we explicitly calculate all relevant

6

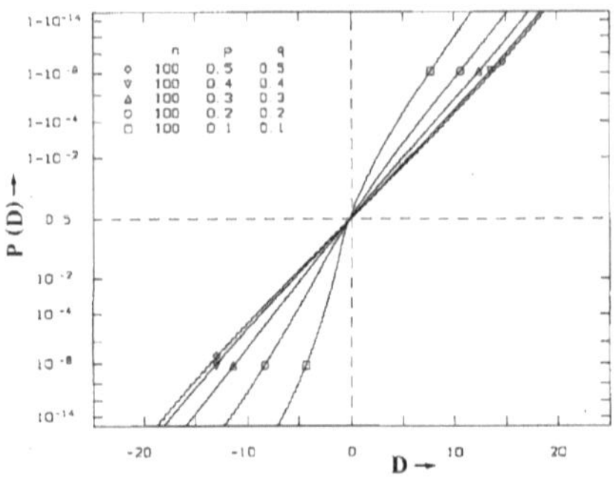

Fig. 3. Examples of cumulative distribution function of correlation measure D for constant number of stimulus presentations n and different single unit firing probabilities $p=q$. The cumulative distribution function $P(D)$ is plotted on Gaussian probability paper

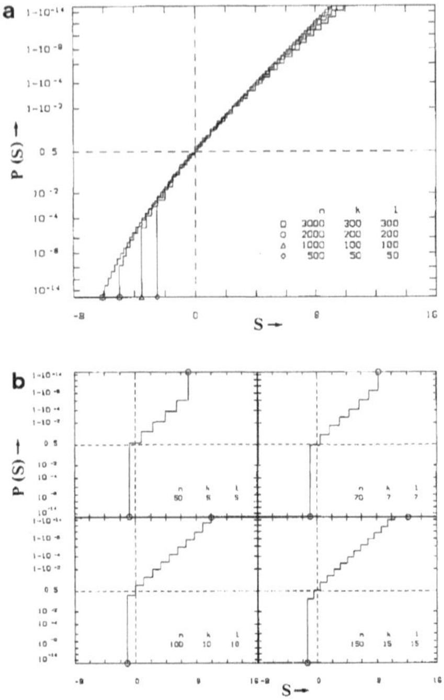

Fig. 4a and b. Examples of cumulative distribution function of standardized correlation S for different numbers of stimulus presentations n and constant spike counts per presentation $\kappa=\lambda$. The cumulative distribution function $P(S)$ is plotted on Gaussian probability paper. **a** Superposition of four $P(S)$-curves for large n. **b** Separate plots of four $P(S)$-curves for smaller n

distributions without resorting to Gaussian approximations in the first place. Also this case is probably the experimentally more relevant one. Of course, for large n, all the measures of correlation will be approximately normally distributed. To the extent that such an approximation is valid, the normal distributions will then be fully characterized by the corresponding mean and variance, as given in (22)–(26). Such mean and variance can then be used directly for significance statements.

In the usual experimental situation, however, the number n of stimulus presentations will not be very large – values of $n \leqq 100$ are a common case. Therefore it is appropriate to study the departure from normality of the various correlation measures as a function of the parameters n, k, and l (or n, p, and q). This can be done by examining the exact distribution of the correlation measures on Gaussian probability paper, i.e. transformed by the inverse of the cumulative normal distribution function (as in Figs. 3 and 4). In such a plot the normal distribution appears as a straight line, going through the origin.

The most commonly used measure of correlation is the correlation coefficient C, referred to as "cross-correlation surface" by Habib and Sen (1985) and as "normalized Joint Peri-Stimulus Time Histogram" in Aertsen et al. (1988). For investigation of the departure from Gaussian, however, it is more convenient to work with the standardized correlation S (as in Fig. 4), because in such a plot all curves should have the same slope, independent of all parameters, in particular of n. When we worked with the firing probabilities p and q, we were unable to calculate the distribution of S; therefore for this case we display the distribution of D (Fig. 3). In such a plot the curves will have different slope for different parameters.

Let us start with the discussion of the expressions in terms of single unit firing probabilities p and q (cf. Sects. 1.1 and 2.1). Figure 3 shows some cumulative distributions of D for different values of $p=q$ and constant $n=100$, plotted on Gaussian probability paper. The frame in Figs. 3 (and 4) has been set to cover the range of cumulative probability values between 10^{-15} and $1-10^{-15}$, corresponding to significance levels of up to 10^{-15}. As should be expected, the curves for higher values of p have a lower overall slope and are closer to a straight line. For $p=0.5$ the curve is symmetric around $D=0$, and virtually indistinguishable from a straight line (i.e. from a Gaussian distribution). For lower values of p, the curves becomes increasingly asymmetric and skewed, with higher average slope for negative D. Altogether, the deviations from the Gaussian distribution appear to be rather small for the relatively low value of $n=100$.

We now turn to the expressions in terms of observed single unit spike counts k and l (cf. Sects. 1.2 and 2.2), which in general show a more pronounced deviation from the Gaussian. Figure 4 shows some cumulative distributions for the standardized correlation S as plotted on Gaussian probability paper. Note that in this figure we varied n, while keeping the fractions κ and λ constant (0.1). Figure 4a shows a superposition of four curves for $n=500$, 1000, 2000, and 3000; Figure 4b shows four separate plots of the cases $n=50$, 70, 100, and 150. Interestingly, the curves in Fig. 4a (i.e. for large n) can be reasonably well approximated by a straight line. There is, however, always a slight curvature to the right: decreasing slope for increasing S. As was to be expected, the departure from Gaussian increases for decreasing n. They are worst at the smallest, but experimentally realistic values of n used in Fig. 4b.

The probability of zero coincidences is given by the size of the first step in the cumulative distribution staircase plot; the probability of maximal number of coincidences is given by the size of the last step. In each case the position of the first step along the S-coordinate is indicated by a symbol. Observe that these positions, as well as the corresponding probabilities, do not reach down to the lower left corner of the graph, especially for smaller n. The position of the last step along the S-coordinate is indicated by a symbol for small n (Fig. 4b), for larger n (Fig. 4a) this last step is well beyond the displayed S-range. The last step in the plot is always much smaller than the first step; in fact it is virtually zero, with the exception of the cases $n=50$ and $n=70$. Furthermore the range of positive S-values is always considerably larger than the range of negative S-values (note the asymmetry of the S-axis). This means that the range of attainable S-values, as well as probability values and, consequently, significance values, for negative interaction is considerably smaller than for positive interaction. In particular, very low probabilities, i.e. very high levels of significance for negative interaction cannot be reached, except for unrealistically large values of n. Note that in the more realistic cases shown in Fig. 4b, cumulative probabilities smaller than 0.5 or negative S-values (i.e. inhibition) can only be found in the extreme case of zero coincidences.

The practical consequence of this positive-negative asymmetry can be formulated in two ways. For a given number of stimulus presentations n, there is a severe limitation to the attainable deviation in the negative direction, and hence to the significance that can be attained for negative interaction ("inhibition"); there is no severe limitation for deviations in the positive direction ("excitation"). Conversely, attaining a required level of significance requires a higher value of n (i.e. more stimulus presentations) for deviations in the negative direction than for deviations in the positive direction. These consequences of the asymmetry have already been noted in cross-correlation analysis of multi-neuron data: for neurons firing at typical rates, inhibition is much harder to detect than excitation (Aertsen and Gerstein 1985).

Figure 4 and the above comments are appropriate for the case of typical parameter values k and l which are less than $n/2$. For k and l around $n/2$ the curves become more symmetric and actually quite similar to a normal distribution. This suggests another strategy for enhancing detectability of inhibition: increase the firing rates of the neurons by some means (e.g. by sensory or pharmacological stimulation). For k and l larger than $n/2$, the plots now become curved in the opposite direction, and the asymmetry reverses: it is now harder to attain significance for positive interaction than for negative interaction.

4 Maximum and Minimum of the Distribution of C, D or S

A simple way of quantifying the difference in sensitivity of the correlation measures for "excitation" (positive correlation) and for "inhibition" (negative correlation) is to calculate their *range*. Let us consider, for example, the correlation coefficient C. Theoretically, C values are allowed to cover the range $[-1, 1]$. For any specific set of numbers k, l, m, n, however, the positive and negative ranges that are actually covered will in general not be the same. For k and l smaller than $n/2$ (the usual physiological case), we show below that the positive range is larger than the negative range. For k and l larger than $n/2$, the relationship between positive and negative ranges is reversed.

Let $\mathrm{Max}(C)$ and $\mathrm{Min}(C)$ denote the maximum and minimum of C. The ratio $a=|\mathrm{Max}(C)/\mathrm{Min}(C)|$ can then be taken as a measure for the asymmetry of the positive and negative ranges. From the known range of the coincidence count Z, the asymmetry a for measure C can be easily calculated. Furthermore, the asymmetries for all measures C, D, R, and S are identical, because these measures differ only in the denominator. Since the range of D is the easiest to calculate, we shall in the following determine a from

$$a=|\mathrm{Max}(D)/\mathrm{Min}(D)|\,. \tag{27}$$

For counts k and l such that $k+l \leq n$, it is clear that Z ranges from 0 to the minimum of k and l. For $k+l>n$ there must be some coincidences: Z is at least $k+l-n$.

8

A more general formulation can be obtained by introducing the notation $\mathrm{Inf}(x, y)$ for the smaller of x and y, and $\mathrm{Sup}(x, y)$ for the larger of x and y. Then for the range of Z we have

$$\mathrm{Min}(Z) = \mathrm{Sup}(0, k+l-n), \tag{28}$$

$$\mathrm{Max}(Z) = \mathrm{Inf}(k, l). \tag{29}$$

To determine the range of D we have to subtract $k \cdot l/n$ from Z, i.e.

$$\mathrm{Min}(D) = \mathrm{Sup}(0, k+l-n) - k \cdot l/n, \tag{30}$$

$$\mathrm{Max}(D) = \mathrm{Inf}(k, l) - k \cdot l/n. \tag{31}$$

Let us assume that $k \leqq l$ and that $k+l \leqq n$. Then the asymmetry is:

$$a = \frac{n-l}{l} = \frac{1-\lambda}{\lambda} \tag{32}$$

and for $k+l \geqq n$ we have:

$$a = \frac{k}{n-k} = \frac{\kappa}{1-\kappa}. \tag{33}$$

For the experimentally relevant parameter range, we always have $k+l \leqq n$. According to (32) this means that for firing rates below 10% the asymmetry is larger than $0.9/0.1 = 9$. The asymmetry of sensitivities for excitatory and inhibitory interactions for the normal experimental situation thus is quite large indeed.

5 Discussion

In this paper we have examined the (delayed) coincidence counts between the firings of two simultaneously recorded neurons as expressed in the (normalized) Joint Peri Stimulus Time Histogram. This is usually expressed as a matrix of values, where each entry (bin) corresponds to a particular pair of time delays of the two neurons with respect to a stimulus time marker. We have derived analytic expressions for the probability density of such coincidence counts, under the null hypothesis of independent neurons with the observed individual PST histograms.

These results allow the evaluation of the coincidence probability densities after various normalization procedures that have previously been introduced in the literature. For each of the normalized correlation measures we have given explicit expressions for the probability density, and from it, the expected value and variance at each bin. The corresponding distributions are non-Gaussian, and have an asymmetric shape and range of values. With physiologically reasonable counts, this asymmetry implies, as has previously been shown, that inhibitory processes are harder to detect than excitatory processes. The given distributions provide the necessary framework for formal evaluation of significance in observed deviations of the coincidence counts from values predicted under the null hypothesis. This is a completely model free approach which results only in a positive or negative statement about the validity of the null hypothesis. Later steps in the analysis process may require the invocation of specific models of neuronal interaction in order to quantify "connectivity" among the neurons (Aertsen et al. 1988).

5.1 Display of Significance Values: Surprise

The usual application of significance testing is to state the probability with which the actual observation could occur under the null hypothesis ($p = 0.05$ or 0.01, for example). The significance value for a given count can be read directly from the graph of the (cumulative) distribution function (cf. Sect. 2.3). However, since we are interested in very small values of the probability, it is appropriate to introduce a logarithmic scale for a more effective display of this graph. Thus, instead of plotting the direct distribution, we will use its negative logarithm; this has been called *entropy* (e.g. Boltzmann 1887; Shannon 1948; Billingsley 1965) or *surprise* (Legéndy 1975; Palm 1981, 1988). The correspondence between the usual significance statement and the surprise measure is a straightforward scale transformation: for example $p = 0.05$ corresponds to a surprise value of 2.996 [i.e. $-\ln(0.05)$]; $p = 0.01$ means a surprise of 4.605.

5.2 Extension of the Surprise to Events in Several Bins

Everything in this paper has so far been based on counts in a single bin. As noted earlier, each such bin corresponds to a particular choice of time delays in the

Fig. 5a and b. Correlation of firing activity among two neurons in the cat's visual cortex (area 17) during stimulation with moving bars (Krüger 1982). **a** Joint PST histogram. *Left:* Matrix of "raw" coincidence counts Z, divided by the number of stimulus presentations *n*; values coded according to grey scale above), together with both single unit PST histograms (along abscissa and ordinate); *right:* histogram representation of values in narrow band along main diagonal of JPSTH (time course of near-coincident firing), and time average of near-coincident firing as a function of relative time delay (upper right histogram across diagonal, obtained by integrating JPSTH along diagonal (comparable to normal cross correlogram). **b** Significance of correlation: difference between surprise for "excitation" and surprise for "inhibition" (display format of **b** as in **a**). Further explanation of the surprise measure is given in the text. [See Aertsen et al. (1988) for a more elaborate discussion on this type of analysis and figures]

9

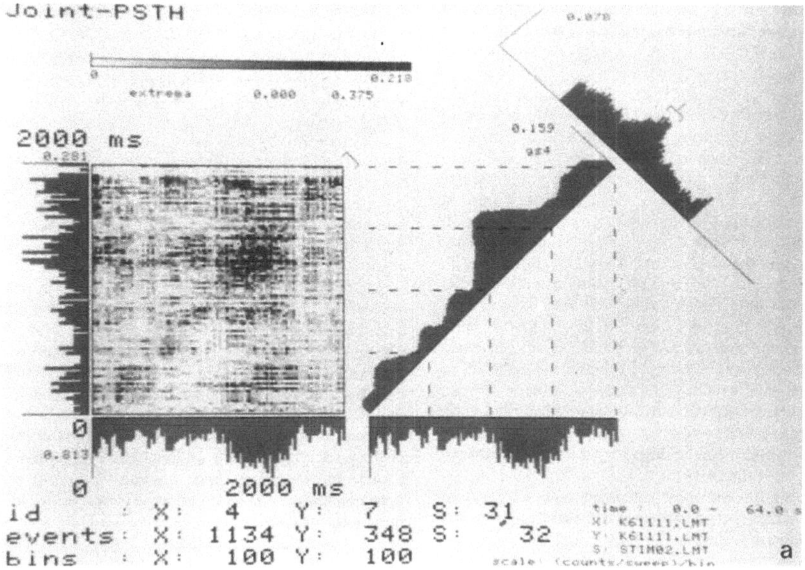
Joint-PSTH
0
0.210
extrema 0.000 0.375
2000 ms
0.159
0
0.813
0
2000 ms
id : X: 4 Y: 7 S: 31
events: X: 1134 Y: 348 S: 32
bins : X: 100 Y: 100
a

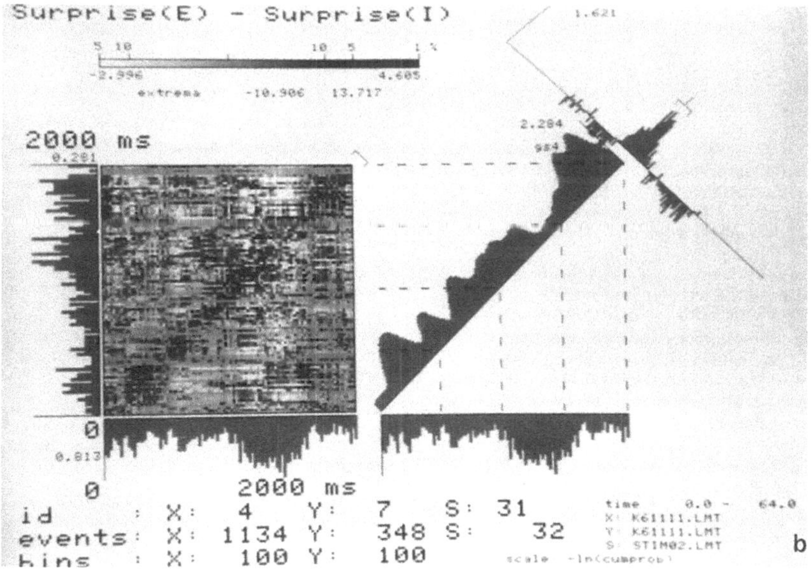
Surprise(E) - Surprise(I)
1.621
-2.996
4.605
extrema -10.906 13.717
2.284
2000 ms
0
0.813
0
2000 ms
id : X: 4 Y: 7 S: 31
events: X: 1134 Y: 348 S: 32
bins : X: 100 Y: 100
b

10

firings of the two neurons with respect to a stimulus marker. Normally, however, we are interested in the complete matrix of such bins; with reasonable choice of time resolution our attention is drawn to coherent deviant regions ("features") in such a matrix (for an example see Fig. 5).

Figure 5 is obtained from spike sequences, simultaneously recorded from two neurons in the cat's visual cortex (area 17) during stimulation with moving bars (Krüger 1982). Each bin in the matrices corresponds to the coincidence counts between spikes in a particular pair of time bins, one for each neuron. Figure 5a displays the "raw" coincidence counts Z; Figure 5b displays the difference between the surprise for "excitation" and the surprise for "inhibition". Both surprise values are calculated from the cumulative distribution obtained from (7), evaluated for the observed counts of X, Y, and Z at each pair of bins. [For a more comprehensive discussion on the use of such matrices and their interpretation, the reader is referred to Aertsen et al. (1988).] In Fig. 5b we note a distinct, broad region of high surprise values along the main diagonal; this is certainly a more "significant" feature than any of the isolated high surprise values, scattered throughout the matrix.

From this viewpoint, isolated bins which show highly significant deviations in the statistical sense, are of less interest. If we require significant deviation over some contiguous region of bins, we are in effect imposing an extended requirement on significance. (compare Palm 1988). Here we can make use of another property of the surprise measure: it makes sense to add the surprise values of different observations, because surprise is a logarithmic measure of improbability. In our case this implies that we can simply add the surprise values of individual bin counts. Such an integration is, of course, also a smoothing process applied to the surprise matrix. Automatic detection of deviant features in the matrix could utilize algorithms similar to those used for the detection of surprising bursts of firing in the spike trains of neurons (Legéndy and Salcman 1985): starting from a bin with high surprise value, one keeps adding adjacent bins which do not lower the average surprise over the combined area.

5.3 *Comparison of the Various Measures of Correlation in Models and Experiments*

Several normalised correlation measures were defined in (14)–(18). Studies with spike trains originating from simulated neuronal circuits show that normalizations C (and S) are the most appropriate, in the sense that they most directly allow recovery of the underlying circuit from the correlation measures (Aertsen et al. 1988). The standardized correlation S differs from the normalized correlation C only by a factor containing the number of sweeps: $\sqrt{(n-1)}$ [cf. (18)].

S has a fixed variance of 1, and, as a consequence, a variable range of values which depends on the number of stimulus presentations. We prefer the correlation measure C, since its values are normalized to the fixed range $[-1, 1]$; this makes comparison across experiments relatively straightforward.

In Fig. 4 we have shown that the coincidence count distribution deviates from Gaussian. However, since in the typical experimental situation these deviations are not dramatic, it is appropriate to examine the applicability of a Gaussian distribution for significance testing. The obvious advantage would be that one can work with the variance alone, rather than with the complete distribution. The necessary variance in this case would be computed in the usual manner, i.e. by averaging the appropriate squared difference over all stimulus presentations. Therefore in this situation we are computing the variance of the coincidence count only over the current realization of the experiment. In contrast, the distributions we have derived in this paper refer to the ensemble of all possible realizations of the coincidence counts under the null hypothesis. Clearly the Gaussian approximation is a more restricted measure, and hence, perhaps to be avoided for these purposes.

The theoretical results given in this paper have been applied in the analysis of correlation of firing in both simulated and physiological spike trains. Such analyses and discussion of their interpretations can be found in (Aertsen et al. 1988; Gerstein 1987; Gerstein et al. 1988). These studies of correlation in multi-neuron firing show that – over a wide range of preparations (both animals and brain areas) – neuronal assembly organization is dynamic on several time scales (milliseconds to seconds), and depends on both stimulus and context.

Acknowledgements. GLG is grateful to Prof. V. Braitenberg for the opportunity of a sabbatical stay at the MPI for Biological Cybernetics during 1986–87. The spike train data for Fig. 5 were kindly provided by Jürgen Krüger. We thank Volker Staiger for skillful assistance in generating the figures.

References

Abeles M (1982) Local cortical circuits. Studies of brain function, vol 6. Springer, Berlin Heidelberg New York

Aertsen AMHJ, Gerstein GL (1985) Evaluation of neuronal connectivity: sensitivity of cross correlation. Brain Res 340:341–354

Aertsen A, Bonhoeffer T, Krüger J (1987) Coherent activity in neuronal populations: analysis and interpretation. In: Caianiello ER (ed) Physics of cognitive processes. World Scientific Publishing, Singapore, pp 1–34

Aertsen AMHJ, Gerstein GL, Habib M, Palm G et al. (1988) Dynamics of neuronal firing correlation: modulation of "effective connectivity" (submitted for publication)

Billingsley P (1965) Ergodic theory and information. Wiley, New York

Boltzmann L (1887) Über die mechanischen Analogien des zweiten Hauptsatzes der Thermodynamik. J Reine Angew Math 100:201–212

Boogaard H van den, Hesselmans G, Johannesma P (1986) System identification based on point processes and correlation densities. I. The nonrefractory neuron model. Math Biosci 80:143–171

Gerstein G (1987) Information flow and state in cortical neural networks: interpreting multi-neuron experiments. In: Seelen W von, Shaw G, Leinhos U (eds) Organization of neural networks: structures and models. VCH Verlagsgesellschaft, Weinheim, pp 53–75

Gerstein GL, Perkel DH (1969) Simultaneously recorded trains of action potentials: analysis and functional interpretation. Science 164:828–830

Gerstein GL, Perkel DH (1972) Mutual temporal relationships among neuronal spike trains. Biophys J 12:453–473

Gerstein G, Bloom M, Espinosa I, Evanczuk S, Turner M (1983) Design of a laboratory for multi-neuron studies. IEEE Trans SMC-13:668–676

Gerstein GL, Aertsen AMHJ et al. (1988) Neuronal assemblies as observed in multi-neuron experiments: dynamic organization depending on stimulus (in preparation)

Glaser EM, Ruchkin DS (1976) Principles of neurobiological signal analysis. Academic Press, New York

Grinvald A (1985) Real-time optical mapping of neuronal activity: from single growth cones to the intact mammalian brain. Ann Rev Neurosci 8:263–305

Habib MK, Sen PK (1985) Non-stationary stochastic point-process models in neurophysiology with applications to learning. In: Sen PK (ed) Biostatistics: statistics in biomedical, public health and environmental sciences. Elsevier/North-Holland, Amsterdam, pp 481–509

Krüger J (1982) A 12-fold microelectrode for recording from vertically aligned cortical neurons. J Neurosci Methods 6:347–350

Krüger J (1983) Simultaneous individual recordings from many cerebral neurons: techniques and results. Rev Physiol Biochem Pharmacol 98:177–233

Kuznetsov PI, Stratonovich RL (1956) A note on the mathematical theory of correlated random points. Izv Akad Nauk SSSR Ser Math 20:167–178; also in: Kuznetsov PI, Stratonovich RL, Tikhonov VI (eds) (1965) Nonlinear transformations of stochastic processes. Pergamon Press, New York, pp 101–115

Legéndy CR (1975) Three principles of brain function and structure. Int J Neurosci 6:237–254

Legéndy CR, Salcman M (1985) Bursts and recurrences of bursts in the spike trains of spontaneously active striate cortex neurons. J Neurophys 53:926–939

Palm G (1981) Evidence, information, and surprise. Biol Cybern 42:57–68

Palm G (1988) Information theory. MIT Press, Cambridge, Mass (in press)

Perkel DH, Gerstein GL, Moore GP (1967) Neuronal spike trains and stochastic point processes. II. Simultaneous spike trains. Biophys J 7:419–440

Perkel DH, Gerstein GL, Smith MS, Tatton WG (1975) Nerve-impulse patterns: a quantitative display technique for three neurons. Brain Res 100:271–296

Shannon C (1948) A mathematical theory of communication. Bell Syst Techn J 27:379–423, 623–656

Received: November 11, 1987

Dr. Ad M. H. J. Aertsen
Max-Planck-Institut für
Biologische Kybernetik
Spemannstrasse 38
D-7400 Tübingen
Federal Republic of Germany

Chapter 9
Spatio-Temporal Spike Patterns

Searching for Significance of Repeating Spike Patterns

Moshe Abeles

I met George for the first time in 1969 at the Second Intensive Study Program organized by Francis O. Schmitt in Boulder Colorado. I was a very young faculty at the Hebrew University of Jerusalem and was invited as such to attend the program. At the registration, there were several bins each with a range of participant names. After getting my badge and program, a young person approached me and with a kind smile asked "how did you like that?" pointing to the registration desk stating "Abeles to Goy." At first, I did not get what was the punch. But, after a short pause I got it. "GOY" is the Hebrew word for a non-Jewish believer and also has a connotation of a simple minded person. That was George who was recruited to help with the organization of the program. His presence, the young physicist who moved into "wet" neurophysiology but used very sophisticated analysis tools, was evident throughout the workshop.

His kind and friendly attitude spanned our entire encounters. Back in Jerusalem, I started to record form auditory cortex using advanced spike sorting techniques for isolating in parallel several spike trains (Abeles and Goldstein 1977). I used extensively the analysis methods devised by Don Perkel and George to quantitatively study the dynamic firing properties of single neurons and pairs of neurons (Perkel et al. 1967a,b; Gerstein and Perkel 1969). However, I felt there is room for studying

M. Abeles (✉)
Hebrew University, Jerusalem, Israel

Bar-Ilan University, Ramat Gan, Israel
e-mail: Moshe.Abeles@biu.ac.il

A. Aertsen et al. (eds.), *Introducing Computation to Neuroscience*, Springer Series in Computational Neuroscience, https://doi.org/10.1007/978-3-030-87447-6_9

more complex interactions. Looking at the literature, the only thing I found was their "snowflake" method for studying 3-fold interactions (Perkel et al. 1975).

When measuring times of 3 spikes, say at t_1, t_2, t_3, there are 3 intervals to be considered: $(t_2 - t_1)$, $(t_3 - t_2)$, and $(t_1 - t_3)$. However, they are not independent as: $(t_2 - t_1) + (t_3 - t_2) + (t_1 - t_3) = 0$. They found an elegant way to present the 3 intervals on a plane by using the Einthoven's Triangle (used in medicine to present the activity vector of the electrocardiogram). If the maximal allowed delay was some fixed value, the dots fall within a hexagonal area looking similar to a "snowflake." This display provided a qualitative impression of the relations among the 3 spike trains. I could convert this display into a quantitative one by tessellating the area with triangular bins, counting the number of points that fell in each bin, and finding various ways of normalization (Abeles 1983).

While studying the 3-fold correlations with this method, I encountered several cases in which 1 or 2 bins stood up with a lot of counts. Such cases represented a situation where 3 spikes were repeating again and again with almost precisely the same intervals (see Abeles (1982 Chapter 6)). While searching the literature for such phenomena, I found the impressive work of Dayhoff and Gerstein (see also Chap. 5 this Volume). They looked on intervals between successive spikes within a spike train as "characters" and searched for "words" that repeated many times.

George was interested in my way of quantifying the "snowflakes," while I was interested in his method of finding repeating structures within a spike train. So, I paid him a visit in Philadelphia to discuss these issues. I felt this to be very useful and asked whether he would host me for a sabbatical. George agreed and very kindly offered to supplement my somewhat inadequate university salary from his grant money.

George was also very kind in helping us to find a house to rent at Haverford just outside Philadelphia. We (my spouse, two little daughters, and me) spent a wonderful year there. My main interest was to find a way for detecting complex spatio-temporal patterns that repeat many times. I had numerous discussions with George in which we examined several alternatives. But, they all seem too cumbersome to implement with the computing power available at the time.

One morning George came up with the following analogy: Suppose we represent the spike times of a neuron as holes along a line in a punched paper tape. If we recorded several neurons in parallel, their spiking times would be represented as holes along parallel lines. Take two copies of such a punched paper tape and slide them past each other. Any repeated patterns of holes will show up at some point of sliding. I coded this magnificently simple idea in Fortran and added ways to estimate what might be found by chance. When data was analyzed this way, it often showed a significant excess of repeating spatio-temporal firing patterns (Abeles and Gerstein 1988). This algorithm serves me for detecting such patterns to this very day (although the Fortran was replaced by C and now by Matlab).

A couple of years later, in an international workshop in Jerusalem, I presented my synfire chain model and the results on repeated spatio-temporal patterns. As supporting evidence, I showed the number of patterns found in surrogate spike trains that were very close to the estimated chance occurrence. At the end of my talk, van Essen said:

– Can you show the table for the surrogate data?
I showed it.
– The numbers you found are too close to the expected, and this cannot be true.
I was stunned! A famous scientist accuses me, a young and unknown researcher, of faking data. It took me a while to recover and then.
– Here the expected was 39 patterns and I found 38. I assume that had I found 31 you would be happier.
– Ah hah.
– But the probability of finding 31 when the expected is 39 is much smaller than the probability of finding 38 when 39 is expected.
– Silence.

A somewhat similar case arose many years later in an international workshop in France where both George and I participated. The organizers scheduled a talk by Roger Lemon to immediately follow mine, without warning me of what is coming. After giving my talk about repeated spatio-temporal patterns in recordings from the cortex of behaving monkeys, Roger described his very elegant experiments on recording pyramidal tract neurons in monkeys performing a precision motor task. In these data, he searched for repeating spatio-temporal patterns and found many with a high number of repetitions. This number was much higher than expected based on surrogate trains generated with Poisson statistics. However, Roger claimed, the spike trains are not Poissonian. When using the appropriate Gamma distribution of inter-spike intervals (ISI), the number of patterns found was much larger than the number in the experimental data. All my "friends" were happily grinning. The fallacy of these results was immediately clear to me. But now, being more mature, I did not want to offend him publicly. So I suggested we discuss his findings over lunch. At lunch, I told him that he showed an expected number of $\approx$ 100 patterns, while in the data there were only 35. The probability of finding 35 or less when 100 is expected is well below 10^{-13}. (Note: I do not recall the exact numbers, but the principle of the above description is correct). Thus, if the surrogates were appropriate, there should be some special mechanism that avoids the repetition of any firing patterns that were already generated.

The moral of these two incidences is: Do not mess with statistics and probabilities unless you have a profound grasp of these issues.

Following that talk, George told me you gave a hell of a lecture. Maybe I am taking an unjustified credit, but following this incident, George took enhanced interest in synfire chains and the precisely repeating spatio-temporal patterns. One of the first studies he made was to find what would be the appropriate statistics to use for surrogate spike trains (Gerstein 2004), followed by work on how to decompose complex patterns into sub-patterns and several works with Markus Diesmann, Sonja Grün, and their students on how to detect repeated patterns in massively parallel spike trains (see also Chap. 14 in this Volume).

For me, the most illuminating conclusion that emerged from these studies was that if multiple synfire chains were embedded in a small cortical volume you would need to record from at least 200 neurons in parallel to detect such synfire activities. Recording from over 100 neurons in parallel is now available, and getting to over 200 is coming. However, such recordings are now either spread over a too large of area or have a too low time resolution. Yet the time that such recordings will be possible from a small volume of cortical tissue is fast approaching. Only then it will be possible to establish or refute the synfire chain model.

References

Abeles M, Goldstein M (1977) The quantification and graphic display of correlations among three spike trains. Proc IEEE BME-30(4):762–773. https://doi.org/10.1109/TBME.1983.325226

Abeles M (1982) Local cortical circuits: an electrophysiological study. In: Studies of Brain Function. Springer, Berlin. https://doi.org/10.1007/978-3-642-81708-3

Abeles M (1983) The quantification and graphic display of correlations among three spike trains. IEEE Trans Biomed Eng BME-30(4):235–239. https://doi.org/10.1109/TBME.1983.325226

Abeles M, Gerstein GL (1988) Detecting spatiotemporal firing patterns among simultaneously recorded single neurons. J Neurophysiol 60(3):909–924. https://doi.org/10.1152/jn.1988.60.3.909

Gerstein GL (2004) Searching for significance in spatio-temporal firing patterns. Acta Neurobiol. Exp. (Wars) 64(2):203–207

Gerstein GL, Perkel DH (1969) Simultaneously recorded trains of action potentials: analysis and functional interpretation. Science 164(3881):828. https://doi.org/10.1126/science.164.3881.828

Perkel DH, Gerstein GL, Moore GP (1967a) Neuronal spike trains and stochastic point processes: I. The single spike train. Biophys J 7(4):391–418. https://doi.org/10.1016/S0006-3495(67)86596-2

Perkel DH, Gerstein GL, Moore GP (1967b) Neuronal spike trains and stochastic point processes: II. Simultaneous spike trains. Biophys J 7(4):419–440. https://doi.org/10.1016/s0006-3495(67)86597-4

Perkel DH, Gerstein GL, Smith MS, Tatton WG (1975) Nerve-impulse patterns: a quantitative display technique for three neurons. Brain Res 100(2):271–296. https://doi.org/10.1016/0006-8993(75)90483-7

JOURNAL OF NEUROPHYSIOLOGY
Vol. 60, No. 3, September 1988. *Printed in U.S.A.*

Detecting Spatiotemporal Firing Patterns Among Simultaneously Recorded Single Neurons

MOSHE ABELES AND GEORGE L. GERSTEIN

Department of Physiology, School of Medicine, Hebrew University, Jerusalem 91010, Israel; and Department of Physiology, School of Medicine, University of Pennsylvania, Philadelphia, Pennsylvania 19104

SUMMARY AND CONCLUSIONS

1. A particular firing pattern among simultaneously observed neurons represents a particular sequence of activity. If any multineuron pattern repeats significantly more than expected by chance, we may be observing a repeated state of a neural assembly as it processes similar units of information.

2. We present here an algorithm that rapidly finds all single or multineuron patterns that repeat two or more times within a block of data, as well as equations for calculating the number of patterns of given length and repetition that would be expected. The complexity of patterns for which it is practical to compute expected numbers is three to six spikes (inclusive).

3. Confidence limits are based on these expected numbers of patterns, so that it is possible to identify groups of patterns that are worthy of further analysis.

4. These methods are tested against simulated multineuron data that has various types of known nonstationarities, with good agreement between observed and expected values.

5. Application to real spike trains shows a large excess of observed repeating patterns, of which some, but not all, are shown to be due to bursts of high frequency firing.

6. It should be possible to apply the new method as a filter in real time in order to search for an association between repeated pattern events and externally observable events (stimulus, behavior, etc.). Any repeated pattern events which cannot be so associated may represent a new indicator of internal events in the nervous system.

INTRODUCTION

In 1949, Hebb (8) suggested that cell assemblies are the building blocks that carry out higher brain functions. According to Hebb, cell assemblies are formed by strengthening of particular synaptic connections between cells, thereby forming circuits in which only some of the neurons in a given region may participate. Such cell assemblies are dynamic entities that become functional when activated by the appropriate spatiotemporal firing patterns at their input. When activated, the cell assembly produces some appropriate spatiotemporal firing patterns at its output.

Although the concept of information processing by such cell assemblies is popular, there has been very little direct experimental research on the level of cell assemblies. The reason for this deficiency is probably the difficulty and relatively primitive capabilities of current experimental methods for identifying and analyzing neural assemblies. Standard recording of single neurons sequentially is of little help because with such data it is impossible to detect the interactions between individual neurons that are part of the assembly process. With such data only the timing relations between a single neuron and external events may be studied, as with the peristimulus time histogram. Even this fails if the process of interest is not accurately time locked to an external event. (This would be the case for internal processes like recalling a memory or planning a movement.)

In principle, a cell assembly could be studied experimentally by monitoring the activity of all its members simultaneously. This is an impractical goal with the present recording techniques because the number of neurons involved in (mammalian) neural assemblies is likely to be very large, perhaps many thousands. Present techniques do allow recording from about 30 single neurons in parallel [see (7) and (11) for reviews]; this is a small and usually a relatively random sample of the putative neuronal assembly.

Let us now assume, whenever an assembly is engaged in a particular computational process (which need not be time locked to any external event), that a unique wave of activity passes among the neurons. If the same information is repeatedly processed, we would expect some particular wave of activity to appear repeatedly, representing some particular repeated "state" of the assembly. Conversely if a particular wave of activity is observed to repeat, we might assume that this is an indicator that the neural assembly is repeatedly in some particular state, corresponding to the repetition of some particular computational process. Note that this one-to-one relationship is probably an oversimplification: We might expect that some *range* of processes will correspond to some *range* of patterns or states, i.e., that these are noisy quantities. Let us assume, however, that these ranges of similarity are small, so that the one-to-one relationship is approximately valid. We are deliberately leaving the nature of the underlying computational process vague at this point; a partial strategy for identifying the type of information that is involved will be proposed in the DISCUSSION.

In the context of practical simultaneous recording, only a small part of the repeated wave of activity could be observed. This sample would take the form of a repeated sequence of firing among some of the observed neurons. Thus by experimentally detecting (possibly noisy) repetitions of multineuronal firing patterns, we mark the occurrence of particular repeated states in the underlying neuronal assembly.

Detection of repeating patterns per se is only the beginning of such investigations. Patterns can and do repeat by chance, so that it is essential to develop a significance test for the experimental detections. If the entire analysis process becomes fast enough, it is in principle possible to search for the repeated sensory or behavioral event which in some cases might be associated with the repeated pattern.

There have been some previous investigations of repeating patterns in both single and parallel spike train recordings in which some repeating patterns were found in each case. Some investigators (4, 5, 9) studied only patterns generated by intervals between successive spikes in a recording of one single unit. In Ref. 9 patterns were defined in terms of relative size of successive intervals rather than in the more usual terms of absolute interval lengths. Other investigators (2, 3) have studied single unit patterns made up only of 3 spikes. Still others (6, 12) have started with simple seed patterns of 2 or 3 spikes (across two neurons) and looked for more complex patterns in which the seeds were embedded. Legendy and Salcman (13) examined firing bursts in each of several spike trains and found that occasionally they contained embedded multineuron patterns. However, the detecting techniques used by most of these investigators required enormously prolonged computations (many hours) and did not even detect all the repeating patterns. Basically, the brute force calculations for this problem are intrinsically unsatisfactory. We note that in parallel to the new work reported here, two other groups have made alternate approaches to the pattern problem (Perkel, private communication of working notes and Johannesma, private communication); it is premature to compare and contrast these different ideas.

In the present paper we describe a new algorithm that detects all possible patterns that repeated twice or more within a section of recorded multineuron data. The algorithm is fast enough to be carried out during an experiment, so that it is possible to consider an interactive search for the events associated with given repeated patterns. We also describe methods for statistical estimation of the number of patterns of a given length and repetition that is expected to appear by chance.

METHODS

The detection algorithm

The detection of all patterns that repeated at least twice may be described by the following metaphor. Suppose the multineuron spike data were represented in the form of a long punched-paper tape. Each row represents the activity of a particular neuron, and each of its spikes is represented by a hole; time runs along the tape. (Old-fashioned paper teletype tape could represent the activities of 8 neurons in this metaphor.) Let us take a second copy of the same tape and slide it past the stationary original tape. At a given displacement we examine the superimposed tapes against a light source. A region which contains *any* pattern that repeats at this particular tape displacement will be much brighter than average. Note that by selecting appropriate thresholds for length (duration) of region and brightness (number of spikes matched), any pattern can be investigated for match to any degree of accuracy, including both extra and omitted spikes. Note also that regions of very fast (burst) activity will tend to match *any* pattern; we will deal with this problem separately below.

A straightforward application of this algorithm with a digital computer will consume a lot of time. As an illustrative example, take a case where 10 single neurons are observed in parallel for 100 s. The average firing rate in each spike train is 5/s, and the spike times are measured with accuracy of 1 ms. (These are typical values for single neurons recorded in mammalian cortex.) The sliding copy of the tape has to be shifted by 100,000 steps (number of ms in 100 s), and after each such shift the position of all 5,000 spikes (10 spike trains $\times$ 5/s $\times$ 100 s) would have to be examined for pattern matches with the stationary copy. This means 500 million comparisons. Even a supermini computer which is dedicated to this process would take many minutes to make all these comparisons.

In order to expedite this search process, we have developed the following algorithm. Let us construct interval histograms for the time between every spike in the stationary tape and the next spike of the same train in the sliding copy. This creates N histograms, one for each of the simultaneously observed spike trains.

Initially, when the two tapes are in exact register, the histograms are identical to ordinary interspike interval histograms. Note that at each subsequent step in the tape sliding process, the entries in each bin remain unchanged, but the time value of each bin is reduced by one bin time. In other words, at each sliding step each current histogram is shifted left by one bin. At a given step, the entries that have just reached zero time (i.e., the leftmost bin) correspond to all those spikes of the stationary tape that fall exactly over a spike (of the same train) in the sliding tape. The final step in updating the histograms after each sliding step is to recalculate interval values only for the spikes to which the zero bin points.

For each such spike in the stationary tape, we measure the time to the next spike of the same train in the sliding tape, and increment the appropriate bin of the corresponding histogram.

Now, instead of just keeping a tally for each bin of a histogram, we also maintain a pointer list. These pointers define the location along the tape (i.e., the time of occurrence) of all the events (intertape spike intervals of a particular length for a particular spike train) that contributed to that tally. The pointers also identify which particular spike train they describe.

As a final step of the algorithm, we merge the N histograms, bin by bin, into a single compound histogram. In this process all pointers belonging to each particular bin of the compound histogram are sorted into a time ordered list. The new pointers created in updating the events in the zero bin must of course be inserted at the correct place into whichever of these lists (bins) is appropriate.

It is now trivial to examine the time ordered pointer list at the zero bin with a sliding time window whose duration is the chosen pattern duration. Whenever the count of pointers falling within such a window exceeds the criterion chosen to define pattern "complexity", we have identified a repeated pattern. Its location and details are spelled out by the pointers.

For the typical data situation described above, on the average only 25 spikes would overlap at each sliding step. Thus only 2.5 million rather than 500 million comparisons are needed. On a Data General MV 10000 it takes ~1 min to detect all the repeating patterns in such typical data, a calculation time comparable to the data collection time. Thus a typical personal computer will not suffice for this purpose.

If one wishes to use bins wider than the time resolution at which the data was collected, one can use one of the following methods: *1*) convert the data into coarser time grain, or *2*) use the spikes that are within the first *n* bins of the interval histograms as pointers to patterns whose match is within *n* time units.

Simulated data

In order to evaluate our detection scheme and to obtain some experience with the numbers of patterns that might be generated by chance, we simulated the activity of 10 spike trains in the following form. The simulation advanced in 1-ms steps. After each step the probability of firing for each cell was determined (so as to generate various types of nonstationarity and correlations, as described in *A* through *G*, below), and then a decision of whether the cell fired at this step was taken by drawing a random number.

It is important to be cautious with random number generators. Using the commercial random number generator supplied with our computer produced huge numbers of repeated patterns. Only when we wrote a special random generator which combines an additive generator (*Eq. 7*, p. 26 of Ref. 10) with randomizing by shuffling

(Algorithm B, p. 32 of Ref. 10) did the results seem random. (Note that we did not experiment extensively with various forms of random number generators; we used the first satisfactory generator we came across.) A recent review article on the merits and testing of random number generators can be found in Ref. 14.

Seven simulations were used for the data described in this paper. All the simulations were run for 100 s of simulated time with 10 parallel spike trains.

A (sim.002a): Spike trains were realizations of independent and stationary Poisson processes. All spike trains were firing at 5/s. *B* (sim.003a): Spike trains were realizations of independent and stationary Poisson processes. Firing rates were between 10 and 1/s. (Averaged over cells, the firing rate was 4.8/s.) *C* (sim.004a): Spike trains were realizations of independent and nonstationary Poisson processes. The firing rates in each spike train fluctuated slowly between 9 and 22 spikes/s (average firing rate 3.8/s and standard deviation of the fluctuations 4.1/s). The time constant of the firing rate fluctuations was 2 s. *D* (sim.005a): Spike trains were realizations of independent but nonstationary Poisson processes. The firing rates in each spike train fluctuated rapidly between 0 and 50 spikes/s (average firing rate 4.0/s and standard deviation of the fluctuations 6.7/s). The time constant of the firing rate fluctuations was 0.05 s. *E* (sim.006a): Spike trains were realizations of correlated and nonstationary Poisson processes. The firing rates in each spike train fluctuated slowly just as in *C* above, but in addition the cells were divided into five pairs (1 and 2, 3 and 4, 5 and 6, 7 and 8, 9 and 10). In each pair the firing rate fluctuations as a function of time were precisely the same. *F* (sim.007a): Spike trains were realizations of correlated and nonstationary Poisson processes. The firing rates in each spike train fluctuated rapidly between 0 and 28 spikes/s (average firing rate 4.3/s and SD of the fluctuations 4.7/s). The time constant of the firing rate fluctuations was 0.05 s. The cells were divided into five pairs (1 and 2, 3 and 4, 5 and 6, 7 and 8, 9 and 10). In each pair the firing rates as a function of time were precisely the same. *G* (sim.007b): Spike trains were realizations of correlated and nonstationary Poisson processes. The spike trains were similar to those in *F* except that the rates of fluctuations were larger, as in *D*.

Experimental data

The detection scheme was applied to four 100-s recordings of spontaneous activity from eight single units measured in parallel. The recordings were made with metal microelectrodes inserted into the primary auditory cortex of a cat anesthetized with nitrous oxide. The eight units were recorded from six electrodes; at most two units could be isolated from a single electrode. The distance between adjacent electrodes was 2.5 mm, and the distance between the two furthest electrodes was 5.6 mm. Detailed description of the recording procedures and stimulus conditions between the periods of recording that are analyzed here are given in Ref. 3.

RESULTS

The expected number of repeating patterns

Even if there were no special brain processes that increase the occurrence of repeated firing patterns in the recorded data, some number of patterns might repeat just by chance. We have developed several equations for estimating the expected number of such repeating patterns, with detailed derivations given in the APPENDIX; here we shall describe some of the final equations and their applicability.

First let us assume a very simple random process in which there are n spike trains each firing at λ spikes per second. The firing times follow a stationary Poisson process and the spike trains are independent. Within this process we look for patterns made out of c spikes with r repetitions of the pattern in the T seconds of recording. Let us denote by w the maximum allowed delay between the first and last spike in a pattern and by Δ the time accuracy with which the firing time is measured (e.g., 1 ms). In this case *Eqs. A6* and *A8* are reduced to

$$N_c^{(r)} = n^c[(w/\Delta)^{c-1}/(c-1)!][T\Delta^{c-1}\lambda^c]^r e^{-T\Delta^{c-1}\lambda^c}/r! \quad (1)$$

where $N_c^{(r)}$ is the expected number of patterns of complexity c that repeat r times.

Although the conditions for which this equation was derived hardly hold in any physiological recordings of real spike trains, it does illustrate several important properties of pattern repetition. The number of expected repetitions is most sensitive to the firing rate λ, since it depends on the firing rate raised to the power $r \cdot c$. (Note that the argument of the exponential is very small, about -10^{-2}, for typical parameter values). Thus if, for instance, the firing rate of the cells increases from 5/s to 8/s, the number of patterns of 4 spikes that repeat 2 times will increase 43-fold ($8/5^8$). We must be extremely careful with data that involve high and variable spike rates.

The next critical parameter in *Eq. 1* is the "bin width" Δ, i.e., the time resolution used for defining spikes. The number of patterns that repeat by chance is proportional to Δ raised to the power of $(c-1)\cdot(r-1)$. This suggests using the highest possible time resolution (minimum Δ) that is physiologically reasonable. In all the examples illustrated in this paper we have used 1 ms.

Real spike trains frequently show firing rate fluctuations that exceed those predicted for a stationary Poisson process. Such large rate fluctuations may be interpreted as the result of nonstationary processes of some appropriate time scale. Note that *Eq. 1* predicts many more chance repeating patterns during periods of high firing rate. For real nonstationary neurons, the use of average firing rates in *Eq. 1* or *A8* would lead to a gross underestimate of the expected number of patterns that repeat by chance. Accordingly, for real spike trains we have developed two alternate ways to estimate the expected number of repeating patterns: the "ad hoc" method for triplet and quadruplet patterns, and the "bootstrap" method for patterns made up of five or more spikes. A brief description of these methods follows.

AD HOC METHOD. Rather than assuming some kind of Poisson process as a basis for calculation, this method actually counts the number of patterns that occur in a given piece of data. If, for example, we are interested in patterns of 3 spikes (2 intervals) we should count the number of patterns composed of 3 firings, respectively from neurons (1, 1, 1), (1, 1, 2), . . . , etc. (i.e., all combinations of three names). For each such set of ordered names we can calculate the number of possible different time combinations (including intervals of zero) that could occur in a pattern window duration w and time resolution Δ. For pattern complexity (length) 3 *Eq. A3* becomes

$$n(\mathcal{D}_3) = (w/\Delta)^2/2$$

where $\mathcal{D}_3$ is the set of all possible time combinations for a particular ordered name list of complexity 3. For the present example, if we have a time resolution of 1 ms and we are interested in triplet patterns that may last up to 500 ms, then we could have 125,000, i.e., $500\cdot500/2$ different possible time delays between firings of, e.g., neurons 1, 2, and 3 in that order.

Assume that the observed counts are distributed uniformly among all the possible different time delay combinations that can occur for each set of names. (The assumption is testable if we wish.) For every list of neuron names S_c, we divide the number of patterns actually observed for these ordered names, $m(S_c)$, by the number of possible different time delays $n(D_c)$. This yields $x(S_c)$, the expected number of occurrences of any one particular delay pattern with these names in this particular data set. (This calculation is similar to sampling with replacement in an urn problem.) For instance, if we observed 1,250 cases where neurons (1, 2, 3) fired in that order within 0.5 s, then the expected number of occurrences for any particular delay pattern of this same firing sequence will be 0.01 (i.e., 1,250/125,000) under the uniform distribution assumption. From this small expected count we can compute the probability of seeing a given pattern r times, using *Eq. A1*. For our example, the probability of seeing a particular pattern twice is 0.00005. By adding up such probabilities over all possible c-fold names, we can find the expected number of all c-spike patterns which repeat r times (again for this particular data set)

$$N_c^{(r)} = \frac{(w/\Delta)^{c-1}}{(c-1)!} \sum_{S \in \zeta_c} e^{-x(S)} \cdot x(S)^r/r! \qquad (2)$$

THE BOOTSTRAP METHOD. The ad hoc method described above requires counting the number of times that every possible pattern S_c occurs. Such a counting process is not practical for patterns which are longer than 4 spikes. With the bootstrap method we calculate the expected number of repeating 5-spike patterns from the expected number of repeating 4-spike patterns. In turn the expected number of 6-spike patterns is calculated from the expected number of 5-spike patterns. Although this method might be carried on also to patterns of ≥ 7 spikes, it is not advisable to do so, because the inaccuracies (introduced by the assumption that the added spike trains are Poissonian) accumulate and yield an underestimate of the expected numbers. Thus for cortical units the bootstrap method is useful only for two steps beyond what is obtained with the ad-hoc method. The APPENDIX describes *Eqs. A18* and *A19* which are used for these computations.

In Table 1, for data taken from control simulations *A–G*, we compare the number of repeating patterns that are actually found

TABLE 1. *Comparison between the number of repeating patterns that were found and the expected number for simulated spike trains*

Complexity:	3				4		5	6
Repetition:	2	3	4	5	2	3	2	2
A found	9,710	48	0		402	0	13	0
A expected	9,515	39.7	0.125		400	0.008	12.5	0.3
B found	21,703	390	4	0	1,367	0	53	2
B expected	21,640	353	5.34	0.073	1,327	0.18	53.1	1.74
C found	11,443	103	4	0	814	0	45	2
C expected	11,970	95.4	0.83	0.0078	877	0.069	23	0.48
D found	4,357	14	0		175	0	3	
D expected	4,171	13.1	0.034		169	0.003	3.64	
E found	8,868	59	0		721	0	51	2
E expected	8,961	68.5	0.60		750	0.076	15.45	0.25
F found	4,120	13	0		150	0	2	0
F expected	4,108	11.7	0.025		150	0.002	3.33	0.059
G found	7,133	28	0		359	0	13	
G expected	6,538	24.7	0.074		294	0.007	7.68	

For patterns of 3 and 4 spikes the expected number was computed by the ad hoc method, whereas for 5 and 6 spikes it was computed by the bootstrap method. The parameters of the simulations *A* through *G* were described in METHODS. Window length was 0.5 s. Data were included only when the expected number of repeating patterns was >0.001 or when patterns were actually found.

with the number of repeating patterns that are expected from the calculations described here. In most cases the estimates are quite accurate, even though various forms of nonstationarity were included in the simulations. Thus we should be able to detect excess numbers of repeating patterns in real data if they are present.

The assumptions used to derive the ad hoc method held only in simulations *A*, *B*, and *C* where the firing rates were either stationary or changed slowly. In the other simulations, the firing rates were changing rapidly, or pairs of spike trains had (various independent) covariation of rate, or both. Nevertheless, the estimations of counts are quite good. The assumptions used to derive the bootstrap method held only in simulations *A* and *B*, where the firing rates were stationary. Indeed, the bootstrap method gives inaccurate prediction in simulations *C* and *E* where the firing rates fluctuated slowly but with large amplitudes (0–22 spikes/s).

Figure 1 gives some insight into the degree of fluctuation and correlation (from covariation of rates) that can be tolerated by these calculations. Satisfactory results were obtained with the degree of cross-correlation (from covariation of rate in pairs of neurons) as in Fig. 1, *A* and *C*, or alternatively, in the absence of cross-correlation with the degree of auto-correlation shown in Fig. 1*B*. Inaccurate (low) predictions of the expected number of repeating patterns occurred if the firing of the individual neurons showed the degree of autocorrelation shown in Fig. 1*B* and in addition there was cross-correlation (from covariation of rates in pairs) of the degree in Fig. 1*D*.

Finding which patterns did not appear by chance

In simulation *G* (Table 1), we found 13 quintuplet patterns that repeated twice, whereas only 7.68 were expected to repeat twice. Is this difference significant? Those quintuplet patterns that did appear twice are an extremely small fraction of the many possible quintuplet patterns (given a time resolution and window duration). For the particular values in this example there are $1.3 \cdot 10^{17}$ possible quintuplet patterns; each of these has an extremely low probability of appearing twice. These are appropriate conditions for using the Poisson formula (*Eq. A1*) to compute the probability of getting ≥13 repeating patterns when 7.68 are expected. The result is 2.5%, so that these numbers represent a not

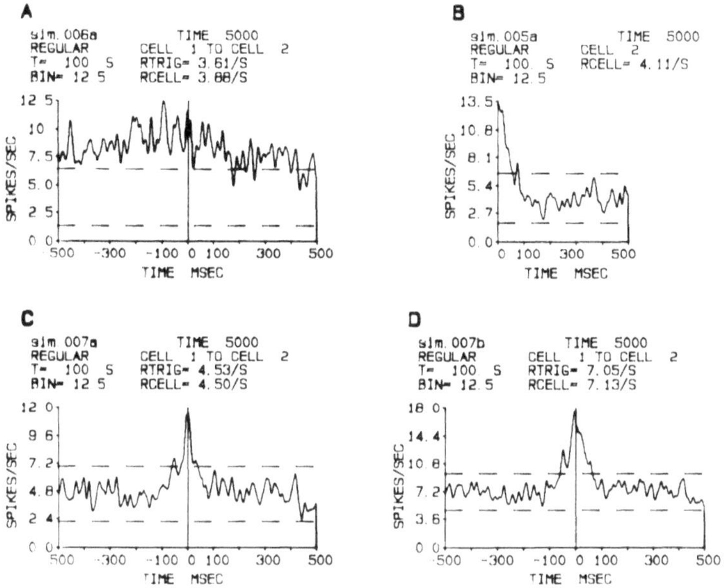

FIG. 1. Auto (*B*) and cross-correlations (*A, C, D*) of spikes in the several simulations with rate fluctuations. The cross-correlations are from simulations in which pairs of "neurons" had covarying rates. The auto-correlation is from a simulation where each neuron had independent rate fluctuations. *Dashed lines* show the 99% confidence limits for a comparable correlogram of a stationary Poisson process of the same mean rate. The expected number of repeating patterns calculated by the ad hoc method agreed with the observed number of all simulations that had less covariation of pair rates than shown in *D*.

very significant excess in the number of repeating quintuplet patterns.

Confidence limits for any expected number of patterns can be obtained by calculating the corresponding probabilities for all possible observed numbers of patterns with the Poisson formula. For 99% confidence limits we would sum the probabilities for low numbers until we achieve a total probability of 0.005; we would then continue summing probabilities for higher counts until we reach a total probability of 0.995. This determines the upper and lower integers that set the 99% confidence limit. When the expected number of patterns is large, the normal approximation to the Poisson formula was used (see Ref. 1 for a more detailed description).

We emphasize that Fig. 2 and all similar figures in this paper compare observed and expected *number of patterns* that meet the stated criteria of repetition, complexity, window, and in some cases specific name or delay list. The expected counts are calculated from the number of single occurrences of the patterns that meet the same stated criteria (except, obviously for the repetition). These figures do not show the observed and expected counts of repetitions for particular patterns; such measure are not appropriate at this stage of the analysis.

Normally we would plot vertical bars (e.g., for 99% confidence limits) of each point in these graphs. If such a bar does not reach the line where observed and expected numbers are equal, then the observation falls outside the stated limits. Since this procedure would produce a rather messy graph we have instead plotted a vertical 99% confidence region

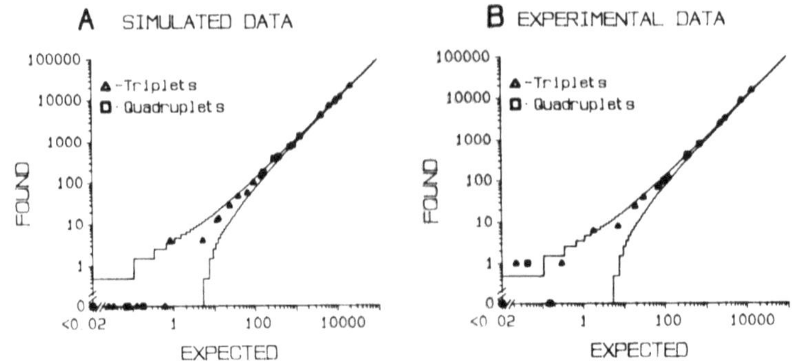

FIG. 2. Comparison of the number of patterns found versus the expected number that repeated two or more times. The triangles describe triplets; the squares describe quadruplets. The jagged lines give the 99% confidence limits. *A*: 7 simulations described in Table 1. *B*: 4 experimental data files described in Table 3.

around the observed equals expected line. The limit values are indicated as next outer half integers, so that there is no doubt when a particular point (which must have an integral observed value) falls beyond the confidence limits. Fig. 2*A* shows in this manner the triplets and quadruplets of the simulation data in Table 1. Only a few points (mostly from Simulation *G*) fall outside the 99% limits. These results suggest that the ad hoc and the bootstrap methods give reasonable estimates for the expected number of repeating patterns in "random" control data. The confidence limits, in turn, can therefore be used on unknown data to focus our attention on those types of repeating patterns that appear in excessively large numbers.

When the number of repeating patterns is small, the confidence limits may be sufficient to point out candidate patterns which do not occur by chance. However, when the number of repeating patterns is very large, the mere knowledge that there is an excess number of such patterns is not enough. For example, in Table 1 we see that in simulation *G* there were 7,133 different triplets that appeared twice, while the expected number was 6,538. This difference is certainly significant, but we wish to know which of the 7,133 triplets are most likely to be nonrandom. This can be done by dividing the 7,133 triplets into subgroups and looking whether there are subgroups in which the number of triplets exceeds the confidence limits. One way for such a division is to separate the triplets by the names of the single units that make up these patterns (e.g., all triplets made of single units 1, 2, and 3; all triplets made of single units 1, 2, and 4; etc.). *Equation A5* can be used for getting an estimate of the expected number of occurrences for any particular list of single unit names. This expected count can then be used to compute confidence limits.

Figure 3*A* shows graphically such a division of all the triplets that repeated twice in simulation *B*. The jagged line gives, as in Fig. 2, the 99% confidence limits. In simulation *B* there were 10 single units, so that there were 1,000 different triplet names [such as (1, 1, 1), (1, 1, 2), (1, 1, 3), . . . , (10, 10, 10)]. When using 99% confidence limits one can expect that 10 out of the 1,000 points will exceed the limits. This is roughly what happens in Fig. 3*A*. Note also, that even the points that fall outside the confidence limits are not placed very far away from the limits.

Experimental data

When experimental data were analyzed it became evident immediately that the number of repeating patterns was far beyond what could be expected by chance. Table 2 shows the data for 100 s of "spontaneous" activity of eight single units which upon shallow ex-

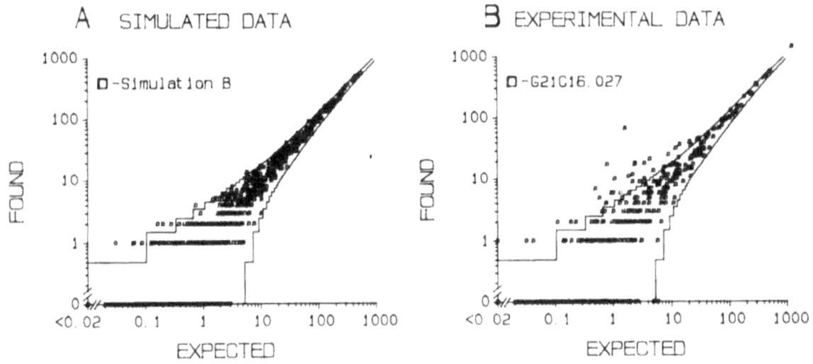

FIG. 3. Breaking triplets into subgroups. All the patterns of 3 spikes that appeared twice are divided into subgroups according to the names of the single units that participated in the patterns. *A*: simulated data (1,000 subgroups). *B*: experimental data (512 subgroups).

amination looked reasonably stationary and not heavily correlated. Among the striking differences between the observed and the expected number of repeating patterns were one pattern of seven spikes that repeated twice, a pattern of four spikes that repeated four times and two patterns of three spikes that repeated seven (!) times.

Figure 3*B* shows the division of 9,330 triplets that occurred twice in this record, into all possible triplet names. Both Fig. 3*B* and Table 2 show great excess of repeating patterns. Close examination of these unusual patterns revealed that they are associated mostly with bursting cells. For instance the two triplets that repeated seven times were (using the nomenclature S_c, D_c of the APPENDIX, in which a pattern is defined by a list of single unit names followed by a list of delays from the first spike): (4, 4, 4), (0, 2, 4) and (4, 4, 4), (0, 2, 6). This means that when single unit 4 emits a high-frequency burst, the detailed

TABLE 2. *Effect of eliminating bursts and high firing rates on the number of patterns found*

Patterns		All spikes		Filtering	
Complex.	Repeat	Found	Expected	Found	Expected
3	2	9,330	8,785	3,134	3,011
	3	263	195	39	30
	4	12	4.7	1	0.29
	5	4	0.1	0	0.003
	6	2	0.0023		
	7	2	<0.0001		
4	2	635	554	94	95.3
	3	8	0.159	0	0.007
	4	1	<0.0001		
5	2	70	15.3	2	2.6
6	2	9	0.337	0	0.05
7	2	1	0.0062		

Data taken from a record containing 100 s of spontaneous activity of 8 single units. Filtering means eliminating all spikes around which (within 500 ms) there were >10 other spikes from the same single unit (i.e., all time sections in which a single unit fired at a high rate) and all the spikes before which another spike from the same single unit occurred within 5 ms (i.e., eliminating all but the first spike in a burst).

structure of the burst tends to repeat. It is possible that such a burst is a highly significant event in the overall activity. Its detailed structure may reflect some interesting organization of the excitation which drives the cell, but it may be also a reflection of the internal changes of excitability in the driven unit itself. If the latter mechanism is true, excessive numbers of patterns associated with high frequency bursts may obscure other patterns which are more likely to represent significant organization of activity in the cortical network.

Periods in which a spike train fired at high rates contributed a large fraction of the repeating patterns produced by these neurons. For instance in another recording from the same single units as described in Table 2 we got >120,000 triplets that repeated twice. About 100,000 of these came from periods in which some single unit(s) fired at rates > 20/s. We can detect that there is an excess number of repeating patterns under the particular conditions, but it is obviously quite difficult to identify which patterns are candidates for further analysis in these circumstances.

We have examined the effect of eliminating bursts and periods of high firing rates from the data. This elimination, which will be called filtering, was carried out in the following manner: a burst was defined as a group of spikes from the same single unit following each other with interspike intervals that did not exceed some minimal time (e.g., 5 ms). When a burst was found, only the first spike was considered in the search for patterns. The low-pass frequency filtering was achieved by setting a window (e.g., 500 ms) around each spike and counting how many more spikes from the same single unit occurred within this window. If the number of spikes exceeded some threshold (e.g., 10) the spike around which the window was set was not considered by the pattern searching routines.

The right half of Table 2 shows what happens when such a filtering scheme is applied to experimental data. The number of observed patterns at each complexity and repetition are very close to what can be expected by chance. Figure 4*B* shows the separation of the 3,134 triplets that occurred twice according to single unit names. Figure 4*A* shows the data from the same record when all the spikes were considered. It is evidence that most of the excessive patterns were removed by this filtering scheme. However some of the cell combinations still tend to produce more patterns than can be expected by chance.

For instance, the single units (3, 1, 4) generated one pattern that appeared twice [at delays (0, 177, 266)], while the expected number of patterns appearing twice by chance was 0.021, or the single units (4, 2, 2) generated two patterns that repeated twice [at delays (0,

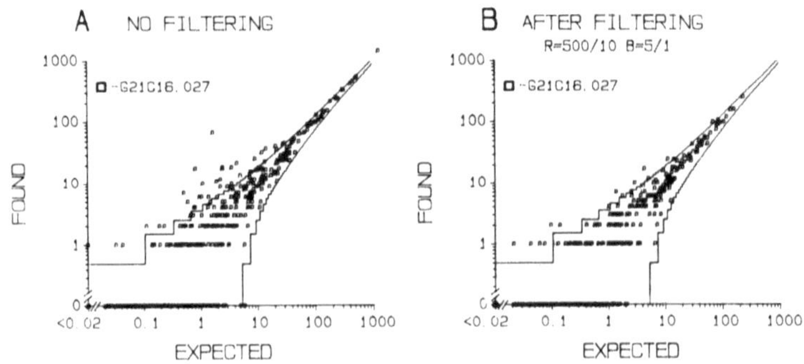

FIG. 4. The effect of filtering on the number of triplets that repeated twice. *A*: all the spikes in the recordings were considered when searching for patterns. *B*: only spikes which were not embedded in periods of high firing rates or which were not embedded in a burst were considered. See legend of Table 2 for the exact definition of the spikes that were filtered out.

TABLE 3. *Observed and expected repeating patterns*

Complexity:	3				4		5	6
Repetition	2	3	4	5	2	3	2	2
1 found	8,209	120	6	1	396	1	13	0
1 expected	7,789	119	1.74	0.023	358	0.044	26.5	1.57
2 found	3,134	39	1	0	94	0	2	0
2 expected	3,011	30	0.29	0.003	95	0.007	2.6	0.05
3 found	14,542	355	8	0	759	0	27	1
3 expected	13,710	328	7.1	0.14	701	0.15	30	1.03
4 found	2,406	24	0	0	69	0	0	
4 expected	2,248	18.5	0.14	0.001	70	0.004	3.74	

Data from 4 records of "spontaneous" activity each of 8 single units in the primary auditory cortex of cat. All records were filtered as described in the legend of Table 2.

160, 182) and (0, 317, 332)], while the expected number of patterns appearing twice by chance was 0.097.

Table 3 and Fig. 2*B* show the number of repeated patterns found in 4 recordings of experimental data after filtering (as described in Table 2). When comparing Table 3 to Table 1, Fig. 2, *A* to *B*, and Fig. 3*A* to 4*B*, it is evident that the experimental data tend often to have more repeated patterns than were found in comparable simulations. One may wonder whether more drastic filtering would cause these patterns to disappear. We found that although one can reduce the number of repeating patterns by more strict filtering of the data, there are always some single-unit combinations that generate excess numbers of repeating patterns.

At this stage one cannot say whether this excess is due to large deviation from the assumptions made while developing the equations used to compute the expected number of patterns that repeat themselves by chance, or due to the presence of some special neural processes that make the cells fire repeatedly in the same pattern. Some clue as to whether the excess of patterns is real may be obtained by merging the list of detected patterns from all four records and looking for patterns that were detected in more than one record. When this procedure was carried out on the data described in Table 3, it was found that indeed some patterns appeared twice (or three times) in more than one record. For example, the pattern (2, 2, 2), (0, 14, 23) appeared three times in record 1 and twice in record 2; the pattern (1, 6, 1), (0, 125, 234) appeared twice in record 1 and twice in record 3. Altogether there were 51 patterns of this sort.

DISCUSSION

We began this paper with the notion that excessively repeating patterns could be used to indicate some sort of repeating state of the nervous system. Since we are examining only a very limited number of neurons, any repeating pattern which we detect probably is a (small) portion of a larger pattern which involves neurons that are not under observation. It is entirely possible that different states (different large patterns) could be associated with the same observed subpattern. This is an intractable sampling problem and ultimately sets limits on the breadth of inference that can be drawn from detection of repeating patterns in the spike trains of a limited number of simultaneously observed neurons.

We have presented algorithms and confidence limit methods that detect all repeating spatiotemporal firing patterns in a given stretch of multineuron data and which subsequently show which pattern classes are appearing in excess. It should be noted that our method seeks exact pattern matches (to within the time resolution). A missing spike in a "repetition" would cause the remaining spikes to be treated as a different pattern, possibly of another class. On the other hand, our matches ignore extra spikes; this means that firing bursts cause excess and possibly meaningless matches. We have demonstrated the effects of filtering bursts out of the data.

It is possible, of course, to take the view that firing bursts are themselves highly significant events whose pattern of occurrence marks particular states of information processing. In this case one can implement a reversed scheme for selection of the data: filter out the isolated spikes and leave the bursts. The methods of this paper can then be used to study patterns made up of or within bursts. Identification of bursts for purposes of elimination, as in this paper, can be made simply on a criterion of spikes following each other with interspike intervals shorter then some minimal delay. If instead the bursts are to be kept and studied, it would probably be more appropriate to use a more rigorous method to define bursts in the context of an ongoing spike train. One excellent approach to this is the Poisson surprise calculation used by Legendy and Salcman (13).

Other approaches to the repeated pattern detection problem can be envisioned. For example, during the detection phase, matches could be made "inexact" according to some continous measure so that missing and excess spikes are both equally penalized (Perkel, private communication of working notes). This approach is computationally expensive. However, once templates have been made out of candidate patterns detected rapidly by the methods presented here, such continuous measures of fit are feasible and appropriate.

When the various algorithms described here were developed, we had the following experimental situation in mind. Suppose that the activity of several single units is being recorded in parallel. Initially the units are studied, possibly under some set of stimulus conditions, for a period of about 100 s. These recorded data are submitted to the pattern searching routines. The computer compares the number of observed repeating patterns of a given description (repetition, complexity, name, or delay list, etc.) with the number of such patterns expected by chance. Presumably repeating patterns of certain descriptions will be singled out as occurring in excess numbers.

If the number of such patterns of a particular description is relatively small, then all of them can be chosen as candidates for further attention. For instance, if we examine the results of expt. *1* in Table 3, we see that only 1.74 patterns of 3 spikes were expected to repeat 4 times while 6 were found. All these 6 patterns are candidates to serve as templates for further search. Similarly, by looking at Fig. 4*B* we see that in this experiment there were 4 points outside the 99% confidence limits at the level of 1. That means that there were 4 different unit combinations [these were (1, 2, 3), (3, 4, 5), (1, 8, 1), and (7, 7, 7)] for which <0.1 patterns are expected to repeat twice but for which 1 pattern actually repeated twice. These 4 patterns are also candidates for templates. The experience of one of us (M. Abeles) from analysis of records obtained from the auditory and the prefrontal cortex of unanesthetized monkeys is that such patterns can be singled out in about half of the records.

If, however, the number of patterns of a particular description is large, it is impractical to designate them all as candidates for further analysis. The experimenter must devise external criteria to make a selection. Alternatively, different subgroupings in the analysis can be attempted (as in the breakdown by name list in Fig. 3) which might single out repeating patterns which are in excess, but which form a smaller group. This selection process is extremely important and deserves considerable future attention.

Once the list of candidate patterns is determined, the computer can generate a corresponding set of templates; all further incoming data from the experiment is examined against the templates with some appropriate criterion of fit. The computer is programmed to announce the further occurrences of any of the candidate patterns (e.g., with a beep). The experimenter may now manipulate the environment (stimuli, behavioral conditions, motor responses, etc.) in a search for the situation, if any, in which some of the candidate patterns reappear.

This type of experiment is analogous to the early experiments on single neurons. Once the single unit was isolated, the experimenter tried to determine the adequate stimulus for firing of that unit. In our case, instead of studying the conditions that result in single unit firing, the experimenter studies the "receptive field" properties of the generator producing the repeated spatiotemporal firing pattern events.

There is no technical difficulty in building several tens of templates that would detect the reappearance of candidate firing patterns in real time, assuming that we solve the selection problem discussed above. The algorithm described here for detecting repeating patterns is very fast and will automatically recognize all patterns that repeat twice or more. The confidence limits developed here should help focus the experimenter's attention on those patterns which are less likely to be generated by chance. The entire process of detecting patterns, sorting them according to spike names, and computing confidence limits for a typical multineuron recording would take ~10 min on a fast minicomputer. This is a vast improvement on past approaches to this problem.

The next step, then, is to install the pattern detecting "machine" as a routine filter in ongoing multineuron experiments. This will allow a continual study of temporal association between repeating patterns and externally observable events. Repeating patterns which continue to occur in an experiment, but for which such an association can not be found, may be a new indicator of repeating internal events in the workings of the observed portion of the nervous system.

APPENDIX

Symbols

- c — Complexity of a pattern. The number of spikes in a pattern. $c = 1, 2, \cdots$
- D, D_c — A list of c delays, $D_c = (0, \tau_2, \tau_3, \ldots, \tau_c)$ The delay τ_n is the time elapsed from the first to the nth spike in a pattern. Therefore τ_1 is always 0
- $\mathcal{D}_c$ — The set of all possible D_c. E.g., $\mathcal{D}_3 = \{(0, 0, 0), (0, 0, \Delta), (0, 0, 2\Delta) \cdots\}$
- Δ — The accuracy of the time measurement in seconds. E.g., 0.001 s
- λ_i — The firing rate of the ith single unit, in spikes per second
- $m(S_c)$ — The total number of cases in which the spike sequence $S_c = (i_1, i_2, \ldots, i_c)$ appeared regardless the delays, except that the total duration of the pattern was less than w
- n — The number of recorded single neurons.
- $n(\mathcal{D}_c)$ — The number of different time delays D_c that can be fitted in a time window w seconds wide
- $N^{(r)}(S_c)$ — The expected number of patterns made from spike trains $S_c = (i_1, i_2, \ldots, i_c)$ which repeated r times
- $N_c^{(r)}$ — The expected number of patterns of complexity c which repeated themselves r times. All possible ways of writing c names are allowed
- r — The number of repeated appearances of the same pattern. $r = 2, 3, \cdots$
- S, S_c — A list of c single unit names. $S_c = (i_1, i_2, \ldots, i_c)$. E.g., $S_4 = (1, 5, 6, 1)$. The i_j can be assigned any names from among the n recorded neurons
- ζ_c — The set of all possible S_c. E.g., $\zeta_3 = \{(1, 1, 1), (1, 1, 2), (1, 1, 3), \cdots\}$
- T — The total duration of single unit recording, in seconds. E.g., 100 s
- t — Time
- τ_i — The delay between the first and the ith spike, in seconds
- $x, x(S)$ — or $x(S, D)$ The expected number of repetitions of a particular pattern S, D
- w — The window duration. The maximal allowed delay between the last and first spikes in a pattern, in seconds. E.g., $w = 0.5$ s

Expected number of repeating patterns

In this section we shall develop equations for evaluating the number of different patterns that are expected to appear repeatedly. We shall assume that for any particular predetermined pattern (e.g., a spike from cell 1 followed, after 0.015 s, by a spike from cell 3, etc.) we can compute the expected number of occurrences x. In the following sections we shall develop several ways to estimate this x; here we assume that x is known.

Let us denote the list of single-unit names that appear in a pattern of c spikes by S_c, $S_c = (i_1, i_2, \ldots, i_c)$, and the list of delays between these spikes and the first spike by D_c, $D_c = (0, \tau_2, \tau_3, \ldots, \tau_c)$, where $i > j$ implies $\tau_i \geq \tau_j$ and $\tau_c < w$. Let us denote the expected number of appearances of the pattern S_c, D_c by $x(S_c, D_c)$.

We can think of the appearances of the sequence S_c, D_c as successes of a trial that is repeated many times in the following way. At time $t = 0$ we look whether this particular pattern occurred, that is whether spike i_1 fired during the time interval $[0, \Delta)$ and spike i_2 fired during the time interval $[\tau_2, \tau_2 + \Delta)$ and spike i_3 fired during the time $[\tau_3, \tau_3 + \Delta)$, etc. Then we try again to find this pattern at times $t = \Delta$ and so on for T/Δ times. The probability of getting these spikes at any one trial is very low, but since we repeat this trial many times, we may succeed sometimes.

Because the few successes that we might get are due to repeating the experiment a great many times with an extremely low probability of success on any one trial, we may compute the probability of getting r successes in this experiment by the Poisson probability density which tells the probability of getting r successes when x successes are expected

$$pr\{r, x\} = e^{-x}x^r/r! \qquad (A1)$$

It is important to note that the validity of *Eq. A1* does not rest on the assumption that the spike trains behave like Poisson processes, but just on the assumption that at any time instance t the probability of getting one prespecified pattern is very low.

Under the conditions of interest to us (low firing rates, small and weak correlations between the cells) this $pr\{r; x\}$ will be always very small. Thus, there is no point in asking what is the probability of finding r repetitions of one particular pattern S_c, D_c. But we can ask what is the expected number of different patterns made out of spikes from cells $S_c = (i_1, i_2, \ldots, i_c)$ that appeared repeatedly r times in our record. This expected number $[N^{(r)}(S_c)]$ can be obtained by adding *Eq. A1* over all possible delays D_c

$$N^{(r)}(S_c) = \sum_{D \in \mathcal{D}_c} [e^{-x(S_c,D)} \cdot x(S_c, D)^r/r!] \qquad (A2)$$

Note that since time is measured with a finite accuracy, Δ, and the maximal delay is limited to w, the total number of terms on the summation is finite and is given by the following considerations: if we combine together all spike patterns S_c starting with the same (i_1) spike we shall have $(c-1)!$ different S_cs included. In this set of patterns each of $\tau_2, \tau_3, \ldots, \tau_c$ may obtain w/Δ different values. There would be $(w/\Delta)^{c-1}$ different delay lists. Thus for each particular S_c there are

$$n(\mathcal{D}_c) = \frac{(w/\Delta)^{c-1}}{(c-1)!} \qquad (A3)$$

different time delays.

If we wish to know what is the expected number of patterns of complexity c (regardless of which particular spikes are involved in the pattern) which appeared r times in our record we would have to sum *Eq. A2* over all possible spike names S_c

$$N_c^{(r)} = \sum_{S \in \mathcal{S}_c} N^{(r)}(S) \qquad (A4)$$

In some situations it may be possible to show that the expected number of occurrences $x(S, D)$ is independent of D. That is, all time delays between the spikes in a pattern are equally probable. In such cases we may write: $x(S, D) = x(S)$, and the sum over D in *Eq. A2* is just this expectation times the number of different time delays that one can have among c cells

$$\sum_{D \in \mathcal{D}_c} x(S, D) = n(\mathcal{D}_c)x(S)$$

By using *Eq. A3* we get from *Eq. A2* that the expected number of patterns from cells S_c that repeated r times is

$$N^{(r)}(S_c) = \frac{(w/\Delta)^{c-1}}{(c-1)!} e^{-x(S_c)} \cdot x(S_c)^r/r! \qquad (A5)$$

and from *Eq. A4* we get that the expected number of patterns of c spikes that repeated r times is

$$N_c^{(r)} = \frac{(w/\Delta)^{c-1}}{(c-1)!} \sum_{S \in \mathcal{S}_c} e^{-x(S)} \cdot x(S)^r/r! \qquad (A6)$$

Equations *A2* and *A4* or *A5* and *A6* give a straightforward way of estimating the number of repeating patterns that one expects to get by chance, but they rely on a valid way for getting $x(S_c, D_c)$ the expected number of occurrences of every specific pattern S_c, D_c. In the following four sections we develop ways to estimate this x.

Stationary Poisson processes

In this section we assume that our record is a realization of n parallel stationary Poisson processes, recorded for T seconds. Let us denote the firing rate of the ith single unit by λ_i. We assume also that the single units fire independently. The time of events is assumed to be quantized by units of Δ seconds. We assume that Δ is small enough so that the probability of getting two spikes from the same train within Δ seconds is negligibly small.

At every time t we can ask what is the probability of getting the pattern S_c, D_c. This probability is given by multiplying the probabilities of getting the first spike exactly at t by the probability of getting the second spike during the interval $[t + \tau_2, t + \tau_2 + \Delta)$ by the probability of getting the third spike during the interval $[t + \tau_3, t + \tau_3 + \Delta)$ and so on. We get the following probability

$$pr\{S_c, D_c, t\} = \lambda_{i_1}dt \cdot \lambda_{i_2}\Delta \cdot \lambda_{i_3}\Delta \cdot \cdots \cdot \lambda_{i_c}\Delta \quad (A7)$$

Note that here we assumed that the pattern started exactly on t and therefore the probability of finding there the first spike (i_1) is $\lambda_{i_1}dt$ and not $\lambda_{i_1}\Delta$.

By integrating *Eq. A7* over all the available time (T) we get the expected number of occurrences of S_c, D_c during T

$$x(S_c, D_c) = x(S_c) = T\Delta^{c-1} \prod_{k \in S_c} \lambda_k \qquad (A8)$$

Note that the expected number of occurrences x is not a function of the delays D_c. This is so because we assumed that the spike trains behaved like independent and stationary Poisson processes. Thus we can substitute the expected count of *Eq. A8* in *A5* to get, for any list of single unit names S_c, the expected number of patterns made of S_c which repeat r times.

Similarly, by substituting *Eq. A8* in *A6* we get a simple way to compute the expected number of patterns of c spikes that repeated r times by chance.

However, these equations are of limited value for physiological spike data, mostly because the firing rates of neurons tend to fluctuate excessively so that stationary Poisson processes are poor approximations to the statistics of neuronal firing times.

Nonstationary Poisson processes

Let us assume that the firing times of the single units are realizations of independent Poisson processes with varying firing rates. Let us denote by $\lambda_i(t)$ the firing rate of the ith single unit at time t. All the other assumptions are as stated in the previous section. Then the probability of getting a specific pattern S_c, D_c at time t is given by

$$pr\{S_c, D_c, t\} = \lambda_{i_1}(t)dt \cdot \lambda_{i_2}(t + \tau_2)\Delta \cdot \cdots \cdot \lambda_{i_c}(t + \tau_c)\Delta \qquad (A9)$$

and the expected number is given by integrating over t.

$$x(S_c, D_c) = \Delta^{c-1} \int_0^T \prod_{j=1}^{c} \lambda_{i_j}(t + \tau_j)dt \qquad (A10)$$

This is not a practical equation to compute for all possible D_c and S_c. If the fluctuations of the firing rates are slow (relative to the maximal pattern duration w) then $\lambda(t + \tau) \approx \lambda(t)$ and *Eq. A10* can be simplified to

$$x(S_c, D_c) = x(S_c) = \Delta^{c-1} \int_0^T \prod_{j=1}^{c} \lambda_{i_j}(t)dt \qquad (A11)$$

Again in this form $x(S_c, D_c)$ is not a function of D_c, or of t. Therefore, *Eq. A11* can be used in *A5* and *A6* to get the expected number of patterns of cells S_c that repeated r times, or the expected number of patterns of c cells that repeated r times.

In most cases the firing rates $\lambda_i(t)$ are not known a priori because they fluctuate in apparently random fashion. Evaluating them from the same record in which patterns are searched will usually result in under estimations of N. Nevertheless these equations can be useful for firing times obtained in simulations (where $\lambda(t)$ are known), and for some cases where the only variations in firing rates are brought about by stimuli, the poststimulus time (PST) histograms for the stimuli are known, and the units respond in a stable fashion.

Ad hoc method

We know from *Eq. A11* that if the firing times behave like independent Poisson processes whose firing rates change slowly with time then $x(S_c, D_c)$ (the expected number of occurrences of the pattern S_c, D_c) is independent of D_c. Usually we cannot use *Eq. A11* because we do not know the firing rate functions $\lambda(t)$. However, we may estimate x by counting all the occurrences of the sequence S_c (with delays that are less than w seconds) in the data. This number $m(S_c)$ is approximately the sum of $x(S_c, D_c)$ over all the possible delays D_c. If we assume that all the $x(S_c, D_c)$ for a given sequence of spike names (S_c) are identical, we have

$$x(S_c) \approx m(S_c)/[(w/\Delta)^{c-1}/(c - 1)!] \qquad (A12)$$

where $m(S_c)$ is obtained by counting how many times S_c appeared in the data. When $m(S_c)$ is large we may assume equality in *Eq. A12* and substitute in *A5* and *A6*. Thus we get an ad hoc estimation of the number of patterns made of S_c that repeated r times and of the total number of patterns of complexity c that repeated r times.

This ad hoc method of estimation was proven to be fairly accurate for both simulated and real data (see RESULTS). However, the process of counting all the occurrences of S_c for all possible combinations of c-spike names becomes very tedious for $c \geq 5$. Thus we found it practical to use the ad hoc method for patterns of 3 or 4 spikes, but had to resort on the next method for patterns of ≥ 5 spikes.

The bootstrap method

We shall develop here estimations for the expected number of patterns of complexity $c + 1$ that appeared repeatedly r times from the expected number of patterns of complexity c that appeared repeatedly r or more times. This kind of estimation is possible because any pattern S_{c+1}, D_{c+1} can be generated from the pattern S_c, D_c by appending a spike name to S_c and a delay to D_c. Furthermore a pattern of complexity $c + 1$ cannot appear r times unless the subpattern S_c, D_c (in which S_c is the list of the first c-spike names of S_{c+1} and D_c is the list of the first c delays from D_{c+1}) appeared also for at least r times.

Therefore, we may compute for every pattern S_c, D_c that appeared k times, the probability that it would produce r ($r \leq k$) repetitions of a particular pattern of $c + 1$ spikes whose names are S_c, i_{c+1} and whose delays are D_c, τ_{c+1}, where $\tau_c \leq \tau_{c+1} < w$. Given that a subpattern S_c, D_c has occurred at time t, what is the probability of finding a spike from i_{c+1} at the time interval $[t + \tau_{c+1}, t + \tau_{c+1} + \Delta)$? To compute this probability we assume that the i_{c+1} single unit is firing according to Poisson statistics with a firing rate of $\lambda_{i_{c+1}}$.

Then the probability that it fired is

$$pr\{i_{c+1} \text{ fired}\} = \lambda_{i_{c+1}}\Delta \qquad (A15)$$

and the probability that it did not fire is

$$pr\{i_{c+1} \text{ did not fire}\} = 1 - \lambda_{i_{c+1}}\Delta. \qquad (A16)$$

We know that the subpattern S_c, D_c occurred k times and we wish to know what is the probability that in exactly r times it was followed by a spike from i_{c+1} at the appropriate time. This probability is given by the binomial term

$$pr\{i_{c+1} \text{ fired exactly } r \text{ times}\} = \binom{k}{r}(\lambda_{i_{c+1}}\Delta)^r(1 - \lambda_{i_{c+1}}\Delta)^{k-r} \qquad (A17)$$

On the average there are $w/c\Delta$ different time delays after the last spike of S_c, D_c at which the i_{c+1} spike might fire, and we expect to have $N^{(k)}(S_c)$ cases in which the subpattern S_c repeated k times, so that the expected number of patterns S that repeated r times is

$$N^{(r)}(S_{c+1}) = \frac{w}{c\Delta}(\lambda_{i_{c+1}}\Delta)^r \sum_{k=r}^{\infty}\left[\binom{k}{r}\cdot(1-\lambda_{i_{c+1}})^{k-r}\cdot N^{(k)}(S_c)\right] \qquad (A18)$$

The term $N^{(k)}(S)$ falls off very fast as k increases, so that there is no practical value for summing up all the way from r to infinity.

The total number of patterns of $c + 1$ spikes that repeated r times is computed as in *Eq. A4*, by adding all the S_{c+1} that repeated r times

$$N_{c+1}{}^{(r)} = \sum_{S\in t_{c+1}} N^{(r)}(S) \qquad (A19)$$

Now that we have $N^{(r)}(S_{c+1})$ and $N_{c+1}{}^{(r)}$ we can apply *Eqs. A18* and *A19* to these estimates to obtain $N^{(r)}(S_{c+2})$ and $N_{c+2}{}^{(r)}$, and so on.

These estimates are based on the assumption that the $c + 1$ neuron in the pattern can be regarded as a Poisson process. As discussed earlier this is a poor approximation. However, when applied for only one or two steps beyond c and if the baseline estimates $N^{(r)}(S_c)$ are good, the errors introduced by the Poisson approximations are not too bad. In the analysis discussed in the results we applied the ad hoc method for patterns of 3 and 4 spikes and the bootstrap method for patterns of 5 and 6 spikes. The agreement with the number of patterns found both in simulated and experimental data were quite good.

ACKNOWLEDGMENTS

We thank Dr. Ad Aertsen for discussions and critical reading of the manuscript.

This work was supported by a grant from the System Development Foundation SDF-00013 (G. L. Gerstein) and by a grant from The Israel Institute for Psychobiology, Charles E. Smith Family Foundation (M. Abeles).

Received 25 June 1987; accepted in final form 5 April 1988.

REFERENCES

1. ABELES, M. Quantification, smoothing, and confidence limits for single-units' histograms. *J. Neurosci. Methods* 5: 317–325, 1982.
2. ABELES, M. *Local Cortical Circuits.* Berlin: Springer Verlag, 1982.
3. ABELES, M., DE RIBAUPIERRE, F., AND DE RIBAUPIERRE, Y. Detection of single-unit responses which are loosely time-locked to a stimulus. *IEEE Trans. Systems, Man and Cybernetics* 13: 683–691, 1983.
4. DAYHOFF, J. AND GERSTEIN, G. L. Favored patterns in spike trains. I. Detection. *J. Neurophysiol.* 49: 1334–1348, 1983.
5. DAYHOFF, J. AND GERSTEIN, G. L. Favored patterns in spike trains II. Application. *J. Neurophysiol.* 49: 1349–1363, 1983.
6. FROSTIG, R., FROSTIG, Z., FRYSINGER, R., AND SCHECHTMAN, V. Multineuron analysis reveals complex patterns of interaction among neurons. *Soc. Neurosci. Abstr.* 11: 1020, 1985.
7. GERSTEIN, G. L., BLOOM, M., ESPINOSA, I., EVANCZUK, X., AND TURNER, M. Design of a laboratory for multi-neuron studies. *IEEE Trans. Systems, Man and Cybernetics* 13: 668–676, 1983.
8. HEBB, D. O. *The Organization of Behavior: a Neurophysiological Theory.* New York: Wiley, 1949.
9. KLEMM, W. R. AND SHERRY, C. J. Serial ordering in spike trains: what's it "trying to tell us"? *Int. J. Neurosci.* 14: 15–23, 1981.
10. KNUTH, D. E. *The Art of Computer Programming* (2nd ed.). Reading, MA: Addison-Wesley, 1981, vol. 2.
11. KRUGER, J. Simultaneous individual recordings from many cerebral neurons: techniques and results. *Rev. Physiol. Biochem. Pharmacol.* 98: 178–233, 1983.
12. LANDOLT, J., REINIS, S., AND WEISS, D. Identification of local neuronal circuits in the visual cortex of the cat. *Soc. Neurosci. Abstr.* 11: 1010, 1985.
13. LEGENDY, C. AND SALCMAN, M. Bursts and recurrences of bursts in the spike trains of spontaneously active striate cortex neurons. *J. Neurophysiol.* 53: 926–939, 1985.
14. WHITNEY, C. A. Generating and testing pseudorandom numbers. *Byte* 9: 128, 1984.

ACTA NEUROBIOLOGIAE EXPERIMENTALIS

Acta Neurobiol Exp 2004, 64: 203-207

Searching for significance in spatio-temporal firing patterns

George L. Gerstein

Department of Neuroscience, University of Pennsylvania, 210 Stemmler Hall, 3450 Hamilton Walk, Philadelphia, PA 19104-6074, USA

Short communication

Abstract. We examine a specific candidate for temporal coding of information by spike trains, the occurrence of a temporal firing pattern among some number of neurons that repeats more often than expected by chance. Methods for detection of repeating patterns have long been available, but there are no analytic methods for calculating the expected numbers of repeating patterns to enable assignment of significance to the results from the experimental data. The expected numbers can be calculated by Monte-Carlo methods by repeatedly modifying the original data spike trains. Ideally the surrogates produced by such changes should destroy all patterns and cross-correlations but preserve other aspects of the trains such as rate, interval structure etc. We present here a novel variant of the "dither surrogate" (Date et al. 1998) and use surrogates generated by this algorithm to evaluate repeating pattern significance in data recorded in monkey motor cortex during behavior. Although we can demonstrate high statistical significance for the excess repetition of some spike patterns, it is not obvious that this has physiological meaning or that such patterns are used for information transfer.

Key words: neuron firing patterns, multi-neuron, synchrony, cortex, synfire-chains, coding, statistical significance

The basic problem. The nervous system certainly uses firing rate coding to represent information, but it is not clear whether and to what extent specific spike timing is used despite many papers on this topic. Perhaps the simplest timing structure would be firing synchrony of two or more neurons. This can have an amplified effect on downstream neurons because of nonlinear addition of the EPSPs. The extent and circumstances in which such "unitary" synchrony events exceed expected rates by significant amounts have been thoroughly investigated in a series of papers by Grün et al. (2001) with the conclusion that excess synchrony can be temporally associated with behavioral events. In this paper we will examine a more complicated type of temporal firing structure consisting of a firing pattern of some length among some number of neurons that repeats more than expected by chance. Such repeating patterns are at least potentially a carrier of information. Also in terms of Abeles' concept of synfire chains (Abeles 1991) they could represent the repeated activation of a particular synfire chain from which we record some number of neurons.

Available methods. Short patterns consisting of two intervals among the firings of one to three neurons (i.e., three successive spikes) have been studied by a variant of the Joint Peri-Stimulus Time Histogram (JPSTH) (Aertsen et al. 1989, Gerstein and Perkel 1972) where the occurrence of a spike in a particular train is used as reference (instead of stimulus) and the delayed occurrence of the next two spikes (of the other trains being analyzed) are represented along the x and y axes of the JPSTH. If a particular interval pattern is repeating more than other interval sequences there will be a hot spot in this variant JPSTH. The method and appropriate significance tests were worked out in a series of papers by Prut et al. (1998). Again it was possible to demonstrate a fairly loose time locking of such excess pattern repetitions to behavioral events. (Note that the "snowflake" (Perkel et al.1975) could be used in a similar way.)

Repetitions of more complex patterns of arbitrary length and membership can be found by a method of Abeles and Gerstein (1988). This is easily visualized by representing the data as an old fashioned punched paper tape. Each line of holes along the tape represents the firing of a particular neuron, quantized time (bins) represented along the tape. Simultaneous firings to within the bin time are represented by more than one hole across the tape at a particular temporal location. Now make a second copy of the tape. With the two tapes superimposed over a light source, start shifting one relative to the other in steps of one bin. We choose a window for maximum pattern length, and at each shift we scan the entire overlaid tapes for superimposed holes. For each tape shift and widow position we keep track of the number and identity of superimposed holes. This identifies patterns of any complexity (number of firings involved) and with any temporal structure within the chosen pattern length window that repeat two or more times in the data. In the original paper significance of counts was obtained by analytic calculation on Poisson and inhomogeneous Poisson trains. This is a poor match to most real data, so that other tests for significance are needed.

Significance tests. Other analytic tests have not yet been developed, so that it is necessary to use surrogate and Monte-Carlo calculation after the original data has been analyzed with the two tapes method. One such approach has been made by Baker and Lemon (2000). For each of the N neurons in the data they obtained firing rate function by convolution of the spike train with a suitable kernel. This rate function was used to generate a spike train with intervals drawn from rate modulated gamma functions of order 1 to 30. For each of these surrogate candidates an interval histogram (IH) was calculated, and the order which best matched the corresponding data IH was used in generating all further surrogates of that spike train. The process was repeated for each spike train in the data set. Unlike Poisson trains this approach replicates the refractory period and lack of short intervals that is typical of most data. The pattern search computation was then calculated for each set of N surrogates, and the process of generating surrogates and pattern search was repeated until adequate statistical significance could be assigned to events in the pattern analysis of the original data. Baker and Lemon's conclusion was that the number of repeating patterns in the surrogate data was the same as or even higher than in the original data, and therefore that patterns could not have any physiological significance (Baker and Lemon 2000).

However, Baker's method of generating surrogates has some problems. When we applied it to artificial data consisting of rate modulated gamma functions, the surrogate generating procedure always produced surrogates at least two orders larger than the original. The order parameter is connected to regularity – the higher the order the more repeating patterns are expected. Thus these were inappropriate surrogates and would produce inappropriately high pattern counts.

A completely different class of surrogates can be generated by dithering the time of each spike. This is an idea first developed by Date et al. (1998), Hatsopoulos et al. (2003) and used in Abeles and Gat (2001). Such a surrogate completely destroys any original patterns, but preserves the overall rate structure of the spike train. In this original formulation the distribution of dithers was flat over some appropriate window (± 8 ms). Such a surrogate however does change the original IH by adding short intervals and lowering the peak. In terms of gamma distributions, such a surrogate is a move to lower order and hence produces an inappropriately low number of patterns.

A modified result can be obtained by defining the dither distribution on the original Joint Interval Histogram (JIH) for adjacent intervals. Dithering a spike on such a 2 dimensional surface corresponds to movement along a trajectory perpendicular to the principal diagonal (for each three spikes as we dither the middle spike the first interval gets shorter and the second interval gets longer, or *vice versa*). The values of the JIH surface along this trajectory can are used to define the actual dither distribution. For typical parameter values the resulting surrogate also destroys any pattern structure, but creates an IH with less short intervals and a higher peak than in the IH of the original data. In terms of gamma distributions such a surrogate is a move to higher order and hence produces an inappropriately high number of patterns (as in the Baker and Lemon surrogate method).

It turns out that the same 2 dimensional procedure carried out on the surface which represents the square root of the JIH produces a surrogate with IH and JIH remarkably similar to the original. This version of dither algorithm has no obvious mathematical foundation, but because of the good IH fits we have used it extensively to make surrogates for pattern searches. The IH perfor-

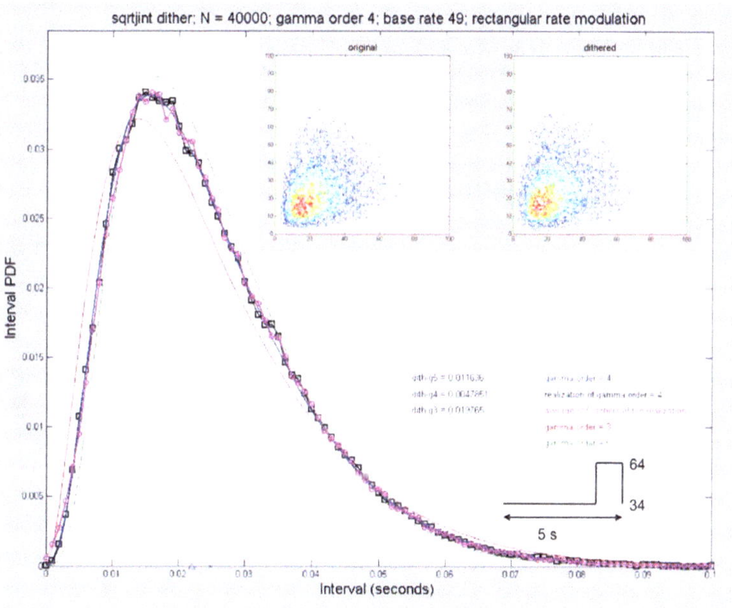

Fig. 1. Performance of the $\sqrt{\text{JIH}}$ dither for data generated by a rate varying gamma process of order 4. The rate modulation cycle is indicated at lower right. Note that the theoretical IH, its realization and its dither all superimpose. Theoretical IH for gamma processes of order 3 and 5 are included for comparison. Joint IH are shown at upper right for original and dithered, and also show little if any distortion.

mance of the $\sqrt{JIH}$ dither method for a data file generated with a rate modulated gamma process of order 4 is shown in Fig. 1. The superposition of theoretical, generated and dithered IH is obvious. The graph also shows theoretical IH for gammas of order 3 and 5 for calibration.

Pattern search results. Before analyzing the original data we filter it to remove bursts of activity. These may be defined as three or more consecutive intervals shorter than some criterion like 10 ms. Such bursts produce many repeating patterns that are essentially uninteresting (the recurrence of bursts *per se* as a potential carrier of information has been studied by Legendy and Salcman 1985). Application of the overlaid paper tape pattern search algorithm to burst filtered data is relatively straightforward, but this is a brute force computation that requires considerable time even with today's fast computers. We have generally processed the burst filtered original data and then made and processed 20 surrogate sets. Results can be summarized in a two dimensional histogram of number of patterns as a function of complexity (how many spikes are involved in the pattern) and of number of repetitions. We call this the pattern spectrum. Details of individual patterns, their repetition score and their location in time are of course also available for further analysis.

The pattern spectra for the 20 surrogates are averaged, and the standard deviation at each point is calculated. We then compute the difference between the pattern spectrum of the original data and this average surrogate spectrum, and express the final result normalized at each point by the corresponding standard deviation of the surrogate average.

We show the analysis of a data set of six neurons simultaneously recorded in M1 cortex of a monkey doing a repetitive thumb-finger grasp task with both hands (Baker laboratory, methods as in Baker and Lemon 2000). Individual spikes within a pattern are defined to 2 ms, and total pattern duration is < 300 ms. The count difference plane in Fig. 2 at the left is shown from above (original pattern counts > average surrogate counts); the view from below the difference plane is at the right. Bar heights are in units of standard deviation of the average count at those coordinates. There are clearly more small differences in the positive direction (original > average of surrogates) than in the negative direction. Among the larger differences there is one bar at 5 SD and one bar at 3 SD in the positive direction; the largest bar in the negative direction is 2 SD

The 5 SD bar represents patterns of 4 spikes that repeat 8 times in the data set. There are 10 such patterns in the original data. We should now examine whether the

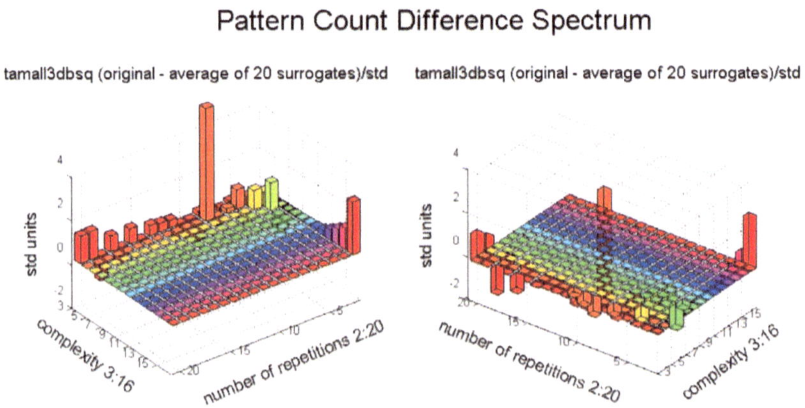

Fig. 2. An example of pattern count difference spectrum in units of standard deviation of the surrogate average. Data consisted of 6 well isolated neurons simultaneously recorded from left M1 of a monkey executing a repeating thumb-finger grasp task with both hands (Baker laboratory). Left panel shows view from above the plane, corresponding to original pattern counts > surrogate average counts; right panel is view from below the plane corresponding to original pattern counts < surrogate average counts. See text.

repeating occurrences of any of these patterns are related to particular behavioral events in the monkey's task.

Another test that compares the pattern occurrences in original and surrogate average is to compute the total number of repeating patterns, i.e., to sum all bars in the original and average surrogate pattern spectra. For the data set used for Fig. 2:

Total repeating patterns in ORIGINAL:	1,624,787
Total repeating patterns in SURROGATE:	1,612,632
DIFFERENCE:	12,155
DIFFERANCE / $\sqrt{\text{SURROGATE}}$:	9.6

Thus the total number of repeating patterns in the original data set would seem to exceed those in the average of the surrogates by a statistically very significant amount.

Discussion. Does the pattern excess in the original data indicate that patterns are used for information transfer? The question of physiological significance is much harder to answer. In fact the numbers are rather discouraging. In the data example shown here there were more than 10^6 total patterns that repeated 2 or more times, with under 1% excess count over the corresponding surrogate average. What physiological mechanism could be used to single out this under 1% of excess patterning in order to use it for communication between neurons or domains?

Another possible interpretation of multi-spike patterns is that some such events are a signature of activity in an underlying synfire chain, while others are just random noise. Information would now lie in whether and which particular synfire chain is activated, not in the pattern *per se*. The numbers might be less daunting, since it is not unreasonable to suppose that many intersecting synfire chains could come through the region where the observed neurons are located.

One possible approach to testing the physiological significance of patterns, either directly or as evidence for a synfire chain, would be to perturb. If spikes are added (or subtracted) from a pattern by local electrical or chemical stimulation will there be any observable behavioral consequence? Certainly appropriate electrically imposed rate changes in cortical area MT can affect behavioral judgment of movement direction in the visual scene (Nichols and Newsome 2002). Related experiments to perturb specific patterns rather than rates have as yet not been carried out, are probably very difficult, and ultimately, because large populations of neurons are involved in typical task behavior, may produce effects far smaller than any observable behavioral changes.

So, for the present, although we have demonstrated a much better surrogate than previously used in such pattern computations, it is as yet impossible to assess the physiological significance of excess patterning. We might just be studying surrogates of different properties, some producing more, some fewer patterns than their original data.

Abeles M, Gerstein GL (1988) Detecting spatio-temporal firing patterns among simultaneously recorded single neurons. J Neurophysiol 60: 909-924.

Abeles M (1991) Corticonics: Neural circuits of the cerebral cortex. Cambridge Universtity Press, Cambridge, 280 p.

Abeles M, Gat I (2001) Detecting precise firing sequences in experimental data J Neurosci Methods 107: 141-154.

Aertsen AMHJ, Gerstein GL, Habib M, Palm G (1989) Dynamics of neuronal firing correlation: modulation of "effective connectivity". J Neurophysiol 61: 900-917.

Baker S, Lemon R (2000) Precise spatiotemporal repeating patterns in monkey primary and supplementary motor areas occur at chance level. J Physiol Lond 84: 1770-1780.

Date A, Bienenstock E, Geman S (1998) On the temporal resolution of neural activity. Technical report, Division of Applied Mathematics, Brown University (Available at: http://www.dam.brown.edu/people/elie/papers/temp–res.ps).

Gerstein GL, Perkel DH (1972) Mutual temporal relationships among neuronal spike trains. Biophysical J 12: 453-473.

Grün S, Diesmann M, Aertsen A (2001) Unitary events in multiple single-neuron spiking activity. I. Detection and significance. Neural Comput 14: 43-80.

Hatsopoulos N, Geman S, Amarasingham A, Bienenstock E (2003) At what time scale does the nervous system operate? Neurocomputing 52-54: 25-29.

Legendy C, Salcman M (1985) Bursts and recurrences of bursts in the spike trains of spontaneously active striate cortex neurons. J Neurophysiol 53: 926-939.

Nichols MJ, Newsome WT (2002) Middle temporal visual area microstimulation influences veridical judgments of motion direction. J Neurosci 22: 9530-9540.

Perkel DH, Gerstein GL, Smith MS, Tatton WG (1975) Nerve impulse patterns: a quantitative display technique for three neurons. Brain Res 100: 271-296.

Prut Y, Vaadia E, Bergman H, Haalman I, Hamutal S, Abeles M (1998) Spatio-temporal structure of cortical activity: Properties and behavioral relevance. J Neurophysiol 79: 2857-2874.

Received 3 February 2004, accepted 4 March 2004

Chapter 10
Cross-Correlations of Unresolved Multi-Neuron Recordings

Does Cross-Correlation Have a Unique Interpretation?

Purvis Bedenbaugh

George Gerstein fearlessly explored how neurons cooperated in assemblies. He began his undergraduate studies in physics at Harvard about the time that "The Organization of Behavior" by Hebb (1949) was published. Hebb introduced the concept of a "cell assembly," a collection of connected neurons, which through repeated coactivation enhance their connectivity, so that activity in one cell enhances activity in another. George was greatly influenced by this work, and he sought to design experiments that would directly observe key features of cell assemblies. In particular, cell assemblies should have stimulus-dependent coordinated activity, and changes in coordinated activity should be stimulus-dependent. The anticipated signature of such assemblies was narrow peaks in the cross-correlation between the activities of simultaneously observed neurons.

George brought his experience as a high-energy physicist (Gerstein et al. 1957), analyzing sequences of events, to the earliest applications of computers (the MIT TX-O and the LINC) to the analysis of neuronal spike trains (Gerstein 1960; Clark 1987) and the search for cell assemblies. With Don Perkel and George Moore, he formalized the analysis of the first- (Perkel et al. 1967a) and second-order (Perkel et al. 1967b) statistics of spike trains, including non-stationarity. The anticipated narrow cross-correlation peaks turned out to be rare, and George and others (Abeles 1982) introduced models that might explain the diversity of observed cross-correlation histograms. George proceeded carefully, from a deep understanding that

P. Bedenbaugh (✉)
Touchspring Technologies, LLC, Tampa, FL, USA
e-mail: pbedenbaugh@gmail.com

A. Aertsen et al. (eds.), *Introducing Computation to Neuroscience*, Springer Series in Computational Neuroscience, https://doi.org/10.1007/978-3-030-87447-6_10

for such a complex enterprise to be successful, that it was necessary to be very clear about what was known and what was not known.

I first encountered George's work as an undergraduate working with Janice Ovelmen-Levitt in the neurosurgery lab of Blaine Nashold (Ovelmen-Levitt et al. 1984). At the time, she was leading a study of how denervation changed the driven and spontaneous activity in neurons in the dorsal horn of the cat spinal cord. Some of these neurons were spontaneously bursting. This lab experience impressed upon me that while the firing rate of sensory neurons played an important role in their signaling, spontaneous activity, and the pattern of firing, it could also be playing an important role in the integration of neuronal signals. Dr. Ovelmen-Levitt gave me a copy of the Perkel, Gerstein, and Moore papers and asked me to assist her with the data collection and the analysis. I was a biomedical engineering student and felt at home when I read the systematic mathematical treatment of neuronal signals. Shortly after I completed my undergraduate degree, a set of papers describing an analysis of patterns in spike trains were published by Judy Dayhoff with George (Dayhoff and Gerstein 1983a,b). I saw the analysis of neuronal signals as an active and evolving area of research and became interested in working with George to unlock the secrets of how groups of neurons integrated patterns of spike trains to generate perception.

After working in George's lab for a time, I was confronted with the reality that interpretation of multiple simultaneous neuronal recordings was not entirely straightforward. I began to conduct simulations with the objective of understanding how changes in how a neuronal network was activated could lead to changes in the cross-correlation between the spike trains of simulated neurons (Gerstein et al. 1989). Around this time, the paperback editions of the Parallel Distributed Processing (PDP) books were published (McClelland and Rumelhart 1989; McClelland et al. 1987), and I considered the problem of how the cross-correlation between the "neurons" in a PDP model might change form as signals were passed from stage to stage within the network. To address the problem, I examined the correlation between the outputs of a pair of sigmoidal processing elements, with correlated shot noise inputs. I found that when the variance was small, so that the inputs were largely confined to the linear portion of the sigmoid, then the output normalized cross-correlation matched the input correlation. When the input variance was large, however, so that the nonlinearity of the sigmoid limited the range of the outputs, then the cross-correlation of the outputs was rectified—positive correlations in the input gave rise to similar correlations in the output, but negative correlations in the input gave rise to output correlations of reduced magnitude.

George often emphasized the importance of normalizing cross-correlation functions in order to ensure that this second-order statistic—relating simultaneously recorded signals—reflected second-order properties of the system under observation. Stimulus modulation, or structured autocorrelation functions in the observed recordings, can give rise to non-zero cross-correlation functions that could be computed from the recordings even if they were derived from sequential recordings. In order to count as evidence for the observation of a neuronal assembly, the statistic had to indicate a relationship that differed from what could have been obtained from recording obtained sequentially. George also emphasized the importance of spike sorting, not just to increase the efficiency of data collection—obtaining more spike trains from each microelectrode recording, but as critical to the interpretation of the experiment. If different cells are recorded in different contexts, changes in cross-correlation cannot be interpreted as changes in the effective connectivity between cells in a neuronal assembly. Sometimes George's concern that it was inappropriate to compute the cross-correlation between multi-unit recordings in the context of the study of neuronal assemblies was dismissed as pedantic, rather than embraced as a guide to separating what was known from what was not known. Multi-unit recordings are much easier to obtain than single-unit recordings. Spike sorting is an inexact science. It is easier to find correlations in multi-unit data than in single-unit data, and it was thought solely because the spike counts were higher.

The scientific milieu of the time was shaped by the landmark paper by Charlie Gray and Wolf Singer, describing stimulus-specific oscillations in the visual cortex (Gray and Singer 1989). There was some controversy about oscillations, and some investigators disputed whether they even existed. Pedro Maldonado in the lab recognized that the oscillations could be an important feature of the architecture of neuronal computation, different from how cell assemblies were conceived up to then. This motivated all of us in George's lab to refine and clarify our thinking and our specific intent when we used cross-correlation measures.

As I scrutinized some data that I had collected as part of my dissertation research (Bedenbaugh 1993), I discovered that if I combined the sorted units obtained from each electrode into something like multi-unit spike trains, the cross-correlation was much greater than the cross-correlation between the pairs of individual sorted spike trains recorded on the same electrode pairs. It was a large effect and a straightforward system, so I set to understand the effect in terms of the asymptotic expectation of a summed process. That is to say, I dashed off a model in terms of expectation without considering the properties of the estimator, neglecting the details of the discretization of time and counts. George reminded me that the analysis is only meaningful if it holds for the properties of the estimator, with all its detailed properties explicit (Bedenbaugh and Gerstein 1997). I continue to be grateful for that lesson.

Later, when George was in residence for a time at a European laboratory, the conclusions of our paper were vigorously discussed. George defended and extended our work (Gerstein 2000) to what he described to me "realistic correlations, and realistic numbers of neurons." In discussion of specific models, he returns to his concern for the cell assemblies popularized by Hebb. Among the models, he highlights, he shows that a narrow, low-latency peak in the cross-correlation histogram, such as would be generated by direct synaptic connection between neurons in an assembly, would be harder to detect in the presence of incomplete spike sorting. The paper shows his consistent attention to clear interpretation of the data. George was concerned that the recorded data could show how reality related not only to the investigator's hypothesis, but also clearly support or falsify alternative interpretations. Such interpretation rests upon an unambiguous provenance of the data.

Throughout his career, George Gerstein sought ways to observe cell assemblies within a quantitative framework that would lay a secure foundation for understanding the operation of neuronal circuits. He had great concern that simultaneous neuronal recordings were sometimes casually misinterpreted. These papers demonstrate a process of formalizing a simple model and then carrying through with analysis to understand what could be observed in an experimental situation. Such a process can show the signatures that would be produced by different models, and processes related to observation can be incorporated into the predictions. As he noted in Gerstein (2000), this forward problem is straightforward and a necessary first step. He reminds us that science advances when we can make a unique interpretation of the backward problem, for example, making an inference of the "effective connectivity between neurons from their spike trains."

References

Abeles M (1982) Local cortical circuits: an electrophysiological study. Studies of brain function. Springer, Berlin. https://doi.org/10.1007/978-3-642-81708-3

Bedenbaugh PH III (1993) Plasticity in the rat somatosensory cortex induced by local microstimulation and theoretical investigations of information flow through neurons. PhD thesis, University of Pennsylvania

Bedenbaugh P, Gerstein GL (1997) Multiunit normalized cross correlation differs from the average single-unit normalized correlation. Neural Comput 9(6):1265–1275. https://doi.org/10.1162/neco.1997.9.6.1265

Clark WA (1987) The LINC was early and small. In: HMI '87 Proceedings of ACM conference on History of medical informatics, pp 51–73. https://doi.org/10.1145/41526.41532

Dayhoff JE, Gerstein GL (1983a) Favored patterns of spike trains. I. Detection. J Neurophysiol 49(6):1334–1348. https://doi.org/10.1152/jn.1983.49.6.1334

Dayhoff JE, Gerstein GL (1983b) Favored patterns of spike trains. II. Application. J Neurophysiol 49(6):1349–1362. https://doi.org/10.1152/jn.1983.49.6.1349

Gerstein GL (1960) Analysis of firing patterns in single neurons. Science 131(3416):1811–1812. https://doi.org/10.1126/science.131.3416.1811

Gerstein GL (2000) Cross-correlation measures of unresolved multi-neuron recordings. J Neurosci Methods 100(1–2):41–51. https://doi.org/10.1016/S0165-0270(00)00226-0

Gerstein G, Niederer J, Strauch K (1957) Elastic scattering of 96-Mev protons. Phys Rev 108(2):427–432. https://doi.org/10.1103/PhysRev.108.427

Gerstein GL, Bedenbaugh P, Aertsen AMHJ (1989) Neuronal assemblies. IEEE Trans Biomed Eng 36(1):4–14. https://doi.org/10.1109/10.16444

Gray CM, Singer W (1989) Stimulus-specific neuronal oscillations in orientation columns of cat visual cortex. Proc Natl Acad Sci 86(5):1698–1702. https://doi.org/10.1073/pnas.86.5.1698

Hebb DO (1949) The organization of behavior: A neuropsychological theory. Lawrence Erlbaum Associates, Inc.

McClelland JL, Rumelhart DE (1989) Explorations in parallel distributed processing: A handbook of models, programs, and exercises. MIT Press, Cambridge

McClelland JL, Rumelhart DE, PDP Research Group, et al (1987) Parallel distributed processing, vol 2. MIT Press, Cambridge, MA

Ovelmen-Levitt J, Johnson B, Bedenbaugh P, Nashold Jr BJ (1984) Dorsal root rhizotomy and avulsion in the cat: a comparison of long term effects on dorsal horn neuronal activity. Neurosurgery 15(6):921–927. https://doi.org/10.1097/00006123-198412000-00030

Perkel DH, Gerstein GL, Moore GP (1967a) Neuronal spike trains and stochastic point processes. I. The single spike train. Biophys J 7(4):391–418. https://doi.org/10.1016/S0006-3495(67)86596-2

Perkel DH, Gerstein GL, Moore GP (1967b) Neuronal spike trains and stochastic point processes. II. Simultaneous spike trains. Biophys J 7(4):419–440. https://doi.org/10.1016/s0006-3495(67)86597-4

Communicated by William Skaggs

Multiunit Normalized Cross Correlation Differs from the Average Single-Unit Normalized Correlation

Purvis Bedenbaugh
Department of Otolaryngology and Keck Center for Integrative Neuroscience, University of California at San Francisco, San Francisco, CA 94143, U.S.A.

George L. Gerstein
Department of Neuroscience, University of Pennsylvania, Philadelphia, PA 19104, U.S.A.

As the technology for simultaneously recording from many brain locations becomes more available, more and more laboratories are measuring the cross-correlation between single-neuron spike trains, and between composite spike trains derived from several undiscriminated cells recorded on a single electrode (multiunit clusters). The relationship between single-unit correlations and multiunit cluster correlations has not yet been fully explored.

We calculated the normalized cross-correlation (NCC) between single-unit spike trains and between small clusters of units recorded in the rat somatosensory cortex. The NCC between small clusters of units was larger than the NCC between single units. To understand this result, we investigated the scaling of the NCC with the number of units in a cluster.

Multiunit cross-correlation can be a more sensitive detector of neuronal relationship than single-unit cross-correlation. However, changes in multiunit cross-correlation are difficult to interpret uniquely because they depend on the number of cells recorded on each electrode and because they can arise from changes in the correlation between cells recorded on a single electrode or from changes in the correlation between cells recorded on two electrodes.

1 Introduction

One of the great challenges of contemporary neuroscience is to understand the distributed patterns of brain activity associated with learned behaviors and remembered stimuli. Increasing evidence supports the view that assemblies of neurons (Hebb, 1949; Gerstein, Bedenbaugh, & Aertsen, 1989) subserving particular percepts and behaviors are embedded in a nonlinear, cooperative network. The collective activity of the cells within a neuronal assembly provides a context for interpreting the activity of each of its component cells. The experimental detection of such neuronal assemblies and study of their properties depends on simultaneously recording from many neurons and on measurement of timing interrelationships among the several neuronal activities. Analysis of the interactions between neuronal assemblies is even more demanding of recording technology and intensive calculations.

For over thirty years, the cross-correlation between two neuronal spike trains (Perkel, Gerstein, & Moore, 1967b), its generalizations (Gerstein & Perkel, 1972; Aertsen, Gerstein, Habib, & Palm, 1989), and its homolog in the frequency domain (Perkel, 1970; Rosenberg, Amjad, Breeze, Brillinger, & Halliday, 1989) have been the principal tools for assessing neuronal relationships. Methods have been developed to normalize the cross-correlation so that the resulting correlation coefficient may be made independent of both firing rate and of modulations in firing rate (Aertsen et al., 1989; Palm, Aertsen, & Gerstein, 1988) and may be fairly compared across different observations. When these methods are applied to single-unit recordings, correlating the spikes of two simultaneously recorded single neurons, the results are unambiguous. It is clear which particular neurons are correlated; changes in correlation are due to a change in the coordination of particular cells, not to changes in recruitment or to poor recording stability.

There are, however, difficulties in using spike train pair correlation to study neural assembly properties. Single-unit cross-correlation is a relatively insensitive measure of neuronal relationship (Bedenbaugh, 1993). Given the low firing rates of most cortical neurons, this makes it difficult to detect a relationship between simultaneously recorded spike trains within a reasonably short recording time. In turn, this translates into fewer data points per experiment, a major limitation for investigations that seek to explore the distributed pattern of neuronal responses and their coordination. The development of spike-shape sorting technology has partially improved the situation with respect to sampling and efficiency problems, since it allows separation of multiunit observations from a single electrode into spike trains that originate from different neurons. However, the price is a complex technology, and also a combinatorial explosion of pairs as the number of separable simultaneously recorded neurons increases. The latter problem has been partially mitigated, although at the cost of intensive computation (Gerstein, Perkel, & Dayhoff, 1985; Gerstein & Aertsen, 1985).

In reaction to these difficulties, a number of investigators have begun to use multiunit recordings without separation of waveforms; that is, they use a composite data stream consisting of several superimposed spike trains from each electrode. The rationale is that cells in localized cortical columns have homogeneous response properties. Changes in the firing rate of a multiunit recording often reflect changes in the activity of the individual neurons in the population. If we assume that possible confounding factors, such as recruitment of additional neurons or recording instability, can be minimized, it is reasonable and increasingly common to study neuronal assembly properties by cross-correlation of multiunit recordings.

We might expect that the normalized cross-correlation (NCC) between two multiunit spike trains would correspond to the average cross-correlation between the single units recorded on different electrodes. However, for our recordings in rat somatosensory cortex, we found that the multiunit normalized cross-correlation was larger than the single-unit normalized cross-correlation. The analysis in this article suggests that this may be because the correlation between units recorded on a single electrode influences the multiunit cross-correlation. This view is also consistent with Kreiter and Singer's (1995) recent report of dependence of multiunit cross-correlation on the local correlation between the single-unit spike trains that comprise the multiunit spike trains.

2 Methods

2.1 Experimental. Extracellular neuronal potentials were recorded from the forepaw region of the somatosensory cortex of seven rats using either three or five tungsten microelectrodes and conventional methods, described elsewhere (Bedenbaugh, 1993; Bedenbaugh, Gerstein, & Bolton, 1997). Recordings from each of the several electrodes used in each experiment were separated into a total of eighty-nine single-unit spike trains with 0.5 millisecond time resolution using hardware principal component spike sorters (Gerstein, Bloom, Espinosa, Evanczuk, & Turner, 1983). Of these, 638 pairs of single-unit trains had been simultaneously recorded and were available for cross-correlation analysis. Finally, a total of twenty-one multiunit spike trains were created by merging of the sorted single-unit spike trains. Each cell pair was recorded under four different somatic stimulation conditions: no stimulus (spontaneous activity), touch to the best receptive field of the cells recorded by one electrode, touch to the best receptive field of the cells recorded by another electrode, and a simultaneous touch to both skin sites (Bedenbaugh, 1993; Bedenbaugh et al., 1997).

The data were blocked into a sequence of stimulus-dependent trials and peri-stimulus-time (PST) histograms and raw cross-correlograms were computed for each stimulus condition. The normalized cross-correlation (correlation coefficient) histogram for a pair of spike trains was calculated by subtracting the cross-correlation of the two PST histograms divided by the number of stimuli (PST predictor) (Perkel et al., 1967a) from the raw cross-correlogram, then normalizing the result by the geometric mean variance of the two point processes (Bedenbaugh, 1993; Bedenbaugh et al., 1997).

2.2 Analytical. Consider the situation illustrated schematically in Figure 1. There are two electrodes, A and B, each recording from m cells so that the spike train recorded on electrode A is $A(t) = a_1(t) + a_2(t) + \cdots + a_m(t)$, and the spike train recorded on electrode B is $B(t) = b_1(t) + b_2(t) + \cdots + b_m(t)$, where t is time, and $m = 2$ in the figure. Discretize time into N_T total bins of equal duration, short enough that any one cell may fire no more than one spike in one bin-width of time. If we assign the value 1 to the occurrence of a spike, and 0 to the absence of a spike in a bin, then the expected number of spikes per unit time of any individual cell, say a_i, is N_i/N_T, where $N_i = \sum_{k=1}^{N_T} a_{ik}$ where k denotes a particular time bin. Since the point process derived from the spike train can only take the values 0 or 1, the expectation (mean value) $E[a_{ik}^2]$ is also N_i/N_T. Further, assuming that all of the individual spike trains contain the same number of spikes and that the number of times any two cells recorded on a single electrode both fire in one time bin is N_{close}, we see then that the mean and variance of a multiunit process, say A, are

$$
\begin{aligned}
\mu_A &= E[A_k] = \sum_{i=1}^{m} E[a_{ik}] = m\frac{N_i}{N_T} \\
\sigma_A^2 &= E[A_k^2] - E^2[A_k] \\
&= E\left[\left(\sum_{i=1}^{m} a_{ik}\right)\left(\sum_{j=1}^{m} a_{jk}\right)\right] - \mu_A^2 \\
&= \sum_{i=1}^{m}\sum_{j=1}^{m} E[a_{ik}a_{jk}] - \mu_A^2 \\
&= m(m-1)\frac{N_{\text{close}}}{N_T} + m\frac{N_i}{N_T} - m^2\frac{N_i^2}{N_T^2},
\end{aligned}
$$

where expectation is taken across the time index k. The assumptions leading to these equations are equivalent to saying that all of the cells recorded by a single electrode have the same firing rate, and the normalized cross-correlation between any two of the single-unit spike trains recorded by a single electrode has the same value, ρ_{close}.

If we further assume that the correlation coefficient between the spike trains of any two cells recorded on different electrodes has the same value, ρ_{distant}, and that the average firing rate of all of the cells recorded on both electrodes is the same, we can find that the covariance of the two multiunit spike trains is

$$
\begin{aligned}
C_{AB} &= E[A_k B_k] - E[A_k]E[B_k] \\
&= E\left[\left(\sum_{i=1}^{m} a_{ik}\right)\left(\sum_{j=1}^{m} b_{jk}\right)\right] - \mu_A\mu_B \\
&= \sum_{i=1}^{m}\sum_{j=1}^{m} E[a_{ik}b_{jk}] - \mu_A\mu_B \\
&= m^2\frac{N_{\text{distant}}}{N_T} - m^2\frac{N_i^2}{N_T^2}.
\end{aligned}
$$

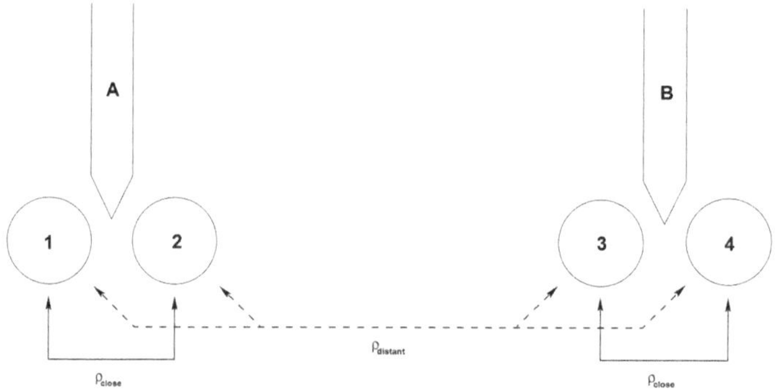

Figure 1: Analytical paradigm. The analysis considers the cross-correlation between two multiunit spike trains recorded by electrode A (recording cells 1 and 2) and electrode B (recording cells 3 and 4). The analysis considers the situation in which all spike trains recorded on a single electrode have one value of normalized cross-correlation, (ρ_{close}, solid lines) and spike trains recorded on different electrodes have another value of NCC (ρ_{distant}, dashed lines).

Normalizing by the geometric mean of the variance of the two multiunit spike trains and canceling terms, we find that the normalized cross-correlation (correlation coefficient) function for the multiunit spike trains is

$$\begin{aligned}\rho_{AB} &= \frac{C_{AB}}{\sqrt{\sigma_A^2 \sigma_B^2}} \\ &= \frac{m\rho_{\text{distant}}}{(m-1)\rho_{\text{close}} + 1}.\end{aligned}$$

Although it might appear that this expression for ρ_{AB} might range beyond the interval $[-1, 1]$, this possibility is excluded by assumptions that the correlation between any two cells recorded on one electrode is the same, and that the correlation between any two cells recorded on different electrodes is the same. Moreover, in the usual situation for neurophysiological recordings, with the number of spikes small compared to the total number of time bins, only positive and very small negative values for cross-correlation are possible.

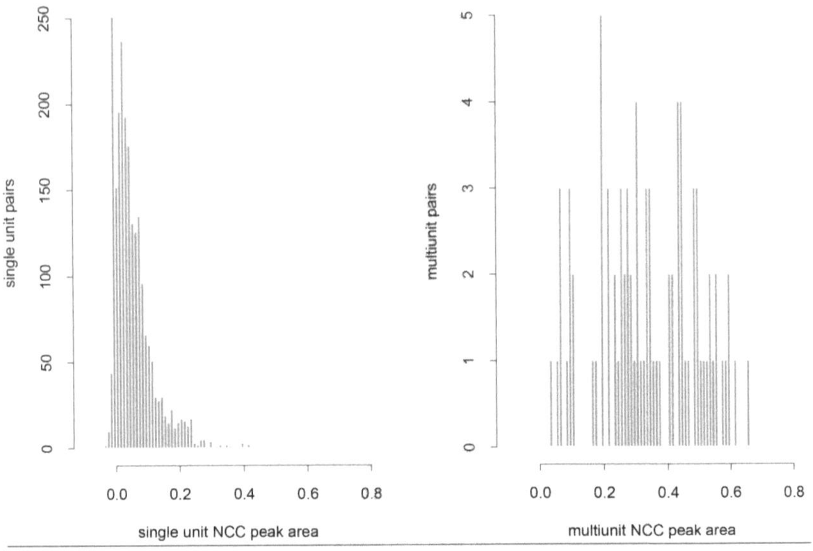

Figure 2: Single- and multiunit peak NCC (normalized cross correlation). The peak NCC for our population of pairs of single-unit spike trains is shown at the left. At the right, the peak NCC for the corresponding population of multiunit spike trains is shown. The multiunit correlation peaks are much larger.

3 Results

3.1 Experimental. Figure 2 shows the area under the central peak ($\tau = \pm 10$ msec, τ the time shift variable) of the normalized cross-correlation for our rat somatosensory system data set (Bedenbaugh, 1993; Bedenbaugh et al., 1997). The normalized cross-correlation for the multiunit spike trains is clearly larger than the normalized cross-correlation for the single-unit spike trains. Note that the scale on the ordinate is different for the two graphs; since only one multiunit spike train was obtained from each electrode, the data set included many more single-unit cross-correlograms.

3.2 Analytical. Figure 3 shows that for our model system, the multiunit normalized cross-correlation ($\rho_{\text{multiunit}}$) is a linear function of the NCC between the single-unit spike trains recorded on different electrodes, ρ_{distant}. The slope of this relationship is determined by the NCC between spike trains recorded on the same electrode. Counting the correlation between each of the cells recorded on one electrode and each cell recorded on the

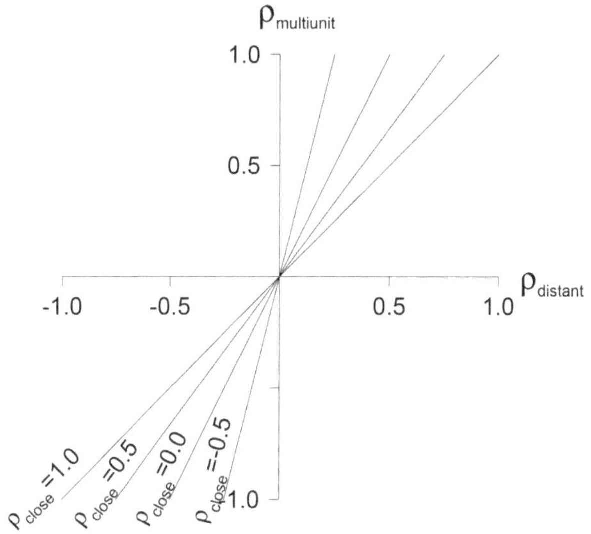

Figure 3: Multiunit NCC as a function of distant NCC ($m = 2$). The multiunit NCC, $\rho_{\text{multiunit}}$, is proportional to the distant NCC, ρ_{distant}. The slope is linearly proportional to m, the number of cells recorded on a single electrode, and is inversely related to the close NCC, ρ_{close}. Note that the minimum slope is 1, when $m = 1$ and $\rho_{\text{close}} = 1$.

other electrode effectively amplifies the multiunit NCC. Each spike train has multiple opportunities to contribute to the multiunit NCC.

Note that the linear dependence of $\rho_{\text{multiunit}}$ on ρ_{distant} with zero intercept implies that a difference in the sign of two $\rho_{\text{multiunit}}$ measurements may be unambiguously interpreted as arising from a difference in the sign of ρ_{distant}. This is also evident in the symmetry about the abscissa of $\rho_{\text{multiunit}}$ as a function of ρ_{close}, as seen in Figure 4.

The special case of $\rho_{\text{close}} = \rho_{\text{distant}} = \rho$ is illustrated in Figure 5. Note that the slope for negative correlation is steeper than for positive correlation, implying that in this circumstance multiunit NCC is a more sensitive detector of negative cross-correlation than of positive cross-correlation. The NCC is plotted as an inset in Figure 5 for the case of two cells recorded on each electrode. The special case of $\rho_{\text{close}} = \rho_{\text{distant}}$ courses a trajectory diagonally across this surface.

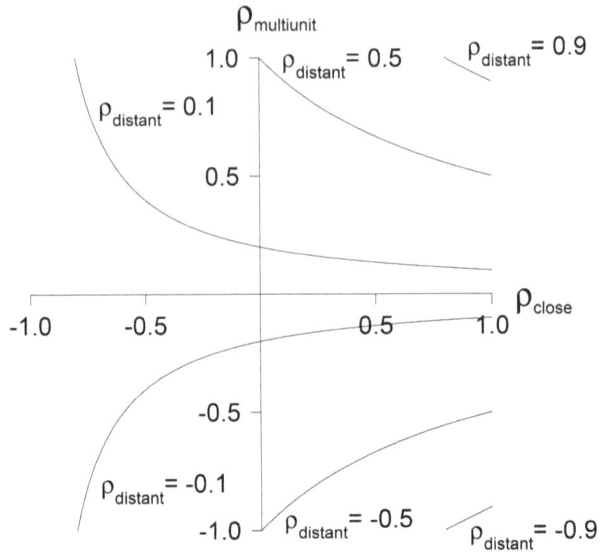

Figure 4: Multiunit NCC as a function of close NCC ($m = 2$). The multiunit NCC, $\rho_{\text{multiunit}}$, is a nonlinear function of close NCC, ρ_{close}. Note that in the situation we considered, positive values for ρ_{distant} lead to positive values for $\rho_{\text{multiunit}}$, and negative values for ρ_{distant} lead to negative values for $\rho_{\text{multiunit}}$.

3.3 Discussion. This study was inspired by the observation that the normalized cross-correlation (correlation coefficient) between multiunit spike trains recorded from the rat somatosensory cortex was significantly larger than the cross-correlation between the single-unit spike trains composing the multiunit recordings. Our analysis suggests why this should be so and suggests that multiple-unit cross-correlation is an intrinsically more sensitive detector of neuronal relationships than is the single-unit cross correlation.

Two technical points should be noted. First, the analysis here is valid for any bin in a normalized cross-correlation histogram, for both exact coincidences and delayed coincidences. Second, the analysis presented here does not take into account the refractory period of the recording system measuring the multiunit spike trains, making it incapable of resolving simultaneous events on a single electrode. This dead time deletes some nearly simultaneous spikes from different cells from the multiunit spike train, and

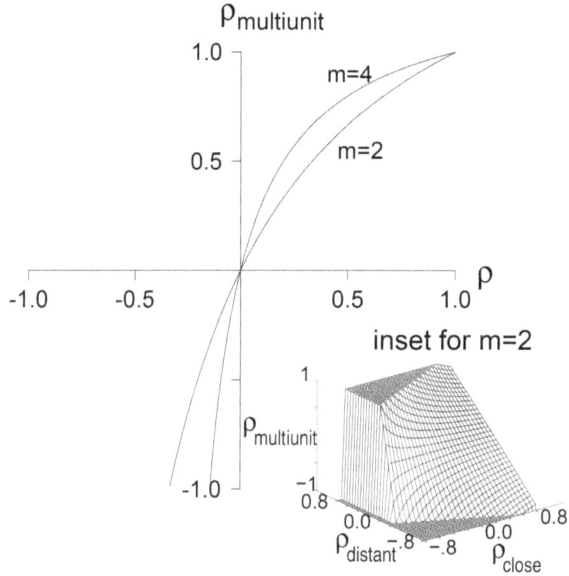

Figure 5: Multiunit NCC when all single-unit NCCs are equal. In the special case where the NCC between any two cells is the same, regardless of location ($\rho_{close} = \rho_{distant} = \rho$), $\rho_{multiunit}$ is a nonlinear function of ρ. Given our assumptions, it is an even more sensitive indicator of negative cross-correlation than of positive cross-correlation. Note, however, that the range of negative correlations that can satisfy the restriction of all recorded cells equally correlated is small. The inset plots the multiunit correlation as a surface parameterized by ρ_{close} and $\rho_{distant}$ for the case of two neurons recorded on each electrode. The large curve for $m = 2$ courses a trajectory diagonally across this surface.

so should reduce the sensitivity of the multiunit cross-correlation to correlations between spike trains recorded on a single electrode.

This analysis considered the case with all observed neurons having the same, constant firing rate. If firing rate changes slowly, these results should still apply. If some of the neurons within a multiunit cluster fire at different rates, we expect less dependence of cross-correlation on the number of neurons in a cluster.

For this model system, the multiunit NCC is larger than the single-unit cross-correlation between cells recorded on different electrodes. This means that although the NCC between two multiunit spike trains is related to the NCC between the individual spike trains recorded by the two electrodes, it is not an unambiguous quantitative measure of that relationship. For example, if we take $\rho_{\text{close}} = 0$, the multiunit NCC increases linearly with the number of cells comprising the multiunit spike trains, so that a change in recruitment may masquerade as a change in multiunit cross-correlation. More generally, a change in multiunit NCC can also result from either a change in the local NCC between cells recorded on a single electrode or the distant NCC between cells recorded on different electrodes, unless the change results in a sign change. These ambiguities limit the utility of comparisons among individual multiunit cross-correlograms. There is a place for comparing populations of multiunit cross-correlograms across treatment groups, though possible changes in the typical number of cells recorded due to the treatment would remain a concern.

On the other hand, multiunit NCC is a more sensitive indicator of a relationship between spike trains than is single-unit NCC, and if the recordings are stable and stationary, a change in multiunit cross-correlation implies some underlying change in the cross-correlation between the single-unit spike trains. Multiunit NCC may be especially useful for detecting negative cross-correlations, which have been observed more rarely than have positive cross-correlations.

The ambiguity with respect to the underlying correlation relationship and the difficulty of assessing the stationarity (uniform recruitment) and stability of multiunit recordings limit the conclusions that can be drawn by comparing multiunit cross-correlation measurements. With appropriate caution, the technical ease with which such measurements can be obtained, and their special sensitivity for detecting cross-correlation, make them useful tools for neuroscientists nonetheless.

Acknowledgments

A preliminary version of this work was presented as a poster at the 1993 annual meeting of the Society for Neuroscience (Bedenbaugh & Gerstein, 1993). It was supported by NIDCD F32-DC00144 and RO1-DC01249 and by NIMH MH R37-467428. Thanks to Michael Brecht and Heather Read for reading the manuscript. Thanks to Hagai Attias for helpful discussions. Portions of this work were performed in the laboratory of Michael M. Merzenich.

References

Aertsen, A. M. H. J., Gerstein, G. L., Habib, M. K., & Palm, G. (1989). Dynamics of neuronal firing correlation: Modulation of "effective connectivity." *Journal of Neurophysiology, 61* (5), 900–917.

Bedenbaugh, P. H. (1993). Plasticity in the rat somatosensory cortex induced by local microstimulation and theoretical investigations of information flow through neurons. Unpublished doctoral dissertation, University of Pennsylvania, Philadelphia.

Bedenbaugh, P. H., & Gerstein, G. L. (1993). Multiunit normalized cross correlation differs from the average single unit normalized correlation in rat somatosensory cortex. *Society for Neuroscience Abstracts, 19* (2), 1566.

Bedenbaugh, P. H., Gerstin, G. L., & Bolton, M. (1997). Intracortical microstimulation in rat somatosensory cortex. Receptive field plasticity and somatic stimulus dependent spike train correlations. Unpublished manuscript.

Gerstein, G. L., & Aertsen, A. M. H. J. (1985). Representation of cooperative firing activity among simultaneously recorded neurons. *Journal of Neurophysiology, 54* (6), 1513–1528.

Gerstein, G. L., Bedenbaugh, P., & Aertsen, A. M. H. J. (1989). Neuronal assemblies. *IEEE Transactions on Biomedical Engineering, 36* (1), 4–14.

Gerstein, G. L., Bloom, M. J., Espinosa, I. E. Evanczuk, S., & Turner, M. R. (1983). Design of a laboratory for multineuron studies. *IEEE Transactions on Systems, Man, and Cybernetics, SMC-13* (5), 668–676.

Gerstein, G. L., & Perkel, D. H. (1972). Mutual temporal relationships among neuronal spike trains. *Biophysical Journal, 12*, 453–473.

Gerstein, G. L., Perkel, D. H., & Dayhoff, J. E. (1985). Cooperative firing activity in simultaneously recorded populations of neurons: Detection and measurement. *Journal of Neuroscience, 5* (4), 881–889.

Hebb, D. O. (1949). *The organization of behavior.* New York: Wiley.

Kreiter, A. K., & Singer, W. (1995). Dynamic patterns of synchronization between neurons in area mt of awake macaque monkeys. *Society for Neuroscience Abstracts, 21* (2), 905.

Palm, G., Aertsen, A. M. H. J., & Gerstein, G. L. (1988). On the significance of correlations among neuronal spike trains. *Biological Cybernetics, 59*, 1–11.

Perkel, D. H. (1970) Spike trains as carriers of information. In F. O. Schmitt (Ed.), *The Neurosciences: A Second Study Program* (pp. 587–596). New York: Rockefeller University Press.

Perkel, D. H., Gerstein, G. L., & Moore, G. P. (1967a). Neuronal spike trains and stochastic point processes ii. simultaneous spike trains. *Biophysical Journal, 7* (4), 419–440.

Perkel, D. H., Gerstein, G. L., & Moore, G. P. (1967b). Neuronal spike trains and stochastic point processes i. The single spike train. *Biophysical Journal, 7* (4), 391–418.

Rosenberg, J. R., Amjad, A. M., Breeze, P., Brillinger, D. R., & Halliday, D. M. (1989). The Fourier approach to the identification of functional coupling between neuronal spike trains. *Progress in Biophysics and Molecular Biology, 53*, 1–31.

Received May 14, 1996; accepted November 12, 1996.

ELSEVIER

Journal of Neuroscience Methods 100 (2000) 41–51

JOURNAL OF NEUROSCIENCE METHODS

www.elsevier.com/locate/jneumeth

Cross-correlation measures of unresolved multi-neuron recordings

George L. Gerstein *

Department of Neuroscience, University of Pennsylvania, A306 Richards Building, 3700 Hamilton Walk, Philadelphia, PA 19104-6085, USA

Received 8 November 1999; received in revised form 17 April 2000; accepted 18 April 2000

Abstract

An increasing number of laboratories are studying population properties of the nervous system using data where the spike activity of more than one neuron is recorded on each electrode and where, accidentally or deliberately, these activities are not resolved into single unit spike trains. We have previously examined the consequences for measurement of cross-correlation between two such electrodes in the limited case where all individual distant (between electrode) correlations are the same and all individual close (on a single electrode) correlations are the same [Bedenbaugh, P.H., and Gerstein, G.L. (1997). Multiunit normalized cross correlation differs from the average single-unit normalized correlation. Neural Computation 9, 1265–1275]. Here, we lift these unrealistic restrictions to allow all values of individual correlation, and examine explicitly the cases of two or three unresolved neurons on each electrode. In these situations, the cross-correlation coefficient measured between the electrodes is a linear sum of the distant correlations, divided by a non-linear function of the close correlations. We then examine in detail the case of a single direct distant correlation and take account of all relevant indirect correlations. The measured interelectrode correlation shows a reduction of this actual distant correlation by a non-linear function of the close correlations on each electrode over most of their possible values. Finally, we examine the consequences of poor waveform sorting for correlation measures; here a supposedly isolated spike train is contaminated by some fraction of the activity of another train, a situation that unfortunately is all too common in experiments. All these distortions become far more serious in the more realistic situation of dynamic firing rates and correlations. This paper is intended as a cautionary note for those who want to draw inferences about neuronal organization and/or coding or representation by using cross-correlation analysis of unresolved recordings. © 2000 Elsevier Science B.V. All rights reserved.

Keywords: Multi-neuron; Extracellular spikes; Waveform sorting; Cross-correlation

1. Introduction

After a lengthy era of single neuron recordings, an increasing number of laboratories are moving towards study of neuronal assemblies and associated population codes. It has become increasingly clear that computation in the nervous system involves populations of neurons as originally proposed by Hebb (1949), and that it is necessary to study the relations among neurons, particularly in dynamic in vivo situations like behavior. Such work requires the simultaneous observation of activity in many single neurons and requires a new level of technology. Initial investigations of this type, some 30 years ago, made use of sharp electrodes that easily allowed isolation of the signal from a single neuron. However, the yield per electrode was low, and experimental access to assembly processes was extremely inefficient and time-consuming. The same time period saw the development of mathematical tools appropriate for assessing timing relationships among spike trains in various contexts. Many of these methods are based on cross-correlation (Perkel et al., 1967), and its various generalizations (Gerstein and Perkel, 1972; Gerstein and Aertsen, 1985; Palm et al., 1988; Aertsen et al., 1989) in the time domain and (Perkel, 1970; Rosenberg et al., 1989) in the frequency domain. Additional material on various experimental and analytic methods for assembly studies is in Nicolelis et al. (1998).

More recently, in an effort to increase experimental efficiency, instead of sharp metal or glass electrodes some number of wire electrodes have been introduced, singly or in combinations like sterotrodes or tetrodes

* Tel.: + 1-215-8988752; fax: + 1-215-5735851.
E-mail address: george@mulab.physiol.upenn.edu (G.L. Gerstein).

0165-0270/00/$ - see front matter
PII: S0165-0270(00)00226-0

(McNaughton et al., 1983; Wilson and McNaughton, 1993; Gray et al., 1995). In these circumstances the signals from more than one neuron will generally appear on each electrode, and it is necessary to employ spike shape sorting methodology to sort out the activities of each neuron. A wide variety of algorithms expressed both in hardware and software has been developed for this purpose; a recent review is in Lewicki (1998). Although all wire electrodes are subject to sorting uncertainties, it was thought that the additional information available from stereotrodes and tetrodes would make vast improvements. Unfortunately considerable residual sorting problems remain, as demonstrated in Wehr et al. (1999), Harris et al. (2000). The intrinsic sorting uncertainties as well as the difficulties in controlling such technologies especially in the pressure of the on-line situation during experiments, all too often lead to the unintended result of poorly separated activities i.e. unresolved multiple neuron recordings.

In addition, given the technical difficulties and combinatorial explosion of neuron pairs, many investigators have made a deliberate choice to use unresolved recordings. The rationale here is that the properties of the population or assembly should be similar to those of the single neurons of which it is composed, i.e. that the response properties are homogeneous. Thus it is increasingly common, sometimes deliberately, sometimes unwittingly, to encounter the use of various cross-correlational and other analytical tools on unresolved neural activity.

Some time ago, we examined cross-correlational measures in the context of unresolved multi-neuron recordings, but with very restrictive and unrealistic assumptions about the nature of activity (Bedenbaugh and Gerstein, 1997) The purpose of the present paper is to extend these results to a number of more general and realistic cases, and to sound a cautionary note about the inferences that may be drawn from cross-correlation based methods applied to such data.

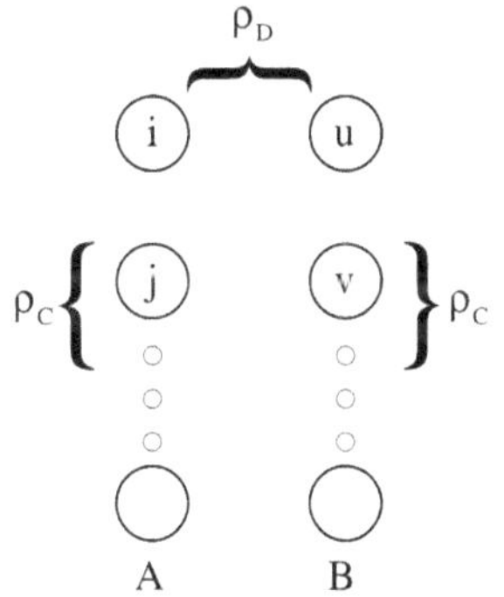

Fig. 1. Arrangement of neurons recorded, respectively, by electrodes A and B. The ρ_D represent correlations between neurons recorded on different electrodes, the ρ_C represent correlations between neurons recorded (and not resolved) on the same electrode.

2. Basic formulation

Suppose we are recording from two electrodes, and that on each, there are some number m of superimposed spike trains. This situation is shown in Fig. 1; the two electrodes are labeled A and B, the neurons recorded on A are i, j, ..., their spike trains are a_j, a_j, ... and those recorded on B are u, v, ..., with spike trains b_u, b_v, ... The correlation coeffcient between spike trains from distant neurons (i.e. one neuron on each electrode) will be called ρ_D with the appropriate further subscripts to identify the neurons; for the correlation coefficient between spike trains from close neurons (i.e. both neurons on the same electrode) we will use ρ_C. The individual spike trains are binned so that we have a 0, 1 process for each neuron; this means that for each neuron's spike train, the expected (average) counts per bin is the same as the expected counts squared per bin, i.e. $E(a_s) = E(a_s^2)$, where s is the neuron identity. There will be a total of N_T bins in the data, and the spikes of each neuron will occupy some N_s of these bins. The composite spike trains from more than one neuron seen on each electrode are of course not 0, 1 processes and there will be bins containing spikes from two (or more) neurons. The number of bins which contains spikes from two particular neurons, will be N_{st}, where s and t identify the neurons. (We assume that practical matters like waveform overlap in a sorting process do not affect these counts).

With these conventions, if m unresolved neurons are recorded on each electrode, and with the further, rather unrealistic assumption that all neurons have the same firing rates, that all close correlation coefficients ρ_C are identical and that all distant correlation coefficients ρ_D are identical, we have previously shown that,

$$\rho_{AB} = \frac{m\rho_D}{(m-1)\rho_C + 1} \tag{1}$$

We note that this is a linear function of the ρ_D but a non-linear function of the ρ_C. A more detailed description and graphs are in Bedenbaugh and Gerstein (1997).

In the present paper, we lift the restrictions on firing rates and correlation coefficients so as to further examine the effects of unresolved spike trains on measurements of correlation. We first examine the case of two unresolved spike trains on each electrode. All naming conventions are as above, and the circuit is shown in Fig. 2.

G.L. Gerstein / Journal of Neuroscience Methods 100 (2000) 41–51 43

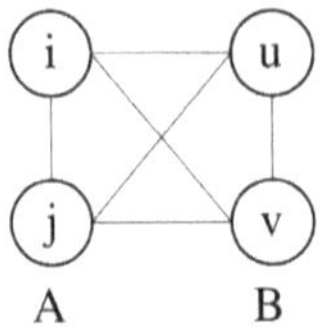

Fig. 2. The general case with two unresolved neurons on each of electrodes A and B. The lines between neurons indicate all possible correlations, close and distant.

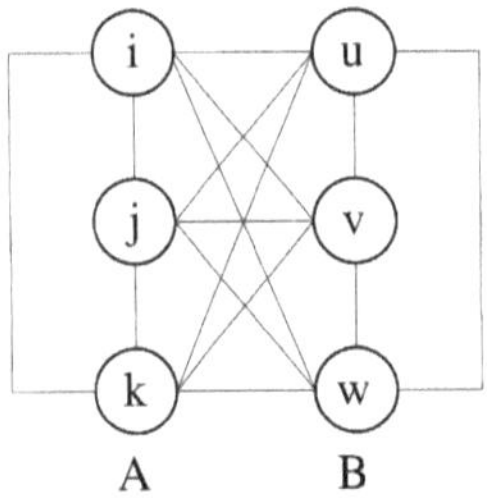

Fig. 3. The general case with three unresolved neurons on each of electrodes A and B. Lines indicate all possible correlations.

To start, the covariance between activities on the two electrodes A and B is,

$$\begin{aligned} C_{AB} &= E(AB) - E(A)E(B) \\ &= E[(a_i + a_j)(b_u + b_v)] - E(a_i + a_j)E(b_u + b_v) \end{aligned} \quad (2)$$

The covariance of the spike trains from any two neurons is

$$C_{Rst} = \frac{N_{st}}{N_T} - \frac{N_s N_t}{N_T^2} \quad (3)$$

where R will represent D (distant) or C (close) neurons and s,t represent the particular neurons under consideration. Using this in Eq. (2) we get,

$$C_{AB} = C_{Diu} + C_{Div} + C_{Dju} + C_{Djv} \quad (4)$$

For the correlation coefficient between electrodes we have

$$\rho_{AB} = \frac{C_{AB}}{\sqrt{\sigma_A^2 \sigma_B^2}} \quad (5)$$

To calculate the denominator of Eq. (5), we note that

$$\begin{aligned} \sigma_A^2 &= E(A^2) - [E(A)]^2 = 2\frac{N_{ij}}{N_T} + \frac{N_i}{N_T} + \frac{N_j}{N_T} - \left(\frac{N_i}{N_T} + \frac{N_j}{N_T}\right)^2 \\ &= 2C_{Cij} + \sigma_i^2 + \sigma_j^2 \end{aligned} \quad (6)$$

and similarly for σ_B^2. Thus we obtain,

$$\rho_{AB} = \frac{C_{Diu} + C_{Div} + C_{Dju} + C_{Djv}}{\sqrt{(2C_{Cij} + \sigma_i^2 + \sigma_j^2)(2C_{Cuv} + \sigma_u^2 + \sigma_v^2)}} \quad (7)$$

Each covariance can be expressed in terms of the corresponding correlation coefficient,

$$C_{Rst} = \rho_{Rst}\sqrt{\sigma_s^2 \sigma_t^2} \quad (8)$$

Again, R represents D (distant) or C (close) and s,t represent the neurons under consideration. Thus Eq. (7) (and Eq. (9) below) can equivalently be expressed in terms of correlation coefficients.

We make the analogous computation for three neurons on each electrode. Neurons i, j, k, are on electrode A and neurons u, v, w, on electrode B, as shown in Fig. 3. With the same notation as above, we obtain,

$$\rho_{AB} = \frac{\sum_{\substack{m=i,j,k \\ n=u,v,w}} C_{D_{mn}}}{\sqrt{\left(2\sum_{\substack{m,n=i,j,k \\ m=n}} C_{C_{mn}} + \sum_{m=i,j,k} \sigma_m^2\right)\left(2\sum_{\substack{m,n=u,v,w \\ m=n}} C_{C_{mn}} + \sum_{m=u,v,w} \sigma_m^2\right)}} \quad (9)$$

The individual quantities in Eqs. (7) and (9) represent the covariances and variances that would be observed if the corresponding two neurons were observed in situ but with perfectly resolved spike trains. The numerators in Eqs. (7) and (9) tell us that the observed correlation between two electrodes with unresolved spike trains is proportional to a linear sum of any distant (across the electrodes) correlations that are present. This means that, if several such distant correlations exist, the observed correlation between electrodes will be larger than for any single resolved recording within the data. As we have previously noted (Bedenbaugh and Gerstein, 1997), unresolved correlation between two electrodes will tend to be a more sensitive indicator than resolved correlation if several different distant individual correlations exist. However, the denominators in Eqs. (7) and (9) are a non-linear function of the individual close (within electrode) correlations, so that if these exist, they will tend to diminish the observed interelectrode correlation. As we demonstrate in the next section, if only a single direct distant correlation exists, then the close correlations will considerably diminish the observed interelectrode correlation to below that which would be observed with a well resolved recording.

3. Specific examples

In this section, we will examine some specific simplified cases of the general equations developed above. We reiterate that the quantities in Eqs. (7) and (9) are the correlations that would be observed if that particular pair of neurons was well isolated. Such correlation is the result of direct connections between the two neurons, but also depends on indirect connections (via other neurons). The latter depends on the specific details of the connectivity.

3.1. Rules for the indirect connectivity

1. If two neurons are (indirectly) connected by two (or more) direct connections in series, the effective correlation is the product of the individual direct correlations.
2. If two neurons share direct (divergent) input from third neuron, the effective correlation between them is the product of the individual direct correlations.
3. If there are two (or more) parallel pathways of connections between two neurons, the effective correlation is the sum of the individual direct correlations.
4. Two otherwise unrelated neurons whose output converges upon a third have zero effective correlation. Note that if one of these direct connections is reversed in direction, we have the situation of rule Eq. (1).

Thus, the indirect correlation quantities to be put into Eqs. (7) and (9) depend upon the assumptions made about the direct connections in the circuit. This, of course, is easy to do in the forward problem, when we assume a circuit and examine the predicted outcome for the unresolved correlation between the two electrodes. However, the backward problem of inferring a

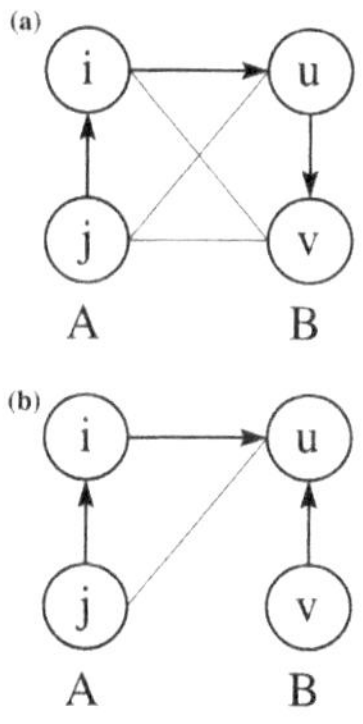

Fig. 4. Two simplified cases with two unresolved neurons on each electrode. In each case, there are direct close correlations but only a single direct distant correlation as indicated by bold lines and arrows. Indirect correlations which must be taken into account are indicated by thin lines.

circuit from the observed unresolved correlation is ambiguous.

3.2. Simplified cases of unresolved recordings

We now examine the cases of two and three unresolved neurons on each electrode so as to clarify some of the parametric relations. Consider the two simplified circuits of Fig. 4a and b. The direct connections are depicted by bold lines and arrows, the indirect connections by the thin lines. On each electrode, the two unresolved neurons are related by a different ρ_C on each electrode but there is only a single direct distant correlation ρ_{Diu}. Suppose that

$Ni = Nj \neq Nu = Nv$ and, therefore that

$$\sigma_i^2 = \sigma_j^2 \neq \sigma_u^2 = \sigma_v^2 \tag{10}$$

In other words, the individual (unresolved) neurons on each electrode have the same firing rates, but these differ on the two electrodes. In the situation of Fig. 4a, Eqs. (7) and (8) lead to

$$\rho_{AB} = \frac{\rho_{Diu}(1 + \rho_{Cij} + \rho_{Cuv} + \rho_{Cij}\rho_{Cuv})}{2\sqrt{(\rho_{Cij} + 1)(\rho_{Cuv} + 1)}} \tag{11}$$

With the circuit of Fig. 4b, the result is somewhat simpler, since there are no indirect correlations between neurons i and v or j and v because of the reversed direction of the uv connection and rule Eq. (4) above. This means only the first two terms in the numerator of Eq. (11) would be non zero.

For the two circuits of Fig. 4a and b, respectively, the correlation coefficient observed between the two electrodes, ρ_{AB}, is shown in Fig. 5a and b; we set the value of $\rho_{Diu} = 0.1$ and show ρ_{AB} as a function of ρ_{Cij} and ρ_{Cuv}. Note that the observed correlations are different non-linear functions of the close correlations in the two circuits, symmetric in Fig. 5a, unsymmetric in Fig. 5b, and that the observed correlations are always less than the 'real' known value of ρ_{Diu}.

A similar but more extreme situation develops with the simplified case of three unresolved neurons on each electrode. We will consider only one of the possible circuits as shown in Fig. 6, with all direct connections running clockwise. Again, we assume that there is only a single direct distant correlation ρ_{Diu}.

Again, we will take the three neurons on each electrode as having the same firing rate, but with different rates on the two electrodes. This means that

$Ni = Nj = Nk \neq Nu = Nv = Nw$ and therefore

$$\sigma_i^2 = \sigma_j^2 = \sigma_k^2 \neq \sigma_u^2 = \sigma_v^2 = \sigma_w^2 \tag{12}$$

We, then, have from Eqs. (8) and (9) a quite complicated expression that is relegated to the Appendix A.

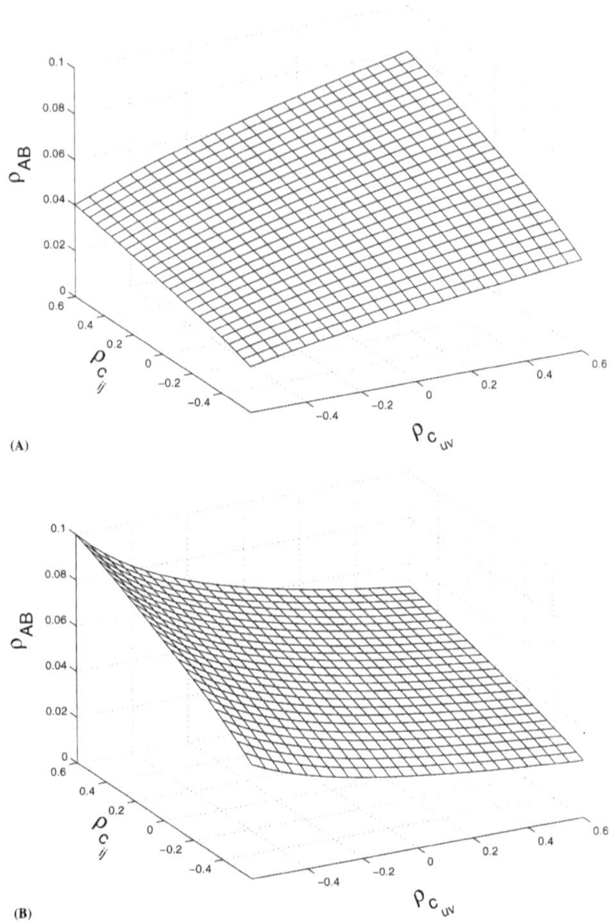

Fig. 5. The observed correlation coefficients between the two electrodes, respectively, for the situations in Fig. 4 as a function of the two close correlations. We have assumed that the single direct distant correlation coefficient has the value 0.1. Note that the observed correlation depends on the two close correlations, and is for all their indicated range below the actual value of the distant correlation.

The observed correlation coefficient ρ_{AB} between the two electrodes is shown in Fig. 7, again for the case that the 'real' direct distant correlation $\rho_{Diu} = 0.1$. Note again that the observed correlation is a non-linear function of the close correlations, and is much depressed below the 'real' value of ρ_{Diu} for all the ρ_C value range. Obviously, with these parameters, we are restricted to the $\rho_C \geq -0.5$.

3.3. *A simplified case of bad waveform sorting*

In all the above, we have assumed the simple unresolved situation, i.e. that all the spike trains present on each electrode are accepted as data. Here we will examine the case where a waveform sorter is used, but where it does an imperfect job of isolating the spike trains on each electrode. We will use a situation similar to that shown in Fig. 4a and the following assumptions.

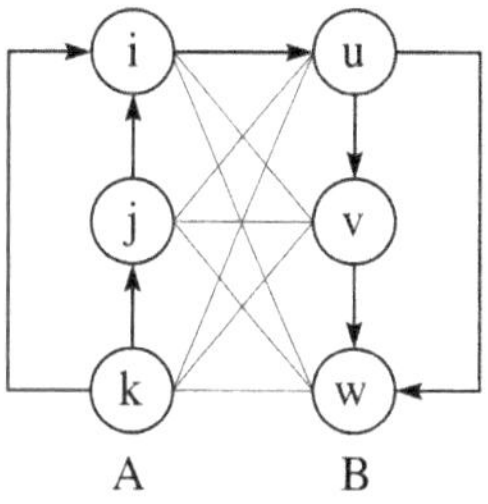

Fig. 6. A simplified case for three unresolved neurons on each electrode. Direct close correlations and the single direct distant correlation are indicated by bold lines and arrows. Indirect correlations are indicated by thin lines. (Indirect correlations between neurons i, k and between u, w are not shown, but are taken into account).

1. There is only a single direct distant correlation ρ_{Diu}.
2. Before sorting, both neurons on a given electrode have the same underlying firing rate. Firing rates and the close correlations ρ_C are different on the two electrodes.
3. The sorting process is imperfect so that on electrode A although all i spikes are accepted, there is an additional fraction f_A of the j spikes. Similarly on electrode B all u spikes are accepted, but there is an additional fraction f_B of the v spikes.
4. The contaminating spikes are randomly selected from the underlying spike train. This assumption may not be realistic for all types of sorting malfunctions.

Assumption Eq. (3) means that the effective number of bins (of the total number N_T occupied) by j spikes is $N_j' = f_A N_j$; similarly for v spikes $N_v' = f_B N_u$. As shown in the Appendix A, with the use of the appropriate form of Eq. (3), the effective variance for such a spike train is

$$\sigma_j^{2\prime} = f_A \sigma_j^2 + f_A(1 - f_A)\frac{N_j^2}{N_T^2} \tag{13}$$

The second term on the right of Eq. (14) is much smaller than the first, and we will ignore it in the subsequent calculations. With similar use of Eq. (3), we can show that the effective close covariance on electrode A is $C_{ij}' = f_A C_{ij}$, and comparably for electrode B using subscript u and v. Thus, for this case of imperfect spike sorting and the above assumptions, we will measure a cross-correlation between the two electrodes

$$\rho_{AB} = \frac{\rho_{Diu}(1 + f_A\rho_{Cij} + f_B\rho_{Cuv} + f_A f_B \rho_{Cij}\rho_{Cuv})}{\sqrt{(2f_A\rho_{Cij} + [1 + f_A])(2f_B\rho_{Cuv} + [1 + f_B])}} \tag{14}$$

Note that this reduces properly to Eq. (11) for $f_A = f_B = 1$ when all spikes on each electrode are accepted. In contrast, with $f_A = f_B = 0$, i.e. perfect sorting, we will measure the actual direct distant correlation between well-isolated spike trains i and u.

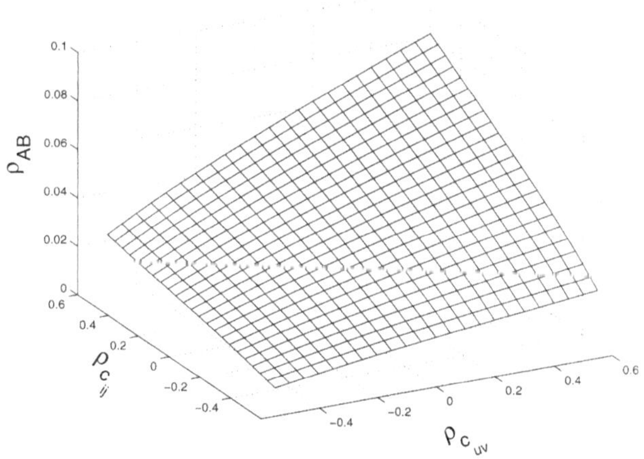

Fig. 7. The observed correlation coefficient between the two electrodes for the situation in Fig. 6 as a function of the close correlations. We have assumed that the single direct distant correlation coefficient has the value 0.1. All individual firing rates and close correlations on a given electrode are assumed to be the same. Note the increased non-linearity compared with Fig. 5a and b in the extent to which the observed correlation depends on the values of the close correlations.

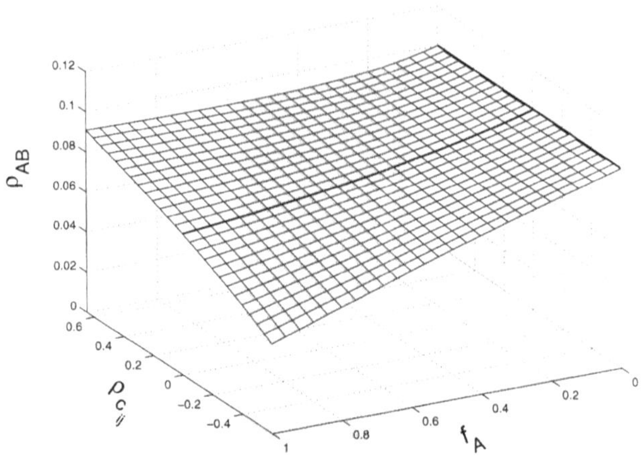

Fig. 8. The observed correlation coefficient between the two electrodes in the situation of Fig. 4a, with imperfect sorting on electrode A, but perfect sorting on electrode B. f_A is the fraction of j spikes erroneously accepted on A; ρ_{Cij} is the direct (close) correlation between the two neurons on electrode A which be measured if both spike trains were fully and correctly sorted out. As before, there is only a single direct distant correlation ρ_{Duv} with a value of 0.1 shown by the bold contour line at right. There is another bold line marking $\rho_{Cij} = 0$; note that even for this value the apparent cross-correlation between electrodes is reduced from its 'true' value.

In order to examine the implications of Eq. (14) graphically, we make the further assumption that the sorting on electrode B has been ideal, with no contamination. For this situation, Fig. 8 shows the measured correlation coefficient between electrodes as a (non-linear) function of the sorting qualify f_A and the close correlation ρ_{Cij} between spike trains on electrode A. As in previous examples, the value of the single direct distant correlation is $\rho_{Diu} = 0.1$, and this value is indicated as a bold contour line at right of the surface in Fig. 8. Over most of the range of sorting quality and close correlation, the measured correlation coefficient ρ_{AB} is lower than the actual direct distant correlation ρ_{Diu}. Even if there is no close correlation (the middle bold line for $\rho_{Cij} = 0$) there is a reduction of the measured correlation; we might call this effect a 'dilution' by bad sorting.

3.4. *Some practical examples*

In this section, by examining cross-correlograms, we will demonstrate the practical consequences of rather dry equations and parametric graphs of the preceding portions of the paper. In order to have available spike trains with well-defined relationships, we used a simple simulator. Four independent and uncorrelated spike trains, each with an interval density described by a Gamma function of order four were generated. The desired correlations were produced by making probabilistic, delayed and temporally noisy copies from a 'pre synaptic' to a 'post synaptic' train. If a spike, copied in this manner, fell within 6 ms of an existing spike in the target train, the copy was aborted. Therefore, the actual number of spikes copied was lower than specified by the copy probability. This kind of simulator is, of course, completely unrelated to physiology, but produces spike trains with appropriate and known characteristics.

Spike trains 1 and 3 were assigned to electrode A, trains 2 and 4 to electrode B. All spike trains had a rate of 20 per s. Correlation across electrodes (only between trains 1 and 2) was produced with a copy probability of 0.2 (effectively about 0.15 because of the restriction of copies that fell closer than 6 ms to an existing spike), and a delay of 5 ± 3 ms. When needed, correlation for use on electrode A (between trains 3 and 1) was produced with the same copy probability and a delay of 3 ± 2 ms. Unresolved trains were produced as needed by superimposing the two trains assigned to each electrode.

Results of this exercise are shown in Fig. 9 and relate to variations of the circuit in Fig. 4a and the corresponding graph of Fig. 5a. The correlations between electrode A and electrode B for various situations are presented in the form of normalized cross-covariances, normalized so that the autocovariances at lag 0 are identically unity. This choice of normalization for the correlograms is closest to the formalism used in the rest of this paper, and the results can readily be compared with the graphical results. (A number of other correlogram normalizations can be used in the experimental situation, and appear frequently in the literature, but are not as useful for the present didactic purpose.)

The top panel of Fig. 9 shows the correlogram corresponding to the perfect, fully resolved situation, and would be the experimental ideal that most closely reflects the (in this case known) relations between the observed neurons. The area of the peak is about 0.15, in keeping with the effective copy probability used in the simulation. In the middle row, we show the results of unresolved activity on electrode A, on the left without any local correlation, on the right with such correlation. Without local correlation on electrode A, the admixture of the second spike train produces a 'dilution', and considerable drop in correlation (peak area). Note that (in agreement with Fig. 5a) the addition of local correlation increases the peak area relative to the pure 'dilution' effects. In the lower row of Fig. 9, on the left we see the results of resolved activity on electrode A and unresolved activity on electrode B (without local correlation). On the right, we have the results of unresolved activity on both electrodes, with local correlation on A, without on B. These panels mainly show the 'dilution' effects.

Thus, in summary, compared with resolved activity, we see that in correlograms the presence of unresolved but uncorrelated local activity on either (or both) electrodes 'dilutes' the apparent strength of the correlation. If in addition, there is local positive correlation, there is an increase of the apparent correlation strength, but not to the level of the resolved situation. Correlograms made with unresolved spike train data in the laboratory

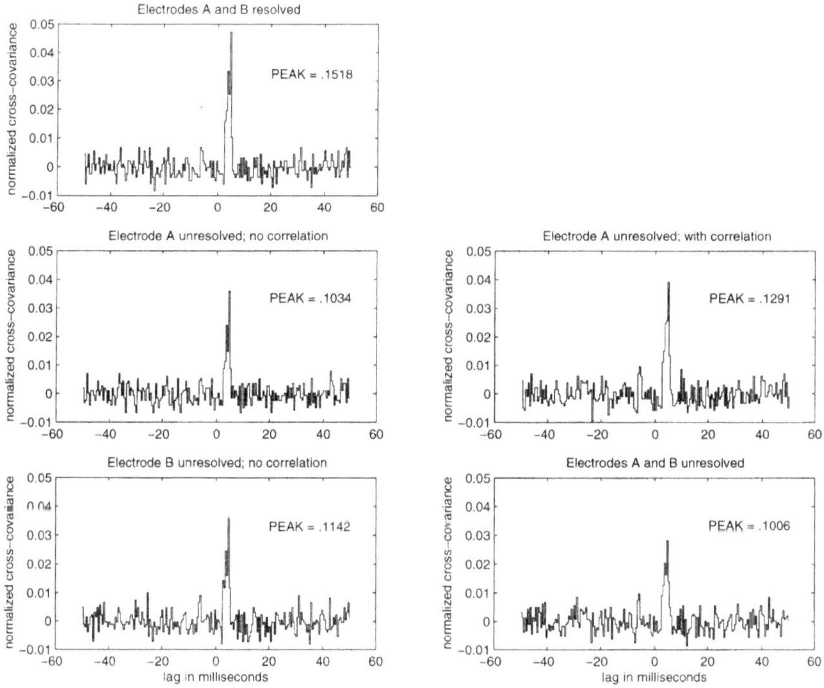

Fig. 9. Normalized cross-covariances for spike trains generated by a simulator mimicking the circuit arrangement in Fig. 4a. In each case the covariance is taken between the total spike trains 'observed', respectively, on electrodes A and B, with the data from electrode A as the 'reference event'. The labels on the individual panels describe the extent to which there are unresolved data on either electrode, and whether these additional spike trains are correlated. The bottom right panel for both electrodes unresolved has an additional spike train on each electrode; trains on electrode A are correlated; trains on electrode B are uncorrelated. See text.

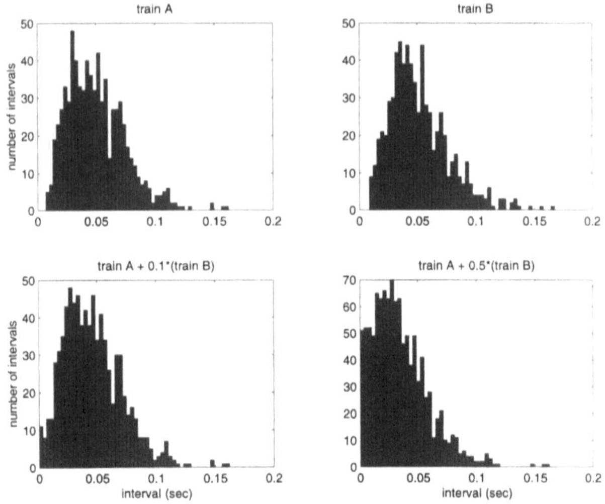

Fig. 10. Interval histograms to demonstrate of effects of spike train superposition in partially unresolved situations. Top row shows interval histograms of two individual independent spike trains from a simulator. These distributions are both Gamma order 4, and are uncorrelated. The bottom row shows interval histograms for, respectively, a 10 and a 50% superposition of the spike trains. Note the appearance of short intervals in the 'unresolved' histogram. Although the presence of short intervals below usual refractory periods signals the unresolved situation, the absence of such small intervals does not necessarily guarantee resolved data. See text.

where we are trying to infer the relations between the observed neurons are, therefore, subject to the dual distortions of 'dilution' and the additional positive or negative effects of local correlation. The strength of these effects depend in detail on the actual circuit among the observed neurons.

An easy experimental test of the extent of contamination in a spike train can be made with the use of the interval histogram. This is demonstrated in Fig. 10. The top panels show the interval histograms for two spike trains, again Gamma order 4 intervals, rate 20 per s. Note that there is an effective refractory time, so that short intervals are improbable. When spike train A is contaminated, on the left with 10% of train B, on the right by 50% of train B, the interval histogram shows many short intervals, and essentially an absence of a refractory period. Unfortunately, even when data pass this simple test, we cannot be absolutely sure that we have isolated a single spike train. We could still be observing two unresolved spike trains from neurons that have a strong mutual inhibitory connection so that they never fire in close temporal juxtaposition. This, however, is probably a rare situation, and the interval histogram test is extremely worth doing routinely.

4. Discussion

We have examined the observed correlation of spike trains as recorded on two electrodes in the situation where there are unresolved multiple neural activities on each electrode, and where all these spikes are accepted. We have treated these unresolved spike trains as contaminants of the 'real' neuronal relationships, and pointed out their considerable effects on the observed correlation. For the simplified situation that there is only a single distant neuronal interaction (as in all the graphs), the effect of accepting all spike of the unresolved neurons is to lower the observed correlation between electrodes. The exact amount of this reduction is a non-linear function of the correlation between close neurons as seen by each electrode. In all these situations, the effect on the interelectrode correlation depends also on the detailed circuit diagram, particularly through the indirect correlation paths. Thus, for example, the cases of Fig. 4a and b, each with two unresolved spike trains on the two electrodes, produced quite different effect of local correlation on the interelectrode correlation. The practical consequences on cross-correlograms of unresolved situations similar to Fig. 4a were demonstrated in Fig. 9, showing the effects of both the dilution and the local correlation. The effect of three unresolved trains on each electrode was shown in Figs. 6 and 7, and one should compare Fig. 5a and Fig. 7.

In the above situations have assumed full acceptance of all spike trains present on each electrode, i.e. multineuron recording. We have also, in Eqs. (13) and (14) and Appendix B, examined the effects of sorting quality, where one spike train is imperfectly isolated on each electrode, with some admixture of other spike trains that are present. A simplified example with variable sorting quality on electrode A and perfect sorting on electrode B was shown in Fig. 8. The circuit was like Fig. 4b, but without neuron v. Again there is a reduction of the observed correlation by a non-linear function of the sorting quality and the close correlations.

It is of interest to note, as demonstrated, that although close and distant correlations both affect the observed interelectrode correlation, they are not symmetric. If there is no distant correlation in an unresolved recording the local correlation will not affect the null interelectrode correlation. Similarly, if the recording is cleanly resolved, local correlations are irrelevant. Thus, we might say that the close correlations are 'lost' in some situations, which are dominated by the distant correlations.

Note that in all these calculations, we are solving the forward problem, i.e. given a set of circuit arrangements, correlations and specification of which spike trains are unresolved, we calculate the observed correlation between electrodes. The backward problem, i.e. the usual laboratory situation where we are trying to infer 'effective connectivity' between neurons from their spike trains, can only be solved uniquely with resolved spike trains.

All the above has been in completely steady state; we have assumed all neurons to have constant firing rates and correlations throughout, and that spike train acceptance is constant. In such a steady state experimental situation, since all values are fixed, the calculations shown in Fig. 5a and b, Figs. 7 and 8 for the simplified circuits shown in Fig. 4a and b, Fig. 6 show that the observed correlation between electrodes will be the 'real' direct distant correlation times a simple scale factor. One might be tempted to suggest this is of no great significance for drawing inferences about network properties.

However, considerable further problems arise if firing rates and correlations are made more realistic and generalized so that they vary in time. Note that such variations may be real, since the observed neurons are presumably participating in dynamic network activity during behavior or sensory stimulation. In addition, even in the case of constant context (repeated stimuli or behavior with repeated identical modulations of the neurons), there may be non-stationarities in the resulting neuronal activity arising from variables like attention or physiological state.

In addition, apparent temporal variations may also originate from technical difficulties with the recording. For example, if a moveable electrode structure creeps slightly, altering the observed spike amplitude and shape distribution, many spike-shaped sorting algorithms behave erratically. In the extreme, a given neuron may (temporarily) drop in or out of what is thought to be well-sorted output. Thus, what may have been a 'clean' and properly sorted set of spike trains may easily and dynamically change into the unresolved situations we have been examining here.

Note that a simple test of sorting quality is to examine the interval histogram of sorted data as was demonstrated in Fig. 10. If there is a sorting problem, short intervals will appear in the interval histogram. If on the other hand, there is little contamination from other neurons, an interval histogram will show no intervals below a few milliseconds, corresponding to the refractory period. (The latter outcome may still conceal problems. Two unresolved but completely anti-correlated neurons would meet the test, but it is likely that this is a very rare case. Further, erroneous misses in spike identification are still possible even if the interval histogram test is met.)

The result of temporal variations, real or technical, in an unresolved recording situation is obvious from Figs. 5, 7 and 8. For constant 'real' ρ_D the point representing observed correlation between electrodes A and B will skitter about on the calculated surfaces depending on variations in the ρ_C and/or the sorting quality f. In addition, with temporal variations in the ρ_D, the entire surface will move up and down. Thus, the observed correlation between the electrodes will represent some unknown mixture of the dynamics of actual long-range correlations with contamination by the dynamics of local correlations and of sorting quality.

Dynamic correlation of activity among neurons is usually interpreted either in terms of changing effective connectivity or of changing coding or representation of sensory or internal variables. As demonstrated in this paper, in the unresolved recording situation, interpretation of neuronal correlation and of temporal structure in such correlation are extremely ambiguous, and should be undertaken only with great caution.

Appendix A

The equation corresponding to the circuit of Fig. 6 and depicted in Fig. 7. This is a particular instance of general Eq. (9). There are one direct (ρ_{Diu}) and eight indirect terms for the numerator. We use the notation 'rs' for each term, r and s denoting the neurons under consideration. Using rules (1–4) for indirect connections, and since all close direct connections on an electrode have the same value, the 'rs' are

$$\mathrm{iu} = \rho_{\mathrm{Diu}}$$
$$\mathrm{iv} = \rho_{\mathrm{Diu}}\rho_{\mathrm{Cuv}}$$
$$\mathrm{iw} = \rho_{\mathrm{Diu}}(\rho_{\mathrm{Cuv}} + \rho^2_{\mathrm{Cuv}})$$
$$\mathrm{ju} = \rho_{\mathrm{Diu}}\rho_{\mathrm{Cij}}$$
$$\mathrm{jv} = \rho_{\mathrm{Diu}}\rho_{\mathrm{Cij}}\rho_{\mathrm{Cuv}}$$
$$\mathrm{jw} = \rho_{\mathrm{Diu}}\rho_{\mathrm{Cuv}}(\rho_{\mathrm{Cij}} + \rho^2_{\mathrm{Cij}})$$
$$\mathrm{ku} = \rho_{\mathrm{Diu}}(\rho_{\mathrm{Cij}} + \rho^2_{\mathrm{Cij}})$$
$$\mathrm{kv} = \rho_{\mathrm{Diu}}\rho_{\mathrm{Cij}}(\rho_{\mathrm{Cuv}} + \rho^2_{\mathrm{Cuv}})$$
$$\mathrm{kw} = \rho_{\mathrm{Diu}}(\rho_{\mathrm{Cij}} + \rho^2_{\mathrm{Cij}})(\rho_{\mathrm{Cuv}} + \rho^2_{\mathrm{Cuv}}) \quad \text{(A.1)}$$

For the denominator, we have for direct and indirect connections

$$\mathrm{ij} = \rho_{\mathrm{ij}} + \rho^2_{\mathrm{ij}}$$
$$\mathrm{ik} = \rho_{\mathrm{ij}} + \rho^2_{\mathrm{ij}}$$
$$\mathrm{jk} = \rho_{\mathrm{ij}}$$
$$\mathrm{uv} = \rho_{\mathrm{uv}}$$
$$\mathrm{uw} = \rho_{\mathrm{uv}} + \rho^2_{\mathrm{uv}}$$
$$\mathrm{vw} = \rho_{\mathrm{uv}} + \rho^2_{\mathrm{uv}} \quad \text{(A.2)}$$

Thus Eq. (9) becomes

$$\rho_{\mathrm{AB}} = \frac{(\mathrm{iu}+\mathrm{iv}+\mathrm{iw}+\mathrm{ju}+\mathrm{jv}+\mathrm{jw}+\mathrm{ku}+\mathrm{kv}+\mathrm{kw})}{\sqrt{(2[\mathrm{ij}+\mathrm{ik}+\mathrm{jk}]+3)(2[\mathrm{uv}+\mathrm{uw}+\mathrm{vw}]+3)}} \quad \text{(A.3)}$$

Appendix B

Derivation of Eq. (13) for the poor sorting situation. The fraction of spikes from neuron j on electrode A that are observed is f_{A}. The number of such spikes is $N'_{\mathrm{j}} = f_{\mathrm{A}}N_{\mathrm{j}}$, where N_{j} is the full actual number of j spikes. When binned this is still a 0, 1 process so that, as before, $E(N'^2_{\mathrm{j}}) = E(N'_{\mathrm{j}})$.

The variance for this spike train is:

$$\sigma^{\ell}_{\mathrm{j}} = \frac{f_{\mathrm{A}}N_{\mathrm{j}}}{N_{\mathrm{T}}} - \frac{(f_{\mathrm{A}}N_{\mathrm{j}})^2}{N^2_{\mathrm{T}}} \quad \text{(B.1)}$$

$$\sigma^{\ell}_{\mathrm{j}} = f_{\mathrm{A}}\left[\frac{N_{\mathrm{j}}}{N_{\mathrm{T}}} - \frac{f_{\mathrm{A}}N^2_{\mathrm{j}}}{N^2_{\mathrm{T}}} + \frac{N^2_{\mathrm{j}}}{N^2_{\mathrm{T}}}\frac{N^2_{\mathrm{j}}}{N^2_{\mathrm{T}}}\right] \quad \text{(B.2)}$$

Collecting the first and fourth terms, we have

$$\sigma^{\ell}_{\mathrm{j}} = f_{\mathrm{A}}\sigma^2_{\mathrm{j}} + f_{\mathrm{A}}\left[\frac{N^2_{\mathrm{j}}}{N^2_{\mathrm{T}}} - \frac{f_{\mathrm{A}}N^2_{\mathrm{j}}}{N^2_{\mathrm{T}}}\right] \quad \text{(B.3)}$$

And finally

$$\sigma^{\ell}_{\mathrm{j}} = f_{\mathrm{A}}\sigma^2_{\mathrm{j}} + f_{\mathrm{A}}[1 - f_{\mathrm{A}}]\frac{N^2_{\mathrm{j}}}{N^2_{\mathrm{T}}} \quad \text{(B.4)}$$

References

Aertsen AHMJ, Gerstein GL, Habib MK, Palm G. Dynamics of neuronal firing correlation: modulation of 'effective connectivity'. J Neurophysiol 1989;61:900–17.

Bedenbaugh PH, Gerstein GL. Multiunit normalized cross correlation differs from the average single-unit normalized correlation. Neural Computation 1997;9:1265–75.

Gerstein GL, Perkel DH. Mutual temporal relationships among neuronal spike trains. Biophys J 1972;12:453–73.

Gerstein GL, Aertsen AMHJ. Representation of cooperative firing activity among simultaneously recorded neurons. J Neurophysiol 1985;54:1513–28.

Gray CM, Maldonado PE, Wilson M, McNaughton B. Tetrodes markedly improve the reliability and yield of multiple single-unit isolation from multiunit recordings in cat striate cortex. J Neurosci Methods 1995;63:43–54.

Harris, K.D., Henze, D.A., Csicsvari, J., Hirase, H., & Buzasaki, G. (2000). The accuracy of tetrode spike separation as determined by simultaneous intracellular and extracellular measurements. *J. Neurophysiol.*, in press.

Hebb DO. The Organization of Behavior. New York: Wiley, 1949.

Lewicki MS. A review of methods for spike sorting: the detection and classification of neural action potentials. Netw Comput Neural Syst 1998;9:53–78.

McNaughton BL, O'keefe J, Barnes CA. The stereotrode: a new technique for simultaneous isolation of several single units in the central nervous system from multiple unit records. J Neurosci Methods 1983;8:391–7.

Nicolelis M, Simon S, Corless J, editors. Methods for simultaneous neuronal ensemble recordings. Boca Raton, FL: CRC Press, 1998.

Palm G, Aertsen AMHJ, Gerstein GL. On the significance of correlations among neuronal spike trains. Biol Cybernetics 1988;59:1–11.

Perkel DH. Spike trains as carriers of information. In: Schmitt FO, editor. The Neurosciences: A Second Study Program. New York: Rockefeller University Press, 1970:587–96.

Perkel DH, Gerstein GL, Moore GP. Neuronal spike trains and stochastic point processes II. Simultaneous spike trains. Biophys J 1967;7:419–40.

Rosenberg JR, Amjad AM, Breeze P, Brillinger DR, Halliday DM. The Fourier approach to the identification of functional coupling between neuronal spike trains. Prog Biophys Mol Biol 1989;53:1–31.

Wehr M, Pezaris JS, Sahami M. Simultaneous paired intracellular and tetrode recordings for evaluating the performance of spike sorting algorithms. Neurocomputing 1999;26–27:1061–8.

Wilson MA, McNaughton BL. Dynamics of hippocampal ensemble code for space. Science 1993;261:1055–8.

Chapter 11
Plasticity in the Receptive Fields by Microstimulation

My Experience Working with George L. Gerstein

Pedro E. Maldonado

Throughout George Gerstein's scientific career, the concept of neuronal plasticity was always a recurrent theme. George became interested in the topic of plasticity about the same time as the seminal work by Eric Kandel on the Aplysia, which was performed in the late 1960s. Together with William Kristan, George published in 1970 an article in science describing a study in Aplysia where they showed that electrical activity from simultaneously recorded neurons can be modified by electrical stimulation over a single nerve (Kristan Jr. and Gerstein 1970). Four years later, George wrote another paper (Gerstein 1974) where he re-introduced his newly developed Joint-PSTH as an analytical tool to examine the plasticity changes in these neuronal systems.

Sixteen years later, Jing Xing who arrived from China and me coming from Chile entered the physiology Ph.D. program at Penn and joined George's "Mulab." As many of you may remember, that was the Internet domain address for George's Multiunit Laboratory. I got strongly attracted by George's teachings where he emphasized the idea of brain functionality in terms of neuronal assemblies. His insistence that many neurons must be recorded at a time to make sense of neuronal operations was novel and new to all my generation, in part because at this time the neuron doctrine (Barlow 1972) was still on a full-blown mode. I was impressed by George's out-of-the-box thinking and his impressive tools to analyze neuronal data. So together with Purvis Bedenbaugh, we worked on our thesis projects. George

P. E. Maldonado (✉)
Department of Neuroscience and BNI, Faculty of Medicine, Universidad de Chile, Santiago, Chile
e-mail: pedro@uchile.cl

A. Aertsen et al. (eds.), *Introducing Computation to Neuroscience*, Springer Series in Computational Neuroscience, https://doi.org/10.1007/978-3-030-87447-6_11

proposed for Jing a modeling study in plasticity, which fitted well her skills in computational neuroscience.

Jing ended up co-authoring five papers with George. In the field of plasticity, they employed a three-layered network model to study basic features of neurons, while the synaptic strengths of lateral excitatory connections were made modifiable by a Hebbian learning rule and a normalization process (Xing and Gerstein 1996a). They concluded that neurons in the cortical network naturally assembled into functional groups, which were highly dynamic in their responses and receptive field properties. In another paper (Xing and Gerstein 1996b), they modeled plasticity using a three-layered neural network model with neuronal groups that were already formed in the cortical layer and simulated then changes in receptive field properties induced by behavioral training or intracortical microstimulation. They showed that their network model was capable of reproducing most characteristics of experiments on cortical reorganization.

In my thesis dissertation, George and I decided to explore the consequences of neural plasticity by electrical intracortical microstimulation (ICMS). By that time, Gregg Recanzone and Michael Merzenich were performing similar experiments in the somatosensory cortex of rats and did show that intracortical microstimulation when applied to a single site in the middle somatosensory cortex layers resulted in an enlarged cortical representation of a restricted skin region spanning several hundred microns. In our laboratory, we studied plasticity onto the receptive field of the auditory cortex of the rat. We found that ICMS induced fast changes in the cortical map and in the receptive field properties of cells at the electrically stimulated and adjacent electrodes. As a result, there was an enlargement of the cortical domain tuned to the acoustic frequency that had been represented at the stimulating electrode. This study is also reproduced here (Maldonado and Gerstein 1996a) with an accompanying paper on neuronal dynamics (Maldonado and Gerstein 1996b). We were also interested in assembly properties in the rat's auditory cortex during plasticity changes. We found that cortical neurons exhibited highly synchronous oscillatory firing patterns that were enhanced by ICMS. Cross-correlation studies between two spike trains showed that statistically significant correlations depended on the anatomical distance between the two neurons. ICMS changed the strength and the incidence of paired cross-correlations between the recorded neurons. Joint peri-stimulus analysis (Gerstein and Perkel 1969) and gravity analysis (Lindsey et al. 1992) showed that the correlation between the neuronal activities varied dynamically at several time scales. We showed that neuronal assemblies in the rat auditory cortex can be established transiently in time and that their membership is dynamic. These studies were followed up by two other studies in George's lab (Kisley and Gerstein 2001; Talwar and Gerstein 2001).

My stay at George's lab was intense both at technical as well at conceptual levels. On one hand, George was indeed a tinkerer. Because of my undergraduate work and former master thesis work with Francisco Varela and Humberto Maturana in Chile, I was already familiar with the need to become a do-it-yourself scientist. That experience was a very good prelude for working with George, who was able to comfortably work in building amplifiers, electrodes, play with surgical techniques, and implement computer programs in primitive and advanced computers. His lab was a wonderful collection of all sorts of paraphernalia, where you can always find something that can solve your technical problem. His experimental lab was located at the basement of the building (unlike the offices who had a proper set of windows), which enabled quite a setting that electrophysiologists learn to love. George was always on the lookout for new technological advances, and on one occasion, we traveled to Tucson, to learn about a new technique Bruce McNaughton and Matt Wilson had been developing. Thus, I learned about the tetrode (Gray et al. 1995), which was going to be a long-term companion in my research.

At the same time, George would engage in deep conceptual discussions about neuronal assemblies and dynamical properties of groups of cells where timing issues were as important as rate. While this may sound a quite familiar concept today, it was quite a novel concept among my generation. This vision was accompanied by George's constant search for statistical tools to analyze neuronal data, especially their timing properties. It has been fascinating to have experienced an early jump into this theoretical wagon. Looking backward, his early vision of how the brain functions in terms of neuronal assemblies and dynamical properties is arguably the most relevant of George Gerstein's contributions to neuroscience.

References

Barlow HB (1972) Single units and sensation: A neuron doctrine for perceptual psychology? Perception 1(4):371–394. https://doi.org/10.1068/p010371

Gerstein GL (1974) Plasticity in small neuronal assemblies. Acta Neurobiol Exp (Wars) 34(1):81–92

Gerstein GL, Perkel DH (1969) Simultaneously recorded trains of action potentials: Analysis and functional interpretation. Science 164(3881):828. https://doi.org/10.1126/science.164.3881.828

Gray CM, Maldonado PE, Wilson M, McNaughton B (1995) Tetrodes markedly improve the reliability and yield of multiple single-unit isolation from multi-unit recordings in cat striate cortex. J Neurosci Methods 63(1-2):43–54. https://doi.org/10.1016/0165-0270(95)00085-2

Kisley MA, Gerstein GL (2001) Daily variation and appetitive conditioning-induced plasticity of auditory cortex receptive fields. Eur J Neurosci 13(10):1993–2003. https://doi.org/10.1046/j.0953-816X.2001.01568.x

Kristan WB Jr, Gerstein GL (1970) Plasticity of synchronous activity in a small neural net. Science 169(3952):1336–1339. https://doi.org/10.1126/science.169.3952.1336

Lindsey BG, Hernandez YM, Morris KF, Shannon R, Gerstein GL (1992) Respiratory-related neural assemblies in the brain stem midline. J Neurophysiol 67(4):905–922. https://doi.org/10.1152/jn.1992.67.4.905

Maldonado PE, Gerstein GL (1996a) Reorganization in the auditory cortex of the rat induced by intracortical microstimulation: A multiple single-unit study. Exp Brain Res 112(3):420–430. https://doi.org/10.1007/BF00227948

Maldonado PE, Gerstein GL (1996b) Neuronal assembly dynamics in the rat auditory cortex during reorganization induced by intracortical microstimulation. Exp Brain Res 112(3):431–441. https://doi.org/10.1007/BF00227949

Talwar SK, Gerstein GL (2001) Reorganization in awake rat auditory cortex by local microstimulation and its effect on frequency-discrimination behavior. J Neurophysiol 86(4):1555–1572. https://doi.org/10.1152/jn.2001.86.4.1555

Xing J, Gerstein GL (1996a) Networks with lateral connectivity. II. Development of neuronal grouping and corresponding receptive field changes. J Neurophysiol 75(1):200–216. https://doi.org/10.1152/jn.1996.75.1.200

Xing J, Gerstein GL (1996b) Networks with lateral connectivity. III. Plasticity and reorganization of somatosensory cortex. J Neurophysiol 75(1):217–232. https://doi.org/10.1152/jn.1996.75.1.217

Exp Brain Res (1996) 112:420–430

RESEARCH ARTICLE

Pedro E. Maldonado · George L. Gerstein

Reorganization in the auditory cortex of the rat induced by intracortical microstimulation: a multiple single-unit study

Received: 9 May 1995 / Accepted: 10 June 1996

Abstract Many manipulations are able to change or perturb various aspects of single neuron properties and interneuronal relationships. Changes of cerebral cortex organization have been observed in different cortical areas and at different time scales in relation to peripheral stimulation, peripheral damage, associative learning, and electrical stimulation. Here we describe studies on separable multineuron recordings in the rat's auditory cortex under two different anesthetics. Acoustic stimuli were used as a normal, physiological input, and weak electrical intracortical microstimulation (ICMS) as a perturbation that forces a rapid cortical reorganization. ICMS induced fast changes in the cortical map and in the receptive field properties of cells at the electrically stimulated and adjacent electrodes. In effect there was an enlargement of the cortical domain tuned to the acoustic frequency that had been represented at the stimulating electrode. ICMS also incremented afterdischarge responses; these consisted of an initial response to the auditory stimulus followed by less intense repetitive activity that was stimulus-time locked and had a period of 8–12 Hz, similar to that of the spontaneous synchronous activity. Cortical activity under ketamine differed from that under pentobarbital sodium, although in both situations we observed that cortical neurons were highly synchronous.

Key words Plasticity · Auditory cortex · Neuronal assemblies · Microstimulation · Cortical maps · Rat

Introduction

A striking feature of sensory cortex is its topographic organization. This refers to the fact that two receptive fields that are adjacent in the sensory surface are likely to be contiguously represented in the corresponding sensory cortex. This topography is usually carried through all intermediate connecting structures. The extension of the adjacent property to the entire appropriate sensory cortical surface constitutes a map. In the rat, the auditory map extends from responses to 500 Hz in the caudal portion of primary auditory cortex to up to 45 kHz in the rostral portion (Sally and Kelly 1988). Neurons along the dorso-ventral axis tend to have similar characteristic frequencies. Although there are some minor variations, the general location and orientation of the frequency maps and isofrequency contours are similar across animals (Kelly 1990).

Until Hubel et al. demonstrated changes in ocular dominance columns in primary visual cortex (Hubel et al. 1977), it was thought that the organization of cerebral cortex, although continuously changing during development, was quite static in adult life. Since then, various manipulations have revealed many instances of reorganization in many different sensory cortices. Merzenich et al. (1983), while mapping the glabrous region of somatosensory cortex in monkey after medial nerve section, discovered that representation of the skin surface along the edges of the denervated region appeared to expand topographically into part of the deprived cortex. Striking topographic changes also have been found in somatosensory cortex after the removal of one or two digits in monkeys (Merzenich et al. 1984) and flying fox (Calford and Tweedale 1988); after the surgical fusion of two contiguous digits (Clark et al. 1988); after small, focal cortical lesions (Jenkins and Merzenich 1987); and after behavioral training (Jenkins et al. 1990). Plasticity of visual maps has been demonstrated after retinal lesions (Kass et al. 1990) and by an "artificial scotoma" (Pettet and Gilbert 1992). Recent studies of deafferentated visual cortical neurons (Chino et al. 1995) showed that 98% of cells in the initially deafferented zone acquired normal receptive-field properties after 3 months of recovery. Changes in the cortical auditory frequency representation have been observed after behavioral conditioning (Bakin and

P.E. Maldonado (✉)[1] · G.L. Gerstein
Department of Neuroscience, School of Medicine,
University of Pennsylvania, Philadelphia, PA 19104, USA

Present address:
[1] Center for Neuroscience, 1544 Newton CT,
University of California, Davis, CA 95616, USA;
Fax: +1-916-757-8827, e-mail: pedro@chaos.ucdavis.edu

Weinberger 1990) and after lesions in the cochlea (Robertson and Irvine 1989, Harrison et al. 1991; Rajan et al. 1993). In these studies, the region of cortex that formerly represented the frequency response of the lesioned cochlear region was now occupied by enlarged representations of frequencies adjacent to the lesioned area. Most of the effects of all these manipulations, are seen only with a long time scale of weeks or months. However, some studies have observed changes immediately following the manipulation, with subsequent further modulation (Calford and Tweedale 1988; Pettet and Gilbert 1992).

Intracortical microstimulation (ICMS) can alter topographic maps in a few hours. Recanzone et al. (1992) showed that the receptive fields of neurons in the somatosensory cortex of adult rats and monkeys exhibit plastic changes in the vicinity of a region of electrical microstimulation. After application of ICMS for 2–6 h, multiunit receptive fields of neurons near the stimulating electrodes were often enlarged and showed new sensitivity to the same skin area to which the cells at the stimulating electrode responded. These changes persisted for some hours after the stimulation period was over. In this paper we present similar studies of the rat auditory cortex under ketamine or pentobarbital sodium anesthesia, where we observed several simultaneously recorded single neurons during change in cortical organization induced by electrical microstimulation.

Materials and methods

Animals and surgical preparation

All experiments followed guidelines published in *Principles of laboratory animal care* (NIH publication No. 86-23, revised 1985) and *Preparation and maintenance of higher mammals during neuroscience experiments* (NIH publication No. 94-3207, 1994). Male Wistar rats (325–450 g) were anesthetized with an induction solution of either ketamine-xylazine (Ketalar) or pentobarbital (Nembutal) plus glycopyrrolate and Ringer's lactate. The induction dose for the ketamine/xylazine experiments (ketamine 70 mg/kg; xylazine 7.2 mg/kg) or pentobarbital experiments (Nembutal 40 mg/kg) was administered via intraperitoneal injection. In all experiments, immediately after induction, 0.1 mg of glycopyrrolate (a synthetic atropine analog) was administrated via intramuscular injection. Then, a catheter was threaded into the femoral vein and a continuous infusion of fluid and anesthetics was delivered by a pump. Infusion solutions (ketamine 20 mg/kg per hour; xylazine 2 mg/kg per hour; glycopyrrolate 0.01 mg/kg per hour or Nembutal 20 mg/kg per hour; glycopyrrolate 0.01 mg/kg per hour) were diluted in Ringer's lactate and delivered at a rate of 1.6–2.0 ml/kg per hour. Ketamine/xylazine (Ketamine)-anesthetized rats were found to yield longer survival times than pentobarbital (Nembutal)-anesthetized rats. Nembutal rats often suffered respiratory stress. In the experiments with Nembutal rats, we performed a tracheotomy and the tracheal tube was frequently cleaned with a suction probe.

Body temperature was maintained at 37°C, with a feedback-controlled, fluid-filled heating pad. The animals were not positioned in a stereotaxic apparatus, since the ears were to be unobstructed. Instead, the hair of the head was clipped, skin reflected, and two (size 0-80-1/16″) machine stainless steel screws were screwed into the medial section of the skull about 2 cm frontal to the bregma point. A third screw (flat head down, size 4-40 3/4″) was positioned nearby and dental cement applied to bone and the heads of the three screws. After hardening, the third screw was held by a stainless steel bar that was firmly attached to the operating table. This arrangement provided a strong and stable support for recording and at the same time ensured accessibility to cortical auditory areas.

Thresholds for evoked auditory field potentials were measured with electrodes positioned on the skull over the auditory cortex to test for auditory impairment. A craniotomy was then made in the dorsotemporal skull, exposing most of the primary auditory cortex, with a diameter of about 6.0 mm. The dura was removed and a video snapshot was taken of the brain's surface vascular pattern. This photo was used as spatial reference for locating electrode penetrations. A linear array of four closely spaced electrodes were positioned along the rostrocaudal axis of the auditory cortex and the cavity was filled with a 1% agar solution. A very thick layer of agar was built to insure thermal insulation as well as to prevent dryness and cortical movements. At the end of the experiments, the animals were killed with a pentobarbital overdose, administered intravenously.

Electrical recordings and data acquisition

Parylene-insulated tungsten microelectrodes (Micro Probe) of 1.4–1.6 MΩ were gold-plated and glued together in a linear array with a 200- to 300-μm spacing. The electrode array was attached to a Burleigh, electronically driven micro-manipulator, oriented along the anteroposterior axis and driven into the cortex. Recordings were sought from the middle cortical layers, typically at a depth between 200 and 600 μm. Although the detailed cortical frequency representation varies from animal to animal, the anteroposterior axis runs across isofrequency lines (Sally and Kelly 1988). The electrodes were attached to a d.c.-coupled head stage amplifier built with LM308 operational amplifiers with 20 MΩ input impedance and a gain of 30. Signals were a.c. coupled with appropriate filtration to a multichannel amplifier and recorded on tape for further off-line analysis.

The taped signal was played back and individual spike trains from each electrode were separated using principal components spike sorters (Abeles and Goldstein 1977). The projections of every incoming data spike waveform onto the first two principal components calculated from a library of similar data were plotted as a dot in a two-dimensional space; well-separated clusters of dots were considered to represent spike waveforms belonging to a single cell.

Auditory stimulation

A PDP-11-based computer and interface system was used to control all aspects of timing, frequency, and amplitude of the auditory stimulus. Continuous sine waves from a Voltage Variable Systems (VVS) model 404 (THD <–90 dB) were shaped by an enveloping device that produced tone bursts of the desired length and with rise-fall time of 5 ms. Overall intensity was controlled by a VVS model 271 dual isolated attenuator (THD <–60 dB). Tone bursts or clicks were fed to a Crown D-75 power amplifier, a post-attenuator, and then to a tweeter located at 22 cm from the bregma point on the contralateral side at an angle of 60°, 0° azimuth. The speakers (Realistic 40-1377) delivered a flat, free field response between 4 and 40 kHz (±2.0 dB) with harmonic distortion (HD) of less than –40 dB below the fundamental. Calibration was performed using substitution of the rat head by a Brüel and Kjaer 12.5-mm microphone. Intensity levels for pure tone were constant to ±2 dB in an 8- by 8-cm area around the rat's head position.

Three types of auditory free-field stimuli were used in the experiments, all presented contralateral to the observed cortex. Initially, clicks were used to determine the intensity threshold of the skull field potential and in order to insure auditory normality. The field potentials at both sides of the head were determined for comparison. Both click and pure tones were used to search for a region of cortex where multiunit activity exhibited maximum response. For frequency response histograms, 15 different frequency tone bursts were randomly presented at 20–40 dB above the threshold for increased multiunit activity; absolute intensity was usually between 30 and 40 dB SPL. Stimulus length was always 50 ms, with an interstimulus time of 600 ms. Normally each frequency was presented 50–60 times. For the study of afterdischarge responses, tone bursts of a frequency that was optimal, or near optimal, for neurons at the electrode that would later be used for microstimulation were presented with a repetition rate of one stimulus per 2 s. Once a set of auditory stimuli was selected, they were presented at various times along the experiment. All experiments began with a multiunit frequency response determination; it was required that this response remain stable for a period of 60–90 min before electrical microstimulation was initiated.

Electrical microstimulation and protocols

The ICMS consisted of trains of 11 biphasic pulses of 0.5 ms duration in 60 ms, repeated at 0.5-s intervals. These were produced by two Tektronix type 161/162 pulse generators and delivered through two stimulus isolating units (model Bioelectric ISA100) with outputs connected in series. The amplitude of the pulses was 4–5 μA. These parameters had previously been successfully used by other laboratories in experiments with ICMS-induced cortical plasticity (Recanzone et al. 1992)

A typical experiment started with the search for a strong auditory response to tones or to clicks. Then a characterization of the multiunit activity's frequency response was performed, and online peristimulus (PST) histograms were computed for reference. After a period of 1–1.5 h, during which it was required that the frequency response remain stable, we recorded the frequency response of the cells, spontaneous activity and the 2-s-long afterdischarges. The latter were elicited by 50-ms-tone bursts at the best frequency of multiunit activity at what was to be the stimulating electrode. All recordings made prior to the ICMS are called PRE. After the PRE characterization, we started the ICMS for 4 h in one of the four electrodes of the array. Immediately after the completion of the ICMS, data were again taken for frequency response, spontaneous activity, and afterdischarges. All data recorded here is referred to as POST. After POST recordings were made, the rat was left unstimulated for four more hours. Then a characterization was again obtained using the same stimulus protocol. Data taken here are referred to as LATER. A number of control experiments was performed in which every step was as described, except that no ICMS current was delivered after the PRE period.

Data analysis

Using amplitude discriminators and three hardware principal components spike sorters, a spike list of up to 20 cells was obtained for each experiment. The sorted data files contained cell identity codes, stimulus codes, and a time stamp for each event with a 0.5-ms resolution. Response planes (arrays of PST histograms arranged according to stimulus frequency) were constructed from the spike data. Frequency response histograms were computed from these by adding all spikes in a time window between 5 and 45 ms after the initiation of the acoustic stimulus at each frequency. In order to quantitatively compare the similarity between two frequency response histograms, we developed a similarity test based on the inner product of the two frequency response histograms treated as vectors. Since the goal was to compare the frequency response histogram shapes regardless of absolute response magnitude, each histogram was first normalized by its maximum amplitude, with an arbitrary threshold to prevent overestimation of the frequency response histograms of cells with very small counts. The variance of the inner product was estimated assuming a Poisson distribution of the counts in each bin. A projection value of 1.0 reflected perfect similarity; a value close to zero reflected extreme dissimilarity. We estimated a 99% confidence limit for response dissimilarity when the inner dot product plus 3 times the estimated standard deviation was smaller than 1.0.

In order to assess rhythmic activity in cell afterdischarges, autocorrelations were computed using unbiased correlograms (Bedenbaugh 1993). The PST predictor was subtracted from each raw correlogram. Significance at the 95% level was calculated from the PST predictor on the assumption of Poisson counts in each bin. Autocorrelograms were then normalized by the product of the square roots of the number of reference and target events contributed to the correlogram by the cell. This subtraction and normalization convert the raw correlograms into normalized covariance functions (NCFS).

Results

Experimental sessions and recording quality control

A total of 21 experiments that met the requirement of recording stability were used for analysis. Of the 21 analyzed experiments, 12 were performed under ketamine/xylazine anesthesia, 3 of which were controls, and 9 were performed under Nembutal, 2 of which were controls. Critical for including a given experiment in the analyzed pool was a constant pattern in the cluster distribution from the spike shape sorting process. Normally a displacement of the electrode array or significant changes in cortical activity substantially modified the clustering. Figure 1A shows patterns of sorted spike shapes, each cluster representing the activity of a single neuron, in a Nembutal experiment, taken from data recorded every 4 h. The overall cluster pattern remained stable, although the number of spikes in each cluster could increase or decrease during the experiment. This pattern would generally change drastically if the population of cells that were being recorded changed. Figure 1B,C shows cluster patterns taken from different experiments for comparison. Normally three to four distinct clusters representing the activity of single units were identified in the data from each electrode. Occasionally after ICMS, some cluster amplitudes would increase or decrease, reflecting that particular cell's activity change under ICMS, but the overall pattern remained unchanged. In this way between 12 and 16 single cells were recorded in every experiment.

In order to make an appropriate evaluation of the effects of ICMS on neuronal assemblies, we used two types of controls. A first set of controls was obtained from experiments in which no ICMS was performed, in order to statistically compare responses of electrically stimulated and unstimulated cells. In addition, we used the characterization of properties of cells located far away from the stimulating electrodes as a second set of controls. Changes of properties of cells only at the stimulating or adjacent site, but not at far away sites, indicate that the changes were a direct consequence of the ICMS manipulation.

Frequency response

Response planes and frequency response histograms (FRH) were, respectively, constructed for multiunit

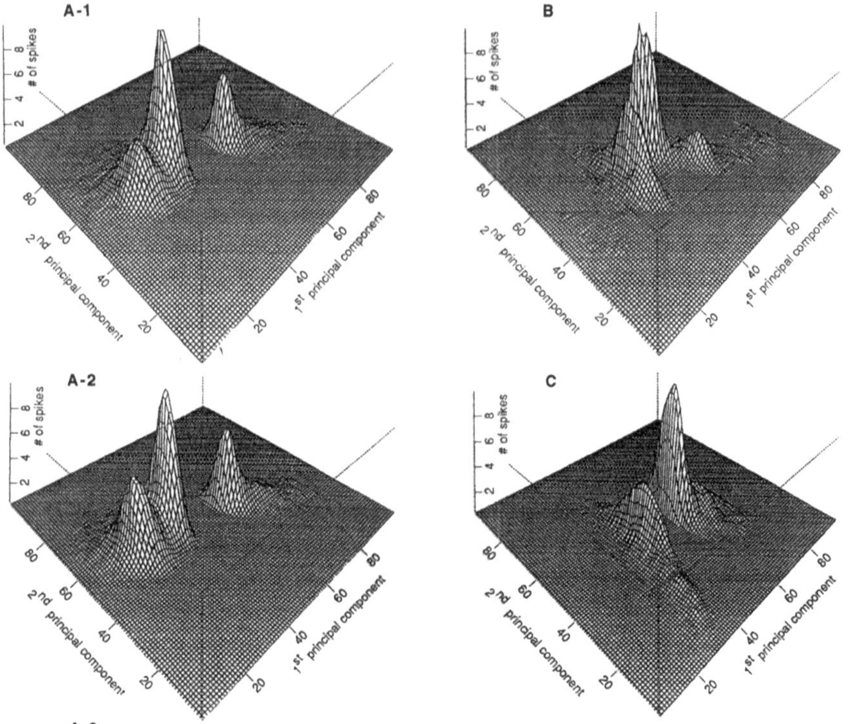

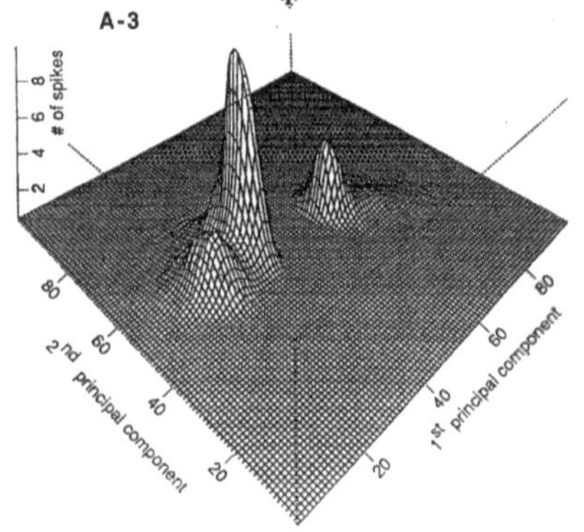

Fig. 1A–C Spike waveform clustering patterns in a Nembutal experiment. Principal components were calculated from an existing library of spike waveforms recorded in many similar experiments. The sorting device projected each incoming spike onto the two pre-calculated principal components and incremented the bin at the appropriate *x-y* location in a two-dimensional space. Similar waveforms are represented at close points in this space. The distribution of waveform projections produced by the sorting device are shown here as three-dimensional perspective plots. **A** *1, 2, 3* are successive recordings taken 4 h apart during a single experiment. Note that the pattern of clusters is stable throughout the recording period, although the number of spike events in each cluster may vary. In **B** and **C**, spike waveform clustering patterns of two other experiments are shown to illustrate that patterns would generally change drastically if the population of cells that are being recorded changed

(*n*=84) and single-unit activity (*n*=183), using spike counts during repeating cycles of acoustic stimulation at a constant intensity. The peak in the response plane or histogram, i.e., the most intense response, was taken as the "best frequency" (BF) of the cell or cells. In most cases, for both single and multiunit activity clear differences in the BF were observed on the several electrodes

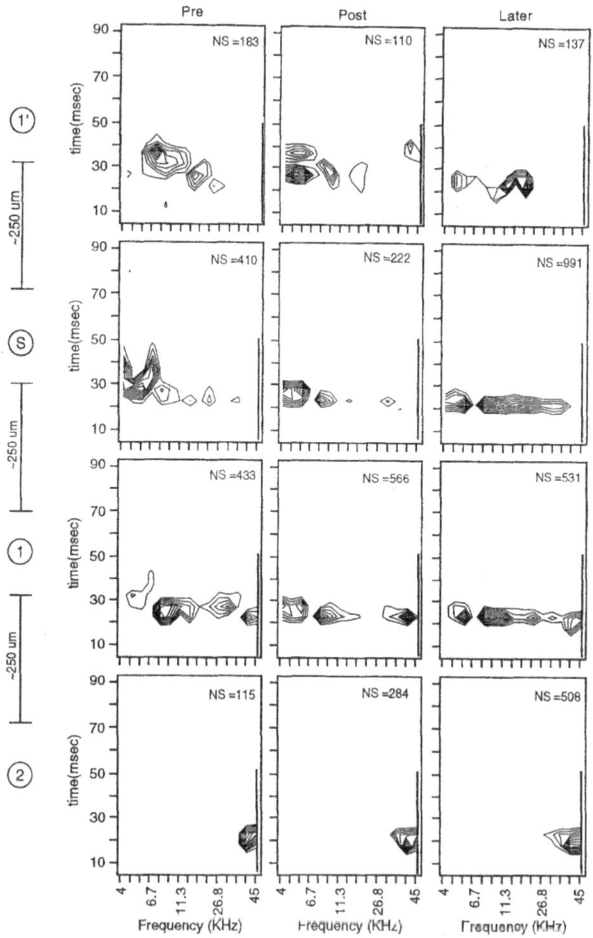

Fig. 2 Multiunit receptive field changes induced by intracortical microstimulation (ICMS) in Nembutal anesthesia. Each panel represents the response planes for the multiunit activity in one of the electrodes numbered on the *left side*. *S* represents the stimulating electrode. In each row the response planes for each of the experimental stages are shown. The *black bar at the right* of each panel represents the duration of the acoustic stimulus. In each response plane, isolevels of spike count are drawn as *continuous lines*. Each panel has been normalized for the maximum spike count. In the *PRE* column, the multiunit activity at different electrodes reflects different "best frequencies" (BFs). After ICMS (*POST*) cells in *1* and *1'* exhibit changes in their response planes that reflect a new sensitivity to the BF at the stimulating electrode. Neurons observed on the stimulating electrode did not significantly change their receptive field properties. Note that the multiunit activity of the farthest electrode also did not change its receptive field. In *LATER*, some degree of reversibility can be observed in *1'*, together with a widening of the frequency response of the stimulating electrode and electrode *1*. Again, no change is observed in the far electrode *2*. The distance between the electrodes is shown on the *left*

in the array. These differences were in accordance with the rat's frequency map: generally cells recorded on the rostral side of the electrode array showed BFs of higher frequency than those located on the caudal side. The range of the BF depended on the location of the electrode array, but because of the characteristic pattern of blood vessels and other landmarks in the rat's auditory cortex, it was usually possible to locate the array in the BF range that matched the frequency capability of the acoustic stimulation system. The mean stimulus intensity used in the ketamine experiments were 47±2 dB SPL, while in the Nembutal experiment the mean was 42±5 dB SPL. These intensity values were chosen to be slightly higher than the threshold values for the multiunit activity.

We observed that in the ketamine-anesthetized animals, the receptive fields or FRHs seemed often to exhibit a broader bandwidth than in the case of Nembutal and were frequently multipeaked. Quantification of this observation is not shown here. In each phase of every experiment, we compared the response planes and FRHs of multiunit activity at each electrode to those at the stimu-

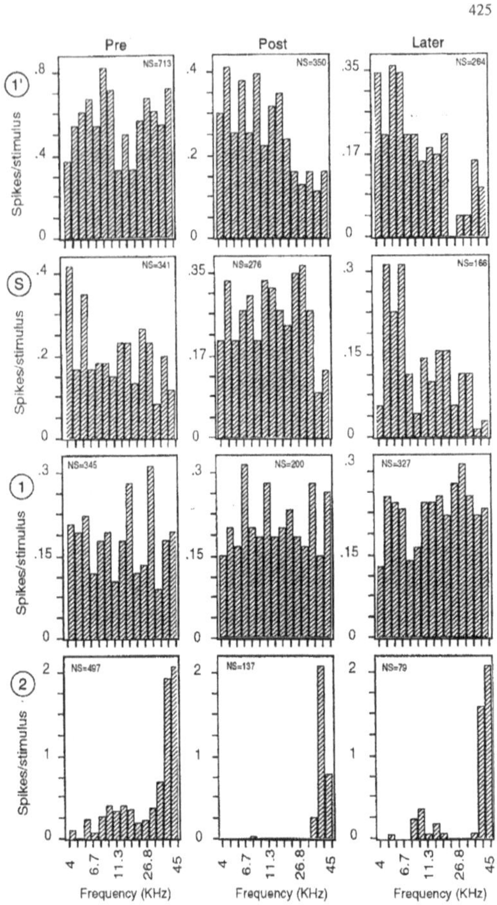

Fig. 3 ICMS-induced changes in single-cell receptive fields in a Nembutal experiment. The histograms represent the frequency response for individual neurons. The electrode from which the cell signal is extracted is shown at the *left*. *S* represents the stimulating electrode and the numbers represent the other electrodes in order of distance from S. The histograms show averaged spikes per presentation of each tone burst frequency. For each stimulus presentation, the spike count was the sum of all spikes within the 5- to 40-ms window after the stimulus onset and represents the primary response to each frequency. Each row shows frequency response histograms for the same cell at the three experimental stages. Note that, in several cases, the frequency response histograms lack smoothness and display multiple peaks, although an estimation of BF and the change in the frequency response histogram can be adequately made. Refer to the Discussion section for further comments. The cell at the stimulating electrode shows a broadening of its frequency response histogram in *POST*. In *LATER* this change is reversed. The cell located at the electrode *1'* develops a new sensitivity to the lower frequencies that correspond to the BF of the cell located at the stimulating electrode. A similar phenomenon can be observed in the cell located at electrode *1*. Finally, note that the cell located at the farthest electrode shows no changes in its frequency response histogram

lating electrode. After the ICMS period, there were often changes in the receptive fields recorded from electrodes at locations adjacent to the stimulation electrode (Fig. 2). However, the acoustic response of neurons at the stimulating electrode did not change substantially. In some cases the multiunit FRH at the stimulating electrode appeared to become less sharp, with an increase in the range of frequency response. At the adjacent electrodes the multiunit activity became newly responsive to the preferred acoustic frequency at the stimulation electrode site. In some cases this took the form of a broadening or additional peak in the FRH, in others there was a shift of the original FRH peak toward the preferred frequency at the stimulation electrode site (Fig. 3). Multiunit recordings from electrodes further away from the stimulating electrode usually did not show any changes in response pattern. These observations were valid both for Nembutal and Ketamine experiments. In the control experiments, where no ICMS was applied, little or no changes were observed at any electrode.

In order to quantify changes in the frequency response after ICMS manipulation, FRHs of 183 single cells were compared with the multiunit FRH at the stimulating electrode using the similarity test described in the Materials and methods section. Although there were fewer controls than ICMS experiments, we shuffled each of the electrodes in turn as the reference electrode in these control recordings, since no ICMS had actually been applied. This maneuver made it possible to obtain comparable numbers in the statistical comparisons with the ICMS experiments. The control curves were constructed by pooling Ketamine and Nembutal control (sham) ICMS experiments. In the control conditions, the similarity index decreased with distance from the reference electrode, signaling different frequency response preferences, and there was no effect on the similarity index after sham ICMS (Fig. 4). In the experimental conditions, more under Nembutal than under Ketamine, similarity decreased with distance, but the values significantly changed after ICMS. For both Ketamine and Nembutal, cells at the stimulating electrode and in the adjacent electrode increase their similarity. The change is clearly significant at the adjacent site for Ketamine and for the stimulating and adjacent sites in Nembutal ($P<0.05$). No such increase in response similarity was observed for cells recorded more than 300 μm from the stimulating electrode.

Afterdischarges

In the construction of FRHs, the acoustic stimulus was delivered with an interstimulus presentation time of 600 ms. A typical response of a single unit to a 50-ms tone burst had a latency of 5–10 ms. If the interstimulus interval was increased significantly, a characteristic afterdischarge could be observed. This consisted of the initial spike discharge response to the auditory stimulus, followed by a silent period and then a series of less intense, periodic but stimulus-time locked responses (Fig. 5), with a silent period that ranged between 80 and 150 ms. We assessed, by autocorrelation, the periodicity of the afterdischarge response of 163 single units; their secondary correlation peaks were tested for statistical significance. Normally, the intensity of the afterdischarge decayed with time after stimulus, and the first afterdischarge was always larger in magnitude than that of the following series. Occasionally the afterdischarge response was observed even when the neurons did not respond or responded poorly to the auditory stimulus, but in general the intensity of the afterdischarges appeared to be related to the amplitude of the acoustic stimulus re-

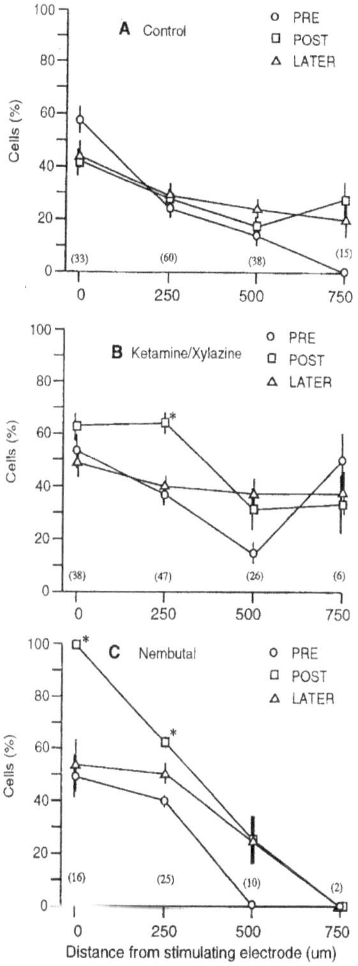

Fig. 4A–C ICMS-induced changes in similarities of single-cell receptive fields. The percentage of total cells that met the similarity test, using the multiunit activity in the stimulating electrode as a reference. The *number at the bottom* refers to the total number of cells compared, pooling all experiments. In **A**, where all electrodes were successively used as reference, it can be seen that the frequency response histograms of cells are more similar the closer they are to the reference electrode. After sham ICMS, in the electrode adjacent to the reference electrode there is no significant change in the percentage of similarity compared with PRE. In the reference electrode there is a decrease in the similarity percentage. Refer to the discussion section for further comments. **B** shows the similarity percentage for the Ketamine ICMS experiments. In POST, there is an increment of similarity among cells at the stimulating electrode as well as between them and cells located in the adjacent electrode, although this change only reached significance level at the adjacent electrode. This change reverses in LATER. **C** shows the similarity percentage for the Nembutal experiments. Note that at a distance of two electrodes none of the frequency response histograms were found to be similar to the reference frequency response histogram. After ICMS, in POST, there is a significant increase in frequency response histogram similarity for the stimulating and adjacent electrode. The increment in more distant number 2 electrode was found not to be significant. The changes reverse in LATER, particularly at the stimulating electrode. The *asterisk* indicates levels that were significantly different from PRE ($P<0.05$. Paired Student's t-test). The *vertical lines* at each point indicate SE

Fig. 5A, B ICMS effect in afterdischarge activity. Peristimulus histograms are shown for multiunit activity on an electrode before (**A**) and after (**B**) ICMS, in a Ketamine experiment. The first, large peak represents the primary response to the acoustic stimulus. The silent period that follows the primary response ranges from 80 to 120 ms. Note the distinctive series of less intense, periodic but stimulus-time locked responses. Each *insert* shows the autocorrelogram of the afterdischarge. Note the secondary peak in the range of 100 to 200 ms. This period is similar to that seen in spontaneous activity

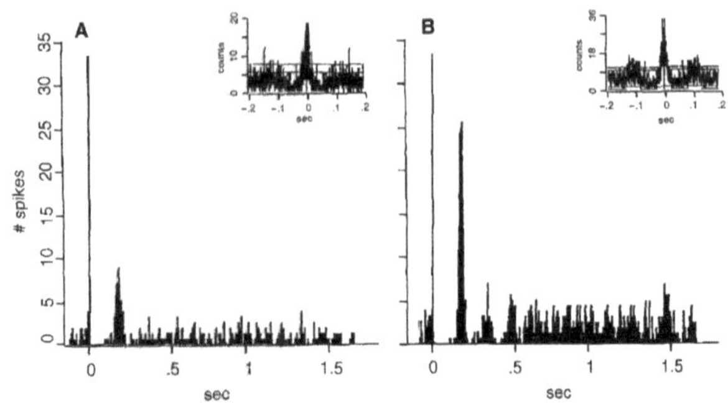

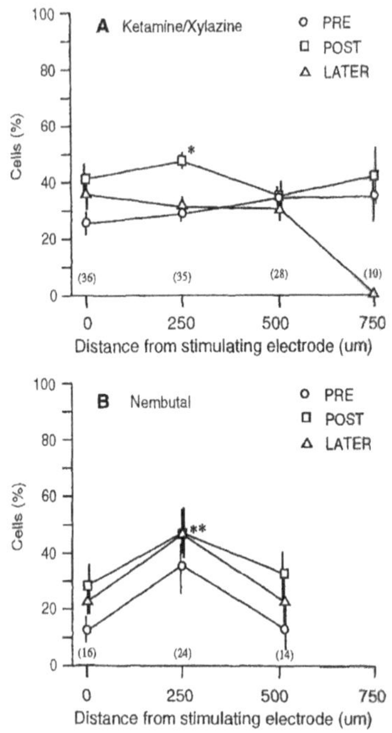

Fig. 6A, B Distribution of significant afterdischarges. The effect of ICMS on the distribution of cells that exhibited correlograms with significant secondary peaks is shown here. **A** shows the distribution for the Ketamine experiments. At PRE there is a spatially constant distribution of cells that show a significant secondary peak. After ICMS, there is a significant increase in the proportion of cells with secondary peaks at the electrode adjacent to the stimulating electrode. Other increases were found not to be significant. The increase is reversed after the waiting period (LATER). In **B** the same analysis is shown for the Nembutal experiments. After ICMS there is a significant increase only at the adjacent site. The *number in parentheses* along the abscissa is the total number of cells analyzed at that electrode distance. The asterisk indicates levels that were significantly different from PRE ($P<0.05$, paired Student's t-test). The *vertical lines* indicate the standard error

sponse. The frequency of the afterdischarges was in the range of 8–12 Hz, similar to that of the spontaneous activity. We found that before ICMS about 27% (n=36) of Ketamine and 12.5% (n=16) of Nembutal recorded cells exhibited significant secondary autocorrelation peaks. After ICMS there was an increase to 41% and 28%, respectively. For the adjacent electrode, comparing PRE and POST, the percentage of cells that exhibited such afterdischarges also increased significantly from 29 to 47% in Ketamine and from 35 to 45% in Nembutal ($P<0.05$) (Fig. 6). No significant increases were seen at larger distances from the stimulating electrode. Not all cells that exhibited a significant secondary peak in their autocorrelogram before ICMS still showed the same peak after ICMS and the waiting period (LATER). Under Ketamine, 68% of cells that showed a secondary peak in PRE showed the peak in POST and 46% showed it in LATER. The corresponding percentages with Nembutal were 65% in POST and 50% in LATER.

Discussion

ICMS-induced changes in receptive fields

For both multiunit and single-cell recordings, we observed that in Ketamine-anesthetized animals the receptive fields or FRHs were much broader than in the case of Nembutal, and in addition that the probability of finding a FRH with a secondary peak was greater.

The FRHs were variable in both shape and quality in our cell population. A principal factor contributing to the variability in quality of the single-cell FRH was the use of a single stimulus intensity value throughout the entire experiment. While the intensity values we used were chosen to be just above threshold levels for the multiunit activity, they were not the optimal stimulus for all cells in the recording. Different cortical cells have different intensity threshold levels, and it is known that measurement with an intense stimulus may produce a much broader receptive field than can be obtained with a lower intensity value. At the same time, for another nearby recorded neuron, the same stimulus intensity could be too low, thus underestimating the cell response. ICMS could result in changes in the excitability of such cells, hence contributing to the altered frequency response.

It is conceivable that by using a click to search the cortex we over-sampled regions with cells having multiple-frequency sensitivities. However, the similarity test we used to compare frequency histograms is independent of the detailed tuning properties because of its vectorial representation of the response. Moreover, we found that the observed FRHs adequately reflect the cortical frequency map. The linear electrode arrays were positioned across the putative isofrequency lines and, indeed, the similarity of the receptive fields decayed with distance between the neurons being compared.

Consistent with the fact that receptive fields measured under Nembutal anesthesia were sharper, we found that the similarity decay was faster under this anesthetic agent; no FRHs were found to be similar beyond the adjacent electrodes. For both multiple- and single-unit activity after the ICMS period, there was a change in the receptive fields, observed at locations adjacent to the stimulation electrode. Most cells located at the adjacent electrodes became sensitive to the BF of the stimulating electrode. This phenomenon is similar to reports of analogous multiunit recording experiments performed on the somatosensory system in rat and monkey (Recanzone et al. 1992). Our results also are consistent with most of the experiments on cortical reorganization that show an expansion of receptive fields that were adjacent to regions that were modified by damaged or stimulated peripheral input. We also observed that a number of FRHs of single cells at the stimulating electrode became broader, a phenomenon that previously has been shown only for multiunit activity (Recanzone et al. 1992).

Afterdischarges

Afterdischarge responses, or oscillatory rebound, were found in about half the cells recorded under Ketamine and in about 20% of the Nembutal neurons. They were observed even when the neurons did not respond or responded poorly to the auditory stimulus, but in general the intensity of the afterdischarges appeared to be related to the amplitude of the acoustic stimulus response. The frequency of the afterdischarges was found to be 8–12 Hz, similar to that of spontaneous activity. We conclude that the afterdischarge responses were entrained, spontaneous oscillatory firing patterns. These cell responses are similar to those of cells recorded in the cat's auditory cortex, where afterdischarge showed a clear dependence on the physical parameters of a wide range of auditory stimuli and are thought to have a functional significance in the cortical processing of auditory stimuli (Altman 1980; Eggermont 1992).

After ICMS, we found that there was an increase in the proportion of cells that exhibited a significant secondary peak in the autocorrelogram of the afterdischarge response. The effect of ICMS in the autocorrelograms appeared to be dependent on distance. For the electrode adjacent to the stimulating electrode (~250 μm), under ketamine anesthesia, the increase was found to be statistically significant ($P<0.05$). For cells at the stimulating site, the increment did reach significant levels only for $P<0.10$, and in cells located at electrodes far away from the stimulating electrode (≥500 μm) none of the changes were significant. It was clear, however, that the effect was more evident on cells that were closer to the stimulating electrode. The changes induced by ICMS on the afterdischarge responses were similar to the changes induced in the spontaneous activity (Maldonado and Gerstein 1996). This is consistent with the view that the afterdischarge responses reflect the entrainment of spontaneous background oscillation. The silent period, seen between the termination of the acoustic stimulus response and the first afterdischarge, could be explained in terms of known postsynaptic potentials in the cells. Intracellular studies in the cat's auditory cortex have shown that some cells showed short-latency (20-ms duration) excitatory postsynaptic potentials (EPSPs) that were usually followed by long-duration (50-160 ms) inhibitory postsynaptic potentials (IPSPs; Ribaupierre et al. 1972). These investigators found that the EPSP/IPSP pattern usually could be deduced from the mean spike output; the PST histogram resembled the poststimulus mean membrane potential. Thus, the onset of the first afterdischarge would coincide with the offset of the cell's inhibition. It is possible that network interactions that are responsible for the spontaneous oscillations could be instrumental in the IPSP duration, thus entraining the cell activity back to oscillation, once the stimulus-related IPSP is over.

Modification of receptive-field properties

The mechanisms by which the cells modify their receptive field properties are yet unknown. A possible explanation can be based on the Hebb model of assemblies. This model proposed in part that two cells or systems of cells that are repeatedly active at the same time will tend to become "associated" so that activity in one cell facilitates activity in the other one (Hebb 1949). During ICMS, cells surrounding the stimulating electrode are being synchronously driven and thus forced into an unusual simultaneous firing pattern. Cells that are weakly associated near the stimulated site will, therefore, increase their association because of Hebb's rule. It is well known that thalamic afferents arborize to a much larger extent than the receptive field properties of cortical cells suggest. If these synaptic connections change strength or, if intrinsic cortico cortical inhibitory connections change strength, the result will be seen as a shift in sensitivity of the receptive fields of cells adjacent to the stimulating site, or as an increase in size of the receptive field of cells at the stimulating site. Both phenomena were observed in our experiments, therefore are consistent with changes only in the cortical circuitry. On the other hand, we cannot rule out that the loci of synaptic changes were at least partly located in subcortical structures. It is possible that cortical microstimulation induced synchronized retrograde and anterograde actuation of subcortical structures such as the thalamus.

The effect of anesthetic agents

There is much evidence that the *N*-methyl D-aspartate (NMDA) receptor is involved in various plasticity and learning processes. The known blocking interaction between ketamine and the NMDA receptor (Rauschecker and Hahn 1987; Albers et al. 1989; Carla and Moroni 1992) led us to consider the use of Nembutal as well in this series of experiments.

Ketamine and Nembutal produce somewhat different effects on the activity of cortical cells. Patel and Chapin (1990) found that ketamine produced a strong inhibition of all somatosensory responsiveness in this area and a tonic excitatory influence expressed heterogeneously on a subgroup of neurons. Auditory cortex also showed reduced response. 2-Deoxyglucose (2-DG) studies in the gerbil central auditory system under ketamine have shown slightly reduced 2-DG uptake, when the animal was exposed to auditory noise stimulation, and slightly decreased uptake in silence (Wang et al. 1987). Nembutal, on the other hand, significantly decreased 2-DG uptake in auditory cortex, whether or not there was auditory stimulation. In addition, Hammer and Herkenham (1983) reported that during subanesthetic ketamine administration, peak 2-DG uptake shifts from cortical layer IV to layer Va. Nembutal also has been found to depress synaptic transmission of excitatory neurons (Richards 1978), while it appears not to prevent inhibition of cortical neurons by GABA (Crawford 1970).

Nonetheless, some receptive field properties of cortical cells do not seem to differ significantly under the effect of these anesthetics. In the somatosensory cortex, Stryker et al. (1987) found that the receptive field defined for small clusters of cells differed little in size or skin locations in comparing alert and pentobarbital-anesthetized or ketamine-anesthetized owl monkeys. We observed, however, a striking difference between ketamine- and Nembutal-anesthetized rats with respect to the overall cortical activity. The auditory cortex in ketamine-anesthetized rats was much more active than under Nembutal. The level of activity under Nembutal depended to a considerable degree on the anesthesia levels. At deep Nembutal-anesthesia levels, the cortex was almost silent. At very light levels, just enough to abolish corneal reflexes, the cortex was much more active and resembled more the Ketamine cortex. Despite all these differences, in our experiments we were able to observe plastic changes associated with the ICMS manipulation using both anesthetics.

Acknowledgement This work was supported by NIH grants MH46428 and DC01249.

References

Abeles M, Goldstein MH (1977) Multispike train analysis. Proc IEEE 65:762–773

Albers GW, Goldberg MP, Choi DW (1989) *N*-methyl-D-aspartate antagonists: ready for clinical trial in brain ischemia? Ann Neurol 25:398–403

Altman IA, Bekhterev NN, Kotelenko LM, Kudriavtseva IN (1980) Afterdischarges of cat medial geniculate body neurons. Fiziol Zh SSSR 66(1):80–88

Bakin JS, Weinberger NM (1990) Classical conditioning induces CS-specific receptive field plasticity in the auditory cortex of the guinea pig. Brain Res. 536:271–286

Bedenbaugh P (1993) ICMS-induced plasticity in the rat somatosensory cortex and theoretical investigations of information flow through neurons. Ph.D. Dissertation Department of Bioengineering, University of Pennsylvania, Philadelphia, Pennsylvania, USA

Calford MB, Tweedale R (1988) Immediate and chronic changes in responses of somatosensory cortex in adult flying-fox after digit amputation. Nature 332(6163):446–448

Carla V, Moroni F (1992) General anesthetics inhibit the responses induced by glutamate receptor agonist in the mouse cortex. Neurosci Lett 146:21–24

Chino YM, Smith EL III, Kaas JH, Sasaki Y, Cheng h (1995) Receptive field properties of deafferentated visual cortical neurons after topographic map reorganization in adults cats. J Neuroscience 15:2417–2433

Clark SA, Allard T, Jenkins WM, Merzenich MM (1988) Receptive fields in the body-surface map on adult cortex defined by temporally correlated inputs. Nature 332:444–445

Crawford JM (1970) Anesthetics agents and the chemical sensitivity of cortical neurons. Neuropharmacology 9:31–46

Eggermont J (1992) Stimulus induced and spontaneous rhythmic firing of single units in cat primary visual cortex. Hear Res 61:1–11

Hammer RP, Herkenham M (1983) Altered metabolic activity in the cerebral cortex of rats exposed to ketamine. J Comp Neurol 220:396–404

Harrison RV, Nagasawa A, Smith DW, Stanton S, Mount RJ (1991) Reorganization of auditory cortex after neonatal high frequency cochlear hearing loss. Hearing Res 54(1):11–19

Hebb DO (1949) The organization of behavior. A neuropsychological theory. Wiley, New York

Hubel D, Wiesel T, LeVay S (1977) Plasticity of ocular dominance columns in monkey striate cortex. Philos Trans R Soc Lond Biol 278:377–409

Jenkins WM, Merzenich MM (1987) Reorganization of neocortical representations after brain injury: a neurophysiological model of the bases of recovery from stroke. Prog Brain Res 71: 249–266

Jenkins WM, Merzenich MM, Recanzone G (1990) Neocortical representational dynamics in adult primates: implications for neurophysiology. Neuropsychologia 28(6):573–584

Kaas JH, Krubitzer LA, Chino YM, Langston AL, Polley E, Blair N (1990) Reorganization of retinocortical maps in adult mammals after lesions of the retina. Science 248:229–231

Kelly JB (1990) Rat auditory cortex. In: Kolb B, Tees R (eds) The cerebral cortex of the rat. MIT Press, Cambridge, MA, pp 381–405

Maldonado PE, Gerstein GL (1996) Neuronal assembly dynamics in the rat auditory cortex during reorganization induced by intracortical microstimulation. Exp Brain Res 112:431–441
Merzenich MM, Kaas JH, Waal J, Nelson JM, Sur M, Felleman M (1983) Topographic reorganization of somatosensory cortical areas 3b and 1 in adult monkeys following restricted deafferentation. Neuroscience 8(1):33–55
Merzenich MM, Nelson RJ, Stryker MS, Cynader MS, Schopmann MS, Zook JM (1984) Somatosensory cortical map changes following digit amputation in adult monkeys. J Comp Neurol 224:591–605
Patel IM, Chapin JK (1990) Ketamine effects on somatosensory cortical single neurons and on behavior in rats. Anesth Analg 70(6):635–644
Pettet MW, Gilbert CD (1992) Dynamic changes in receptive-field size in the primary visual cortex. Proc Natl Acad Sci USA 89:8366–8370
Rajan R, Irvine DR, Wise LZ, Heil P (1993) Effect of unilateral partial cochlear lesions in adult cats on the representation of lesioned and unlesioned cochleas in primary auditory cortex. J Comp Neurol 338(1):17–49
Rauschecker JP, Hahn S (1987) Ketamine-xylazine anesthesia blocks consolidation of ocular dominance changes in kitten visual cortex. Nature 326:183–185
Recanzone GH, Merzenich MM, Dinse HR (1992) Expansion of the cortical representation of a specific skin field in primary somatosensory cortex by intracortical microstimulation. Cereb Cortex 2:181–196
Ribaupierre F, Goldstein MH, Yeni-Komshian G (1972) Intracellular study of the cat's primary auditory cortex. Brain Res 48:185–204
Richards CD (1978) The action of anesthetics on synaptic transmission. Gen Pharmacol 9:287–293
Robertson D, Irvine D (1989) Plasticity of frequency organization in auditory cortex of guinea pig with partial unilateral deafness. J Comp Neurol 282:456–471
Sally SL, Kelly JB (1988) Organization of the auditory cortex in the albino rat: sound frequency. J Neurophysiol 59:1627–1638
Stryker MP, Jenkins WM, Merzenich MM (1987) Anesthetic state does not affect the map of the hand representation within area 3b somatosensory cortex in owl monkey. J Comp Neurol 258:297–303
Wang Z, Ryan A, Woolf NK (1897) Pentobarbital and ketamine alter the pattern of 2-deoxyglucose uptake in the central auditory system of the gerbil. Hear Res 27:145–155

Exp Brain Res (1996) 112:431–441

RESEARCH ARTICLE

Pedro E. Maldonado · George L. Gerstein

Neuronal assembly dynamics in the rat auditory cortex during reorganization induced by intracortical microstimulation

Received: 9 May 1995 / Accepted: 10 June 1996

Abstract Single neurons, acting alone, cannot account for the complex and rapid computations that are routinely accomplished by the behaving nervous system. Recent studies with separable multineuron recordings are showing that neuronal assemblies can indeed be detected and that their organization is very dynamic, depending on variables such as time, physical stimulus, and context. Here we explore both single-neuron and assembly properties in the rat's auditory cortex. Acoustic stimuli are used as a normal, physiological input, and weak electrical intracortical microstimulation (ICMS) as a perturbation that forces a rapid cortical reorganization. In this setting, various aspects of neuronal interactions are changed by the ICMS. We found that cortical neurons exhibited highly synchronous oscillatory firing patterns that were enhanced by ICMS. Cross-correlation studies between two spike trains showed that statistically significant correlations depended on the anatomical distance between the two neurons. ICMS changed the strength and the local number of such correlations. Joint petri-stimulus analysis and gravity analysis showed that the correlation between neuronal activities varied dynamically at several time scales. We have identified neuronal assemblies in two ways, defined through similarity of receptive field properties and defined through correlated firing. Close anatomical spacing between neurons was conducive to, but not sufficient for membership in, the same assembly with either definition. ICMS changed cortical organization by altering assembly membership. Our data show that neuronal assemblies in the rat auditory cortex can be established transiently in time and that their membership is dynamic.

P.E. Maldonado (✉) · G.L. Gerstein
Department of Neuroscience, School of Medicine,
University of Pennsylvania, Philadelphia, PA 19104, USA

Present address:
[1] Center for Neuroscience, 1544 Newton Ct.,
University of California, Davis, CA 95616, USA;
Fax: +1-916-757-8827; e-mail: pedro@chaos.ucdavis.edu

Key words Plasticity · Auditory cortex · Neuronal assemblies · Microstimulation · Cortical maps · Rat

Introduction

It is customary to characterize the activity of sensory (cortical) cells, by presenting the stimulus numerous times in order to obtain mean responses: these are considered an adequate estimation of the response properties. However, we normally find large variations in the response of a cell to which we are repeatedly presenting the same stimuli. For example, in primary visual cortex, the individual responses to identical visual stimuli vary between one spike and bursts of more than 12 spikes (Boch 1986). Moreover, the same response can be obtained to the presentation of different stimuli. The reason for such variability in the response is probably due to the large number of connections to any given neuron and a profusion of other, unrelated processes. Thus, the activity of that particular neuron is governed by the integration of a large number of synaptic activities, of which only a subset is directly related to the pathway we are stimulating (Sheperd 1990). Yet, our perceptual world is constructed upon single exposures to stimuli. How does the brain solve the problem of response ambiguity and how does it do so with a great deal of constancy even under very different conditions of stimulation?

One possible solution to the problem is that there is a hierarchical information process in the brain: single cells in other cortical regions integrate information from many of the more primary neurons and are thus more reliable in detecting some particular stimuli (White 1989; Felleman and Van Essen 1991). Another solution could be based on cells that are highly selective to particular stimulus configurations such as the "grandmother neurons" (Lettvin et al., 1959) or "face cells" (Gross et al. 1972; Young and Yamane 1992). In many of these cases, ambiguity can be overcome by pooling a large number of such cells. This, however, only translates the ambiguity problem to the next layer or level of processing. However, it is unlikely that there are a sufficient number of neurons in the higher levels of the cortex to represent the multiple possible combinations presented in the sensory world (Ballard et al. 1983; Malsburg and Singer 1988). Even cells that are found to be selective to a particular object or collection of objects, are highly variable in their response, in contrast to certainty at the perceptual level.

Another way to surmount the ambiguity problem is to consider as the functional unit not a single cell but a set of single neurons that work together to subserve a common behavior or cognitive process. One of the possible properties that can be used to define a neural assembly is the similarity of receptive fields. Another possibility is correlation of firing (Gerstein et al. 1989; Singer 1991; Deppisch et al. 1994; Lindsey et al. 1994). In this case the members of an assembly are identified by the mutual, preferred timing relationships in their firing. The latter definition of neuronal assembly will be used throughout most of this paper.

In the accompanying paper (Maldonado and Gerstein 1996) we have shown that intracortical microstimulation (ICMS) changes the organization of the rat's auditory cortex. ICMS in one of an array of electrodes rapidly induces changes in the receptive field properties of nearby cortical cells: (1) cells located at the stimulating electrode show a broadening of their receptive fields; (2) neurons located at electrodes adjacent to the stimulated electrode exhibit a moderate increase in responsiveness to the best frequency of cells at the stimulating electrode. In these experiments we could define assemblies of neurons based on the similarity of their frequency response. Close anatomical distance was not a requisite for such similarity of tuning characteristics, nor sufficient for it. ICMS changed frequency response structures, and, therefore, on this definition of assembly, changed neuronal memberships of particular assemblies.

By defining neuronal assemblies through preferred firing time relationships, we can study the assemblies' dynamics by looking at changes in the temporal interactions within the observed population of neurons. Such characterization of a neuronal assembly may provide insight into computational function. In this paper, we present studies in which the activity of several neurons in auditory cortex of the rat is recorded simultaneously, in a situation where a change in cortical organization is induced by electrical microstimulation. We use several analytical tools to examine the activity of ensembles of neurons and compare our results with different conceptual models for neuronal assemblies.

Materials and methods

Most of our methods are described in the accompanying paper (Maldonado and Gerstein 1996), where the details of surgical preparations, recording technology, and experimental protocol can be found. Here we emphasize only methodological aspects specific to these experiments. All experiments followed guidelines published in: *Principles of laboratory animal care* (NIH publication No. 86-23, revised 1985) and *Preparation and maintenance of higher mammals during neuroscience experiments* (NIH publication No. 94-3207, 1994).

Stimulus and protocol

Parylene-insulated tungsten microelectrodes (Micro Probe) of 1.4–1.6 MΩ were gold-plated and glued together in a lineary array with a 200- to -300-μm spacing. The electrode array was attached to a Burleigh, electronically driven micro-manipulator, oriented along the anteroposterior axis and driven into the cortex. Recordings were sought from the middle cortical layers, typically at a depth between 250 and 600 μm. A PDP-11-based compuster and interface system was used to control all aspects of timing, frequency, and amplitude of the auditory stimulus. The principal type of acoustic stimulus used in these experiments was 50-ms-tone bursts, presented free-field contralateral to the cortical recording site. A characterization of the multiunit activity frequency response was performed using 15 different frequency tone bursts presented repeatedly in random order, and on-line peristimulus (PST) histograms were computed for reference. After a period of 1–1.5 h during which it was required that the frequency response remained stable, we recorded neural activity during 5 min of silence (spontaneous activity). We then recorded afterdischarges to 50-ms-tone bursts presented every 2 s. These were at the best frequency for the multiunit activity recorded from the electrode that would become the ICMS stimulating electrode. All recordings made prior to the ICMS are called PRE. After the PRE characterization, we started the ICMS for 4 h in one of the four electrodes of the array. Immediately after the completion of the ICMS, data were again taken with the above stimulus protocol. All data recorded here is referred to as POST. Subsequently the rat was left unstimulated for four more hours and another recording obtained again using the same stimulus protocol. Data taken here are referred to as LATER. A number of control experiments were performed in which every step was a described, except that no ICMS current was delivered after the PRE period. When presented, the intracortical microstimulation consisted of trains of 11 (4–5 μA) biphasic pulses of 0.5 ms duration in 60 ms, repeated at 0.5-s intervals.

Data analysis

Autocorrelations of spontaneous activity and cross-correlations of cell afterdischarges were, respectively, computed using unbiased correlograms, i.e., correlation coefficients (Eggermont 1992; Bedenbaugh 1993). The cross-correlation of the PST histograms of the contributing cells (PST predictor) was subtracted from the raw correlogram to remove that part of the correlation that could be explained by the stimulus-locked modulation of firing rate (Perkel et al. 1967). Significance at the 95% level was calculated from the PST predictor on the assumption of Poisson counts in each bin and is shown on each correlogram. Cross-correlograms and autocorrelograms were then normalized by the product of the square roots of the number of reference and target evens (Glaser and Ruchkin 1976). To determine statistically significant differences in correlation strength for the afterdischarge data, after the correlograms had been normalized and corrected as above for firing rates and their stimulus-related modulations, a bin-by-bin subtraction was computed between the PRE and POST cross-correlogram of a given cell pair. The variance of the area under the correlogram was estimated upon the assumption of Poisson distribution of the counts in each bin. Significant changes were those differences in excess of 3 times the estimated standard deviation of the area (99% confidence level). Positive and negative differences reflected, respectively, an increment or decrement in correlation strength.

In order to study the changes in stimulus-locked interactions, cell pairs that were consistently recorded throughout the experiment and stimulated with their preferred frequency were further analyzed with joint PST histograms (JPST; Aertsen et al. 1989).

This is a time-dependent measure of correlation, averaged over many repetitions of the same stimulus and time-locked to the onset of the stimulus. A statistical measure, "surprise" (Palm et al. 1988), tests the significance of the excess or deficit of joint firing expected for each bin. The surprise is defined as the negative natural logarithm of the probability of finding that bin value or a more deviant one, given the marginal counts and number of stimuli. See Aertsen et al. (1989) for a full description.

In addition, for many sets of several neurons recorded simultaneously, we examined the temporal relationships between their spike trains using the "gravity" analysis (Gerstein and Aertsen 1985; Gerstein et al. 1985). Each of the N neurons is represented by a particle moving in an n-dimensional space. Each particle is given a charge that reflects the firing of the neurons it represents; the resulting attractive or repulsive force causes the particles to move from their initial equidistant configuration. The final result is aggregation of particles (possibly into several clusters) which represent those neurons whose spike trains have an excess of near-synchronous firing. Appropriate correction can be made for stimulus-related modulations of the individual neuron firing rates, just as in the calculation of normalized cross-correlograms. Results of the calculation are presented either as particle-pair distances compared with time or as projections of particle trajectories from the n-space to a 2-space. Only the latter is shown here. In either case, possible time-varying aggregation of particles is immediately apparent and can be interpreted as time-varying organization and assembly structure among the observed neurons. The advantage of this procedure is that it analyzes all available spike trains in a single calculation, rather than by pairs as in cross-correlation analysis.

Results

Spontaneous activity

We recorded a total of 163 single units from 15 experiments (ketamine 109 units; pontobarbital sodium, Nembutal, 54 units). Most recordings were obtained from the middle layers of the cortex, ranging in depth from 250 to 630 (mean 480 µm) as measured from the cortical surface (ketamine) and ranging from 340 to 570 µm (mean 409 µm) for Nembutal. The first noticeable difference between ketamine- and Nembutal-anesthetized rats was the overall spontaneous activity of the cells in the cerebral cortex. The auditory cortex in ketamine rats was much more active in terms of firing rate than in the Nembutal rats. As soon as the electrode array was deeper than 150 µm in the cortex, we saw acoustic evoked and spontaneous activity that increased as the array was advanced deeper into the cortex. In the Nembutal experiments, little or no activity could be observed until the array reached at least 300 µm; still the overall activity was less than in the ketamine condition. The most striking feature of the spontaneous activity was the presence of a bursting firing pattern, which was observed in the multiunit activity on all electrodes. Bursting was particularly pronounced in the animals under ketamine anesthesia. The envelope of the temporal pattern was almost identical in all electrodes even when the distance between them reached as much as 1 mm, indicating a highly synchronized activity across cortex. The frequency of the rhythmic burst pattern was in the range of 0.8–2 Hz, differing from that of cardiac (typically 3–4 Hz) or respiratory activity (0.4–0.6 Hz). Within each burst, a fine structure could be observed, which was much more evident under ketamine anesthesia. Typically spike trains from two or three neurons contributed to the burst, each with frequency in the range of 8–10 Hz. Figure 1 shows traces for spontaneous activity under ketamine anesthesia. Autocorrelation measurements of single units demonstrated the existence of a significant secondary peak that corresponded to the rhythmicity of the intraburst temporal pattern (Fig. 2).

The ICMS manipulation resulted in an increment in the number of cells that exhibited a significant secondary peak in their autocorrelograms. This effect appeared to depend on the distance of the cells from the stimulating electrode as well as one the anesthetic used (Fig. 3). At the stimulating electrode, the increment under ketamine conditions was not statistically significant; under Nembutal the increment was larger, although it did not quite reach significance level. In the adjacent electrode, located about 250 µm away, the increment was significant under both ketamine and Nembutal. In farther electrodes, the changes were not significant. After the waiting period (LATER), with both anesthetics, measurements were statistically indistinguishable from PRE levels, indicating some reversal from POST. Most of the cells that exhibited a significant secondary peak before ICMS (PRE), still showed the peak after ICMS. Under ketamine, 87% of the cells with a peak in PRE showed a secondary peak

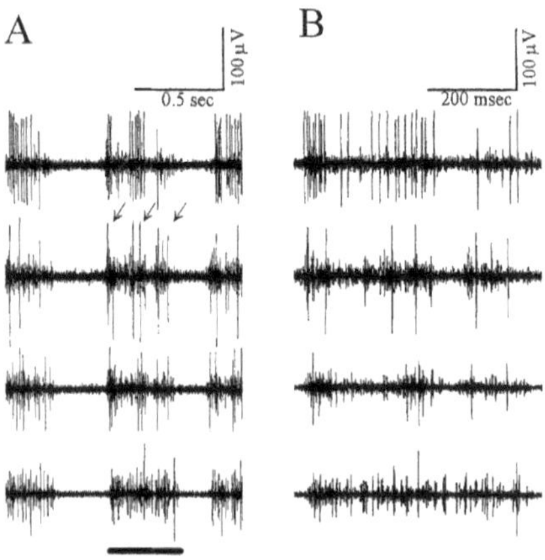

Fig. 1A, B Four-channel recording of spontaneous activity from a ketamine experiment. **A** shows traces at a low time resolution in which the burst and interburst time structure is indicated by *arrows*. All neuronal activity in the four electrodes, separated by about 250 µm, appeared to be synchronized. **B** shows the recording time marked by a *bar at the bottom of* **A**, at a higher time resolution. Note that, in the intraburst time structure, synchronous spike activity between different electrodes is less apparent

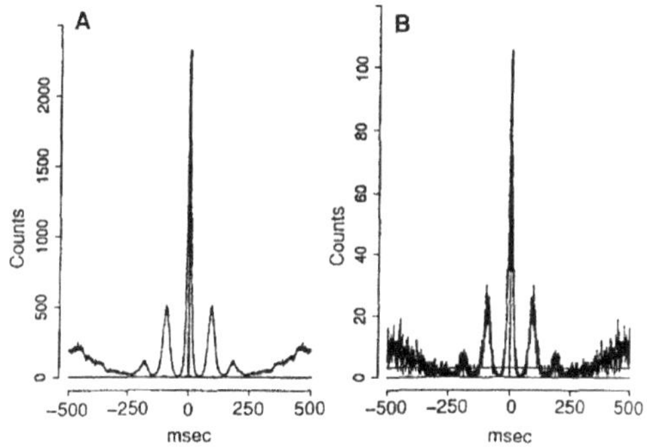

Fig. 2A, B Autocorrelation measurement of spontaneous activity. **A** shows a correlogram of the multiunit activity of ketamine experiment. **B** shows the autocorrelogram for a single cell extracted from the multiunit activity. Note the significant secondary peak at ~100 ms delay, another small peak at 200 ms. These peaks reflect the intraburst rhythmic activity that oscillates with a frequency of 8–12 Hz. A wide peak with a delay of about 450 ms can also be distinguished. This peak is associated with the 2- to 4-Hz burst structure. The *horizontal lines* indicate the 95% confidence limit for significance level. Bin size 0.5 ms. The central bin has been deleted

in their autocorrelograms at POST and 67% were still showing the peak at LATER. Under Nembutal, or POST and LATER, 62% of the cells with a peak at PRE were found to conserve the secondary peak in their autocorrelogram.

Afterdischarge cross-correlations

Cross-correlograms were computed for 1780 pair combinations in the PRE, POST and LATER conditions. Statistical analysis of the cross-correlograms showed that about half of all correlograms exhibited peaks that were over the 95% confidence level. There was no significant change in the number of peaked correlograms throughout the several phases of the experiments (Table 1). As examination of all correlogram shapes showed that about 5% showed narrow peaks of about 5 ms width with amplitude at least 3 times larger than the background. Most correlograms exhibited wide peaks, multiple narrow peaks, or a combination of both as shown in the samples of Fig. 4. The number of correlograms with significant peaks observed under ketamine anesthesia showed no dependence on the distance between the neurons in each pair. Under Nembutal, the probability of a correlogram with a significant peak appeared to be dependent on distance between the neurons (Fig. 5).

Although the total number of correlograms that exhibited a significant peak did not change during successive phase of the experiments, we observed changes in cross-correlogram area after the ICMS period. The integral of the correlogram was used as a measure of the strength of

Fig. 3A, B Effect of intracortical microstimulation (ICMS) on spontaneous oscillations. The number of autocorrelograms of spontaneous activity with secondary peaks increased after ICMS. **A** shows the percentage of total correlograms in the ketamine experiments that exhibited a significant secondary peak. The increment produced by ICMS only reached statistical significance in the adjacent (to stimulating) electrodes and appeared to be reversible. In **B**, similar effects are seen in the Nembutal experiments. Note that, in the PRE stage with ketamine, an overall higher percentage of cells with autocorrelograms with significant secondary peaks was observed. In **A** and **B** the *asterisks* indicate levels that were significantly different from PRE ($P<0.05$, paired Student's *t*-test). The *vertical line* at each point indicates SE. Population for each condition is indicated in *parentheses*

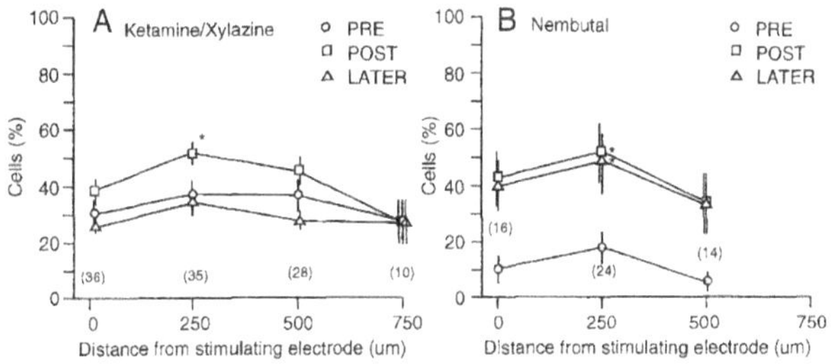

Table 1 Significant cross-correlations. The total proportion of correlograms that exhibit peaks above the 95% significance level. No significant changes were found in the total counts for either the ketamine or the Nembutal experiments

	PRE	POST	LATER	Cells (*n*)
Ketamine	56%	57%	50%	1402
Nembutal	41%	43%	56%	378

the interactions between the two cells of a given pair and was compared across the phases of an experiment. Only those cell pairs that were consistently recorded during the entire experiment were included (n=1180). Under ketamine anesthesia, 26.6% of all correlograms changed in area after ICMS. Nineteen percent of all cell pairs showed an increase in the area of the correlogram, corresponding to an increase in the strength of their interaction; only 7.6% decreased the strength of their interaction. Changes in the correlogram areas were found to be dependent on the distance between cells of the pair (Fig. 6). For instance, the probability of an increase in correlation strength for the spike trains of a cell located at the stimulating electrode and any other cell was inversely proportional to distance. At the same time, correlations between activity of neurons recorded at the same electrode increased in all sites. Under Nembutal anesthesia, about the same percentage (28.4%, n=322) of the cell pairs exhibited changes in the correlogram area after ICMS, but increases and decreases were about equally likely: 12.4% showed an increase, while 16% showed a decrease. Unlike the ketamine observations, these area changes did not appear to be particularly dependent on the distance between neurons.

Joint peristimulus time histograms

From the large number of cell pairs in the afterdischarge sample, a subset of cell pairs was selected for analysis with JPST histograms (Aertsen et al. 1989). The selection criterion required preservation of a significant peak in the ordinary cross-correlogram throughout the experimental protocol. JPST histograms measure the near-coincidence firing at different times from the initiation of the trial (stimulus presentation), thus producing a temporal dissection of a cross-correlogram. Correction is made for the correlation produced by covariation of neuron firing rates, and the significance of the excess (or deficit) of near-coincidence firing is established.

A total of 311 neuron pairs recorded under ketamine were examined. In PRE, 11.5% (36) of the JPST histograms exhibited excess joint firing (above the 99% confidence limit) at the time of the acoustic stimulus re-

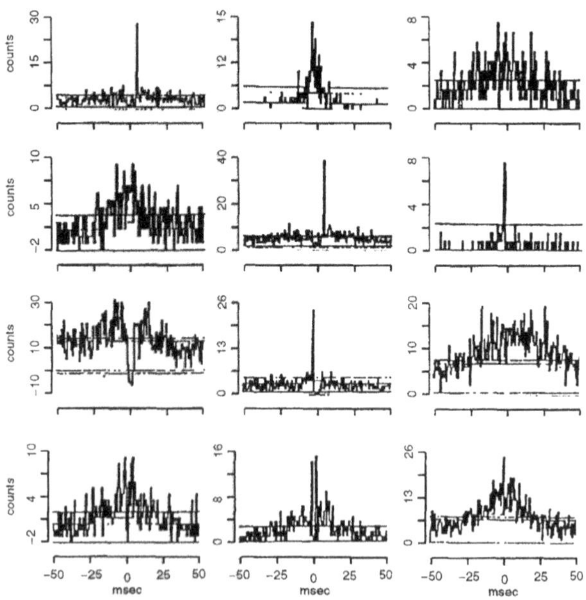

Fig. 4 Samples of cross-correlations between isolated neurons, taken from various experiments under ketamine and Nembutal anesthesia. The *horizontal lines* reflect the 95% confidence limit for significance

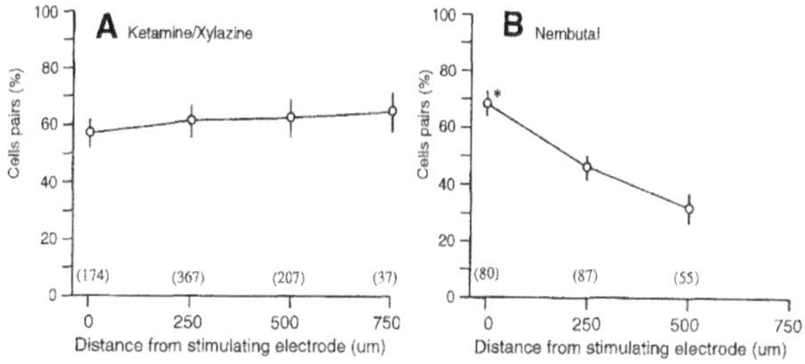

Fig. 5A, B Distribution of cross-correlations with significant peaks versus distance between neurons recorded during the POST afterdischarge activity. For ketamine (**A**), there is a uniform distribution. For Nembutal (**B**), the probability of finding a significant cross-correlation is dependent on the distance between the two cells. Refer to the Discussion for further comments. The *number in parentheses at the bottom* represents the total number of cells analyzed at that location. In **B**, the percentage of cells at the stimulating electrode (*asterisk*) is significantly higher ($P<0.05$, paired Student's *t*-test) than in the adjacent and farther electrodes. The *vertical lines* indicate the standard error

Table 2 Significant excess joint firing in neuron-pair joint peristimulus time histograms (JPST). Proportions of corrected JPST that exhibit peaks along the diagonal, indicating excess joint firing. *STIM* refers to the time of the primary response to the acoustic stimuli and *AD* refers to the afterdischarge period. Following ICMS in ketamine as well as in Nembutal experiments, the incidence of excess joint firing appeared to be unaffected during the stimulus response time. During AD time, we observed an increase in incidence of significant joint firing. At the LATER stage, a reversal to original values or a decrement occurred

	PRE		POST		LATER		cells (n)
	STIM	AD	STIM	AD	STIM	AD	
Ketamine	11.5%	10.9%	9.0%	18.6%	4.8%	6.7%	311
Nembutal	4.8%	3.2%	3.2%	9.6%	0%	3.2%	62

sponse, spanning between 5 and 50 ms. About the same percentage was found to have excess joint firing in the afterdischarge response during the interstimulus time (Table 2). After ICMS, there was no change in the number of JPST histograms with excess joint firing during the stimulus response time. However, we found an increase in the percentage (from 10.9 to 18.6) of JPST histograms that showed excess joint firing during the afterdischarge response, suggesting that ICMS increased the probability of joint firing during this time, but not during the direct stimulus response time. After the waiting period (LATER), the total number of JPST histograms with significant joint firing decreased during the afterdischarge response time, reflecting some degree of reversibility. An example of a neuron-pair JPST histogram measurement, at different times during the experiment, is shown in Fig. 7.

Fig. 6 The changes in correlation strength computed in the POST relative to the PRE stages for ketamine experiments. Increment in strength is noted by an *upward arrow* and the opposite is used for decrement. *Loop arrows* refer to cross-correlations measured between cells isolated from the same electrode. There are more increases in strength than decreases. Changes depend on both separation of neurons in a pair and on the distance from the stimulating electrode (*n* total number of correlograms analyzed in that situation)

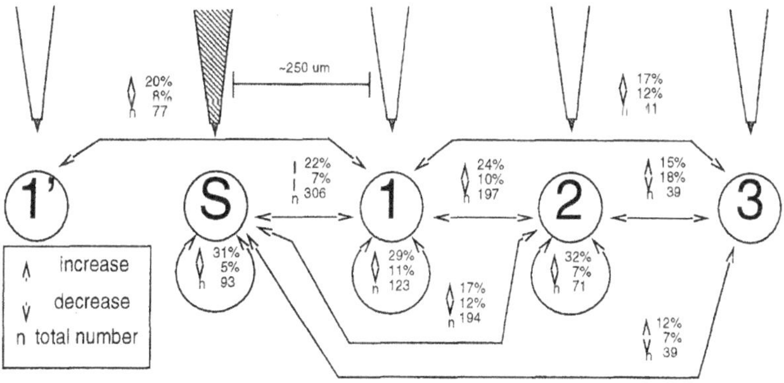

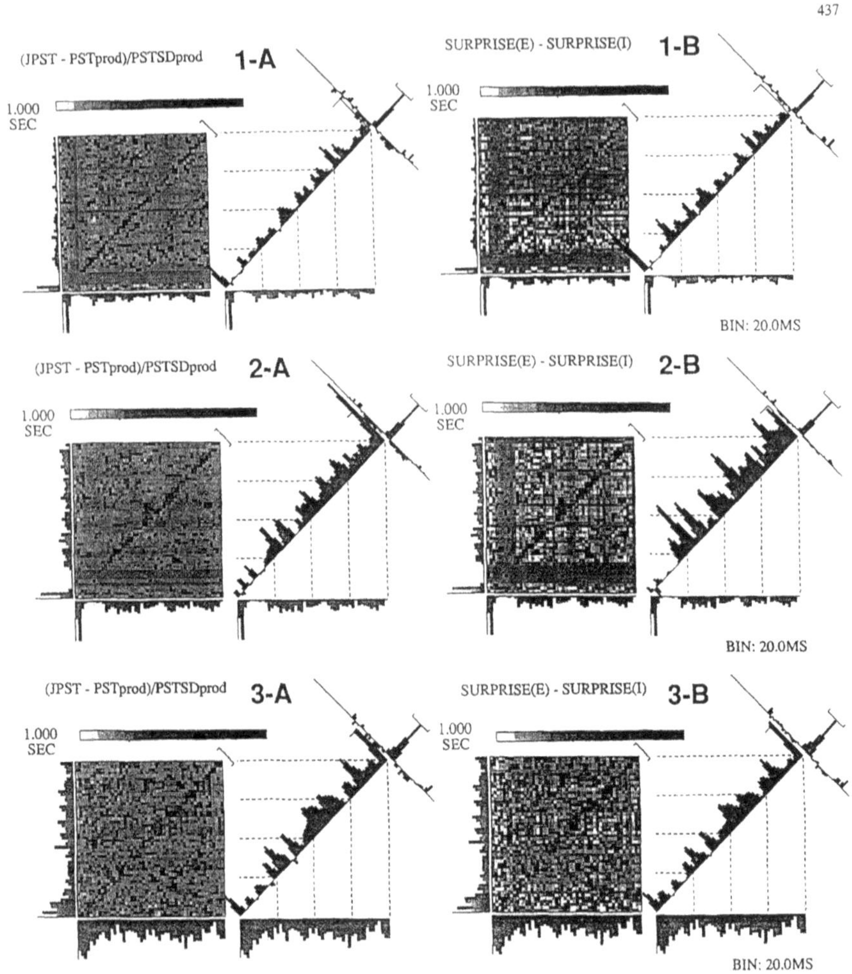

Fig. 7 Example of a joint peristimulus time histogram (*JPST*) analysis for a cell pair, one located at the stimulated electrode and one at an adjacent electrode. *1-A* shows the JPST corrected for the effects of stimulus modulation of the firing. At the *upper right* is the cross-correlogram constructed by summation within paradiagonal stripes of the corrected JPST in the matrix at *left*, and normalized for the number of small squares along that diagonal. The remaining diagonal histogram represents the normalized mean stimulus-time locked course of near-coincidence firing of the two neurons in the same sense that PST histograms show the mean stimulus-time locked course of the individual neuron firing. The PST histograms for the two neurons are, respectively, at the *left* and *bottom* of the JPST box. *1-B* shows the significance measurement of the joint firing. Surprise matrices were scaled for display between −2.996 (corresponding to a significance level of 95% for less joint firing than expected) and +4.605 (similarly of 99% of more joint firing than expected) as indicated along the *gray wedge*. Here, there is a significant, excess joint firing during the stimulus response time, but not during the subsequent interstimulus time. *2-A* and *2-B* show JPST of the same cell pair following ICMS. Note that excess joint firing is no longer significant at stimulus response time, but highly significant at times during the afterdischarge period. Also note that the cross-correlation central peak is larger, although its profile appeared unchanged, as compared with that of the previous experimental stage. *3-A* and *3-B* show JPST of the same cell pair in LATER. Note that excess joint firing is no longer significant at stimulus response time, nor during the afterdischarge period. Again, note that the cross-correlation appeared unchanged as compared with that of the previous experimental stages

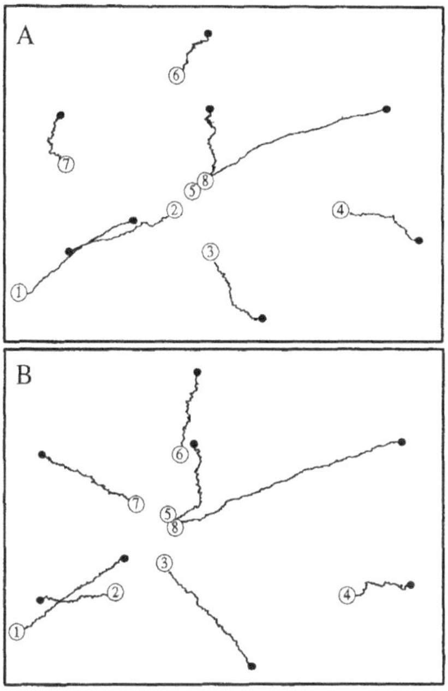

Fig. 8A, B A gravity run on a group of eight cells from a Nembutal experiment. Each neuron was represented as a particle in an 8-space. Particle trajectories in an 8-space projected down to a 2-space with initial position defined by a *black point* drawn at the beginning of each trace. Axes represent projected location of particles on an arbitrary scale. Cells *5* and *8* were recorded at the stimulating electrode; cells *2*, *3*, *4*, *6*, and *7* were recorded at adjacent electrodes and cell *1* was at the farthest electrode. Cells 2, 5, and 8 interact strongly before ICMS (**A**), while 3, 4, 6, and 7 interact weakly. Cell 1 does not interact at all with any of the other cells, and its particle is seen moving away from the other particles. Following ICMS (**B**), cells 3, 5, 6, 7, and 8 show stronger interactions, while cells 2 shows a weaker interaction with cells 5 and 8. Cells 1 and 4 exhibit little or no change in their interactions

From the Nembutal recordings, 62 neuron pairs were taken for analysis. The percentage of JPST histograms with significant excess joint firing during both stimulus response time and interstimulus response time was smaller than in the ketamine experiments (Table 2), but showed the same effects of ICMS. After ICMS, there was no change in the number of JPST histograms with significant excess joint firing during stimulus response time, but an increment in the number with excess joint firing during the afterdischarge response. With the additional waiting period of the LATER condition, the numbers returned to values comparable with those observed in the PRE condition.

Gravity analysis

Gravity analysis detects temporal relationships between spike trains as a clustering of particles in a multidimensional space. Particles positioned in a multidimensional space represent neurons and they attract or repel each other, depending on the coincidence of neuronal firing (Gerstein and Aertsen 1985). Clustering of a set of particles implies a functional relationship among the neurons the particles represent, in other words, a neuronal assembly based on preferred timing relationships. The analysis is particularly suited to demonstrate dynamics in the neuronal organization.

Although a quantitative comparison between different experiments has not yet been developed, a number of consistent features appeared in the data analyzed by the gravity method. First, distinct condensations, that is, aggregation of particles, and formation of clusters were observed in all experiments. In 80% (n=15) of the analyzed cases, a change in the condensation was observed after ICMS was presented, implying a change in the underlying neuronal organization. In all data examined, cells that were located more distant to the stimulating electrode did not cluster or clustered more slowly (Fig. 8). ICMS increased the overall rate of condensation, thus reflecting an increased level of interaction between cells. A close inspection of the condensation steps revealed that the changes in particle pair distances reflect the rhythmicity seen in spontaneous activity. Indeed, an autocorrelation of the distance steps (not shown here) revealed, in most cases, a large secondary peak similar to those seen in the autocorrelograms of spontaneous activity. In control experiments, condensations did not exhibit any changes through the sham ICMS protocol.

Discussion

Spontaneous activity and rhythmic firing patterns

The most striking feature of the spontaneous activity was the presence of repetitive temporal firing patterns with both high- and low-frequency components. The 8- to 12-Hz oscillatory activity was observed in all ketamine experiments, as well as in light Nembutal anesthesia. This firing pattern occurred in bursts rather than in a continuous fashion, the onset of the bursts coinciding with the low-frequency activity structure. The low-frequency firing pattern (in the 0.8- to 2-Hz range) was prominent in multiunit activity, but more irregular than the 8- to 12-Hz oscillations.

Complex temporal firing patterns have been seen in various circumstances in other brain structures. Synchronized rhythmic activity has been reported for neuronal groups of similar response characteristics within columns, in neighboring hypercolumns, in different visual cortical areas, and even between different hemispheres in the cat visual cortex (Gray and Singer 1989; Engel et al. 1991; Gray et al. 1992; Whittington et al. 1995). Recently, oscillations with the same characteristics also have been found in the primary visual cortex of awake monkeys (Eckhorn et al. 1993). Such rhythmic activity has recently received attention because it involves spatial coordination of temporal activity patterns; this may be involved in stimulus integration (binding) and emergence of perceptions (Singer and Gray 1995). It has been suggested that these oscillations arise in the cortex, get reorganized in the thalamus, and then reenter the cortex completing a cortico-thalamo-cortical resonant loop (Llinas et al. 1991). Indeed, Silva et al. (1991) demonstrated that intrinsic oscillations in the range of 5–12 Hz were generated by layer V pyramidal cells in rat neocortex slices. Recently, it has been demonstrated that the cat's primary auditory cortex has stimulus-induced and spontaneous rhythmic behavior of single units in the frequency range of 5–15 Hz (Eggermont 1992) and 20–60 Hz (McCormick and Gray 1995). Spontaneous oscillations appeared less frequently at depths corresponding to layer IV in comparison with both the upper and deeper layers. In our experiments, we found rhythmic firing patterns similar to those reported in cat's auditory cortex.

We observed that, overall, ICMS increased the rhythmic firing activity. The artificial joint firing induced during ICMS in the affected region produced a change in the proportion of cells that subsequently exhibited a significant secondary peak in their autocorrelogram. Although there was a consistent increase in cells with these characteristics in all but the farthest recording sites, it was statistically significant only in the adjacent electrodes, located about 250 μm away. The observed change in the proportions of cells with significant 8- to 12-Hz oscillatory behavior could be seen under both ketamine and Nembutal anesthetics.

The mechanism by which cortical cell interactions and properties change with ICMS is yet unknown. There is even the possibility that the loci of the changes may not be located in the cortex itself, but that the changes are reflections of ICMS-induced changes in thalamus caused by anterograde and retrograde stimulation. In addition, it is unclear how these changes depend on the ICMS parameter choice. In all our experiments, only one set of parameters such as current amplitude and frequency was used. Although the effect on the proportion of cells with oscillatory behavior reached no further than the adjacent electrode, during the stimulation we were able to observe ICMS-evoked activity in cortical cells that were located as far as 750 μm from the stimulating electrode.

Afterdischarge cross-correlation studies

Afterdischarge responses show a clear dependence on the physical parameters of a wide range of auditory stimuli and may be instrumental in the cortical processing of auditory stimuli (Altman et al. 1980). Statistical analysis of all cell pair cross-correlograms showed that about half of all correlograms exhibited peaks that were over the 95% confidence level. The shapes of the correlogram peaks suggested that most of the correlations were probably caused by shared input. Only about 5% of the cross-correlograms exhibited short-delay, narrow peaks that would reflect direct synaptic interactions. In the Nembutal but not in the ketamine experiments, the probability of finding significant cross-correlations was inversely proportional to the distance between the two neurons. Several reports have previously indicated that the probability of long-distance interactions decreases as a function of the distance between the recorded units (Toyama et al. 1981; Vaadia et al. 1991; Hata et al. 1993). Other parameters besides distance that may affect the incidence of significant correlations may be the receptive field properties. Cells that share receptive field properties may be more likely to exhibit significant correlation of activity even when they are located far apart (Engel et al. 1991; T'so 1991; Lowel and Singer 1992).

The total number of cross-correlograms that exhibited a significant peak did not change after our weak microstimulation (ICMS). However, we observed that there were changes in correlation strength, seen as changes in the area under the cross-correlogram. After ICMS, 19% of all cross-correlations increased in strength, while 7.6% showed a decrease in strength. These changes were distributed differently, depending on the location relative to the stimulating electrode and on distance between the cells. The largest increases in correlation strength occurred between cells located near the same electrode, while the probability of a correlation between a cell in the stimulating electrode and one in another electrode decreased with distance from the stimulating electrode (Fig. 6). In the Nembutal experiments, a similar but much weaker tendency in the changes was seen. These results are consistent with the Hebbian notion of changes in synaptic strength associated with frequent coactivation.

Joint peristimulus histograms

PST histograms characterize neuron-pair relationships as cross-correlation does, but as a function of time relative to the initiation of the stimulus, thus detecting significant joint firing that may be transitory and variable in time, but locked to the stimulus. Significant excess joint firing was shown by a smaller fraction of the JPST histograms than of the ordinary cross-correlograms. In the JPST histograms cell pairs could show significant joint firing only during the stimulus response time, only during the interstimulus time, or in both. The excess joint firing during the afterdischarge (interstimulus time) often exhibited a temporal pattern that resembled that of the spontaneous activity, but time-locked to the stimulus. After ICMS, we found an increment in the number of JPST histograms with significant joint firing during the interstimulus time; the number with significant joint firing during stimulus response time did not change. We interpret this finding as a reflection of an increment in the corticocortical connections among cells, rather than an increment in the shared input coming from subcortical structures. The fact that the proportion of significant JPST histograms is larger under ketamine may also provide some support for this assertion, in light of the possible observation of different populations of cells under the two anesthetics as we have discussed above.

In recent years, the study of oscillatory firing patterns and cross-correlations has been explored because it has been hypothesized that oscillations and synchronic activity are instrumental in perceptual integration. The idea is that an assembly of detectors that respond to coherent features of a sensory input becomes distinguished from detectors responding to noncoherent features, by synchronous and in-phase oscillatory responses (Gray et al. 1989; Singer 1991). However, in our experiments, we observed increased stimulus-locked synchronous oscillations only during the afterdischarge responses. Electrophysiological recordings of oscillatory activity in monkey somatosensory cortex have also shown suppression immediately after sensory stimulation (Ahissar and Vaadia 1990). Thus different oscillatory activities may be present during, in comparison with after, the stimulus presentation. Apparently, the 40- to 60-Hz synchronous oscillations in the visual system are only significant during presentation of the stimulus and are not time-locked to it, while lower-frequency oscillations such as the ones observed in our experiments are observed only during the afterdischarges that follow a stimulus presentation.

Gravity analysis

Although more quantitative aspects of the gravity method have yet to be developed, there were some consistent qualitative features in these data. We observed that the distance between particles mostly decreased at a constant rate, indicating a constant interaction among the observed cells. Occasionally, a short-duration, transient increase was observed in the condensation rate. Such special events are an indication of transient, highly correlated activity that would normally be lost by a standard cross-correlogram owing to averaging. We suggest that one factor that may significantly contribute to the mostly constant particle condensation rate in these experiments was the fact that most of the data was taken from interstimulus time, that is, during the afterdischarge activity. It is possible that fast, transient, high-correlation events tended to occur more frequently during stimulus response time, as was observed in visual cortex (Gray et al. 1989; Singer 1991).

The gravity analysis corroborated that cells spatially located near each other, i.e., recorded on the same electrode, had the tendency to interact strongly, and cells located at large anatomical distances frequently did not interact at all. However, a small anatomical distance was not a requisite for strong interaction, nor sufficient for it. Indeed, we found that the formation of a functional cluster of neurons did not necessarily relate to the anatomical distance between the cells that belonged to that cluster. After ICMS, in most cases, we observed that the rate of condensation was increased, implying an increased level of interaction between the cells. This was consistent with the results obtained from cross-correlation measurements as well as the JPST histograms. Again, these results suggest an increased corticocortical interaction during the synchronous afterdischarge activity.

Neuronal assemblies

The afterdischarge consists of strong, low-frequency synchronous activity that follow sensory stimulation and exhibits a rhythmicity similar to that seen in the ongoing spontaneous activity. Most of the cells shared these characteristics, thus possibly forming the basis for a high-level assembly; the coexistence of different levels of assemblies has been proposed (Gerstein et al. 1989). Application of ICMS produced a change in the afterdischarge response (Maldonado and Gerstein 1996), increasing the number of cells that exhibit significant synchrony and increasing the entrainment of the low-frequency oscillatory firing pattern. The proportion of cells that showed entrainment of oscillatory activity was found not to be dependent on the location of the cell in the cortex, but rather on the distance of the cell from the stimulating electrode. This is consistent with a Hebbian-like mechanism, which would dependent on the spatial extent of the cofiring effects of the ICMS.

The Hebbian and the correlation assembly models are both defined through interactions among the member neurons. Our cross-correlation studies show that assemblies that are identified by criteria based on correlation of firing are also related to assemblies based on receptive field properties (Maldonado and Gerstein 1996). The horizontal extent of the membership decays similarly with distance both in the case of assembly membership based on frequency response histograms or on correlation of firing. In addition, the effects of ICMS is similar in the two types of assembly.

Cross-correlogram analyses only give an average estimation of the interactions among cells in the assembly; the JPST histogram and gravity analyses, on the other hand, showed that even before ICMS interactions and assembly membership could be transitory and related to the auditory stimulus delivery time. Clearly ICMS modifies assembly organization: the strength of many neuronal interactions either increased or decreased after a period of ICMS. At the same time we have shown that many of the changes induced by ICMS showed a degree of reversibility. Thus weak electrical stimulation is able to modify synaptic interactions temporarily and reversibly and, therefore, assembly membership.

Acknowledgements This study was supported by NIH grants MH46428 and DC01249.

References

Aertsen AM, Gerstein GL, Habib MK, Palm G (1989) Dynamics of neuronal firing correlation: modulation of "effective connectivity". J Neurophysiol 61:900–917

Ahissar E, Vaadia E (1990) Oscillatory activity of single units in a somatosensory cortex of an awake monkey and their possible role in texture analysis. Proc Natl Acad Sci USA 87:8135–8139

Altman JA, Bekhterev NH, Kotelenko LM, Kudriavtseva IN (1980) Afterdischarges of cat medial geniculate body neurons. Fiziol Zh SSSR 66(1):80–8

Ballard DH, Hinton GE, Sejnowski TJ (1983) Parallel visual computation. Nature 306:21–26

Bedenbaugh P (1993) Plasticity in the rat somatosensory cortex induced by local microstimulation and theoretical investigations of information flow through neurons. PhD dissertation, Department of Bioengineering, University of Pennsylvania

Boch R (1986) Behavioral modulation of neuronal activity in monkey striate cortex: excitation in the absence of active central fixation. Exp Brain Res 64:610–614

Deppisch J, Pawelzik K, Geisel T (1994) Uncovering the synchronization dynamics from correlated neuronal activity quantifies assembly formation. Biol Cybern 71(5):387–99

Eckhorn R, Frien A, Bauer R, Woelbern T, Kehr H (1993) High frequency (60–90 Hz) oscillations in primary visual cortex of awake monkeys. Neuroreport 4:243–246

Eggermont J (1992) Stimulus induced and spontaneous rhythmic firing of single units in cat primary visual cortex. Hear Res 61: 1–11

Engel AK, Kreiter AK, Koenig P, Singer W (1991) Interhemispheric synchronization of oscillatory neuronal responses in car visual cortex. Science 252:1177–1179

Felleman DJ, Van Essen DC (1991) Distributed hierarchical processes in the primary cerebral cortex. Cereb Cortex 1(1):1–47

Gerstein GL, Aertsen AM (1985) Representation of cooperative firing activity among simultaneously recorded neurons. J Neurophysiol 54:1513–1528

Gerstein GL, Perkel DH, Dayhoff JE (1985) Cooperative firing activity in simultaneously recorded populations of neurons: detection and measurements. J Neurophysiol 5:881–889

Gerstein GL, Bedenbaugh P, Aertsen AM (1989) Neuronal assemblies. IEEE Trans Biomed Engl 36(1):4–14

Glaser EM, Ruchkin DS (1976) Principles of neurobiological signal analysis. Academic, New York

Gray C, Singer W (1989) Stimulus-specific neuronal oscillations in orientation columns of cat visual cortex. Proc Natl Acad Sci 86:1698–1702

Gray C, Koenig P, Engel AK, Singer W (1989) Oscillatory responses in cat visual cortex exhibit inter-columnar synchronization which reflects global stimulus properties. Nature 338: 334–337

Gray C, Engel AK, Koenig P, Singer W (1992) Synchronization of oscillatory neuronal responses in cat striate cortex. Temporal properties. Vis Neurosci 8:337–347

Gross CG, Rocha-Miranda CE, Bender BE (1972) Visual properties of neurons in the inferotemporal cortex of the macaque. J Neurophysiol 35:96–112

Hata Y, Tsumoto T, Sato H, Hagihara K, Tamura T (1993) Development of local horizontal interactions in the cat visual cortex studied by cross-correlation analysis. J Neurophysiol 69:40–56

Lettvin JY, Maturana HR, McCulloch WS, Pitts WH (1959) What the frog's eye tells the frog's brain. Proc IRE 47(11):1940–1951

Lindsey BG, Segers LS, Morris KF, Hernandez KM, Saporta S, Shannon R (1994) Distributed actions and dynamic associations in respiratory-related neuronal assemblies of the ventrolateral medulla and brain stem midline: evidence from spike train analysis. J Neurophysiol 72(4):1830–51

Llinas R, Grace AA ,Yarom Y (1991) In vitro neurons in mammalian cortical layer 4 exhibit intrinsic oscillatory activity in the 10- to 50-Hz frequency range. Proc Natl Acad Sci USA 88: 897–901

Lowel S, Singer W (1992) Selection of intrinsic horizontal connections in the visual cortex by correlated neuronal activity. Science 255:209–212

Maldonado PE, Gerstein GL (1996) Reorganization in the auditory cortex of the rat induced by intracortical microstimulation: a multiple single-unit study. Exp Brain Res 112:420–430

Malsburg C von der, Singer W (1988) Principles of cortical network organization. In: Radik P, Singer W (eds) Neurobiology of neocortex. Wiley and Sons, New York, pp 69–99

McCormick DA, Gray CM (1995) Physiologically identified cell groups have distinct receptive field and morphological properties in cat striate cortex. Soc Neurosci Abstr 21(2):1506

Palm G, Aertsen AM, Gerstein GL (1988) On the significance of correlations among neuronal spike trains. Biol Cybern 59:1–11

Perkel DH, Gerstein GL, Moore GP (1967) Neuronal spike trains and stocastic point processes. II. Simultaneous spike trains. Biophys J 7(4):419–440

Sheperd G (1990) The synaptic organization of the brain. Oxford University Press, Oxford

Silva LR, Amitai Y, Connors BW (1991) Intrinsic oscillations of neocortex generated by layer 5 pyramidal neurons. Science 251:432–435

Singer W (1991) The formation of cooperative cell assemblies in the visual cortex. In: Kruger J (ed) Neuronal cooperativity. Springer, Berlin Heidelberg New York, pp 165–183

Singer W, Gray CM (1995) Visual feature integration and the temporal correlation analysis. Annu Rev Neurosci 18:555–586

Toyama K, Kimura M, Tanaka K (1981) Cross-correlation analysis of interneuronal connectivity in cat visual cortex. J Neurophysiol 46:191–201

Tso D (1991) Connectivity and functional organization in the mammalian visual cortex. In: Kruger J (ed) Neuronal cooperativity. Springer, Berlin Heidelberg New York, pp 249–279

Vaadia E, Ahissar E, Bergman H, Lavner Y (1991) Correlated activity of neurons: a neural code for higher brain functions? In: Kruger J (ed) Neuronal cooperativity. Springer, Berlin Heidelberg New York, pp 249–279

White EL (1989) Cortical circuits. Birkhauser, Boston

Whittington MA, Traub RD, Jefferys JG (1995) Synchronized oscillations in interneuron networks driven by metabotropic glutamate receptor activation. Nature 373 (6515):612–615

Young MP, Yamane S (1992) Sparse population coding of faces in the inferotemporal cortex. Science 256 (5061):1327–1331

Chapter 12
Plasticity of Receptive Fields in Auditory Cortex

Auditory Experiments and Modeling

Michael A. Kisley

Auditory cortex function, like that of other cortical areas, depends importantly upon the following principles that characterize the papers included in this section: (i) operational units are not single neurons, but neuronal *populations* and (ii) cortical organization is *dynamically* regulated. George Gerstein helped pioneer these ideas, and throughout his career, he contributed to the relevant theoretical and experimental literatures that have validated and refined them.

The studies described here were completed, while I was a neuroscience graduate student working under George's supervision. Although this research formed the backbone for my dissertation, it would not have been possible, nor would it have been as rigorous or important, without George's insight, experience, and intellectual generosity. Reading back through these papers, I can see the influence of George's guiding hand and high scientific standards. These papers in particular remind me that working with George meant tackling the most important scientific questions without regard for the particular method or measures to be utilized. We would ask questions first, only afterward learn the most appropriate methods, and build the necessary equipment. The practical skills I gained working with George were invaluable. But even more impactful to me was George's contagious scientific curiosity and his inventive spirit.

Among the valuable lessons I learned from George was the wisdom of designing a study based on a current scientific controversy. All three papers included in

M. A. Kisley (✉)
Department of Psychology, University of Colorado, Boulder, CO, USA
e-mail: mkisley@uccs.edu

A. Aertsen et al. (eds.), *Introducing Computation to Neuroscience*, Springer Series in Computational Neuroscience, https://doi.org/10.1007/978-3-030-87447-6_12

this section stemmed from a "hot topic" of the time. Although these studies were conducted almost 20 years ago, all of the topics are *still* hot and associated research activity has only increased. For "The continuum of operating modes for a passive model neuron" (Kisley and Gerstein 1999), we employed neural modeling to assess whether single neurons respond to synaptic inputs as temporal integrators or as coincidence detectors. Answering this question is vital to understanding how and when neurons work together to form neuronal assemblies. An insight from this paper concerning the slope of a neuron's membrane potential immediately prior to action firing continues to influence this field (e.g., Koutsou et al. 2012).

The other two papers mentioned in this section describe experimental studies conducted with rodents designed to deepen our understanding of *dynamic* cortical organization, specifically learning-induced plasticity and brain state-dependent modulation. "Daily variation and appetitive conditioning-induced plasticity of auditory cortex receptive fields" (Kisley and Gerstein 2001) describes the first study to demonstrate that conditioned pairing of a tone with a pleasant stimulus, as opposed to an unpleasant stimulus such as a foot shock, remaps auditory cortex receptive fields. Recent follow-up work by other researchers has shown that this effect is generalizable to other types of pleasant stimuli (we utilized electrical stimulation of the medial forebrain bundle), is behaviorally relevant, and occurs in humans as well (Puschmann et al. 2013). "Trial-to-trial variability and state-dependent modulation of auditory-evoked responses in cortex" (Kisley and Gerstein 1999) describes research that demonstrates a violation of the assumption that neuronal responses to an auditory stimulus are combined linearly with ongoing background activity. This observation is important because it suggests that computing an average response profile through simple stimulus-locked averaging may not be appropriate or accurate depending upon current brain state. This work continues to inform studies of cortical function conducted under different types and levels of anesthesia (e.g., Schölvinck et al. 2015). It has also been influential in human event-related potential research for which findings depend critically upon response averaging (e.g., Ouyang et al. 2017). The most innovative experimental aspect of this study was that we utilized an analysis of ongoing EEG activity in an anesthetized animal in order to trigger the presentation of auditory stimuli in real time.

George Gerstein's contributions to the study of cortical function are invaluable and enduring. In addition to his impact on the scientific literature, he continues to exert a lasting, positive influence through those of us with whom he has collaborated and mentored. I'm grateful to count myself among them.

References

Kisley MA, Gerstein GL (1999) Trial-to-trial variability and state-dependent modulation of auditory-evoked responses in cortex. J Neurosci 19(23):10451–10460. https://doi.org/10.1523/jneurosci.19-23-10451.1999

Kisley MA, Gerstein GL (1999) The continuum of operating modes for a passive model neuron. Neural Comput 11(5):1139–1154. https://doi.org/10.1162/089976699300016386

Kisley MA, Gerstein GL (2001) Daily variation and appetitive conditioning-induced plasticity of auditory cortex receptive fields. Eur J Neurosci 13(10):1993–2003. https://doi.org/10.1046/j.0953-816X.2001.01568.x

Koutsou A, Christodoulou C, Bugmann G, Kanev J (2012) Distinguishing the Causes of Firing with the Membrane Potential Slope. Neural Comput 24(9):2318–2345. https://doi.org/10.1162/NECO_a_00323

Ouyang G, Hildebrandt A, Sommer W, Zhou C (2017) Exploiting the intra-subject latency variability from single-trial event-related potentials in the P3 time range: a review and comparative evaluation of methods. Neurosci Biobehav Rev 75:1–21. https://doi.org/10.1016/j.neubiorev.2017.01.023

Puschmann S, Brechmann A, Thiel CM (2013) Learning-dependent plasticity in human auditory cortex during appetitive operant conditioning. Hum Brain Mapp 34(11):2841–2851. https://doi.org/10.1002/hbm.22107

Schölvinck ML, Saleem AB, Benucci A, Harris KD, Carandini M (2015) Cortical state determines global variability and correlations in visual cortex. J Neurosci 35(1):170–178. https://doi.org/10.1523/jneurosci.4994-13.2015

The Journal of Neuroscience, December 1, 1999, 19(23):10451–10460

Trial-to-Trial Variability and State-Dependent Modulation of Auditory-Evoked Responses in Cortex

Michael A. Kisley and George L. Gerstein

Department of Neuroscience, University of Pennsylvania School of Medicine, Philadelphia, Pennsylvania 19104-6074

Recent experimental work has provided evidence that trial-to-trial variability of sensory-evoked responses in cortex can be explained as a linear superposition of random ongoing background activity and a stationary response. While studying single trial variability and state-dependent modulation of evoked responses in auditory cortex of ketamine/xylazine-anesthetized rats, we have observed an apparent violation of this model.

Local field potential and unit spike trains were recorded and analyzed during different anesthesia depths—deep, medium, and light—which were defined by the pattern of ongoing cortical activity. Estimation of single trial evoked response was achieved by considering whole waveforms, rather than just one or two peak values from each wave. Principal components analysis was used to quantitatively classify waveforms on the basis of their time courses (i.e., shapes).

We found that not only average response but also response variability is modulated by depth of anesthesia. Trial-to-trial variability is highest under medium levels of anesthesia, during which ongoing cortical activity exhibits rhythmic population bursting activity. By triggering the occurrence of stimuli from the spontaneously occurring burst events, we show that the observed variability can be accounted for by the background activity. In particular, the ongoing activity was found to modulate both amplitude and shape (including latency) of evoked local field potentials and evoked unit activity in a manner not predicted by linear superposition of background activity and a stereotyped evoked response. This breakdown of the linear model is likely attributable to rapid transitions between different levels of thalamocortical excitability (e.g., spike-wave discharges), although brain "state" is relatively fixed.

Key words: variability; modulation; local field potential; principal components analysis; anesthesia; ketamine; xylazine; auditory cortex; spike-wave discharge; high-voltage spindle

On a trial-by-trial basis, cortical activity evoked by sensory stimulation is extremely variable. This has been shown for both single units (Schiller et al., 1976; Whitsel et al., 1977; Heggelund and Albus, 1978; Rose, 1979; Tolhurst et al., 1983; Scobey and Gabor, 1989; Vogels et al., 1989; Softky and Koch, 1993) (but see Gur et al., 1997) and macropotentials, especially human evoked potentials (for review, see Childers et al., 1987) (also see Thomas et al., 1989; Liberati et al., 1991). The traditional method for dealing with such variability is averaging over many trials [units: Gerstein (1960); evoked potentials: Dawson (1951); for review, see Aunon et al. (1981)]. Such a manipulation requires the assumption of linear superposition between basically random ongoing background activity and a highly stereotyped, repeatable evoked response. Recently, voltage-sensitive dye (Arieli et al., 1996) and intracellular (Azouz and Gray, 1999) recordings from the visual cortex of anesthetized cats have provided support for the validity of this assumption. Nevertheless, nonstationarity of general brain state can lead to nonstationarity of the so-called "repeatable" response (for review, see Coenen, 1995) (also see Discussion), thus violating the assumptions necessary for averaging (Coppola et al., 1978; Möcks et al., 1987). In the present report we demonstrate a violation of linearity between ongoing and evoked activity in rat auditory cortex even at a stable plane of anesthesia.

Traditionally, trial-to-trial variability studies of auditory evoked potentials in subhuman species have involved characterizing each individual waveform with only one or two points per waveform (Tunturi, 1959; Worden et al., 1964; Horvath, 1969). Such simple measures of single trial evoked responses, such as the amplitude between two peaks, might not always be sufficiently sensitive to detect modifications of waveform time course. Principal components analysis (PCA), on the other hand, provides a method for quantifying the shape of an entire single trial waveform. This type of analysis has seen increasing application in neuroscience, including sorting of extracellularly recorded action potential waveforms (Abeles and Goldstein, 1977), analysis of evoked potentials (for review, see Chapman and McCrary, 1995), synaptic potentials (Astrelin et al., 1998), evoked extracellular field potentials (Musial et al., 1998), and current source density waveforms (Di et al., 1990). In the present study we use PCA to quantify waveform variability, and further to reduce variability by numerically classifying single trial evoked field potential responses into groups of similarly shaped waveforms.

By recording evoked local field potential and unit activity from the auditory cortex of rats anesthetized with ketamine/xylazine, we have endeavored to (1) quantify single trial evoked responses and their variability using whole waveforms rather than just one or two points, (2) examine the modulation of response and response variability by the different brain states available with our anesthetic regimen, (3) investigate the impact of ongoing activity (as the potential cause of variability), and (4) compare the variation in local field potential evoked responses with that of neu-

Received May 17, 1999; revised Aug. 30, 1999; accepted Sept. 13, 1999. This work was supported by Grants MH 46428 and DC 01249 from National Institutes of Health. We gratefully acknowledge Dr. Jeff Keating, Sanjiv Talwar, and Dr. Stuart Baker for technical assistance and helpful discussions, and Drs. Pawel Musial and Larry Palmer for comments on an earlier version of this manuscript.

Correspondence should be addressed to Michael A. Kisley, Department of Neuroscience, 215 Stemmler Hall/6074, University of Pennsylvania, Philadelphia, PA 19104-6074. E-mail: mike@mulab.physiol.upenn.edu.

rons. Our results are discussed in the context of thalamocortical rhythms and spike-wave discharges.

This work has been published previously in abstract form (Kisley and Gerstein, 1998).

MATERIALS AND METHODS

Female albino rats (Charles River, Wilmington, MA) weighing 250–350 gm were used for this study. Recordings were taken from both acutely prepared and chronically implanted animals. Results are presented as from a single population. The chronically implanted animals were also being used for a separate study involving recording during the performance of various auditory discrimination tasks. Detailed differences in procedures are noted below. All procedures are in accordance with the Institutional Animal Care and Use Committee at the University of Pennsylvania and the Society for Neuroscience's policy on the use of animals in neuroscience research.

Surgical procedures. Animals were initially anesthetized with an intraperitoneal injection of ketamine (70 mg/kg) and xylazine (8 mg/kg). Additionally, smaller doses were administered to maintain surgical anesthesia. Glycopyrrolate (0.06 mg/kg), a synthetic atropine antagonist, was injected subcutaneously to prevent respiratory tract secretions. Body temperature was monitored and maintained at 37°C with a fluid-filled heating pad.

Once surgical anesthesia was achieved (checked with pedal-withdrawal reflex), the head was shaved and fixed in a stereotaxic apparatus. Skin, muscle, and connective tissue were then cleared from the top and left temporal portion of the skull. Small stainless-steel screws were driven into holes drilled with a dental drill on the top of the skull. For acute experiments, the head of a long screw was affixed to the skull screws with dental acrylic. This "upward-facing" screw would later be attached, through a holder, to the operating table so that the stereotaxic ear bars could be removed. In this condition the head was held rigidly, but the ears were clear for auditory stimulation. A small craniotomy, ~1 mm in diameter, was made with a dental drill over stereotaxic coordinates 5.0 mm posterior and 4.0 mm ventral of bregma, corresponding to left primary auditory cortex (Paxinos and Watson, 1997). After the dura was cut with a fine needle, microwire electrodes were rapidly inserted into the brain, then slowly (~10 μm/min) lowered to the infragranular layers (0.8–1.2 mm). Once at their desired depth, the electrodes were fixed in place using dental acrylic. Recordings for acute experiments were begun once the extracellular action potential waveforms appeared stable. For chronically implanted animals, dental acrylic was then liberally applied over the exposed skull and around a connector (Microtech, Boothwyn, PA), the skin on the head was sutured, and the animal was allowed to recover in isolation for 1 week. These animals were later reanesthetized with the same regimen to acquire data for the present study.

Electrophysiology and sensory stimulation. Both single wire electrodes and tetrodes were used in this study. Single wires were either stainless steel or gold-plated Ni-Ch, 50-μm-diameter microwires (California Fine Wire, Grover Beach, CA). Tetrodes consisted of four 25 μm tungsten microwires twisted together. Impedance at 1 kHz for all wires was <500 kΩ.

Signals were buffered near the head with operational amplifier follower circuits, then passed through preamplifier, amplifier, and digital signal processing hardware (Plexon, Dallas, TX). Before the preamplifier, signal lines were split into two channels: extracellular action potentials ("units") and local field potentials. Unit data were filtered between 300 and 3000 Hz, and candidate spike-waveforms were sampled at 40 kHz. These waveforms were stored for later spike sorting analysis. Field potentials were bandpass-filtered between 0.5 and 300 Hz and sampled continuously at 1 kHz. All signals were referenced to a ground wire securely attached to the skull screws.

Clicks (0.6 msec square waves) were presented from a speaker located 22 cm from the right ear, 45° to the right of the animal's horizontal body axis, 0° azimuth. For some analyses, clicks were presented randomly with an interclick interval between 1.5 and 3.0 sec. Clicks were also presented in a manner time-locked to the ongoing local field potential activity as follows: for each single trial, one of several preselected delays was randomly selected. Once the ongoing field potential activity crossed an amplitude threshold, the delay was imposed and a click was presented. After 1.5 sec, another delay was randomly selected and so on. The timing of stimulus presentation was controlled with a Lab-PC board (National Instruments, Austin, TX) and accompanying software routines. All experiments were performed in a double-walled, sound-attenuated room.

Spike sorting and analysis. Bullock's (1997) definition of "multi-unit" activity was used in this report. All threshold crossings, regardless of waveform size or shape, were considered part of the multi-unit activity. Threshold levels were arbitrary but held fixed for an entire experiment. Generally, it appeared that between 5 and 10 single units contributed to the multi-unit recordings.

Often, several single units could be sorted out for individual analysis. For tetrode recordings, sorting was accomplished by making a scatter plot of the peak-to-peak amplitude of an action potential on one electrode versus the peak-to-peak amplitude of the same action potential recorded on another electrode of the tetrode (McNaughton et al., 1983; Gray et al., 1995). For single wires, sorting was accomplished by making a scatter plot of a waveform's projection onto the first principal component taken from a library of prerecorded waves versus the waveform's projection onto the second principal component (Abeles and Goldstein, 1977). For both sorting techniques, well isolated single units will tend to form well isolated clusters in the scatter plots.

Unit data were analyzed with peristimulus time histograms. The "centroid" of these histograms was computed as the mean latency of all spikes that occurred between 5 and 30 msec from the onset of the click. Rhythmicity of ongoing activity was assessed by autocorrelogram.

Analysis of evoked local field potentials. Each single trial evoked waveform, U_i, can be represented as a series of discrete values:

$$U_i = u_{ij}\{j = 1, 2, \ldots, m\},$$

where time between samples is 1 msec, and m is the number of points per waveform (either 100 or 128). In the first part of the study, the amplitude of single trial evoked responses was estimated using an "average-as-template" method. The average wave is computed from all n relevant waveforms using the standard method:

$$\bar{U} = \bar{u}_j\{j = 1, 2, \ldots, m\}$$

$$\bar{u}_j = \frac{1}{n}\sum_{i=1}^{n} u_{ij},$$

and then normalized (by the square-root of average power per point) to give the template waveform:

$$\bar{V} = \bar{v}_j\{j = 1, 2, \ldots, m\}$$

$$\bar{v}_j = \frac{1}{b}\bar{u}_j \qquad b = \left[\frac{1}{m}\sum_{j=1}^{m}\bar{u}_j^2\right]^{1/2}.$$

Single trial amplitude, A_i, is the scalar result of taking the inner product of this normalized template and a single trial waveform:

$$A_i = \frac{1}{m}\sum_{j=1}^{m}(u_{ij}\cdot\bar{v}_j)$$

Thus not only larger deflections from zero, but also greater similarity to the average waveform, will lead to a larger single trial amplitude. For comparison with a more traditional method, single trial amplitudes were also computed as the magnitude between temporally restricted minima and maxima (peak-to-peak method).

In the second part of this report, analysis of waveforms was performed using PCA. For this analysis, each single trial waveform was upsampled (using interpolation) by a factor of 5. After the mean (DC offset) was removed from each wave, the principal components were computed from the eigenvectors of the signal covariance matrix by the method described in Glaser and Ruchkin (1976). Respective eigenvalues describe the amount of data variance accounted for by each principal component.

By its very nature, PCA allows partial reconstruction of a waveform from those basis vectors that account for the greatest amount of variance present in the data. Because the first two components, F_1 and F_2, generally accounted for >90% of the data variance in our study, all single trials were represented with only these components:

$$F_1 = f_{1j}\{j = 1, 2, \ldots, m\} \qquad F_2 = f_{2j}\{j = 1, 2, \ldots, m\}$$

J. Neurosci., December 1, 1999, 19(23):10451–10460 10453

First and second "scores" were computed by projecting each single trial waveform onto the first and second principal components, respectively:

$$S_{1i} = \frac{1}{m}\sum_{j=1}^{m}(f_{1j}\cdot u_{ij}) \qquad S_{2i} = \frac{1}{m}\sum_{j=1}^{m}(f_{2j}\cdot u_{ij}).$$

Each score squared represents the signal power accounted for by that component. The sum-of-squares of all component scores is equivalent to the total power of the waveform, whereas the sum-of-squares of only the first two scores represents the power of the single trial waveform that can be fit by the shape of the first two components:

$$P_i = S_{1i}^2 + S_{2i}^2.$$

This quantity was taken as an estimate of the signal power for each waveform (the impact of this estimation is considered in Results). Because the first and second components are shaped differently, a quantitative estimate of shape for purposes of comparison between single trials is the fraction of waveform power accounted for by only the first component (S_{1i}^2/P_i).

All data analysis was performed with custom-written programs and Matlab analysis package (Math Works, Natick, MA). All statistical estimations and hypothesis tests were performed under the assumption of normal distributions.

RESULTS

Definition of anesthesia depth

One way to assess state-dependent modulation of both evoked responses and response variability is to record such responses under different depths of anesthesia. In the present study we define three depths of ketamine/xylazine anesthesia: deep, medium, and light. This classification is based on our empirical observations of these relatively stable, easily defined states. During an experiment, these states can be cyclically repeated by giving small (approximately one-fourth the amount used for induction), supplemental doses of ketamine/xylazine every 90 or so min. Although the time course of transition between these states varies from rat to rat, deep anesthesia is generally stable by 20 min after a supplemental dose of ketamine/xylazine, medium from ~50 to 80 min, and light after ~80 min.

Definition of anesthesia depth is achieved by a constellation of experimental parameters, the most important and reliable being the ongoing multi-unit activity. During deep anesthesia, multi-unit activity exhibits low spontaneous rates and infrequent bursts or very slow (<1.2 Hz) rhythmic bursting. Medium anesthesia is characterized by higher spontaneous rates (usually ~50% higher than deep anesthesia) and very clear rhythmic bursting activity at ~1.6 Hz. Under light anesthesia, multi-unit activity exhibits the highest spontaneous rates (often >100% higher than deep anesthesia) and usually tonic firing patterns, but very occasionally rhythmic bursting (>2 Hz). Breathing patterns are also correlated with anesthesia depth. Breathing is very regular and slow (~50 breaths per min) under deep anesthesia. In medium anesthesia, breathing is a little faster (50–80 per min) but still regular. Breaths are significantly faster (100–160 per min) under light anesthesia and quite irregular. The transition from medium to light anesthesia is also accompanied by the beginnings of spontaneous whisker movement and a pedal-withdrawal reflex. The spectral composition of ongoing field potential activity is also somewhat correlated with depth of anesthesia, but generally depended too strongly on cortical depth to be useful in our recordings. With regard to the classic definition of anesthesia depth, all anesthetic levels are within stage III, "surgical," anesthesia (Thurmon et al., 1996). We do not necessarily intend our definitions to align directly with past attempts to delineate deep, medium, and light surgical anesthesia, which are based on many more physiological variables than we monitored.

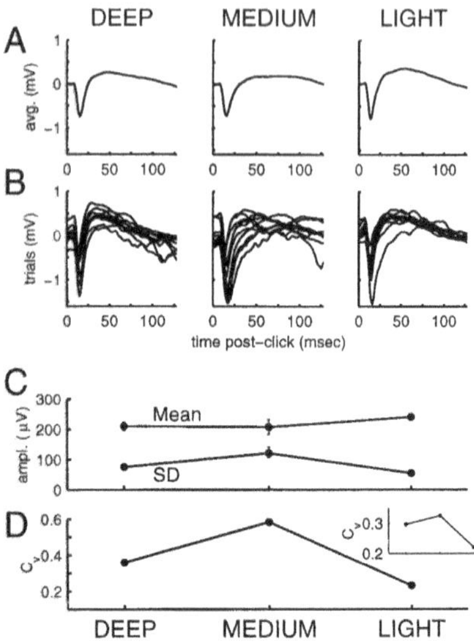

Figure 1. Depth of anesthesia (deep, medium, light) modulates click-evoked response and response variability. *A*, Average waveforms computed from 100–120 clicks at each anesthesia depth, presented randomly with interclick intervals between 1.5 and 3.0 sec. Click occurs at time = 0 msec. Positive deflections of field potential are plotted upward. *B*, Ten example single trials for each state. Note the increased spread of waveforms at medium depth of anesthesia. *C*, Mean and SD of single trial amplitudes as estimated by taking the inner product with the normalized average waveform for each anesthesia depth. Error bars indicate 95% confidence intervals. *D*, Coefficient of variation (C_v = SD/mean) as a function of anesthesia depth. *Inset* shows C_v as calculated using the traditional peak-to-peak method for estimating single trial amplitude. Although the general shape of the plot is similar, the coefficient of variation at the medium depth of anesthesia is substantially lower for this method.

Modulation of response and response variability by anesthesia depth

We found that average evoked responses differ slightly between the three anesthesia depths, but the single trial variability differs quite dramatically. This is shown for one rat in Figure 1. During this experiment, at least 100 click-evoked responses (random interstimulus intervals between 1.5 and 3.0 sec) were recorded under each anesthesia depth: deep, medium, and light. The general shape of these average evoked responses is quite similar to those recorded intracortically by Hall and Borbely (1970) in awake and sleeping rats. It can be seen from example single trials (Fig. 1*B*) that the evoked response is more *consistent* under deep and light, compared with medium, anesthesia. Not only does the peak amplitude vary widely under medium anesthesia, but even the overall time course (i.e., shape) varies significantly from one trial to the next.

Table 1. Size and variability of evoked responses for individual rats

Rat	Mean: medium/light	Variance: medium/light	Mean: deep/light	Variance: deep/light
A	0.86[a]	2.60[b]	0.95	2.56[b]
B	1.17	1.96[b]		
C			0.99	1.06
D	0.96	1.78[b]	0.93	1.34
E	0.88	4.17[b]	0.95[a]	2.16[b]

To allow direct comparison between rats, mean evoked response sizes and corresponding variances for medium and deep levels of anesthesia were normalized by equivalent variables under light anesthesia.

[a]Mean under medium or deep anesthesia is significantly *less* than mean under light anesthesia (large sample z test, $\alpha = 0.01$).

[b]Variance under medium or deep anesthesia is significantly *more* than variance under light anesthesia (F distribution test, $\alpha = 0.01$).

To quantify individual evoked field potential responses and their variability, estimates were made of each single trial amplitude by taking the inner product of each trial with a normalized average waveform (see Materials and Methods). This approach, rather than using just one or two points from each waveform, takes the shape and size of the entire wave into account. Statistics can then be estimated from the distribution of these single trial amplitudes (Fig. 1*C*, Table 1). Although evoked responses were generally largest under light anesthesia, only one of four rats exhibited significantly smaller evoked responses under medium and deep anesthesia (Table 1). On the other hand, the variance of evoked response was significantly higher under medium than under light anesthesia for all rats tested. In general, the variance increased from light to deep to medium anesthesia. This sensitivity of response *variability* to anesthesia depth can also be visualized with the coefficient of variation (C_v = SD/mean) (Fig. 1*D*). Calculating single trial amplitudes with a peak-to-peak method yields basically similar results (*inset*). However, the coefficient of variation under medium anesthesia was quite a bit less than that computed using the inner product method. This confirms the notion that the peak-to-peak method is not very sensitive to variability of waveform shape.

PCA of evoked responses under medium depth of anesthesia

To characterize and even quantify the variability in waveform shape seen especially under medium depths of anesthesia, we used PCA. The goal of PCA is the same as Fourier analysis: to represent a waveform as a linear combination of a set of basis waveforms. In Fourier analysis the basis waveforms are predetermined (sine waves of different frequencies); in PCA the basis waves are computed from the data to be represented. In particular, the principal component bases explain the maximum amount of variance present in the data with the least number of waveforms. This aspect of PCA allows each waveform to be relatively well characterized by only a few parameters [for further discussion, see Glaser and Ruchkin (1976)].

Figure 2 shows the results of PCA on the 100 single trial evoked local field potentials recorded during a single session under medium anesthesia (the same set of waveforms shown in Fig. 1). For this set of data, the first component accounts for 63.7% of the variance of the data set, and the second for 27.0%. Thus >90% of the power in the population of waveforms can be described using the projection of each waveform onto only these two components. This percentage agrees with Musial et al. (1998) who recorded somatosensory evoked field potentials from barrel cortex of rats anesthetized with urethane.

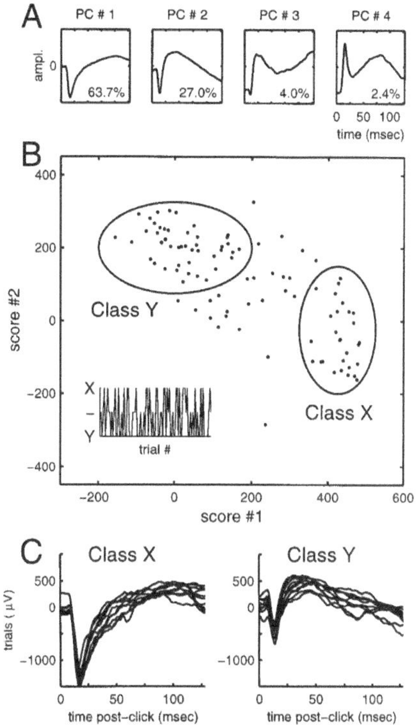

Figure 2. PCA allows arbitrary quantitative separation of single trials into response classes (*X* and *Y*). *A*, First four principal components for the medium depth of anesthesia data shown in Figure 1. The value in the *bottom right* of each plot corresponds to the percentage of total data variance accounted for by that component. *B*, Scatter plot of the first component score versus the second component score for each single trial. Selection of trials into either class *X* or *Y* is made with the ellipses. *Inset* shows chronological order of which single trials belong to *Class X*, *Class Y*, or neither (−). *C*, Ten example single trials for each of the two classes. Note the striking difference between the two classes and the relative consistency within each class.

The first and second component scores for each single trial waveform are the scalar results of taking the inner product of that waveform and the first and second principal components, respectively. A typical way of displaying component scores is a scatter plot, such as that shown in Figure 2*B*. Although not clearly distinct, there is some suggestion of two clusters of response types. We made these response types explicit by selecting responses as either class X or class Y with the illustrated ellipses. The *inset* of Figure 2*B* shows that the variability of response shape, as summarized by class, was not caused by some slowly changing brain state, but rather a dynamic ongoing process.

Although the exact placement of the elliptical boundaries was somewhat arbitrary, Figure 2C shows that classification succeeded in two ways. First of all, the general time course of the two response types differs quite substantially. Second, within each response class, the repeatability of the waveforms is relatively good. This latter point can be summarized quantitatively by comparing the average normalized inner product (i.e., correlation coefficient) between single trial waveforms before and after classification. The average normalized inner product between all 100 single trial waves is 0.43. In contrast, this similarity measure is 0.94 for only the 28 class X waves, and 0.74 for only the 47 class Y waves. Therefore, PCA has allowed a numerically based separation into classes that reduced the observed time course variability of evoked field potential responses. It should be noted that the qualitative aspects of these results do not depend on the precise shape or location of the selection boundaries. In fact, separation into classes X and Y—and the resulting reduction of variability—can be achieved nearly as well with a single straight line that roughly divides the two clusters of points (data not shown).

Evoked unit activity also shows quite dramatic differences between responses after PCA-based sorting of evoked field potentials. The click-evoked multi-unit activity, recorded simultaneously with the waveforms shown in Figure 2, is summarized in Figure 3. Figure 3C is a peristimulus time histogram of the multi-unit activity for all waves. The time course of this multi-unit response generally matches the negative-going peak of the average evoked local field potential. This is expected because a negative deflection of the field potential represents negative extracellular currents, which corresponds to depolarization of nearby neurons. Figure 3D shows the evoked multi-unit activity after PCA-based sorting into the two response classes. The later peak of the multi-unit activity for class X compared with class Y is correlated with the later peak in the evoked field potential waveform for class X. Also, the sustained unit discharge (>30 msec after the click) for class X responses could have been predicted from the sustained negative deflection of the class X field potential average. However, the difference between classes in spontaneous multi-unit activity directly preceding the clicks was quite unexpected. Note that there is virtually no spontaneous activity preceding a click when a class X evoked field potential occurs, compared with quite significant spontaneous activity preceding class Y evoked waveforms. This result suggests that the occurrence of a class X or a class Y evoked response depends on, and thus can be predicted from, the ongoing activity immediately preceding the click.

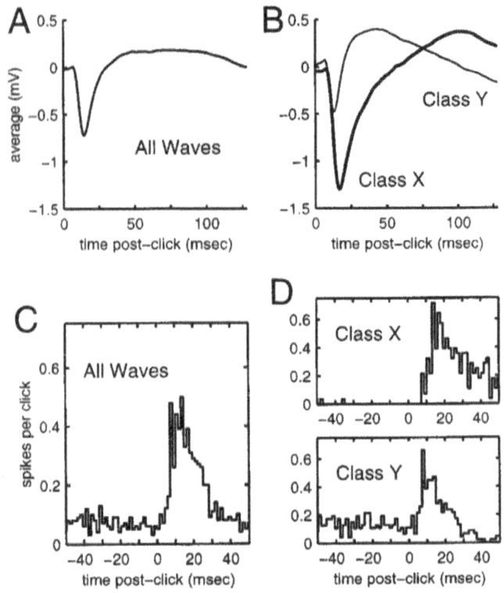

Figure 3. Evoked multi-unit activity differs for the response classes as defined by PCA. *A*, Average click-evoked field potential computed from all single trials for the data from medium depth of anesthesia in Figures 1 and 2. *B*, Average click-evoked field potentials for single trials as separated into *Class X* (*thick line*) and *Class Y* (*thin line*) by PCA. *C*, Multi-unit peristimulus time histogram for all single trials. *Vertical axis* shows the average number of spikes per click in each bin (bin size = 1.5 msec). Note that time scale is different than for field potentials. Click occurs at time = 0 msec. *D*, Multi-unit peristimulus time histograms after single trials have been sorted via PCA of corresponding field potential waveforms. Later peak latency and sustained negative deflection of class X field potentials correlates with later onset and sustained multi-unit discharges. Also note difference in spontaneous activity preceding clicks of the different response classes.

Modulation of evoked response by ongoing activity

To test the hypothesis that the observed variability in waveform-shape was caused by, or at least correlated with, the ongoing cortical activity, we presented clicks at fixed delays from spontaneously occurring events. Figure 4 shows an average of 30 such events, which are very common during medium depths of anesthesia. From this figure it can be seen that a large negative spike in the field potential is accompanied by a large burst of multi-unit activity. In this particular recording, the burst events are occurring rhythmically at ~1.6 Hz (Fig. 4C).

During an experiment, an amplitude threshold would be set on the ongoing field potential activity. When the threshold was crossed by a burst event, a fixed delay would be imposed and a click presented. The delay was determined randomly for each single trial and could be either 150, 300, or 450 msec (Fig. 4A). Responses were grouped together, or classified, on the basis of these delays.

We found that evoked responses are quite dramatically modulated by the spontaneously occurring population burst events in a manner that cannot be accounted for by linear superposition of a response and ongoing background activity. It can be seen from the single trials (Fig. 5A) that increasingly long delays from spontaneous threshold crossings lead to larger evoked responses and different time courses. There is no single evoked waveform that could be added to the average background waveform (Fig. 4A) at each of the different delays that would produce the observed time course distortions. Once again, these differences in waveform shape can be quantified with PCA. Figure 5B shows the scatter plot of first and second component scores for all single trial evoked responses. Notice how the different symbols, corresponding to the different response classes, segregate quite well. Recall that these response classes are defined not by the analysis itself but rather by the relationship between each click and ongoing cortical activity.

Evoked multi-unit activity is also quite different depending on

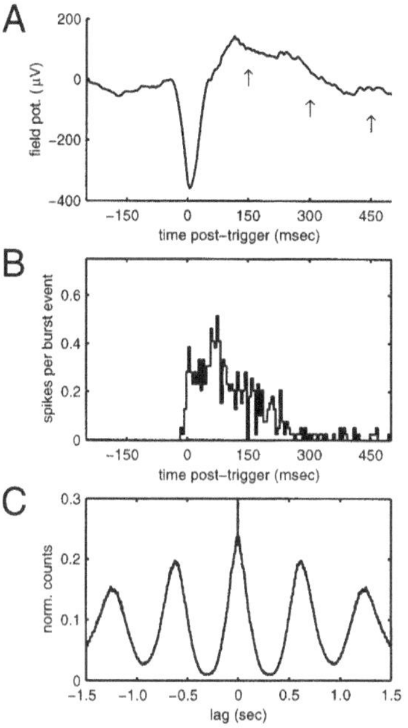

Figure 4. Population burst events occur spontaneously at medium depth of anesthesia. *A*, Average (field potential) of 30 spontaneous burst events, aligned on a threshold crossing (0 msec). In subsequent figures, responses are sorted into classes that correspond to clicks presented at delays of 150, 300, and 450 msec (indicated with *arrows*) from the burst event trigger crossing. *B*, Histogram of multi-unit activity locked to the same threshold crossing (bin size = 1.5 msec). *C*, Autocorrelogram of spontaneous multi-unit activity (bin size = 5 msec), normalized by height of central bin. Presence of secondary and tertiary peaks indicates rhythmic activity. Secondary peak occurs at a lag of ~625 msec, which corresponds to a frequency of ~1.6 Hz.

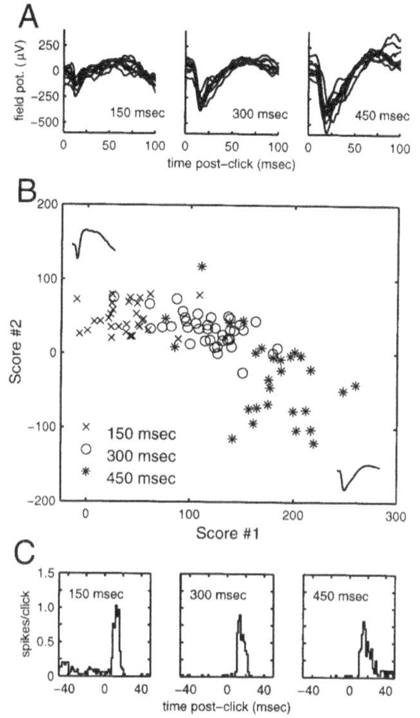

Figure 5. Spontaneous population burst events dynamically modulate evoked responses. *A*, Ten examples of single click-evoked field potentials (DC offsets removed) for the three different values of delay (indicated in *bottom right* of each plot) between spontaneous burst event and click. Note the change in both waveform size and shape with increasing delay. *B*, Scatter plot of first versus second principal component score for each single trial. Note how the different response classes (indicated by different *symbols*) segregate nicely in the principal component space. The shape of the first two components is shown near their respective axes. *C*, Multi-unit peristimulus time histograms for the three response classes (bin size = 1.5 msec). Increasing delay from the burst event leads to reduced spontaneous activity preceding the click, slower response onset, and longer response duration.

how much time elapses between a population burst event and a click. In particular, notice the spontaneous activity preceding the clicks delayed by 150 msec and the relative lack before the other clicks. In addition, there is a tendency for multi-unit activity to be sustained for a longer period of time after the click when it is delayed longer from the burst event.

The modulation of evoked field potential and multi-unit responses can be summarized quantitatively. For example, Figure 6*A* shows a monotonic and statistically significant (large sample *z* test, $\alpha = 0.01$) increase in mean power of evoked field potential response as a function of delay from the population burst event. Note that, as opposed to response amplitudes calculated above, response power is computed as the sum-of-squares of the first two principal component scores for each wave (computing response power as the sum-of-squares of all component scores, i.e., total waveform power, yields a nearly identical relationship between mean power and delay from burst event: correlation between the two methods of computing power is >0.99 in this instance). Differences in shape can be quantified by computing the percentage of a waveform's power accounted for by only one of the components, e.g., the first component. Plotting this value as a function of delay from a spontaneously occurring burst event (Fig. 6*B*) shows that longer delays (300 and 450 msec) lead to significantly modified evoked response shapes than a short delay (150 msec). Figure 6*C* indicates that mean latency of the negative-going peak increases monotonically and significantly with increas-

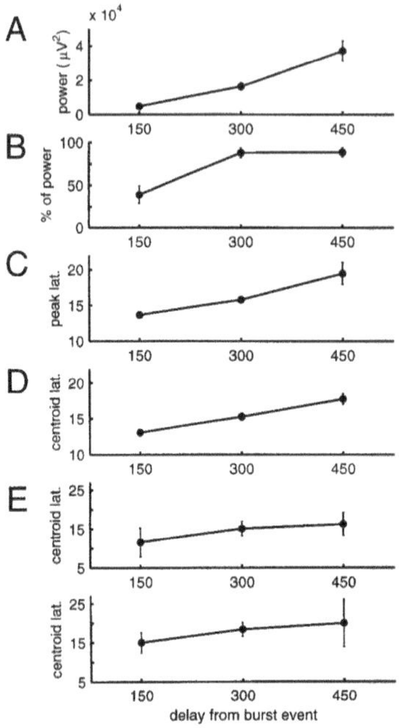

Figure 6. Summary of response modulation for data shown in Figure 5. *A*, Mean waveform power, estimated as the sum-of-squares of the first two principal component scores, as a function of click's delay from spontaneous burst event. *B*, Mean percentage of waveform power accounted for by the first principal component. *C*, Mean latency of peak (between 5 and 30 msec) for single-trial evoked field potentials. *D*, Centroid latency (between 5 and 30 msec) for multi-unit histograms. *E*, Centroid latency for two single units recorded from different tetrodes (~300 μm apart). Note change in vertical-axis scale. The trends for these units are the same as the multi-unit data, but there is large overlap of the confidence intervals because of relatively low spike counts. For all plots, error bars indicate 95% confidence intervals.

ing duration of delay from the spontaneous burst events, where the peak is the minimum value between 5 and 30 msec after a click. Related to this value, the centroid of the multi-unit peristimulus time histogram is computed by summing the latency of all spikes in the histogram between 5 and 30 msec after the click, then dividing by the number of spikes. This temporal measure of multi-unit activity also increases steadily with larger delays from the burst events (Fig. 6*D*).

There are at least two ways to account for the change in latency of multi-unit response as a function of increasing delay from the population burst events. (1) The responses are composed of a combination of two separate neural populations that have different latencies of response (Wróbel et al., 1998). As delay from a burst event increases, the relative contribution of evoked activity to the multi-unit histogram shifts from being dominated by the earlier responding population to the later responding population. (2) Only one population of neurons constitutes the evoked multi-unit response. As delay from a burst event increases, the actual latency of response of these neurons shifts. We believe the second explanation to be the most likely one in this particular situation because single units, isolated from the multi-unit population using tetrode-sorting techniques (see Materials and Methods), individually showed an increase in the centroid of their peristimulus time histograms as a function of delay from the burst events (Fig. 6*E*). This conclusion must be made cautiously, however, because the spike counts for single units were quite low, and some well isolated single units were not responsive to clicks. This is not surprising because auditory evoked single unit activity is somewhat suppressed under ketamine anesthesia (Zurita et al., 1994).

Although the spontaneously occurring population burst events were found to modulate click-evoked responses in all rats tested, the strength of modulation varied between rats and was usually but not always monotonically related to delay from the bursts. All five rats tested showed, with increasing delay from the burst events, increased mean latency of negative-going field potential peak (mean correlation coefficient = 0.78), increased signal power (0.88), increased percentage of power accounted for by the first principal component (0.73), and increased latency of multi-unit centroid (0.87; data only available for three rats). It should be noted that some rats were tested with different sets of delays. For example, in one rat, delays of 100, 200, 300, and 400 msec from the burst events were used for the analysis. Using different delays in this manner did not yield qualitatively different results. Significantly longer delays could not be used because the burst events typically occurred rhythmically with mean inter-burst intervals of ~600 msec under medium depths of anesthesia.

Although not shown in the present report, we also found that responses to other types of auditory stimuli, e.g., tones and tone sweeps, are also dramatically modulated by these spontaneously occurring population burst events. Also, changing stimulus intensity does not particularly affect the character of the results. Finally, we found that burst events occurring during deep and light depths of anesthesia also modulate click-evoked responses. However, such burst events occur less frequently in deep and light anesthesia, and the degree of modulation is generally less than for similar events occurring during medium anesthesia. This observation explains why variability of evoked response was highest under medium levels of anesthesia.

DISCUSSION

We investigated trial-to-trial variability and state-dependent modulation of auditory cortex evoked responses in ketamine/xylazine-anesthetized rats. We found that not only *average* responses but also response variability is modulated by the depth of anesthesia. In particular, it was shown that trial-to-trial variability is usually lowest under light anesthesia and highest under medium anesthesia. To quantify the observed variability in evoked waveform shapes, we used PCA. Such an analysis allowed classification of single trial evoked field potential responses into groups. This classification reduced single trial variability, and evoked multi-unit activity was found to differ substantially among these groups.

We next examined whether the observed variability of waveform time course, as quantified with PCA, can be accounted for by the ongoing cortical activity. Indeed, by triggering the occurrence of auditory stimuli from the ongoing local field potential, it was found that large population burst events common to ketamine/xylazine anesthesia dynamically modulate both the size and shape of single trial evoked responses. In particular, under moderate depths of anesthesia, increasingly longer delays between the occurrence of a burst event and the presentation of a click lead to larger amplitude responses, modified response shapes, and later peak latencies. This modulation cannot be explained as a linear combination of ongoing activity and stereotyped evoked response.

Potential cause(s) of observed variability

The extreme variability of evoked responses observed under moderate levels of anesthesia could possibly be caused by the large-scale changes occurring throughout thalamocortical circuits during typical rhythms associated with sleep and anesthesia (for review, see McCormick and Bal, 1997; Steriade, 1997). For example, it is known that the responsiveness of both cortical and thalamic neurons varies with the phase of slow anesthesia rhythms (Contreras et al., 1996; Timofeev et al., 1996). The frequency of such rhythmicity in ketamine-anesthetized cats, typically <1 Hz (Steriade et al., 1993), is lower than that which we observed (down to ~1.2 Hz under deep anesthesia), but this discrepancy might be explained as a difference between species.

The presently observed rhythmic population bursting under medium levels of anesthesia more closely resembles spike-wave discharges, also known as high-voltage spindles (for review, see Coenen, 1995). These events, which can grow out of and likely share neural circuits with slow anesthesia rhythms (Steriade et al., 1998), are thought to be associated with certain forms of epilepsy (Snead, 1995). The basic phenomenology we observed roughly matches that shown by other researchers recording from rat cortex (Coenen, 1995; Kandel and Buzsáki, 1997; Seidenbecher et al., 1998). Although the frequency of spike-wave rhythmicity is typically >2 Hz, it has been previously observed down to 1.6 Hz in rats (Buzsáki, 1991) and 1.5 Hz in cats (Steriade et al., 1998). Ketamine, the anesthetic used in the present study, can induce spike-waves in cats (Black et al., 1980) and is often used to study neural activity during spike-waves (Steriade et al., 1998). Furthermore, ketamine has been shown to regulate the frequency of high-voltage spindle rhythmicity in rats (Buzsáki, 1991). Finally, it has previously been shown that presenting a visual stimulus in the interspike-wave interval causes a significantly different evoked potential than presenting a stimulus during the wave portion of a spike-wave, in a manner quite similar to our results [compare Fig. 5 of the present report with Fig. 5 of Pellegrini et al. (1986)].

Regardless of terminology, we believe the presently observed modulation of evoked responses to be primarily a reflection of oscillating excitability in both thalamic and cortical neurons. Immediately after a burst event there will be an accumulation of inhibitory currents that will cause a reduction in amplitude and even a modification in shape of evoked response. These might include $GABA_B$ mediated (Destexhe, 1998) and outward potassium currents (Steriade et al., 1998). After a sustained hyperpolarization, neurons would become primed to exhibit an enhanced (i.e., larger amplitude and extended time course) response to a sensory stimulus. This priming could be caused by deinactivation of voltage-gated sodium and perhaps also low-threshold calcium channels. These special calcium channels have been found to be involved in both thalamocortical rhythms and epilepsy [succinctly reviewed by Huguenard (1998); see de la Peña and Geijo-Barrientos (1996) for cortical distribution of these channels]. It is also possible that the sustained inhibitory currents could activate hyperpolarization-activated cation currents (Pape, 1996) that would further prime the neurons.

The linear model of ongoing background activity and evoked response

One very critical issue to processing and interpreting neurophysiological data, and even to general cortical function, is whether relatively stereotyped evoked responses are linearly superimposed with ongoing brain activity. Such a condition is requisite for computing traditional average responses. Our results at first seem to contradict the optical recording study of Arieli et al. (1996) who found well behaved superposition of responses and background activity. However, the signals recorded are very different: local field potentials represent extracellular *population currents* (Bullock, 1997), whereas voltage-sensitive dye signals correspond to *localized* changes in neuronal membrane *potentials* (Arieli et al., 1995). More importantly, the stability of thalamocortical responsivity (i.e., excitability) apparently differed between our studies. Both Arieli et al. (1996) and Azouz and Gray (1999) indirectly assumed constancy of intrinsic responsivity and demonstrated sufficiency of the linear model under such circumstances. We have demonstrated conditions under which there is a systematic variation of responsivity and for which the linear model is deficient. The present results are probably most applicable to brain states associated with population bursting, rhythmic or not, such as natural sleep [Amzica and Steriade (1998); in humans: Achermann and Borbély (1997)], anesthesia, and paroxysmal episodes (see above).

Variability and state-dependent modulation of evoked responses

Our finding that variability of click-evoked response is lowest for light levels of anesthesia and highest under medium anesthesia is in direct contrast to Horvath (1969) who found that increasing depth of dial/urethane anesthesia leads to monotonically decreasing variability of surface-recorded auditory evoked potentials in cats. This discrepancy is most likely attributable to the use of different anesthetics, particularly because dial is a barbiturate. Barbiturates are known to cause generalized depression of central nervous activity (Thurmon et al., 1996) and have been specifically shown to reduce spontaneous activity in auditory cortex (Zurita et al., 1994). In contrast, as discussed above, moving from light to medium levels of ketamine/xylazine anesthesia increases the strength and prevalence of population burst events, each of which strongly modulates the shape and size of evoked responses, thus increasing the *apparent* variability of response.

Variability of long-latency (>80 msec) auditory evoked responses has previously been investigated, especially in the human evoked potential field. For example, Zerlin and Davis (1967) showed that the peak-to-peak amplitude of the scalp-recorded N1-P2 complex (110–190 msec) is stochastic and obeys a Gaussian distribution. Especially pertinent to the present study, it has been shown that evoked potentials can be modulated by the *phase* of an ongoing rhythm during which they are presented [for review and alternative viewpoint see Rudell (1980)]. For example, Pfurtscheller (1976) found that an ongoing slow oscillation rhythmically modulated the amplitude of N90 waves recorded from awake subjects.

State-dependent modulation of shorter-latency evoked responses has previously been investigated in subhuman species. For example, the modulation of evoked potentials by different behavioral conditions (e.g., sleep and waking) has been examined in cat auditory (Herz, 1965), visual (Sigüenza et al., 1984), and somatosensory cortex (Howe and Sterman, 1973), and rat auditory (Hall and Borbely, 1970; Knight et al., 1985) and visual cortex (Bringmann and Klingberg, 1995) (for review, see Coenen, 1995). Also, modification of evoked potentials by spike-wave discharges has been studied in monkey (Mirsky et al., 1973), cat (Pellegrini et al., 1986), and rat visual cortex (Inoue et al., 1992; Meeren et al., 1998). In all of these studies only average waveforms were analyzed. By analyzing single trial waveforms we found that not only the average response but also response variability is modulated by brain state. Furthermore, in previous studies, average waveforms were computed from *all* evoked responses that occurred during a given behavioral state [see Pellegrini et al. (1986) for single exception]. We have shown that even during a fixed state, such as a particular depth of anesthesia, one evoked response can be significantly different than another because of very rapid transitions between *sub*-states.

The finding that ongoing activity can modulate evoked responses in a ketamine/xylazine-anesthetized rat was not unexpected. Previously, Eggermont and Smith (1995) showed that ongoing activity, caused by a click, has a nonlinear impact on evoked field potentials generated by a subsequent click in ketamine/xylazine-anesthetized cats. Similarly, Kenmochi and Eggermont (1997) found that certain rhythmic click rates produce larger field potential deflections than others, suggesting again that activity evoked by preceding clicks can affect subsequent evoked responses. Furthermore, Barth and Di (1991) showed that directly stimulating rat somatosensory cortex leads to oscillatory changes between higher and lower levels of excitability, as measured with field potentials. One novel aspect of the present study is that the responses recorded were modulated by spontaneously occurring cortical activity, rather than by stimulation-evoked activity.

REFERENCES

Abeles M, Goldstein MH (1977) Multispike train analysis. Proc IEEE 65:762–773.

Achermann P, Borbély AA (1997) Low-frequency (<1 Hz) oscillations in the human sleep electroencephalogram. Neuroscience 81:213–222.

Amzica F, Steriade M (1998) Cellular substrates and laminar profile of sleep K-complex. Neuroscience 82:671–686.

Arieli A, Shoham D, Hildesheim R, Grinvald A (1995) Coherent spatiotemporal patterns of ongoing activity revealed by real-time optical imaging coupled with single-unit recording in the cat visual cortex. J Neurophysiol 73:2072–2093.

Arieli A, Sterkin A, Grinvald A, Aertsen A (1996) Dynamics of ongoing activity: explanation of the large variability in evoked cortical responses. Science 273:1868–1871.

Astrelin AV, Sokolov MV, Behnisch T, Reymann KG, Voronin LL (1998) Principal component analysis of minimal excitatory postsynaptic potentials. J Neurosci Methods 79:169–186.

Aunon JI, McGillem CD, Childers DG (1981) Signal processing in evoked potential research: averaging and modeling. CRC Crit Rev Bioeng 5:323–367.

Azouz R, Gray CM (1999) Cellular mechanisms contributing to response variability of cortical neurons *in vivo*. J Neurosci 19:2209–2223.

Barth DS, Di S (1991) Laminar excitability cycles in neocortex. J Neurophysiol 65:891–898.

Black JA, Golden GT, Fariello RG (1980) Ketamine activation of experimental corticoreticular epilepsy. Neurology 30:315–318.

Bringmann A, Klingberg F (1995) Behavior-dependent and drug-induced changes of rat visual evoked potential: relation to the EEG spectral power. Neuropsychobiology 31:89–97.

Bullock TH (1997) Signals and signs in the nervous system: the dynamic anatomy of electrical activity is probably information-rich. Proc Natl Acad Sci USA 94:1–6.

Buzsáki G (1991) The thalamic clock: emergent network properties. Neuroscience 41:351–364.

Chapman RM, McCrary JW (1995) EP component identification and measurement by principal components analysis. Brain Cogn 27:288–310.

Childers DG, Perry NW, Fischler IA, Boaz T, Arroyo AA (1987) Event-related potentials: a critical review of methods for single-trial detection. CRC Crit Rev Biomed Eng 14:185–200.

Coenen AML (1995) Neuronal activities underlying the electroencephalogram and evoked potentials of sleeping and waking: implications for information processing. Neurosci Biobehav Rev 19:447–463.

Contreras D, Timofeev I, Steriade M (1996) Mechanisms of long-lasting hyperpolarizations underlying slow sleep oscillations in cat corticothalamic networks. J Physiol (Lond) 494:251–264.

Coppola R, Tabor R, Buchsbaum MS (1978) Signal to noise ratio and response variability measurements in single trial evoked potentials. Electroencephalogr Clin Neurophysiol 44:214–222.

Dawson GD (1951) A summation technique for detecting small signals in a large irregular background. J Physiol (Lond) 115:2P–3P.

de la Peña E, Geijo-Barrientos E (1996) Laminar localization, morphology, and physiological properties of pyramidal neurons that have the low-threshold calcium current in the guinea-pig medial frontal cortex. J Neurosci 16:5301–5311.

Destexhe A (1998) Spike-and-wave oscillations based on the properties of $GABA_B$ receptors. J Neurosci 18:9099–9111.

Di S, Baumgartner C, Barth DS (1990) Laminar analysis of extracellular field potentials in rat vibrissa/barrel cortex. J Neurophysiol 63:832–840.

Eggermont JJ, Smith GM (1995) Synchrony between single-unit activity and local field potentials in relation to periodicity coding in primary auditory cortex. J Neurophysiol 73:227–245.

Gerstein GL (1960) Analysis of firing patterns in single neurons. Science 131:1811–1812.

Glaser EM, Ruchkin DS (1976) Principles of neurobiological signal analysis. New York: Academic.

Gray CM, Maldonado PE, Wilson M, McNaughton B (1995) Tetrodes markedly improve the reliability and yield of multiple single-unit isolation from multi-unit recordings in cat striate cortex. J Neurosci Methods 63:43–54.

Gur M, Beylin A, Snodderly DM (1997) Response variability of neurons in primary visual cortex (V1) of alert monkeys. J Neurosci 17:2914–2920.

Hall RD, Borbely AA (1970) Acoustically evoked potentials in the rat during sleep and waking. Exp Brain Res 11:93–110.

Heggelund P, Albus K (1978) Response variability and orientation discrimination of single cells in striate cortex of cat. Exp Brain Res 32:197–211.

Herz A (1965) Cortical and subcortical auditory evoked potentials during wakefulness and sleep in the cat. Prog Brain Res 18:63–69.

Horvath RS (1969) Variability of cortical auditory evoked response. J Neurophysiol 32:1056–1063.

Howe RC, Sterman MB (1973) Somatosensory system evoked potentials during waking behavior and sleep in the cat. Electroencephalogr Clin Neurophysiol 34:605–618.

Huguenard JR (1998) Low-voltage-activated (T-type) calcium-channel genes identified. Trends Neurosci 11:451–452.

Inoue M, Van Luijtelaar ELJM, Vossen JMH, Coenen AML (1992) Visual evoked potentials during spontaneously occurring spike-wave discharges in rats. Electroencephalogr Clin Neurophysiol 84:172–179.

Kandel A, Buzsáki G (1997) Cellular-synaptic generation of sleep spindles, spike-and-wave discharges, and evoked thalamocortical responses in the neocortex of the rat. J Neurosci 17:6783–6797.

Kenmochi M, Eggermont JJ (1997) Autonomous cortical rhythms affect temporal modulation transfer functions. NeuroReport 8:1589–1593.

Kisley MA, Gerstein GL (1998) Variability of auditory cortex click responses as a function of preceding rhythmic activity. Soc Neurosci Abstr 24:1878.

Knight RT, Brailowsky S, Scabini D, Simpson GV (1985) Surface auditory evoked potentials in the unrestrained rat: component definition. Electroencephalogr Clin Neurophysiol 61:430–439.

Liberati D, Bertolini L, Colombo DC (1991) Parametric method for the detection of inter- and intrasweep variability in VEP processing. Med Biol Eng Comput 29:159–166.

McCormick DA, Bal T (1997) Sleep and arousal: thalamocortical mechanisms. Annu Rev Neurosci 20:185–215.

McNaughton BL, O'Keefe J, Barnes CA (1983) The stereotrode: a new technique for simultaneous isolation of several single units in the central nervous system from multiple unit records. J Neurosci Methods 8:391–397.

Meeren HKM, Van Luijtelaar ELJM, Coenen AML (1998) Cortical and thalamic visual evoked potentials during sleep-wake states and

spike-wave discharges in the rat. Electroencephalogr Clin Neurophysiol 108:306–319.
Mirsky AF, Bloch S, Tecce JJ, Lessell S, Marcus E (1973) Visual evoked potentials during experimentally induced spike-wave activity in monkeys. Electroencephalogr Clin Neurophysiol 35:25–37.
Möcks J, Gasser T, Pham DT, Köhler W (1987) Trial-to-trial variability of single potentials: methodological concepts and results. Int J Neurosci 33:25–32.
Musial P, Kublik E, Wróbel A (1998) Spontaneous variability reveals principal components in cortical evoked potentials. NeuroReport 9:2627–2631.
Pape H-C (1996) Queer current and pacemaker: the hyperpolarization-activated cation current in neurons. Annu Rev Physiol 58:299–327.
Paxinos G, Watson C (1997) The rat brain in stereotaxic coordinates, Ed 3. San Diego: Academic.
Pellegrini A, Zanotto L, Ermani M, Chemello R, Testa G (1986) Laminar field potentials and unit activity in the cortex during visual evoked potentials in feline generalized penicillin epilepsy. Exp Neurol 94:455–468.
Pfurtscheller G (1976) Variability of cortical evoked responses in man related to slow wave activity. Pflügers Arch 362:193–199.
Rose D (1979) An analysis of the variability of unit activity in the cat's visual cortex. Exp Brain Res 37:595–604.
Rudell A (1980) The rhythm of cortical excitability. Electroencephalogr Clin Neurophysiol 49:125–134.
Schiller PH, Finlay BL, Volman SF (1976) Short-term response variability of monkey striate neurons. Brain Res 105:347–349.
Scobey RP, Gabor AJ (1989) Orientation discrimination sensitivity of single units in cat primary visual cortex. Exp Brain Res 77:398–406.
Seidenbecher T, Staak R, Pape H-C (1998) Relations between cortical and thalamic cellular activities during absence seizures in rats. Eur J Neurosci 10:1103–1112.
Sigüenza JA, De Andres I, Ibarz JM, Reinoso-Suarez F (1984) Cortical visual evoked potentials during the sleep wakefulness cycle of the freely moving cat. Characterization and statistical comparisons. Electroencephalogr Clin Neurophysiol 59:165–171.
Snead OC (1995) Basic mechanisms of generalized absence seizures. Ann Neurol 37:146–157.
Softky WR, Koch C (1993) The highly irregular firing of cortical cells is inconsistent with temporal integration of random EPSPs. J Neurosci 13:334–350.
Steriade M (1997) Synchronized activities of couple oscillators in the cerebral cortex and thalamus at different levels of vigilance. Cereb Cortex 7:583–604.
Steriade M, Nuñez A, Amzica F (1993) A novel slow (<1 Hz) oscillation of neocortical neurons *in vivo*: depolarizing and hyperpolarizing components. J Neurosci 13:3252–3265.
Steriade M, Amzica F, Neckelmann D, Timofeev I (1998) Spike-wave complexes and fast components of cortically generated seizures. II. Extra- and intracellular patterns. J Neurophysiol 80:1456–1479.
Thomas DG, Neer CM, Price JM (1989) Analyses of single-trial N1 amplitude and latency variability and their influence on the average evoked potential. Electroencephalogr Clin Neurophysiol 74:228–235.
Thurmon JC, Tranquilli WJ, Benson GJ (1996) Lumb and Jones' veterinary anesthesia, Ed 3. Baltimore: Williams & Wilkins.
Timofeev I, Contreras D, Steriade M (1996) Synaptic responsiveness of cortical and thalamic neurones during various phases of slow sleep oscillation in cat. J Physiol (Lond) 494:265–278.
Tolhurst DJ, Movshon JA, Dean AF (1983) The statistical reliability of signals in single neurons in cat and monkey visual cortex. Vision Res 23:775–785.
Tunturi AR (1959) Statistical properties of near threshold responses to brief sounds in the MES auditory cortex of the anesthetized dog. Am J Physiol 196:1168–1174.
Vogels R, Spileers W, Orban GA (1989) The response variability of striate cortical neurons in the behaving monkey. Exp Brain Res 77:432–436.
Whitsel BL, Schreiner RC, Essick GK (1977) An analysis of variability in somatosensory cortical neuron discharge. J Neurophysiol 40:589–607.
Worden FG, Marsh JT, Abraham FD, Whittlesey JRB (1964) Variability of evoked auditory potentials and acoustic input control. Electroencephalogr Clin Neurophysiol 17:524–530.
Wróbel A, Kublik E, Musial P (1998) Gating of the sensory activity within barrel cortex of the awake rat. Exp Brain Res 123:117–123.
Zerlin S, Davis H (1967) The variability of single evoked vertex potentials in man. Electroencephalogr Clin Neurophysiol 23:468–472.
Zurita P, Villa AEP, de Ribaupierre Y, de Ribaupierre F, Rouiller EM (1994) Changes of single unit activity in the cat's auditory thalamus and cortex associated to different anesthetic conditions. Neurosci Res 19:303–316.

European Journal of Neuroscience, Vol. 13, pp. 1993–2003, 2001

Daily variation and appetitive conditioning-induced plasticity of auditory cortex receptive fields

Michael A. Kisley and George L. Gerstein
Department of Neuroscience, University of Pennsylvania School of Medicine, Philadelphia, PA 19103, USA

Keywords: classical conditioning, ketamine, learning, pitch perception, plasticity, rat, receptive field

Abstract

Long-term modification of cortical receptive field maps follows learning of sensory discriminations and conditioned associations. In the process of determining whether appetitive – as opposed to aversive – conditioning is effective in causing such plastic changes, it was discovered that multineuron receptive fields, when measured in rats under ketamine-sedation, vary substantially over the course of a week, even in the absence of classical conditioning and electrode movement. Specifically, a simple correlation analysis showed that iso-intensity frequency response curves of multiunit clusters and local field potentials recorded from auditory cortex are nonstationary over 7 days. Nevertheless, significant plastic changes in receptive fields, due to conditioned pairing of a pure tone and electrical stimulation of brain reward centres, are detectable above and beyond these spontaneous daily variations. This finding is based on a novel statistical plasticity criterion which compares receptive fields recorded for three days before and three days after conditioning. Based on a more traditional criterion (i.e. one day before and after conditioning), the prevalence of learning-induced changes caused by appetitive conditioning appears to be comparable to that described in previous studies involving aversive conditioning.

Introduction

In the past decade it has become increasingly clear that sensory receptive fields, and corresponding sensory-topic maps, can be modified throughout adulthood in mammals. This plasticity is believed to be adaptive such that neural resources devoted to processing sensory information can be re-allocated to those particular features of the environment, determined by past experience, to be most behaviourally relevant. Such learning-induced changes have been observed in auditory, somatosensory and visual systems (for recent reviews see Scheich *et al.*, 1997; Buonomano & Merzenich, 1998; Edeline, 1999).

Frequency receptive fields of neurons at both cortical and subcortical levels of the mammalian auditory system can be rapidly modified by many learning paradigms, including conditioned pairing of tone and footshock (Bakin & Weinberger, 1990; Edeline & Weinberger, 1991; Gao & Suga, 1998), aversive differential conditioning (Edeline & Weinberger, 1993; Ohl & Scheich, 1996), instrumental avoidance conditioning (Bakin *et al.*, 1996) and habituation (Condon & Weinberger, 1991). Modifications can also be induced by direct experimental manipulations, including application of pharmacological agents (Ashe *et al.*, 1989; McKenna *et al.*, 1989; Metherate & Weinberger, 1989; Zhang *et al.*, 1997), conditioned pairing of tone and electrical stimulation of subcortical nuclei (e.g. Gonzalez-Lima & Scheich, 1984, 1986; Bakin & Weinberger, 1996), direct electrical stimulation of auditory cortex (Cruikshank & Weinberger, 1996; Maldonado & Gerstein, 1996; Yan & Suga, 1998) and presentation of intense acoustic stimuli (Willott & Lu, 1982; Calford *et al.*, 1993).

Several chronic manipulations (applied for weeks) have been shown to induce long lasting plastic changes in auditory receptive fields. These include sudden (Robertson & Irvine, 1989) and progressive (Willott *et al.*, 1993) cochlear lesions, stimulation of cochlea with implanted electrodes (Dinse *et al.*, 1997), discrimination training (Recanzone *et al.*, 1993) and pairing of tones with direct stimulation of nucleus basalis (Kilgard & Merzenich, 1998).

Only one short-term (< 1 h) manipulation has been demonstrated to produce long-term (up to 8 weeks) plastic changes of auditory cortex receptive fields; conditioned pairing of tone and footshock (Weinberger *et al.*, 1993). For the present investigation we ask whether appetitive conditioning (pairing of tone and reward) has a similarly powerful and enduring effect on the receptive fields of auditory cortical neurons as this aversive conditioning paradigm. 'Reward' consists of direct electrical stimulation of the medial forebrain bundle (MFB) in the region of lateral hypothalamus and ventral tegmental area (Olds, 1962). As we set out to answer the above question we found that the frequency response curves (FRCs) of multiunit clusters and local field potentials in rat auditory cortex vary from day to day, even in the absence of conditioned pairing. In order to interpret properly any observed modification of FRCs after training, it is important to understand the nature and magnitude of these 'spontaneous' daily variations. However, this phenomenon has not been closely investigated. Therefore, in the first part of this study we employ a simple correlation analysis to show that these variations do not represent random, or stochastic, variability about an otherwise stationary FRC, but instead are a progressive change of the neuronal receptive fields. In the second part we compare the relative frequency of observable receptive field modification induced by appetitive conditioning to the relative frequency of observable changes due to

Correspondence: Michael A. Kisley, Department of Psychiatry, University of Colorado Health Sciences Center, 4200 E. 9th Avenue, Box C268-71, Denver, CO 80262, USA.
E-mail: Michael.Kisley@UCHSC.edu

Received 15 August 2000, revised 9 March 2001, accepted 15 March 2001

these spontaneous daily variations. We argue that observed changes can be more confidently attributed to learning-induced plasticity when such a comparison to the spontaneous 'plasticity' is undertaken. This work has appeared previously in abstract form (Kisley & Gerstein, 1999a).

Materials and methods

Surgical preparation and implantation of chronic electrodes

All procedures were in accordance with the Institutional Animal Care and Use Committee at the University of Pennsylvania and the Guidelines of the United States National Institutes of Health for use of animals in biomedical research. Female albino rats weighing 250–350 g (~ 10–15-week-old) were anaesthetized with an intraperitoneal injection of ketamine (70 mg/kg) and xylazine (8 mg/kg). Periodic supplemental doses of anaesthetic were administered to maintain surgical anaesthesia during implantation of electrodes. Glycopyrrolate (0.06 mg/kg), a muscarinic antagonist, was injected subcutaneously at the beginning of surgery to reduce respiratory tract secretions. Body temperature was monitored with a rectal probe, and maintained at 37 °C with a heating pad.

Once a surgical plane of anaesthesia was reached (assessed by pedal-withdrawal reflex), the rat's head was shaved and fixed in a stereotaxic frame. Skin, muscle, and connective tissue were removed from the top and left temporal portion of the skull. Small stainless-steel screws, to serve as recording reference and anchors for the dental acrylic cap, were inserted into drilled holes at several locations on the top of the skull. A small craniotomy, ~1 mm in diameter, was centred 5.0 mm posterior and 1.0 mm lateral (to the animal's right) of bregma. A twisted pair of 100 µm stainless steel microwires (California Fine Wire, Grover Beach, CA, USA) were lowered through the craniotomy and to 8 mm directly ventral from dura. This location corresponds to the MFB at the level of the lateral hypothalamus and ventral tegmental area (Nieuwenhuys *et al.*, 1982; Paxinos & Watson, 1997). Self-stimulation through electrodes at these coordinates has been shown to cause extremely powerful 'approach' behaviour, with virtually no associated 'avoidance' behaviour (Olds & Olds, 1963). These electrodes were fixed in place with dental acrylic.

A 2 × 1 mm craniotomy for the recording electrodes was centred at 5 mm posterior and 4 mm ventral of bregma, corresponding to Te1, left primary auditory cortex (Paxinos & Watson, 1997). After cutting the dura, a linear array of four or eight 50 µm stainless steel microwires (impedance < 500 kΩ at 1 kHz; interelectrode spacing ~200 µm) was rapidly inserted into the brain, then slowly (~10 µm/min) lowered to the infragranular layers (0.8–1.0 mm below dura). Before insertion, these electrodes were orientated perpendicular to the surface of the brain. Dental acrylic was used to fix the array in place, and then to make a protective cap over the entire exposed skull and around the connectors for the stimulation and recording channels. The skin at the edges of the acrylic cap was sutured, and the animal was placed in an isolated recovery cage. Recording sessions began two weeks after surgery.

One week after surgery, and again after all recording was completed, physiological/behavioural verification of stimulating electrode placement in the intended brain reward centre was made. Specifically, rats were allowed to activate a nose-poke and consequently receive an electrical stimulus identical to that used for conditioned pairing (see below). In all cases, after no more than four trials, the animals would begin poking rapidly and repeatedly. Although they attempted to continue poking, we removed them from the cage after a maximum of 15 trials. Localization of stimulating electrodes to the intended brain areas was also visualized with gross histological examination in half of the conditioned animals. Briefly, animals were anaesthetized with an intraperitoneal injection of ketamine (70 mg/kg) and xylazine (8 mg/kg). DC current of 50 µA was passed through the stimulating electrodes for 20 s. Animals were perfused intracardially with saline, followed by 10% formalin. Brains were removed and placed in 10% formalin for 24 h, then 30% sucrose–formalin for 1 week. Frozen sections (40 µm) were examined with light microscopy (not shown).

Daily acquisition of frequency response curves

Each day the animals were sedated with just enough ketamine (~65 mg/kg i.p.) to allow head-restraint without any struggling or signs of distress. This also provided a consistent behavioural state from day to day. The precise dose differed between rats, but once an effective dose was established it was used every day for each individual animal. Care was taken to avoid using doses large enough to cause synchronized bursting of cortical neurons, which has been shown to strongly modulate evoked responses in this system (Kisley & Gerstein, 1999b). Typically, animals were just starting to move their hindlimbs at the end of the 35 min recording session. Other than head-restraint, no painful or distressing procedures took place during these recording sessions.

At 10, 20, and 30 min after intraperitoneal injection of the ketamine, iso-intensity FRCs were obtained from multiunit clusters and local field potentials recorded from the microwires chronically implanted in auditory cortex. Auditory stimulation consisted of ten pure tones, logarithmically spaced from 5–31.06 kHz. Tones were produced by computer-controlled operation of a Voltage Variable Systems model 404 sine wave generator and model 271 attenuator (Philadelphia, PA, USA). Free-field stimulation was achieved with a speaker (#40-1377, Radio Shack, Fort Worth, TX, USA) located directly in front of the rat, 10 cm from the ears (head held by connection between a 10-pin plug and connector on head). After amplification (D-75 power amplifier, Crown, Elkhart, IN, USA) and postattenuation (#31-1957, Optimus, Fort Worth, TX, USA), tone intensity at the ear was 50 dB sound pressure level (SPL), which is well above the albino rat's behavioural threshold across this frequency range (< 20 dB SPL: Fay, 1988). For this speaker, intensity variation for 50 dB tones across a 15-cm by 15-cm square region was previously determined to be ± 5 dB for all frequencies (calibrated with a 6.25-mm microphone, Brüel and Kjaer, Denmark). Since the rats' heads were fixed to a location in this sound field to within 0.25 cm each day, daily variation of sound intensity was very small. Tones were 100 ms in duration, with 5 ms rise and fall time. For each FRC, ≈ 50 tones of each frequency were presented in random order. A total of 500 ms elapsed between the end of one tone and the beginning of the next. All experiments were carried out in a double-walled, sound-attenuated room.

Cortical signals were buffered near the head with follower circuits, then passed to preamplifier, amplifier and digital signal processing hardware (Plexon, Dallas, TX, USA). Before the preamplifier, the signal lines were split into extracellular action potential ('unit') and local field potential channels. Unit data were filtered between 300 and 3000 Hz, and candidate spike-waveforms crossing a fixed threshold were sampled at 40 kHz. Action potential waveforms were stored for off-line analysis. Field potentials were filtered between 0.5 and 300 Hz, and sampled continuously at 1 kHz. All signals were referenced to a ground wire attached to the skull screws.

For multiunit FRCs, response amplitude at each frequency was computed as the average number of spikes arising between 5 and

40 ms after tone onset. The multiunit cluster recorded from each electrode was considered to consist of all threshold crossings, regardless of waveform size or shape. Threshold levels were arbitrary (typically around three standard deviations beyond noise), but held fixed across days for each individual electrode. In order to ensure that we were recording from the same population of neurons from day to day, electrode stability was assessed by examining spike waveforms from the first and last days of a 7-day recording block. This was done by projecting the action potential waveforms onto the first and second principal components that had been computed from a library of action potential recordings from rat auditory cortex. This computation is expressed graphically as a scatter plot where each point represents a single spike waveform. It has been shown that electrode movement alters these scatter plots (Abeles & Goldstein, 1977; Snider & Bonds, 1998). About 5% of 7-day recording blocks were determined to be contaminated by electrode instability, and, thus, excluded from further analysis. An example of multiunit activity and two examples of stability assessment are shown in Fig. 1. Average firing rates of multiunit clusters across days were also examined to ensure that no progressive changes (which might indicate tissue damage) were occurring.

FRCs were also constructed from tone-evoked local field potentials recorded from one electrode in six of the eight rats. Response amplitude was calculated as the difference between the peak of the initial positive deflection (6–10 ms after tone onset) and the trough of the primary negative wave (18–28 ms after tone onset) of the average evoked field potential for each frequency. Subsequent analysis was identical to that carried out for multiunit clusters.

Classical Conditioning

At least 3 weeks after surgery, six animals were trained with a conditioned pairing paradigm. No recording took place on the day of conditioning. While freely moving about a small shielded cage the animal was presented with 30 trials – random intertrial intervals between 50 and 120 s – consisting of a 2.5 s tone (conditioned stimulus) immediately followed by 0.5 s of electrical stimulation (unconditioned stimulus) through the electrodes in the lateral hypothalamus/ventral tegmental area, contralateral to the recording electrodes in the auditory cortex. The tone was fixed at one of the ten frequencies used for generating the FRCs. The intensity of the tone was 50 dB SPL (± 5 dB, see above for calibration) throughout the wire-mesh cage. The electrical stimulation was a 200-Hz sine-wave, with a peak amplitude of ± 100 μA. The electrical stimulation parameters (duration, frequency and amplitude) are within the range of parameters previously found to be effective in eliciting self-stimulation behaviour in rats (reviewed by Olds, 1962).

All animals, within five to ten conditioned pairing trials, exhibited a clear and consistent conditioned emotional response during the conditioned stimulus (i.e. tone). This reaction, alternatively labelled 'freezing' or conditioned suppression, was comparable to the description of that used to verify the establishment of conditioned associations in studies of aversive conditioning-induced neural plasticity (e.g. Bakin & Weinberger, 1990; LeDoux & Muller, 1997).

In the same cage, and before conditioned pairing took place, the animal was presented with a random tone series consisting of all ten frequencies, in a manner identical to that used for acquiring the FRCs. This preconditioning acoustic stimulation was intended to allow time for habituation to the physical and auditory environments. Three of six animals experienced the conditioned pairing paradigm twice. In these instances, the conditioning sessions were separated by at least 2 weeks, and different tone frequencies were used.

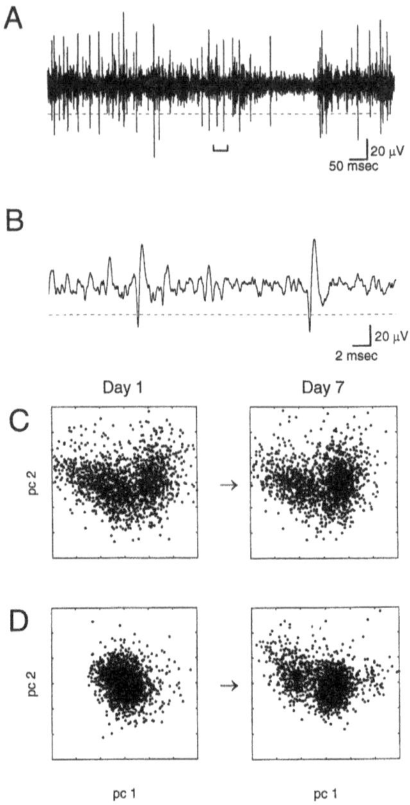

FIG. 1. Example of multiunit activity and assessment of electrode stability. (A) One second of continuous data, with threshold indicated by dashed line. (B) Close-up of area indicated in A to show typical action potential waveforms and brain noise. (C) Scatter plots of action potential waveform projections onto first ('pc 1') and second principal components ('pc 2') at days one and seven of a 7-day block. Because the overall shape of the scatter plot is consistent between these two days, this electrode was considered stable. (D) An example taken from a different animal which was considered unstable, and, therefore, not analysed further.

In order to assess whether any observed learning-induced plasticity required both the tone and the electrical stimulation, we repeated the exact same conditioning paradigm with only tones or only MFB stimulation. These control experiments were carried out on the same animals. Half of the animals experienced conditioned pairing first, then tone-alone, then MFB stimulus-alone sessions. The other half of the animals experienced a tone-alone session first, then MFB stimulus-alone, then conditioned pairing of tone and stimulus.

Long-term analysis of frequency response curves

All FRCs presented in this report were normalized. Specifically, the amplitude of response at each frequency was divided by the mean response amplitude for all ten frequencies of a given FRC. This allowed analysis of daily fluctuation of relative, rather than absolute, response magnitude at each sound frequency (frequency selectivity, rather than overall firing rate, is the variable of interest). For daily analysis, an average of the three normalized FRCs – taken 10, 20, and 30 min after ketamine injection – was computed. For two of eight rats, recordings were made at 15 and 45 min after ketamine injection, in which case the mean curves were computed from these two time-points. To exclude frequency receptive fields that varied excessively, even during the course of a 30 min recording session, we excluded those multiunit clusters that had mean intraday correlation coefficients (i.e. between FRCs taken at 10, 20, and 30 min of the same day) below 0.8. Five of 43 multiunit clusters were removed because of this restriction.

To assess whether observed variation in FRCs was due to random fluctuation about a fixed curve or due to progressive receptive field changes we performed a simple correlation analysis on all available 7-day 'blocks' of tuning curves taken from the multiunit clusters and field potentials of individual electrodes. The correlation coefficient was computed between all pairs of averaged tuning curves (day 1 vs. day 2, day 1 vs. day 3,... day 2 vs. day 3,... etc.). These coefficients were plotted as a function of 'lag' between curves (1 day, 2 days... 6 days), and a straight line was fit to the data. The slope of this line, hereafter referred to as the correlation slope, indicates whether a FRC was fixed or whether it was nonstationary across days; if the curve was changing then the correlation coefficient should go down as the number of days between curves increases, thus, leading to a negative correlation slope. However, if the receptive field was stationary, and exhibiting only random (i.e. stochastic) fluctuations, the correlation between any two FRCs in the 7-day block should be the same on average, leading to a correlation slope of zero. To ensure that the analysis itself was not biased towards negative correlation slopes, it was repeated using the same 7-day blocks after random scrambling (without replacement) of the daily FRCs. If not biased, the analysis should provide an estimate of zero correlation slope for these randomly shuffled data.

Learning-induced plasticity of FRCs was approached in a different manner, still using 7-day blocks. For each day of conditioning, and for each electrode, the magnitude of response at the conditioned frequency was compared between the 3 days before and the 3 days after conditioned pairing. In particular, 'plasticity' required that the response magnitude to the conditioned frequency increased significantly according to a two-distribution, lower-tailed t-test ($P < 0.01$) between the 3 pre- and postconditioning days. This analysis requires the assumption that the two sampled populations consist of FRCs recorded from the multiunit cluster before and after conditioning. In the event that three consecutive days of recorded FRCs were not available, the data from the nearest day was used (this occurred in four of 33 cases involving conditioned pairing of tone and electrical stimulation of brain reward centres).

To determine if the likelihood of observing plasticity after classical conditioning was greater than that observed during normal daily experience, the same plasticity criterion was applied to every available 7-day block recorded in the absence of conditioning. Since these data were not associated with any particular tone frequency, the number of plasticity tests performed was increased by using all ten frequencies in each FRC. To further increase confidence in the estimate of spontaneously occurring plasticity, all overlapping 7-day blocks were analysed. Assuming the resulting proportion is a reliable estimate of the likelihood of plastic changes in the absence of conditioned pairing, normal-curve approximation of the binomial distribution can be used to assess the significance of observed prevalence of plastic changes after conditioning.

Results

Of 50 functional microwires implanted in the auditory cortex of eight rats, 38 were utilized for further analysis. These 38 exhibited obvious multiunit activity (e.g. Fig. 1) and FRCs that satisfied a within-session consistency criterion (see Materials and methods). The latter requirement prevented the inclusion of frequency receptive fields that exhibited extreme variability even over the course of 30 min. Such highly variable curves were generally associated with multiunit clusters that exhibited peak onset firing rates only two or three times background rates. Peak onset firing rates of more stable tuning curves were between five and 20 times average background firing rates. Best frequencies (i.e. frequencies of maximal response for each FRC) for multiunit clusters varied across the entire range of frequencies presented (5–31.06 kHz), but were most commonly found between 6.13 and 13.79 kHz. This frequency range corresponds to the albino rat's highest behavioural sensitivity (reviewed by Fay, 1988). Because only ten logarithmically spaced frequencies were used, best frequencies could not be precisely resolved. We occasionally observed FRCs with two clear peaks. Since these tuning curves were obtained from multiunit clusters, it could not be determined whether such double-peaks were due to contributions from two differently tuned populations of neurons, or from individual neurons with 'multipeaked' tuning curves (e.g. Sutter & Schreiner, 1991).

Daily variation of frequency response curves

Receptive fields of multiunit clusters were analysed in 7-day blocks. As discussed in the Materials and methods section and as illustrated in Fig. 1, if an electrode moved during the 7-day period, the data obtained from that electrode were not analysed. A total of 54 multiunit cluster (there were more 7-day blocks than electrodes because several animals were recorded from for two 7-day blocks) and nine local field potential 7-day blocks were available for analysis.

A rather extreme example of daily variation of a multiunit receptive field across 7 days is demonstrated in Fig. 2. Note that there were three different best frequencies for this multipeaked tuning curve over the course of 1 week. The plot in the upper right of Fig. 2A suggests that the three tuning curves taken within a single day were generally more consistent than tuning curves taken across days. Quantitatively only a trend was observed; the average correlation coefficient was 0.944 between curves taken during a single recording session at 10 and 20 min after ketamine injection, vs. 0.919 between curves taken at 10 min one day and 20 min the subsequent day (t-test, $P < 0.1$). Note that, across days, the response magnitude to some frequencies (e.g. 9.19 kHz) changed progressively and others changed more suddenly (e.g. 11.26 kHz).

To assess quantitatively whether these variations represented stochastic variability about an otherwise stationary tuning curve, or progressive changes from day to day, we employed a simple idea; if the receptive fields were nonstationary over 7 days, then the average correlation between response curves should decrease as the number of intervening days increases (i.e. they should gradually become less similar to each other). If the receptive field was fixed and the observed variation is simply due to random variability, then the average correlation between curves should be independent of the

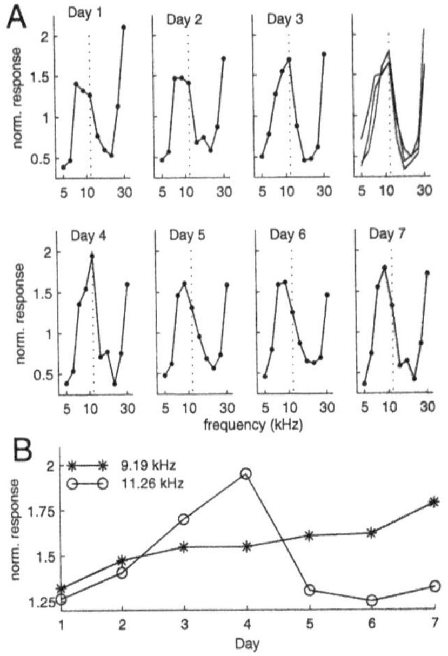

FIG. 2. Multiunit cluster receptive fields vary during the course of 1 week in the absence of conditioned pairing. (A) FRCs of a stable multiunit cluster during a 7-day block. For a given curve, response at each frequency was normalized by the mean response magnitude for all ten frequencies. Each response curve represents an average of the three curves taken each day (at 10, 20 and 30 min after ketamine injection). The plot in upper right shows these three individual curves for day three. Dotted vertical lines at 11.26 kHz are for comparison purposes only. (B) Normalized response magnitude for two frequencies as a function of day.

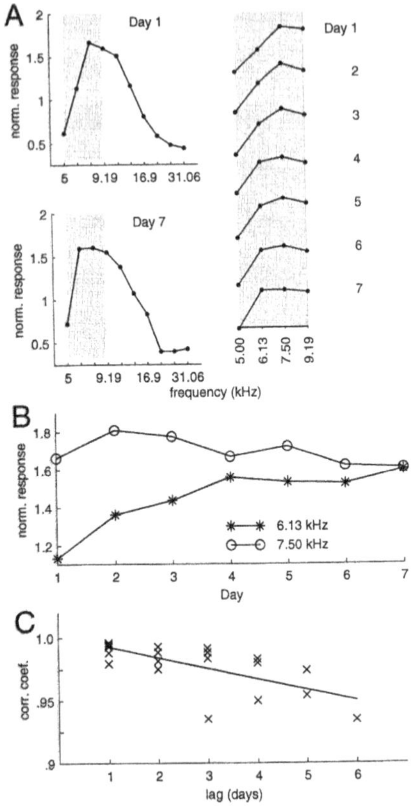

FIG. 3. Systematic change in FRC leads to decreasing correlation as a function of days between curves. (A) FRCs of a multiunit cluster from days one and seven (left) and close-up of response magnitudes for the first four frequencies to show gradual progression of centre of gravity from 7.50 to 6.13 kHz (right). (B) Plot of response magnitude as a function of day for these two frequencies. (C) Correlation coefficient between FRCs as a function of days between curves. Slope of line fit to data (correlation slope) = -8.41×10^{-3}.

number of intervening days. The analytical implementation of this idea is exemplified in Fig. 3 (see Methods for details). It can be seen that, for this more typical multiunit cluster during this 7-day block, there were progressive changes across days in response magnitude at some tone frequencies. As expected, the correlation between tuning curves decreased as the lag (in days) between curves increased (Fig. 3C). This is summarized by considering the slope of a line fit to the scatter-plot of correlation coefficients ('correlation slope'). A negative slope indicates a nonstationary receptive field, whereas, a slope of zero indicates random variability about a fixed curve. In this instance the slope was -8.40×10^{-3}. A population of such slopes must be considered in order to make a statistical assessment.

The correlation slopes for all 54 multiunit clusters are shown in Fig. 4A. This histogram clearly shows that, for the population of 7-day blocks, the correlation slope was negative (one-distribution, upper-tailed t-test, $P < 10^{-10}$). Therefore, the average correlation between FRCs decreased as the number of intervening days increased, supporting the idea that the observed variations of receptive fields were progressive and not simply random. The mean correlation slope for all 7-day blocks was -4.98×10^{-3} (SD = 4.92×10^{-3}).

Although not shown, the population of 7-day blocks taken from evoked local field potentials showed basically the same behaviour as the data from multiunit clusters. The mean correlation slope for these FRCs was -5.30×10^{-3} ($n = 9$, SD = 4.23×10^{-3}). Like the population of multiunit clusters, the mean correlation slope was

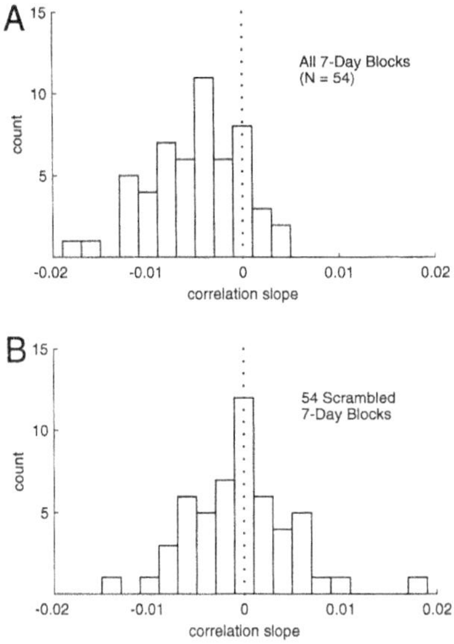

FIG. 4. Multiunit FRCs are nonstationary over the course of 1 week. (A) Histogram of correlation slopes taken from all available multiunit cluster 7-day blocks. Dotted line indicates slope of zero, which would be expected if daily variations were simply due to random variability about an otherwise stationary FRC. Mean of distribution = -4.98×10^{-3}, SD = 4.92×10^{-3}. (B) Histogram of correlation slopes from the same 7-day blocks shown in A after random shuffling of the FRCs within each block. As expected, this data is centred around a slope of zero, indicating that the analysis is not biased towards generating negative correlation slopes.

found to be significantly less than zero (one-distribution, upper-tailed t-test, $P < 0.001$).

In order to ensure that our correlation analysis was not biased towards negative correlation slopes, we repeated it after randomly shuffling the multiunit FRCs within each 7-day block. Randomly scrambling each 7-day block results, as expected, in a population of correlation slopes centred around zero (Fig. 4B). Since there are thousands of random permutations for the FRCs within each 7-day block, we can repeat the shuffling many times to generate a more reliable estimate of the actual mean correlation slope. Scrambling each 7-day block 200 times leads to a total of 10800 correlation slopes, the mean of which is 0.00×10^{-3} (SD = 5.80×10^{-3}).

Learning-induced plasticity of frequency response curves

How do learning-induced modifications of frequency receptive fields compare to the changes described above which occurred in the absence of conditioning? To answer this question, FRCs of multiunit clusters were again analysed in 7-day blocks, with conditioned pairing occurring on day four. As in the previous section, only data for which the electrode did not appear to move were analysed.

An example of learning-induced plasticity is illustrated in Fig. 5. It can be seen that, for the multiunit cluster recorded from this electrode, conditioned pairing of a 6.13-kHz tone and electrical stimulation of brain reward centres produced a clear enhancement of response to the conditioned frequency. The significance of this enhancement was assessed with a lower-tailed, two-distribution t-test ($P < 0.01$) between the response magnitude to the conditioned frequency for the 3 days before and the 3 days after conditioning (see Materials and methods for assumptions regarding populations and sampling). In this case, the response to the conditioned frequency did increase significantly, and is, thus, considered to be 'plasticity'. The response magnitude to the frequency immediately above the conditioned frequency was significantly decreased as a result of conditioned pairing (Fig. 5A and B). The local field potential evoked response recorded from the same electrode was also significantly enhanced at the conditioned frequency (Fig. 5C and D).

Another example of significant learning-induced plasticity of a multiunit FRC is shown in Fig. 6. In this case the conditioned frequency became the best frequency after conditioning. An adjacent frequency band was also enhanced, though not significantly. Note that the sudden and sustained change in the FRC caused by the conditioned pairing is obvious even in the presence of day-to-day variability. Although there appears to be further increase in response magnitude at the conditioned frequency from day one to day three postconditioning (i.e. putative 'consolidation'), it is difficult to attribute this change to the training event; as shown above, FRCs often show this degree of modification simply as a result of everyday experience.

To determine whether the observed learning-induced plasticity occurred more often after conditioned pairing of tone and reward than it did during normal everyday experience, the plasticity criterion was applied to all available 7-day blocks acquired in the absence of conditioned pairing. For this analysis, day four was considered pseudo-conditioning day. Of the 2200 available t-tests (see Materials and methods), 2.6% exhibited significant enhancement that qualified as plasticity. Because this proportion was based on so many trials, we assumed that it reliably estimated the actual likelihood of observing plastic changes with our criterion in the absence of conditioning. Of the 33 available t-tests (across six rats) involving conditioned pairing of tone and electrical brain stimulation, 18.2% qualified as plasticity. The difference between these two proportions is significant (normal-curve approximation of binomial distribution, $P < 0.01$). Thus, appetitive conditioning does induce significant changes in the FRCs above and beyond that which is observed in the absence of conditioning. Incidentally, utilizing an upper-tailed t-test criterion – effectively considering significant decreases at the conditioned frequency as 'plasticity' – did not uncover a significant effect of training (3.0% after training compared to 2.6% spontaneously, $\alpha = 0.05$).

To determine whether the induction of receptive field plasticity required both the tone and electrical brain stimulation, we repeated the conditioning paradigm but with either tones only or MFB stimulation only (half of the animals experienced these nonassociative tasks before the conditioned pairing task). For tone-alone, 6.7% of the 30 t-tests passed the plasticity criterion. For stimulus-alone, 2.1% of the 240 available t-tests were significant (normalized response magnitude at all ten frequencies was used for analysing the stimulus-alone paradigm). Neither of these proportions were significantly different from the likelihood of observing plastic changes in the absence of conditioned pairing ($\alpha = 0.05$).

In order to compare the prevalence of plastic changes induced by conditioned pairing of tone and reward to that of tone and

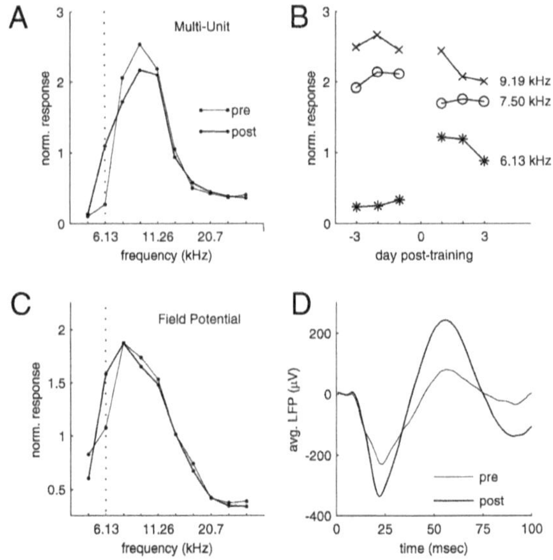

FIG. 5. Example of learning-induced plasticity after conditioned pairing of tone and electrical stimulation of brain reward centres. (A) FRCs, averaged across the 3 days before ('pre') and 3 days after ('post') conditioning. Conditioned frequency, 6.13 kHz, is indicated with dotted vertical line. (B) Response magnitudes for select frequencies as a function of day. Increase of response at conditioned frequency was statistically significant (see Results), and, thus, qualifies as 'plasticity'. (C) Average FRCs computed from local field potentials recorded from same electrode as multiunit cluster shown in A. Field potential response magnitude at the conditioned frequency also passed the plasticity criterion. (D) Evoked local field potential taken 30 min after ketamine injection, averaged across the 3 days before and 3 days after conditioning, in response to the conditioned frequency, 6.13 kHz. Tone onset is at 0 ms.

punishment, the present data were re-analysed using a slightly modified version of the criterion employed by Weinberger *et al.* (1993). This criterion states that learning-induced plasticity has taken place if the difference between the normalized response to the conditioned frequency on the day after conditioning, and that taken just before conditioning, is greater than 0.1 for a majority of sound intensities. Whereas, our criterion involves comparison of response magnitudes for multiple days before and after learning (and, thus, allows a statistical assessment), that of Weinberger *et al.* (1993) utilizes responses for only 1 day pre- and postconditioning. The other important difference is that, whereas, Weinberger *et al.* (1993) normalized the response to the conditioned frequency by the response to the preconditioning best frequency, we normalized by the mean response strength for all frequencies in the FRC. Unlike the study of Weinberger *et al.* (1993) we did not record evoked neural activity for multiple tone intensities. Therefore, we cannot employ the exact same criterion. Nevertheless, applying a 1-day pre- and postconditioning criterion with normalization by the best frequency to our data, we find that conditioned pairing of tone and reward produced detectable plastic changes 36.4% of the time ($n = 33$). Conditioned pairing of tone and punishment (footshock) produced plasticity 45% of the time (Weinberger *et al.*, 1993; 24 h after training, $n = 20$). These proportions are not significantly different at any reasonable significance level (two-proportion, binomial distribution).

It is useful to know whether the likelihood of observing learning-induced plasticity depends upon the relationship between a conditioned frequency and the particular shape of a FRC. To address this, we assessed correlation between changes in response amplitude at the conditioned frequency and (i) the preconditioning response amplitude at the conditioned frequency as compared to (and normalized by) the mean response amplitude for all frequencies; and (ii) the octave distance (and direction) between the conditioned frequency and the preconditioning best frequency. Regarding the first comparison, it was found that conditioned frequencies with relatively low response magnitudes before conditioned pairing were more likely to exhibit large learning-induced increases in response. Figure 7 shows how an increase in response to the conditioned frequency depends on the normalized response at that frequency before conditioning (averaged over 3 days). The correlation between these values was –0.59 ($P < 0.001$). It can be seen in this figure that the 1-day pre- or postconditioning criterion classified as plastic increases two points which actually had decreased mean responses (as computed from 3-day averages) at the conditioned frequency. Since this criterion is only based on response amplitudes one day before and one day after conditioning, daily variability can cause such misclassifications.

We did not find a similar linear correlation between the octave distance from the preconditioning best frequency to the conditioned frequency and the magnitude of observed learning-induced plasticity (correlation coefficient = 0.0091). Although a nonlinear dependence between octave distance and the size of aversive conditioning-induced changes has been demonstrated (Gao & Suga, 1998), we do not have sufficient data to determine whether a similar relationship applies to the appetitive paradigm. Finally, we note that the average activity across all frequencies of the FRCs did not change significantly as a result of appetitive conditioning (paired t-test between 3 days pre- and postconditioning, $P < 0.1$).

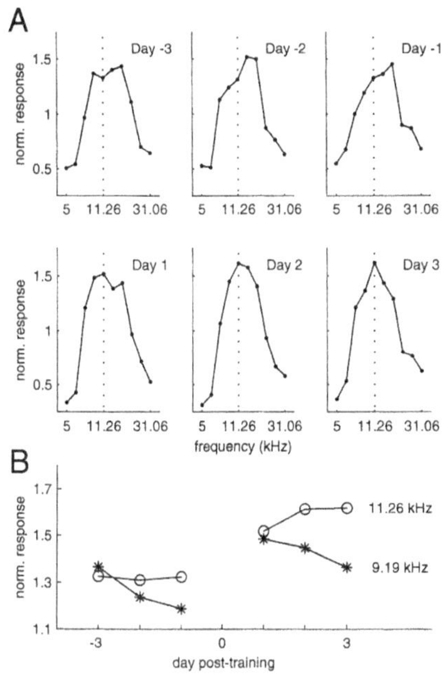

FIG. 6. Example of significant learning-induced plasticity for which conditioned frequency becomes best frequency after conditioned pairing. (A) FRCs for the 3 days before and 3 days after conditioning. Conditioned frequency, 11.26 kHz, is indicated with dotted line. (B) Response magnitude as a function of day for select frequencies. Increase in response magnitude at 11.26 kHz was significant.

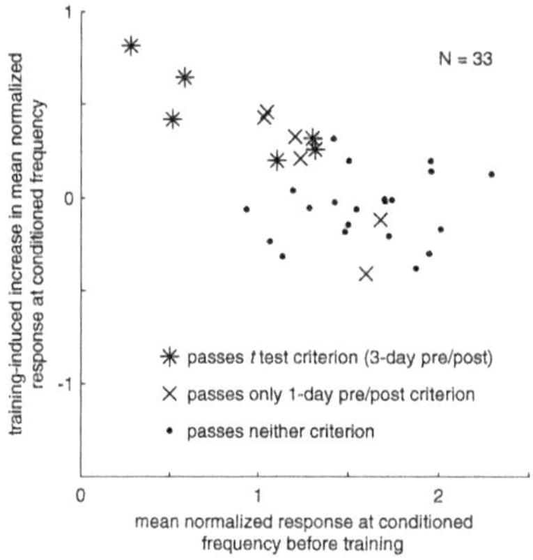

FIG. 7. Potential for plasticity depends upon relative response magnitude before conditioned pairing. Magnitude of learning-induced change in mean normalized response for the conditioned frequency as a function of mean normalized response magnitude before conditioning. Mean responses were computed from the 3 days before and 3 days after conditioning. *Cases which passed the t-test criterion of the present report. ×Cases which failed the above criterion but still passed the 1-day pre- and postconditioning criterion (with normalization by the preconditioning BF). •All other cases. For all 33 points, correlation coefficient = −0.59 ($P < 0.001$).

Discussion

Of necessity, the present study consisted of two parts. In the first part it was demonstrated that multiunit cluster and local field potential receptive fields in rat primary auditory cortex are nonstationary when measured daily (under Ketamine sedation) for a period of 1 week. Classical conditioning, however, was shown in part two to induce changes in receptive fields that are detectable above and beyond these spontaneous variations. The latter point was shown with two statistical tests. First, paired t-tests were used to compare the mean response strength at the conditioned frequency for the three days before and three days after classical conditioning. Then, a test of binomial probabilities was implemented to demonstrate that significant plastic changes, as assessed by t-test, occur significantly more often after classical conditioning than as a result of the spontaneous daily variations described in part one. For the first time it has been shown that conditioned pairing of an acoustic stimulus and reward produces enduring (at least 3 days) modification of auditory receptive fields, and that it does so about as often as conditioned pairing involving punishment.

Daily variation of frequency receptive fields

Before discussing the significance of the finding that receptive fields change progressively and 'spontaneously', it is necessary to consider potential caveats associated with our methodology. First, and foremost, this conclusion depends upon having obtained recordings from the same multiunit clusters from day to day. We ensured that only stable 7-day blocks of FRCs were analysed by assessing the relative consistency of principal component scatter plots for the action potential waveforms recorded from each electrode. It has previously been shown that electrode movement of only 10 μm can have an effect on these scatter plots (fig. 9 of Abeles & Goldstein, 1977; see also Snider & Bonds, 1998). It is also relevant that FRCs taken from evoked field potentials varied nonrandomly over the course of a week. Since these signals represent summation over a relatively large volume (up to 1 mm from the tip of the electrode, Bullock, 1997), it is unlikely that small electrode movements could produce such significant alterations in the field potential receptive fields.

All analysis in the present report was applied to normalized FRCs. That is, for each neuron population recording, the evoked response to each tone frequency was divided by the average response to all frequencies. This manipulation was employed because the relative, rather than absolute, firing rate of a neuron (or neurons) determines pitch-selectivity, or 'tuning'. While reducing the day-to-day random variability of the FRCs, normalization by itself does not introduce nonrandom variations. Since all frequencies in an FRC are normalized by the same value, tuning curve shape is not affected. Further, although normalization can be problematic when firing rates are very low, this should not have been an issue in the present study since multiunit clusters with low tone-evoked rates and high within-session variability were excluded from further analysis (see Results).

It is important to remember that all signals used in the present study were taken from populations of neurons. This restriction was of necessity, not of choice. In our experience, extracellular action potential waveforms recorded acutely with microwires from rat primary auditory cortex all exhibit relatively similar shapes, making sorting by principal components analysis somewhat difficult, though quite tractable. Difficulty becomes near impossibility in chronic recordings where the signal to noise ratio is reduced (e.g. Figure. 1C and D), presumably because of gliosis around the electrode (e.g. Rousche & Normann, 1998; Turner *et al.*, 1999). Even when sophisticated single-wire sorting techniques are applied to acutely recorded waveforms from other cortical areas of larger species, classification errors can occur (Gray *et al.*, 1995). Although we have had significant success isolating single units from acute recordings with tetrodes, similar attempts with fixed-electrode chronic recordings have failed, perhaps because of long-term tissue damage caused by the large, twisted bundle of wires.

It might be argued that the gradual changes in receptive fields observed in the present study were due, in fact, to damage caused by the electrodes, and subsequent reorganization of the tonotopic map. This is unlikely because we observed such systematic changes even in rats that had been chronically implanted for 7 months (not shown). Further, we would expect to see changes in the principal component scatter plots of action potential waveforms if the physical health and/or organization of the cortex surrounding the electrode tip was changing during a 7-day block.

Unfortunately we cannot be certain that giving daily doses of ketamine is not in fact what caused the auditory receptive fields to be nonstationary. However, in order to be certain that learning-induced changes observed in the receptive fields were due to plastic changes, rather than simply selective attention to a particular tone frequency, it was necessary to record from a basically unconscious animal. This condition also prevented the animal's general behavioural state from increasing the variability between days. Because anaesthesia depth can modulate sensory receptive fields in general (Duncan *et al.*, 1982; Armstrong-James & George, 1988; Wörgötter *et al.*, 1998; Friedberg *et al.*, 1999) and auditory cortex evoked activity in particular (Kisley & Gerstein, 1999b), we were careful to give the same dose of ketamine every day, to record at fixed postinjection intervals and then average together FRCs taken at these time points.

Although not quantified, we observed many different types of receptive field changes. For example, some FRCs exhibited drifting best frequencies (both upwards and downwards), while others showed relatively stable best frequencies and systematically changing response magnitudes in the side-bands. Because of this, and because cortical FRCs display an incredible diversity of shapes, we chose to compare receptive fields across days with correlation coefficients rather than using more traditional auditory neurophysiology measures (e.g. characteristic frequency and bandwidth). Any type of nonrandom variation in a FRC's shape will result in a negative correlation slope. Further studies will be necessary to understand in detail precisely what types of changes occur and their relevance for the frequency selectivity of cortical neurons and neuron populations.

The finding of nonstationary receptive fields depends upon the total duration of measurement used for the correlation analysis. It is possible that, over longer time-scales, the FRCs are basically stable. This would be the case if, for example, the receptive fields oscillate or follow a random walk on a time-scale of months. We certainly do not expect that the pitch-selectivity of a cortical neuron continues to change monotonically throughout the life of an animal. Regardless, it can be concluded from our study that detectable changes occur over the course of 24 h (or less), and they do not cycle during seven or fewer days. Although not shown, we analysed several 14-day blocks of data, and came to basically the same conclusion. Here we present data from 7-day blocks only because increasing the length of analysis block increases the likelihood of electrode instability (Liu *et al.*, 1999).

Caveats aside, we believe the observation of nonrandom changes in FRCs was due to actual modification of the functional organization of auditory cortex receptive fields. It cannot yet be determined whether these changes happened spontaneously or as a result of the rats' daily experiences. Nor can the behavioural relevance (if any) of these changes be appreciated from the present data. Nevertheless, these findings have implications for past and future studies of long-term receptive field plasticity. For example, analysing several days before conditioning allows a more reliable estimate of each multiunit cluster's receptive field, and its natural rate of variation. As discussed above, a criterion using only one day's receptive fields pre- and postconditioning doubled the estimated number of multiunit clusters classified as exhibiting significant plastic changes by our criterion (which is statistical and compares curves from multiple days). This suggests that the day-to-day variability of FRCs could lead to errors when estimating the prevalence of significant and detectable learning-induced changes.

Learning-induced plasticity of frequency receptive fields

Given the very strong behavioural significance associated with electrical stimulation of the MFB (Olds, 1962), we were not surprised to find that conditioned pairing involving such a reward would produce detectable changes in the receptive field of cortical neurons. Previously it has been shown that conditioned pairing of sound and stimulation of lateral hypothalamus leads to modified evoked responses in the auditory system (Woody *et al.*, 1994). However, it was not clear until the present study whether: (i) receptive fields, in particular, can be modified by such appetitive conditioning and (ii) the modifications endure for days, as opposed to hours, beyond the training event. Our results are consistent with the idea that the modification of evoked response is due to actual plastic changes rather than selective attention (during the postconditioning measurement phase of the experiment), because the FRCs were altered even when assessed while the animals were under sedation. However, attentional modulation cannot be completely ruled out since the animals were generally beginning to move near the end of the recording period. Previous studies have shown definitive plastic changes after aversive conditioning (Lennartz & Weinberger, 1992; Weinberger *et al.*, 1993) and discrimination training (Recanzone *et al.*, 1993) where receptive fields were subsequently measured in fully anaesthetized animals. One very important issue that remains to be addressed – for which the present data are too scant – is the detailed relationship between the shape of an FRC before appetitive conditioning, the frequency of the conditioning tone and the resulting receptive field changes.

In order to determine whether the learning-induced changes observed in the present study required both a tone and stimulation of brain reward centres, we repeated the training but with stimulus-alone and tone-alone paradigms. Although self-stimulation behaviour has been shown to produce structural changes in several brain regions including hippocampus and cortex (Shankaranarayana Rao *et al.*, 1999), we found that the prevalence of significant modifications in auditory cortex frequency receptive fields after stimulus-alone training was essentially the same as that after no training. Training with a pure tone in the absence of brain stimulation, however, did produce a slight – though not significant – increase in the percentage of receptive fields that exhibited plastic changes. While still unexplained, this finding is consistent with previous studies demonstrating elevated probability of plastic changes after tone-alone training (Gao & Suga, 1998; Yan & Suga, 1998). However, as with these aversive conditioning studies, the prevalence of plasticity was clearly highest after associative training. The effects of strict 'nonassociative' training (i.e. presentation of both tones and brain stimulation during a single session but without temporal pairing) were not assessed in this study.

From the present results we cannot determine the exact mechanism of appetitive conditioning-induced plasticity of auditory receptive fields. Further experiments are required to decide whether this type of learning-induced plasticity involves some or all of the same neural circuits, still under debate, believed to be responsible for aversive conditioning (LeDoux & Muller, 1997; Gao & Suga, 1998; Weinberger, 1998). In addition, it is too early to tell whether conditioned pairing of tone and reward in general, rather than pairing of tone and direct electrical stimulation of the brain in particular, is the necessary element for inducing plasticity. The MFB contains fibers from many different modulatory brain centres (Nieuwenhuys *et al.*, 1982), some of which might be directly or indirectly involved in learning-induced plasticity of cortical receptive fields (e.g. cholinergic nuclei of the basal forebrain: Bakin & Weinberger, 1996; Kilgard & Merzenich, 1998).

Regardless of the precise mechanism of learning-induced plasticity, we would ultimately like to characterize the behavioural relevance of the additional neural resources apparently devoted to processing stimuli at the frequency of the conditioning tone after the pairing described in this and previous reports. A recent study from our laboratory (S. K. Talwar & G. L. Gerstein, unpublished observation) showed that expansion of a given sound frequency's representation by intracortical microstimulation does not improve an animal's frequency-discrimination abilities (but other abilities associated with cortical function including spatial-discrimination were not assessed). Is the same true of plasticity evoked with this more behaviourally relevant appetitive conditioning paradigm? The answer to this question has implications for the physiological basis of perceptual learning, and perhaps, for the future of noninvasive compensatory treatment of individuals with selective brain damage.

Acknowledgements

This work was supported by Grants MH 46428 and DC 01249 from the National Institutes of Health. For helpful discussion and suggestions we gratefully acknowledge Drs Leif Finkel, Jeff Keating, Larry Palmer, David Perkel, James Saunders, and Sanjiv Talwar.

Abbreviations

FRC, frequency response curve; MFB, medial forebrain bundle.

References

Abeles, M. & Goldstein, M.H. (1977) Multispike train analysis. *Proc. IEEE*, **65**, 762–773.

Armstrong-James, M. & George, M.J. (1988) Influence of anesthesia on spontaneous activity and receptive field size of single units in rat Sml neocortex. *Exp. Neurol.*, **99**, 369–387.

Ashe, J.H., McKenna, T.H. & Weinberger, N.M. (1989) Cholinergic modulation of frequency receptive fields in auditory cortex: II. frequency-specific effects of anticholinesterases provide evidence for a modulatory action of endogenous ACh. *Synapse*, **4**, 44–54.

Bakin, J.S., South, D.A. & Weinberger, N.M. (1996) Induction of receptive field plasticity in the auditory cortex of the guinea pig during instrumental avoidance conditioning. *Behav. Neurosci.*, **110**, 905–913.

Bakin, J.S. & Weinberger, N.M. (1990) Classical conditioning induces CS-specific receptive field plasticity in the auditory cortex of the guinea pig. *Brain Res.*, **536**, 271–286.

Bakin, J.S. & Weinberger, N.M. (1996) Induction of a physiological memory in the cerebral cortex by stimulation of the nucleus basalis. *Proc. Natl. Acad. Sci. USA*, **93**, 11219–11224.

Bullock, T.H. (1997) Signals and signs in the nervous system: the dynamic anatomy of electrical activity is probably information-rich. *Proc. Natl. Acad. Sci. USA*, **94**, 1–6.

Buonomano, D.V. & Merzenich, M.M. (1998) Cortical plasticity: from synapses to maps. *Annu. Rev. Neurosci.*, **21**, 149–186.

Calford, M.B., Rajan, R. & Irvine, D.R.F. (1993) Rapid changes in the frequency tuning of neurons in cat auditory cortex resulting from pure-tone-induced temporary threshold shift. *Neuroscience*, **55**, 953–964.

Condon, C.D. & Weinberger, N.M. (1991) Habituation produces frequency-specific plasticity of receptive fields in the auditory cortex. *Behav. Neurosci.*, **105**, 416–430.

Cruikshank, S.J. & Weinberger, N.M. (1996) Receptive-field plasticity in the adult auditory cortex induced by hebbian covariance. *J. Neurosci.*, **16**, 861–875.

Dinse, H.R., Reuter, G., Cords, S.M., Godde, B., Hilger, T. & Lenarz, T. (1997) Optical imaging of cat auditory cortical organization after electrical stimulation of a multichannel cochlear implant: differential effects of acute and chronic stimulation. *Am. J. Otol.*, **18**, S17–S18.

Duncan, G.H., Dreyer, D.A., McKenna, T.M. & Whitsel, B.L. (1982) Dose- and time-dependent effects of ketamine on SI neurons with cutaneous receptive fields. *J. Neurophysiol.*, **47**, 677–699.

Edeline, J.-M. (1999) Learning-induced physiological plasticity in the thalamo-cortical sensory systems: a critical evaluation of receptive field plasticity, map changes and their potential mechanisms. *Prog. Neurobiol.*, **57**, 165–224.

Edeline, J.-M. & Weinberger, N.M. (1991) Subcortical adaptive filtering in the auditory system: associative receptive field plasticity in the dorsal medial geniculate body. *Behav. Neurosci.*, **105**, 154–175.

Edeline, J.-M. & Weinberger, N.M. (1993) Receptive field plasticity in the auditory cortex during frequency discrimination training: selective retuning independent of task difficulty. *Behav. Neurosci.*, **107**, 82–103.

Fay, R.R. (1988). *Hearing in Vertebrates: a Psychophysics Databook*. Hill-Fay Associates, Winnetka.

Friedberg, M.H., Lee, S.M. & Ebner, F.F. (1999) Modulation of receptive field properties of thalamic somatosensory neurons by the depth of anesthesia. *J. Neurophysiol.*, **81**, 2243–2252.

Gao, E. & Suga, N. (1998) Experience-dependent corticofugal adjustment of midbrain frequency map in bat auditory system. *Proc. Natl. Acad. Sci. USA*, **95**, 12663–12670.

Gonzalez-Lima, F. & Scheich, H. (1984) Neural substrates for tone-conditioned bradycardia demonstrated with 2-deoxyglucose. I. activation of auditory nuclei. *Behav. Brain Res.*, **14**, 213–233.

Gonzalez-Lima, F. & Scheich, H. (1986) Neural substrates for tone-conditioned bradycardia demonstrated with 2-deoxyglucose. II. auditory cortex plasticity. *Behav. Brain Res.*, **20**, 281–293.

Gray, C.M., Maldonado, P.E., Wilson, M. & McNaughton, B. (1995) Tetrodes markedly improve the reliability and yield of multiple single-unit isolation from multi-unit recordings in cat striate cortex. *J. Neurosci. Methods*, **63**, 43–54.

Kilgard, M.P. & Merzenich, M.M. (1998) Cortical map reorganization enabled by nucleus basalis activity. *Science*, **279**, 1714–1718.

Kisley, M.A. & Gerstein, G.L. (1999a) Long term variation of frequency response curves recorded from neuronal populations of auditory cortex: random variability or plasticity? *Soc. Neurosci. Abstr.*, **25**, 392.

Kisley, M.A. & Gerstein, G.L. (1999b) Trial-to-trial variability and state-dependent modulation of auditory-evoked responses in cortex. *J. Neurosci.*, **19**, 10451–10460.

LeDoux, J.E. & Muller, J. (1997) Emotional memory and psychopathology. *Phil. Trans. R. Soc. Lond. B Biol. Sci.*, **352**, 1719–1726.

Lennartz, R.C. & Weinberger, N.M. (1992) Frequency-specific receptive field plasticity in the medial geniculate body induced by pavlovian fear conditioning is expressed in the anesthetized brain. *Behav. Neurosci.*, **106**, 484–497.

Liu, X., McCreery, D.B., Carter, R.R., Bullara, L.A., Yuen, T.G.H. & Agnew, W.F. (1999) Stability of the interface between neural tissue and chronically

implanted intracortical microelectrodes. *IEEE Trans. Rehab. Eng.*, **7**, 315–325.
Maldonado, P.E. & Gerstein, G.L. (1996) Reorganization in the auditory cortex of the rat induced by intracortical microstimulation: a multiple single-unit study. *Exp. Brain. Res.*, **112**, 420–430.
McKenna, T.M., Ashe, J.H. & Weinberger, N.M. (1989) Cholinergic modulation of frequency receptive fields in auditory cortex: I. frequency-specific effects of muscarinic agonists. *Synapse*, **4**, 30–43.
Metherate, R. & Weinberger, N.M. (1989) Acetylcholine produces stimulus-specific receptive field alterations in cat auditory cortex. *Brain Res.*, **480**, 372–377.
Nieuwenhuys, R., Geeraedts, L.M.G. & Veening, J.G. (1982) The medial forebrain bundle of the rat. I. general introduction. *J. Comp. Neurol.*, **206**, 49–81.
Ohl, F.W. & Scheich, H. (1996) Differential frequency conditioning enhances spectral contrast sensitivity of units in auditory cortex (field AI) of the alert mongolian gerbil. *Eur. J. Neurosci.*, **8**, 1001–1017.
Olds, J. (1962) Hypothalamic substrates of reward. *Physiol. Rev.*, **42**, 554–604.
Olds, M.E. & Olds, J. (1963) Approach-avoidance analysis of rat diencephalon. *J. Comp. Neurol.*, **120**, 259–295.
Paxinos, G. & Watson, C. (1997). *The Rat Brain in Stereotaxic Coordinates*. 3rd edn. Academic Press. San Diego.
Recanzone, G.H., Schreiner, C.E. & Merzenich, M.M. (1993) Plasticity in the frequency representation of primary auditory cortex following discrimination training in adult owl monkeys. *J. Neurosci.*, **13**, 87–103.
Robertson, D. & Irvine, D.R.F. (1989) Plasticity of frequency organization in auditory cortex of guinea pigs with partial unilateral deafness. *J. Comp. Neurol.*, **282**, 456–471.
Rousche, P.J. & Normann, R.A. (1998) Chronic recording capability of the Utah intracortical electrode array in cat sensory cortex. *J. Neurosci. Methods*, **82**, 1–15.
Scheich, H., Stark, H., Zuschratter, W., Ohl, F.W. & Simonis, C.E. (1997) Some functions of primary auditory cortex in learning and memory formation. *Adv. Neurol.*, **73**, 179–193.
Shankaranarayana Rao, B.S., Raju, T.R. & Meti, B.L. (1999) Increased numerical density of synapses in CA3 region of hippocampus and molecular layer of motor cortex after self-stimulation rewarding experience. *Neuroscience*, **91**, 799–803.
Snider, R.K. & Bonds, A.B. (1998) Classification of non-stationary neural signals. *J. Neurosci. Methods*, **84**, 155–166.
Sutter, M.L. & Schreiner, C.E. (1991) Physiology and topography of neurons with multipeaked tuning curves in cat primary auditory cortex. *J. Neurophysiol.*, **65**, 1207–1226.
Turner, J.N., Shain, W., Szarowski, D.H., Andersen, M., Martins, S., Isaacson, M. & Craighead, H. (1999) Cerebral astrocyte response to micromachined silicon implants. *Exp. Neurol.*, **156**, 33–49.
Weinberger, N.M. (1998) Physiological memory in primary auditory cortex: characteristics and mechanisms. *Neurobiol. Learn. Mem.*, **70**, 226–251.
Weinberger, N.M., Javid, R. & Lepan, B. (1993) Long-term retention of learning-induced receptive-field plasticity in the auditory cortex. *Proc. Natl. Acad. Sci. USA*, **90**, 2394–2398.
Willott, J.F., Aitkin, L.M. & McFadden, S.L. (1993) Plasticity of auditory cortex associated with sensorineural hearing loss in adult C57BL/6J mice. *J. Comp. Neurol.*, **329**, 402–411.
Willott, J.F. & Lu, S.-M. (1982) Noise-induced hearing loss can alter neural coding and increase excitability in the central nervous system. *Science*, **216**, 1331–1332.
Woody, C.D., Wang, X.-F. & Gruen, E. (1994) Response to acoustic stimuli increases in the ventral cochlear nucleus after stimulus pairing. *Neuroreport*, **5**, 513–515.
Wörgötter, F., Suder, K., Zhao, Y., Kerscher, N., Eysel, U.T. & Funke, K. (1998) State-dependent receptive-field restructuring in the visual cortex. *Nature*, **396**, 165–168.
Yan, W. & Suga, N. (1998) Corticofugal modulation of the midbrain frequency map in the bat auditory system. *Nature Neurosci.*, **1**, 54–58.
Zhang, Y., Suga, N. & Yan, J. (1997) Corticofugal modulation of frequency processing in bat auditory system. *Nature*, **387**, 900–903.

Chapter 13
Improvements in the Analysis of Neuronal Interactions

Thinking About Analysis Methods with George in Philadelphia, Cambridge, and Newcastle

Stuart N. Baker

I started my PhD in Roger Lemon's laboratory in 1992, developing approaches for multiple single-unit recording from awake behaving monkeys. At that time, neuroscience was starting to ask questions about the nature of the neural code. Since the time of Hubel and Wiesel (1959) and Evarts (1966), these questions had not seemed worth asking to most people. If an experimenter found the right stimulus, or the animal performed a certain behaviour, it made the cell fire more spikes. Clearly then, neurones coded the presence or absence of their preferred stimulus by their firing rate. Experimenters typically averaged discharge over multiple trials, and over relatively wide bins (maybe tens of milliseconds) to remove the noise inevitably introduced by coding with a series of discrete firing events. It was assumed that the brain could do something similar: long membrane time constants within neurons would provide temporal smoothing, and averaging across multiple cells carrying the same signal would have the same effect as averaging across trials. But in the 1990s, many investigators provided evidence that single neurons might code information by the precise timing of their action potentials, and especially by synchronisation between neurons. Could it be that single spikes, measured with millisecond precision, mattered? This was a big question, fundamental to how we think about neural computation, and there were good arguments on both sides—see, for example, the papers by Softky and Koch (1992, 1993) and Shadlen and Newsome (1994, 1998).

S. N. Baker (✉)
Institute of Neuroscience, Medical School, Newcastle University, Newcastle Upon Tyne, UK
e-mail: stuart.baker@ncl.ac.uk

A. Aertsen et al. (eds.), *Introducing Computation to Neuroscience*, Springer Series in Computational Neuroscience, https://doi.org/10.1007/978-3-030-87447-6_13

From his very earliest work, George had thought about these issues. He had certainly considered firing rate codes. The peri-stimulus time histogram (PSTH), which is the most basic analysis tool in the rate coder's armoury, was first introduced by Gerstein and Kiang (1960). He was also a pioneer in the analysis of neural synchrony, and his early work beautifully described the factors that could influence the cross-correlation histogram (Perkel et al. 1967; Moore et al. 1970). His subsequent work was ahead of its time. Already in 1975, George was considering the problem of quantitative display of synchrony between three neurons (Perkel et al. 1975); by the mid-eighties, he was working on detecting temporal patterns across multiple cells (Gerstein et al. 1985).

In 1989, George published one of his seminal papers (Aertsen et al. 1989), which set the scene for our subsequent interactions. Against the background described above, it was a logical progression, for it looked at stimulus-locked changes in synchrony—a combination of the PSTH and cross-correlation histogram. This chapter introduced two innovations. First, it presented a new way of displaying the data, which cleverly showed all the important information in one simple composite figure (the joint peri-stimulus time histogram, JPSTH); this had evolved from George's earlier work (Gerstein and Perkel 1972). Second, an approach to statistical significance was developed. This relied on forming a prediction of the JPSTH surface from the single-cell PSTHs; by comparing actual and predicted coincidence counts, it was possible to determine whether there were more synchrony than expected by chance.

The JPSTH was an undoubted advance, but it made one key assumption. Because the prediction of the JPSTH was derived from the PSTHs, it was a requirement that the PSTHs (averaged across trials) accurately represented the single-trial firing of the neurons. This might be reasonable when recording a stimulus-evoked response in anaesthetised animals, and the two example datasets presented by Aertsen et al. (1989) were both of this type. My own experimental recordings made in the Lemon Laboratory were from the motor cortex of an awake monkey performing a precision grip task. In my data, there was clear trial-by-trial variability in the behavioural performance, and not unsurprisingly this led to variation in the neural firing profile. The PSTH was a smeared-out average representation of how a cell changed its rate during the task; in reality, no trial had a rate profile that matched the PSTH. Brody (1998, 1999a,b) showed that changes in amplitude and latency of responses from trial-to-trial could disrupt the statistical assessment of correlation and lead to false positives. His suggestion was to shift and scale the PSTH for each trial to provide a better estimate of the single-trial rate.

With Quentin Pauluis from the Université de Louvain, we investigated an alternative strategy, whereby we would generate an estimate of the single-trial firing rate by smoothing the raw data (Pauluis and Baker 2000). The problem of deriving density estimates for point processes was well studied in mathematics (Silverman 1986). We produced a method that smoothed rate over a reasonable timeframe but detected rapid rate changes and did not smooth across these boundaries—in this way, we hoped to reduce noise where possible, but not over-smooth. We tested our approach on unitary event analysis, which was an alternative way of analysing synchrony introduced by Grün (1996); Riehle et al. (1997). As expected, in non-stationary data, there was spurious detection of synchrony when relying on the PSTH, which could be avoided by using single-trial density estimates. I subsequently used this general approach to analyse task-related changes in synchrony and higher-order patterns (Baker and Lemon 2000; Baker et al 2001); Quentin Pauluis applied the method to his own experimental recordings from cat superior colliculus (Pauluis et al. 2001).

By early 1999, the manuscript that would become Pauluis and Baker (2000) had been submitted to Neural Computation and was under review. I was spending that year writing up my experimental work, prior to setting up my own laboratory in Cambridge. I was keen to take some time out to visit a group with strengths in developing novel analysis methods and naturally contacted George, who agreed to host me for 3 months in his laboratory in Philadelphia. Funding for my travel and stay was generously provided by Christ's College in Cambridge. On arriving, George declared himself as the reviewing editor of our paper, which led to some interesting discussions (the keen-eyed will spot "Communicated by George Gerstein" at the top of the paper). Although he was in agreement with the overall idea that trial-by-trial variation would generate spurious correlations, he was less keen on our solution. He intuitively felt that bootstrapping the spike density from the data might lead to issues, although at that stage he could not be more specific. In the subsequent years, he worked on this further and demonstrated some problems with this method applied to pattern detection (Gerstein 2004). We therefore agreed that during my time in Philly, I would work on alternative approaches. This ultimately led to a joint paper, also published in Neural Computation (Baker and Gerstein 2001).

We first discussed ways in which a dataset could be tested to see whether a trial-by-trial variation was really a problem. It soon became clear that simply estimating the variance across trials of the firing rate (determined by Gaussian kernel convolution) could provide a powerful diagnostic (see Fig. 1 of Baker and Gerstein 2001). A consistent rate profile across trials led to a variance that was a scaled version of the mean rate. Amplitude variation led to inconsistencies that marked out the period of concern. Jitter in the response latency led to "rabbit ears" on the variance plot, which again identified the region of interest.

This observation led to the first of three methods reported in our paper to detect response latency. We shifted each trial by a small amount and recalculated the rate variance time course; an optimisation algorithm then found the best shifts to minimise the total variance. We also developed methods based on Bayesian change point detection and simple threshold crossing of a kernel density estimate. Parametric analysis on different simulated datasets showed that there was no overall best method: each was optimal for a particular range of parameters (Fig. 5 of Baker and Gerstein 2001). As expected, recalculation of the JPSTH after correcting for latency jitter effectively removed spurious correlation.

As well as the figures that we obtained by simulation to address the core problem, this paper was notable in my career as the first and only time in which I have contributed a mathematical proof. As part of our analysis, we needed an expression for the autocorrelation of a gamma process. This is easily expressed as an infinite sum over all orders of inter-spike interval distributions; our Appendix derived a closed-form solution as a finite sum. I well recall thinking long and hard about this problem during my time in Philadelphia and excitedly finding a solution over one weekend. On announcing this to George on the Monday, he was not hugely impressed. Practically, the form of the autocorrelation could easily be determined numerically to high precision by truncating the infinite sum. George was always interested first and foremost in the brain; novel analytical methods were just a means to the end of understanding neural function. Maths for its own sake held little interest for him. Still, the Appendix is there if ever you need this particular result! To my knowledge, this is given nowhere else in the literature.

Around halfway through my time in his lab, this paper was largely completed, and George suggested that I turn my attention to another problem. He had always been frustrated that even if we were able to record many cells simultaneously, most analysis methods simply considered the correlation between pairs. One interesting method that he had developed to extract higher-order structure from neural recordings was gravitational clustering (Gerstein et al. 1985). In this, cells are represented as points in a multi-dimensional space and allowed to move according to simulated forces generated by their firings. Synchronous firing tends to lead to clusters of points. George was aware that his previous formulation of this approach had been somewhat arbitrary: he had defined forces between particles in the simplest way he could think of, and the method had worked. In the paper that arose from our work together (Baker and Gerstein 2000), we looked at the best way to determine the forces, showing that introduction of a non-linearity could increase the separation of synchronous from asynchronous cells using a given length of data. We also showed that the optimal setup would depend on the regularity of the inter-spike intervals. As part of this work, we generated a new way to produce simulated spike trains with arbitrary levels of synchrony and inter-spike interval regularity. This involved merging multiple Poisson processes and then "decimating" (selecting every nth event) to produce a gamma process of order n.

At the same time as I was working on these two problems, George had realised that the trend towards recording from increasing numbers of electrodes should bring more benefits than simply an increase in the size of recorded population. Electrophysiological recordings are always corrupted by electrical noise, most of which is common across electrodes (e.g., mains hum, movement artefact, or EMG contamination resulting from the animal chewing). Pavel Musial, Jeff Keating, and I looked into ways of exploiting the multiple recordings to improve signal-to-noise ratio, inspired by older papers on signal processing for RADAR arrays. This issue consumed many hours of discussion in George's office and eventually led to the third paper to arise during my 3 months in Philadelphia (Musial et al. 2002).

Modern technology is currently allowing electrode channel counts to rise rapidly into the thousands (Jun et al. 2017; Fiáth et al. 2018; Angotzi et al. 2019), providing quantities of data undreamt of when George started his career in the 1960s. Many groups are working on automated approaches to spike discrimination from these large-scale recordings but often ignore the signal-to-noise ratio gains that can be made relatively easily from simple data pre-processing (Rossant et al. 2016; Chung et al. 2017). This is clearly something that should be included in new algorithms.

On my return from Philadelphia, I set up my own group in Cambridge; within a short time, George formally retired. Freed from his teaching and administrative commitments, his greatest desire was to travel to his friends and collaborators internationally to work on the problems so dear to his heart. Accordingly, we arranged for George to visit Cambridge in 2002, reversing the former roles of host and guest. His stay coincided with a very talented undergraduate student Ronnie Davies doing a project in my lab. When simulating a gamma spike train, the two necessary input parameters are the firing rate and the regularity. Having worked on measurement of firing rate, I now wished to look at regularity, since improved ways to estimate this could allow us to simulate better surrogate data for significance estimation of correlation measures. The problem, of course, was that measures had to be made over single trials, as trial-to-trial rate fluctuation would otherwise artificially inflate estimates of firing irregularity. After many false starts and much discussion with George, Ronnie came up with a very simple measure and showed that this had many attractive properties.

At the end of Ronnie Davies' project (just 24 weeks long!), much progress had been made, but the work was not yet ready for publication. As soon as possible, I invited George back for a further visit; by this time (2004), my group had moved to Newcastle. Ronnie, now a clinical medical student, was able to join us for part of this time; the work was completed, leading to my final paper in collaboration with George (Davies et al. 2006). George especially enjoyed his time in Newcastle and took full advantage of the many cultural opportunities on offer, being a frequent visitor to the Theatre Royal and Sage Concert Hall in the city.

As well as these collaborative visits, George visited Newcastle twice to attend conferences that I organised as part of the UK Spike Train Analysis Network. These were conceived of as highly interactive workshops, where attendees would set out problems in neural data analysis which they thought important, and then work on them with others of like mind. George's participation on each occasion was critical to the success of these meetings; he had the ability to ask the difficult questions and to probe and push until there were some answers. These long workshops suited his style. George was not one to give quick responses; he liked to listen. He then preferred to think, read, and check something with a piece of computer code before making a suggestion, which was invariably to the point.

George was a hugely influential figure in our field, but he was also a wonderful colleague and friend. Aside from the many papers that are reproduced in this volume, his gentle encouragement of younger scientists has spread his approach to the brain far beyond his direct work. During his visits to my lab, the students were in awe that such a senior figure was sitting at a desk in the open plan area. The awe was soon dispelled, as George talked through their current projects and gave suggestions and kindly support. Likewise, I remember on the Sunday before one of our conferences, two of our Institute secretaries nervously welcoming Prof. George Gerstein at the conference registration desk, and doing their best to make sure that all went smoothly for someone so eminent. The nervousness was dispelled when George produced a model airplane from his bag, gave it to my six year old son who was sitting bored in the corner, and suggested that the two of them went and tested how it flew on the field outside. No more bored six year old, and I was left free to help with the rest of the registration queue. George took a genuine interest in everyone.

Neuroscience has lost one of its founding fathers; we are profoundly grateful for all that he gave. In his mother tongue:

БОЛЬШОЕ СПАСИБО, ГЕОРГИЙ!

References

Aertsen AM, Gerstein GL, Habib MK, Palm G (1989) Dynamics of neuronal firing correlation: modulation of "effective connectivity". J Neurophysiol 61(5):900–917. https://doi.org/10.1152/jn.1989.61.5.900

Angotzi G, Boi F, Lecomte A, Miele E, Malerba M, Zucca S, Casile A, Berdondini L (2019) SiNAPS: an implantable active pixel sensor CMOS-probe for simultaneous large-scale neural recordings. Biosensors Bioelectron 126:355–364. https://doi.org/10.1016/j.bios.2018.10.032

Baker SN, Gerstein GL (2000) Improvements to the sensitivity of gravitational clustering for multiple neuron recordings. Neural Comput 12(11):2597–2620. https://doi.org/10.1162/089976600300014863

Baker SN, Gerstein GL (2001) Determination of response latency and its application to normalization of cross-correlation measures. Neural Comput 13(6):1351–1377. https://doi.org/10.1162/08997660152002889

Baker SN, Lemon RN (2000) Precise spatiotemporal repeating patterns in monkey primary and supplementary motor areas occur at chance levels. J Neurophysiol 84(4):1770–1780. https://doi.org/10.1152/jn.2000.84.4.1770

Baker SN, Spinks R, Jackson A, Lemon RN (2001) Synchronization in monkey motor cortex during a precision grip task. I. Task-dependent modulation in single-unit synchrony. J Neurophysiol 85(2):869–885. https://doi.org/10.1152/jn.2001.85.2.869

Brody CD (1998) Slow covariations in neuronal resting potentials can lead to artefactually fast cross-correlations in their spike trains. J Neurophysiol 80(6):3345–3351. https://doi.org/10.1152/jn.1998.80.6.3345

Brody CD (1999a) Disambiguating different covariation types. Neural Comput 11(7):1527–1535. https://doi.org/10.1162/089976699300016124

Brody CD (1999b) Correlations without synchrony. Neural Comput 11(7):1537–1551. https://doi.org/10.1162/089976699300016133

Chung JE, Magland JF, Barnett AH, Tolosa VM, Tooker AC, Lee KY, Shah KG, Felix SH, Frank LM, Greengard LF (2017) A fully automated approach to spike sorting. Neuron 95(6):1381–1394.e6. https://doi.org/10.1016/j.neuron.2017.08.030

Davies RM, Gerstein GL, Baker SN (2006) Measurement of time-dependent changes in the irregularity of neural spiking. J Neurophysiol 96(2):906–918. https://doi.org/10.1152/jn.01030.2005

Evarts EV (1966) Pyramidal tract activity associated with a conditioned hand movement in the monkey. J Neurophysiol 29(6):1011–1027. https://doi.org/10.1152/jn.1966.29.6.1011

Fiáth R, Raducanu BC, Musa S, Andrei A, Lopez CM, van Hoof C, Ruther P, Aarts A, Horváth D, Ulbert I (2018) A silicon-based neural probe with densely-packed low-impedance titanium nitride microelectrodes for ultrahigh-resolution in vivo recordings. Biosensors Bioelectron 106:86–92. https://doi.org/10.1016/j.bios.2018.01.060

Gerstein GL (2004) Searching for significance in spatio-temporal firing patterns. Acta Neurobiol Exp (Wars) 64(2):203–207

Gerstein GL, Kiang NS (1960) An approach to the quantitative analysis of electrophysiological data from single neurons. Biophys J 1(1):15–28. https://doi.org/10.1016/s0006-3495(60)86872-5

Gerstein GL, Perkel DH (1972) Mutual temporal relationships among neuronal spike trains: statistical techniques for display and analysis. Biophys J 12(5):453–473. https://doi.org/10.1016/S0006-3495(72)86097-1

Gerstein GL, Perkel H, Dayhoff E (1985) Cooperative firing activity in simultaneously recorded populations of neurons: detection and measurement. J Neurosci 5(4):881–889. https://doi.org/10.1523/JNEUROSCI.05-04-00881.1985

Grün S (1996) Unitary joint-events in multiple-neuron spiking activity. PhD thesis, Ruhr University Bochum

Hubel DH, Wiesel TN (1959) Receptive fields of single neurones in the cat's striate cortex. J Physiol 148(3):574–591. https://doi.org/10.1113/jphysiol.1959.sp006308

Jun JJ, Steinmetz NA, Siegle JH, Denman DJ, Bauza M, Barbarits B, Lee AK, Anastassiou CA, Andrei A, Aydin Ç, Barbic M, Blanche TJ, Bonin V, Couto J, Dutta B, Gratiy SL, Gutnisky DA, Häusser M, Karsh B, Ledochowitsch P, Lopez CM, Mitelut C, Musa S, Okun M, Pachitariu M, Putzeys J, Rich PD, Rossant C, Sun Wl, Svoboda K, Carandini M, Harris KD, Koch C, O'Keefe J, Harris TD (2017) Fully integrated silicon probes for high-density recording of neural activity. Nature 551:232–236. https://doi.org/10.1038/nature24636

Moore GP, Segundo JP, Perkel DH, Levitan H (1970) Statistical signs of synaptic interaction in neurons. Biophys J 10(9):876–900. https://doi.org/10.1016/S0006-3495(70)86341-X

Musial PG, Baker SN, Gerstein GL, King EA, Keating JG (2002) Signal-to-noise ratio improvement in multiple electrode recording. J Neurosci Methods 115(1):29–43. https://doi.org/10.1016/S0165-0270(01)00516-7

Pauluis Q, Baker SN (2000) An accurate measure of the instantaneous discharge probability, with application to the unitary joint-event analysis. Neural Comput 12(3):647–669. https://doi.org/10.1162/089976600300015736

Pauluis Q, Baker SN, Olivier E (2001) Precise burst synchrony in the superior colliculus of the awake cat during moving stimulus presentation. J Neurosci 21(2):615–627. https://doi.org/10.1523/jneurosci.21-02-00615.2001

Perkel DH, Gerstein GL, Moore GP (1967) Neuronal spike trains and stochastic point processes. II. Simultaneous spike trains. Biophys J 7(4):419–440. https://doi.org/10.1016/s0006-3495(67)86597-4

Perkel DH, Gerstein GL, Smith MS, Tatton WG (1975) Nerve-impulse patterns: a quantitative display technique for three neurons. Brain Res 100(2):271–296. https://doi.org/10.1016/0006-8993(75)90483-7

Riehle A, Grün S, Diesmann M, Aertsen A (1997) Spike synchronization and rate modulation differentially involved in motor cortical function. Science 278(5345):1950–1953. https://doi.org/10.1126/science.278.5345.1950

Rossant C, Kadir S, Goodman D, Schulman J, Hunter M, Saleem A, Grosmark A, Belluscio M, Denfield G, Ecker A, Tolias A, Solomon S, Buzsáki G, Carandini M, Harris K (2016) Spike sorting for large, dense electrode arrays. Nature Neurosci 19:634–641. https://doi.org/10.1038/nn.4268.10.1038/nn.4268

Shadlen MN, Newsome WT (1994) Noise, neural codes and cortical organization. Curr Opin Neurobiol 4(4):569–579. https://doi.org/10.1016/0959-4388(94)90059-0

Shadlen MN, Newsome WT (1998) The variable discharge of cortical neurons: implications for connectivity, computation, and information coding. J Neurosci 18(10):3870–3896. https://doi.org/10.1523/JNEUROSCI.18-10-03870.1998

Silverman BW (1986) Density estimation. Chapman & Hall/CRC, London

Softky WR, Koch C (1992) Cortical cells should fire regularly, but do not. Neural Comput 4(5):643–646. https://doi.org/10.1162/neco.1992.4.5.643

Softky WR, Koch C (1993) The highly irregular firing of cortical cells is inconsistent with temporal integration of random EPSPs. J Neurosci 13(1):334–350. https://doi.org/10.1523/JNEUROSCI.13-01-00334.1993

LETTER — Communicated by Ad Aertsen

Improvements to the Sensitivity of Gravitational Clustering for Multiple Neuron Recordings

Stuart N. Baker
Department of Anatomy, University of Cambridge, Cambridge, CB2 3DY, U.K.

George L. Gerstein
Department of Neuroscience, University of Pennsylvania, Philadelphia, PA 19104, U.S.A.

We outline two improvements to the technique of gravitational clustering for detection of neuronal synchrony, which are capable of improving the method's detection of weak synchrony with limited data. The advantages of the enhancements are illustrated using data with known levels of synchrony and different interspike interval distributions. The novel simulation method described can easily generate such test data. An important dependence of the sensitivity of gravitational clustering to the interspike interval distribution of the analysed spike trains is described.

1 Introduction

There is an increasing realization among neuroscientists that neurons can code information by more complex means than just the mean firing rate of their discharge (Rieke, Warland, de Ruyter van Steveninck, & Bialek, 1997). One important additional neural code appears to be the simultaneous discharge of two or more cells. Such short-term synchrony has been proposed to underlie processing in a wide range of neural systems (Espinosa & Gerstein, 1988; Lindsey, Hernandez, Morris, & Shannon, 1992; Singer & Gray, 1995; Vaadia et al., 1995; Baker, Olivier, & Lemon, 1997). The detection of above-chance synchrony between cells from data limited in recording length and number of spikes is therefore an important and challenging analysis issue (Strangman, 1997).

Gravitational clustering (Gerstein, Perkel, & Dayhoff, 1985; Gerstein & Aertsen, 1985) has been proposed to permit rapid detection of synchronous cell assemblies in simultaneous recordings from large numbers of neurons. Each cell is represented as a particle in a higher-dimensional space. Action potentials produce transient "charges" on the particle corresponding to the cell that fired. The particles move according to forces caused by attraction or repulsion of the charges. Particles relating to cells that fire synchronously thereby aggregate together. Following a gravitational analysis, aggregated groups of particles indicate the presence of synchronous cell assemblies. The dynamic nature of the method allows it to visualize changes in groupings of neural synchronization over the duration of the recording. The gravitational clustering approach has been successfully applied to a range of experimental data (Gochin, Gerstein, & Kaltenbach, 1990; Lindsey, Morris, Shannon, & Gerstein, 1997; Lindsey, Shannon, & Gerstein, 1989; Aertsen, Bonhoeffer, & Kruger, 1987; Aertsen et al., 1991).

When compared with conventional cross-correlation-based approaches, the principal advantage of gravitational clustering is that it scales simply with the number of neurons whose firing has been recorded and multiple pairwise interactions can be simultaneously detected. The sensitivity of gravitational clustering has been shown to be similar to cross-correlation approaches in detecting synchrony with limited data (Strangman, 1997). An important difference between the two methodologies is that while cross-correlation represents a model-free approach to synchrony detection, gravitational clustering is more restricted. In particular, the time course of charge increment on a given particle following a spike limits the detection of synchrony to that which has a comparable jitter in synchronization. A similar limitation is present in the unitary event method (Grün, 1996; Pauluis & Baker, 2000). Such constraints prevent detection of synchronies that do not conform to the implied template. However, in theory, focusing on only those events deemed of interest a priori could have the capability to improve the sensitivity of detection.

In this article we present several modifications to the gravitational clustering approach, which are designed to improve its sensitivity in detecting neural synchronization. The desired improvement is demonstrated for simulated data having a magnitude and time course of synchrony in accord with that seen in experimental data. In addition, an important dependence of the sensitivity of the gravitational approach on the interspike interval distribution of the analyzed units is described.

2 Simulation of Spike Trains with Different Interspike Interval Statistics and Fixed Synchrony Magnitude

In order to quantify our improvements, we require a means of producing simulated spike trains in which there is a prespecified level of synchronization above that expected by chance. This can be relatively straightforwardly achieved if the spikes are simulated as a Poisson process, which have an exponential distribution of interspike intervals (Aertsen & Gerstein, 1985; Strangman, 1997). However, real recordings of neurons in the central nervous system usually show a relative refractory period (Rodieck, Kiang, & Gerstein, 1962), and their interspike interval histograms can be better approximated by a gamma distribution (Kuffler, Fitzhugh, & Barlow, 1957; Stein, 1965). The following is a method by which specified synchrony in gamma distributions may be achieved.

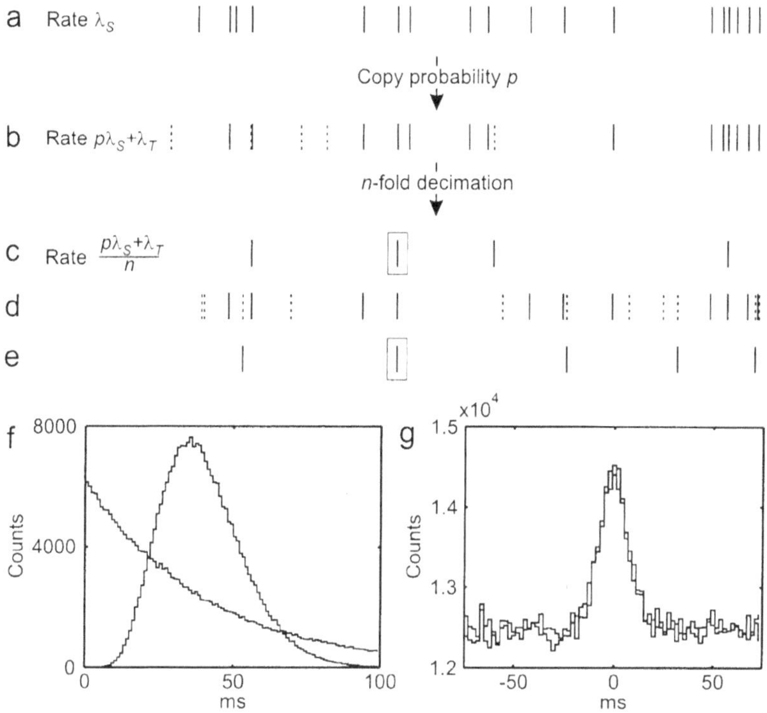

Figure 1: Method of synchronous spike train simulation. (a) Poisson spike train S, which serves as the source of synchronous spikes. (b) These are copied with probability p to the destination train T, which also has independent spikes (dotted lines). (c) Decimation produces a spike train with interspike intervals following a gamma distribution. (d, e) The process can be repeated to produce another such spike train. The spike trains of c and e will show synchronization above the chance level (boxed spikes), since in some instances spikes may result from the same S spike. (f) Interspike interval histograms for simulated trains with Poisson and gamma-(order 4) distributed intervals. Firing rate in each case was 25 Hz. (g) Cross-correlation histograms for two pairs of spike trains. One pair had Poisson interval statistics, the other gamma statistics (fourth order). The shape and magnitude of the cross-correlation peak can be fixed independently of the interval statistics.

We begin by simulating two event trains, denoted S and T, which are Poisson processes having rates λ_S (see Figure 1a) and λ_T (the dotted lines in Figure 1b). A randomly chosen fraction p of the events from the first process are copied to the second, producing a Poisson process that has a rate of $p\lambda_S + \lambda_T$. This composite process is then decimated by selecting every nth event (see Figure 1c). A gamma process of order n is defined as the waiting time until the nth event of a Poisson process. The resultant simulated spike train R_1 of Figure 1c will therefore have an interspike interval distribution that follows a gamma distribution, order n, and a rate given by

$$\lambda_R = \frac{p\lambda_S + \lambda_T}{n}. \tag{2.1}$$

The process can then be repeated, using the same source spike train as used previously (see Figure 1a), but with a different seed for the random number generator used to generate T and determine which spikes will be copied from S (see Figure 1d). After decimation, this will result in a second spike train R_2, having the same rate and interspike interval statistics as R_1.

Because R_1 and R_2 were generated from the same spike train S, they will show an excess of synchronous discharges above the level expected by chance for two independent spike trains firing at a rate λ_R, since on occasion the same spike will be copied into each spike train from S and also survive the n-fold decimation (boxed spikes in Figures 1c and 1e). This can be quantified as follows. Consider the central region of the cross-correlation histogram, with R_2 acting as the trigger and R_1 the target neuron. If a spike from R_2 occurs at a given time, we require the expected number of spikes from R_1, which will be observed in a time window Δt wide around the time at which R_2 fired. This is given by:

$$\begin{aligned}
E(R_1 \text{ within } \Delta t | R_2) &= P(R_2 \text{ originated from } S) \\
&\quad \times \begin{pmatrix} P(\text{same } S \text{ spike was copied to } R_1 \\ \quad \text{and survived decimation}) \\ + E(\text{other } S \text{ spikes occurred in } \Delta t, \text{ were} \\ \quad\quad \text{copied to } R_1, \text{ and survived} \\ \quad\quad \text{decimation}) \\ + E(T_1 \text{ spikes occurred in } \Delta t \text{ and} \\ \quad\quad \text{survived decimation}) \end{pmatrix} \\
&\quad + P(R_2 \text{ originated from } T_2) \\
&\quad \times \begin{pmatrix} E(S \text{ spikes occurred in } \Delta t, \text{ were copied} \\ \quad \text{to } R_1, \text{ and survived decimation}) \\ + E(T_1 \text{ spikes occurred in } \Delta t \text{ and} \\ \quad\quad \text{survived decimation}) \end{pmatrix} \\
&= P(R_2 \text{ originated from } S) \\
&\quad \times P(\text{same } S \text{ spike was copied to } R_1 \\
&\qquad \text{and survived decimation}) \\
&\quad + E(S \text{ spikes occurred in } \Delta t \text{ were copied} \\
&\qquad \text{to } R_1, \text{ and survived decimation}) \\
&\quad + E(T_1 \text{ spikes occurred in } \Delta t \text{ and survived} \\
&\qquad \text{decimation}).
\end{aligned} \tag{2.2}$$

Since

$$P(R_2 \text{ originated from } S|R_2) + P(R_2 \text{ originated from } T_2|R_2) = 1. \quad (2.3)$$

The sum of the last two terms of equation 2.2 is simply the expected number of spikes from neurone R_1 over a time Δt. This is therefore the expected number of counts in the central part of the cross-correlation histogram that would be seen if R_1 and R_2 were independent. The first term is the excess counts in this part of the cross-correlation, due to the fact that R_1 and R_2 are synchronized together more than expected by chance. Both counts are normalized to the firing rate of R_2, the unit chosen to trigger the cross-correlation histogram. Denoting the strength of excess synchrony by s (Abeles, 1991):

$$\begin{aligned} s &= P(R_2 \text{ originated from } S)P(\text{same } S \text{ spike was copied to } R_1 \text{ and survived decimation}) \\ &= \frac{\lambda_S p}{\lambda_S p + \lambda_T} \cdot \frac{p}{n} \\ &= \frac{\lambda_S p^2}{n(\lambda_S p + \lambda_T)}. \end{aligned} \quad (2.4)$$

In general, we know the values of s and λ_R, which are desired for the simulated spike trains. The three unknowns (p, λ_S, and λ_T) may then be adjusted to produce these in accordance with equations 2.1 and 2.4. Note that the solution to this problem is not unique. Varying p, λ_S, and λ_T over ranges that maintain the same s and λ_R will change the correlation of the resultant spike trains R with S. This is of no interest, since S will not be used in any subsequent analysis, and hence any valid combination of parameters can be used.

Valid solutions to equations 2.1 and 2.4 will have $0 \leq p \leq 1$, and all firing rates positive. This leads to a constraint on the maximum level of synchrony, which can be simulated for a particular gamma order n:

$$s < 1/n. \quad (2.5)$$

In practice when simulating realistic levels of physiological synchrony and interval distributions (typically $s < 0.1$, $n < 10$), this constraint is not a serious limitation.

Excess synchrony between real neurons does not occur at exactly zero lag, but exhibits a jitter (typically some tens of milliseconds; Smith & Fetz, 1989). Such jitter can be easily incorporated into the simulation method described by jittering the spikes copied from S to R by a time distributed according to a given random distribution. This will not change the level of excess synchrony s produced, so long as it is quantified over a sufficiently long section of the cross-correlation histogram. The shape of the peak in the cross-correlation histogram will be given by the autoconvolution of the jitter distribution function.

Figure 1f shows the interspike interval histograms for two such simulated spike trains, having $n = 1$ (Poisson process) and $n = 4$. The rate λ_R was 25 Hz for both simulations, and $s = 0.07$. The higher-order gamma process produced an interspike interval histogram with a realistic relative refractory period and therefore more closely approximated real neuron spiking. Figure 1g shows the correlation between pairs of simulated neurons for these two cases. The copying process included a jitter that was gaussian distributed with a standard deviation of 5 ms. The overlaid cross-correlation histograms are identical to within estimation noise. It was verified that the shape of the peak was equal to the autoconvolution of the gaussian jitter function. The peak area above baseline divided by the number of trigger spikes was very close to 0.07.

The above method thereby allows the simulation of any number of spike trains with preset synchronization and interspike interval characteristics, using methods that are highly computationally efficient to implement. While the description above has assumed that the parameters of the spike trains (firing rate and synchrony strength) are stationary throughout the simulation, the method is not restricted to this case. Using equations 2.1 and 2.4, the firing rates λ_S and λ_T may be computed as a function of time once the desired rate $\lambda_R(t)$ and synchrony strength $s(t)$ are known. Spike trains S and T can then be simulated as inhomogeneous Poisson processes, and mixed and decimated as described above to produce nonstationary simulated spike trains R.

3 Improvements to the Sensitivity of Gravitational Clustering

3.1 Formalism of Gravitational Technique. N neurons are assumed to be simultaneously recorded. Following previous work on gravitational clustering (Gerstein et al., 1985; Aertsen & Gerstein, 1985), we represent these cells as N particles in N-dimensional space. The initial coordinate along the jth axis of the ith particle is given by:

$$\mathbf{x}_i^j = \begin{matrix} 1 & i = j \\ 0 & i \neq j. \end{matrix} \tag{3.1}$$

The charge on particle i at time t is given by:

$$q_i(t) = \sum_k K(t - T_k) - \lambda_i, \tag{3.2}$$

where T_k are the times of the spikes fired by this cell and $K(t)$ is a kernel function that is convolved with the spike train. The kernel function is assumed to have unit area. The firing rate of the ith neuron is given by λ_i; this is subtracted to ensure that the mean charge on a given particle is zero. In this article, only data with stationary firing rates are considered. However, it is possible to extend the method to nonstationary rates, in which case a time-varying rate $\lambda_i(t)$ is subtracted in equation 3.2. The rate is normally estimated over a short window of data (Gerstein & Aertsen, 1985). The enhancements to the gravity method here proposed should also be usable when neurons change their firing rates by making this modification.

In the usual gravitational clustering analysis, particle positions are adjusted according to the pairwise attractions between particles. Thus, using a discrete time step of δt, the position update rule is:

$$\mathbf{x}_i(t + \partial t) = \mathbf{x}_i(t) + \kappa \partial t \sum_{\forall j \neq 1} Q_{i,j} \frac{\mathbf{x}_j - \mathbf{x}_i}{|\mathbf{x}_j - \mathbf{x}_i|}, \tag{3.3}$$

where

$$Q_{i,j} = q_i q_j, \tag{3.4}$$

The constant κ controls the rate of aggregation of synchronized neurons. The time step δt was fixed at 1 ms throughout this article. The denominator term within the summation prevents a dependence on the current separation of particles; an alternative version, where this is omitted, has been previously described (Dayhoff, 1994).

3.2 Effect of Moving-Window Smoothing on Charge Product. As a first potential improvement to the method of gravitational clustering, we propose increasing the sensitivity of the method to repeated instances of synchrony between the same two cells within a particular observation window. The inspiration for this approach comes from the intuitive consideration that whereas two neurons may fire once in synchrony simply by chance, two such occurrences within a short space of time are considerably less likely unless both are firing in bursts. Such an extension may be straightforwardly generated by calculating $Q_{i,j}$ as a moving average of the charge product over a window extending back in time for a duration w:

$$Q_{i,j}(t) = \frac{1}{w} \int_{t-w}^{t} q_i(u) q_j(u) du. \tag{3.5}$$

Figure 2a shows the effect such a moving-window smoothing approach has on the distribution of $Q_{i,j}$ computed between two independently firing neurons. The two spike trains were simulated to have a firing rate of 25 Hz and interspike intervals that followed a gamma distribution, order 4. The kernel function $K(t)$ was an exponential:

$$K(t) = \begin{matrix} \exp(-t/\tau) & t > 0 \\ 0 & t < 0. \end{matrix} \tag{3.6}$$

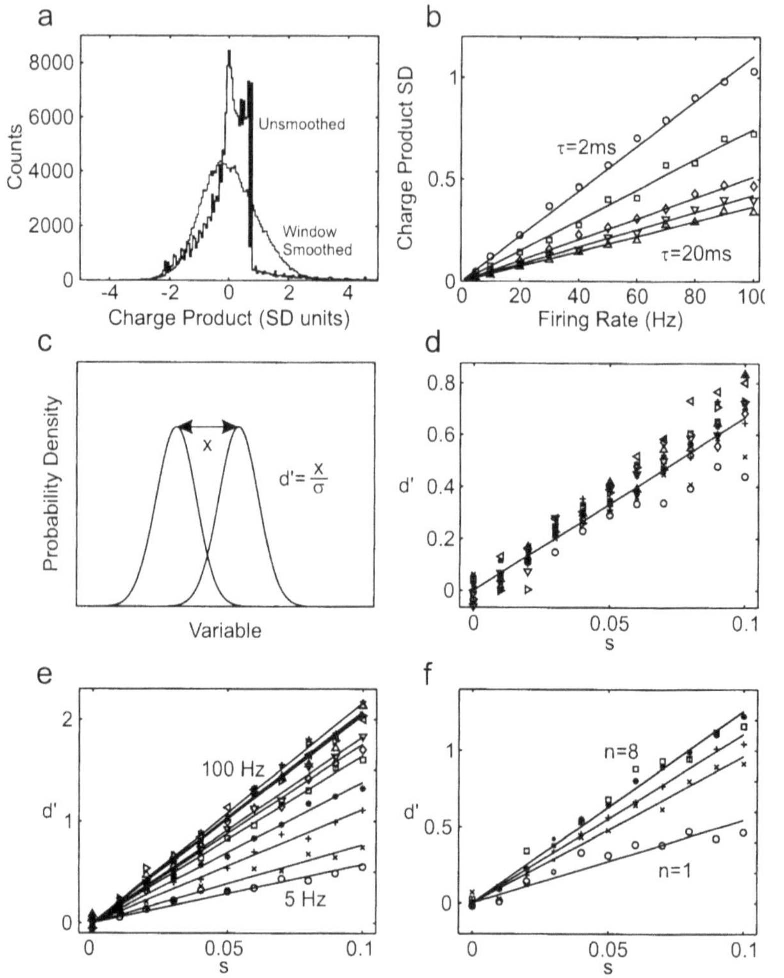

Calculation of $Q_{i,j}$ on single points as equation 3.4 led to a distribution that was sharply peaked. By contrast, using a window length $w = 1$ s long, the distribution was approximately gaussian. The distribution histograms of Figure 2a have been plotted with the x-axis scaled by the standard deviation of the distributions. In addition to changing the form of the charge product distribution function, the standard deviation was reduced by a factor of 12.7 using the 1 s window. The conventional gravity analysis method achieves smoothing and the consequent noise reduction implicitly through the development of the particle trajectory. However, as will be shown below, it can be advantageous to perform this explicitly according to equation 3.5.

3.3 Detection of Synchrony Using Charge Products. Figure 2b shows the effect of different firing rates on the variability of the charge products. The lines show a linear regression relating the standard deviation of the $Q_{i,j}$ calculated with a 1 s window to the cell firing rate; each line is calculated for runs using different values of the time constant τ in the kernel function $K(t)$. The variability of the charge product increases with the increasing firing rate of the cells, as more transient increases in the individual charges occur each second. The slope of this relationship is much steeper when small values of τ are used. Small τ leads to a single charge being flat except just after a spike, when a high-amplitude sharp transient occurs (the kernel amplitude will be large, since it is brief but constrained to have unit area). Large τ results in a much more even, and hence less variable, charge profile. These changes in single charge variabilities are reflected in the variance of the charge product $Q_{i,j}$ as shown in Figure 2b.

The variability in the charge product $Q_{i,j}$ shown in Figure 2a when two neurons fire independently means that detection of a mean increase in $Q_{i,j}$ due to slight excess synchrony between the neurons is not trivial. The extent to which the synchronous and independent cases are separated can be measured by the statistic d' (Green & Swets, 1966). This is illustrated by Figure 2c, and calculated according to

$$d' = \frac{\overline{Q_{i,j}^{SYNC}} - \overline{Q_{i,j}^{INDEP}}}{stdev\left(Q_{i,j}^{INDEP}\right)}. \tag{3.7}$$

The mean charge for an individual neuron is fixed to zero by equation 3.2; hence for independent neurons, the charge product will have zero mean, and the term $\overline{Q_{i\,j}^{INDEP}}$ can be ignored. Figure 2d plots d' versus the amount

Figure 2: *Facing page.* (a) Distribution of the charge product $Q_{i,j}$ using no smoothing, or a window of 1 s. (b) Variation of the standard deviation of the charge product $Q_{i,j}$ for independently firing neurons with different firing rates. The lines indicate regression fits to the results using kernel time constants τ of 2, 5, 10, 15, and 20 ms (from highest to lowest line). All results are based on simulated spike trains with Poisson interval distributions. (c) Definition of d'. (d) Variation of discriminability d' with strength of synchrony s, for spikes having a Poisson interval distribution. Different symbols mark different firing rates. The line is a linear regression fit calculated across all data points. Regardless of firing rate, the measurements are approximately fitted by this single line. (e) The same, but using spikes with a fourth-order gamma distribution of interspike intervals. Firing rate now has an important effect on discriminability. Firing rates of 5 Hz, plus 10–100 Hz in steps of 10 Hz were used in *d* and *e*. (f) Variation of d' with s for different orders of the gamma distribution n. Regression lines are for $n = 1, 2, 3, 4$, and 8 (lines for $n = 4$ and $n = 8$ are indistinguishable).

of synchrony between the neurons, quantified by the measure s defined above. This is shown for simulated spike trains with Poisson distributed interspike intervals. Different superimposed symbols show the results for spike trains with firing rates varying from 5 to 100 Hz. The points appear to lie approximately on the same line, indicating that the detectability of synchrony depends only weakly on the unit firing rates in this case. This is in agreement with previous observations (Strangman, 1997).

The rate of occurrence of excess coincidences will be directly proportional to the neuron firing rate (where the constant of proportionality is s as defined above). Hence, the numerator of equation 3.7 must show firing-rate dependence. The finding that the d' measure is essentially independent of firing rate indicates that the changes in the denominator (the standard deviation; see Figure 2b) balance those in the numerator when working with spikes having Poisson distributed interspike intervals.

Figure 2e shows a similar plot of d' versus s, but now calculated using spike trains simulated to have interspike intervals distributed according to a fourth-order gamma process. The difference from Figure 2d is readily apparent. Firing rate now has a profound effect on the slope of the relationship between d' and s, as shown by the fitted regression lines for each tested frequency. The changes in the numerator of equation 3.7 are independent of the interval statistics of the spike trains (not shown). However, gamma-distributed intervals have lower variance than a Poisson train of the same firing rate (Shadlen & Newsome, 1998); the standard deviation of Q_{ij} in the denominator of equation 3.7 thus rises less steeply with firing rate than the increase in the mean of Q_{ij} in the presence of excess synchrony. If the neurons have a refractory period with a gamma-like interval distribution, synchrony of a given strength is more easily detected when the cells fire at higher rates.

Figure 2f shows in more detail the effect of changing n, the order of the gamma process underlying the simulated spike trains, on the slope of d' versus s. The simulations displayed in this example had a constant firing rate of 25 Hz. There is a sharp increase in the slope as the order is increased above $n = 1$; thereafter, the slope rises only slowly. For an order of $n = 4$, the fitted regression line is indistinguishable from $n = 8$. It can be seen that detection of synchrony is improved by more than twofold if neurons show a relative refractory period in their interspike intervals histograms, compared with the detectability obtained from Poisson spike trains with the same-size cross-correlation peak. A similar result was obtained using spike trains constructed as Poisson processes with a dead time to model the refractory period (data not shown). The effect of interspike interval distribution on the detection of correlation has also been noted by Davey, Ellaway, and Stein (1986) in a different context.

3.4 Use of a Nonlinearity to Improve Detection. At high noise levels, when the measure d' is small, detection of synchrony (i.e., a nonzero mean

$Q_{i,j}$) can occur only by averaging over sufficient data to overcome the noise. However, at higher levels of d', it is possible to enhance detection by making the forces between particles depend nonlinearly on $Q_{i,j}$, yielding the following update rule:

$$\mathbf{x}_i(t + \partial t) = \mathbf{x}_i(t) + \kappa \partial t \sum_{\forall j \neq i} f(Q_{i,j}) \frac{\mathbf{x}_j - \mathbf{x}_i}{|\mathbf{x}_j - \mathbf{x}_i|}, \tag{3.8}$$

where $f(u)$ is a nonlinear function with a slope that increases its magnitude with increasing distance from $u = 0$. This will have the effect of suppressing small, noisy fluctuations and enhancing the large excursions in $Q_{i,j}$, which are unlikely to have been caused by chance. The simplest suitable form of nonlinearity is:

$$f(u) = A|u|^m sign(u), \tag{3.9}$$

where the power m can be varied to change the functional form; use of this form permits even and nonintegral m while ensuring that the sign of $f(u)$ always equals the sign of u. The constant A permits scaling of the function. For practical application of the algorithm, we have found it convenient first to scale $Q_{i,j}$ by its standard deviation:

$$Q'_{i,j}(t) = \frac{Q_{i,j}(t)}{stdev(Q_{i,j})}. \tag{3.10}$$

Such a scaling makes the variability of all charge products the same, regardless of cell firing rate or interspike interval statistics. By normalizing for this first, the choices of constant κ in equation 3.8 can be standardized, and the same value used across different data sets. This avoids the need to tune performance of the algorithm by testing multiple values of κ. The standard deviation of the charge product can be calculated over the entire length of the data if the firing rate of the neurones is stationary or by using a moving window if not (compare with the calculation of mean charge using a moving window in nonstationary data; Gerstein & Aertsen, 1985). Normalization by standard deviation as a means to correct for nonstationarity has also been used by Aertsen et al. (1987; Aertsen, Gerstein, Habib, & Palm, 1989).

Assuming that $Q'_{i,j}$ is normally distributed (a reasonable approximation if a moving-window smoothing process is used, due to the central limit theorem; see Figure 2a), the variance of $f(Q'_{i,j})$ will then be

$$\sigma^2 = \frac{1}{\sqrt{2\pi}} \int_{-\infty}^{\infty} A^2 u^{2m} e^{-u^2/2} du. \tag{3.11}$$

The parameter A can be adjusted to make $\sigma^2 = 1$ for the chosen value of m. This permits experimentation with different m without causing the severe changes to the excursions of the random walks undergone by particles that relate to independent spikes.

A measure of the improvement of discrimination achieved by using $f(Q'_{i,j})$ rather than $Q'_{i,j}$ is given by

$$\frac{\overline{f(Q'_{i,j})}}{\overline{Q'_{i,j}}} = \frac{\int_{-\infty}^{\infty} f(u)e^{\frac{-(u-d')^2}{2}}du}{\int_{-\infty}^{\infty} ue^{\frac{-(u-d')^2}{2}}du}. \tag{3.12}$$

Figure 3a shows the behavior of this measure for different values of d', using various powers m of nonlinearity. It can be seen that for d' as low as 0.5, the ratio of equation 3.12 is below one for all powers higher than unity. This indicates that for such a low signal-to-noise ratio, no improvement in discriminability can be obtained above and beyond the linear case. Use of a nonlinearity simply enhances noise fluctuations and degrades discrimination. However, for larger d', the curves rise above one, indicating that a nonlinearity can enhance performance. The peak in these curves, indicating the optimal power to use, occurs at greater powers the larger d'.

Figure 3b shows this best power as a function of d' (solid line). It can be seen that even for quite modest values of d' (corresponding to a mean $Q_{i,j}$ just over one standard deviation away from zero), use of a nonlinearity is expected to improve the performance of the gravitational clustering algorithm. The dotted line in Figure 3b indicates the point at which curves similar to those of Figure 3a recross an ordinate value of one. This indicates the highest power that can be safely used to yield performance that is no worse than the usual gravity algorithm with $f(u) = u$. From this perspective, use of a low power function (e.g., cubic, $m = 3$) will enhance, or at least not degrade, the performance of gravitational clustering for all but the very weakest correlations.

Moving-window smoothing of the charge product $q_i q_j$ as described by equation 3.5 is an essential prerequisite for application of a nonlinearity as described above. Without it, the standard deviation of the charge product will be so great that d' will never exceed the region of Figure 3b where no better performance than a linear $f(u)$ can be achieved. Conversely, window smoothing cannot achieve any performance gains without a nonlinear f. This can readily be understood upon reexamination of equations 3.3 through 3.5. With a linear f, the position of a particle at a given time after starting the analysis will depend on the integral of the products between its charge and its neighbors up to that time. This will be the same whether the integration is performed entirely implicitly by the movement of the particles themselves (equations 3.3 and 3.4), or effectively in two stages by equation 3.5 followed by particle movements.

3.5 Performance Improvement by Triple Pair Enhancement. An important feature of the gravity technique is its ability to detect multiple cells

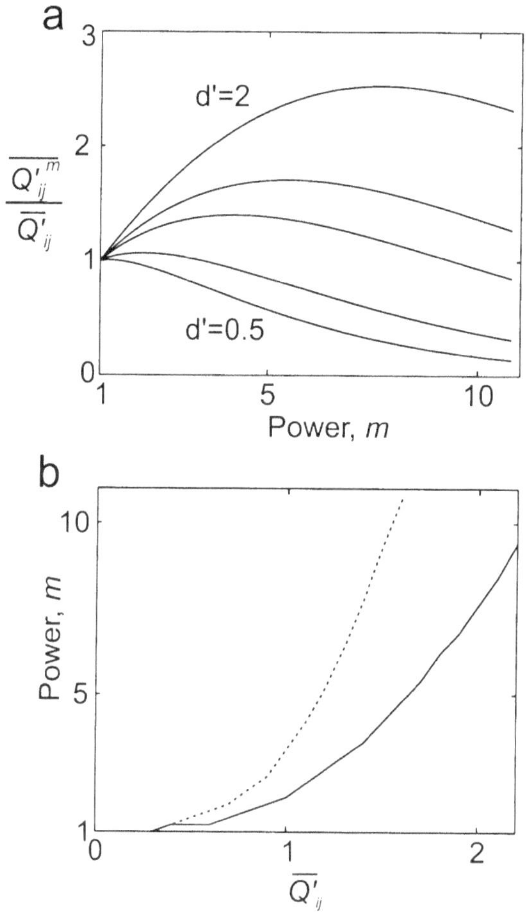

Figure 3: (a) Effect of the power used in the nonlinearity of equations 3.8 and 3.9 on discriminability, relative to the discriminability with a power of one (linearity). Lines are shown for $d' = 0.5, 1, 1.5, 1.7, 2$. (b) Optimal power of nonlinearity to use for the best discrimination as a function of d' (solid line) and the highest power that will be no worse than linearity (dotted line).

that are members of the same synchronous assembly; such assembly membership is indicated by the clusters into which particles coalesce. This feature of the method suggests a further possible enhancement. Suppose three neurons, *i*, *j*, and *k*, are members of the same assembly. The pairwise synchronous firings (i, j), (i, k), and (j, k) will each occur at above-chance rates, and this will lead the particles to move toward each other in the ordinary gravity computation. However, since all three neurons are part of the same assembly, we might expect an excess temporal clustering of the several pair coincidences within a moving window of length w. Such counts of multiple pair coincidences are highly unlikely to occur by chance. The occurrence of triple pairs will be a much more robust indicator of assembly membership than any one of the individual pairwise synchronies alone.

The above observation inspires the following modification to the particle movement rule of equation 3.8:

$$\mathbf{x}_i(t+\partial t) = \mathbf{x}_i(t) + \kappa\,\partial t \sum_{\forall j \neq i} f(Q'_{i,j})\, g\left(\sum_{\forall k \neq i,j} Q'_{i,k} Q'_{j,k}\right) \frac{\mathbf{x}_j - \mathbf{x}_i}{|\mathbf{x}_j - \mathbf{x}_i|}. \tag{3.13}$$

The function $g(u)$ is intended to scale the force experienced by the particle. The following are desirable features of this function. When the argument of g is zero, no augmentation or diminution of the force should occur; hence $g(0) = 1$. We have found that a sigmoid function is not suitable, since the maximal slope around $u = 0$ enhances small chance fluctuations in u. The slope of the function g should instead be zero at $u = 0$ but increase in magnitude to either side. A suitable function among the many that are possible, and which we have found effective, is:

$$g(u) = \begin{matrix} 0 & u < -2 \\ 1 + \frac{u^3}{8} & -1 \leq u < 2\sqrt[3]{3} \\ 4 & u \geq 2\sqrt[3]{3}. \end{matrix} \tag{3.14}$$

This function is thresholded at $g = 0$; it can thus never become negative, so that force directions cannot be reversed. It is likewise thresholded at $g = 4$; this prevents extreme outliers of the summed charge products from excessively disrupting the gravity condensation.

If no other neuron fires synchronously with cells i and j during the window period w, the summation over k in equation 3.13 will be close to zero; g will be nearly one, and the algorithm will be unaffected. However, if another cell k does participate in the assembly in the period w, the summation will on average be above zero, since both $Q'_{i,k}$ and $Q'_{j,k}$ will have nonzero, positive mean. The force causing aggregation between i and j will therefore be enhanced. By symmetry, a similar effect will occur for the force between particle pairs (i, k) and (j, k), and all three particles will therefore condense together more rapidly than had the term in $g(u)$ not been included. The practical success of this modification is demonstrated in the next section.

It would be possible to extend this approach to groups larger than three neurons; however, this would increase the number of combinations that need to be examined and is expected to be helpful only if large assemblies are present. Use of groups of three is expected to cause assemblies with larger numbers of members to aggregate considerably faster, since such populations have several subgroups of size three. We therefore have not investigated further than the triple pair approach outlined.

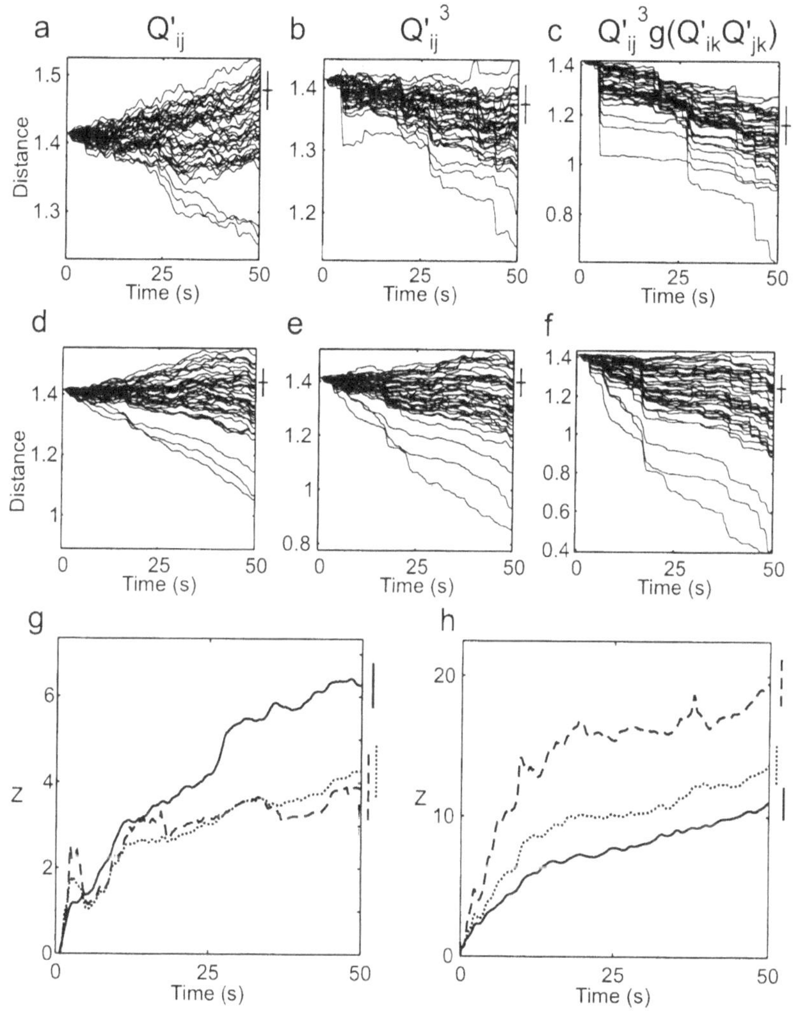

4 Quantification of Improvements to Gravitational Clustering

Figure 4 illustrates the results of gravitational analysis using the modifications described above. This was performed on simulated data. Ten neurons were used, all with a constant firing rate of 25 Hz. Three of these cells were synchronized together, with a synchrony strength $s = 0.1$; the copying process used to produce the synchronized spikes incorporated a temporal jitter that was normally distributed, with 5 ms standard deviation (as in Figure 1g). Figures 4a through 4c present the results when the simulated spike trains were Poisson processes (order $n = 1$); Figures 4c, 4d, and 4f are the results for a different data set, simulated so that the interspike intervals were gamma distributed, order $n = 4$. The figures are plots of the pairwise distances between particles as a function of time, for the 50 s duration of the simulation. A smoothing window $w = 1$ s was used, and the kernel time constant was $\tau = 10$ ms.

Figures 4a and 4d show the results for a conventional gravity analysis, as defined by equations 3.3 and 3.4. Three of the distance traces, corresponding to the distances between the particles representing the three cells synchronized together, condensed and became clearly separated from the other lines. (The crossed line to the right of each graph indicates the mean and standard deviation of distance trajectories for pairs where neither neuron is part of the synchronized assembly.) The separation occurs earlier and is somewhat more pronounced at the end of the run, for the data with gamma order 4 interspike intervals (see Figure 4d) compared with the Poisson case (see Figure 4a); this agrees with the results of Figure 2, which suggested that synchrony is more easily detected in a gamma spike train.

Figures 4b and 4e show the performance of the algorithm modified to use a cubic nonlinearity ($m = 3$ in equation 3.9). Particle condensation is considerably impaired for the Poisson spike trains, such that only two traces appear clearly separated from the noise excursions by the end of the analysis. Separation is adequate for the gamma distributed spike train.

Figures 4c and 4f show the performance when using the further modification described by equations 3.13 and 3.14. Once again, separation of the three distance traces relating to synchronized pairs is poorly achieved for the Poisson simulations (see Figure 4c), but good separation occurs for the spikes with gamma intervals (see Figure 4f).

Figure 4: *Facing page.* Results of gravitational analysis using different algorithms. (a–c) Pairwise distances between particles as a function of time, using Poisson distributed spike trains. (d–f) Using spike trains having a gamma distribution of interspike intervals (fourth order). (a,d) Conventional linear gravity method; (b,e) with cubic nonlinearity; (c,f) cubic nonlinearity and triple pair enhancement. The cross on the right of each plot shows the mean plus or minus standard deviation of the interpair distance at $t = 50$ s for all pairs where neither neuron was part of the synchronized assembly. The ordinate scales have been chosen such that the height of this marker is the same in *a–c* and *d–f*, permitting visual comparison of the magnitude of the particle aggregation relative to the aggregation noise. (g,h) Relative performance of the linear, cubic, and cubic plus triple enhancement algorithms, shown by solid, dotted, and dashed lines respectively. The lines show the mean of 20 runs using different simulated data in each case. The bars to the right indicate the standard error of the mean computed over the 20 runs at time 50 s. Spike trains were Poisson in *g* and had fourth-order gamma interval statistics in *h*.

While some differences between the performance of the different algorithms can be assessed by eye, it is desirable to develop a quantitative measurement of the degree to which particles representing synchronized neurons aggregate. We have chosen to use the following measure for this purpose:

$$Z = \frac{\overline{d_{i,j\in I}} - \overline{d_{i,j\in S}}}{stdev(d_{i,j\in I})}, \tag{4.1}$$

where $d_{i,j}$ is the distance between the *i*th and *j*th particle, *S* represents the set of synchronized neurons, and *I* is the set of independently firing neurons in the same simulation. This measure *Z* has analogies with d' as defined in equation 3.7; it is the difference between the mean separation of the particles representing synchronous and independently firing neurons, expressed in terms of the variability seen in the independently firing interparticle distances. A similar measure would be appropriate for real, nonsimulated data where neuron synchrony properties are not known a priori. Here we would identify the synchronized neurons in terms of condensing trajectories.

Figures 4g and 4h show how *Z* varies during the gravity analysis for the different situations illustrated in the earlier part of the figure. The lines represent the mean of *Z* calculated over 20 data sets simulated with different random number seeds. The variability is indicated by the bars to the right of each graph, which represent one standard error either side of the mean, assessed from the 20 separate analyses at time 50 s. The solid lines show the behavior of *Z* using conventional gravity analysis (see Figures 4a and 4d), the dotted line using a cubic nonlinearity (see Figures 4b and 4e), and the dashed line using the triple pair enhancement (see Figures 4c and 4f). Figure 4g presents the results for analysis of Poisson spike trains. Conventional gravity analysis without modification here clearly gives the best result, with a discriminability some 50% higher than either of the two more complex algorithms. By contrast, Figure 4h shows the opposite effect when gamma-distributed spike trains are analyzed. All three lines are consistently above their corresponding plots in Figure 4g, confirming the increased detectability of synchrony for a spike train with gamma compared with Poisson interval statistics. A modest but worthwhile improvement over the conventional algorithm is achieved by using the cubic nonlinearity. A considerable improvement (almost a twofold rise in *Z*) can be achieved by combining the nonlinear and triple pair enhancement approaches. Since experimentally recorded spike trains will usually have interspike intervals closer to a gamma than a Poisson distribution, a similar improvement should be seen in experimental data.

Figure 5 presents data showing the robustness of this result. Figure 5a plots the final value of *Z* after a 50 s analysis, similar to that in Figure 4, for the three algorithms, as a function of the order of the gamma distribution of interspike intervals of the simulated spike trains. The same line style

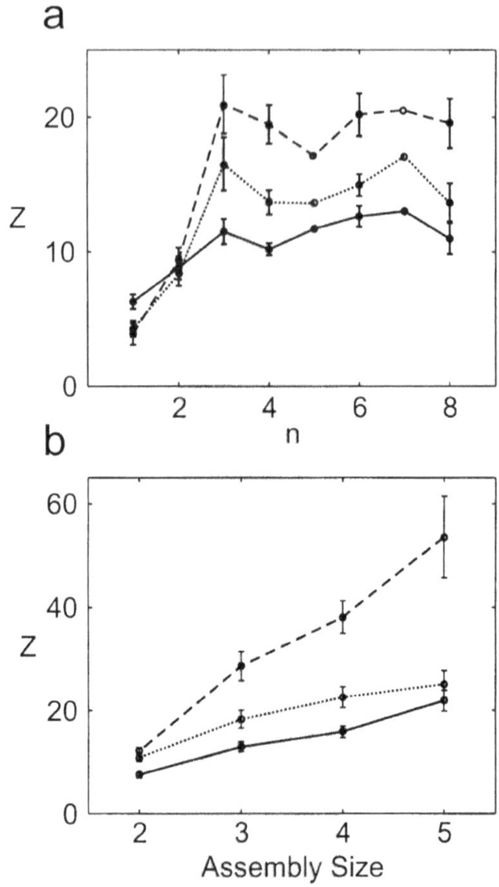

Figure 5: (a) Efficiency of gravitational algorithms (measured by Z at the end of a 50 s analysis run) depending on the order of the gamma distribution that their interspike intervals follow (n). Results with conventional linear gravity, a cubic nonlinearity, and the nonlinearity combined with triple pair enhancement are shown by solid, dotted, and dashed lines, respectively. Each point marks the mean from the analysis of 20 runs; error bars mark the standard error calculated over the same sample. (b) Variation of algorithm efficiency with the number of neurons in the synchronized assembly. Display conventions as in *a*.

conventions as in Figures 4g and 4h have been used to denote the three algorithm variants under investigation. The usual gravitational method is superior only for Poisson spike trains ($n = 1$). Performance of the three techniques is the same for $n = 2$; at all higher orders, there is a clear advantage in using the modifications detailed in this article.

The triple pair enhancement algorithm was specifically designed to enhance assemblies with three members; it is important to be certain that its performance in other situations is adequate. Figure 5b accordingly indicates the effect of the size of the assembly on the relative performance of the three methods. Use of a cubic nonlinearity provided a fairly consistent increment over the usual, linear algorithm at all assembly sizes tested. Values of Z using triple pair enhancement overlap those produced without it when only two cells are synchronized, indicating that this method introduces no appreciable extra noise in the small assembly case. However, the benefits of triple pair enhancement grow rapidly with larger assemblies, presumably due to the combinatorial rise in the number of triple pairs that can be formed.

5 Gravitational Clustering Averaged Relative to an External Event

The techniques described allow detection of synchronous cell assemblies when analysis can be performed over a relatively long period of time. The enhancements described rely on moving-window smoothing. The width of this window must be sufficient to produce a reasonable probability of more than one excess coincidence between a given pair of cells within it. At the firing rate (25 Hz) and strength of synchrony ($s = 0.1$) considered above, there is a probability of 0.71 of two excess synchronous events occurring within a 1 s window (computed from the Poisson counting distribution). Reducing the window duration would decrease the probability of more than one excess synchronous event falling within the window. The improvements in performance described above would thus be considerably degraded, due to the reduction in d'.

However, in many experiments, the aim is to investigate the modulation of synchronous firing over much shorter timescales (typically tens to hundreds of milliseconds). The small number of spikes in such short sections means that reliable conclusions cannot be reached on the basis of a single trial. The usual solution is to average the time course of synchrony over multiple trials with reference to an external event such as a stimulus occurrence or behavioral marker (Aertsen et al., 1989; Grün, 1996; Pauluis & Baker, 2000). The increased amount of data then available permits increased statistical power.

The gravitational clustering approach is also amenable to such across-trial averaging. In order to use the enhancements described above, this averaging is performed as follows. The pairwise, smoothed charge products $Q_{i,j}(t)$ are found for each available trial using equation 3.5. These are then averaged across trials, to obtain a set of mean charge product time courses $\overline{Q_{i,j}}(t)$. The standard error of the mean, computed across trials, can then be used to scale the charge product time courses according to equation 3.10, obtaining $\overline{Q'_{i,j}}(t)$. The gravitational analysis is carried out by using these values in equations 3.3, 3.8, or 3.13. By averaging over multiple trials, the noise level of $Q'_{i,j}$ is reduced, increasing d' and permitting the use of a nonlinearity and triple pair enhancement.

Figure 6 shows an example of the application of such trial-averaged gravitational clustering. One hundred trials, each 3 s long, were simulated in which 10 neurons discharged at 25 Hz with interspike intervals following a fourth-order gamma process. For the first 1.5 s of each trial, neurons 1, 2, and 3 were synchronized together; for the last 1.5 s of each trial, these cells ceased to show excess correlation, and cells 4, 5, and 6 became synchronized instead ($s = 0.1$ in all cases). Figures 6a, 6b, and 6c show the results of gravitational clustering using the linear, nonlinear (quintic, $m = 5$ in equation 3.9), and triple-pair enhancement algorithms, respectively. A considerably shorter smoothing window, $w = 100$ ms, was used in comparison to previous analyses to retain temporal resolution. The higher power of nonlinearity compared with that used earlier (quintic compared with cubic) takes advantage of the improved signal-to-noise ratio available by averaging across trials (see Figure 3b). The pairwise distances plotted here have been adjusted by subtracting from each distance its value at $t = 1.5$ s. This renders the changes that occur on either side of this point more easily seen.

In each case, the three pairs relating to the synchronized cells show clear separation from the remaining pairs. It should be noted that while the plots appear symmetrical, this is in fact not the case, since a different set of pairs condenses before and after the normalization point in the middle of the analysis. In addition to a clustering of the distances between particles representing the three synchronized neurons, a second group can be seen to separate from the remaining distances to a lesser extent. These are distance pairs that include one of the synchronized cells and one of the independently firing ones. This small condensation of distance is expected from the geometry of the situation, as illustrated schematically in Figures 6d and 6e. If the two filled particles condense together, but there are no net attractive forces between them and the white particle, this will still result in a narrowing of the distance between the white and black particles. Such a small apparent condensation of pairs including only one synchronized neuron is often not noticeable in gravity analysis due to a high noise level (e.g., see Figure 4).

The improved algorithms described in this article clearly provide for a reduced level of noise in the distance plots of Figures 6b and 6c compared with Figure 6a. This is quantified in Figures 6f and 6g using the measure Z defined in equation 4.1. Figure 6f presents the Z calculation for the (1,2,3) assembly, and Figure 6g for the (4,5,6) one. The calculation of Z has again been carried out over 20 separate simulations (each of 100 trials); the thin lines show the mean, plus or minus one standard error. The same line styles have been used as in Figures 4 and 5 to code the different analysis types. The order of relative efficiency of the three versions of the algorithm is as expected from the previous analysis (linear worse than nonlinear, which is

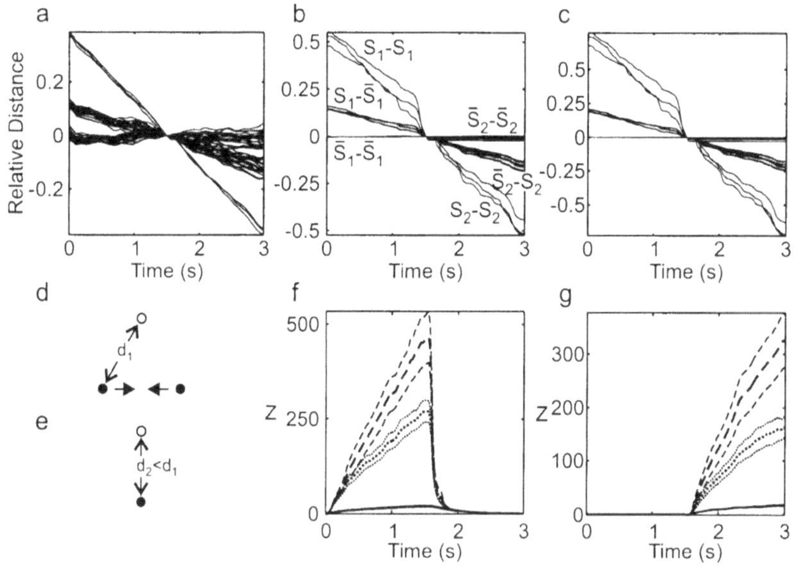

Figure 6: Results using trial-averaged gravitational analysis. (a–c) The results using conventional linear gravity, a quintic (fifth power) nonlinearity, and the nonlinearity combined with triple pair enhancement, respectively. Traces are the interpair distances as a function of time relative to the trial onset. Distances have been normalized by subtraction of their value at time 1.5 s for clarity of display. Distance plots have been labeled in *b* according to whether they relate to pairs in which both neurons were part of the synchronized assembly (*S*-*S*), only one was in the assembly (S-$\bar{S}$) or neither were ($\bar{S}$-$\bar{S}$). Note that the two different assemblies S_1 and S_2, which were active before and after time 1.5 s, respectively, are separated before and after this time. (d) Schematic showing how apparent condensation can occur simply due to the geometry of the system. The black particles attract each other due to excess synchronization; the white particle, by contrast, represents an independently firing cell and experiences no net forces. After the black particles have condensed together (e), the distance between them and the white particle is nevertheless reduced. This explains the apparent condensation of the S-$\bar{S}$ pairs in *a*–*c*. (f) Efficiency of the three algorithms computed for assembly (1,2,3), which was synchronized for the first half of each trial. Line-style conventions are as in Figures 4g and 4h. (g) The same for assembly (4,5,6), which was active in the second half of each trial. Traces show the mean and standard deviation of *Z* computed over 20 simulation runs.

worse than nonlinear with triple pair enhancement); the magnitude of the performance enhancement (greater than a 17-fold increase in *Z*) is, however, considerably greater than in the previous examples, which did not use trial averaging.

6 Conclusion

The two enhancements to the gravitational clustering algorithm presented in this article are capable of a considerable improvement in its sensitivity for detecting synchronous cell assemblies. While not universally applicable, the range of parameters over which it becomes worthwhile to use the two adjustments is wide and encompasses much of that seen physiologically. Implementation of the improvements to the gravity analysis will carry with it the cost of somewhat lengthened computation time. While this was of some concern only a few years ago (Dayhoff, 1994), rapid advances in computer power have rendered this of less consequence.

An interesting observation made in the course of the work presented here is that a given level of synchrony is harder to detect (i.e., requires more data before reaching significance) for a Poisson event train compared with events having a relative refractory period, as simulated by the gamma distribution. Detection of synchrony by neurons has some analogues to the gravity method (e.g., compare exponentially decaying kernels with cortical excitatory postsynaptic potentials), suggesting that this result may also hold for synchrony detection by neurons. If true, it would imply that the relative refractory period can be viewed as an important adaptation to permit accurate synchrony detection. This hypothesis, currently speculative, clearly requires further work.

Acknowledgments

This work was supported by NIH MH46428, DC01249, and NSF to G. L. G. S. N. B. was funded by Christ's College, Cambridge.

References

Abeles, M. (1991). *Corticonics*. Cambridge: Cambridge University Press.

Aertsen, A. M. H. J., Bonhoeffer, T., & Kruger, J. (1987). In E. R. Caianiello (Ed.), *Physics of cognitive processes* (pp. 1–34). Singapore: World Sciences Publishers.

Aertsen, A. M. H. J., & Gerstein, G. L. (1985). Evaluation of neuronal connectivity: Sensitivity of cross correlation. *Brain Research, 340*, 341–354.

Aertsen, A. M. H. J., Gerstein, G. L., Habib, M. K., & Palm, G. (1989). Dynamics of neuronal firing correlation—modulation of effective connectivity. *Journal of Neurophysiology, 61*, 900–917.

Aertsen, A., Vaadia, E., Abeles, M., Ahissar, E., Bergman, H., Karmon, B., Lavner, Y., Margalit, E., Nelken, I., & Rotter, S. (1991). Neural interactions in the frontal-cortex of a behaving monkey—signs of dependence on stimulus context and behavioral state. *Journal für Hirnforschung, 32*, 735–743.

Baker, S. N., Olivier, E., & Lemon, R. N. (1997). Task dependent coherent oscillations recorded in monkey motor cortex and hand muscle EMG. *Journal of Physiology, 501*, 225–241.

Davey, N. J., Ellaway, P. H., & Stein, R. B. (1986). Statistical limits for detecting change in the cumulative sum derivative of the peristimulus time histogram. *Journal of Neuroscience Methods, 17,* 153–166.

Dayhoff, J. E. (1994). Synchrony detection in neural assemblies. *Biological Cybernetics, 71,* 263–270.

Espinosa, I. E., & Gerstein, G. L. (1988). Cortical auditory neuron interactions during presentation of 3-tone sequences—Effective connectivity. *Brain Research, 450,* 39–50.

Gerstein, G. L., & Aertsen, A. M. H. J. (1985). Representation of cooperative firing activity among simultaneously recorded neurons. *Journal of Neurophysiology, 54,* 1513–1528.

Gerstein, G. L., Perkel, D. H., & Dayhoff, J. E. (1985). Cooperative firing activity in simultaneously recorded populations of neurons: Detection and measurement. *Journal of Neuroscience, 5,* 881–889.

Gochin, P. M., Gerstein, G. L., & Kaltenbach, J. A. (1990). Dynamic temporal properties of effective connections in rat dorsal cochlear nucleus. *Brain Research, 510,* 195–202.

Green, D., & Swets, J. (1966). *Signal detection theory and psychophysics.* New York: Wiley.

Grün, S. (1996). *Unitary joint-events in multiple-neuron spiking activity.* Thun, Frankfurt am Main: Harri Deutsch.

Kuffler, S. W., Fitzhugh, R., & Barlow, H. B. (1957). Maintained activity in the cat's retina in light and darkness. *Journal of General Physiology, 40,* 683–702.

Lindsey, B. G., Hernandez, Y. M., Morris, K. F., & Shannon, R. (1992). Functional connectivity between brain stem midline neurons with respiratory-modulated firing rates. *Journal of Neurophysiology, 67,* 890–904.

Lindsey, B. G., Morris, K. F., Shannon, R., & Gerstein, G. L. (1997). Repeated patterns of distributed synchrony in neuronal assemblies. *Journal of Neurophysiology, 78,* 1714–1719.

Lindsey, B. G., Shannon, R., & Gerstein, G. L. (1989). Gravitational representation of simultaneously recorded brain-stem respiratory neuron spike trains. *Brain Research, 483,* 373–378.

Pauluis, Q., & Baker, S. N. (2000). An accurate measure of the instantaneous discharge probability, with application to unitary joint-event analysis. *Neural Computation, 12,* 647–669.

Rieke, F., Warland, D., de Ruyter van Steveninck, R., & Bialek, W. (1997). *Spikes: Exploring the neural code.* Cambridge, MA: MIT Press.

Rodieck, R. W., Kiang, N. Y.-S., & Gerstein, G. L. (1962). Some quantitative methods for the study of spontaneous activity of single neurons. *Biophysical Journal, 2,* 351–368.

Shadlen, M. N., & Newsome, W. T. (1998). The variable discharge of cortical neurons: Implications for connectivity, computation, and information coding. *Journal of Neuroscience, 18,* 3870–3896.

Singer, W., & Gray, C. M. (1995). Visual feature integration and the temporal correlation hypothesis. *Annual Review of Neuroscience, 18,* 555–586.

Smith, W. S., & Fetz, E. E. (1989). Effects of synchrony between corticomotoneuronal cells on post-spike facilitation of primate muscles and motor units. *Neuroscience Letters, 96,* 76–81.
Stein, R. B. (1965). A theoretical analysis of neuronal variability. *Biophysical Journal, 5,* 173–194.
Strangman, G. (1997). Detecting synchronous cell assemblies with limited data and overlapping assemblies. *Neural Computation, 9,* 51–76.
Vaadia, E., Haalman, I., Abeles, M., Bergman, H., Prut, Y., Slovin, H., & Aertsen, A. (1995). Dynamics of neuronal interactions in monkey motor cortex in relation to behavioural events. *Nature, 373,* 515–518.

Received May 14, 1999; accepted February 2, 2000.

ELSEVIER

Journal of Neuroscience Methods 115 (2002) 29–43

JOURNAL OF NEUROSCIENCE METHODS

www.elsevier.com/locate/jneumeth

Signal-to-noise ratio improvement in multiple electrode recording

P.G. Musial [a,b], S.N. Baker [c], G.L. Gerstein [a,*], E.A. King [a], J.G. Keating [a]

[a] *Department of Neuroscience, University of Pennsylvania, A306 Richards Building, 3700 Hamilton Walk, Philadelphia, PA 19104-6085, USA*
[b] *Department of Neurophysiology, Nencki Institute of Experimental Biology, Warsaw, Poland*
[c] *Department of Anatomy, University of Cambridge, Cambridge CB2 3DY, UK*

Received 6 July 2001; received in revised form 26 November 2001; accepted 26 November 2001

Abstract

Recordings of spike trains made with microwires or silicon electrodes include more noise from various sources that contaminate the observed spike shapes compared with recordings using sharp microelectrodes. This is a particularly serious problem if spike shape sorting is required to separate the several trains that might be observed on a particular electrode. However, if recordings are made with an array of such electrodes, there are several mathematical methods to improve the effective signal (spikes) to noise ratio, thus considerably reducing inaccuracy in spike detection and shape sorting. We compare the theoretical basis of three such methods and evaluate their performance with simulated and real data. © 2002 Elsevier Science B.V. All rights reserved.

Keywords: Signal/noise; Multi-electrode; Multi-neuron; Extracellular spikes; Waveform sorting

1. Introduction

Increasing numbers of laboratories are now making recordings of multiple neurons simultaneously, frequently in awake behaving animals. The appropriate technology has evolved from single sharp electrodes that allow clean observation of single unit activity to arrays of blunter electrodes, either small diameter insulated wires or silicon arrays. Each such electrode generally records action potentials from more than one neuron, thus forcing the use of spike shape sorting to separate the several activities for further analysis. Many computational techniques are available for shape sorting, and have recently been reviewed by Lewicki (1998). All such methods involve mapping some measure of individual spike shape to a point in multi-dimensional property space. The clustering of points in this space is then used to identify and differentiate among the several different spike trains present in the recording. Sometimes closely spaced arrays of electrodes are used—stereotrodes, triodes, tetrodes (McNaughton et al., 1983; Gray et al., 1995)—so that the action potential from a given neuron is recorded by more than one electrode. The signal seen by each such electrode is different because of the different relative neuron/electrode geometry. When the measured values to be used in the property space are obtained from different electrodes, waveform sorting can be more effective.

Whatever measures are used for the clustering process, it is ultimately necessary to define boundaries between the different clusters obtained. It is at this stage that the effects of signal to noise ratio become critically apparent. If the signal to noise ratio is favorable, the clusters will be well isolated with few outliers, and the separation process will be free of errors. If on the other hand the spike waveform does not rise much above the background noise, and particularly if this background noise has similar spectral content to the desired spikes, the clusters in the property space will overlap. It is then impossible to define clean boundaries, and hence there are inevitably errors of inclusion or exclusion in the assignment of an individual action potential to a particular neuron. Such poor sorting can render subsequent analysis at best inaccurate, and at worst invalid. Errors can be especially serious for cross-correlation and related analyses (Gerstein, 2000).

Obviously the best way to avoid problems in sorting is to have a good signal to noise ratio. Little can be done to increase the size of recorded spikes with a given

* Corresponding author. Tel.: +1-215-898-8752; fax: +1-215-573-5851.
E-mail address: george@mulab.physiol.upenn.edu (G.L. Gerstein).

PII: S0165-0270(01)00516-7

type and size of electrode; the relevant and usually fixed factors involve the impedance and shape of the electrode tip, and its placement relative to the nearby active neuron. The noise component of the recording however offers some possibilities for improvement, and that will be the topic of this paper.

We assume that the low frequency local field potentials have been filtered out of the recordings, so that our pass band is ≈ 300 Hz–8 kHz. By definition, noise in this residual recording is everything except the (relatively) large spikes. Such noise can come from a number of sources: (1) Johnson noise from individual electrode impedances, (2) Johnson or other noise from the reference electrode, (3) high frequency components of field potentials from distant active populations of neurons, (4) electromyogram (EMG) from muscles in the scalp, jaws, neck, and sometimes body, (5) electrical artifacts generated in the wiring harness by abrupt movements, and (6) other types of electrical artifact as the animal moves and touches various portions of the apparatus. Fortunately, the spatial gradient across an electrode array of the potentials from sources (2) to (6) differs from that of action potentials from a particular neuron, and makes possible the several approaches to be examined in this paper.

To make these differences clearer, consider a single electrode that is successively moved in 100 μm steps. At each location, the magnitude of the Johnson noise will be the same. The large spikes recorded will differ at each site. However, potentials from noise sources (2) to (5) will change little from site to site. Thus if a closely spaced electrode array is used, these potentials from distant sources will appear approximately the same on each electrode. This is amply illustrated in Fig. 3 of Bierer and Anderson (1999): for 50 μm electrode spacing, a particular spike train is essentially visible only on a single channel while the non-spike activity is almost identical across four channels and 150 μm. The same can be seen in our Fig. 4. It is the shared nature of the noise across multiple electrodes that will be used in this paper to produce an improvement in the signal to noise ratio. These ideas are only applicable to recordings made with an array of similar electrodes, and with inter-electrode spacing no more than several hundred micrometers.

One approach to this form of noise reduction is based on signal processing used in radar (Applebaum, 1976) and was tested on spike data by Bierer and Anderson (1999). Here we present two other approaches appropriate for arrays of both single electrodes and tetrodes. Using simulated data we carry out a parametric analysis of the performance of each method, and show that the three methods produce almost identical results. The theoretical basis for this is explored in Appendix A. Finally we demonstrate the improvement which can be obtained on real data.

2. Methods

2.1. Simulation of test data

In order to allow parametric study of the noise reduction methods explored here it is necessary to have independent knowledge of the data properties. This is impossible with real data, so that the main part of this paper uses simulated data with known parameters. We choose to mimic many of the parameters of our experimental recording system. Twelve 'recording' channels are simulated mimicking an electrode array with 100 μm spacing, sampled at 25 kHz and with a duration of 1 s. In total, 15 independent noise voltages are simulated as Gaussian random noise bandwidth limited to 300 Hz–5 kHz. The amplitudes of each are specified in advance, and can vary from 0 to 1. A different independent noise waveform is added to each one of the 12 recording channels, mimicking Johnson noise for that electrode. Another noise waveform is added equally to all recording channels, mimicking effects from an indifferent or reference electrode. Finally, two other noise waveforms are added to all channels, but with an amplitude gradient that varies linearly across the array from 1.0 at one side to 0.45 at the other. The side with maximum noise amplitude is different for each of these two common noise sources; they mimic high frequency components of field potentials generated by two distant neuronal populations. The choice of gradient parameter is reasonable for mimicking a 1.2 mm wide array, but in fact turns out to be irrelevant for the results; we have even tried random amplitude of these shared noise distributions across the array with identical final performance. For simplicity, we did not include simulation of the short noise events originating from scratching/chewing EMG or electrical artifact; such events are however apparent in the real data analyzed in Fig. 4.

Finally, we generate a spike train of specified rate and with a gamma interval distribution of specified order (order 1 would be a Poisson train, higher orders mimic interval distributions with refractory time; Baker and Gerstein, 2000). A single experimentally recorded spike waveform is then added to the 12 channels at each time location specified by the generated spike train. For this process we treat the 12 channels as if they were sequential rather than simultaneous, using sequential segments of the generated spike train. This is justified since we are using a steady state spike train without rate or pattern modulations by a stimulus event.

Thus the final output of the simulator is a 1 s 'recording' of 12 channels, with independent parameters setting a specified set of independent and shared noises, and with an independently timed spike train on each channel (Fig. 2A). (However, the waveform of the spike is the same on each channel, unlike real data; this again is justified for our particular usage in this paper of the simulated data). Examination of individual spike waveforms in the data at this stage shows the clean original waveform badly distorted by the combined noise voltages; the distortions are of course different for each occurrence of the spike on any channel (Fig. 2C).

2.2. Algorithms for improvement of signal:noise ratio

2.2.1. General notation

Denote the data to be processed as matrix $\mathbf{D}(t,c)$, where t indexes the successive data samples in time (rows) and c indexes channels (columns). Let $\max(c) = K$ and $\max(t) = T$. For our particular simulated data sets $K = 12$ and $T = 25000$, representing 12 channels recorded for 1 s at 25000 samples per second.

2.2.2. The PCA cleaner

The general idea here is to use principal components analysis to predict as much of the noise contribution to each electrode as possible. For each electrode in turn, we find the two or three principal components that best represent the data from all the *other* electrodes. This representation mostly reflects the noise component of the data, and is only slightly dependent on the spike trains. An iterative procedure (Stage 2, described below) can subsequently be used to reduce the influence of the spike trains on the calculation. We project the data from the electrode being processed onto these principal components, finding thereby the coefficient that should multiply each principal component when we use them to predict the noise on this channel. This prediction is then subtracted from the data on the electrode being processed, yielding a first order estimate of the cleaned signal. The entire process is carried out successively for each electrode, i.e. 12 times for the 12 channels of data provided by the simulator.

In more detail, in order to clean the data from electrode i, we form a reduced data matrix $\mathbf{Dnoti} = \mathbf{D}(t,c_{\text{noti}})$ which contains the data from all channels *except* for i, the channel being processed.

The covariance $\mathbf{Cnoti} = \text{cov}(\mathbf{Dnoti})$ has dimensions $(K-1)$ where $K = \max(c)$. We obtain the eigenvectors $\mathbf{V}_n(\text{noti})$ corresponding to the three largest eigenvalues using Matlab's 'eig' function. These three $(n = 1-3)$ eigenvectors also have dimension $(K-1)$. Each represents a set of weighting factors with which to sum the $(K-1)$ data channels (columns of $\mathbf{Dnoti}$) in computing the corresponding principal component.

Thus the principal components are $\mathbf{pc}_n(\text{noti}) = \mathbf{Dnoti}\mathbf{V}_n(\text{noti})$, have a dimension of T where $T = \max(t)$ and give an optimal representation of the data from all but the ith channel.

We now project the data of the channel being processed, $\mathbf{D}(t,c_i)$, onto each of these principal components, obtaining a normalized coefficient $b_n(i)$. Letting the superscript T denote transpose of a vector or matrix, we have:

$$b_n(i) = \mathbf{D}(t,c_i)^{\mathrm{T}}\mathbf{pc}_n(\text{noti})/(\mathbf{pc}_n(\text{noti})^{\mathrm{T}}\mathbf{pc}_n(\text{noti})) \quad (1)$$

These coefficients are the amplitude fraction of the corresponding principal component that should be used for representing the noise on the ith channel.

Finally, to obtain $\mathbf{D}_{\text{clean}}(t,c_i)$, the estimate of the ith data channel without noise, we take the raw data values and subtract the contributions from each principal component specified by the $b_n(i)$.

$$\mathbf{D}_{\text{clean}}(t,c_i) = \mathbf{D}(t,c_i) - b_1(i)\mathbf{pc}_1(\text{noti}) - b_2(i)\mathbf{pc}_2(\text{noti}) - b_3(i)\mathbf{pc}_3(\text{noti}) \quad (2)$$

This entire process is carried out for each electrode in turn, producing the first stage cleaned data. Calculations were done in Matlab (MathWorks Inc.).

2.2.3. The regression cleaner

The general idea here is to use multiple linear regression to predict as much of the noise contribution to each electrode as possible. We find the weighted sum of the data on all other electrodes that best represents the data on the electrode being processed, and carry out this calculation on each electrode in turn. This representation mostly reflects the noise components of the data, and is only slightly dependent on the spike trains. An iterative procedure (Stage 2, described next) can subsequently be used to reduce the influence of the spike trains on the calculation. In a common use of regression analysis the regressors, i.e. the functions for which we seek optimal weights for the representation are a constant and a linear slope, and we seek two coefficients. In our case the regressors are the $K-1$ data vectors on the *other* electrodes, and we seek $K-1$ coefficients. The criterion for calculating these coefficients is a least squares fit between the weighted sum of regressors and the data vector on the electrode being processed. Once the coefficients are determined, we have an optimal representation of the noise, and can subtract it from the data vector on the electrode being processed. The result of this subtraction is called the residual, and is the first order cleaned signal. The entire process is carried out successively for each electrode.

In more detail, a weighted sum of the $K-1$ regressors $\mathbf{D}(t,c_{\text{noti}})$ will be used to approximate $\mathbf{D}(t,c_i)$:

$$\mathbf{D}(t,c_i) = \sum_{\substack{k=1 \\ k \neq i}}^{K} g_k \mathbf{D}(t,c_k) + \varepsilon_i(t) \quad (3)$$

where $\varepsilon_i(t)$ is a normal random variable with mean 0 and some variance σ^2.

The expectation of $\mathbf{D}(t,c_i)$ is then

$$E\{\mathbf{D}(t,c_i)\} = \sum_{\substack{k=1\\k\neq i}}^{K} g_k \mathbf{D}(t,c_k) \quad (4)$$

Thfle error in the approximation is $\varepsilon_i(t)$:

$$\varepsilon_i(t) = \mathbf{D}(t,c_i) - \sum_{\substack{k=1\\k\neq i}}^{K} g_k \mathbf{D}(t,c_k) \quad (5)$$

The g_k coefficients will be determined by minimizing the error in the least squares sense. Thus we seek

$$\min[\varepsilon_i^2(t)] = \min\left\{\mathbf{D}(t,c_i) - \sum_{\substack{k=1\\k\neq i}}^{K} g_k \mathbf{D}(t,c_k)\right\}^2 \quad (6)$$

The $K-1$ coefficients are calculated by taking and setting equal to zero the $K-1$ partial derivatives of Eq. (6) with respect to each of the g_k, and solving the resulting equations (Winer, 1971; Blum and Rosenblatt, 1972).

The optimal representation of the noise component of the data on electrode i is then given by Eq. (4), and the cleaned data on electrode i is the residual:

$$\mathbf{D}_{\text{clean}}(t,c_i) = \mathbf{D}(t,c_i) - \sum_{\substack{k=1\\k\neq i}}^{K} g_k \mathbf{D}(t,c_k) \quad (7)$$

The above procedure is carried out successively for each electrode in turn, producing the first order cleaned data. Calculations were done in Matlab (MathWorks Inc.) using the multiple regression function *regress*.

We note that successful application of the regression algorithm may not always be possible. Calculation of the $K-1$ coefficients a_k depends (Blum and Rosenblatt, 1972) on the existence of the inverse matrix:

$$\{\mathbf{D}(t,c_{\text{noti}})\mathbf{D}(t,c_{\text{noti}})^{\text{T}}\}^{-1} \quad (8)$$

where the superscript T indicates transpose. If the regressors (i.e. columns of $\mathbf{D}(t,c_{\text{noti}})$ representing the data on the *other* electrodes) are linearly independent the existence of the inverse matrix is assured. However, in our application the regressors have some degree of shared noise as well as individual noise (and spikes). If the individual noise components are sufficiently small relative to the shared components, linear independence of the regressors is violated, and the inverse matrix may not exist. However, we note that in all instances of real data analysis we have never encountered this problem.

2.2.4. The array cleaner

The general idea here is to find a linear combination of *all* channels that optimizes the ratio of signal to noise, where for our application the signal consists of spike waveforms and the noise is everything else in the recording. This approach is based on a processing method for an array of radar antennas (Applebaum, 1976) and has been implemented for multi-electrode recordings by Bierer and Anderson (1999). Note that the immediate objective here is different from the two algorithms described earlier. In those calculations we derived a representation of the noise components from the *other* channels and subtracted this from the channel being processed. The identity of the K channel recordings was preserved through both stages of the cleaning process. Here, in contrast, the channel identity is generally preserved only in the first order application of the algorithm; as before this allows better detection of whatever spikes may be in the data. The output of a second order application of the algorithm will, depending on parameters, consist of vectors that may still represent channels, or alternatively may represent a weighted sum of transformed data on the original channels. In the latter situation the output vectors are no longer necessarily associated with a particular channel but might be associated with a particular neuron. In spite of these apparently fundamental differences all three algorithms produce essentially the same improvement in the data (see Section 3) and are shown in Appendix A to be closely related.

The required basic linear transformation equation (derived in Applebaum, 1976) is:

$$\mathbf{Z}(t,u) = \mathbf{D}(t,c)\mathbf{C}_{\text{N}}^{-1}\mathbf{w}(c,u) \quad (9)$$

where $\mathbf{Z}(t,u)$ is the output of the algorithm, with u indexing the signal sources (individual neurons in the data for our application), and $\max(u) = U$. $\mathbf{D}(t,c)$ is the raw data which includes signal (spikes) and noise, $\mathbf{C}_{\text{N}}$ is the covariance matrix of *only* the noise in the data, $\mathbf{C}_{\text{N}}^{-1}$ is its inverse, and $\mathbf{w}$ is a weighting matrix described further. The dimensions of $\mathbf{Z}$ are T rows, U columns, $\mathbf{D}$ is T rows (sample time points) and K columns (channels), while $\mathbf{C}_{\text{N}}$, $\mathbf{C}_{\text{N}}^{-1}$ are $K \times K$, and $\mathbf{w}$ is K rows and U columns. Note that Eq. (9) can be interpreted as a projection of the data $\mathbf{D}$ onto a set of vectors $\mathbf{C}_{\text{N}}^{-1}\mathbf{w}(c,u_i)$, $i = 1,\ldots, U$. Each such vector is the optimal projection choice to maximize signal:noise for the ith neuron.

In the first order application of this algorithm we obviously cannot yet parse the raw data into signal and noise. Therefore, even though biased, we compute the covariance of the raw data. Thus initially we have $\mathbf{C}_{\text{N}} = \text{cov}(\mathbf{D}(t,c))$ and can usually obtain its inverse $\mathbf{C}_{\text{N}}^{-1}$.

In the original radar application there was only a single signal, and $\mathbf{w}$ was a column vector representing the distribution of signal amplitude across the K channels. In the present application we potentially have U signals (spikes from U different neurons), and we take $\mathbf{w}$ as a matrix where each of the U columns describes the amplitude distribution across the K channels of the spikes from each recorded neuron. Obviously, we do not have the information to determine such a $\mathbf{w}$ at this stage, and therefore choose it to be unitary, i.e. ones on the diagonal, zeros elsewhere. This choice corresponds to the assumption that each channel of data contains spikes from only one neuron, i.e. that no two electrodes record the same neuron's spikes and that neuron identity corresponds to channel identity. Both the calculation of $\mathbf{C}_N$ and the choice of $\mathbf{w}$ can be improved in the subsequent second order application of the algorithm as indicated below.

Note that with the stated conditions, the first order $\mathbf{Z}(t,u)$ is not necessarily equivalent to the first order $\mathbf{D}_{clean}(t,c)$ obtained from the PCA and regression algorithms. One difference is that each column of $\mathbf{Z}$ generally has a *different* scaling factor, a problem that does not arise in $\mathbf{D}_{clean}$. It is convenient to normalize each column of $\mathbf{Z}$ so that $\mathbf{Z}(t,c_i)$ is divided by $\mathbf{C}_N^{-1}(i,i)$.

The algorithm expressed in Eq. (9) is somewhat unfamiliar in neuroscience, so that it is appropriate to examine its origin. Here we present a derivation of the array processing algorithm using the method of discriminant analysis. The data to be processed are contained as before in matrix $\mathbf{D}(t,c)$. There are K channels and each has T time points. Now consider a new K-dimensional vector space in which individual vectors will represent the data values at consecutive time points across all K channels, i.e. the rows of $\mathbf{D}$. At each time point we assume there is a linear sum of the signal and noise. For the present purpose we will assume that $U=1$, i.e. that the signal consists of spikes *only from one particular neuron* (but observed with different amplitude across the K channels). The noise comes from shared and individual sources, so that it may be variously correlated across the K channels. If we examine the T noise and signal vectors separately, they will form two clusters in our vector space.

Let $s(t)$ be the time course of the signal (spikes) and let its amplitude be distributed across the K channels with weights described by column vector $\mathbf{w}_i$ ($i = 1,\ldots,K$). Thus, *at each time point* t ($t = 1,\ldots,T$) the signal across all channels can be described as $\mathbf{S}(t) = s(t)\mathbf{w}^T$; both $\mathbf{S}(t)$ and $\mathbf{w}^T$ are row vectors of dimension K. Analogously, the noise is described *at each time point* t by $\mathbf{N}(t)$, also a K-dimensional row vector. For purposes of illustration, let us take $K=2$, i.e. two channels only. As seen on Fig. 1, all signal points (i.e. $\mathbf{S}(t)$ for each of the T values of t) lie on the direction defined by $\mathbf{w}^T$. To make Fig. 1 clearer, we assume the signal has a DC component larger than the noise variance so that the signal and noise clusters are clearly separated. The noise component of the data (i.e. $\mathbf{N}(t)$ for each of the T values of t) forms an ellipsoid-like cloud whose shape depends on the pattern of noise correlations between channels (Fig. 1, lower panel). In the case of zero correlation between channels, the noise cloud is spherical (Fig. 1, upper panel). However, the noise cluster is always centered at the origin because we assume it has no DC component.

The problem of finding the channel mix that will maximize the signal to noise ratio is equivalent to a search for the one-dimensional projection from our K-space that yields maximal separation of the projected signal and noise clusters, i.e. a discriminant analysis. If the noise is not correlated across channels, the optimal direction for the projection is parallel to the signal vectors (see projection marked by **o** in Fig. 1, upper panel). If the noise is correlated, the optimal projection is a trade off between projections onto the signal direction **s** and the direction **n** that has minimal variance of the noise (see Fig. 1, lower panel).

Let us define the separation (Δ) of signal and noise clusters *after* projection as

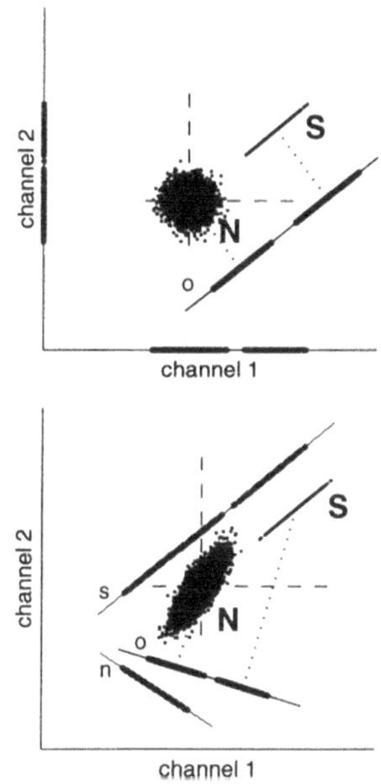

Fig. 1. Illustration of the optimal projection for maximal signal to noise ratio in the array algorithm. This is a two-dimensional caricature of the K-dimensional data space (i.e. simulation of two channel recording, with each axis representing one channel). The signal and noise clusters (**S**, **N**, respectively) are well separated for illustrative purposes by assuming a DC component for the signal. Upper figure: uncorrelated noise. The variance of the noise cluster is the same in all possible projections. Thus the optimal projection (marked by **o**) yielding maximal separation (Δ; see text) is the one leading to maximal signal amplitude, i.e. projection onto a vector parallel to the signal vectors. Lower figure: noise correlated across channels. The optimal projection (**o**) is determined by $\mathbf{C}_N^{-1}\bar{\mathbf{S}}$ and is a trade-off between the signal direction **s** and the minimal noise variance direction **n**.

$$\Delta = \frac{d^2}{V_S + V_N} \tag{10}$$

where d is the difference between mean values of the projected signal and the projected noise; V_S, V_N are variances of the projected noise and signal, respectively.

Let the direction of the optimal one-dimensional projection be defined by the vector **a**. For any time point the vector $\mathbf{S}(t)$ projected onto **a** is $\mathbf{a}^T\mathbf{S}(t)$ (analogously, for the noise vectors: $\mathbf{a}^T\mathbf{N}(t)$). These are both inner products. Thus, given the signal (**S**) and noise (**N**) clusters the separation Δ can be expressed as a function of projection vector **a**:

$$\Delta = \frac{(\mathbf{a}^T\bar{\mathbf{S}} - \mathbf{a}^T\bar{\mathbf{N}})^2}{\frac{1}{T}\sum_{t=1}^{T}(\mathbf{a}^T\mathbf{S}(t) - \mathbf{a}^T\bar{\mathbf{S}})^2 + \frac{1}{T}\sum_{t=1}^{T}(\mathbf{a}^T\mathbf{N}(t) - \mathbf{a}^T\bar{\mathbf{N}})^2} \tag{11}$$

where $\bar{\mathbf{N}}$, $\bar{\mathbf{S}}$ are the mean values of **S**, **N**, respectively, over the time of analysis.

But

$$\frac{1}{T}\sum_{t=1}^{T}(\mathbf{a}^T\mathbf{S}(t) - \mathbf{a}^T\bar{\mathbf{S}})^2 = \mathbf{a}^T\left(\frac{1}{T}\sum_{t=1}^{T}(\mathbf{S}(t) - \bar{\mathbf{S}})(\mathbf{S}(t) - \bar{\mathbf{S}})^T\right)\mathbf{a} = \mathbf{a}^T\mathbf{C}_S\mathbf{a} \tag{12}$$

where $\mathbf{C}_S$ is covariance matrix of the signal. After applying the same operation to the noise cluster **N** and denoting the noise covariance matrix as $\mathbf{C}_N$ we can rewrite Eq. (11) in the following way:

$$\Delta = \frac{(\mathbf{a}^T\bar{\mathbf{S}} - \mathbf{a}^T\bar{\mathbf{N}})^2}{\mathbf{a}^T(\mathbf{C}_S + \mathbf{C}_N)\mathbf{a}} \tag{13}$$

Projection of the clusters on vector **a** is intended to maximize separation. Vector **a** can be found by setting to zero the partial derivative of Eq. (13) with respect to **a**. This gives (Rao, 1973, p. 60):

$$\mathbf{a} = (\mathbf{C}_S + \mathbf{C}_N)^{-1}(\bar{\mathbf{S}} - \bar{\mathbf{N}}) \tag{14}$$

Since we assume that there is no DC component of noise on any channel we can neglect the $\bar{\mathbf{N}}$ term in Eq. (13). Also, since $\mathbf{S}(t) = s(t)\mathbf{w}$, we get $\bar{\mathbf{S}} = \bar{s}\mathbf{w}$, where $\bar{s}$ is the mean value of $s(t)$ over time:

$$\bar{s} = \frac{1}{T}\sum_{t=1}^{T} s(t)$$

We are not interested in the norm of **a**, but only in its direction, so we can write

$$\mathbf{a} = (\mathbf{C}_S + \mathbf{C}_N)^{-1}\mathbf{w} \tag{15}$$

This can be further simplified. Multiply Eq. (15) by $(\mathbf{C}_S + \mathbf{C}_N)$ from the left giving:

$$\mathbf{C}_S\mathbf{a} + \mathbf{C}_N\mathbf{a} = \mathbf{w} \tag{16}$$

But

$$\mathbf{C}_S = \frac{1}{T}\sum_{t=1}^{T}(\mathbf{S}(t) - \bar{\mathbf{S}})(\mathbf{S}(t) - \bar{\mathbf{S}})^T = \frac{1}{T}\sum_{t=1}^{T}(s(t) - \bar{s})\mathbf{w}(s(t) - \bar{s})\mathbf{w}^T = v(s)\mathbf{w}\mathbf{w}^T \tag{17}$$

where $v(s)$ is the variance of $s(t)$:

$$v(s) = \frac{1}{T}\sum_{t=1}^{T}(s(t) - \bar{s})^2$$

Multiplying Eq. (17) by **a** from the right and rearranging terms:

$$\mathbf{C}_S\mathbf{a} = v(s)\mathbf{w}(\mathbf{w}^T\mathbf{a}) = v(s)(\mathbf{w}^T\mathbf{a})\mathbf{w} = \beta\mathbf{w} \tag{18}$$

The rearrangement in Eq. (18) is allowed since $\mathbf{w}^T\mathbf{a}$ is an inner product of vectors **w** and **a**, and hence a scalar. The new constant β is the product of that scalar with $v(s)$. Eq. (18) states that $\mathbf{C}_S\mathbf{a}$ is parallel to **w**. This is not surprising because, as seen in Fig. 1, all points in the signal cluster are aligned with vector **w**. This means that $\mathbf{C}_S$, the covariance matrix of the signal, has only one eigenvector which is parallel to **w**. Now, using Eq. (18) we can transform Eq. (16) into $\mathbf{C}_N\mathbf{a} = \alpha\mathbf{w}$, where $\alpha = (1 - \beta)$, a scalar. Thus the direction on which to project the data for optimal signal:noise separation is defined by:

$$\mathbf{a} = \mathbf{C}_N^{-1}\mathbf{w} \tag{19}$$

which is the same as shown in Eq. (9).

2.2.5. Iteration of the algorithms

After the first application of any of the earlier three cleaning algorithms there is a considerable reduction of all common mode noise on each channel, leaving largely the individual Johnson like noise component, and of course all the large spikes. However, for each channel, the noise representation produced by the PCA and regression cleaning algorithms was based on original data from the $K - 1$ *other* channels, including their spikes. All these spikes will appear as small events in the noise representation with amplitude of about $1/(K - 1)$ of the original spike amplitudes. When the subtraction of noise representation from original data on the channel being processed is carried out, the residual, besides showing a reduction of noise, will in addition show these small spike waveforms (inverted). A similar situation occurs with the array algorithm. The initial calculation of the 'noise' covariance is necessarily made using the raw data, and is therefore biased by whatever spikes are present. It is of course desirable to eliminate the artifactual 'mixing' across channels. The following procedure is used.

In the raw data with its large amount of noise, detection of spike events would generally not be reliable. However, after the first stage of cleaning by any of the algorithms there is considerable noise reduction, and we can readily and accurately detect the spike occurrences on each channel. Such spike detection can be done either by an amplitude threshold or a time derivative (slope) criterion. We now go back to the original data and remove a small section around each of the now well defined spike occurrences. We could substitute zeros (or the mean) for each of these edited sections, but there is a better choice. Consider the matrix formed by the original raw data minus the first order cleaned:

$$\mathbf{F}(t,c) = \mathbf{D}(t,c) - \mathbf{D}_{\mathrm{clean}}(t,c) \quad (20)$$

In this difference matrix, spikes are removed, while noise components remain. We use this matrix as a source for substitution into the edited chunks of the original data. By this procedure we have edited the original data so that spikes are removed and substituted by a close approximation of the residual noise at each time point where there had been a spike. Call this edited data $\mathbf{D}_{\mathrm{mod}}(t,c)$.

Now we repeat the entire cleaning algorithm procedure. For the PCA and regression algorithms we process each channel with the noise representation calculated from the *other* $K-1$ channels, but always using as input the appropriate $\mathbf{D}_{\mathrm{mod}}(t,c_{\mathrm{noti}})$ from the composite spike free modification of the original data. After this second stage cleaning, common mode noise is much reduced and spike shapes therefore much improved, all as in the first stage cleaning. But now there is no artifactual channel mixing with the small inverted spike waveforms from all other channels. For the array algorithm the 'noise' covariance is now calculated from the spike free $\mathbf{D}_{\mathrm{mod}}(t,c)$, which eliminates the mixing. For K channels this cross-contamination has relative amplitudes of $1/(K-1)$; it is more noticeable therefore when the number of electrodes is smaller than in the present 12 channel example, and the second stage cleaning procedure as described earlier is more necessary. Thus the first stage of cleaning is used only for accurate spike detection and substitution. The spike free data are then used in a second stage of cleaning.

2.2.6. Tetrode recordings, 'unitary events' and near-coincidences

When dealing with recordings made with multiple stereotrodes or tetrodes, the same spikes are observed at *exactly* the same time on more than one channel within a tetrode because spacing is under 50 μm. This multiplicity gives additional possibilities for spike sorting, but at first glance causes problems with the cleaning algorithms as described earlier. The PCA and regression algorithms, besides reducing noise on the processed channel, would then have an amplitude reducing effect on its spikes since they are shared across some channels. In stage one of the calculation, this reduction could reach a relative factor of $3/(K-1)$ for a recording made with $K/4$ tetrodes (the total number of channels is K). This would likely make the identification of spikes more difficult. Even if spike identification is successful this amplitude reduction would contaminate the composite spike free modification of the original data needed for Stage 2, and consequently distort the final spike waveform. The simple solution for PCA and regression algorithms, although it reduces the effective K, is to exclude *all* channels of the stereotrode or tetrode being processed from the computations both in the initial and second stage cleaning procedure (not shown in this paper). For the array algorithm it is necessary to make an appropriate choice of the weighting matrix $\mathbf{w}$ for the second stage procedure, as described by Bierer and Anderson (1999).

A related problem occurs if the objective is to study 'unitary spike events' (Grün et al., 1999), i.e. excess near coincidences among different neurons on different channels. If the time coincidence is *exact*, even though different waveforms are involved, the cleaning algorithms will indeed reduce amplitude and distort waveforms (see next paragraph). However, usually a tolerance of 1–5 ms is allowed to search for and define 'unitary events'. If the several spikes comprising such an event have as little as 300 μs of time offset or jitter among them, the cleaning algorithms have almost no effect on final spike amplitude or waveform (not shown). The computations to remove shared signal are on a time point by time point basis, and apparently with even a minimal offset the interference between near simultaneous waveforms on different channels is low. Thus the use of the cleaning algorithms should have little effect on detection of 'unitary spike events'.

The interference applies to coincidences between spike trains (within 300 μs) and will cause the algorithms to reduce amplitude and distort waveforms on such events. For a pair coincidence, the distortion of a given spike waveform will essentially be by subtraction of $1/(K-1)$ of the near coincident spike waveform on some other channel. In the case of equal amplitude waveforms on the two channels, and with 12 channels as in our data, this is less than a 10% distortion, and in detail depends on the exact time offset between the near coincident spikes. In our simulations the individual channel noise (which is not affected by the algorithms) is at least the same fraction of spike amplitude, and is frequently larger in real data. Thus, the waveform distortions contribute under half of the residual spike waveform variability, and in practice can be neglected. The rate of near coincidences of course depends on the individual spike rates, but since the interference effects are small, the rate dependence can also be neglected. None of these effects have any noticeable effect on individual spike detection.

2.3. Performance assessment

There are a number of ways to demonstrate the noise reduction performance of these several cleaning algorithms. We could, before and after the cleaning process, show a superposition of the spike waveform over all channels (a single spike waveform is used for all channels in the simulated data). The noise-induced distortions of the waveform will be smaller after cleaning, as can easily be shown by a variance calculation. Another method appropriate for simulated data would be to project each cleaned waveform onto the known elemental waveform (an inner product of two vectors). This is a one-dimensional goodness of fit measure, and we can examine its mean and variance.

However, it is preferable to use a measure that is demonstrable in terms of clustering in a shape property space. For this we again use principal components. We have available a library of waveforms recorded in similar experiments, and have calculated the principal components that best describe this set of waveforms. Recall that *all* spikes from our simulated data set consist of the same waveform plus noise. Each such waveform is projected onto the first and second of the library PCA components, thus defining a point that is plotted on a PC1–PC2 plane. This process is carried out over all spikes in the raw and cleaned data, and directly demonstrates in the relative cluster sizes the extent of improvement by the cleaning process. Since the clusters are roughly circular for the simulated data, we can draw a bounding circle that takes in 95% of the points by examining the distribution of distances from each cluster mean. The ratio of circle diameters for the raw and cleaned data gives a good single number measure of the performance of the cleaning algorithm. The same process carried out on real data rarely results in circular clusters because the occurrences of each underlying waveform have some degree of amplitude or shape variability. Furthermore there will be different clusters, and possibly more than one cluster from the spikes recorded on each electrode. For real data we simply examined the raw and cleaned clusters on each electrode, using longer data sets to obtain reasonable statistics.

For simulated data we used the assessment method above to study the cleaning results from the different algorithms as a function of noise properties and spike amplitudes. For real data we simply compared the results of the different algorithms.

3. Results

3.1. A typical generated data set

The simple simulator described above allowed full control of the relative amplitudes of spikes, individual channel noise and three independent sources of shared noise. In addition, the spike rate and inter-spike interval distribution (order of the gamma distribution), and the waveform used for the spikes, could be varied. All parameters were constant within a given simulation, so that for example there were no spike rate modulations that would mimic stimulus presentation. However, it should be noted that the algorithms we use for shared noise reduction are very insensitive to spike rate modulation (not shown but described in the section on tetrode recordings and near coincidences). Fig. 2A shows a typical one second data set. The spike amplitude was 0.4 (for the negative peak) and individual channel noise amplitude 0.1. The three shared noise sources each had an amplitude 0.4. The upper panel of Fig. 2C shows a superposition of many noise-contaminated waveforms from the raw data set of Fig. 2A.

3.2. The cleaning of simulated data

We applied all three algorithms to a wide variety of data from our simulator, examining a range of individual and shared noise levels relative to a constant spike amplitude. Fig. 2B and the lower panel of Fig. 2C give an example of the cleaning process using the PCA algorithm, for the data shown in Fig. 2A. There is an impressive reduction in noise, such that individual spikes are much more clearly evident and have much smaller variance of shape.

The noise reduction is demonstrated more quantitatively in Fig. 2D, which shows projections of the spike waveforms before (circles) and after (crosses) the cleaning procedure onto a PC1 and PC2 calculated from a library of spike waveforms. A 95% bounding circle has been drawn about each cluster, and for these data the ratio of diameters is 0.19 (cleaned/raw). For data with the same levels of individual noise but zero shared noise both circles have the same diameter as the inner circle in Fig. 2D (not shown). This means that the algorithm removes essentially all of whatever shared noise is present.

Full parametric studies for the three algorithms are shown in Fig. 3, plotting the bounding circle ratio (cleaned/raw) as determined from results like Fig. 2D over a range of individual (0–0.3) and shared (0–0.4) noise amplitudes for a fixed spike amplitude (1.0). We show the full performance surface for the PCA algorithm in Fig. 3A. Fig. 3B and C shows the *difference* between that surface and those produced by the regression and array algorithms, with vertical gain multiplied by 10. The differences in the performance of the three algorithms are small and apparently randomly distributed over the investigated parameter space; all three algorithms give equally good results. This 'experimental' result is corroborated in the Appendix where we

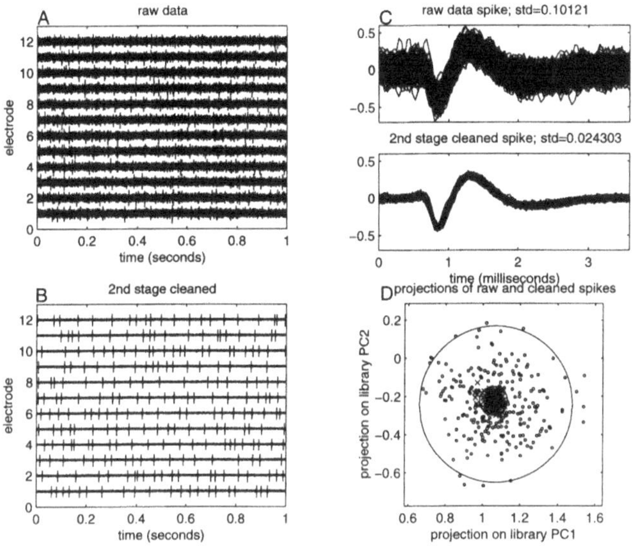

Fig. 2. (A) Simulated data mimicking 1 s of 12-channel recording sampled with 25 kHz frequency. There is a different independent noise on each channel with amplitude 0.1 and bandwidth 300 Hz–8 kHz; one source of shared noise is distributed equally across all channels, amplitude 0.4 and bandwidth 300 Hz–8 kHz there are two additional sources of shared noise distributed across all channels, each with a different spatial pattern, maximum amplitude 0.4 and bandwidth 300 Hz–5 kHz. An experimentally recorded spike waveform (scaled to negative amplitude equal to 0.4) is added to noise on all channels. Interspike intervals follow a fourth order gamma distribution; spike rate 30 per second. See text for further details. (B) Data from (A) after second stage of the cleaning process using the PCA algorithm. Note the extent of noise reduction. The cleaning process removes noise shared across channels and leaves spike waveforms with individual (i.e. channel specific) noise. (C) Upper panel: superposition of noise-contaminated spike waveforms from all channels of the raw data shown on (A). Lower panel: the same after second stage cleaning with the PCA algorithm. Standard deviations are given above each panel. (D) Projections of spike waveforms onto the first two principal components calculated from a library of experimentally recorded spike waveforms. '○': raw data, ' × ': after cleaning. Circles bounding 95% of the points in each cluster quantify the reduction of noise caused variance in the spike waveforms.

demonstrate the underlying mathematical similarity of the three algorithms. For each algorithm the performance surface given by the set of cluster diameter ratios is somewhat curved as a function of individual and shared noise amplitudes. Smaller values of the ratio indicate better cleaning, and occur for smaller values of individual noise but also for larger values of the shared noise amplitude. The latter result is deceptive: the algorithms are able to remove most of the shared noise no matter how big it is. Therefore the improvement seems larger when this shared noise is larger, since it forms a greater fraction of the total noise.

3.3. The cleaning of real data

Application of the regression cleaner to a noisy real data set is shown in Fig. 4, together with the corresponding projections of spike waveforms from one electrode onto a library PC1–PC2 plane. Arrows in Fig. 4A indicate shared spike like events that are removed by the cleaning process. Note also that much of the other 'noise' is also shared, since it gets removed.

With real data there may be different spike waveforms in each of the several electrodes, so that it is inappropriate to lump waveforms across channels as we did for the simulated data. Instead we process a longer stretch of data, the PCA calculation being applied over each consecutive second. Then waveforms from any particular channel can be collected and projected on the library principal components. The bounding circle is no longer appropriate. Accordingly Fig. 4C and D shows waveform projection planes before and after cleaning for a particular electrode. In this case there is apparently only a single waveform that forms a diffuse cluster with a widespread background. As in the simulated data, the cluster size after cleaning is considerably smaller and denser than in the raw data. However, a considerable background scattering of points remains, so that although there is improvement, isolation of these identified spike events by a boundary on the plane remains imperfect. Clearly data that initially had a better spike/noise ratio would fare better; the cleaning methods described here are useful but can only do so much. For these data the other two cleaning algorithms produce very similar results to those in Fig. 4 (not shown).

PCA cleaning of another recording (made with only three electrodes) is shown in Fig. 5 as waveform projection planes. Panel A shows the waveform clusters for the raw data when spikes are detected by a threshold amplitude criterion in the raw data. There are many outlier waveforms, many of which turn out to be shared and removable; the total number of points $n = 2267$. Panel B shows the waveform clusters for the raw data, but now the spikes are detected by threshold *after* cleaning. Most of the outliers are removed compared to Panel A since spike detection is now much more reliable so that $n = 1699$. Finally Panel C shows the waveform clusters in the cleaned data with spikes detected in the cleaned data (as for Panel B). The difference between Panels B and C demonstrates the removal of shared noise from the spike waveforms since they show the same spike events before and after the cleaning process. Panel C has denser clusters with better separation. Note that there are three outliers in the lower part of Panel C which were not evident in Panel B. In the raw data (Panel B) these were lumped into the main clusters. After removal of the shared noise, however, these three events were sufficiently atypical that they became outliers. Below each panel there is a histogram

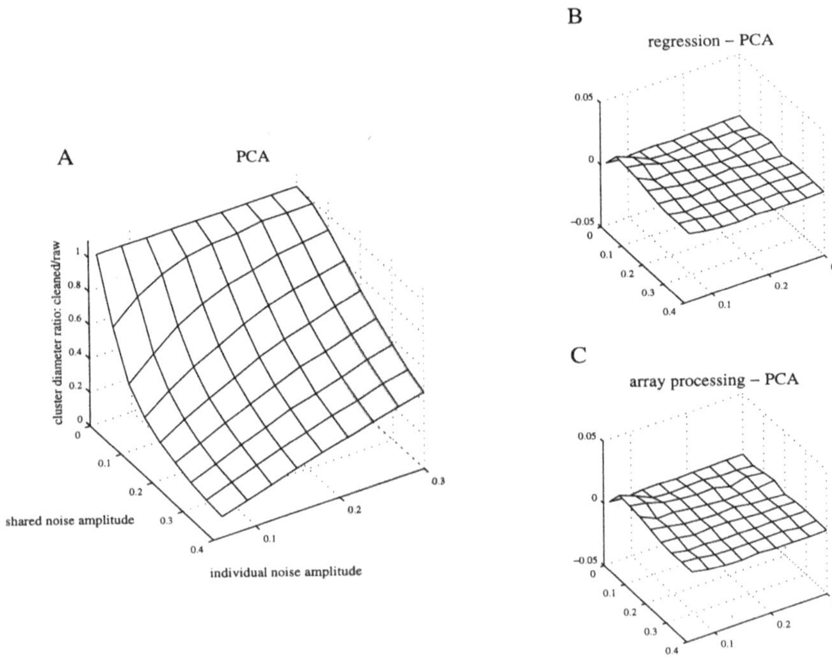

Fig. 3. Parametric performance surfaces for the three algorithms. Bounding circle diameter ratios (cleaned/raw) are determined as in Fig. 2D over a range of individual and shared noise amplitudes for a fixed spike amplitude of 1. The lower the ratio, the more effective the cleaning. (A) Surface for the PCA cleaner; (B) and (C) *differences* between the PCA surface and the regression and array surfaces. Note that the vertical gain in (B) and (C) is ten times that in (A). All three algorithms gave the same results within the investigated parameter range.

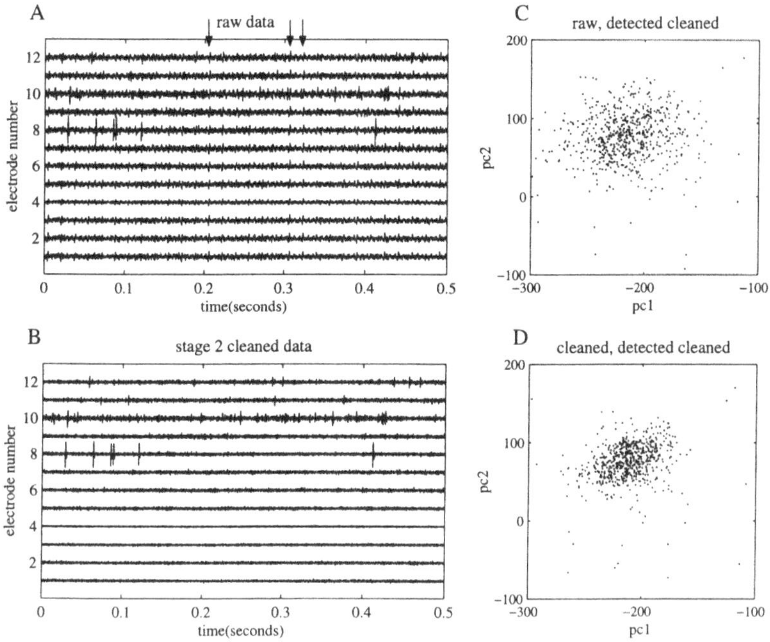

Fig. 4. Application of the regression cleaner to a real noisy data set obtained from multielectrode recordings in the rat auditory cortex. (A) Raw data. Arrows indicate shared noise events present on all channels. These artifacts (removed after cleaning, see the lower panel) would probably have been detected as neuronal spikes. (B) Same data after cleaning. Shared noise events and general background have been removed. Note that spikes on channels 11 and 12, hardly visible in the raw data, are easy to detect after cleaning. (C) Projections of spike waveforms from one electrode (channel 8 in (A)) onto the plane of first two principal components from the library. (C) Waveforms in the raw data but detected in the cleaned data; (D) waveforms in the cleaned data detected in the cleaned data. Note the cluster shrinkage although there are also some points left in a widespread background.

of the counts between PC2 = ± 100, demonstrating more quantitatively the successive reduction of outliers and improvement in waveform cluster density and separation.

4. Discussion

We propose two new algorithms for removing correlated noise from multichannel recordings made with arrays of micro-wires or other electrodes. We have compared these methods to the array processing algorithm (Applebaum, 1976) previously implemented for spike detection and sorting in multichannel data (Bierer and Anderson, 1999; Rebrik et al., 1999; only for detection). The first new method is based on principal component analysis. In this approach we subtract from the data in a given channel that part of its noise content that is correlated with the principal component representation of noise on the remaining channels. It is necessary to edit these remaining channels to remove spikes before computing the principal components. The second method is based on multiple linear regression. First we find a linear combination of the data on the other channels that best expresses (in terms of least squares) the noise in a given channel. Then, using these calculated regression coefficients, we subtract that linear combination of data on other channels from the one being processed. Although requiring different procedures, both new algorithms appear to be as effective as the array processing algorithm in simulation studies over a broad range of shared and individual noise parameters. As we show in the Appendix A, all three shared-noise cleaning methods, despite being based on different concepts, are in fact mathematically equivalent. Each method however has some specific advantages that we mention in more detail later in this section. We have also given an alternate derivation of the array processing formula. This makes use of linear discriminant analysis to separate maximally signal and noise in multichannel data, and leads to the same recipe for optimal mixing of the data. Although this derivation does not lead to a new method we believe it helps to understand the topic.

All three algorithms have some common features. All of them require two steps. Initially the raw data spikes are often masked and distorted by the noise. Therefore, in a first stage, we apply the particular method to the raw data set in order to detect spikes better. At this stage the noise removal process is biased because its parameters were based on data that included spikes in the other channels (for PCA and regression) or all channels (for array). In addition the biased first stage process introduces small cross-channel artifacts. In a second stage, we then again process each channel, using edited data in which we substitute noise for the spikes detected in the first stage on the other channels (for PCA and regression) or all channels (for array). The noise used in these short spike-centered substitution windows is obtained by subtracting the first stage cleaned data from the original raw data. (In principle the entire process could be iterated, using successively improved estimates of the spike waveforms to estimate the noise substitution for the next stage. We found little improvement beyond the second stage as used throughout this paper.)

In experimental recordings, it is highly likely that the shared noise characteristics and distribution across channels can change in time. This suggests that noise processing should be done dynamically in a short and moving time window, typically about 1 s in duration. There is a trade off for all three methods: for short time windows, even in case of stationary noise, the results are more sensitive to individual channel noise and thus less reliable (although taking less computation time).

If the number of shared noise sources is small, the PCA method offers some reduction of computation time and memory usage (due to the more economic noise representation; see Appendix A) while maintaining high accuracy. There will be some relation between the number of independent shared noise sources and the number of PCs that must be used in the cleaning process. For our simulated data with three shared noise sources and for all the real data we have examined, the amplitude of PC3 is about an order of magnitude smaller than PC1 and PC2, and therefore has little influence on the cleaning process. It would however be a good rule-of-thumb to examine the relative magnitudes of PCs for the data being processed, and to carry the computation out to as many terms as appropriate.

With the recent development of independent component analysis (ICA; Lee, 1998; Brown et al., 2001 for a tutorial review) we attempted to use this method to denoise multichannel data but our attempts were not successful. One of the limitations of the ICA method at the current stage of development is that it requires the availability of at least as many recorded channels as the number of sources (Lee, 1998, p. 63). This is easily obtained in optical recording, but is rarely possible with

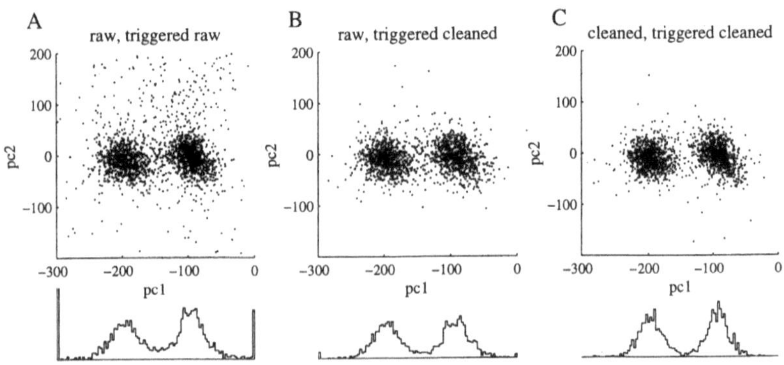

Fig. 5. Projections of spike waveforms from one electrode in a 3-channel recording. (A) Waveforms in the raw data detected in the raw data by amplitude threshold; (B) waveforms in the raw data but detected in the cleaned data; (C) waveforms in the cleaned data detected in the cleaned data. Each panel also has a histogram of the point counts between PC2 = ± 100. Note the narrowing of clusters in (C) after cleaning; this leads to better separation and therefore improvement of spike sorting.

electrodes. For N electrodes (channels) in a typical experiment we might expect $N/2$ (or more) spike trains, N individual noise sources, and several shared noise sources. The number of sources clearly outnumbers the N available channels and separation does not work.

Whereas array processing and PCA methods are both linear computations, the regression method can easily be generalized. Instead of a linear combination in the regression equation, it would be possible to incorporate nonlinear dependencies between channels. A possible candidate for such nonlinear effects would be capacitance between electrodes. A further enhancement to the procedure would be to include time delays into the regression equation. This would allow the solution of deconvolution problems, where there are shared sources of noise which are differently filtered by electrodes that have slightly varied characteristics. That variant of the regression method could also be helpful in the case of data multiplexed during the sampling process.

We have not explored non-linear methods like wavelets for either spike detection or noise reduction. Such application may be difficult when the spectral characteristics of noise and spikes are similar, as in the real and simulated data considered in this paper.

If there are fewer shared noise sources than the number of channels and if individual channel noise amplitude is negligible, there is a problem with redundant representation of noise. In this situation the covariance matrix becomes singular which means that array processing and the regression method are no longer valid. The PCA approach can still work assuming we use only as many principal components as the number of shared noise sources. However, this problem occurred only in simulated data when individual noise amplitudes were made considerably smaller than the shared noise amplitudes. We never faced the problem of covariance matrix singularity in experimental data because the individual channel noise was high enough.

Recall that both PCA and regression algorithms require a time-point by time-point subtraction. Accuracy in calculating the waveforms for this process requires a high sampling rate. A rule-of-thumb for satisfactory performance with our data is about 2.5 times the Nyquist limit at the highest frequency of the noise bandpass. Thus for an upper noise bandpass of 5 kHz we use a 25 kHz sampling rate (i.e. 40 μs between samples).

It should be noted that the cleaning algorithms described in this paper do not alter the underlying spike waveforms in any way as would for example a filter. The only effect of the cleaning procedure is to reduce whatever shared noise contribution there is to the spike shape variance, and thereby to improve performance of any subsequent shape sorting process. It should also be noted that since the second stage algorithms eliminate the influence of spikes on the shared noise estimate, and then operate on each time point individually, the cleaning performance is not affected by rate modulations of the spike trains or by amplitude non-stationarities of the shared noise. Individual channel noise, stationary or not, is of course unaffected by these algorithms.

The real and simulated examples given in this paper have all assumed recording arrangements as for typical spike train studies, with a bandpass of 300 Hz–5 kHz. However, such limitation is not fundamental or necessary. If real recordings include *shared* low frequency signal or noise, as might happen in a study of evoked or field potentials, the algorithms will be effective in eliminating them. Low frequency signal or noise that are individual to each channel will not be affected, and would cause serious problems if these same data are also to be used in subsequent spike shape sorting efforts. Adequate sorting requires appropriate preliminary high pass filtering.

In real spike data we do often observe high individual channel noise in comparison to shared noise. None of the three cleaning methods will then offer much improvement in the quality of spike waveforms. However, they will still get rid of unwanted shared events that may even mimic spikes; an example is abrupt movement artifacts present simultaneously on more than one channel. Routine application of cleaning algorithms to recorded data can only improve quality, costs little in computation time, and would seem highly worthwhile.

Acknowledgements

Supported by NIH Grants MH 46428 and DC 01249 to G.L.G. and a Wellcome Trust Fellowship to S.N.B. A program implementing the PCA cleaning algorithm in MATLAB 5.3 is available by sending an email request to software@mulab.physiol.upenn.edu.

Appendix A

We show in the following section that for the case of multichannel single wire recordings (spikes of any particular neuron present in one channel only) the array processing algorithm is equivalent to the two methods we propose in this paper (principal component analysis and multiple linear regression). For simplicity let us assume that the signal spikes come from only one neuron, and are present only on channel 1 (this may require renumbering of the K channels). Thus, in Eq. (9) we have $U = 1$, and $\mathbf{w}$ is a K-dimensional column vector. Its first component is one, and all remaining components are zeros ($\mathbf{w}(c_1) = 1$ and $\mathbf{w}(c_i) = 0$ for $i = 2,\ldots,K$). Eq. (9) then simplifies to:

$$\mathbf{Z}(t) = \mathbf{D}(t,c)\mathbf{C}_N^{-1}\mathbf{w}(c) \tag{A.1}$$

At this point let us simplify the notation. Set $Z = \mathbf{Z}(t)$, $D_i = \mathbf{D}(t,c_i)$, $\mathbf{C}^{-1} = \mathbf{C}_N^{-1}$, and $w_i = \mathbf{w}(c_i)$. Then

$$Z = \sum_{i,j=1}^{K} (\mathbf{C}^{-1})_{ij} D_i w_j = \sum_{i=1}^{K} (\mathbf{C}^{-1})_{i1} D_i$$

$$= (\mathbf{C}^{-1})_{11} D_1 + \sum_{i=2}^{K} (\mathbf{C}^{-1})_{i1} D_i \tag{A.2}$$

But

$$(\mathbf{C}^{-1})_{ij} = (-1)^{i+j} \frac{\det \mathbf{M}_{ji}}{\det \mathbf{C}} \tag{A.3}$$

where $\mathbf{M}_{ji}$ is a submatrix of $\mathbf{C}$ obtained after deleting the jth row and ith column (Bretscher, 1997, p. 259).

$$Z = D_1 \frac{\det \mathbf{M}_{11}}{\det \mathbf{C}} + \sum_{i=2}^{K} D_i (-1)^{i+1} \frac{\det \mathbf{M}_{1i}}{\det \mathbf{C}}$$

$$= \frac{\det \mathbf{M}_{11}}{\det \mathbf{C}} \left(D_1 - \sum_{i=2}^{K} D_i (-1)^i \frac{\det \mathbf{M}_{1i}}{\det \mathbf{M}_{11}} \right) \tag{A.4}$$

Let us use the Laplace expansion of $\det \mathbf{M}_{1i}$ along the first column of $\mathbf{M}_{1i}$ (Bretscher, 1997, p. 285):

$$\det \mathbf{M}_{1i} = \sum_{k=2}^{K} c_{k1} (-1)^k \det \mathbf{N}_{ki} \tag{A.5}$$

where c_{k1} are elements of the covariance matrix $\mathbf{C}$ and $\mathbf{N}_{ki}$ is a submatrix of $\mathbf{M}_{1i}$ obtained after deleting the $(k-1)$th row and the first column (note that the rows of $\mathbf{M}_{1i}$ are numbered from 2 to K).

Denote the normalization factor $(\det \mathbf{M}_{11}/\det \mathbf{C})$ by η. Thus

$$Z = \eta \left(D_1 - \sum_{i=2}^{K} D_i \frac{(-1)^i}{\det \mathbf{M}_{11}} \sum_{k=2}^{K} c_{k1} (-1)^k \det \mathbf{N}_{ki} \right)$$

$$= \eta \left(D_1 - \sum_{i,k=2}^{K} c_{k1} (-1)^{i+k} \frac{\det \mathbf{N}_{ki}}{\det \mathbf{M}_{11}} D_i \right)$$

$$= \eta \left(D_1 - \sum_{i,k=2}^{K} c_{k1} (\mathbf{M}_{11}^{-1})_{ik} D_i \right) \tag{A.6}$$

Using matrix notation and choosing to express the D_i as a row vector and the c_{k1} as a column vector (the reverse choice is also possible but less convenient):

$$Z = \eta \left\{ D_1 - [D_2 \cdots D_K] \mathbf{M}_{11}^{-1} \begin{bmatrix} c_{21} \\ \vdots \\ c_{K1} \end{bmatrix} \right\} \tag{A.7}$$

Since $\mathbf{M}_{11}$ is in fact a covariance matrix of noise in channels $2,\ldots,K$, the expression subtracted in Eq. (A.7) from D_1 is a multiple linear regression formula for approximation of noise in channel 1 by noises in channels $2,\ldots,K$ (Winer, 1971). Note that while D_1 consists of spikes and noise, there are no spikes present in $D_2,\ldots,D_K$ so after the subtraction channel 1 is left with the spikes and individual noise. In other words, within a normalization factor η, and for the case that spikes of a given neuron are only present on channel 1, we have transformed Eq. (9) for the array algorithm into Eq. (A.7) which is the regression algorithm proposed in this paper.

Let us now define a new matrix $\mathbf{B}$ fulfilling the condition that $\mathbf{BB} = (\mathbf{M}_{11})^{-1}$.

Eq. (A.7) can be rewritten as:

$$Z = \eta \left\{ D_1 - [D_2 \cdots D_K] \mathbf{BB} \begin{bmatrix} c_{21} \\ \vdots \\ c_{K1} \end{bmatrix} \right\} \tag{A.8}$$

The term $[D_2 \cdots D_K]\mathbf{B}$ represents data in channels $2,\ldots,K$ after decorrelating by $\mathbf{B}$. Let us denote this transformed data by $[\tilde{D}_2 \cdots \tilde{D}_K]$. Note that the covariance matrix of $[\tilde{D}_2 \cdots \tilde{D}_K]$ is unity (Applebaum, 1976) which means that transformed vectors are orthogonal to each other and each of unitary variance.

Since there is no DC component in noise the covariance coefficient c_{k1} is

$$c_{k1} = \frac{1}{T} \sum_{t=1}^{T} D_k(t) n_1(t)$$

where $n_1(t)$ is the noise on channel 1. Thus, the term

$$\mathbf{B} \begin{bmatrix} c_{21} \\ \vdots \\ c_{K1} \end{bmatrix}$$

can be transformed in the following way:

$$\mathbf{B} \frac{1}{T} \sum_{t=1}^{T} \begin{bmatrix} D_2(t) n_1(t) \\ \vdots \\ D_K(t) n_1(t) \end{bmatrix} = \frac{1}{T} \sum_{t=1}^{T} \left(\mathbf{B} \begin{bmatrix} D_2(t) \\ \vdots \\ D_K(t) \end{bmatrix} \right) n_1(t)$$

$$= \frac{1}{T} \sum_{t=1}^{T} \begin{bmatrix} \tilde{D}_2(t) \\ \vdots \\ \tilde{D}_K(t) \end{bmatrix} n_1(t)$$

$$= \begin{bmatrix} \tilde{c}_2(t) \\ \vdots \\ \tilde{c}_K(t) \end{bmatrix} \tag{A.9}$$

where $\tilde{c}_k$, $k = 2,\ldots,K$, is a set of covariance coefficients of noise on channel 1 with noises on the other channels after being transformed by $\mathbf{B}$. Now we can rewrite Eq. (A.8) as

$$Z = \eta \left(D_1 - \sum_{k=2}^{K} \tilde{D}_k \tilde{c}_k \right) \tag{A.10}$$

Since vectors $\tilde{D}_k$, $k = 2,\ldots,K$, have unitary variances (and zero mean values) Eq. (A.10) can be interpreted as subtraction from D_1 of that part of its noise content that is correlated with vectors constituting an orthogonal representation of noise in the remaining channels. However, this procedure can not be done unambiguously based just on raw data in channels 2,...,K because there may be correlation of the raw data across channels. The outcome would depend on the order in which the channels were used in processing. The procedure would favor channels taken first while reducing contributions of the later ones. The orthogonal representation of noise determined by **B** as in Eq. (A.8) is only one of many possibilities. The representation based on the principal components analysis proposed in this paper is different but also orthogonal. Thus the PCA and regression algorithms are conceptually related insofar as both subtract a (different) orthogonal representation of the noise on the other channels from the channel being processed. The PCA approach is particularly advantageous (see also Section 4) since it offers reduction in dimensionality of the noise representation. This usually means economy in terms of memory usage and computational time.

References

Applebaum SP. Adaptive arrays. IEEE Trans Antennas Propag 1976;AP-24:585–98.

Baker SN, Gerstein GL. Improvements to the sensitivity of gravitational clustering for multiple neuron recordings. Neural Computat 2000;12:2597–620.

Bierer SM, Anderson DJ. Multi-channel spike detection and sorting using an array processing technique. Neurocomputing 1999;26–27:947–56.

Blum JR, Rosenblatt JI. Probability and statistics. Philadelphia, PA: W.B. Saunders, 1972.

Bretscher O. Linear algebra with applications. Upper Saddle River, NJ: Prentice-Hall, 1997.

Brown GD, Yamada S, Sejnowski TJ. Independent component analysis at the neural cocktail party. Trends Neurosci 2001;24:54–63.

Gerstein GL. Cross-correlation measures of unresolved multi-neuron recordings. J Neurosci Methods 2000;100:41–51.

Grün S, Diesmann M, Grammont F, Riehle A, Aertsen A. Detecting unitary events without discretization of time. J Neurosci Methods 1999;94:67–79.

Gray CM, Maldonado PE, Wilson M, McNaughton B. Tetrodes markedly improve the reliability and yield of multiple single-unit isolation from multi-unit recordings in cat striate cortex. J Neurosci Methods 1995;63(1–2):43–54.

Lee T-W. Independent component analysis. Theory and applications. Dordrecht: Kluwer Academic, 1998.

Lewicki MS. A review of methods for spike sorting: the detection and classification of neural action potentials. Comput Neural Syst 1998;9:R53–78.

McNaughton BL, O'Keefe JO, Barnes CA. The stereotrode: a new technique for simultaneous isolation of several single units in the central nervous system from multiple unit records. J Neurosci Methods 1983;8:391–7.

Rao CR. Linear statistical inference and its applications. New York: Wiley, 1973.

Rebrik SP, Wright BD, Emondi AA, Miller KD. Cross-channel correlations in tetrode recordings: implications for spike-sorting. Neurocomputing 1999;26–27:1033–8.

Winer BJ. Statistical principles in experimental design. New York: McGraw-Hill, 1971:69–71.

Chapter 14
Higher-Order Correlations and Synfire Chains

Ongoing Research Initiated by George

Sonja Grün and Markus Diesmann

I first met George when I did my diploma thesis at the Max Planck Institute for Biological Cybernetics in Tübingen. He was visiting Ad Aertsen, my thesis supervisor. From the first moment, George was easy to approach. When I moved with Ad to Bochum University, George also came for a research stay, as well as he did when later Markus and I followed Ad to the Weizmann Institute. On that occasion, he also invited me to dinner, and he cooked one of his famous lentils/beans stew, which he said, gets better and better the longer it stays. At the third day, it is best but lasts for a week! George celebrated every day and every meal as a gift. Lessons in humbleness for a young researcher to live up to. Also when I was a postdoc in Moshe Abeles' lab in Jerusalem, he came to visit Moshe's lab—they were already old friends since the sixties. When I then started in the lab of Singer in 1998 as a senior fellow, George visited *me*. I was very proud of that, and Markus (at that time at the Max Planck Institute for Nonlinear Dynamics and Self-Organization in Göttingen) and I showed him around Frankfurt and he had the famous green sauce. From there on, we visited George regularly, at Penn University or at the summer house in Up State New York (see photo, on the porch Fig. 14.1) to discuss science and enjoy life together. He preferred to take us to parks with nice plants, to museums, or to classical concerts. Surely when the two got lost in the exhibition, they would find me back in the museum shop. When we were at RIKEN Brain Science Institute, he stayed for 2 months with us in our lab. We discussed lots of

S. Grün (✉) · M. Diesmann (✉)
Institute of Neuroscience and Medicine (INM-6 and INM-10), Jülich Research Centre, Jülich and RWTH Aachen University, Aachen, Germany
e-mail: s.gruen@fz-juelich.de; diesmann@fz-juelich.de

A. Aertsen et al. (eds.), *Introducing Computation to Neuroscience*, Springer Series in Computational Neuroscience, https://doi.org/10.1007/978-3-030-87447-6_14

Fig. 14.1 George, Sonja, and Markus discuss science on the porch in the summer house in Jay

science but also explored Japanese peculiarities. In 2011, Markus and I moved to Jülich Research Center, and also here he visited us for 2 months in 2013. Still he called us "children" in these years if something was not exactly according to plan. George retired in 2003, but he still kept his office at the university and went there almost every day to work. When he finally gave up his office at Penn in 2015, we were allowed to move his science library (a treasure) and integrated it into our institute's library. According to George's wish, all paper works were destroyed. George always supported us all these years, as a mentor, wrote letters of reference for us, and advised us on politics and university games, although he hated them.

The work of George deeply influenced me. Some of his papers led to projects of my own, for example, the study on the influence of imperfect spike sorting on the Unitary Events analysis is inspired by Bedenbaugh and Gerstein (1997) and Gerstein (2000) (see also Chap. 10 this Volume). Another important influence on my work was the idea to compare the measure of interest (e.g., time-resolved correlation as analyzed by the JPSTH) to a prediction based on the firing rates of the neurons (e.g., Aertsen et al. (1989); Abeles and Gerstein (1988)).

George was thinking of neuronal processing in the form of active cell assemblies, i.e., groups of neurons that interact and process information (Gerstein et al., 1989). This line of thinking led also to intense interactions with George. Thus, here I would like to emphasize joint publications with George, which also led to further routes of my own research (see for illustration Fig. 14.2). At around 2002, I started to think about the detection of correlations in massively parallel spike trains, as it was clear that data sets with the order of 100 simultaneously recorded spike trains

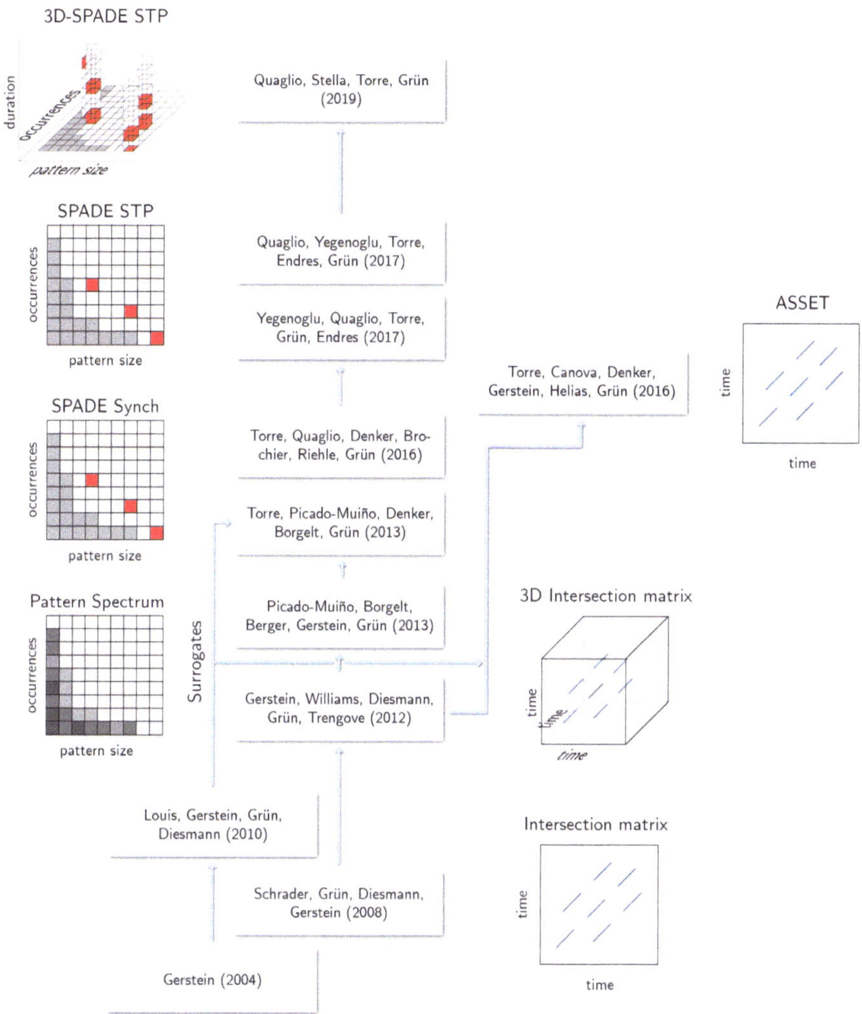

Fig. 14.2 History of methods developed with George. Schrader et al.'s work (2008) is the root for developing analysis methods with use of network simulations and for using the intersection matrix for visualization of repeating sequences of synchronous activity. This work was continued in Gerstein et al. (2012) with an extension of the intersection method to 3D. But that paper was the root for using the pattern spectrum as a display for patterns and was continued in the development of SPADE. The intersection method was further developed into a quantitative method (ASSET) in Torre et al. (2016). The development of good surrogate methods found entry in that work but was also crucial for the SPADE method for the significance test. SPADE was later extended into a method for detection of significant spatio-temporal spike patterns (3d-SPADE)

from different neurons would become available soon. I was initially hoping that such recordings would more easily enable the detection of active cell assemblies. However, the statistics to detect higher-order patterns in such high-dimensional data is tricky—many different synchronous spike patterns can occur and thereby massive multiple testing problems. So new ideas were required. “One step forward, two steps back” as George used to say.

In 2006, George and I met in Freiburg when he was with Ad for a research stay and thought about using a concrete model implementation to detect such active assemblies. We decided to analyze a spiking network model with 50 embedded synfire chains (SFC) as reference data. These data served as ground truth data in so far as we knew the detailed connectivity structure and stimulated each synfire chain with 1Hz Poisson trains. To our surprise, the resulting synfire activity in that network was not trivially visible, so we had to invent a new analysis method to detect the active SFCs. From Diesmann et al. (1999), we knew that not all neurons of a SFC group need to be active to propagate the synchronous activity reliably through the network, but only 60–80% of the neurons of each SFC group need to be active. From this, we concluded that if the same SFC is active twice, a 100% overlap of active neurons of each group cannot be expected, but rather less. Thus we had the idea to discretize the time axis into bins of a width comparable to the latency group activations (about 3ms) and enter the overlap of active neurons in any two bins in an intersection matrix with the time axis running along both axes. Repetitive activation of the same SFC then leads to diagonal structures of elevated entries. When the binning of the time axis did not fully correspond to the delays between the groups, the diagonal structures became wiggly and we called them “worms.” We found that we need to observe at least 200 arbitrary neurons from the network to detect the active SFCs in the network; otherwise, the sensitivity is too low (Schrader et al., 2008).

George continued to work on the “worms” analysis and teamed up with Chris Trengove, Liz Williams, and Markus to analyze even more complex networks with interacting SFCs (Gerstein et al. 2012). In that work, the authors extended the previous analysis to deal with higher firing rates and noise levels and developed two additional tools. One is a three-dimensional version of the intersection matrix, which detects triple re-occurrences of a SFC and thus is much more sensitive in case of higher rates. George here also introduced another display—the pattern spectrum, which continued to be an important element in our later data analysis (the Spike PAttern Detection and Evaluation (SPADE) method). The pattern spectrum collectively counts patterns of a certain signature, i.e., the number of spikes involved and the occurrence count of the pattern. This pooling helps to display large amounts of data in a compact way. George compared the entries in the pattern spectrum to pattern spectra of independent data to get an understanding of what pattern size and occurrence number are expected by chance and what is due to coordinated processes.

In Picado-Muiño et al. (2013), we continued to work on the pattern spectrum approach and explored the usage of frequent item set mining (FIM) for the efficient search and counting of synchronous spike patterns and compared the detectability of patterns to George’s accretion method already proposed in Gerstein et al. (1978). We found that the accretion method may miss higher-order patterns, because it is based on a pairwise correlation analysis. In Torre et al. (2013), we used FIM for pattern counting and included a significance test for the pattern spectrum. This is done by generating pattern spectra from surrogate data, i.e., data in which we intentionally destroy the synchronous correlation by dithering each spike randomly within a small time interval (Louis et al. 2010). From these independent data, we derived a p-value spectrum that showed how probable it is to get patterns by chance for each particular signature. The comparison of empirical entries to the corresponding p-value enables the detection of significant patterns. This method was called SPADE.

SPADE was first applied to experimental data from motor cortex of monkeys involved in a reach to grasp task (Torre et al. 2016). The recordings were performed with a 10×10 Utah array (Blackrock) resulting in up to 150 single units in parallel. The analysis revealed the presence of synchronous spike patterns that are (a) above chance, (b) specific to the behavior (grip type), and (c) have a particular spatial orientation (medio-lateral) on the cortical surface. What we did not expect was that the number of different neurons involved in these patterns barely went beyond 2. This may be due to the distance of the electrodes (400 µm) or that just looking for synchronous patterns is not enough. Therefore, we extended SPADE to also enable the detection and evaluation of spatio-temporal spike patterns (STPs), i.e., patterns between neurons that also include a temporal delay between the spike times (see also Chap. 9 this Volume). This is even a bigger challenge since the number of patterns that need to be considered does not only grow with the number of neurons considered (2^N), but with $2^{N \cdot K}$ with K being the maximal duration of the patterns searched for. So we started out with new methods for dealing with the mined patterns (Yegenoglu et al. 2016) but then realized in later work (Quaglio et al. 2017) that formal concept analysis (FCA) and FIM, after reformatting the data, are identical. Thus we are able to use the same framework as before to analyze STPs, namely SPADE. Very recently, we had to extend SPADE further for the case of patterns of different durations since patterns of different temporal extents require a different null hypothesis (Stella et al. 2019). Patterns of short duration are less probable to occur by chance than longer ones, since the number of possible pattern compositions increases with pattern duration. Therefore, we extended the pattern spectrum to 3d capturing the signatures: a number of neurons involved, occurrence count, and pattern duration. This method is currently applied to experimental data, and indeed, we find many STPs, also of higher neuron counts.

Another line of research continued also from the joint work with George on the intersection matrix published in Schrader et al. (2008). The method as such left the task of identifying these diagonal structures to visual inspection rather than to a quantitative analysis. Equipped with the intersection matrix, we started to analyze experimental data to search for indications of SFCs therein. But we had to learn that there were no clear "worms" visible in the data. This could have several reasons: (a) the sample size of the data (about 100 neurons) is too small, (b) non-stationary firing rates generate large matrix entries drowning any patterns, (c) visual inspection may not be sensitive enough so that we overlook indications of SFCs, and (d) there are no SFCs. At that moment, we could not change the sample size, but we decided to improve the method to quantitatively distinguish coordinated firing from "noise" in an automated fashion. So we worked out a method that we called ASSET (Analysis of Sequences of Synchronous EvenTs) that consists of a sequence of steps. The first tests which entries in the matrix potentially belong to a diagonal structure, i.e., occur more often than expected by the firing rate. This can be estimated bin by bin as the product of trial-by-trial rate estimations. Because we specifically aim to identify SSEs that should be expressed as diagonal structures, we clustered in a second step the significant entries into individual diagonal structures. In a third step, the neurons composing of the SSEs are identified and extracted with their neuron ids and timings. ASSET was published as Torre et al. (2016). Therein, the method was calibrated using parallel point processes generated by stochastic simulations as test data of various kinds to demonstrate the performance of the method under a wide range of realistic scenarios, including different types of non-stationarity of the spiking activity and different correlation structures. We also demonstrated in that work the ability of the method to discover SSEs on the data from the network simulations we performed in 2008 with embedded synfire chains. This was the last paper of George, and Janet had to read it to him, since his eyesight got already very bad at that point.

In summary, George was a great mentor for me. In a world of sharks, he continued his work without need to show off. He supported me in that the work I am doing is worth it even if the large crowd is continuing on evaluating firing rates. "Up the opposition" was one of his favorite toasts!

References

Abeles M, Gerstein GL (1988) Detecting spatiotemporal firing patterns among simultaneously recorded single neurons. J Neurophysiol 60(3):909–924. https://doi.org/10.1152/jn.1988.60.3.909

Aertsen AM, Gerstein GL, Habib MK, Palm G (1989) Dynamics of neuronal firing correlation: modulation of "effective connectivity". J Neurophysiol 61(5):900–917. https://doi.org/10.1152/jn.1989.61.5.900

Bedenbaugh P, Gerstein GL (1997) Multiunit normalized cross correlation differs from the average single-unit normalized correlation. Neural Comput 9(6):1265–1275. https://doi.org/10.1162/neco.1997.9.6.1265

Diesmann M, Gewaltig MO, Aertsen A (1999) Stable propagation of synchronous spiking in cortical neural networks. Nature 402:529–533. https://doi.org/10.1038/990101

Gerstein GL (2000) Cross-correlation measures of unresolved multi-neuron recordings. J Neurosci Methods 100(1–2):41–51. https://doi.org/10.1016/S0165-0270(00)00226-0

Gerstein GL, Perkel DH, Subramanian KN (1978) Identification of functionally related neural assemblies. Brain Res 140(1):43–62. https://doi.org/10.1016/0006-8993(78)90237-8

Gerstein GL, Bedenbaugh P, Aertsen AMHJ (1989) Neuronal assemblies. IEEE Trans Biomed Eng 36(1):4–14. https://doi.org/10.1109/10.16444

Gerstein GL, Williams ER, Diesmann M, Grün S, Trengove C (2012) Detecting synfire chains in parallel spike data. J Neurosci Methods 206(1):54–64. https://doi.org/10.1016/j.jneumeth.2012.02.003

Louis S, Gerstein GL, Grün S, Diesmann M (2010) Surrogate spike train generation through dithering in operational time. Front Comput Neurosci 4:1–16. https://doi.org/10.3389/fncom.2010.00127

Picado-Muiño D, Borgelt C, Berger D, Gerstein G, Grün S (2013) Finding neural assemblies with frequent item set mining. Front Neuroinform 7:9. https://doi.org/10.3389/fninf.2013.00009

Quaglio P, Yegenoglu A, Torre E, Endres DM, Grün S (2017) Detection and evaluation of spatio-temporal spike patterns in massively parallel spike train data with SPADE. Front Comput Neurosci 11:41. https://doi.org/10.3389/fncom.2017.00041

Schrader S, Grün S, Diesmann M, Gerstein GL (2008) Detecting synfire chain activity using massively parallel spike train recording. J Neurophysiol 100(4):2165–2176. https://doi.org/10.1152/jn.01245.2007

Stella A, Quaglio P, Torre E, Grün S (2019) 3d-SPADE: significance evaluation of spatio-temporal patterns of various temporal extents. BioSystems 185(104022):1–10. https://doi.org/10.1016/j.biosystems.2019.104022

Torre E, Picado-Muiño D, Denker M, Borgelt C, Grün S (2013) Statistical evaluation of synchronous spike patterns extracted by frequent item set mining. Front Comput Neurosci 7:132. https://doi.org/10.3389/fncom.2013.00132

Torre E, Canova C, Denker M, Gerstein G, Helias M, Grün S (2016) ASSET: analysis of sequences of synchronous events in massively parallel spike trains. PLoS Comput Biol 12(7). https://doi.org/10.1371/journal.pcbi.1004939

Torre E, Quaglio P, Denker M, Brochier T, Riehle A, Grün S (2016) Synchronous spike patterns in macaque motor cortex during an instructed-delay reach-to-grasp task. J Neurosci 36(32):8329–8340. https://doi.org/10.1523/JNEUROSCI.4375-15.2016

Yegenoglu A, Quaglio P, Torre E, Grün S, Endres D (2016) Exploring the usefulness of formal concept analysis for robust detection of spatio-temporal spike patterns in massively parallel spike trains. In: Haemmerlé O, Stapleton G, Faron Zucker C (eds) Graph-based representation and reasoning, vol 8577, Springer, Berlin, pp 3–16. https://doi.org/10.1007/978-3-319-08389-6

J Neurophysiol 100: 2165–2176, 2008.
First published July 16, 2008; doi:10.1152/jn.01245.2007.

Detecting Synfire Chain Activity Using Massively Parallel Spike Train Recording

Sven Schrader,[1] Sonja Grün,[2] Markus Diesmann,[1,2] and George L. Gerstein[3]
[1]*Bernstein Center for Computational Neuroscience, Albert-Ludwigs-University, Freiburg, Germany;* [2]*Theoretical Neuroscience Group, RIKEN Brain Science Institute, Wako City, Saitama, Japan; and* [3]*Department of Neuroscience, University of Pennsylvania, Philadelphia, Pennsylvania*

Submitted 12 November 2007; accepted in final form 14 July 2008

Schrader S, Grün S, Diesmann M, Gerstein GL. Detecting synfire chain activity using massively parallel spike train recording. *J Neurophysiol* 100: 2165–2176, 2008. First published July 16, 2008; doi:10.1152/jn.01245.2007. The synfire chain model has been proposed as the substrate that underlies computational processes in the brain and has received extensive theoretical study. In this model cortical tissue is composed of a superposition of feedforward subnetworks (chains) each capable of transmitting packets of synchronized spikes with high reliability. Computations are then carried out by interactions of these chains. Experimental evidence for synfire chains has so far been limited to inference from detection of a few repeating spatiotemporal neuronal firing patterns in multiple single-unit recordings. Demonstration that such patterns actually come from synfire activity would require finding a meta organization among many detected patterns, as yet an untried approach. In contrast we present here a new method that directly visualizes the repetitive occurrence of synfire activity even in very large data sets of multiple single-unit recordings. We achieve reliability and sensitivity by appropriately averaging over neuron space (identities) and time. We test the method with data from a large-scale balanced recurrent network simulation containing 50 randomly activated synfire chains. The sensitivity is high enough to detect synfire chain activity in simultaneous single-unit recordings of 100 to 200 neurons from such data, enabling application to experimental data in the near future.

INTRODUCTION

Although a tremendous amount of detail exists about properties and connectivity of cortical neurons (Braitenberg and Schüz 1991; Douglas and Martin 2004; Petersen and Sakmann 2000; Song et al. 2005; Yoshimura et al. 2005) the neuronal organization that underlies computational processes is poorly understood. Neurons usually do not act alone and, since Hebb (1949), the idea of a neuronal assembly has become widely accepted as a functional unit. There is, however, a continuing debate in the literature over whether representation of information in and transmission of information between such assemblies is in the form of firing rates or in the form of spatiotemporal firing patterns. One particularly intriguing suggestion for assembly organization is the synfire chain first proposed by Abeles (1991). The original version was a chain of length l composed of groups of w neurons each (the synfire chain width), with full unidirectional connectivity between successive groups. A group together with its output connections can interchangeably be called a link in the synfire chain. Such a chain structure propagates near-synchronous activity in each group along successive groups like a row of dominoes. Detailed properties of various versions of such systems have received considerable theoretical attention; stability properties (Diesmann et al. 1999) have been studied as well as structural variations such as partial group to group connectivity or such as feedback when a given neuron occurs in more than one group. There also have been studies of partially interconnected synfire chains to define the conditions in which activity can jump between chains (Hayon et al. 2004, 2005). Recent theoretical studies suggest that coupled synfire chain structures can demonstrate compositionality, the hierarchical representation of complex entities in terms of parts and their relations (Bienenstock 1995; Hayon et al. 2004; Schrader et al. 2007).

Intriguing though the synfire concept may be, to date there has been no direct way to examine whether such structures occur in real brains. One approach might be to look for the repeating multineuron firing patterns that would be the signature of synfire activity. There are good methods (Abeles and Gerstein 1988; Gerstein 2004) to detect and test for significance any repeating spatiotemporal firing patterns that exceed the number expected by chance. Such patterns have been found in repeating behavioral task situations (Baker and Lemon 2000; Baker et al. 2001; Kohn and Smith 2005; Prut et al. 1998) and, in several papers, claimed as evidence for synfire chain activity (Abeles et al. 1993; Ikegaya et al. 2004). With parallel recording of the order of 10 neurons the likelihood that the small sample includes several neurons either from the same or successive links of a particular chain is vanishingly small, so that such patterns are more likely to directly represent information coding and flow.

However, as massively parallel recordings become available (electrode methods: Csicsvari et al. 2003; Donoghue 2002; Euston et al. 2007; Nicolelis and Ribiera 2002; Rolston et al. 2007; Warren et al. 2001; optical methods: Ohki et al. 2005, 2006; Sasaki et al. 2007) the sampling problem is eased. The pattern program (Abeles and Gerstein 1988) can easily deal with the larger number of neurons and will detect millions of candidate repeating patterns, some of which will be significant by comparison to surrogate data (proper choice of surrogate is a nontrivial problem in its own right). At this stage synfire chains can only be inferred (Abeles et al. 1993; Ikegaya et al. 2004), since they are a possible but not unique source for patterns. It is then necessary to sort, collate, and variously compare the large number of significant detected patterns to

Address for reprint requests and other correspondence: S. Schrader, HONDA Research Institute Europe GmbH, Carl-Legien-Str. 30, 63073 Offenbach/Main, Germany (E-mail: sven.schrader@honda-ri.de).

The costs of publication of this article were defrayed in part by the payment of page charges. The article must therefore be hereby marked "*advertisement*" in accordance with 18 U.S.C. Section 1734 solely to indicate this fact.

gradually build up a meta organization like several synfire chains. The process would be like assembling the pieces of a jigsaw puzzle, in this case making appropriate matches and interleavings of pattern occurrence time, neuron identities, and spike time interval structure. Such programs have not yet been tried and will have unknown and probably data dependent computing time requirements.

In this study we develop a new method that can rapidly and clearly detect synfire chain activity and identify the participating neurons in a massive recording of even thousands of neurons, but that also is sensitive enough so that it continues to detect and identify even when ≤200 neurons are available. We demonstrate the method's comparative run-time requirements and its effectiveness for detection and for eliciting some aspects of chain connectivity using data from a large-scale cortical network model with and without embedded synfire chains.

METHODS

Simulation

Development and evaluation of a new tool for spike train analysis can only be done using data with well-defined properties. The data need to resemble experimental recordings as closely as possible and contain the effects for which the analysis method searches. In our case this requires the simulation of a local cortical network representing roughly a cubic millimeter of cortex on which we can perform virtual experiments. This simulation is only briefly described here since this report is primarily about a new analysis tool for synfire chain activity.

The basic network architecture (Fig. 1) is the well-studied balanced random network model (Amit and Brunel 1997; Brunel 2000; Vreeswijk and Sompolinsky 1996). The dynamics of the network is adjusted to the asynchronous irregular (AI) regime to reproduce relevant aspects of cortical dynamics: low-rate firing of the individual neurons with Poisson process like interval statistics and large fluctuations of the membrane potential (see Fig. 1 for the full set of parameters). The model consists of 40,000 excitatory and 10,000 inhibitory integrate-and-fire point model neurons in the network.

The neurons establish a random number (drawn from a binomial distribution) of dendritic synapses with the excitatory population of neurons. The mean of the distribution is a fraction $\varepsilon = 0.1$ of the size of the excitatory population. This connection scheme permits multiple synapses between pairs of neurons. The scheme is also applied for establishing the dendritic synapses with the inhibitory population of neurons. The postsynaptic currents are described by alpha-functions with a rise time of 0.2 ms for the excitatory synapses and 0.6 ms for the inhibitory synapses. The excitatory postsynaptic potentials (EPSPs) have an amplitude of 0.1 mV; the inhibitory postsynaptic potentials (IPSPs) are ($g = 6$)-fold larger. The synaptic delays are randomly chosen from a uniform distribution between 0.5 and 3 ms. In addition to the explicitly modeled synaptic input all neurons receive a constant excitatory current of 350 pA describing the excitatory input from outside the local cortical volume. Thus the irregular activity is exclusively generated by the dynamics of the network and not imposed by an external source.

In contrast to the standard balanced random network architecture, in our network the excitatory-to-excitatory connections are excluded from the construction scheme described in the previous paragraph. This leaves the AI state largely unaffected because it is essentially generated by the interplay of the external excitatory suprathreshold drive and the local inhibition. The excitatory-to-excitatory subnetwork is, instead, purely composed of a superposition of $m = 50$ synfire chains. These chains are consecutively constructed. First, we randomly select $w \times l$ neurons (here: $w = 100$, $l = 20$) from the

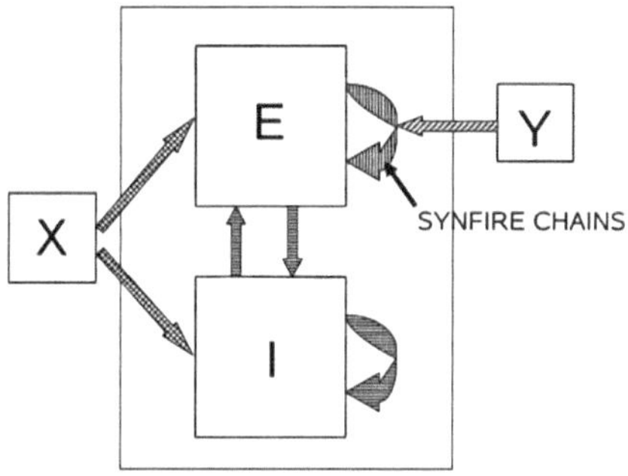

FIG. 1. Block diagram of the cortical network model with all excitatory–excitatory connections formed by synfire chains; 40,000 excitatory (E) and 10,000 inhibitory (I) neurons are recurrently connected (horizontally striped arrows): each neuron establishes a random number of synapses drawn from a binomial distribution with the mean given by 10% of the size of the respective target population. Neurons are integrate-and-fire point models (time constant 20 ms, capacity 250 pF, threshold 20 mV, absolute refractory period 2 ms, excitatory postsynaptic potential [EPSP] amplitude 0.1 mV generated by alpha-function current with 0.2-ms rise time, inhibitory postsynaptic potential [IPSP] amplitude 6-fold larger, alpha current rise time 0.6 ms). Synaptic delays are drawn from a uniform distribution between 0.5 and 3 ms. Each neuron receives (cross-hatched arrows) an excitatory DC drive of 350 pA (X). Excitatory neurons are randomly chosen to form 50 completely connected synfire chains (vertically striped arrow) of length 20 with 100 neurons in each link. A neuron can occupy only one position in each chain but may contribute to many chains. Intrachain synapses are 5-fold stronger than other EPSPs. Intrachain delays are drawn from the same distribution as others but are identical for synapses connecting a particular pair of neuron links. Chains are stimulated with independent Poisson sources (Y) at 1 Hz each. A stimulus consists of 100 spike times drawn from a Gaussian distribution with SD 1 ms sent to all the neurons of the first link in the chain via the same synapse type as used in the chain.

population of excitatory neurons without repetitions. In the next step, these neurons are connected into a feedforward subnetwork of l successive links of w neurons. Each neuron, except the neurons in the first link, establishes a dendritic synapse with all the neurons in the preceding group. This arrangement is called a complete divergent and convergent connectivity (Abeles 1982; see also Griffith and Horn 1963). The connection procedure is repeated m times. Thus an excitatory neuron may participate in several chains but occurs at most once in any given chain. The synaptic delays are drawn from the distribution specified earlier for the other synapses but is kept fixed for all synapses connecting two subsequent groups (links) of neurons. The amplitudes of postsynaptic potentials of synapses in the chains are fivefold larger than those of other excitatory connections.

In the absence of specific stimuli the synfire chains are not active and the excitatory neurons spike irregularly at a low rate. For the data presented herein we apply independent 1-Hz Poisson trains of stimuli to the synfire chains. A stimulus consists of a volley of 100 spikes. The individual spike times are drawn from a Gaussian distribution centered at the time of stimulation with SD of 1 ms. This volley is sent to all the neurons of the first group in a chain using synapses as strong as the intrachain synapses.

The dynamics are integrated on a discrete time grid of 0.1 ms using the exact integration method (Rotter and Diesmann 1999; see Morrison et al. 2007 for a detailed discussion of the application of the method to integrate-and-fire dynamics). We verified our results using a newly developed simulation technique capable of treating spike times and delays in continuous time in the framework of a globally time driven simulation scheme (Morrison et al. 2007). All simulations were carried out using the parallel computing capabilities (Plesser et al. 2007) of the NEST simulation tool (Gewaltig and Diesmann 2008) on an Infiniband compute cluster with 24 nodes with four compute cores and 8 GB of RAM per node.

The average spike rate in this network is 2.2 Hz for the excitatory neurons and 1 Hz for the inhibitory. The probability of synfire activity to survive until the end of the chain (>80% active neurons within 6 ms around the expected firing time) is 0.7–0.8.

Previous studies (Aviel et al. 2003; Mehring et al. 2003; see also the discussion in Vogels et al. 2005) have reported considerable difficulties in embedding even a single feedforward subnetwork capable of stable propagation of synchronous activity into the balanced random network. Our network differs from previous approaches in two essential properties. First, there is heterogeneity in the number of dendritic synapses. Second, the excitatory-to-excitatory subnetwork is purely composed of synfire chains. In fact, Tetzlaff et al. (2005) showed that the former condition is sufficient for stable propagation of a single synfire chain in an otherwise random balanced network.

However, in the present manuscript the network model is only employed as the generator of cortical spike data mimicking synfire activity. The details of the network dynamics and a parametric analysis of performance will be presented elsewhere.

Data overview

Neuronal activity is commonly visualized as a raster display where each tick represents the spike activity of a particular neuron. In the neural net simulations we deliberately randomized the neuron number (identity) with respect to the embedded synfire structures to mimic the random sampling of real recordings. It is instructive, however, to remap some data so that adjacent lines in the raster are associated with particular links and chains. Thus (Fig. 2*A*) raster lines 8,001–8,100 represent the activity of neurons in the first link and lines 8,001–10,000 constitute the activity of the entire fifth (arbitrarily chosen) synfire chain. The strong, nearly vertical lines in the raster of Fig. 2*A* are firings of this synfire chain; other, background neuron firings in the raster represent either spontaneous activity or the activity of other chains, since each neuron in this chain is also a member of two to three other chains. Figure 2*B* shows a portion of this raster with high time resolution revealing details of several chain runs. The offset between short vertical segments corresponds to activations of successive links; the offset is the interlink propagation or synaptic delay. Note that both link synchrony and interlink propagation time are variable and that there is a failure of propagation in the middle run of the chain.

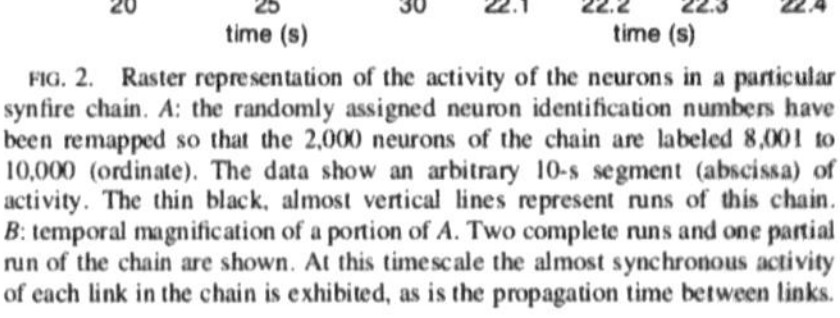

FIG. 2. Raster representation of the activity of the neurons in a particular synfire chain. *A*: the randomly assigned neuron identification numbers have been remapped so that the 2,000 neurons of the chain are labeled 8,001 to 10,000 (ordinate). The data show an arbitrary 10-s segment (abscissa) of activity. The thin black, almost vertical lines represent runs of this chain. *B*: temporal magnification of a portion of *A*. Two complete runs and one partial run of the chain are shown. At this timescale the almost synchronous activity of each link in the chain is exhibited, as is the propagation time between links.

Without the remapping of neuron identities as in Fig. 2 synfire chain sequences are not visible in rasters. Figure 3*A* shows a 5-s segment of the activity of 150 excitatory neurons randomly selected (and not remapped according to chain membership) from a synfire simulation as described earlier (50,000 neurons, 50 fully connected synfire chains independently stimulated by low-frequency Poisson trains); the raster has no particular structure. The raster from a larger set of 10,000 excitatory neurons over the same time range is shown in Fig. 3*B*. Again, since the neurons are chosen randomly as they would be in a real recording, there are no structures like those in Fig. 2*B* that are directly ascribable to activations of individual synfire chains. However, there is an obvious broader vertical striping that connotes a coordinated fluctuation in the population activity; this striping becomes increasingly visible when the raster display shows a large number of neurons (contrast Fig. 3, *A* vs. *B*). It turns out that these coordinated rate fluctuations are the result of fluctuations in the number of synfire chains that have been activated at any time. Individual neurons, however, experience no rate increase. Figure 4 juxtaposes *1*) the times of chain stimulations with *2*) their (low-pass) temporal representation (i.e., the number of chains activated at any time) and *3*) the fluctuations of total population rate. These are clearly strongly, although not completely, related. We note that raster displays of simulation data from random nets, which contain no explicit synfire chains, also exhibit similar vertical striping (i.e., population rate fluctuations), but presumably caused by a different mechanism (Kriener et al. 2008).

Analysis method

The basic data source is a large file specifying the firing times of each of many neurons. The exact format is a matter of laboratory preference and irrelevant here. For the work in this study networks were simulated with a temporal precision of 0.1 ms, the observed number of neurons $n = 50{,}000$, and the total observed time $T = 100$ s. The first step in the analysis is to define a bin size for time based on the expected interlink propagation delay (3 ms in Figs. 6–8 and 11–13; 2 ms in Fig. 10*C* for greater clarity). The number of bins is $K = T/\text{bin size}$, so that a bin is specified by the index i lying between 1 and K. For each such bin along the entire length of the data we find the set $S(i)$ of all the neurons that fired in that (3-ms) interval. For large networks the $S(i)$ can be of the order of ≥ 100. The size of each $S(i)$ defines the population firing rate at the time of that bin. Figure 4 illustrates a smoothed version of $S(i)$ size over a selected epoch.

The fundamental calculation we carry out is to compare the net activity at time bins i and j. This can be done by two conceptually different approaches. The first uses the $S(i)$ sets and calculates the size of the intersection of $S(i)$ and $S(j)$—i.e., the number of neuron identifications that appear in both sets. The appropriate normalization here is the size of the lesser of the two sets, thus producing a number between 0 and 1. With these definitions

$$M_S(i, j) = |S(i) \cap S(j)|/\min(|S(i)|, |S(j)|) \qquad (1)$$

The second, more traditional approach, defines a 0/1 vector $V(i)$ of length N at each bin i, where the entries of 1 indicate the identities of neurons active at that time. Now we take the dot product of the vectors and normalize by the product of their magnitudes (norms). This defines the cosine of the angle between the two vectors $V(i)$ and $V(j)$ and is a frequently used measure of similarity (or difference) between the vectors. Thus

$$M_V(i, j) = V(i) \cdot V(j)/\sqrt{|V(i)|^2|V(j)|^2} \qquad (2)$$

Note that the numerators of the two expressions are numerically precisely the same (for 0/1 vectors), but the normalizations make M_S generally larger and more nonlinear with respect to the number of neuron identity overlaps (higher contrast) than M_V if the two sets $S(i)$ and $S(j)$ differ in size. Clearly $M(i, i) = 1$ and $M(j, i) = M(i, j)$.

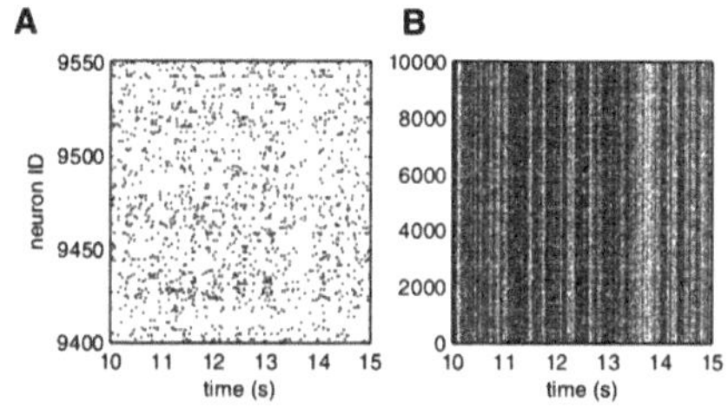

FIG. 3. Fluctuations in network activity. *A*: a raster of 150 randomly selected neurons (ordinate) during a period of 5 s (abscissa). The activity of the neurons appears uncorrelated, irregular, and fairly stationary in time. *B*: a raster of 10,000 neurons during the same time period as in *A*. The activity exhibits prominent vertical bands.

Therefore it suffices to inspect only the triangular matrix $i \leq j$. With the stated bin structure, i and j are both discrete measures of time; we decided to locate the origin at the lower left corner for visualization as in the traditional two-dimensional correlation displays like the JPSTH (joint peristimulus time histogram; Aertsen et al. 1989). Here the diagonal of $i = j$ extends from the lower left to the upper right and we are inspecting the lower triangular part of the intersection matrix. The programs have been designed to allow examination of any time ranges for i and j with $i < j$, allowing details in any portion of the triangular matrix to be visualized. For massively parallel recorded data with large N the set approach is computationally less demanding and was used throughout this study. With either approach we plot $M(i, j)$ as a matrix, the color or gray level indicating the normalized activity intersection (or vector similarity) at the two time bins.

The measurement defined earlier for spike trains is a version of the "recurrence plot" of the mathematical literature (see Marwan et al. 2007 for a review). This was originally developed for following the states of a dynamical system and detecting approximately repeating states of the system. The formal definition of a recurrence plot involves a Heaviside function at a threshold value, so that matrix entries are limited to 0 or 1, and the emphasis is on detecting occurrence of states at times i and j that are more similar than the selected threshold value. In our application the state of the system is defined by the set of neurons active within a time window at time i, and we compute a matrix with continuous, not 0/1, values representing the degree to which the memberships of the active sets overlap at times i and j. Using adjustment of a gray or color scale we can emphasize either high or low values in the matrix.

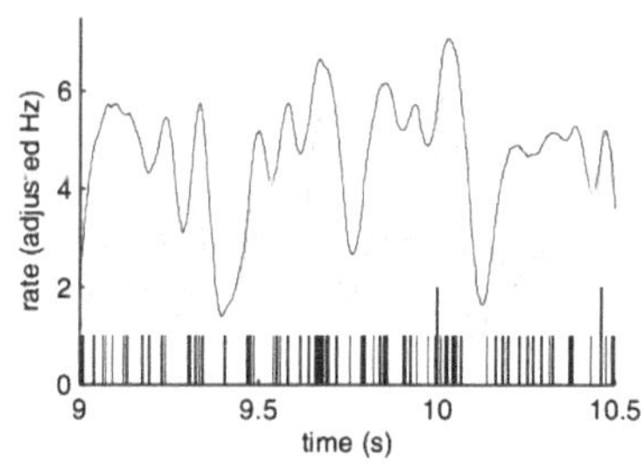

FIG. 4. Relation of population firing rate and synfire activation rate. The black curve represents the smoothed population activity (triangular filter, full width 51 ms); the gray curve represents smoothed stimulation times (same filter) of all synfire chains. The actual stimulation times are shown as vertical bars at the *bottom*.

Implementation and run-time comparisons

All analysis and display programs were implemented in Matlab 2007a (The MathWorks) on a 3.1-GHz laboratory PC with 2-GB memory running the Windows XP system. Using this implementation we compared run times of the traditional pattern program (Abeles and Gerstein 1988) and the intersection matrix method for detecting and identifying members of synfire chains.

For run-time comparisons additional test data sets were made by choosing the activity of a random 500 or 100 neurons from the full 50,000 neuron network with 50 embedded synfire chains subsequently analyzed in Fig. 7. Data durations were 100 or 50 s. Except for the smaller number of neurons these test data had general characteristics similar to those described in *Data overview*.

The pattern program was used with a clock resolution of 2 ms and a time step of 2 ms. Repeating patterns <100 ms were accepted. Since many patterns arise by chance, a minimum of 21 iterations were required (original and 20 surrogates) to calculate significance. The intersection matrix program was used with a clock resolution of 0.1 ms, a 3-ms time bin, and window width of 1.5 s so that 66 or 33 time-stepped iterations were required to analyze the full data of 50 or 100 s. Direct significance testing is not necessary since, as subsequently demonstrated in *Control* and *Signal/noise, detection sensitivity*, the short diagonal lines in the matrix that are the signature of synfire chains do not arise by chance.

Timings are given in Table 1. Note that the time given for the pattern program does not include the as yet unknown time for sorting, collating, and comparing significant patterns to detect the meta structure and neuron identities (IDs) of synfire chains. In contrast the intersection matrix analysis is not only faster but directly detects synfire activity and requires only a simple additional computation to determine the IDs of the neurons in any chain. More detail is given in the next section.

Identities of neurons in a synfire chain

The pattern programs sort spike patterns into groups according to complexity and repetition number and identify those groups whose count significantly exceeds corresponding counts found in surrogate data. All repeating patterns are consistent with an underlying synfire chain but could be generated by various other mechanisms. To resolve this issue it is necessary to examine various combinations of candidate patterns of various complexities and repetition number, searching for

TABLE 1. *Computing time comparisons*

N; T	Matrix, min	Pattern, h
50,000; 100 s	58	
500; 100 s	32	7.4
500; 50 s	16	1.9
100; 100 s	14	0.8
100; 50 s	7	0.2

Five different subsets of neurons from a single network simulation are analyzed. N, number of neurons in data; T, data duration (simulated seconds); analysis durations in minutes or hours; Matrix, intersection matrix method; Pattern, pattern search (Abeles and Gerstein 1988). Application of the pattern search to the largest data set (50,000 neurons) is computationally infeasible. For the pattern program the computing time changes as the square of data length and changes by about ×1.9 the multiplication factor of N, the number of neurons. For the matrix program the computing time is linear data length, for $N < 500$, changes at about ×0.46 the multiplication factor of N for $N < 500$ (and much slower for $N > 500$), and is linear in matrix width. (The individual matrix computation time increases as the square of matrix width, but then fewer iterations are needed to step along the entire data length.)

those that always interleave temporally in the same way (as would be expected from a synfire chain). Since there are very many candidate patterns (likely in the millions) this becomes a difficult combinatorial computation. We have not tried to write such programs as yet and can only guess at time estimates, but it is likely they would add considerably to the run time of the pattern programs.

By contrast, as shown in RESULTS, the intersection matrix programs directly identify the repeated runs of a synfire chain by short diagonal stripes, each of which represents a chain run repetition pair. A simple additional step is needed to actually identify the neurons that are involved in a given chain. Consider a chain that runs at times t_1, t_2, t_3, t_4 . . . in the data set, with analysis as shown, for example, in Fig. 6. The short diagonal stripes representing the various run combinations are arranged in a vertical and horizontal array as in that figure. Contributions to each pixel located in any particular diagonal stripe represent the intersection of both chain and nonchain (accidental) neuron IDs active at those two times.

To identify the chain neurons we intersect pixel by pixel the identities in the diagonal stripe for t_1, t_2 and the diagonal stripe for t_1, t_3, and then t_2, t_3, and so forth. The IDs of accidental firings at those times will disappear, in the multiple intersection process, because their activity is random with respect to the chain runs. The core IDs remaining in this process are precisely members of the chain. We have not yet written such programs, but their computational requirements are obviously small and would not impinge on the computation time comparisons of Table 1.

RESULTS

Synfire chain signature

TWO FIRINGS. We first consider the matrix signature of a single synfire chain that fires twice, once beginning at (binned) time t_1, once at (binned) time t_2, as shown in the cartoons of Fig. 5. If for the moment we simplify by taking the interlink propagation time exactly equal to the chosen bin width, at t_1 we observe a set $S(t_1)$ of neurons that includes all active members of the first link in the chain. The set $S(t_2)$ similarly includes all members of the first link in the chain that are active at t_2. Both sets can additionally include whatever other neurons in the net were active at their respective times. At $t_1 + 1$ and $t_2 + 1$ the two sets include all active members of the second link in each chain, and so on. Thus the matrix M defined earlier has high values (large intersection) at t_1, t_2; $t_1 + 1$, $t_2 + 1$; . . . $t_1 + l$, $t_2 + l$, where l is the chain length. A visualization of the matrix shows a short diagonal stripe of pixels with length l (~20 in our cases) starting at time t_1, t_2. If for some reason activity does not run through the entire set of l links in one of the two chain firings, the diagonal stripe will be shorter as determined by the link at which propagation fails. If the link-to-link propagation time is not equal to the selected bin width, the short diagonal stripes will show some irregularity, which takes the form of shorter diagonal segments with 1-pixel offsets. (The situation is analogous to drawing an arbitrarily inclined line on a pixeled computer screen.) Finally, if the link-to-link propagation time is not constant—so that the two activations of the chain differ in detailed timing structure—there may also be gaps or deviations from 45° along the short diagonal stripe. All these distortions will depend in part on the relationship between selected bin width and the noisiness of the interlink delays. Thus a repeated synfire chain activation always produces a short approximately diagonal stripe in the intersection matrix, but its detailed appearance will be affected by minor moiré-like effects. Examples are shown in Figs. 6 and 8*A*.

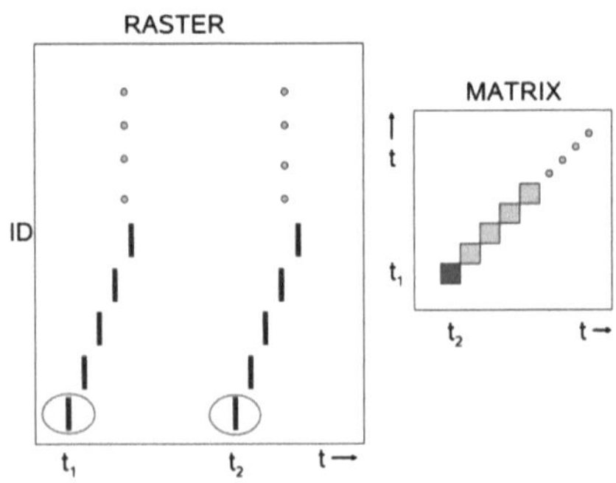

FIG. 5. A cartoon showing 2 runs of a particular synfire chain starting at t_1 and t_2. The identical active first links (*left*, circled) have a high value of intersection in the matrix at location (t_2, t_1) as shown (*right*). The subsequent repeated activity of successive links produces high intersection values along a 45° diagonal in the matrix as shown.

MANY FIRINGS. Now suppose our single synfire chain fires $k > 2$ times within our analysis time window. We now observe short diagonal stripes at $k \times (k - 1)/2$ locations; if the successive firings are numbered, these will be at (binned) times 1, 2; 1, 3 . . . 1, k; 2, 3; 2, 4 . . . 2, k; . . . $k - 2$, $k - 1$; $k - 2$, k; $k - 1$, k. This situation is shown in Fig. 6 for $k = 4$. Finally, if there are several active chains, with each producing its appropriate set of short diagonal stripes, we obtain Fig. 7, which analyzes an arbitrarily chosen 1.5 s of the 100-s data set. Figure 7 is calculated from a full data set, in effect observing all its 50,000 neurons; the result clearly detects the existence and activity of some number of the embedded synfire chains. Although we have not done so in Fig. 7, we could obviously identify and label all diagonal stripes that originate from the activity of a particular synfire chain. Note that besides the prominent short diagonal stripes, Fig. 7 also shows some faint, nonuniform background that will be discussed later. Typical details of the short diagonal stripes together with chain labeling are shown in Figs. 6 and 8*A*, which are enlargements of portions of Fig. 7.

Signal/noise, detection sensitivity

In the context of simulated data we may examine the performance limits of the M_S measure by reducing the number of observed neurons. The neurons are chosen randomly with no relation to the synfire chain structures. The analysis time window is constant for Fig. 8, *A–D*. When observing only 500 randomly chosen neurons out of the 40,000 excitatory network, the short diagonal stripes are somewhat noisier, so that not all elements of a particular stripe appear. However, comparing Fig. 8, *A* and *B*, most if not all stripes remain clearly visible. When only 200 randomly selected neurons are observed, we get Fig. 8*C*. The stripes are now much more noisy, with many omissions, and not all stripes can be seen. This situation probably represents the current limit of sensitivity for detection of possible synfire chains in a data set. Such numbers, how-

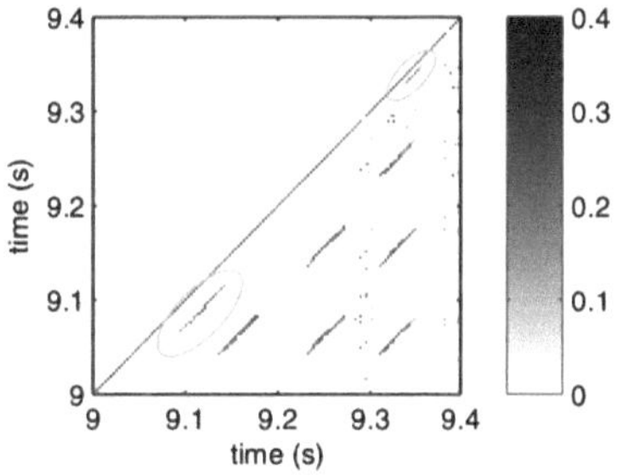

FIG. 6. With several repeated runs of a particular chain the matrix shows diagonal stripes starting at locations corresponding to each pair of repeat times; these are necessarily displaced both vertically and horizontally. The displayed time segment of 400 ms exhibits 6 diagonal stripes from 4 repeats of one particular chain (and also 2 diagonal stripes from one repeat each of 2 other chains indicated by ellipses). Stripes from different run repetition pairs are not identical because of the detailed variability of synfire chain runs as shown in Fig. 2*B*. The bar (*right*) indicates the mapping of activity intersection to gray values.

ever, are just becoming possible in experiments, both those carried out by multiple electrodes or wires and with optical methods using Ca^{2+} indicators.

When small numbers of neurons are observed, considerable visual enhancement for detecting the diagonal-stripe signature of synfire chain activity can be achieved by passing an appropriate short 45° diagonal filter (6–8 pixels along the diagonal, rectangular) over the noisy M matrix. When this is done for the data shown in Fig. 8*C* we get the matrix shown in Fig. 8*D*. All short diagonal stripes visible from the large data set (Fig. 8*A*) can now clearly be seen with the $N = 200$ analysis, so that detection of at least some synfire chains should be possible with even smaller N. Note that even with the noisy matrix of Fig. 8*C* the filtration has produced only a single spurious diagonal stripe (circled).

A more rigorous depiction of the signal/noise situation can be made by varying the number of observed neurons and comparing the average and SD of matrix values on and off a small diagonal stripe that had been detected with the full large data set. To have reasonable statistics we calculate over 40 disjoint sets of random choices of each number of neurons and in each iteration sum over 15 pixels along 45° both on and off the short diagonal stripe location. The results are shown in Fig. 9*A* and give a synfire chain detection threshold of about 50 observed neurons.

Another approach to the signal/noise separation does not depend on knowing about the location of short diagonal stripes, as mentioned earlier, and uses the entire matrix. Here we examine a histogram (probability density function [pdf]) of the pixel population as a function of filtered pixel value, comparing the results of both a 45 and a 135° narrow (6 pixel) filter. It is convenient to examine the results as a "survivor function," i.e., (1 − cdf), where cdf is the cumulative distribution function. For the data shown in Fig. 8, *B* or *C* ($N = 500$ or 200) the survivor functions of the two filtered matrices all show a steep drop at a filtered pixel value 0.166 on a scale of 0–1 (Fig. 9*C*). This drop is an appropriate threshold value for comparing the pixel population in the two filter situations but is data and filter shape dependent, and must be examined for each application of such a calculation. Values larger than this threshold correspond to "signal" (the short diagonal stripes or equivalent pixel values in the 135° filtered case) and values below that threshold correspond to "noise" events. Again calculating over 40 disjoint sets of random choices of neuron at each number of neurons the comparison of pixel counts with values >0.166 after the 45 and 135° filters is shown in Fig. 9*B*. Thus our second type of signal/noise analysis yields a threshold for detecting synfire chain activity at about 100 observed neurons. Note that the validity of this second analysis does depend on

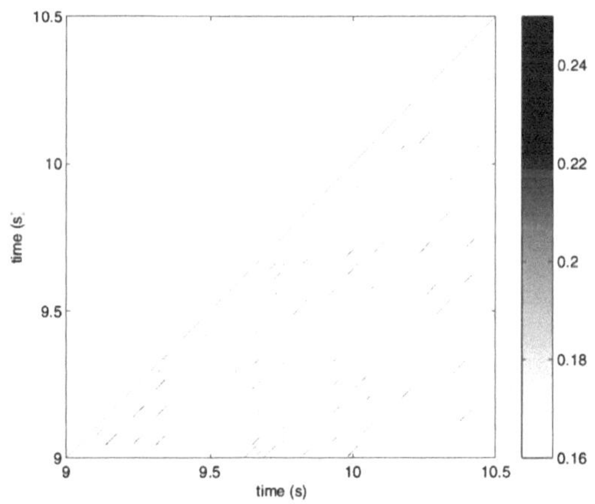

FIG. 7. Normalized M_S intersection matrix calculated with a time window of 1.5 s at (arbitrarily) 9 s of a 100-s simulation. The diagonal stripes correspond to different run repetition pairs of many different chains. Some of the stripes are much shorter; these result from a synfire chain run repetition where at least one of the 2 runs fails in the middle.

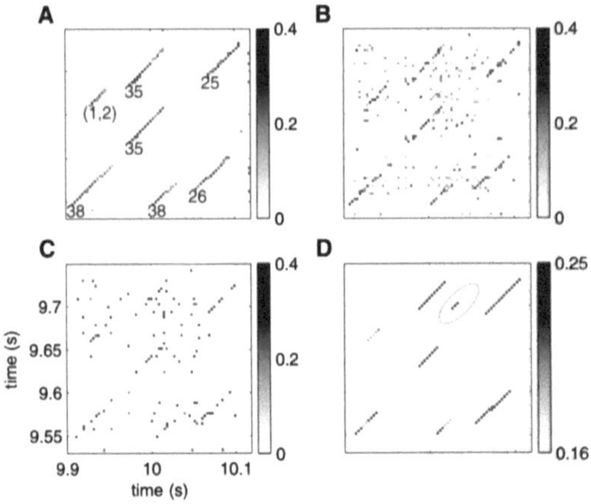

FIG. 8. Detail from the intersection matrix of Fig. 7 calculated with a time window of 220 ms. *A*: matrix using data from all 40,000 excitatory neurons in the simulation. Chain identities are indicated. The strip labeled (1, 2) results from a crossing of these 2 chains; see the text section on chain crossings. There are 2 stripes (run repetition pairs) for each of chains 35 and 38; this corresponds to the presence of 3 runs of each of these chains in the selected time range. *B*: the same matrix as in *A* but calculated using data from only an arbitrary 500 neurons chosen from the 40,000-neuron excitatory network. Although noisy, the diagonal stripes of *A* are still easily visible. *C*: the same matrix again but calculated using data from only an arbitrary 200 neurons. The diagonal stripes are now almost obscured by the noise. *D*: application of a sliding 6-pixel rectangular 45° filter to *C*. All diagonal stripes of *A* are reconstructed by this procedure. There is one spurious stripe (circled) that has emerged from the noise.

using a matrix window that contains 5–10 of short diagonal stripes as in Fig. 8.

In this study we have used data from simulations of fully connected synfire chains; i.e., each neuron in a given link receives input from all neurons of the preceding link but with other parameter values set so there were about 30% failures of complete chain propagation. If we increase the link width but keep the number of converging inputs to each neuron the same as before, we have a diluted chain with partial connectivity. The detection of chain activity in such a case will be easier than that in the fully connected case. During a run of this effectively wider chain the set of active neurons in each link will be larger and the set of intersections between any two chain activations will also be larger (the numerator of *Eq. 1*). The denominator of *Eq. 1*, however, includes the amount of ongoing background activity at that time bin and thus increases by a smaller fraction than does the numerator. Therefore the normalized matrix pixel value corresponding to a particular link activity repetition will increase. One could also use just the unnormalized pixel value (i.e., just the numerator of *Eq. 1*), although variability in firing may make this measure more noisy. Other parametric chain changes such as reduction of the converging input number would cause many more propagation failures that in turn would shorten many of the signature diagonal lines in the matrix and might make detection of synfire activity more difficult.

In the signal-to-noise analyses of Fig. 9 we repeatedly sampled (without replacement) various numbers of excitatory neurons from a total pool of 50,000. Each excitatory neuron in the pool was involved in several of the 50 synfire chains and a given chain comprised about 4% of the total pool. Consider dilution of the data so that we leave intact the pool size and the neurons in some smaller number of chains, but replace the rest of the data by similar rate neurons that do not participate in chain activity. The only result in the intersection analysis matrix (like Fig. 7) will be a diminution of the number of short diagonal stripes, but not their intensity or contrast. If on the other hand the dilution process also replaces neurons in the remaining chains, the detection of synfire activity becomes more difficult and depends on the fraction of pool size represented by each remaining (now only partially observable) synfire chain. With a dilution such that the observable part of a given synfire chain constitutes about 1% of the neuron pool, detection requires simultaneous observation of 200–500 neurons (not shown).

Intersecting synfire chains

In the case of intersecting synfire chains we might have some or all neurons in a given link of chain *A* also appear in a given link of chain *B*. In such a situation activity propagating along either chain could begin to propagate in the other chain, starting at the shared link. A cartoon of this situation is illustrated in Fig. 10*A*. Figure 10*B* is the associated raster with remapped identities showing activities of chain 1 in raster lines 1–2,000 and chain 2 in raster lines 2,001–4,000. Link 10 constitutes the same 100 neurons in both chains and it is clear that activity in chain 2 begins when link 10 of chain 1 is active (or vice versa if chain 2 is stimulated). In terms of the sets of active neurons in the appropriate time bins, this means after the 10th links are active that there will be an increase of active set size used as the denominator of *Eq. 1* (less than doubling since there is also background activity). There will also be a doubling of intersection value to a repetition of the (double) chain run (used as the numerator in *Eq. 1*). The increase in pixel value in the M_S matrix (*Eq. 1*) is clearly visible in the gray levels of Fig. 10*C*, which shows an enlarged and filtered image of the diagonal stripes corresponding to the three 2–1 chain-firing repetitions shown in Fig. 10*B*. For this purpose we could also have used the unnormalized matrix of intersection values, i.e., the numerator of *Eq. 1* alone; however, this may be more noisy if activities are variable between runs.

If instead link 10 of chain 1 were identical with link 5 of chain 2 we would expect a longer stripe with a profile along the stripe that increases and then returns to the single values. Thus

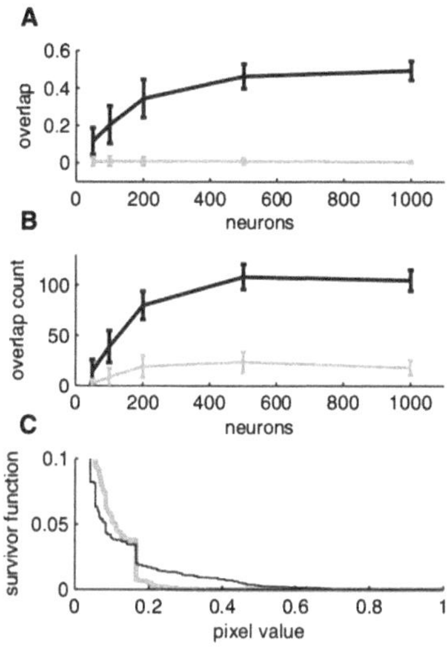

FIG. 9. Signal-to-noise analysis of the intersection matrix as a function of the number of neurons used in the calculation. *A*: sum of pixel values (ordinate) within a particular diagonal repetition stripe (black) compared with sum of values from an equivalent set of pixels off the stripe (gray) for an increasing number of neurons used in the calculation (abscissa). The range of ±SD (vertical bars) is obtained from 40 repetitions of the calculation, each with a different random choice of the included neurons. *B*: analysis of the full matrix; 45 and 135° filters (rectangular 6 pixels) are respectively passed over the matrix. The pixels with value >0.166 (vertical) are shown as a function of the number of neurons (abscissa) used in the calculation (45°, black; 135°, gray). SDs are computed as in *A*. *C*: pixel value histograms (as survivor functions) resulting from the 2 filter operations (gray levels as in *B*). The sharp break at value = 0.166 is used as a lower threshold for pixel values in making the counts in *B*.

within the limitations of count statistics it is possible in simple cases to infer details of intersecting synfire chains, using either normalized or unnormalized versions of the matrix.

More generally, detailed examination of the identities and numbers contributing to successive pixels of one or more short diagonal stripes could be used *1*) to determine the membership of a particular chain (described in more detail in the last section of METHODS) and *2*) to detect partial coupling between chains with resulting probabilistic ignition. Propagation failures would shorten the corresponding diagonal stripe and could be assigned to the responsible chain. However, general methods for appropriate comparisons of identity lists and for interpretation of such comparisons constitute a meta problem that remains to be addressed.

Another aspect of intersecting chains occurs in Fig. 8*A* in the short stripe labeled (1, 2). This stripe represents the neuron identity intersection of the second half of a full run of chain 2 with an earlier run of the second half of chain 2, which was caused by a full run of chain 1. Thus unlike most of the diagonal stripes, this does not represent a repetition of a directly stimulated synfire chain. Such a detailed explanation is possible here only because we know all about the data; it would not be possible in the case of experimental data.

Background and blobs

The small diagonal lines in the intersection matrix that are the signature of synfire chain activity have pixel values near 0.6 and the gray scale of the figures in this study has been chosen to emphasize these larger values. As shown in Fig. 11, a change of gray scale in Fig. 7 allows examination of the many small values in the intersection matrix; these take the form of a general nonuniform background haze. There are broad, diffuse vertical and horizontal bands with blobs of somewhat higher density at their intersections. With high temporal resolution (not shown) individual pixels are spaced and have roughly the same low values, whether in or out of a broad band or blob; it is the density of pixels that creates these features. Referring back to Figs. 3*B* and 4, we note that the fluctuations of overall population firing rates have a similar time structure. During periods of higher population rate the set of active neurons in each time bin is larger and, correspondingly, the probability of a small identity overlap between two close times within a rate surge is higher, thus yielding a higher density of small pixels in the matrix. There is little overlap of neuron identities in the events leading to different broad bands or different blobs. We have demonstrated earlier (Fig. 4) that much of the population rate fluctuation in these data is caused by overlap of synfire chain stimulations. Similar population rate fluctuations in randomly connected nets without any synfire structures remain to be investigated by the active population intersection methods used herein.

Controls

We use data from a number of simulated networks specified in detail in the following text. For each such data set the intersection matrix is featureless, with randomly distributed darker and lighter pixels. An example from a stimulated random network is shown in Fig. 12. For all other control cases the intersection matrices are similar and are therefore not shown.

1) The first data set consists of 40,000 independent inhomogeneous spike trains, with a gamma interval distribution of order 4, and a noisy rate profile matching the average population firing rates in the synfire simulations. Individual neuron firing rates are partially correlated so that the population firing rate is a noisy oscillation similar to that shown in Fig. 4.

2) The second data set consists of spike trains from a randomly connected network, each neuron with the same general parameters and degree of inward and outward connectivity as in the synfire networks. Random activity is without specific external stimulation.

3) The third data set consists of spike trains from the same randomly connected network but with additional independent Poisson stimulation of 50 arbitrarily chosen groups of 100 neurons, matching the stimulation conditions used in the synfire simulations (Fig. 12).

4) The fourth data set consists of spike trains from a network with connectivity for 50 synfire chains but without the external stimulation (Y in Fig. 1) that was used to start chain activity. The network had only its intrinsic "spontaneous" activity.

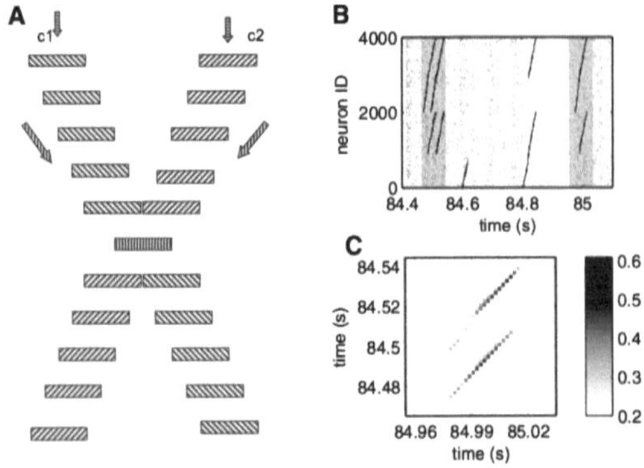

FIG. 10. Crossing synfire chains. *A*: cartoon showing 2 chains of 20 links each, which cross at link 10 (i.e., link 10 is common to both chains). *B*: remapped raster of 2 crossing chains. Chain 1 is mapped to neuron identities 1–2,000; chain 2 mapped to 2,001–4,000. There are 3 examples of chain 2 triggering the second half of chain 1, one example of chain 1 triggering the second half of chain 2, and one failed run of chain 1. The gray shadings denote the time frames compared in *C*. *C*: normalized intersection matrix for the 3 repetitions of the chain 2 triggering chain 1 runs in *B*. The second half of each diagonal pixel stripe has higher values than the first portion. For greater detail here a smaller pixel size is used (2 instead of 3 ms as in the rest of this report) with subsequent filtering by a 6-pixel 45° rectangle.

In an additional control procedure we have increasingly disorganized the orderly timing progression of activity along a synfire chain. This was done by dithering—i.e., randomly shifting within some time window the roughly synchronous firing of all neurons in each link of each chain. For several values of dithering time window Fig. 13 shows a raster display of a typical run of dithered chain data and also shows a portion of the corresponding intersection matrix. Note that this procedure leaves intact the original approximate synchrony of activity in each 100-neuron link of the chain, but destroys the systematic sequential timing between successive links. Because the sequential nature of the link firing timing is increasingly disorganized (Fig. 13, *top* to *bottom*) the corresponding short diagonal lines in the matrix blur and dissolve. With a dither time window of 50 ms, which exceeds the approximately 45-ms duration of a normal synfire sequence, the intersection matrix has no remaining structure and shows only a random scattering of high-value pixels.

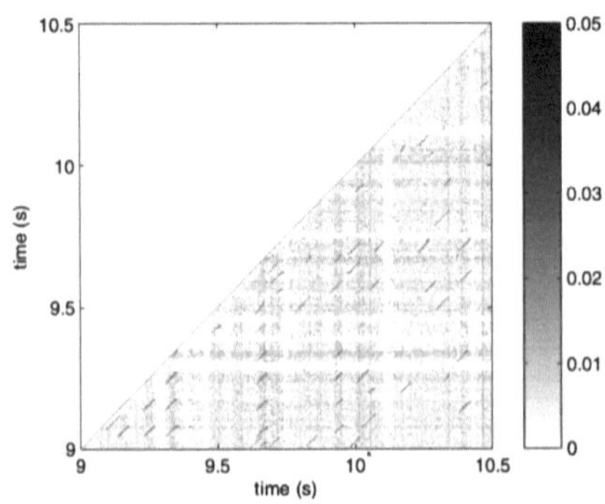

FIG. 11. Intersection matrix for the same data shown in Fig. 7 but with a gray scale that emphasizes the general background of bands and blobs of low-value pixels.

A final control calculation has to do with the possible consequences of bad spike sorting in multineuron recording. Spike-shape sorting is required when more than one neuron is recorded on a given (multi-) electrode. Although the literature is full of methods, sorting continues to be somewhat problematic and frequently results to some extent in multineuron rather than single-neuron spike trains (Harris et al. 2000). Spike train measures related to synchrony, auto- and cross-correlation, and time structure in general are considerably influenced by bad sorting (Gerstein 2000; Pazienti et al. 2006), although population measures are scarcely affected. It is therefore appropriate to examine the sensitivity of the intersection matrix to bad spike sorting. Data sets mimicking this situation were produced by overlaying sets of 5 or 10 randomly selected spike trains in the original 50,000 neuron synfire chain simulation file ana-

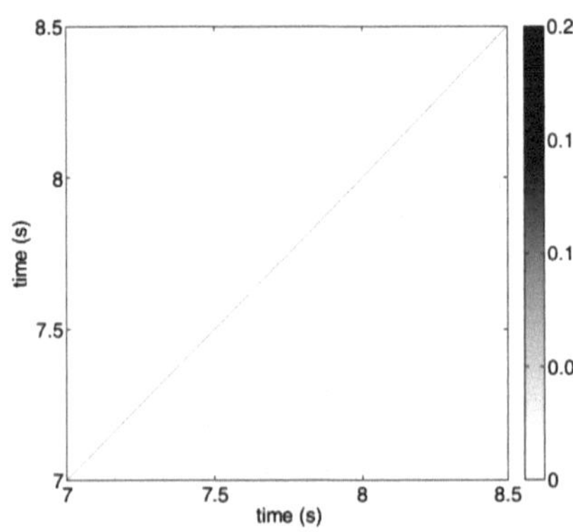

FIG. 12. Intersection matrix for data from control case (c) consisting of a randomly connected network with random groups of 100 neurons stimulated as for synfire chain activation. Other net parameters as for the synfire nets. There are high-value (black) pixels, but they are at random points in the matrix and do not form into any spatial configurations.

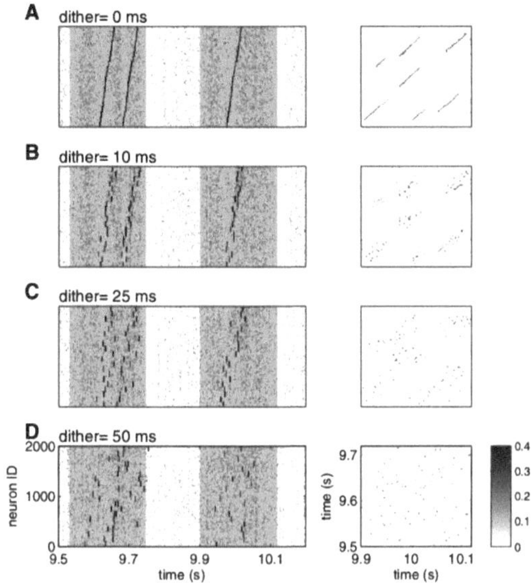

FIG. 13. Consequences of dithering the firing time of each link of a chain. This preserves the near synchrony of firing in each link but destroys the orderly sequence during a run of the chain and disorganizes the short diagonal lines in the intersection matrices. *Left column*: rasters; *right column*: corresponding intersection matrix. *Rows* from the *top*: dither windows (*A*) 0 ms (i.e., no dither), (*B*) 10 ms, (*C*) 25 ms, (*D*) 50 ms.

lyzed in Fig. 7, thus producing data sets with either 10,000 or 5,000 "mixed" spike trains. The resulting intersection matrices have somewhat larger values in the background gray pixels than in Fig. 7, but all the short diagonal lines continue to be dark and prominent. Thus even in this rather worst case of bad sorting scenario, the intersection matrix calculation is robust and continues to deliver the same information as in a clean, well-isolated neuron data situation.

These several control computations demonstrate clearly that the short diagonal structures seen in the intersection matrices of Figs. 6–8 and 11–13 do not arise accidentally from various other types of network activity and are, in fact, the unique signatures of the orderly and sequential activities of groups of neurons arranged in synfire chains.

DISCUSSION

In this study we have presented a new method related to "recurrence plots" for detection and visualization of synfire chain activity. The calculation examines the normalized intersection of identities for neurons firing at any two distinct (binned) times *i* and *j*. With a binning resolution of milliseconds, the rapid and orderly sequence of link activation in a synfire chain produces a signature of short diagonal lines in the intersection matrix $M(i, j)$.

We have shown that two related approaches are possible, one based on set intersections and one on vector representation of firing. Both types of measures are normalized to 0/1 but behave somewhat differently. The set intersection approach is computationally more convenient with very large numbers of neurons as in the simulated nets used herein. Compared with the vector approach, it also offers an improvement in contrast between high- and low-identity overlaps when the number of active neurons (set sizes) at times *i* and *j* differ. We have demonstrated that limited random samples of the order of 100 to 200 neurons from the large network are sufficient to detect synfire chains if they are present. This level of sensitivity should in the fairly near future allow experimental testing and either verification or rejection of the synfire idea. Current recording methods, electrode and optical, are allowing simultaneous observation of some 100 neurons in vivo, and rapid improvements can be expected.

We note that in the simulated data used here all excitatory neurons were involved in several synfire chains and that the neurons from any one synfire chain comprised about 4% of the total pool of observed neurons. Detection required a sample of about 100 neurons from the total pool. In experimental data from real nervous systems it may be that each synfire chain will constitute a smaller percentage of the observed neuron pool. This form of dilution would require larger random sampling numbers: if a single synfire chain constitutes only 1% of the total pool, it is necessary to sample 300–500 neurons for detection.

The vector representation of neuronal firing activity in some time window appears in the literature in a number of ways. It has been used in comparing simultaneously or sequentially recorded responses from several neurons to different and/or repeating stimuli (DiLorenzo 1989; Georgopoulos et al. 1983; Schild 1988). In these studies the response of each of the *N* neurons in the observed population is characterized by a spike count over some appropriate single time window for each repetition of the stimulus. Clustering in the N-dimensional population response space is used to predict the stimulus parameters from the neural activity. More detailed "activity vectors" (the time pattern of spikes after a stimulus) have been used to examine variability of single-unit responses to repeated stimulus presentations (Sasaki et al. 2007; Schreiber et al. 2003). Vector representation in the sense used here (an N length 0/1 vector representing data with millisecond time resolution) also underlies unitary-event analysis (Grün et al. 2002; Riehle et al. 1997). In that work such vectors are compared with what is expected from the firing rates of the individual neurons. Comparison of such vectors at two different times as in the present study (but with very low time resolution >1 s and, consequently, vectors that may not be 0/1) has been used to follow the slowly varying state of an observed neuronal network both by a matrix representation and with a principal components analysis (Sasaki et al. 2007). Although computationally similar in comparing vectors (or active neuron sets) at two times, in the present study we have used a time resolution of 3 ms, which is sufficiently small to make vectors that are always 0/1. Such time resolution is essential to detect the firing of synfire chains with the expected timescale of their ordered activity. A similar display has independently been developed in the field of music processing (Goto 2006).

With a time-bin resolution at the value of interlink delays expected in a synfire chain, the short diagonal stripes appearing in the intersection matrix each represent a particular repetition of activation of a particular chain. Repeated activity of a particular chain therefore appears as multiple short diagonal stripes arranged in rows and columns corresponding to the chain activation times. Different chains form separate series of short diagonal stripes.

Examination of the details in a particular short diagonal stripe yields considerable information about the synfire chain properties. For example, if in one repetition the activity of a synfire chain dies out before running through all links, all signature diagonal stripes involving that repetition will be shorter than the full chain length. If there are partially shared links between two chains this will be evident in the intersection identities and by modulation of the intersection numbers along the signature short diagonal stripe. Partial or full ignition of one chain by another can be extracted from the intersection identities. Link propagation times and their variation can be estimated from the way the signature diagonal is made up of short offset diagonal line segments.

The precision of synchronization in a synfire chain link (temporal jitter in spike coincidences) and the delay between the activation of successive neuron links are two distinct biophysically determined parameters in the simulation. Synchronization is mainly determined by the rise time of the postsynaptic potential (Gödeke and Diesmann 2008), whereas interlink delay is determined by conduction time or synaptic delay. Such parametric properties and issues will be examined in a subsequent study about the network simulation. In the present application, the bin size of our analysis needs to be adjusted to a compromise between the effects of these two parameters. When the synaptic delay is larger than the precision of synchronization and varies between synfire links, the signature of repeated synfire activity in the matrix remains a dark stripe at 45°. However, in this situation, the stripe may contain holes with values of background activity or may consist of several contiguous shorter diagonal stripes offset by one bin. For this reason processing the matrix with a 45° diagonal filter generally enhances the detection of synfire activity. Fluctuations in background activity may also scatter the matrix elements constituting the stripe away from the diagonal; a 45° diagonal filter extending a few bins perpendicular to the orientation of the diagonal may then be useful.

One may wonder why the simple analysis presented here may perform better than the elaborate statistical tools previously developed to detect the presence of individual spatiotemporal spike patterns. A conceptual difference is that our method automatically solves the puzzle of which patterns to consider as subpatterns of a larger one. We directly obtain a prominent diagonal matrix structure if synfire activity is present. Previously developed methods applied to such data detect many individual repeating patterns of neuron firings that do mostly originate from individual chains. However, such detected patterns must then be combined to determine the overall pattern of synfire activity, a computationally daunting project in a blind situation of real rather than simulated data. By intersecting identities of neurons firing at two different times our method integrates over the space of recorded neurons. By visual or filtered detection of diagonal structures that signify synfire activity, the method integrates over the temporal domain. This spatiotemporal integration provides sensitivity and robustness to the method.

ACKNOWLEDGMENTS

We thank M. Abeles for urging development of a sensitivity test based on the full intersection matrix. Major parts of this work were conducted during a sequence of three scientific stays: August 2005, S. Grün and M. Diesmann with G. Gerstein at Jay, New York; September to December 2006, G. Gerstein with S. Schrader and M. Diesmann at Bernstein Center for Computational Neuroscience, Freiburg, Germany; and March 2007, G. Gerstein with S. Grün and M. Diesmann at RIKEN Brain Science Institute, Saitama, Japan.

GRANTS

This work was partially supported by German-Israeli Project Cooperation Grant F1.2, European Union Fast Analog Computing with Emergent Transient States Grant 15879, and German Federal Ministry of Education and Research Grants 01GQ0413 and 01GQ0420 to Bernstein Center for Computational Neuroscience Berlin and Freiburg.

REFERENCES

Abeles M. *Local Cortical Circuits: An Electrophysiological Study*. Berlin: Springer-Verlag, 1982.

Abeles M. *Corticonics: Neural Circuits of the Cerebral Cortex*. Cambridge, UK: Cambridge Univ. Press, 1991.

Abeles M, Bergman H, Margalit E, Vaadia E. Spatiotemporal firing patterns in the frontal cortex of behaving monkeys. *J Neurophysiol* 70: 1629–1638, 1993.

Abeles M, Gerstein GL. Detecting spatiotemporal firing patterns among simultaneously recorded single neurons. *J Neurophysiol* 60: 909–924, 1988.

Aertsen A, Gerstein GL, Habib M, Palm G. Dynamics of neuronal firing correlation: modulation of "effective connectivity." *J Neurophysiol* 61: 900–917, 1989.

Amit D, Brunel N. Model of global spontaneous activity and local structured activity during delay periods in the cerebral cortex. *Cereb Cortex* 7: 237–252, 1997.

Aviel Y, Mehring C, Abeles M, Horn D. On embedding synfire chains in a balanced network. *Neural Comput* 15: 1321–1340, 2003.

Baker S, Lemon RN. Precise spatiotemporal repeating patterns in monkey primary and supplementary motor areas occur at chance level. *J Neurophysiol* 84: 1770–1780, 2000.

Baker S, Spinks R, Jackson A, Lemon RN. Synchronization in monkey motor cortex during a precision grip task. I. Task-dependent modulation in single-unit synchrony. *J Neurophysiol* 85: 869–885, 2001.

Bienenstock E. A model of neocortex. *Network Comput Neural Syst* 6: 179–224, 1995.

Braitenberg V, Schüz A. *Anatomy of the Cortex: Statistics and Geometry*. Berlin: Springer-Verlag, 1991.

Brunel N. Dynamics of sparsely connected networks of excitatory and inhibitory spiking neurons. *J Comput Neurosci* 8: 183–208, 2000.

Csicvari J, Henze DA, Jamieson B, Harris KD, Sirota A, Bartho P, Wise KD, Buzsáki G. Massively parallel recording of unit and local field potentials with silicon-based electrodes. *J Neurophysiol* 90: 1314–1323, 2003.

Diesmann M, Gewaltig MO, Aertsen A. Stable propagation of synchronous spiking in cortical neural networks. *Nature* 402: 529–533, 1999.

DiLorenzo PM. Across unit patterns in the neural response to taste: vector space analysis. *J Neurophysiol* 62: 823–833, 1989.

Donoghue J. Connecting cortex to machines: recent advances. *Nat Neurosci* 5: 1085–1989, 2002.

Douglas RJ, Martin KAC. Neural circuits of the neocortex. *Annu Rev Neurosci* 27: 1419–1451, 2004.

Euston DR, Tatsuno M, McNaughton BL. Fast forward playback of recent memory sequences in prefrontal cortex during sleep. *Science* 138: 1147–1150, 2007.

Georgopoulos AP, Caminiti R, Kalaska JF, Massey JT. Spatial coding of movement: a hypothesis concerning the coding of movement direction by motor cortical populations. *Exp Brain Res Suppl* 7: 327–336, 1983.

Gerstein GL. Cross-correlation measures of unresolved multi-neuron recordings. *J Neurosci Methods* 100: 41–51, 2000.

Gerstein GL. Searching for significance in spatio-temporal firing patterns. *Acta Neurobiol Exp (Warsz)* 2: 203–207, 2004.

Gewaltig MO, Diesmann M. NEST (Neural Simulation Tool) [Online]. *Scholarpedia* 2: 1430, 2008.

Goedeke S, Diesmann M. The mechanism of synchronization in feed-forward neuronal networks. *New J Phys* 10: 015007, 2008.

Goto M. A chorus section detection method for musical audio signals and its application to a music listening station. *IEEE Trans Audio Speech Language Processing* 14: 1783–1794, 2006.

Griffith J, Horn G. Functional coupling between cells in the visual cortex of unrestrained cat. *Nature* 199: 893–895, 1963.

Grün S, Diesmann M, Aertsen A. "Unitary events" in multiple single-neuron spiking activity. I. Detection and significance. *Neural Comput* 14: 43–80, 2002.

Harris KD, Henze DA, Csicsvari J, Hirase H, Buzsáki G. Accuracy of tetrode spike separation as determined by simultaneous intracellular and extracellular measurements. *J Neurophysiol* 84: 401–414, 2000.

Hayon G, Abeles M, Lehmann D. Modeling compositionality by dynamic binding of synfire chains. *J Comput Neurosci* 17: 179–201, 2004.

Hayon G, Abeles M, Lehmann D. A model for representing the dynamics of a system of synfire chains. *J Comput Neurosci* 18: 41–53, 2005.

Hebb D. *The Organization of Behavior: A Neuropsychological Theory*. New York: Wiley, 1949.

Ikegaya Y, Aaron G, Cossart R, Aronov D, Lampl I, Ferster D, Yuste R. Synfire chains and temporal songs: temporal modules of cortical activity. *Science* 304: 559–564, 2004.

Kohn A, Smith M. Stimulus dependence of neuronal correlations in primary visual cortex of the macaque. *J Neurosci* 25: 3661–3673, 2005.

Kriener B, Tetzlaff T, Diesmann M, Aertsen A, Rotter S. Correlations and population dynamics in cortical networks. *Neural Comput* 20: 2185–2226, 2008.

Marwan N, Romano MC, Thiel M, Kurths J. Recurrence plots for the analysis of complex systems. *Physics Reports* 438: 237–329, 2007.

Mehring C, Hehl U, Kubo M, Diesmann M, Aertsen A. Activity dynamics and propagation of synchronous spiking in locally connected random networks. *Biol Cybern* 88: 395–408, 2003.

Morrison A, Straube S, Plesser HE, Diesmann M. Exact subthreshold integration with continuous spike times in discrete time neural network simulations. *Neural Comput* 19: 47–79, 2007.

Nicolelis M, Ribeiro S. Multielectrode recordings: the next steps. *Curr Opin Neurobiol* 12: 602–606, 2002.

Ohki K, Chung S, Ch'ng YH, Kara P, Reid RC. Functional imaging with cellular resolution reveals precise micro-architecture in visual cortex. *Nature* 433: 597–603, 2005.

Ohki K, Chung S, Kara P, Hübener M, Bonhoeffer T, Reid RC. Highly ordered arrangement of single neurons in orientation pinwheels. *Nature* 442: 925–928, 2006.

Pazienti A, Grün S. Robustness of the significance of spike correlation with respect to sorting errors. *J Comp Neurosci* 21: 329–342, 2006.

Petersen CCH, Sakmann B. The excitatory neuronal network of rat layer 4 barrel cortex. *J Neurosci* 20: 7579–7586, 2000.

Plesser HE, Eppler JM, Morrison A, Diesmann M, Gewaltig MO. Efficient parallel simulation of large-scale neuronal networks on clusters of multiprocessor computers. In: *Euro-Par 2007: Parallel Processing. Lecture Notes in Computer Science*, edited by Kermarrec A-M, Bougé L, Priol T. Berlin: Springer-Verlag, 2007, vol. 4641, p. 672–681.

Prut Y, Vaadia E, Bergman H, Haalman I, Slovin H, Abeles M. Spatiotemporal structure of cortical activity: properties and behavioral relevance. *J Neurophysiol* 79: 2857–2874, 1998.

Riehle A, Grün S, Diesmann M, Aertsen A. Spike synchronization and rate modulation differentially involved in motor cortical function. *Science* 278: 1950–1953, 1997.

Rolston JD, Wagenaar DA, Potter SM. Precisely timed spatiotemporal patterns of neural activity in dissociated cortical cultures. *Neuroscience* 18: 294–303, 2007.

Rotter S, Diesmann M. Exact digital simulation of time-invariant linear systems with applications to neuronal modelling. *Biol Cybern* 81: 381–402, 1999.

Sasaki T, Matsuki N, Ikegaya Y. Metastability of active CA3 networks. *J Neurosci* 27: 517–528, 2007.

Schild D. Principles of odor coding and a neural network for odor discrimination. *Biophys J* 54: 1001–1011, 1988.

Schrader S, Morrison A, Diesmann M. A composition machine for complex movements. *Proc 31st Göttingen Neurobiol Conf* TS18-1C, 2007. Available online at http://www.neuro.uni-goettingen.de/archiv/2007/pdf/Proceedings-Goettingen2007.pdf

Schreiber S, Fellous JM, Whitmer D, Tiesinga P, Sejnowski TJ. A new correlation-based measure of spike timing reliability. *Neurocomputing* 925–931, 2003.

Song S, Sjöström PJ, Reigl M, Nelson S, Chklovskii DB. Highly nonrandom features of synaptic connectivity in local cortical circuits. *PLoS Biol* 3: 0507–0519, 2005.

Tetzlaff T, Morrison A, Timme M, Diesmann M. Heterogeneity breaks global synchrony in large networks. *Proc 30th Göttingen Neurobiol Conf* 206B, 2005. Available online at http://www.neuroanatomie.uni-goettingen.de/neurobio_archiv/2005/pdf/Proceedings-Goettingen2005.pdf

Vogels T, Rajan K, Abbott LF. Neural network dynamics. *Annu Rev Neurosci* 28: 357–376, 2005.

Vreeswijk C, Sompolinsky H. Chaos in neuronal networks with balanced excitatory and inhibitory activity. *Science* 274: 1724–1726, 1996.

Warren DJ, Fernandez E, Normann RA. High-resolution two-dimensional spatial mapping of cat striate cortex using a 100-microelectrode array. *Neuroscience* 105: 19–31, 2001.

Yosimura Y, Dantzker JL, Callaway EM. Excitatory cortical neurons form fine-scale functional networks. *Nature* 433: 868–873, 2005.

Journal of Neuroscience Methods 206 (2012) 54–64

Contents lists available at SciVerse ScienceDirect

Journal of Neuroscience Methods

journal homepage: www.elsevier.com/locate/jneumeth

Basic Neuroscience

Detecting synfire chains in parallel spike data

George L. Gerstein[a,*], Elizabeth R. Williams[b], Markus Diesmann[c,d], Sonja Grün[d,e,f], Chris Trengove[d]

[a] Department of Neuroscience, University of Pennsylvania, Philadelphia, PA, USA
[b] Institute of Neuroscience, Newcastle University, Newcastle upon Tyne, UK
[c] Institute of Neuroscience and Medicine (INM-6), Research Center Jülich, Jülich, Germany
[d] RIKEN Computational Science Research Program, Wako City, Saitama, Japan
[e] Theoretical Systems Neurobiology, RWTH University, Aachen, Germany
[f] RIKEN Brain Science Institute, Wako City, Saitama, Japan

ARTICLE INFO

Article history:
Received 22 September 2011
Received in revised form 1 February 2012
Accepted 3 February 2012

Keywords:
Multiple spike trains
Synfire chains
Detection and analysis algorithms
Repeating spike patterns
Active identity intersection matrix

ABSTRACT

The synfire chain model of brain organization has received much theoretical attention since its introduction (Abeles, 1982, 1991). However there has been no convincing experimental demonstration of synfire chains due partly to limitations of recording technology but also due to lack of appropriate analytic methods for large scale recordings of parallel spike trains. We have previously published one such method based on intersection of the neural populations active at two different times (Schrader et al., 2008). In the present paper we extend this analysis to deal with higher firing rates and noise levels, and develop two additional tools based on properties of repeating firing patterns. All three measures show characteristic signatures if synfire chains underlie the recorded data. However we demonstrate that the detection of repeating firing patterns alone (as used in several papers) is not enough to infer the presence of synfire chains. Positive results from all three measures are needed.

1. Introduction

Ever since Hebb's (1949) book there is general agreement that neurons do not act alone and that they join dynamically into functional assemblies both for computation and representation. Despite vast amounts of information about anatomy and connectivity in the real nervous system, the formation of such assemblies, their function and their interaction remain obscure. One interesting theoretical suggestion for the organization of assemblies is the "synfire chain" as proposed by Abeles (1982, 1991). In its original and simplest form the synfire chain of length l consists of l sub-assemblies or links of w neurons each (the synfire chain width), with full unidirectional excitatory connectivity between successive links. Such a chain structure can propagate near-synchronous activity in the first or input link to successive links like a row of dominoes. With suitable conditions the synchrony of firing in a link increases as the activity passes down the chain.

Theoretical studies have defined detailed properties of various versions of such systems. Stability properties and temporal behavior have been examined as well as structural variations such as partial link to link connectivity or feedback when a given neuron occurs in more than one group (Aertsen et al., 1996; Diesmann et al., 1999; Mehring et al., 2003; Aviel et al., 2003; Goedeke and Diesmann, 2008; Gewaltig et al., 2001; Tetzlaff et al., 2005, 2009; Trengove et al., 2010b, 2012). There also have been studies of partially interconnected or overlapping synfire chains to define the conditions in which activity can be transmitted between chains (Hayon et al., 2004, 2005). Further theoretical studies have shown that coupled synfire chain structures can produce compositionality, the hierarchical representation of complex entities in terms of parts and their relations (Bienenstock, 1995; Hayon et al., 2004; Schrader et al., 2007, 2011).

As yet there has been little search for and no convincing demonstration of synfire chains in real nervous systems. Such searches require new analysis methods for multi-neuron spike train recordings and identification of the corresponding synfire chain signatures. One appropriate method is our previously described algorithm (Schrader et al., 2008) that directly detects and shows many details about synfire chains if they are present in the recorded data. The computation is based on intersecting the identities of neurons active within any one time window (like 3 ms) with those at all other time windows. The resulting matrix $M(\mathbf{i},\mathbf{j})$ of intersection values at times $\mathbf{i},\mathbf{j}$ shows the orderly progression of firing in a synfire chain run as a short diagonal stripe of high value pixels. Details, controls and some sensitivity studies are in Schrader et al. (2008). With low background activity, that work showed that if a synfire chain comprises about 5% of the large (40k) neuronal net, detection would require a random sample of about 100–200

* Corresponding author at: Department of Neuroscience, 215 Stemmler Hall, University of Pennsylvania, Philadelphia, PA 19104, USA. Tel.: +1 215 898 8752; fax: +1 215 573 0833.
E-mail address: george@mulab.physiol.upenn.edu (G.L. Gerstein).

0165-0270 © 2012 Elsevier B.V. Open access under CC BY-NC-ND license.
doi:10.1016/j.jneumeth.2012.02.003

neurons. One purpose of the present paper is to extend the intersection algorithm so as to achieve comparable sensitivity in the presence of higher background activity.

A second purpose of this paper is to examine the use of repeating spatio-temporal patterns as a signature of synfire chains. Existing pattern programs (Abeles and Gerstein, 1988; Abeles and Gat, 2001; Gerstein, 2004; Shmiel et al., 2006; Patnaik et al., 2008; Louis et al., 2010a) sort spike patterns whose duration is less than some criterion into groups according to pattern complexity and repetition number and identify those groups whose count significantly exceeds corresponding counts found in surrogate data (Gerstein, 2004; Louis et al., 2010b) or by using conditional probabilities (Sastry and Unnikrishnan, 2010). Such repeating patterns have been suggested as evidence for synfire chains (Abeles et al., 1993; Oram et al., 1999; Ikegaya et al., 2004; Shmiel et al., 2005; Luczak et al., 2007), and indeed synfire chains will generate many repeating patterns. However other mechanisms can be envisioned also as demonstrated toward the end of this paper. To uniquely identify the presence of synfire chains from repeating patterns requires further demonstration that the various repeating patterns have the interleaving temporal structures expected in synfire chain activity. Using data from simulated nets with embedded synfire chains we describe tools for this purpose when parallel recordings from some hundreds of neurons are available.

2. Test data generation and description

Much of the test data for the spike train analyses described in this paper were originally generated by a large scale simulation program which contained 40 thousand excitatory and 10 thousand inhibitory point neurons arranged into 50 synfire chains each 100 wide and 20 long. Each neuron participated in an average of 2.5 different chains (Schrader et al., 2008). For some of the present purposes we edited such spike train data so that the identities and connectivity of only one synfire chain remained while all other neurons were replaced by independent spike trains with gamma 4 interval distribution and either high (8 Hz) or low (1.7 Hz) firing rates as an unstructured background. Individual firing rates were modulated to mimic the average rate fluctuations of the original data file. Further editing in some cases involved random choice of smaller populations out of these single chain data sets. Since neuron identities had been randomly assigned, such smaller populations always contained a sample of the chain neurons.

Additional test data for parts of Figs. 6 and 7 came from two different simulations with 80 thousand excitatory and 20 thousand inhibitory point neurons. In the "111" data set there were 379 chains each 130 wide and (on average) 100 long. Each neuron participated in 62 chains. In addition each chain was connected to two "successor" chains in which it could probabilistically initiate activity, thus forming a large, complicated and probabilistic tree in which the activity could propagate. In effect this produced activity in synfire chains of varying length depending on when (probabilistically) a particular activation would die out (Trengove et al., 2010b). In the "558959" data set there were 914 chains each 148 wide and 32 long. Each neuron participated in 52 chains. All but one of these chains received 3 stimulations during the run. A single selected chain received 240 stimulations, in effect making it the only active chain in the data. Background rates were 10 Hz. This is a variant of a recent model demonstrating large-scale embedding of synfire chains (Trengove et al., 2010a, 2012).

Test data for Fig. 8 came from a simulation of a single chain with 20 links of 100 neurons. The data were edited so that the near-synchronous (within 3 ms) firing of each link was maintained, but so that the activity each group was independently dithered by ±100 ms relative to the orderly sequence of the original synfire activity. An additional 4000 background neurons fired with gamma 4 interval distributions and rates of 1.7 Hz.

3. Signal to noise and firing rate in the intersection algorithm

We have previously described a sensitive algorithm for detecting synfire activity if present in a multi-neuron spike train recording (Schrader et al., 2008). The basic computation detects repetitive activations of a synfire chain, by searching for repeated sequences of synchronous activities. In comparing any two runs of a particular synfire chain similar sets of neurons from link 1, then link 2, etc. would be active. We search for repeats of such activity by first binning the data on a scale that corresponds to the expected inter-link delay (usually 3 ms), and list the identities of all neurons whose spikes fall into a given time bin along the data; call the list $\boldsymbol{S(i)}$ at time bin $\boldsymbol{i}$. We then compare the activities at time bins $\boldsymbol{i}$ and $\boldsymbol{j}$ by calculating the size of the intersection of $\boldsymbol{S(i)}$ and $\boldsymbol{S(j)}$, i.e. the number of neuron identifications that appear in both sets. After appropriate normalization the result is shown as a matrix, called the intersection matrix, with shading proportional to the normalized sizes of the $\boldsymbol{ij}$ list intersections. This is also essentially equivalent to treating each list as a (0,1) vector $\boldsymbol{V(i)}$ at time bin $\boldsymbol{i}$ and computing the angles within all possible vector pairs. The normalizations in both formulations are essential to deal with possible rate variations along the data.

The matrix entry at times $\boldsymbol{ij}$ with appropriate normalizations can therefore be formally expressed in two ways:

$$M_S(i,j) = \frac{|\boldsymbol{S(i)} \cap \boldsymbol{S(j)}|}{\min(|\boldsymbol{S(i)}|, |\boldsymbol{S(j)}|)} \tag{1}$$

$$M_V(i,j) = \frac{\boldsymbol{V(i)} \cdot \boldsymbol{V(j)}}{\sqrt{|\boldsymbol{V(i)}|^2 |\boldsymbol{V(j)}|^2}} \tag{2}$$

Each pair of runs of any particular synfire chain involves a repetition of an ordered sequence of identity lists, and therefore leads to a short diagonal stripe of high values in the intersection matrix. Length of the stripe corresponds to the number of links in the chain. Note that this result is independent of different time delays along the synfire chain as long as the two chain runs have the same time and identity properties. The short diagonal stripe might have moire-like variations of value if the original data binning does not quite correspond to the average inter-link delay in the synfire chain. If the two runs of the chain differ in timing values, the corresponding short diagonal stripe may shift among several adjacent diagonals and appear somewhat crooked. For N runs of the synfire chain there will be $N(N-1)/2$ run-pairs and therefore a systematic array of that number of short diagonal stripes in the intersection matrix. Detailed explanations and examples are given in Schrader et al. (2008).

In that paper we also examined signal to noise performance, i.e. the extent to which such short diagonal stripes stand out over the general background values in the matrix which result from the random (non-synfire) intersections of activity. We showed that the signal to noise increases with the ratio of sample to total population while its variability decreases. With a single synfire chain of 2000 neurons out of a total population of 50,000 neurons (i.e. 4%) and low background firing rates (1–2 Hz) reliable detection of the synfire activity requires observation of about a 100–200 neuron random sample.

Now consider two 50,000 neuron data sets with identical runs of one synfire chain (as above), but with different firing rates in all the non-synfire neurons. For pixels on one of the short diagonal stripes the numerator of Eq. (1) will differ little since most of the contribution to these list intersections will come from the synfire activity.

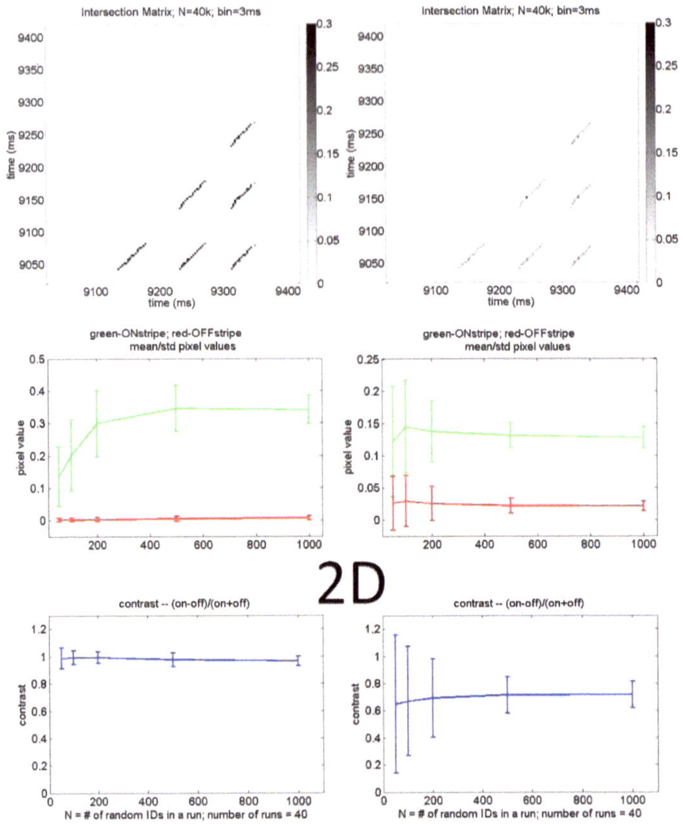

Fig. 1. Detection of synfire chain activity is harder with higher background firing rates. Two-dimensional analysis of simulation data that contains a single synfire chain and 38 thousand independent neurons with gamma 4 interval distributions. The left half data has independent neuron firing rate of 1.7 Hz, right half 8 Hz. Top panels are the intersection matrices for an arbitrarily chosen 0.4 s in the full data sets. The short diagonal stripes at the times of pairs of synfire runs are much fainter relative to background at the 8 Hz background rate. Middle panels are signal to noise analysis using total pixel value sum on or off a known short diagonal stripe. Each data point represents 40 random samples of the indicated size. On-stripe pixel values are lower and off-stripe values higher at the 8 Hz background rate. (Note that the *y*-axis calibration differs in the left and right middle panels.) Bottom panels are the corresponding contrast and show the reduction in contrast and increase in variability as the background rate increases.

But the length of lists in the denominator will depend directly on the background firing rates, and if this is high will reduce the pixel values on the short diagonal stripe. In addition, with higher background rates there will be more random intersections, and pixels *not* on the short diagonal stripes will have increased values. Both these effects of background rate are demonstrated in the matrices of the top two panels of Fig. 1, comparing background rates of 1.8 and 8 Hz for data with four closely spaced runs (and hence 6 run-pairs) of a synfire chain.

A more rigorous depiction of the signal/noise situation can be made by repeat calculations using varying numbers of randomly chosen neurons and comparing the average and standard deviation of intersection matrix values on and off a small diagonal stripe which had been detected with the full large data set. In order to have reasonable statistics we calculate over 40 disjoint random choices of each sample number of neurons and in each iteration calculate the average value over 15 pixels along 45 degrees both on and off the known short diagonal stripe location. Comparisons of results for 1.8 and 8 Hz background rates are in Fig. 1. The middle panels show mean pixel values on and off a (typical) short diagonal stripe (note the different ordinate scale at the different background rates), and the lower panels show the corresponding contrast. At the higher rate the mean pixel value on the short diagonal stripe is about 1/3 of that for the low rate, and the off-stripe values are about twice as high; the corresponding contrast drops from 1.0 to 0.6. Reliable detection of the synfire activity requires only 50 neurons with the low background rate, but 200 neurons at the high rate.

A simple way to improve the detection is to calculate intersections of activity lists at three instead of two times. The formal calculation is given by Eq. (3)

$$M_S(i,j,k) = \frac{\left|\boldsymbol{S}(\boldsymbol{i}) \cap \boldsymbol{S}(\boldsymbol{j}) \cap \boldsymbol{S}(\boldsymbol{k})\right|}{\min(|\boldsymbol{S}(\boldsymbol{i})|, |\boldsymbol{S}(\boldsymbol{j})|, |\boldsymbol{S}(\boldsymbol{k})|)} \tag{3}$$

Any three runs of a synfire chain will produce repeating sequences of identity lists, and hence a short diagonal stripe of high pixel values in the 3-dimensional matrix. For the same data set with 4 runs of one chain as used above, the top panels of Fig. 2 show the short diagonal stripes for the corresponding 4 run-triples in a 3-dimensional display. (Note that pixels in these displays do not have gray levels as in the 2-dimensional matrix display; all pixels above a threshold are shown at the same intensity.) The increased number of background pixels when data have higher background rate is clearly visible.

The quantitative analysis of signal to noise is in the lower panels of Fig. 2. As in the 2-dimensional analysis the pixel values on the short diagonal stripe are reduced at high background rates, but the off-stripe pixels remain low. As a result the contrast stays at 1.0 even at the high background rate, and the synfire chain detection level remains at samples of 50 neurons. Thus the 3-dimensional analysis is more sensitive than the 2-dimensional analysis in the face of higher background rates.

With the network and synfire chain parameters used above, approximately 5% of the neurons in each analyzed sample were members of the same chain. Thus inclusion of 3–5 chain neurons in a sample is needed for detection. Although not shown, if the data are diluted so that less than 5% of the overall network are chain neurons, detection requires a compensating increase in the sample size so that the necessary 5 or so chain neurons are included.

In the above study of sensitivity we compared pixel values on and off a short diagonal stripe whose location was determined from analyzing the full data set of 50,000 neurons (see the upper panels in Figs. 1 and 2). Obviously this is not applicable to experimental data limited by current technology to less than 200 neurons. However as shown previously (Schrader et al., 2008) the 2-dimensional short diagonal stripes become visually detectable (possibly with the aid of appropriate spatial filtering of the matrix) at around sample sizes of 100 neurons at low background firing rates. High background rates require larger samples of the population for synfire detection.

4. Repeating pattern spectrum richness with chains

We have previously described algorithms for finding all repeating temporal patterns among simultaneous recordings of multiple spike trains (Abeles and Gerstein, 1988). The basic idea is to represent the multiple observed spike trains as holes in an old time paper computer tape, different neuron identities across the tape, time along its length. Two copies of this tape are shifted in time (along the length) and regions where holes superimpose represent repeating spatial-temporal patterns. Our implementation extracts the longest repeating patterns that fall within a specified duration. It does not also count the repetitions of sub-patterns of the detected patterns as do the published methods based on bottom up data mining algorithms (Shmiel et al., 2005; Patnaik et al., 2008).

The results of our programs are presented as a pattern repetition spectrum as shown in Fig. 3. Here we have a bar plot of the logarithm of the number of detected patterns as a function of complexity and repetition number. Complexity is the number of spikes that are involved in a given pattern; this is not a measure of how many individual spike trains (neurons) are involved.

Examination of significance of counts in the pattern repetition spectrum can be based on multiple repetition of the spectrum computation with surrogate data. The generation of appropriate surrogates requires matching the original rate and inter-spike interval structures but destroying only patterning. Useful ways to do this depend on dithering of each individual spike, i.e. moving it some relatively small distance in time (Date et al., 1998; Hatsopoulos et al., 2003). However, additional details of the dither can produce better preservation of rate and interval. We have shown that a dither probability shape based on the square root of the joint interval histogram (JIH) of that spike train is very effective in preserving interval structure (Gerstein, 2004) although the rate structure is somewhat distorted. If there is a repeating underlying rate structure even better results can be obtained if the joint interval histogram is used in operational time (Louis et al., 2010b). Other types of significance test for patterns have been suggested using conditional probabilities (Sastry and Unnikrishnan, 2010). Some work using different surrogate generating methods (Baker and Lemon, 2000; Baker et al., 2001) concludes that repeating patterns in some experimental data occur at chance levels.

In Fig. 3A we show the pattern repetition spectrum for 500 independent spike trains, each with inter-spike interval distributions of gamma order 4 and with a rate of 1.7 Hz and slow rate modulation mimicking our original simulation with synfire chains. Note that all repeating patterns are clustered at low complexity and low repetitions. In Fig. 3B we show the pattern spectrum for 500 randomly chosen neurons from the original full data set with 50 synfire chains. There is now an extraordinary richness of patterns of high complexity and repetition. A similar sample in Fig. 3C from the data set that contains only one synfire chain of the 50 with the rest of the net independent, shows fewer but still very many patterns at high complexity and repetition. Finally in Fig. 3D we show the pattern spectrum from the same 50 chain data set as in Fig. 3B but dithered with a flat distribution in a ±5 ms window. With these data the pattern richness has reverted to clustering at low complexity and repetition similar to the result from independent neurons.

The comparison of Fig. 3B and D also demonstrates that the high pattern counts at complexity >3 and repetition >2 in 3B are highly significant, since a dither surrogate almost completely eliminates them. This result might depend on data length or rates, so that it is always wise to make the simple dither test; if met it is generally not necessary to go through the procedure of many repeated surrogate calculations to assess significance.

It is interesting to examine what neurons participate in patterns of any particular complexity and repetition. From the data used in Fig. 3B we arbitrarily chose the five patterns at complexity 8 and repetition 19. These patterns are analyzed in the five panels of the left two columns of Fig. 4. Time structure of each pattern in milliseconds is indicated above its panel; neuron identities are different for each pattern but are omitted for space reasons. Recall that these data contained 50 synfire chains. In each panel we show how many neurons of the analyzed pattern (indicated in the title of the panels) belong to a particular chain. Since in these data each neuron participated in 2–3 different synfire chains we expect that the neurons in each pattern would be members of various chains (the low bars). However for every pattern *all* 8 neurons participating in that pattern are members of one particular chain, different for each pattern (the high bar). The panels in the right two columns of Fig. 4 show the same analysis for the data used in Fig. 3C with a single synfire chain. Here we arbitrarily chose the 7 patterns of complexity 5 and repetition 11. For each pattern *all* 5 neurons were members of chain 35 (the only one in these data). Although not shown, when the detected pattern includes intervals <2 ms some of the neurons are members of the same link in the chain.

Similar analyses (not shown) of the same data for the neuron identities in patterns represented at the edges and near the

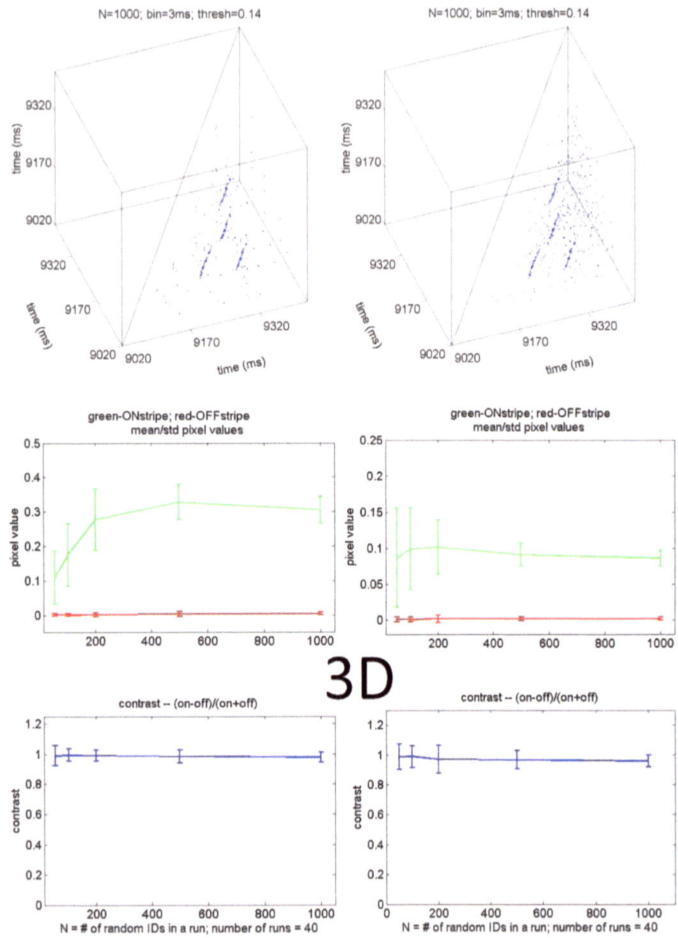

Fig. 2. Analysis of the same simulated data carried out with the 3-dimensional version shows that detection is less affected by background firing rate. Pixels in the 3-dimensional matrix indicate overlap of active neuron identities at three times. The short diagonal stripes remain strong at the higher background rate. The signal to noise analysis in the middle panels differs from the 2-dimensional results mainly in that the off-stripe pixel count remains low with the 8 Hz background rate. (Note that the y-axis calibration differs in the left and right middle panels.) The contrast remains at 1.0 even with the high background rate.

origin of the plane, i.e. with complexity 3 and/or repetition 2 (i.e. low complexities and repetition), would usually include or be entirely composed of non-synfire neurons. As in the examples above, the neuron identities in patterns towards the center of the plane (complexity >3, repetition >2) are entirely from the synfire chain.

These data had low spontaneous firing rates of the non-chain neurons. As will be shown below, increasing these firing rates results in patterns at complexity >3 and repetition >2 that also include non-chain neurons.

The total number of repeating patterns depends both on analysis parameters and on properties of the data. Relevant analysis parameters include maximum allowed pattern length, bin width or time resolution, and the random sample size drawn from the total population. Relevant data properties include data length, the percentage of synfire neurons in the

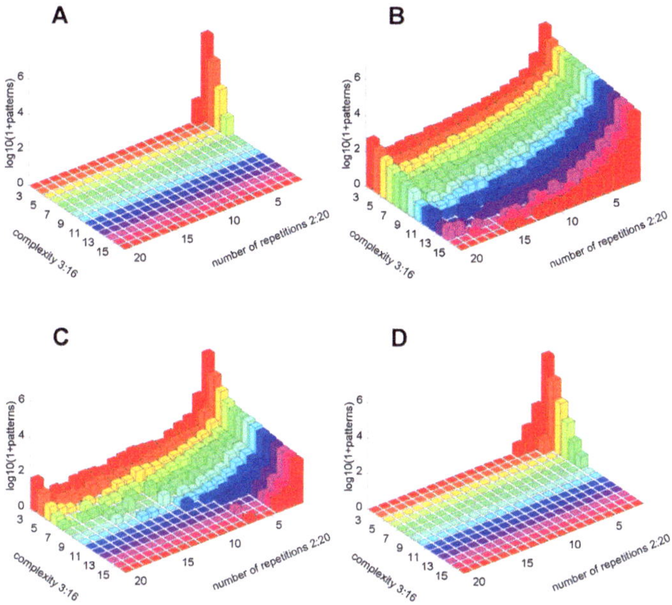

Fig. 3. Pattern repetition spectra for 500 randomly chosen neurons from each of four different simulations. Synfire chain data shows a rich set of patterns at high complexity and repetition. (A) Independent firing of 40 thousand neurons each with gamma 4 interval distribution and mean rate of 1.7 Hz. (B) 40 thousand neurons arranged as 50 synfire chains. Each neuron participated in an average of 3.5 chains. Chains were stimulated by independent Poisson sequences at a rate of 1 Hz. (C) One synfire chain and 38 thousand independently firing neurons as in (A). (D) The same data as in (B) but with each spike time dithered by ±5 ms.

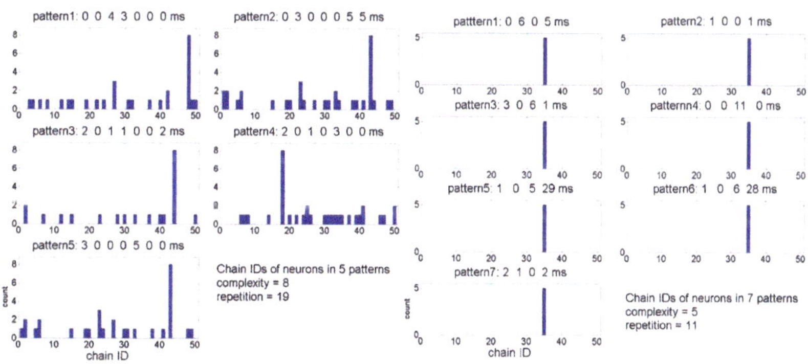

Fig. 4. Chain membership of neurons that participate in patterns arbitrarily chosen from the regions of complexity >3 and repetition >2 in Fig. 3. Each panel shows for one particular pattern how many neurons (*y*-axis) of the pattern are members of a particular synfire chain (*x*-axis: chain ID). Left panels: 5 different patterns of complexity 8 and repetition 19 from data with 50 synfire chains shown in Fig. 3B. The assignment of each neuron to several different chains produces the low bars; but in each panel all eight neurons come from some single chain (the high bar). Right panels: the same analysis for the 7 patterns of complexity 5 and repetition 11 from Fig. 3C. All five neurons in each different pattern come from the single synfire chain in the data.

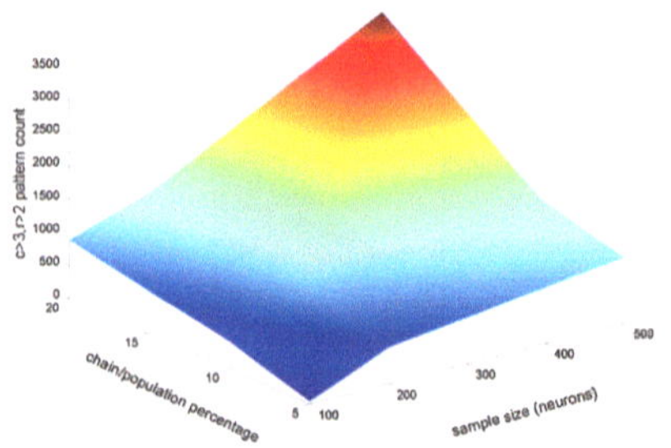

Fig. 5. Dependence of the total number of patterns for complexity >3 and repetition >2 (i.e. the center region of the complete pattern spectrum) on sample size (number of neurons) and the fraction of chain to non-chain neurons in the data.

total population (and in the random sample) and the properties of the non-synfire neurons (like rates, ISIs, temporal modulation, correlated firing, etc.). Using data sets similar to that used for Fig. 3C, Fig. 5 demonstrates the dependence of the total number of complexity >3, repetition >2 patterns on sample size and percentage of synfire neurons in the data set. Increase of either parameter produces a monotonic increase of the total number of patterns in the center part of the plane. Reciprocal changes of these parameters can keep the number of patterns constant.

Although not shown, we have examined pattern properties when the two parameters of rate and regularity are varied. Intuitively the number of patterns should increase with regularity, going to extremely high numbers for a clocklike set of neurons. For this investigation we used simulations with a single synfire chain as in Fig. 3C, and edited the properties of the 38,000 non-chain neurons. Regularity was specified by the order of the gamma function used to generate their inter-spike intervals. Higher gamma corresponds to a narrower ISI distribution and therefore higher regularity. For the non-chain neurons we varied gamma from 4 to 12, and rates from 1.7 Hz (same as in the single synfire chain) to 7 Hz.

Gamma order in this range produces smaller changes in total pattern counts or $c > 3$, $r > 2$ pattern counts than the variability produced by random selection of 500 neurons from the data. It is likely that much higher values of the gamma order would have a visible effect. By contrast increasing the rate of the non-chain neurons in the range examined increases the total pattern count by a factor of approximately 400, but cuts the $c > 3$, $r > 2$ counts in half. This last observation can be explained by recalling that at the original low rate of 1.7 Hz and gamma order 4 most of the $c > 3$, $r > 2$ patterns originate from the synfire chain as shown in Fig. 4. As the firing rate of the non-chain neurons is increased there is an increasing likelihood that additional (non-chain) spikes will appear in the existing synfire patterns and that this will occur repetitively. When this happens the assignment of the pattern moves on the plane to a higher complexity and a lower repetition (usually only $r = 2$). Thus, as also shown in Figs. 1 and 2, the effects of higher rates and/or very high regularity in the non-chain neurons tend to obscure the signature of synfire activity, and make its detection that much harder.

4.1. Pattern timing issues

In the above we have demonstrated that networks containing synfire chains produce a rich spectrum of patterns with high complexity and repetition. The converse, however, is not true. Some repeating patterns with complexity >3 and repetition >2 do occur randomly. Furthermore many such repeating patterns can be generated by other means than synfire chains. An example using approximately synchronous firing within each of several independent neural groups (cortical columns?) is given towards the end of the paper. Thus finding repeating patterns is by itself not enough to infer the presence of synfire chains. Additional measurement of the time structure of the repeating patterns is essential in order to make that inference. In order to develop these requirements we examine the detailed time structures of patterns observed in simulated nets with synfire chains as above.

For this investigation we have used four different large scale simulations (40,000 or 80,000 excitatory neurons) with one or more embedded synfire chains. (Parameter details for these simulations are in the section on data generation.) The pattern spectra of our four example simulations are shown in Fig. 6, with the left column for the simulations with many chains and the right column for the simulations with only a single chain. In each case a random sample of 100 neurons (repeated to check stability but repetitions not shown) was used to mimic the current limits of recording technology. All four spectra show many patterns at complexity >3 and repetition >2, i.e. the middle of the plane excluding the upper edge and right edge values. The effect of sample sizes on pattern counts is seen by comparing Fig. 3B with Fig. 6A and Fig. 3C with Fig. 6B and is explicitly shown in Fig. 5. The pattern richness is characteristic (but not unique) for synfire chains; simulations with only independent spike trains show spectra with patterns mainly at the edges and origin of the plane (see Fig. 3A).

For each such spectrum calculation we have examined temporal characteristics of all patterns of complexity >3 and repetition >2 by constructing (1) a raster of when each pattern (represented as a point) occurs in the data set; (2) each full pattern displayed as a raster and as a spike time histogram (like a PST histogram) in time lock (i.e. alignment) to the known synfire chain stimuli; (3) each full pattern displayed as a raster and as a spike time histogram both aligned on the first spike of the pattern.

The first of these measures, the overall raster, is useful to assess temporal heterogeneity, including for real data the possible modulation associated with behavior or the existence of recording problems. However this is of little use here since we are using simulations that are fairly steady state. The second measure, locking detected patterns to synfire chain initiating stimuli, indeed demonstrates that most patterns in these data are associated with runs of the synfire chain(s). However this measure depends on knowledge of the stimuli times which would generally not be available in an experimental situation. The third measure, aligning patterns on the first spike, turns out to be sensitive to the delay between links in the synfire chain. This should be applicable to experimental data. With simulated data this measure also clearly identifies some patterns which in part include firing of non-chain neurons.

For each of the four data sets in Fig. 6 the corresponding spike time histogram of complexity >3, repetition >2 aligned patterns is shown in Fig. 7. In each case there are prominent peaks which tend to fall in bins (1 ms width) with spacing of about 3 ms. This is the time for propagation between successive links in the chain models used in these simulations so that patterns arising from synfire chain activity would have multiples of this basic timing in their interval structure. The background in the histograms below the peaks represents the patterns which involve the activity of non-chain neurons. Note also that the large peaks come early in the pattern except in case C. All these simulations have variability built into the link to link delays. These temporal variations build up during each run of a chain, so that exactly repeating patterns (to 1 ms time resolution) in two or more runs of the chain can originate only from a limited portion of the full chain length and thus have a duration less than that of a full chain run. The spaced peak structures in each panel of Fig. 7 are therefore an additional signature of synfire chain activity.

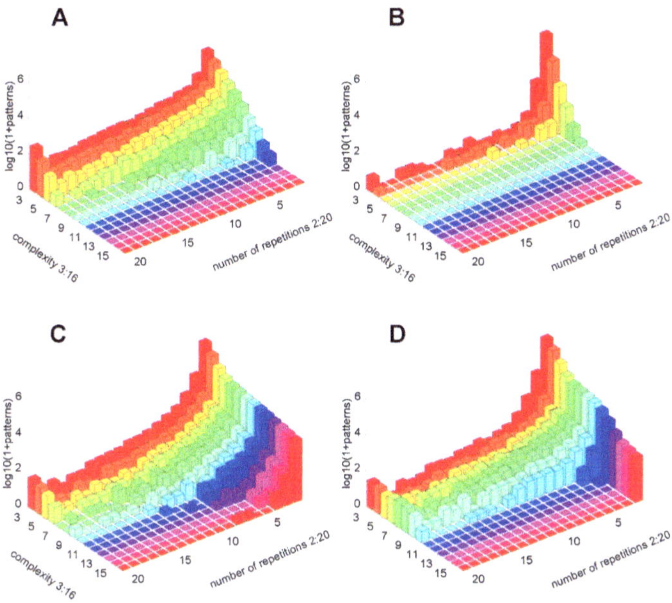

Fig. 6. Pattern repetition spectra for 100 randomly chosen neurons from four different simulations. (A) Same data as Fig. 3B. (B) Same data as Fig. 3C. These two panels demonstrate the effects of sample size on pattern numbers. (C) 100 thousand neurons in a multiply branching tree of synfire chains. (D) 100 thousand neurons arranged as many overlapping chains; in effect only one chain is stimulated. See section on data simulation for details of C and D. All four panels show the rich pattern repetition spectrum at complexity >3 and repetition >2 characteristic of synfire chains.

5. Group synchrony without chain structure

Several papers in the literature have claimed evidence for synfire chain activity based on repeating patterns or sequences (Euston et al., 2007; Ikegaya et al., 2004; Yosimura et al., 2005; Prut et al., 1998; Shmiel et al., 2005, 2006; Abeles et al., 1993). In some of these papers various surrogates were used to assess significance of the pattern counts. In this section we analyze simulation data that produces many complexity >3, repetition >2 patterns, but contains no synfire chains. Such data can be distinguished from synfire-containing spike trains by the two additional measurements set out in this paper: (1) the matrix of neuron identity intersections and (2) the time structure of the detected repeating patterns.

The data consist – as the above synfire chain simulations – of activity from 20 groups of 100 neurons (same parameters as used in many of the above synfire chain simulations) with the neurons in each group firing in near synchrony (within 3 ms) but without the groups being sequentially activated. The data were generated from the simulated data containing one synfire chain but here the synchronous spikes from the individual links were together dithered in time within a window of ±100 ms. A raster display of a 0.7 s segment of the firing activity of the 20 groups is shown in Fig. 8A. There is no ordered sequential firing of the several groups as would be the case for links in a synfire chains. (Comparable rasters of simulated orderly synfire chain activity are in Schrader et al., 2008). The data also contained 4000 independent neurons firing at 2.7 Hz with gamma 4 interval distributions.

The neuron identity intersection matrix is for an arbitrary 700 ms segment of the data is shown in Fig. 8B. Unlike the matrices in the top panels of Figs. 1 and 2, there are no orderly short diagonal stripes of pixels, just an apparently random distribution of single pixels. This of course is expected, since these data have no repeating sequential ordering to the activity of the several groups.

The pattern spectrum of a random 100 neurons chosen from this activity is in Fig. 8C, and shows considerable repeating pattern counts for complexity >3, repetition >2. Since we have full knowledge of this simulation data, examination of the neuron identities contributing to these patterns (as in Fig. 4) shows that most patterns originate within some single group (not explicitly shown). There are few patterns with members spread among one or more groups and background activity as is the case in the ordered sequences of synfire activity. This prevalence of single group origin of patterns is made obvious by the histogram of spike times in aligned patterns shown in Fig. 8D. Unlike the synfire data in Fig. 7, many patterns here have durations of less than 3 ms, indicating their origin in a single group. Note that the other repeating patterns have a continuous distribution of inter-spike times. This contrasts with the spaced peaks in the histograms of spike times in aligned patterns shown in Fig. 7 which correspond to the delay structure of a synfire chain.

62 G.L. Gerstein et al. / Journal of Neuroscience Methods 206 (2012) 54–64

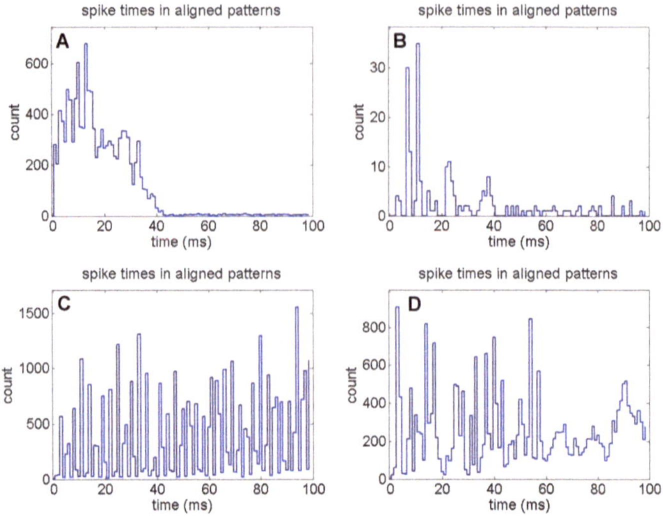

Fig. 7. Histograms of spike times in patterns aligned to first spikes. Data and arrangement are identical to Fig. 6, but only patterns of complexity >3 and repetition >2 are considered. Bin width of the histograms is 1 ms. Each panel shows various numbers of prominent peaks with 3 ms spacing. This corresponds to the interlink delays built into the several synfire simulations.

Thus with at least this one counterexample, repeating patterns alone are not enough to infer synfire activity. It is also essential to examine the matrix of active neuron identity intersections as well as the time structure of the detected repeating patterns.

6. Discussion

The synfire chain is an intriguing theoretical concept with interesting properties that should make it very applicable to understanding many aspects of brain function. Unfortunately there has to date been no convincing experimental demonstration of synfire chains due partially to the difficulties of multi-neuron recording but also to the lack of appropriate analytic tools. We have previously published one such analysis tool (Schrader et al., 2008). In the present paper, using simulated data, we have extended that work and in addition examined two new tools for detecting the presence of synfire chain activity in neuronal spike train recordings. Such analyses require data from at least one hundred simultaneously recorded spike trains, although larger samples would be preferred in order to improve reliability and sensitivity.

There are serious methodological difficulties in obtaining the necessary data. Of the currently available recording methods the multiple electrode approaches all suffer to some degree from the problem of clean separation of spike waveforms (Warren et al., 2001; Csicvari et al., 2003) from different neurons detected on the same electrodes (Harris et al., 2000). Such mixing of spike trains seriously degrades and makes ambiguous any inference from measurements of spike timing (Gerstein, 2000; Pazienti and Grün, 2006), and depending on the degree of contamination, would impact all three methods developed in the present paper. The older dye based optical methods have either spatial or temporal resolution difficulties as well as problems with bleaching and opto-kinetic damage to the neurons. There is hope however in the newer techniques with 2-photon calcium dye methods as well as with the rapidly developing genetically engineered fluorescent protein approaches (Knöpfel et al., 2003; Ohki et al., 2005, 2006; Miri et al., 2011; Miyawaki, 2005).

In a previous paper (Schrader et al., 2008) we described a simple analysis of the active neuron populations in the data that yields a unique signature of synfire activity if it exists. In the present paper we have further studied the signal to noise properties of this analysis and have extended it to the detection of triples rather than pairs of synfire chain runs. This has provided a considerable gain in sensitivity and particularly reduced problems with relatively high rate background neurons. Our signal to noise studies compared matrix pixel values on and off a short diagonal stripe whose location was determined by analyzing the large full set of simulated neurons. This obviously cannot be done in an experimental situation. However, results from the procedure should be applicable to the threshold for visual detection of short diagonal stripes (possibly aided by appropriate spatial filtering of the matrix) for experimental data. We have previously shown that visual threshold is about 100 neurons in the 2-dimensional pair analysis at low background rates. The 3-dimensional analysis of chain run triples with its better signal to noise properties should maintain this performance even at higher background firing rates. Nevertheless the sampling problem remains critical especially with high background rates.

Our second approach to detecting synfire chain activity is based on finding repeating patterns of firing. Methods for finding all such repeating patterns are well known (Abeles and Gerstein, 1988; Shmiel et al., 2005; Patnaik et al., 2008) and can be top down or bottom up. Data from simulations containing multiple synfire chains or even with a single synfire chain in a great sea of background neurons (Figs. 3 and 6) show a great richness of patterns of complexity

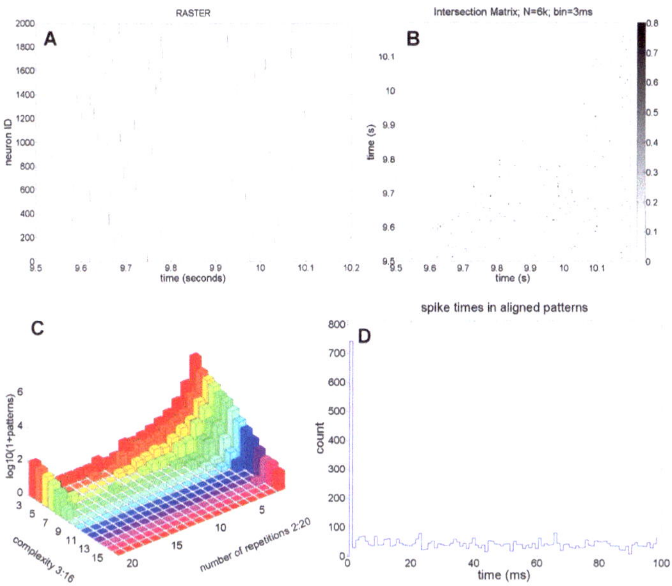

Fig. 8. Analysis of simulation data containing synchronous groups (3 ms precision), that were not active sequentially in time. The data were generated from simulated data containing one synfire chain. Same synfire chain data as in Fig. 3C but here the synchronous spike responses of the 20 individual groups of 100 neurons are dithered in time within a window of ±100 ms instead of being sequentially ordered as in the original synfire chain. (A) Raster display showing the activity of the 20 groups over an arbitrary 0.7 s interval. Activity of the remaining 4000 gamma 4 background neurons is not shown. (B) Activity intersection matrix of an arbitrary 700 ms segment of the data. This is a similar time scale to that used in Fig. 1, but shows only randomly arranged individual pixels rather than short diagonal stripes. (C) Corresponding pattern repetition spectrum for a random sample of 100 neurons. (D) Corresponding time histogram of spikes in aligned complexity >3 repetition >2 patterns. The single peak at 1–2 ms indicates that members of a given repeating pattern were mostly in some single group. Repeating patterns with longer intervals do not show the discrete structure corresponding to synfire chain delays. Thus repeating patterns per se are not necessarily an indicator of synfire activity.

>3 and repetition >2. We have demonstrated in Fig. 4 for synfire data that the neurons involved in each such pattern are usually members of some single chain. Although directly unverifiable, one would hope this would also be true in patterns high complexity, high repetition data when found in experiments. Comparable multi-neuron spike trains from simulations of independent neurons do show repeating patterns but mostly at complexity 3 and/or repetition 2.

Our third approach to detecting synfire chain activity is based on examining the detailed time structure of any repeating patterns. The most useful such measure is to align the repeating patterns to their first spike. With synfire data a histogram of spike times of such aligned patterns tends to show peaks with spacing corresponding to the interlink delays of a synfire chain (Fig. 7).

The detection methods presented here also apply to a generalization of synfire chains known as braids (Bienenstock, 1995) or polychrony (Izhikevich, 2006) which are characterized by the rule that all direct and indirect paths between any two neurons have the same delay. In these feed-forward structures the neurons need not be grouped into discrete links, and they support reproducible, precisely timed spike sequences that need not be grouped into synchronous packets. The delays on the feed-forward connections preserve the key property of synfire chains that, during spike sequence re-activation, each neuron receives synchronously arriving spikes from upstream neurons. Repeated activations of braids, just as for chains, will give rise to diagonal stripes in the intersection matrix and will produce temporal patterns with high complexity and repetition. However, the comb-like structure seen in Fig. 7, being specific to chains, would not generally be found in braids.

Although synfire chains produce many high complexity, high repetition patterns so can other mechanisms. We have demonstrated one such situation with a simulation of near synchronous firing within neural groups but with each such group active at random times. Such data (Fig. 8) show rich patterning at complexity >3 and repetition >2, but as expected the intersection matrix has only randomly arranged pixels rather than short diagonal stripes. The histogram of spike times in aligned patterns reflects the prevalence of patterns limited to the near synchronous firing within each group. Patterns with longer intervals have continuously distributed intervals rather than preferred discrete values which would correspond to the chain delay structure.

Finally we suggest that experimental detection of synfire chains cannot rest exclusively on the demonstration of repeating firing patterns. It is necessary in addition to demonstrate an activity intersection matrix with short diagonal stripes, and a histogram of spike times in aligned patterns that shows that many patterns have a structure appropriate for expected inter-link delays in a synfire chain. We have so far had access to four multiple neuron recording data sets from four different laboratories and preparations. All of them show repeating patterns, but none of them have intersection matrices with short diagonal stripes or appropriate spaced peaks in the spike times of aligned patterns. The search continues.

Acknowledgments

Partially funded by EU Grant 269921 (BrainScaleS), DIP F1.2, the Helmholtz Alliance on Systems Biology (Germany), and the Next-Generation Supercomputer Project of MEXT (Japan). ERW was supported by the CARMEN project, funded by the Engineering and Physical Sciences Research Council, UK (www.carmen.org.uk).

References

Abeles M. Local cortical circuits: an electrophysiological study. Berlin: Springer-Verlag; 1982.

Abeles M. Corticonics neural circuits of the cerebral cortex. Cambridge, UK: Cambridge Univ. Press; 1991.

Abeles M, Bergman H, Margalit E, Vaadia E. Spatiotemporal firing patterns in the frontal cortex of behaving monkeys. J Neurophysiol 1993;70:1629–38.

Abeles M, Gat I. Detecting precise firing sequences in experimental data. J Neurosci Methods 2001;107:141–54.

Abeles M, Gerstein GL. Detecting spatiotemporal firing patterns among simultaneously recorded single neurons. J Neurophysiol 1988;60:909–24.

Aertsen A, Diesmann M, Gewaltig M. Propagation of synchronous spiking activity in feed-forward networks. J Physiol (Paris) 1996;90:243–7.

Aviel Y, Mehring C, Abeles M, Horn D. On embedding synfire chains in a balanced network. Neural Comput 2003;15:1321–40.

Baker S, Lemon RN. Precise spatiotemporal repeating patterns in monkey primary and supplementary motor areas occur at chance level. J Neurophysiol 2000;84:1770–80.

Baker S, Spinks R, Jackson A, Lemon RN. Synchronization in monkey motor cortex during a precision grip task. I. Task-dependent modulation in single-unit synchrony. J Neurophysiol 2001;85:869–85.

Bienenstock E. A model of neocortex. Network Comput Neural Syst 1995;6:179–224.

Csicvari J, Henze DA, Jamieson B, Harris KD, Sirota A, Bartho P, et al. Massively parallel recording of unit and local field potentials with silicon-based electrodes I. Neurophysiol 2003;90:1314–23.

Date A, Bienenstock E, Geman S. On the temporal resolution of neural activity. Technical report, Division of Applied Mathematics, Brown University 1998 (Available at: http://www.dam.brown.edu/people/elie/papers/temp-res.ps).

Diesmann M, Gewaltig MO, Aertsen A. Stable propagation of synchronous spiking in cortical neural networks. Nature 1999;402:529–33.

Euston DR, Tatsuno M, McNaughton BL. Fast forward playback of recent memory sequences in prefrontal cortex during sleep. Science 2007;138:1147–50.

Gerstein GL. Cross-correlation measures of unresolved multi-neuron recordings. J Neurosci Methods 2000;100:41–51.

Gerstein GL. Searching for significance in spatio-temporal firing patterns. Acta Neurobiol Exp (Warsz) 2004;2:203–7.

Gewaltig M-O, Diesmann M, Aertsen A. Propagation of cortical synfire activity: survival probability in single trials and stability in the mean. Neural Netw 2001;14:657–73.

Goedeke S, Diesmann M. The mechanism of synchronization in feed-forward neuronal networks. New J Phys 2008;10:015007, http://iopscience.iop.org/1367-2630/10/1/015007.

Harris KD, Henze DA, Csicsvari J, Hirase H, Buszaki G. Accuracy of tetrode spike separation as determined by simultaneous intracellular and extracellular measurements. J Neurophysiol 2000;84:401–14.

Hatsopoulos N, Geman S, Amarasingham A, Bienenstock E. At what time scale does the nervous system operate? Neurocomputing 2003, 25–29, 52–54.

Hayon G, Abeles M, Lehmann D. Modeling compositionality by dynamic binding of synfire chains. J Comput Neurosci 2004;17:179–201.

Hayon G, Abeles M, Lehmann D. A model for representing the dynamics of a system of synfire chains. J Comput Neurosci 2005;18:41–53.

Hebb D. The organization of behavior: a neuropsychological theory. New York: Wiley; 1949.

Ikegaya Y, Aaron G, Cossart R, Aronov D, Lampl I, Ferster D, et al. Synfire chains and cortical songs: temporal modules of cortical activity. Science 2004;304:559–64.

Izhikevich EM. Polychronization: computation with spikes. Neural Comput 2006;18:245–82.

Knöpfel T, Tomita K, Shimazaki R, Sakai R. Optical recordings of membrane potential using genetically targeted voltage-sensitive fluorescent proteins. Methods 2003;30(May (1)):42–8.

Louis S, Borgelt C, Grün S. Complexity distribution as a measure for assembly size and temporal precision. Neural Networks 2010a;23:705–12.

Louis S, Gerstein G, Grün S, Diesmann M. Surrogate spike train generation through dithering in operational time. Front Neurosci 2010b;4(September) (article 127).

Luczak A, Barthóf P, Marguet SL, Buzsáki G, Harris KD. Sequential structure of neocortical spontaneous activity in vivo. Proc Natl Acad Sci U S A 2007;104: 347–52.

Mehring C, Hehl U, Kubo M, Diesmann M, Aertsen A. Activity dynamics and propagation of synchronous spiking in locally connected random networks. Biol Cybern 2003;88:395–408.

Miri A, Dale K, Bardine R, Aksay E, Tank D. Regression based identification of behavior encoding neurons during large-scale optical imaging of neural activity at cellular resolution. J Neurophysiol 2011;105:964–80.

Miyawaki A. Innovations in the imaging of brain functions using fluorescent proteins. Neuron 2005;48:189–99.

Ohki K, Chung S, Ch'ng YH, Kara P, Reid RC. Functional imaging with cellular resolution reveals precise micro-architecture in visual cortex. Nature 2005;433:597–603.

Ohki K, Chung S, Kara P, Hübener M, Bonhoeffer T, Reid RC. Highly ordered arrangement of single neurons in orientation pinwheels. Nature 2006;442:925–8.

Oram MW, Wiener MC, Lestienne R, Richmond BJ. Stochastic nature of precisely timed spike patterns in visual system neuronal responses. J Neurophysiol 1999;81:3021–33.

Patnaik D, Sastry PS, Unnikrishnan KP. Inferring neuronal network connectivity from spike data: a temporal data mining approach. Sci Program 2008;16:49–77.

Pazienti A, Grün S. Robustness of the significance of spike correlation with respect to sorting errors. J Comp Neurosci 2006;21:329–42.

Prut Y, Vaadia E, Bergman H, Haalman I, Slovin H, Abeles M. Spatiotemporal structure of cortical activity: properties and behavioral relevance. J Neurophysiol 1998;79;2857–74.

Sastry PS, Unnikrishnan KP. Conditional probability based significance tests for sequential patterns in multi-neuronal spike trains. Neural Comput 2010;22:1025–59.

Schrader S, Grün S, Diesmann M, Gerstein GL. Detecting synfire chain activity with massively parallel spike train recording. J Neurophysiol 2008;100: 2165–76.

Schrader S, Morrison A, Diesmann M. A composition machine for complex movements. In: Proceeding of the 31st Göttingen neurobiol conf TS18-1C; 2007, Available online at http://www.neuro.uni-goettingen.de/archiv/2007/pdf/Proceedings-Goettingen2007.pdf.

Schrader S, Diesmann M, Morrison A. A compositionality machine realized by a hierarchic architecture of synfire chains. Front Comput Neurosci 2011;4:154.

Shmiel T, Drori R, Shmiel O, Ben-Shaul Y, Nadasdy Z, Shemesh M, et al. Neurons of the cerebral cortex exhibit precise inter-spike timing in correspondence to behavior. Proc Natl Acad Sci U S A 2005;102:18655–7.

Shmiel T, Drori R, Shmiel O, Ben-Shaul Y, Nadasdy Z, Shemesh M, et al. Temporally precise cortical firing patterns are associated with distinct action segments. J Neurophysiol 2006;96:2645–52.

Tetzlaff T, Morrison A, Timme M, Diesmann M. Heterogeneity breaks global synchrony in large networks. In: Proceeding of the 30th Göttingen neurobiol conf 206B; 2005, Available online at http://www.neuroanatomie.uni-goettingen.de/neurobio_archiv/2005/pdf/Proceedings-Goettingen2005.pdf.

Tetzlaff T, Einevol G, Diesmann M. Synchronization and rate dynamics in embedded synfire chains: effect of network heterogeneity and feedback. BMC Neuroscience 2009;10(Suppl. 1):P258, doi:10.1186/1471-2202-10-S1-P258.

Trengove C, Leeuwen CV, Diesmann M. High storage capacity of synfire chains in large-scale cortical networks of conductance-based spiking neurons. Nineteenth Annual Computational Neuroscience Meeting: CNS*2010. BMC Neuroscience 2010a;11(Suppl. 1):F1, doi:10.1186/1471-2202-11-S1-F1.

Trengove C, Leeuwen CV, Diesmann M. Random compositional networks of synfire chains dynamically self-tune to the critical state for ongoing percolation of activity. In: Front Neurosci Conference Abstract Neuroinformatics: 2010b., doi:10.3389/conf.fnins.2010.13.00041.

Trengove C, van Leeuwen C, Diesmann M. High-capacity embedding of synfire chains in a cortical network model. J Comput Neurosci 2012, in press.

Warren DJ, Fernandez E, Normann RA. High-resolution two-dimensional spatial mapping of cat striate cortex using a 100-microelectrode array. Neuroscience 2001;105:19–31.

Yosimura Y, Dantzker JL, Callaway EM. Excitatory cortical neurons form fine-scale functional networks. Nature 2005;433:868–73.

PLOS | COMPUTATIONAL BIOLOGY

RESEARCH ARTICLE

ASSET: Analysis of Sequences of Synchronous Events in Massively Parallel Spike Trains

Emiliano Torre[1]*, Carlos Canova[1], Michael Denker[1], George Gerstein[2], Moritz Helias[1,3], Sonja Grün[1,4]

1 Institute of Neuroscience and Medicine (INM-6) and Institute for Advanced Simulation (IAS-6) and JARA BRAIN Institute I, Jülich Research Centre, Jülich, Germany, **2** Department of Neuroscience, University of Pennsylvania, Philadelphia, Pennsylvania, United States of America, **3** Department of Physics, RWTH Aachen University, Aachen, Germany, **4** Department of Biology, RWTH Aachen University, Aachen, Germany

* e.torre@fz-juelich.de

Citation: Torre E, Canova C, Denker M, Gerstein G, Helias M, Grün S (2016) ASSET: Analysis of Sequences of Synchronous Events in Massively Parallel Spike Trains. PLoS Comput Biol 12(7): e1004939. doi:10.1371/journal.pcbi.1004939

Editor: Olaf Sporns, Indiana University, UNITED STATES

Received: October 15, 2015

Accepted: April 24, 2016

Published: July 15, 2016

Data Availability Statement: The stochastic data employed in the paper have been generated using Python. The code and data to generate the figures of the manuscript are available from the DRYAD database (http://datadryad.org/) under the doi:10.5061/dryad.cb360.

Funding: This work was partly supported by the Helmholtz Portfolio Supercomputing and Modeling for the Human Brain (SMHB, [http://www.fz-juelich.de/JuBrain/EN/Helmholtz%20Portfolio.html]), the Helmholtz young investigator group VH-NG-1028 [https://www.helmholtz.de/en/jobs_talent/funding_programs/helmholtz_young_investigators_groups/],

Abstract

With the ability to observe the activity from large numbers of neurons simultaneously using modern recording technologies, the chance to identify sub-networks involved in coordinated processing increases. Sequences of synchronous spike events (SSEs) constitute one type of such coordinated spiking that propagates activity in a temporally precise manner. The synfire chain was proposed as one potential model for such network processing. Previous work introduced a method for visualization of SSEs in massively parallel spike trains, based on an intersection matrix that contains in each entry the degree of overlap of active neurons in two corresponding time bins. Repeated SSEs are reflected in the matrix as diagonal structures of high overlap values. The method as such, however, leaves the task of identifying these diagonal structures to visual inspection rather than to a quantitative analysis. Here we present ASSET (Analysis of Sequences of Synchronous EvenTs), an improved, fully automated method which determines diagonal structures in the intersection matrix by a robust mathematical procedure. The method consists of a sequence of steps that i) assess which entries in the matrix potentially belong to a diagonal structure, ii) cluster these entries into individual diagonal structures and iii) determine the neurons composing the associated SSEs. We employ parallel point processes generated by stochastic simulations as test data to demonstrate the performance of the method under a wide range of realistic scenarios, including different types of non-stationarity of the spiking activity and different correlation structures. Finally, the ability of the method to discover SSEs is demonstrated on complex data from large network simulations with embedded synfire chains. Thus, ASSET represents an effective and efficient tool to analyze massively parallel spike data for temporal sequences of synchronous activity.

the EU Grant 269921 (BrainScaleS [https://brainscales.kip.uni-heidelberg.de/]), and the EU Grant 604102 (Human Brain Project, HBP [https://www.humanbrainproject.eu/]), the Priority Program SPP 1665 of the DFG (GR 1753/4-1 and DE 2175/1-1) and the Ger-Jpn Comput Neurosci Project BMBF Grant 01GQ1114. The funders had no role in study design, data collection and analysis, decision to publish, or preparation of the manuscript.

Competing Interests: The authors have declared that no competing interests exist.

Author Summary

Neurons in the cerebral cortex are highly interconnected. However, the mechanisms of coordinated processing in the neuronal network are not yet understood. Theoretical studies have proposed synchronized electrical impulses (spikes) propagating between groups of nerve cells as a basis of cortical processing. Indeed, animal studies provide experimental evidence that spike synchronization occurs in relation to behavior. However, the observation of sequences of synchronous activity has not been reported so far, presumably due to two fundamental problems. First, the long-standing lack of simultaneous recordings of large populations of neurons, which only recently were enabled by advances in recording technology. Second, the absence of proper tools required to find these activity patterns in such high-dimensional data. Addressing the second issue, we introduce here a fully automatized mathematical method that advances an existing visual approach to identify sequences of synchronous events in large data sets. We demonstrate the efficacy of our method on a range of simulated test data that capture the characteristics and variability of experimental data. Our tool will serve future studies in their search for spike time coordination at millisecond precision in the brain.

Introduction

Synchronous input spikes to a receiving neuron are considered most effective in generating an output spike, as predicted by theoretical studies coining the term coincidence detector [1]. The argument rests on the premise that excitatory post-synaptic potentials (EPSPs) in the cortex are typically small in relation to the firing threshold, so that many EPSPs need to overlap to produce an output spike. Due to leak currents in the neuronal membrane, the firing threshold is reached with fewer spikes when these arrive synchronously at the post-synaptic neuron rather than sparsely, thus making the neuron behave like a coincidence detector. Experimental studies provide evidence for the existence of coincidence detectors (e.g., [2]) and relate them to various mechanisms of spike-timing dependent plasticity [3, 4] as well as to different encoding and decoding schemes [5, 6].

Cortical anatomy supports such considerations. Individual neurons receive synaptic connections from a large number of neurons (on the order of 10,000 in the human cortex, see e.g. [7]) and project to a similar number of other cells. Such a connectivity structure combined with suitable synaptic delays may produce spatio-temporal spike patterns, as proposed in [8–11]. A simple case is represented by a temporal sequence of synchronous events (SSE), each event consisting of synchronous spikes from a group of neurons.

The synfire chain model [8] is a neural network model that has been proposed to exhibit such activity through a suitable wiring of the neurons in the network in successive groups interconnected in a highly divergent and convergent manner. Each neuron of one group projects to several neurons of the next group, thus forming a chain structure. Under the assumption that the synaptic transmission delays from neurons of one group to neurons of the next group are identical, the synchronous stimulation of neurons in the first group leads to robust propagation of synchronous spiking activity through the chain [10] even in the presence of noise. The activation of a synfire chain would thus lead to the occurrence of an SSE.

In order to assess whether SSEs are indeed observed in the brain and have a functional role, the spiking activity of several neurons needs to be recorded simultaneously, analyzed for temporal correlation and related to behavior. Under the assumption that an SSE occurs sparsely in a given data set, for instance in relation to a specific behavior, pairwise correlations between

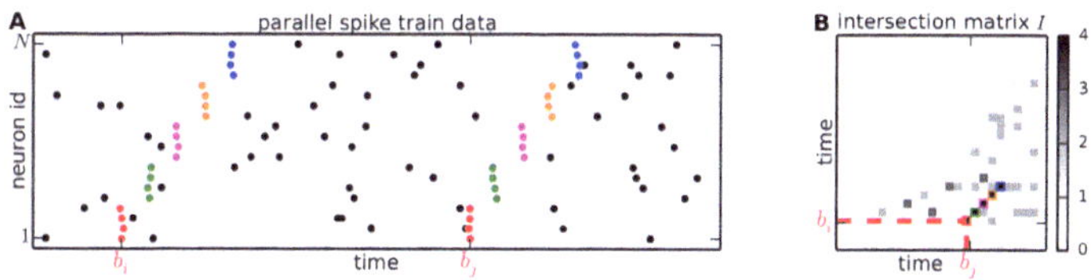

Fig 1. From spike trains to the intersection matrix. (A) Raster plot of parallel spike trains of multiple neurons (vertical axis) over time (horizontal axis). Dots in each row correspond to the spike times of one neuron. Time is discretized into adjacent bins (marked by white and blue shaded backgrounds) to define synchronous events. Synchronous spikes forming an SSE repeating twice are indicated by colored dots (one color per event). **(B)** Intersection matrix I. Each matrix entry I_{ij} (values encoded by gray levels) contains the degree of overlap of neurons active in time bins b_i and b_j. Only the entries I_{ij} with $i < j$ are shown due to the symmetry of the matrix.

doi:10.1371/journal.pcbi.1004939.g001

neurons engaged in the activity are weak and will not be detected by means of pairwise correlation analyses. For this reason, an analysis method was proposed in [12] that directly searches for SSEs in massively parallel spike data.

The basic idea of the method presented in [12] is the following: after discretizing time in bins of a few ms (see Fig 1A), a synchronous event which repeats at two different time bins leads to a large number of neurons that are active in both bins. Building an intersection matrix I where each entry I_{ij} represents the number of neurons active in both bins i and j, a group of synchronous events occurring in bins i and j results in a large overlap I_{ij} compared to other entries of the matrix. An SSE which occurs twice produces in the matrix I a sequence of high-valued entries, which we name *diagonal structures* (DS), aligned parallel to the main diagonal (Fig 1B). A diagonal filter applied to the matrix I enhances the contrast between the DS and surrounding entries, mapping I into a filtered matrix F.

The method was calibrated using data from large-scale synfire chain network simulations. When the neurons of a full chain or a large portion of it are observed, the method reveals the associated DS in the intersection matrix. However, when only a few hundreds of neurons (comparable to the number of cells that can be recorded simultaneously *in vivo* with modern electrophysiological techniques) are randomly sampled from the full network, the DSs are less visible and less likely to be continuous, making it impossible to isolate them from the surrounding entries. Indeed, this method leaves the judgment of whether individual entries are large or small, as well as the grouping of proximal entries into DSs, to visual inspection. As a consequence, the results are prone to subjectiveness and the procedure is not open to automation.

In this paper we present a method, named ASSET (Analysis of Sequences of Synchronous EvenTs), which improves the approach proposed in [12] by providing a mathematical and fully automated detection of SSEs in parallel spike train data. The analysis features i) a statistical assessment of membership of individual entries of the intersection matrix to a DS, ii) a rigorous construction of individual DSs and associated SSEs by clustering, and iii) the reconstruction of the neuronal composition and occurrence times of each event in the found SSEs.

The manuscript is organized as follows. In "*Methods*" we derive a statistical assessment of the membership of matrix entries to a DS based on two statistical tests for the significance of each individual entry and the joint significance of an entry and its neighbors, respectively. Entries passing both tests are then grouped together into individual DSs depending on their reciprocal distance by a clustering procedure. Once the DSs are identified, it is readily possible to reconstruct their neuronal composition.

In "*Results*" we assess the performance and robustness of ASSET on various types of simulated data that replicate typical features of electrophysiological recordings. These include firing rate heterogeneity across neurons, variability over time and different types of correlation structure in the spiking activity. In addition, we demonstrate the ability of the ASSET analysis to find repeated SSE activity in simulated data generated by a neural network model of overlapping synfire chains as introduced in [12].

We conclude by discussing benefits and limitations of the ASSET method, open problems and future plans for the analysis of electrophysiological data.

Methods

In this section we provide a formal definition of the intersection matrix as introduced in [12], introduce statistical tests to determine whether individual entries of the intersection matrix are part of any DS, and propose a clustering technique to group significant entries into the different DSs. After the last step, we can determine the occurrence times and neuronal composition of all events of the repeated SSEs associated to the DSs found. We then formally describe the stochastic models employed in "*Results*" to assess the performance of the method in a variety of scenarios.

The intersection matrix

Our approach builds on the notion of intersection matrix defined in [12]. Here we introduce this concept mathematically and provide definitions. Given a set of N parallel spike trains observed in the time interval $[0, T]$, synchronous spike events across neurons are determined by discretizing the time interval into B adjacent time bins $b_1, b_2, \ldots, b_B$ of identical width $\Delta = T/B$ (typically of a few ms), as illustrated in Fig 1A. Each set S_i of spikes falling in time bin b_i forms a synchronous event. The value $|S_i \cap S_j|$ represents the number of neurons being active at both time bins b_i and b_j. The *intersection matrix* I is defined by

$$I_{ij} := |S_i \cap S_j| \quad \forall i, j = 1, 2, \ldots, B.$$

In this setting I is a symmetric matrix, whose main diagonal contains the population histogram, i.e. the time histogram of the number of neurons simultaneously active in each time bin. A synchronous spike event occurring at two bins i and j, e.g. as a result of the repeated activation of the same synfire chain, results in a larger value for I_{ij} compared to the chance level. A repeated SSE composed of l_{SSE} successive synchronous events, each of which occurs at time bins (b_{i_r}, b_{j_r}), $r = 1, 2\ldots, l_{SSE}$, determines a sequence of large-valued entries I_{i_r, j_r} in I—which we term a *diagonal structure* (DS)—as sketched in Fig 1B.

To account for the variability of firing rates over time when comparing different entries of I, in [12] each entry of I_{ij} was normalized by the number of neurons active in each bin b_i and b_j. After normalization, the entries take value 1 for a complete overlap and value 0 for no overlap. To enhance the contrast between high-valued entries belonging to the same diagonal structure and the surrounding entries, the matrix was further filtered by a linear kernel having an orientation parallel to the main diagonal. We take here a different approach, as outlined in the following section.

Statistical significance of individual entries in the intersection matrix

To derive a measure of overlap that is independent of firing rates, we first need the probability mass function (pmf) $p_{ij}(\cdot)$ of each individual entry I_{ij} in the intersection matrix, under the null hypothesis H_0 that the spike trains under consideration are realizations of mutually independent Poisson processes. If the null hypothesis is rejected, i.e. if the observed value ξ taken by I_{ij}

is too large to be interpreted as chance, we classify the overlap $S_i \cap S_j$ as a statistically significant repeated synchronous event. In "*Results*" we show that the statistics of the method are robust to deviations from Poissonianity as well as to the presence of various types of correlations other than repeated SSE activity.

Under the stated hypotheses, the distribution of I_{ij} is determined by the firing rate of each neuron k at the time bins b_i and b_j. I_{ij} represents the (stochastic) number of neurons firing simultaneously in both bins and can thus be expressed as

$$I_{ij} = \sum_{k=1}^{N} I_{ij}^{(k)}, \tag{1}$$

where $I_{ij}^{(k)}$ is a Bernoulli random variable taking value 1 if neuron k fires in both bins i and j, and 0 otherwise. The probability parameter $p_{ij}^{(k)}$ of $I_{ij}^{(k)}$ is related to the local firing rate $\lambda_i^{(k)}$ of neuron k at bin b_i and $\lambda_j^{(k)}$ at bin b_j by

$$p_{ij}^{(k)} = (1 - e^{-\lambda_i^{(k)}\Delta})(1 - e^{-\lambda_j^{(k)}\Delta})$$

for each $i \neq j$, where each factor $1 - e^{-\lambda_i^{(k)}\Delta}$ is the probability for neuron k to emit at least one spike in the time bin i. The knowledge of the firing rate profiles of all neurons is therefore a pre-requisite for the exact computation of the pmf $p_{ij}(\cdot)$ of I_{ij}, $i, j = 1, \ldots, N$.

In light of Eq 1, I_{ij} is a Poisson Bernoulli random variable [13]. Its pmf $p_{ij}(\cdot)$ is analytically given by

$$p_{ij}(\xi) = \sum_{A \in \mathcal{P}_{N;\xi}} \left\{ \prod_{k \in A} p_{ij}^{(k)} \prod_{h \in A^C} (1 - p_{ij}^{(h)}) \right\}, \tag{2}$$

where $\mathcal{P}_{N;\xi}$ is the family of all possible subsets of ξ elements that can be extracted from the set $\{1, \ldots, N\}$, A is one such subset and $A^C = \{1, \ldots, N\} \setminus A$ is the complement of A. Thus $p_{ij}(\xi)$ is a summation of $\binom{N}{\xi}$ addenda, for a total of 2^N terms needed to compute $p_{ij}(\cdot)$. The computation is feasible for small N, but soon becomes prohibitive as N increases beyond a few dozens, as in the applications to large parallel recordings we are interested in.

By use of Le Cam's theorem [13], we approximate $p_{ij}(\cdot)$ by a Poisson density function $p_{(\lambda)}$ with rate parameter $\lambda = \Sigma_k \lambda_k$

$$p_{ij}(\xi) \simeq p_{(\lambda)}(\xi) = \frac{\lambda^{\xi} e^{-\lambda}}{\xi!}. \tag{3}$$

The approximation error grows quadratically with the $p_{ij}^{(k)}$'s and stays low if the $p_{ij}^{(k)}$'s are sufficiently small, as shown in Fig 2.

Given the firing rate profile (i.e. the rate at each time bin) of each neuron, we can thus calculate the pmf $p_{ij}(\cdot)$ and its cumulative distribution function (cdf) $\mathbb{P}_{ij}(\cdot) := \sum_{\xi<\cdot} p_{ij}(\xi)$, either exactly from Eq 2 or approximately relying on Le Cam's approximation in Eq 3. Transforming each entry I_{ij} by its respective cdf $\mathbb{P}_{ij}(\cdot)$, we map the observed overlaps I_{ij} to cumulative probabilities

$$P_{ij} := \mathbb{P}_{ij}(I_{ij}) = \sum_{\xi < I_{ij}} \frac{\lambda^{\xi} e^{-\lambda}}{\xi!}, \tag{4}$$

obtaining the *probability matrix* $P := (P_{ij})_{ij}$, as illustrated in Fig 3A.

PLOS | COMPUTATIONAL BIOLOGY

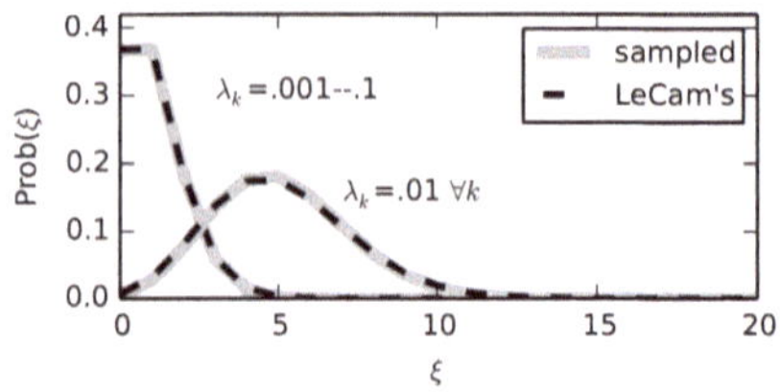

Fig 2. Le Cam's approximation. The continuous gray curves show two Poisson Bernoulli distributions resulting from the sum of 100 Bernouilli random variables, having rates $\lambda_k = 0.01\ \forall k$ (right curve) and $\lambda_k = 0.001, 0.002, \ldots, 0.1$ (left curve), respectively. The distributions are drawn as histograms from samples of size 10^8 (calculating their exact analytical expression as in Eq 2 exact is computationally prohibitive). In both cases, Le Cam's approximation given by a Poisson distribution with parameter $\lambda = \sum_k \lambda_k$ (dashed lines) yields a good match (integrated absolute error <0.04).

doi:10.1371/journal.pcbi.1004939.g002

If the null hypothesis holds, then $\mathbb{P}_{ij}(\cdot)$ is the true probability distribution of the amount of overlap between bins b_i and b_j, and I_{ij} is a realization from that probability distribution. If so, $P_{ij} := \mathbb{P}_{ij}(I_{ij})$ takes $N + 1$ values $x \in [0, 1)$ (as many as the intersection values from 0 to N), and its cdf is the identity function over the domain of this set of values: $\Pr(P_{ij} < x) = x$ for any x. If P_{ij} is large (close to 1), the null hypothesis is rejected in favor of the alternative hypothesis that the observed overlap reflects active synchronization between the involved neurons at time bins b_i and b_j. After setting a significance threshold α_1 (e.g. $\alpha_1 = 0.99$), we classify all entries I_{ij} for which $P_{ij} > \alpha_1$ as statistically significant, along with the associated repeated synchronous events.

Note that the sum in Eq 4 includes only values of ξ strictly lower than I_{ij}. This choice, which assigns the observed value I_{ij} to the critical region of the hypothesis test, ensures that $1 - P_{ij}$ retains the property of a p-value, namely that $\Pr(1 - P_{ij} \leq y) = y$ for each y. $1 - P_{ij}$ reflects the probability, computed under the null hypothesis, that I_{ij} would take a value equal to or exceeding the observed value. We reject the null hypothesis if this probability is lower than $1 - \alpha_1 = 0.01$.

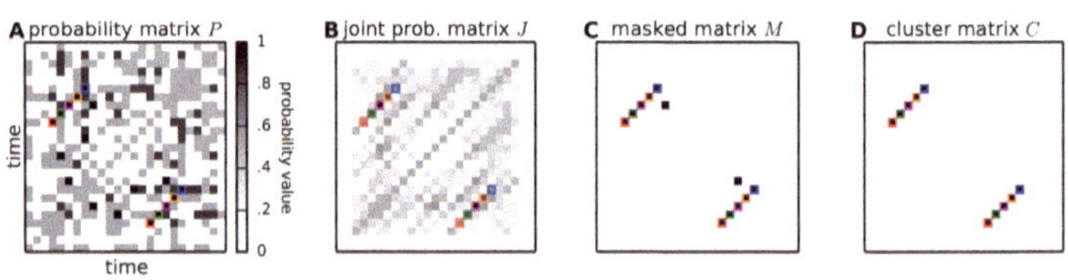

Fig 3. From spike trains to the cluster matrix. (A) Given parallel spike train data and their intersection matrix I as in Fig 1, for each entry I_{ij} its cumulative probability P_{ij} is calculated analytically under the null hypothesis $\mathcal{H}_0$ that the spike trains are independent and marginally Poisson. **(B)** The l largest neighbors of I_{ij} in a rectangular area extending along the 45 degree direction are isolated by means of a kernel and their joint cumulative probability is assigned to the joint probability matrix J at position J_{ij}. **(C)** Chosen a significance threshold α_1 for the probability of individual entries (i.e. for entries P_{ij}) and a significance threshold α_2 for the joint probability of the neighbors of an entry (i.e. for entries J_{ij}), each entry I_{ij} for which $P_{ij} > \alpha_1$ and $J_{ij} > \alpha_2$ is classified as statistically significant. Significant entries of I are retained in the binary masked matrix M_{ij}, which takes value 1 at positions (i, j) where I is statistically significant and 0 elsewhere. **(D)** 1-valued entries in M falling close-by are clustered together (or discarded as isolated chance events) by means of a DBSCAN algorithm, which thus isolates diagonal structures.

doi:10.1371/journal.pcbi.1004939.g003

Joint significance of neighbors of I_{ij}

A DS resulting from a repeated SSE, as sketched in Fig 1B, differs from the surrounding entries of the intersection matrix I due to not just one, but a sequence of entries with large values aligned around a diagonal. We devise a statistical test that exploits this feature to detect DSs in the intersection matrix. In the previous section we have already derived the probability matrix P as a transformation of I that normalizes raw intersection values by the local neuronal firing rates. We can now look for DSs in P rather than in I.

Neighbors extracted by rectangular kernel. A visual technique to enhance DSs in the intersection matrix was proposed in [12], consisting in averaging each entry I_{ij} in the matrix (previously normalized by estimates of the instantaneous rates) with surrounding entries falling in the same diagonal, and thus building a filtered version F of the normalized intersection matrix. F_{ij} is large if the neighbors $I_{i+h,\,j+h}$ of I_{ij} are jointly large, and small otherwise.

Inspired by this approach, we aim at mapping the matrix I into a joint probability matrix J whose entries J_{ij} reflect the joint probability of entries surrounding I_{ij}. We first note that the DS does not have to be composed of adjacent entries in I, and that these entries do not necessarily lie on the same off-diagonal. For instance, if any r-th entry of the DS, $r \in \{1, 2, \ldots, l_{SSE}\}$, is such that $i_r > i_{r-1}+1$ and/or $j_r > j_{r-1}+1$, then (i_{r-1}, j_{r-1}) and (i_r, j_r) will not be adjacent. If $i_r - i_{r-1} \neq j_r - j_{r-1}$, then (i_{r-1}, j_{r-1}) and (i_r, j_r) will not lie on the same diagonal, as sketched in Fig 4. The number of parallel diagonals spanned by the DS, or its "wiggliness", is given by

$$w_{DS} := \max_r\{\Delta i_r - \Delta j_r\} - \min_r\{\Delta i_r - \Delta j_r\} + 1,$$

where $\Delta i_r := i_r - i_1$ and $\Delta j_r := j_r - j_1$ are the lags (in number of bins) of the r-th synchronous event from the first one, for the first and second occurrence of the SSE, respectively.

To select the entries around I_{ij} for which to calculate the joint probability, we center a rectangular kernel of length l_K and width w_K around I_{ij}, aligned along the diagonal I_{ij} belongs to. If I_{ij} belongs to a DS, the kernel should optimally cover all the entries of the same DS. A rectangular kernel having a width $w_K > 1$ allows capturing wiggly DSs, contrarily to a linear kernel. We call the set of matrix entries covered by the kernel the *neighborhood* of I_{ij}. If the intersection matrix between two different data segments is calculated, the neighborhood covers $n = l_K \cdot w_K - \lfloor \frac{w_K}{2} \rfloor \left(\lfloor \frac{w_K}{2} \rfloor + 1 \right)$ entries in total. However, for the intersection matrix between one and the same stretch of data, the main diagonal contains the population histogram and is therefore excluded from the analysis, as is the lower triangular matrix $i > j$ that contains redundant information due to the symmetry of the matrix in this case. The total number n of

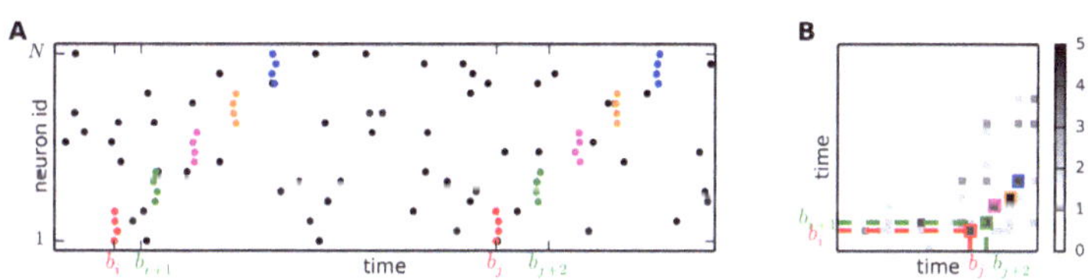

Fig 4. Origin of holes and wiggliness of a DS. Spiking activity containing a repeated SSE **(A)** and associated intersection matrix **(B)**. The SSE comprises 5 successive events. If two successive events of the SSE do not fall into adjacent bins (e.g. the second and third events in the first SSE repetition) a hole appears in the associated DS. If the distance (in number of bins) of two consecutive events is not identical for the two SSE repetitions, the corresponding entries belong to different off-diagonals of the intersection matrix.

doi:10.1371/journal.pcbi.1004939.g004

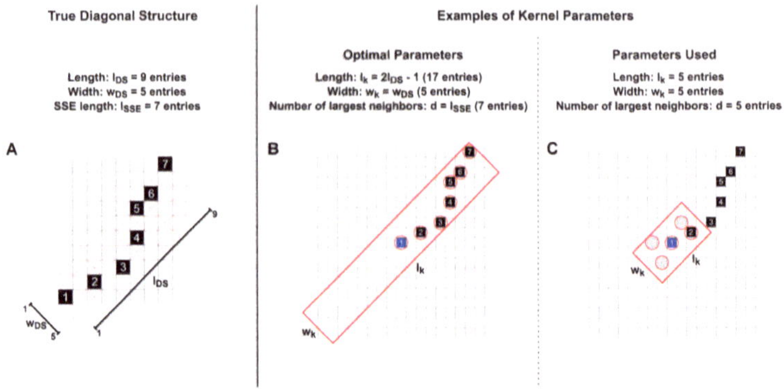

Fig 5. Kernel covering a diagonal structure. (A) Illustration of a DS in the intersection matrix composed of $l_{SSE} = 7$ entries, numbered 1 to 7. The DS has length $l_{DS} = 9$ (larger than l_{SSE}, because some entries are separated by holes) and wiggliness parameter $w_{DS} = 5$ (because the entries are spread over 5 different diagonals). **(B)** Kernel centered around the first DS entry (in blue) with optimal parameters $l_K = 2 \cdot l_{DS} - 1 = 17$ and $w_K = w_{DS} = 5$, ensuring coverage of the full DS. The largest $d = 7$ entries inside the kernel (circled) are considered for joint significance, in this case corresponding to the 7 DS entries. **(C)** Kernel with sub-optimal parameters $l_K = 5$, $w_K = 5$ and $d = 5$ (as employed in the validation, see "Results"), centered around the first DS entry. The kernel covers only 2 DS entries, and additional background entries are considered for joint significance.

doi:10.1371/journal.pcbi.1004939.g005

entries considered in this case is re-calculated accordingly. A similar argument holds for entries I_{ij} close to the edge of the matrix.

Even when I_{ij} belongs to a DS fully covered by the kernel, only some of the n neighbor entries belong to the DS. For this reason we select only the d largest (i.e. most significant) entries in the neighborhood of P_{ij} and evaluate their joint statistical significance. The parameters l_K, w_K and d reflect the putative length l_{DS} of the DS, its wiggliness w_{DS} and the length l_{SSE} of the SSE, respectively. As a rule, we set l_K and w_K odd so that the kernel is symmetric around I_{ij}. The optimal choice of the parameters would be $l_K = 2l_{DS} - 1$ and $w_K = w_{DS}$, such that the kernel fully covers the DS also when placed at one end of the DS itself (see Fig 5B). If the kernel length l_K is larger than this optimal value, the kernel would cover all entries forming the DS plus additional background entries. However, as only the d largest entries are considered for the joint significance, the background entries would not lower the joint significance and thus would not reduce the power of the test. Conversely, a smaller value for l_K results in a kernel that covers only part of the DS, thus yielding decreased joint probability values in the matrix J (see Fig 5C). However, the method tolerates sub-optimal values for l_K because the joint significance is already very high when the kernel covers just a few (e.g. 2 or 3) entries. This is indeed the case we investigated in our validation (see "Results"), where we set the filter length to half of the optimal length. These considerations show that the performance of the method is robust with respect to departures of the value of l_K from the optimal choice.

The kernel width w_K reflects the putative wiggliness w_{DS} of the DS, and $w_K = 5$ corresponds to a value which accounts for highly wiggly DSs, as the one illustrated in Fig 5. Indeed, the events of the SSE are captured even if their delay from the first event changes between the two repetitions of the SSE by up to ±2 time bins. We set $w_K = 5$ in our validation (see "Results") in order to capture this bad case scenario. This value was larger than the real wiggliness of the DSs we actually generated in our test data, $w_{DS} \in \{1, 2\}$, so that the kernel spanned more diagonals than the DS. The performance was nevertheless close to maximal: as for the parameter l_K, the method tolerates excessively large values of w_K because only the d largest entries falling into the kernel are considered.

Finally, setting $d = l_{SSE}$ would ensure that only the l_{SSE} largest entries (likely to be those forming the DS, if any) are considered, and not all the n entries covered by the kernel. Deviations from this optimal value amount to include additional or fewer entries than those forming the DS. In "*Results*" we show that, however, non optimal values are also tolerated. Indeed, as soon as 2 to 3 DS entries are considered among the $d = 5$ that are selected to determine the joint significance in our validation, the statistical test crosses significance threshold in almost all the cases (FN rate close to 0).

Significance of the *d* largest kernel entries. Under the null hypothesis H_0 that the spike trains are independent and Poisson, we calculate the joint cumulative distribution of those d neighbors of I_{ij} whose corresponding cumulative probabilities $P^{(1)}, P^{(2)}, \ldots, P^{(d)}$ (already calculated and stored in the probability matrix P) are the largest among the n neighbors covered by the kernel. We consider $P^{(1)}, P^{(2)}, \ldots, P^{(d)}$, of which we aim to evaluate the joint significance, as independent samples from the same cumulative probability distribution $F(x)\colon = \Pr(P^{(k)} < x) = x$. Indeed, under H_0 each entry P_{ij} is the cumulative probability of the corresponding intersection value, and its distribution is therefore the identity function over the discrete domain of the possible intersection values from 0 to N. These entries are not exactly independent if they share an index, e.g. (i, j_1) and (i, j_2). The approximation as independent samples, however, yields high performance, as shown in the validation (see "Results").

For a sample $(X_1, X_2, \ldots, X_n)$ of n independent realizations extracted from a probability distribution F defined on the interval [0, 1], the joint survival function $\overline{F}$ of the d largest order statistics $X^{(n-d+1)} < X^{(n-d+2)} < \ldots < X^{(n)}$ is defined by

$$\overline{F}_{n-d+1,\ldots,n}(x_1, x_2, \ldots, x_d) := Pr(X^{(n-d+1)} \geq x_1, X^{(n-d+2)} \geq x_2, \ldots, X^{(n)} \geq x_d)$$

and represents the probability that at least d samples are larger than or equal to x_1, at least $d-1$ samples are larger than or equal to $x_2, \ldots$, and at least 1 sample is larger than or equal to x_d.

Since the $x_1 \leq \ldots \leq x_d$ are ordered, the distribution vanishes if at least d samples are not $\geq x_1$. Hence the number i_1 of samples $\geq x_1$ can take values $i_1 = d, d+1, \ldots, n$ and the remaining $n - i_1$ samples are below x_1. Of the i_1 samples $\geq x_1$, there must be at least $d-1$ samples $\geq x_2$. Of the i_2 samples $\geq x_2$, $i_2 = d-1, d, \ldots, i_1$, there must be at least $d-2$ samples $\geq x_3$ and so on. Continuing this reasoning we arrive at the expression

$$\begin{aligned}
&\overline{F}_{n-d+1,\ldots,n}(x_1, x_2, \ldots, x_d) \\
&= \sum_{i_1=d}^{n} \sum_{i_2=d-1}^{i_1} \cdots \sum_{i_d=1}^{i_{d-1}} \Pr(\text{exactly } i_k \text{ X's are } \geq x_k,\ k = 0, 1, \ldots, d) \\
&= \sum_{i_1=d}^{n} \sum_{i_2=d-1}^{i_1} \cdots \sum_{i_d=1}^{i_{d-1}} \Pr(\text{exactly } i_k - i_{k+1} \text{ X's are in } [x_k, x_{k+1}),\ k = 0, 1, \ldots, d) \\
&= \sum_{i_1=d}^{n} \sum_{i_2=d-1}^{i_1} \cdots \sum_{i_d=1}^{i_{d-1}} n! \prod_{k=0}^{d} \frac{[F(x_{k+1}) - F(x_k)]^{i_k - i_{k+1}}}{(i_k - i_{k+1})!},
\end{aligned}$$

for any $x_1 \leq x_2 \leq \ldots \leq x_d$, and 0 otherwise, where $x_0 = 0$, $x_{d+1} = 1$, $i_0 = n$ and $i_{d+1} = 0$.

Substituting to $F(x_r) \to P^{(r)}$ and denoting for simplicity $\overline{F}_{n-d+1,\dots,n}$ by $\overline{F}$ yields

$$\overline{F}(P^{(1)}, P^{(2)}, \dots, P^{(d)}) = n! \sum_{i_1=d}^{n} \sum_{i_2=d-1}^{i_1} \cdots \sum_{i_d=1}^{i_{d-1}} \prod_{k=0}^{d} \frac{(P^{(k+1)} - P^{(k)})^{i_k - i_{k+1}}}{(i_k - i_{k+1})!}, \tag{5}$$

which represents the upper-tail probability of the joint event

$$(X^{(n-d+1)} \geq P^{(1)}) \cap (X^{(n-d+2)} \geq P^{(2)}) \cap \dots \cap (X^{(n)} \geq P^{(d)})$$

under H_0.

The upper-tail probability is small (that is, reflects high significance) either if $P^{(1)}, P^{(2)}, \dots, P^{(d)}$ are jointly large (for instance, all >0.99) or if just one of them (e.g. the largest, $P^{(d)}$) is extremely large (e.g. 0.99999999). The first scenario indicates the presence of a possible DS, while the second corresponds to the presence of an isolated entry which is individually so large (tail probability very close to 1) that its joint significance with neighboring entries taking values at chance level still crosses the significance threshold α_2. To avoid the second scenario we impose an upper bound on the values of P by replacing each $P^{(r)}$, $r = 1, \dots, d$, with $P_*^{(r)} = \min(P^{(r)}, p_{\max})$ (e.g. $p_{\max} = 0.999$) to obtain

$$\overline{F}(P^{(1)}, P^{(2)}, \dots, P^{(d)}) = n! \sum_{i_1=d}^{n} \sum_{i_2=d-1}^{i_1} \cdots \sum_{i_d=1}^{i_{d-1}} \prod_{k=0}^{d} \frac{(P_*^{(k+1)} - P_*^{(k)})^{i_k - i_{k+1}}}{(i_k - i_{k+1})!}. \tag{6}$$

By doing so, statistically highly significant values of $\overline{F}$ are possible only in the presence of jointly large neighbors of P_{ij}, and not, as in the original expression Eq 5, when isolated entries take extremely high values while neighboring entries are at chance level.

The complementary function $1 - \overline{F}(\cdot)$ returns the probability of *not* having the joint event, and can be used to map the probability matrix P into a *joint probability matrix* J

$$P_{ij} \to J_{ij} = 1 - \overline{F}(P^{(1)}, P^{(2)}, \dots, P^{(d)}). \tag{7}$$

Fig 3B illustrates the joint probability matrix J derived from the probability matrix P in panel A.

After setting a significance threshold α_2 (e.g. $\alpha_2 = 0.99999$), we classify entries I_{ij} in the raw intersection matrix (Fig 1B) as having significantly jointly large neighbors if $J_{ij} > \alpha_2$.

Clustering entries of *I* into DSs

Each entry I_{ij} in the raw intersection matrix is tested for its individual significance and for the joint significance of its neighbors. If both tests pass, i.e. if $P_{ij} > \alpha_1$ and $J_{ij} > \alpha_2$, the entry is classified as belonging to one DS. Such entries are collected in a binary *masked matrix* M that takes value 1 if both tests pass, and 0 otherwise,

$$M_{ij} := 1_{\{I_{ij} > \alpha 1\}} \cdot 1_{\{J_{ij} > \alpha_2\}}, \tag{8}$$

as illustrated in Fig 3C. It remains to be established which entries belong together to the same DS. Intuitively, entries in the masked matrix that take value 1 belong to the same DS if close-by, and to different DSs if far apart. The masked matrix in Fig 3C shows for instance two clearly separated DSs.

A suitable notion of "distance" for matrix entries should make entries falling in the same diagonal, i.e. aligned along the natural direction of a DS, closer together than entries aligned along the anti-diagonal. We introduce the following elliptic distance between any two matrix entries (i_1, j_1) and (i_2, j_2):

$$d_\rho((i_1, j_1), (i_2, j_2)) := \left[1 + (\rho - 1) \cdot \sin\left(\left|\theta - \frac{\pi}{4}\right|\right)\right] \cdot \sqrt{(i_2 - i_1)^2 + (j_2 - j_1)^2} / \sqrt{2},$$

where $\theta = \arctan\left(\frac{j_2 - j_1}{i_2 - i_1}\right)$ is the angular coefficient of the line intersecting (i_1, j_1) and (i_2, j_2), the first square root factor is the Euclidean distance between the two points and $\rho \geq 1$ is a stretching factor for angular coefficients deviating from $\pi/4$. $d(\cdot)$ grows as θ approaches $3\pi/4$ or $-\pi/4$, i.e. the anti-diagonal orientation. For instance, $d_\rho((i, j), (i+k, j+k)) = k$ for any ρ, $d_\rho((i, j), (i+k, j-k)) = \rho k$ and $d_\rho((i, j), (i, j+k)) = \left[1 + \frac{\sqrt{2}}{2}(\rho - 1)\right]k$.

We set $\rho = 5$ and, based on the distance $d_5(\cdot, \cdot)$ of their positions in the matrix, group all entries I_{ij} with $M_{ij} = 1$ (see Eq 8) into clusters via a density based scanning (DBSCAN) algorithm [14]. The algorithm considers two entries as part of the same neighborhood if their distance is not larger than a maximum value ε. Neighborhoods sharing an entry are joined together and eventually classified as a cluster if they contain a minimum number l_0 of entries. We set $\varepsilon = 3.5$, thus allowing for a maximum of $\lfloor\varepsilon\rfloor - 1 = 2$ holes between two consecutive entries of a DS along the main diagonal, and $l_0 = 3$, thus requiring a DS to reflect at least 3 repeated synchronous events. The elliptical neighborhood used when clustering should be contained into the kernel used to build the joint probability matrix J. The reason is that the first defines the "immediate" neighbors of the entry, while the second is meant to cover all entries that may belong to the same DS. Thus, the parameter ε should not be chosen larger than the kernel length l_K, while the stretching coefficient ρ should be chosen such that the shorter axis of the ellipse fits into the kernel width w_K. The values $\varepsilon = 3.5$ and $\rho = 5$ we set for the validation satisfy these requirements. Fig 6 illustrates the matrix entries falling inside the ellipse (red dots) for various choices of the parameters ρ and ε, and the kernel (gray area) centered around the same entry. Entries in the matrix M not belonging to any cluster are discarded as events that do not reflect repeated SSE activity.

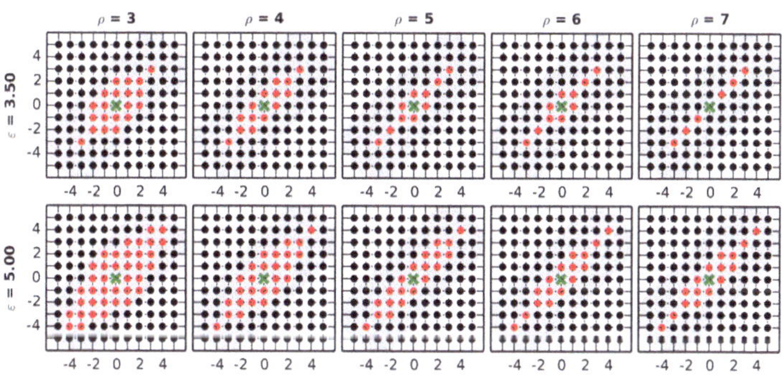

Fig 6. Neighborhood of SSE entries in DBSCAN clustering. Each panel shows, for a different choice of the stretching coefficient ρ and of the maximum distance ε, the entries falling inside (red dots) and outside (black dots) the ellipse defined by the elliptical distance. The ellipse is centered around the entry M_{ij} (green cross, here marked at position (0, 0)). Entries inside the ellipse are considered neighbors of M_{ij} by the DBSCAN clustering algorithm. The gray shaded area represents the kernel, centered around the position of M_{ij}, used to build the joint probability matrix J. The ellipse should be contained within the kernel centered around the same position.

doi:10.1371/journal.pcbi.1004939.g006

Fig 3D shows the *cluster matrix C* assigning value 1 (colored in black) to entries belonging to a cluster, and 0 (white) to the others. The matrix contains two clusters composed of 5 entries each, corresponding to the 5 synchronous events highlighted in Fig 1A.

Rate estimation methods

Calculating the entries of the probability matrix as given by Eqs 2 and 3 requires the knowledge of the firing rate of each neuron over time. Firing rate estimation is a problem that has been targeted by a number of studies.

The peri-stimulus time histogram (PSTH, [15]) is an estimate of the firing rate performed by discretizing time into adjacent bins and by counting the number of spikes falling into each bin. The bin width for the PSTH is typically larger (tens of ms) than the one used to define synchrony, as firing rates change on a slower time scale. The larger the bin width, the coarser but less biased the estimate. Normalizing the spike count in each bin by the bin width yields the rate of the process, i.e. the number of spikes per time unit.

Kernel convolution [16] replaces each spike with a kernel (a probability density function) centered around the spike time and estimates the firing rate by the sum of these distributions. Formally, this is done by representing the spike time by a Dirac delta function centered around the spike time t^* and by convolving it with the kernel. Following a standard choice, we specify the kernel as a normal distribution with assigned standard deviation σ, truncated at $\pm 2.7\sigma$ to yield a finite support. When setting the kernel width $w^* = 5.4\sigma$ to a fixed value, we employ $w^* = 200$ ms.

Both PSTH and kernel convolution can be applied to the case when multiple independent, identically distributed trials of the activity of a neuron are available, by averaging the estimates obtained for each trial. For identical bin and kernel widths, the PSTH typically better represents sharp changes in the firing rate from one bin to the next, while kernel convolution yields smoother curves.

Both estimates are parametric and require the choice of a bin- (kernel-) width. Methods have been recently proposed to determine the optimal bin- or kernel-width by minimizing the error between the true (unknown) rate and its estimate in some statistical sense (see [17, 18]). These methods have been shown to outperform their fixed width variants and are particularly helpful when analyzing parallel spike train data from different neurons with different rates, where the optimal bin width varies across neurons.

In the "*Results*" we compare the performance of ASSET employing either PSTH or kernel convolution or optimized-width kernel convolution [18] estimates of the firing rate profiles over an increasing number of trials.

Results

We propose here an extension of the method presented in [12] that visualizes repeated temporal sequences of synchronous events (SSE) as diagonal structures (DS, a sequence of large values along a diagonal) within an intersection matrix I. I_{ij} represents the number of neurons which have spikes both in time bin i and time bin j (see Fig 1 and Eq 1 for the formal definition). We map the intersection matrix I into a probability matrix P which contains at each position P_{ij} the probability of an overlap lower than I_{ij} between the spike trains at the two time bins (Fig 3A). The calculation is derived analytically under the null hypothesis H_0 that the spike trains are Poisson and independent. We then further compute the joint probability matrix J whose entries J_{ij} represent the cumulative joint probability of neighbors of I_{ij}, again under H_0 (Fig 3B). Set two statistical thresholds $\alpha_1 < \alpha_2$, each entry such that both $P_{ij} > \alpha_1$ and $J_{ij} > \alpha_2$ hold is classified as a potential member of a DS. Technically, this is done by building a masked matrix M such that $M_{ij} = 1$ if the two conditions hold, and $M_{ij} = 0$ otherwise (Fig 3C). Entries in M that take value 1 are finally clustered into individual DSs or discarded as isolated entries based on their reciprocal distance, thus yielding a cluster matrix C (Fig 3D).

The positions (i, j) of entries composing each DS, which can be obtained from the matrix C, allow us to reconstruct the synchronous events forming the associated repeated SSE as the intersection of the neurons active at bins i and j.

We investigate the performance of the proposed method on simulated test data. The data are generated by stochastic simulations of N parallel spike trains (here we chose $N = 100$) over a time period of $1s$. We define 10 types of background spiking activity, differing by the marginal properties of the spike trains (firing rate profile and ISI distribution) and the correlations among the spike trains. We use these data to determine the false positive (FP) rate of the method in cases where SSEs are not included in the data. Then we enrich the data with two occurrences of an SSE composed of $l_{SSE} = 7$ synchronous events, each event involving $\xi_{SSE} = 5$ neurons with IDs $1-5$, $6-10, \ldots, 31-35$, and investigate therein the true positive (TP) and FP rates. In data containing SSE activity, a found SSE is considered as an FP if it is not composed of (or it is only partly composed of) the events forming the embedded SSE. Details are given below. For the diversity of data types we test the power of the method, further identify critical cases, and suggest solutions.

Test data

To assess the quality of the method, we first measure its performance in the case when all assumptions entering the derivation of Eq 2 (alternatively Eq 3) and Eq 6 are met. In addition, we test the robustness of the statistics of the method with respect to deviations from these assumptions which are typically found in experimental data. In particular, we investigate how the following features of the data affect the performance of the method and test if these lead to false positive (FP) outcomes. The first four features relate to aspects of firing rates, such as various types of non-stationarities and rate correlations, i.e. correlations between the spike trains on a slower time scale than SSEs. The last two features relate to spike synchrony, however with a different organization than in SSEs.

1. Variability of firing rates over time by means of a sudden rate jump that is coherent across all neurons (Fig 7A).
   Firing rate changes over time are the basic observation of experimental data, in particular in response to an external stimulus or in relation to behavior. Neurons reacting to a common stimulus often exhibit a co-modulation of their rates. Ignoring or mis-estimating such rate changes is a typical generator of false positives in synchrony analyses [19–21]. We thus investigate a worst case scenario of coherent and instantaneous rate changes.
2. Heterogeneity of the firing rates across neurons (Fig 7B).
   The firing rates of simultaneously observed neurons typically differ. Although we estimate the firing rates on a neuron by neuron basis, we still want to ensure that the method can cope with this rate variability. Such variability combined with cross-trial variability was shown to be a strong generator of FPs (see e.g. [21]).
3. Increased regularity of the spiking activity compared to the Poisson assumption, by means of Gamma-distributed ISIs (Fig 7C).
   Deviations from Poisson statistics in terms of ISI regularity, which is observed in experimental data [22–24], was shown to bias synchrony analysis methods [20, 25] that, like ASSET, assume Poisson spike trains in the null hypothesis.
4. Short lasting, simultaneous rate jumps in a group of neurons, propagating to other groups in a sequence; this "chain" of rate jumps re-occurs at a later time (Fig 7D).

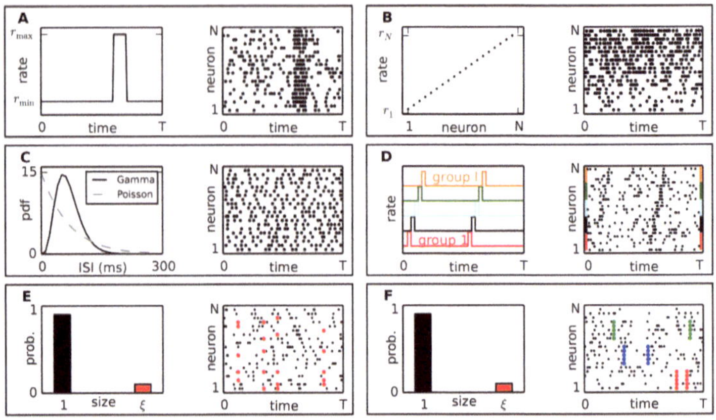

Fig 7. Different features of the data. Each box illustrates one feature (from A to F) of the test data (left column) and an associated exemplifying raster plot (right column). The dots in the raster plots mark the occurrence time of each spike. **(A)** Variability of firing rates over time. The underlying rate profile is shown on the left. **(B)** Heterogeneity of the firing rates across neurons: the different stationary firing rate levels are marked on the left. **(C)** Gamma-distributed ISIs: the underlying Gamma distribution (solid line) is contrasted to a Poisson distribution (dashed). **(D)** Short lasting, sequential rate jumps (different colors) coherent within groups of neurons, repeat identically at two time points. **(E)** Population synchronization: the occurrence probability of the size of the synchronous, but randomly selected spike events, is shown in red, the occurrence probability of independent background spikes (size 1) in black. **(F)** Intra-group synchronization: the complexity distribution is identical to the case of population synchronization in panel E, but the events involve specific groups of neurons (each group is marked in the dot display by one color).

doi:10.1371/journal.pcbi.1004939.g007

This model mimics a rate propagation model, instead of spike synchrony propagation as represented by SSEs. In the ongoing debate on rate coding vs temporal coding [26–28] it was proposed that coherent short-lasting firing rate changes at the input of neurons would be as efficient in bringing neurons to emit a spike as synchronous input. Whether rate correlation and spike synchrony can be distinguished mathematically is being debated (see e.g. [29] vs [30]). Here we ensure that ASSET can distinguish sequences of coherent rate changes from SSE activity when using suitable statistical thresholds.

5. Population synchronization, represented by the occurrence of synchronous spike events at random time points, each involving a random selection of neurons (Fig 7E).
ASSET operates under the null hypothesis of spike train independence given the estimated firing rate of each neuron. Thus, fine temporal correlations are not incorporated in the null hypothesis. However, studies of recurrent neuronal networks show the presence of weak temporal correlations [31], as those caused for instance by the recurrent connectivity in the network. To study their effect on the performance of our method, we generate test data that contain population correlations with unspecific neuronal compositions. As a correlation model we choose the compound Poisson process, which inserts synchronous spikes at a predefined occurrence rate in randomly selected sets of neurons, rather than in specific groups of neurons as for the SSEs.

6. Intra-groups synchronization, where multiple disjoint groups of neurons exhibit synchronous spiking within the group, at independent points in time for each group (Fig 7F). This model represents a second type of correlation structure at fine temporal scale differing from SSE activity. Here, synchronous spike events of specific groups of neurons exist in the same number (7) and size (5) as for the embedded SSE, but in contrast to the SSEs are not emitted in a temporal sequence. This type of correlation was explored already in [32], were it was shown to produce in the intersection matrix isolated high-valued entries, but no DSs. We test here whether this holds true for our more advanced method.

For the concrete test cases we formulate 10 different stochastic models of spiking activity (see Fig 8 as examples of the realizations), each including only one or a combination of the above-mentioned features, as summarized in Table 1. We use these models to generate background activity into which we subsequently embed the spiking activity corresponding to a repeated SSE. We provide here the definition of each model.

Model 0—Independent Poisson. $N = 100$ independent Poisson spike trains having identical, stationary firing rates $\lambda = 15$ Hz.

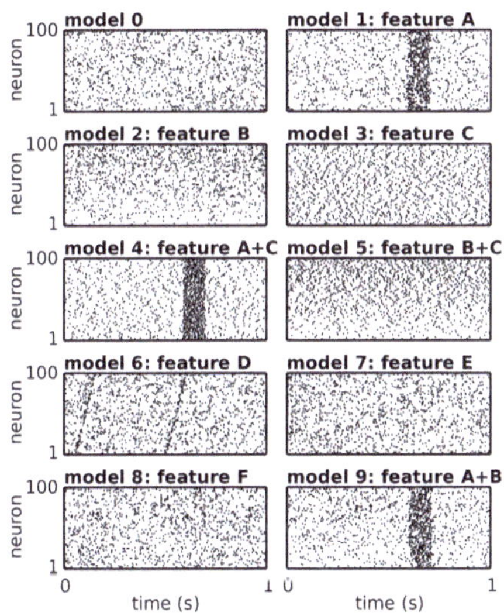

Fig 8. Spiking activity of the stochastic models. Examples of population raster plots of the spiking activity of 100 neurons (vertical axis) over time (horizontal axis), for each stochastic model 0 to 9. Model 0 consists of independent Poisson spike trains with firing rates that are stationary over time and identical across neurons. The other models destroy one or more of these aspects by including some of the features A to F, as listed in Table 1.

doi:10.1371/journal.pcbi.1004939.g008

Table 1. Features embedded in the stochastic models. The crosses mark the features (A to F) included and combined in each model (0 to 9) to simulate the background activity of the test data.

	model id									
	0	1	2	3	4	5	6	7	8	9
A: firing rates varying over time		x			x					x
B: heterogeneous rates across neurons			x			x				x
C: high regularity of ISIs				x	x	x				
D: rate propagation							x			
E: population synchronization								x		
F: intra-groups synchronization									x	

doi:10.1371/journal.pcbi.1004939.t001

Model 1—Poisson with simultaneous rate jump. $N = 100$ Poisson spike trains having an identical, time-varying firing rate profile $\lambda(t) = 10\text{ Hz} + 50\text{ Hz} \cdot 1_{\{600\text{ms}<t<700\text{ms}\}}$. Thus, all spike trains undergo a simultaneous sudden rate excursion of +50 Hz between times $t_1 = 600$ ms and $t_2 = 700$ ms. The average rate over the 1 s simulation period is 15 Hz, as in model 0.

Model 2—Heterogeneous, time-stationary Poisson. $N = 100$ independent Poisson spike trains with heterogeneous firing rates across neurons ranging from 5 Hz to 25 Hz. Neuron k has stationary firing rate $\lambda_k = 5 + \frac{20}{99} k$ Hz, $k = 0, 1, \ldots, 99$. The population-averaged firing rate is 15 Hz, identical to models 0 and 1.

Model 3 / 4 / 5—High inter-spike interval regularity. Same as models 0 / 1 / 2 respectively, but the spike trains have marginally Gamma distributed inter-spike intervals (ISIs) with shape factor $\alpha = 5$ ($\alpha = 1$ for the Poisson case) and mean parameter $\mu = 1/15$ s

$$\mathbb{P}(\text{ISI} = \tau) = \left(\frac{\alpha}{\mu}\right)^{\alpha} \frac{\tau^{\alpha-1} e^{-\alpha\tau/\mu}}{\Gamma(\alpha)}.$$

The Gamma distribution determines ISIs having mean length μ (firing rate: $1/\mu = 15$ Hz), standard deviation $\sigma = \mu/\sqrt{\alpha}$ and thus a coefficient of variation $\sigma/\mu = 1/\sqrt{\alpha}$. The larger the α, the more regular the ISIs compared to their mean value.

Model 6—Fast rate-jump propagation. $N = 100$ independent Poisson spike trains, organized in 20 groups of 5 spike trains each. The first group experiences a sudden, simultaneous rate jump from $\lambda_1 = 14$ Hz to $\lambda_2 = 100$ Hz at two times $t_1 = 50$ ms and $t_2 = 500$ ms, lasting 5 ms (one time bin) each time. Then the second group undergoes the same rate jump 5 ms (one bin) later, and so on. The l-th group experiences the rate jump at time $t_1 + 5(l - 1)$ ms and $t_2 + 5(l - 1)$ ms, $l = 1, \ldots, 20$.

Model 7—Population synchronization. $N = 100$ spike trains having all stationary firing rates $\lambda = 15$ Hz. From time to time, 5 randomly selected neurons fire synchronous spikes, with a frequency that yields a mean pairwise correlation coefficient among any two spike trains of $\rho = 0.01$. The rest of the spiking activity is mutually independent. Formally, the model is described by a Compound Poisson process (CPP; [33, 34]) with a two-peak amplitude distribution $A(\cdot)$ such that $A(1) = 0.938$, $A(5) = 0.062$ and $A(\xi) = 0$, $\forall \xi \neq 1, 5$. Each neuron has an average participation rate in synchronous events of 4.63 Hz.

Model 8—Intra-group synchronization. Multiple Single-Interaction Process (mSIP, see [32, 35]; cf. [36] for details on the basic SIP model) comprising 65 independent neurons plus 7 groups (SIPs) of 5 neurons each, for a total of $N = 100$ neurons. Each SIP exhibits synchronous activity in two randomly selected time bins, independent for each group. Thus, the resulting data contains 7 disjoint repeated synchronous events of size 5. This is the same as for the case when we inject a repeated SSE into the data, with the difference that the synchronous events do not form a temporal sequence here. Each neuron has a total firing rate $\lambda = 15$ Hz, stationary over time.

Model 9—Time-varying heterogeneous Poisson. $N = 100$ independent, marginally Poisson spike trains whose firing rates are heterogeneous across neurons and non-stationary over time. Spike train k has rate profile

$$\lambda_k(t) = \left(5 + \frac{10 \cdot k}{99}\right)\text{Hz} + 50\,\text{Hz} \cdot 1_{\{600\text{ ms}<t<700\text{ ms}\}},$$

$k = 0, 1, \ldots, 99$. Thus, the spike trains have baseline firing rates ranging from 5 Hz to 15 Hz and experience a coherent rate jump of +50 Hz between 600 ms and 700 ms. The time- and population-averaged firing rate is 15 Hz. This model mixes non-stationarity and heterogeneity of firing rates individually characterizing models 1 and 2, respectively.

Fig 8 shows example raster plots of the activity associated to each of these stochastic models, without additional injection of SSE activity.

Validation of the method on test data

Significance of the two statistical tests. Eq 3 provides an analytical expression of the cumulative probability P_{ij} of the intersection value I_{ij} under the null hypothesis H_0 of independent, marginally Poisson spike trains. The complementary test p-value $1 - P_{ij}$ represents the probability that the intersection between bins b_i and b_j is larger than or equal to I_{ij}. Analogously, Eq 6 provides the joint tail probability $1 - J_{ij}$ of neighbors of P_{ij}. For positions (i, j) of the intersection matrix corresponding to a real DS, J_{ij} should take values which are orders of magnitude lower than P_{ij}, as it represents the joint significance of multiple rather than individual repeated synchronous events. To confirm this, we simulate data from each background activity model 0 to 9 (see Fig 8) for 100 realizations. For each realization, we additionally inject at two points in time, chosen at random, an SSE composed of 7 links and 5 neurons/link, resulting in a real DS composed of a set of 7 entries $\mathcal{S}_{\text{DS}} := \{(i_k, j_k),\ k = 1, 2, \ldots, 7\}$ in the intersection matrix. We estimate the firing rate profile of each neuron from single spike trains by kernel convolution (boxcar kernel, kernel width: 200 ms), segment the data in 5 ms bins and calculate the matrices P and J. Given the large number of neurons involved, we need to rely on Le-Cam's approximation of the probability mass function of the intersection values (see Eq 3) to compute P. We then extract from each matrix the tail probabilities $1 - P_{i_k, j_k}$ and $1 - J_{i_k, j_k}$ of the entries forming the embedded DS, and consider the largest (i.e. least significant) ones, $\max_k(1 - P_{i_k, j_k})$ and $\max_k(1 - J_{i_k, j_k})$. Fig 9 shows for each model a cloud of the 100 points $(\max_k(1 - J_{i_k, j_k}), \max_k(1 - P_{i_k, j_k}))$, one per simulation. As expected, for entries belonging to the true DS the joint significance J_{ij} is orders of magnitude higher than the individual significance P_{ij}. For this reason it makes sense to set the respective statistical thresholds α_2 and α_1 such that $\alpha_2 > \alpha_1$. In particular, on the basis of the scatter plots in Fig 9, which show that most tail probabilities $1 - P_{i_k, j_k}$ and $1 - J_{i_k, j_k}$ are lower than 10^{-2} and 10^{-5} respectively, we set $\alpha_1 = 0.99$ and $\alpha_2 = 0.99999$.

True positive and false positive DSs. For each simulation of models 0 to 9 we identify statistically significant entries (i, j) in the intersection matrix. We then cluster these entries into diagonal structures using the DBSCAN algorithm, as explained in "*Methods*". Table 2 summarizes the parameter values used in the analysis.

We define a DS found in stochastic simulations containing SSE activity as a TP if it contains at least 50% of the entries of the true DS, and if at least 50% of its entries belong to the true DS. If either requirement is not met, or if the data do not contain embedded SSE activity in the first place, the found DS is classified as an FP. Fig 10, left, illustrates the entries composing a true

PLOS | COMPUTATIONAL BIOLOGY

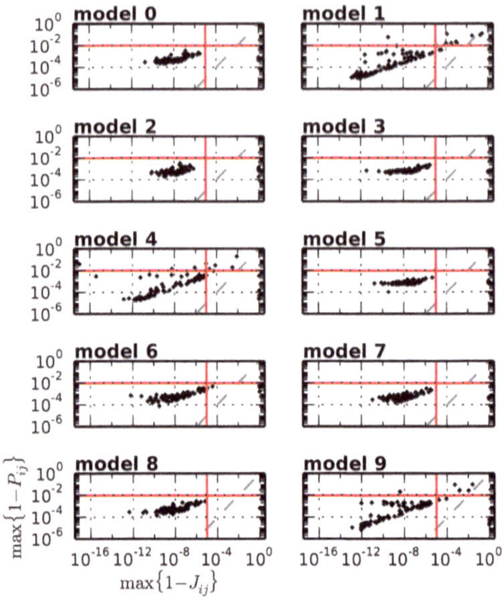

Fig 9. Scatter of the upper-tail probabilities of the two tests in the presence of a repeated SSE. The upper tail probability of the first test (y-axis) in relation to that of the second test (x-axis) for all models 0, . . ., 9. Each model is repeated 100 times and in each repetition an SSE composed of 7 links and 5 neurons/link is injected at two random points in time. The represented values $\max(1 - P_{i_k, j_k})$ and $\max(1 - J_{i_k, j_k})$ are taken each as the maximum (i.e. least significant) significance value across all entries $S_{DS} := \{(i_k, j_k), k = 1, 2, \ldots, 7\}$ of the intersection matrix composing the embedded DS. The figure shows for each model the cloud of the 100 points $(\max(1 - J_{i_k, j_k}), \max(1 - P_{i_k, j_k}))$, one per simulation. The red lines mark the significance levels $1 - \alpha_1$ (horizontal) and $1 - \alpha_2$ (vertical) for individual tail probability values $1 - P_{ij}$ and tail joint probability values $1 - J_{ij}$, respectively.

doi:10.1371/journal.pcbi.1004939.g009

Table 2. Analysis parameters. Parameters of the ASSET method employed for the analysis of stochastic and network data. Each column shows the parameters employed for the corresponding step of the method.

intersection matrix		statistical assessments		DBSCAN clustering	
parameter	value	parameter	value	parameter	value
Δ: bin size	5 ms	α_1: 1st threshold	0.99	ρ: stretch coeff.	5
l: kernel length	5	α_2: 2nd threshold	0.99999	ε: max distance	3.5
w: kernel width	5			s: min cluster size	3
d: # largest neighbors	5				

doi:10.1371/journal.pcbi.1004939.t002

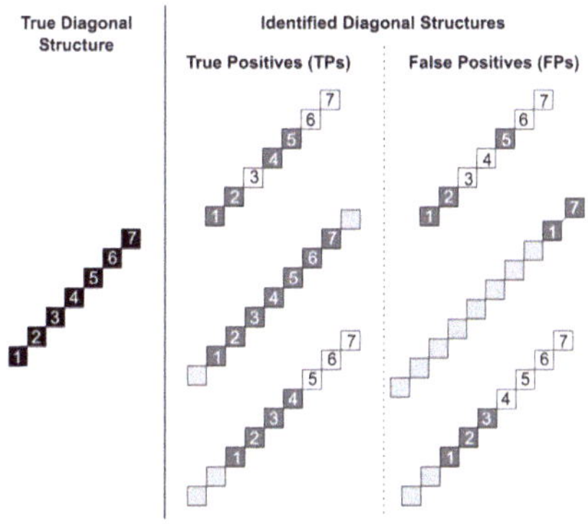

Fig 10. Illustration of true positive and false positive DSs. Left: DS composed of 7 entries, resulting from the two occurrences of the SSE embedded in the test data. **Middle, right:** A found DS in the cluster matrix can contain entries that belong to the true DS (dark gray) or not (light gray), and may miss entries of the true DS (white). **Middle column:** Different types of TP DSs found in data: contained in (*top*), containing (*middle*) and partially overlapping with (*bottom*) the true DS. **Right column:** Different types of FP DSs found in data: contained in (*top*), containing (*middle*) and partially overlapping with (*bottom*) the true DS.

doi:10.1371/journal.pcbi.1004939.g010

DS in data with embedded SSE activity. Fig 10, right, illustrates FPs arising from the fact that the found DS contains less than 50% of the entries composing the true DS (top, first requirement not met), or contains more than 50% of entries not belonging to the real DS (middle, second requirement not met). The central column in the figure similarly illustrates different types of TPs, when both criteria are satisfied. Note that both TP and FP DSs can be contained in the real DS (top), contain it (middle) or partially overlap with it (bottom).

Fig 11, *top*, shows the FP rate (i.e. average number of FPs over the 100 simulations) in each test data model introduced above, for the case where no SSE is embedded in the data (therefore, each found DS is an FP). The middle and bottom panels show the TP and FP rate for the corresponding cases where two occurrences of an SSE are embedded in the data. Optimal performance is achieved in SSE-free data when the FP rate, which can take values between 0 and $+\infty$, is 0. In data with embedded SSEs the TP rate, ranging between 0 and 1, is optimally 1.

Assumptions of null hypothesis met. When the null hypothesis of Poisson and independent spike trains holds, and the firing rates are stationary over time (model 0, identical rates across neurons, and model 2, different rates) the performance is high both in terms of TP rate (= 1.00) and FP rate (= 0.00, both in test data with and without embedded SSEs). Cross-neuron heterogeneity of firing rates does not affect the performance. These results demonstrate the adequacy of kernel convolution, used to estimate the rate profiles on a single spike train basis, when the rates are stationary. The high performance also indicates the efficacy of Le Cam's approximation (Eq 3) of the true but intractable probability distribution of intersection values (Eq 2).

ISI regularity. The optimal TP and FP levels observed in model 0 and 2 are maintained when the spike trains exhibit higher ISI regularity than Poisson (model 3 and 5, respectively). This shows that the method is robust to deviations from the Poisson assumption employed to derive the null distribution.

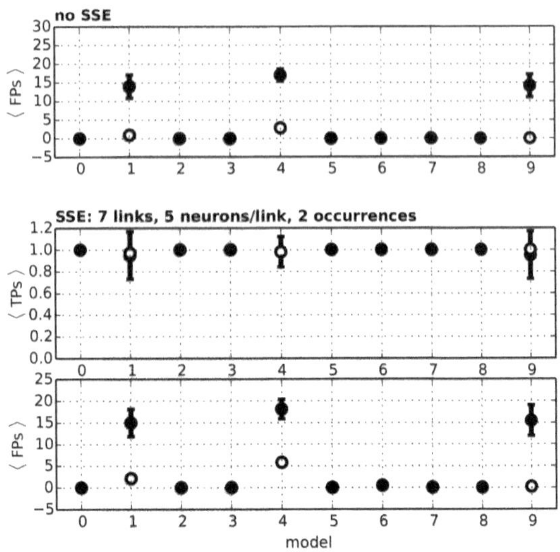

Fig 11. Performance of the method for different models of background activity. The 10 different models of background (SSE-free) activity, numbered from 0 to 9, are simulated 100 times each. **Top panel:** Average number of FPs found in the data when estimating the firing rates of individual spike trains by kernel convolution (filled circles) and when using the theoretical rate profiles that underlie the generation of the test data (empty circles). Bars indicate the standard deviation. **Center and bottom panel:** Average number of TPs and FPs found in the data containing 2 repetitions of an SSE of $l_{SSE} = 7$ synchronous events involving $\xi_{SSE} = 5$ neurons each.

doi:10.1371/journal.pcbi.1004939.g011

Rate propagation. We further evaluate whether high and short-lasting rate increases affecting a group of neurons simultaneously and propagating from one group to the next in a sequence would be interpreted by the method as the occurrence of an SSE. The main difference between rate propagation and SSE activity is that the first causes neurons to fire stochastically (probability <1) rather than reliably (probability $= 1$). The first model converges to the second as the rate increase grows to infinity, so that ASSET should find SSE activity in this case. It is otherwise understood as a different model of information coding, which should not yield SSEs.

Using model 6, we simulate an intermediate scenario in which each of 20 groups of neurons successively increases its firing rate from a baseline of 14 Hz to 100 Hz for a period of 5 ms, with an inter-group delay of 5 ms matching the analysis bin width. This fast rate change propagation occurs twice, thus resembling a repeated SSE (see Figs 7D and 8, model 6), but with the firing probability of each neuron of 0.39 during the high-rate regime (instead of 1 for SSE activity and 0.067 during the baseline regime). With these parameters the method yields an FP rate of 0.48 (equivalent to less than 1 FP every 2 model iterations), despite the unrealistically high and sudden rate jump, its coherence across neurons and the fact that the high-rate state perfectly falls into adjacent analysis bins. Deviations from these three features would decrease the false discovery rate substantially. This indicates that the method distinguishes rate propagation from SSE activity as long as the two models are substantially different, but would identify the two when the two models converge.

Correlated spike trains. The presence of synchrony involving different groups of neurons each time (model 7) does not harm the performance of the method: the TP and FP rates stay at levels 1.00 and 0.00, respectively. The same results are obtained in data characterized by specific groups of neurons that synchronize their activity within the group, however independently among groups (model 8). These results taken together indicate that, for the range of parameters employed here, the presence of synchronous activity, at the population level or even involving specific groups of neurons, does not affect the performance of the method.

In general, however, one may explicitly include such correlations in the null-hypothesis. To this end we suggest here a Monte-Carlo approach to estimate the probability matrix J while taking the observed repeated synchronous events into account. The basic idea is to estimate J from surrogates obtained by manipulations of the intersection matrix I by destroying the arrangement of rows and columns of I. Using this approach, we aim to distinguish synchrony that leads only to isolated high-valued entries in the intersection matrix (non-SSEs) from synchrony that leads to DSs. Concretely, the steps are:

1. Generate S (e.g. $S = 1000$) surrogates of the binned discretized data by shuffling the bins randomly.
2. For each surrogate s, $s = 1, 2, \ldots, S$, compute the corresponding intersection matrix $\tilde{I}_s$ and derive the associated probability matrix $\tilde{P}_s$ by Eq 2 or Eq 3, using the original firing rate profiles. This operation preserves all synchronous events in the original data but not their temporal order, and effectively corresponds to an identical random shuffling of rows and columns of P. Any DS originally present in P is thus destroyed in $\tilde{P}_s$, thereby implementing the null hypothesis $\mathcal{H}_{00}$ that the observed repeated events do not form temporal sequences.
3. Transform each surrogate probability matrix $\tilde{P}_s$ into a filtered probability matrix $\tilde{F}_s$ as follows:
   - apply to each entry $\tilde{P}_{s:ij}$ a rectangular kernel
   - for each $\tilde{P}_{s:ij}$ extract its d largest (most significant) neighbors $\tilde{P}^{(k)}_{s:ij}$, $k = 1, 2, \ldots, d$ and set $\tilde{F}_{s:ij} := 1 - \prod_k (1 - \tilde{P}^{(k)}_{s:ij})$.
4. In the same way transform the original probability matrix P into a filtered matrix F.
5. Compute the significance of each entry F_{ij} in F by comparison with the sample $\{\tilde{F}_{s:ij},\ s = 1, 2, \ldots, S\}$, i.e. set

$$\tilde{J}_{ij} = \frac{1}{S} |\{s:\ \tilde{F}_{s:ij} \le F_{ij}\}|.$$

The matrix $\tilde{J}$ is a Monte-Carlo estimate of the joint probability matrix under $\mathcal{H}_{00}$. Its entries J_{ij} are statistically significant if $\tilde{J}_{ij} > \alpha_2$. In summary, if a repeated SSE was present in the data, the surrogate destroys the corresponding DS in I and thereby leads to higher (more significant) values in $\tilde{J}$ than if there was no such arrangement in the first place, as e.g. for non-SSE synchrony.

Fig 12 shows the performance on models 0 to 9 using this Monte-Carlo estimate (and thus working under $\mathcal{H}_{00}$). The FP rate is comparable to that obtained using the analytical approach under H_0. Importantly, this holds also for models 6 – 8 where correlations are present,

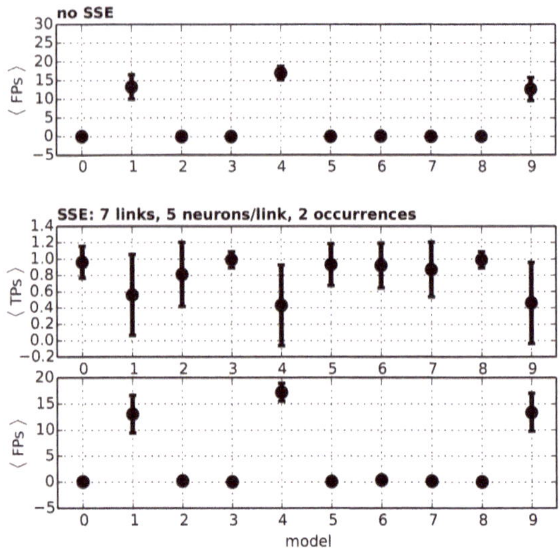

Fig 12. Performance of the Monte-Carlo approach. Top panel: Average number of FPs (vertical axis) found over 100 simulations of each model from 0 to 9 (horizontal axis) using the Monte-Carlo estimate $\tilde{J}$ of the joint probability matrix under $\mathcal{H}_{(0)}$. **Middle panel:** Average number of TPs found for each model from 0 to 9 after injecting two repetitions of an SSE (equivalent to one DS in the intersection matrix) with 7 events and 5 neurons/event. **Bottom panel:** Average number of FPs for the same data as in the middle panel.

doi:10.1371/journal.pcbi.1004939.g012

demonstrating that the correlations embedded in the data do not bias the analytical estimates. The Monte Carlo approach shows also decreased TP rate, especially for the cases where firing rates are non stationary (models 1, 4 and 9, TP rate <0.6). The reason is that time bin shuffling effectively destroys the rate profiles and thus ignores rate changes when building the null distribution. Nevertheless, scenarios may be thought of where correlations need to be taken into account by means of the Monte-Carlo approach illustrated above.

Time-varying firing rates. We further investigate how time-varying firing rates affect the performance of the method. Model 1 contains independent Poisson spike trains, as in model 0, with the difference that the firing rate profiles exhibit a coherent rate jump from 10 Hz to 60 Hz at time $t_1 = 600$ ms and back to 10 Hz at time $t_2 = 700$ ms (see Figs 7A and 8). Similarly, models 4 and 9 combine this rate non-stationarity with Gamma-distributed ISIs (Fig 7C) and with cross-neuron rate heterogeneity (Fig 7B), respectively. In all cases the FP rate increases beyond 14.0 (*filled circles* in Fig 11). The problem resides in the inability to estimate the instantly varying firing rates by means of a fixed-width kernel convolution, which smooths the rate jump over a larger window. Indeed, when using the true rate profiles to compute the statistics, the FP rate drops back to values lower than 2.0 for models 1 and 9, and as low as 5.7 for model 4 which features highly regular ISIs (*empty circles*).

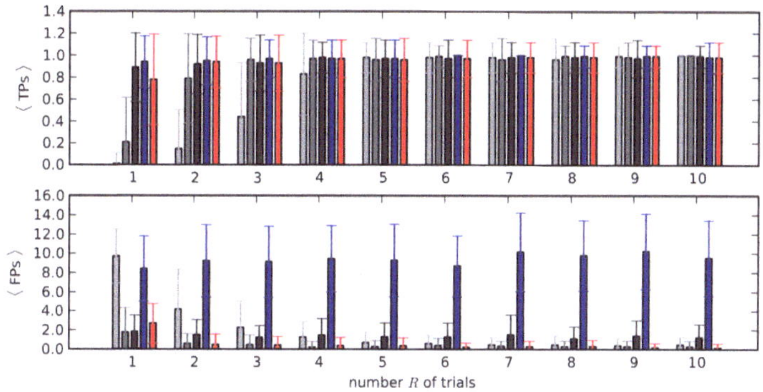

Fig 13. Rate profiles estimated via 3 different methods computed over an increasing number of trials. Average number of TPs (*top*) and FPs (*bottom*) resulting from different firing rate estimation methods, as a function of the number of trials used to estimate the firing rates of individual neurons (horizontal axis). The data are defined by model 4 and enriched with two repetitions of an SSE with l_{SSE} = 7 links and ξ_{SSE} = 5 neurons/link. *Gray bars*: PSTH (dark: 20 ms bin width, medium-dark: 10 ms, light: 5 ms). *Blue bars*: kernel convolution with fixed kernel width (200 ms). *Red bars*: optimized-bandwidth kernel convolution.

doi:10.1371/journal.pcbi.1004939.g013

In realistic scenarios of analysis of experimental data firing rate profiles are not known. However multiple trials, i.e. repetitions of the same stimulus or behavioral condition related to a given trigger event, are often available that make the estimation of the firing rate possible and more reliable. Common tools for firing rate estimation across trials are the peri-stimulus trial histogram (PSTH, [15]) or trial-averaged kernel estimates. We use model 4, which represents a worst case scenario (sudden co-modulation of rates, non-Poisson spiking), to investigate whether cross-trial estimation of the firing rates is accurate enough for our needs since it comprises a worst case scenario (coherent rate jump, non-Poisson). To this end we estimate the firing rate profile of each neuron by simulating its activity R times ("trials"), R = 1,2,. . .,10 and computing its time-resolved average rate over these trials using three different estimation techniques (for details, see "*Rate estimation methods*"): PSTH with a bin width of 5, 10 or 20 ms, kernel convolution [16] with fixed kernel width (200 ms) and kernel convolution with an optimized kernel width [18]. Fig 13 illustrates the resulting performance for each method as a function of the number R of trials considered for the estimation. The PSTH (gray bars) performs well for a properly predetermined bin width (here the best being 10 ms, resulting in a TP rate >0.9 and an FP rate <0.2 when 3 or more trials are available). The same problem affects kernel convolution with a fixed kernel width, which is here chosen too large, resulting in large numbers of FPs. Instead, the optimized-width kernel convolution solves the problem by determining the optimal kernel width in a statistical sense, and yields here a TP rate >0.9 and an FP rate <0.2 as soon as 3 or more trials are considered. Thus, this proves to be a suitable method for the rate estimation for experimental data, which are typically characterized by strong and fast rate changes, provided that multiple trials are available.

Number of neurons and size of the DS. We finally investigate how the power of the method is affected by a reduced number of neurons involved in SSE activity in the total data set. To this end, we repeatedly simulate background activity as in model 0 varying the total number

N of neurons from 50 to 500, and add two occurrences of an SSE varying the number l_{SSE} of links (from 3 to 7) and the number ξ_{SSE} of synchronous spikes per link (from 2 to 5). For each value of N, l_{SSE} and ξ_{SSE} we generate the data 100 times and analyze each of them with ASSET employing the parameters reported in Table 2. Note that the kernel length employed ($l_K = 5$) is suboptimal for values of l_{SSE} higher than 3, as in those cases the kernel centered at the ends of the DS does not cover the full DS and a longer kernel would yield higher performance.

The results are shown in Fig 14. The TP rate (column A) increases for a larger size ξ_{SSE} of the synchronous events and a smaller total number N of spike trains, as a larger fraction of the total neurons are involved in the SSE. However, the TP rate does not increase substantially with the number l_{SSE} of synchronous events composing the embedded SSE, because the kernel has fixed length and does not cover the additional events, thus resulting in stationary joint significance values J_{ij} of the second statistical test. For low l_{SSE}, low ξ_{SSE} or large N the TP rate drops considerably to values lower than 0.5. This is partly due to our strict definition of a TP, which does not include found SSEs containing less than 50% of the synchronous events composing the embedded SSE (true positive events). Fig 14, column B, shows the average fraction of true positive events composing each found SSE. The FP rate (column C) is close to 0 (and always lower than 0.1; note the different scale of the color map) for all investigated values of l_{SSE}, ξ and N. Importantly, it does not increase with N, which is a relevant feature of the method for applications to large-scale data. For $l_{SSE} = 7$ and $\xi_{SSE} \geq 3$ we observe occasionally jumps in the FP rate. Almost all of these FPs consist of SSEs which partially overlap with the embedded one, but not enough to be classified as TPs. As a comparison, Fig 14, column D, shows the rate of FP SSEs that are completely disjoint from the true one. This rate is indeed close to 0 for all parameter choices, showing that the increased FP rates shown in column C result from partially true discoveries.

We then investigate the special case of SSEs composed of 1 spike per group only ($\xi_{SSE} = 1$), to test if our method is able to detect spatio-temporal patterns involving no spike synchronization. We inject as before two repetitions of a spatio-temporal pattern, now composed of l_{SSE} = 7 spikes with a time delay of 5 ms, into independent data (model 0, composed of $N = 10$ neurons). Because the previous analysis indicates low performance already for $\xi_{SSE} = 2$ for the parameters employed so far, we increase the kernel length to $l_K = 7$.

As shown in Fig 15, the tail probability of individual entries in the intersection matrix corresponding to the real DS and the joint tail probability of their $d = 5$ largest neighbors are both too large (weakly significant) to yield sufficient test power. A larger number of repetitions of the spatio-temporal pattern may increase the statistical significance to acceptable levels, as commented in "Discussion". Thus, we conclude that ASSET is not suitable for the detection of spatio-temporal patterns involving no spike synchrony.

Analysis of a synfire chain network simulation

Synfire chain data. Synthetic data from simulations of a balanced neuronal network with embedded overlapping synfire chains (SFCs, [8]) were produced in [12]. The full network comprises 40,000 excitatory and 10,000 inhibitory neurons, with biologically realistic connectivity. 50 synfire chains are embedded, each obtained by selecting 2000 neurons of the network and randomly organizing them in 20 successive groups of 100 neurons each. All neurons in one group are connected to all neurons in the next group in a feed-forward fashion, resulting in high synaptic convergence and divergence. Individual neurons participate in up to 3 SFCs, which therefore overlap. The SFCs are stimulated by injection of a current pulse in the first link of the chain, leading to synchronous spiking activity in the first group of the chain which propagates downstream. The propagation delay between one group and the next is ~2 ms. The

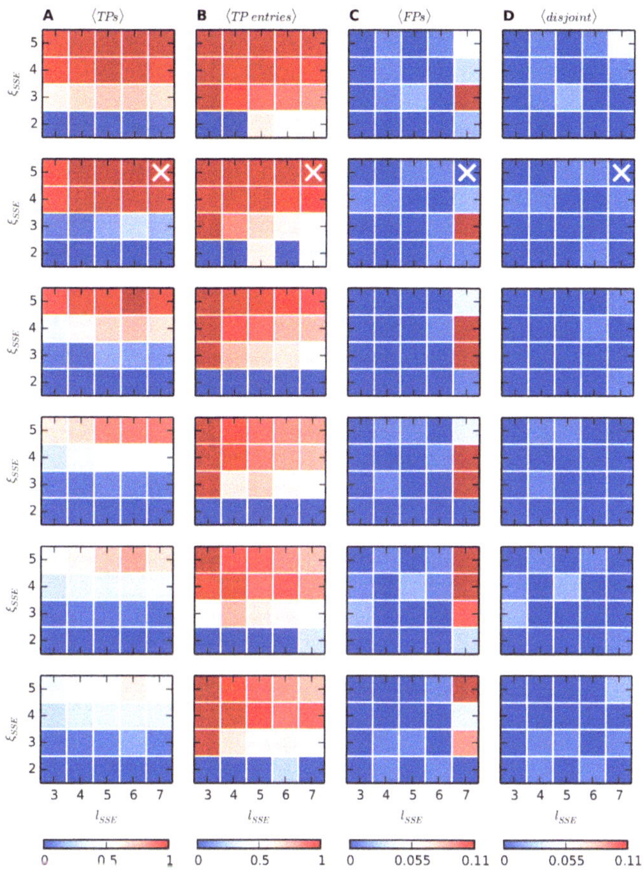

Fig 14. *Performance as a function of l_{SSE}, ξ_{SSE} and N.* **(A)** Average number of TP SSEs (over 100 model simulations) found in data where an SSE with varying number of links (*horizontal axis*, from 3 to 7) and neurons per link (*vertical axis*, from 2 to 5) is injected twice in the activity of *N* otherwise independent spike trains (model 0). *N* varies from 50 to 500 (top to bottom panels). For *N* = 100 the cross at position ($l_{SSE} = 7$, $\xi_{SSE} = 5$) marks the parameters used in the previous calibrations. **(B)** Average number of events of the embedded SSE contained in each found SSE ("true positive events"). **(C)** Average number of FP SSEs (note the different scale of the color map). A FP can either be completely disjoint from the true embedded SSE, or contain part (or all) of its member events (see definition above and Fig 10). **(D)** Average number of found SSEs completely disjoint from the embedded one.

doi:10.1371/journal.pcbi.1004939.g014

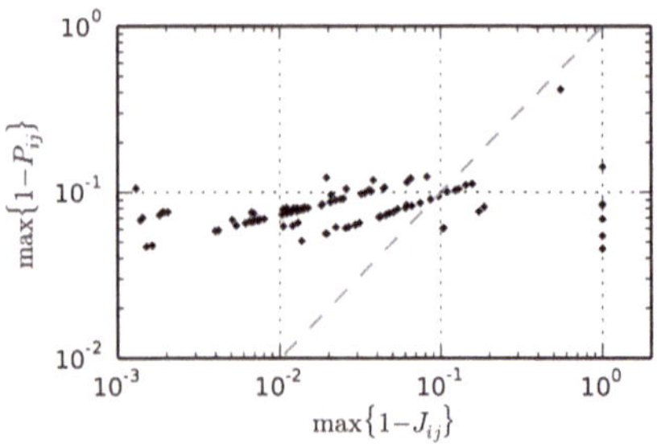

Fig 15. Scatter plot of the upper-tail probabilities of the two tests in the presence of a repeated spatio-temporal pattern. The upper tail probability of the first test (vertical axis) over the upper-tail probability of the second test (horizontal axis) for a reduced version of model 0. The model consists of parallel activity from $N = 10$ independent Poisson spike trains, plus two repetitions of a temporal sequence of $l_{SSE} = 7$ spikes from 7 different neurons ($\xi_{SSE} = 1$). The repeated spatio-temporal pattern generates a DS in the intersection matrix composed of 7 entries $S_{DS} := \{(i_k, j_k), k = 1, 2, \ldots, 7\}$, and corresponding values P_{i_k, j_k} and J_{i_k, j_k} in the probability and joint probability matrices, respectively. The data are generated 100 times. For each data set and corresponding DS in the intersection matrix, the figure shows a point $(\max_k(1 - J_{i_k, j_k}), \max_k(1 - P_{i_k, j_k}))$, corresponding to the least significant entries in the matrices J and P among those composing the DS. The figure shows the cloud of the 100 such points, one per simulation.

doi:10.1371/journal.pcbi.1004939.g015

propagation is robust to the presence of noise and does not require all neurons in a link to be active for propagation to the next link [10].

We consider the activity of 2000 neurons of the network, having identities 8001 to 10,000, over a time segment of 300 ms. The considered neurons compose the entirety of one of the 50 SFCs. In particular, neurons 8001 to 8100 compose the first link, neurons 8101 to 8200 the second link and so on until neurons 9901 to 10,000, which compose the last link of the chain. Fig 16 shows the activity of these neurons over a time stretch of 10 s (A) and of 300 ms (B,C). Panels A and B are adapted from Figure 2 in [12]. As apparent from panel B, in the considered time window the selected synfire chain is active three times. The first and last time the activity runs through all the 20 links, the second time it propagates only until the 13-th link. Panel C reproduces the same data as in panel B, but with a random shuffling of the neuron identities along the vertical axis. Here, the SFC activity misleadingly appears as a co-modulation of firing rates across neurons.

In addition, some of these neurons also belong to 3 successive groups of a second SFC. The 3 groups comprise 14, 12 and 14 neurons of the population analyzed, respectively. In the considered time period, this second SFC is active twice. Contrarily to the first SFC, the activity of the second SFC does not become apparent in the raster plot of Fig 16B, because the sorting of neuron identities along the vertical axis of the plot does not place neurons belonging to the same group close-by. Neither does this activity reflect in the plot as population co-modulation, because it involves a relatively small fraction (2%) of the population.

Analysis of synfire chain data. We analyze the stretch of data illustrated in Fig 16B with ASSET to demonstrate that the method is able to discover repeated synfire chain activation. The activation of each SFC generates an SSE, and each pair of SSEs corresponds to a DS in the intersection matrix. Therefore, the triple activation of the first SFC generates 3 diagonal structures that the method should find in the cluster matrix: one (DS 1, composed of 20 links) is given by the overlap of the two complete runs of the SFC (which involve 20 synchronous events

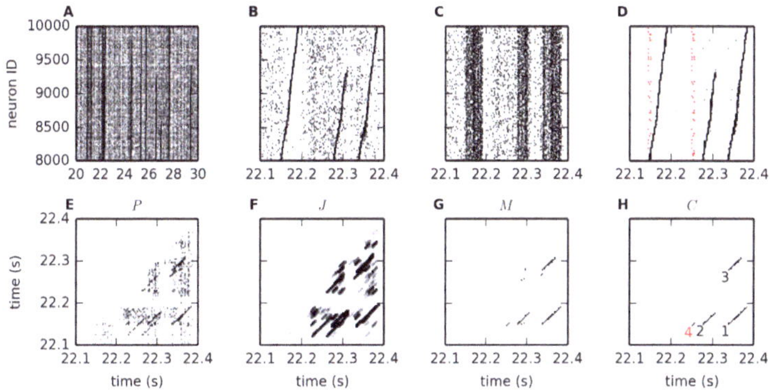

Fig 16. Synfire chain network data and analysis results. (A-D) Activity of 2000 neurons (numbered 8001 to 10,000) forming the entirety of an active SFC in a balanced random network [12] composed of 20 groups. Neurons 8001 – 8100 form the first group of the chain, neurons 8101 – 8200 the second group and so on. A subset of the neurons also participate in 3 successive groups of a second SFC in the network. **(A)** Activity over a time window of 10 s (replicated from [12], Fig 2A). **(B)** Enlargement of (A) showing 300 ms of data (replicated from [12], Fig 2B). The SFC under consideration is activated three times in this time interval. **(C)** Same as in panel (A), but with random sorting of the neuron IDs on the vertical axis. **(D)** SSE activity detected by ASSET. In the time interval considered, the two SFCs are stimulated three times (black dots) and two times (red dots), respectively. **(E-H)** Probability matrix *P*, joint probability matrix *J*, masked matrix *M* and cluster matrix *C*. Numbers 1 to 3 and 4 in the cluster matrix mark the significant DSs found by ASSET and associated to the activations of the first and second SFCs, respectively.

doi:10.1371/journal.pcbi.1004939.g016

each), and the other two (DSs 2 and 3) are composed of 13 links each, resulting from the comparison of each complete activation with the one stopping at the 13-th group. The double activation of the second SFC, composed of 3 successive synchronous events each, should analogously yield 1 significant DS (DS 4) of length 3. Indeed, the method finds all of these DS, as shown in Fig 16H. The raster plot in Fig 16D shows the SSEs generated by the activation of the two SFCs (black and red dots, respectively; spikes not belonging to the SSEs not shown), which ASSET reconstructs in their entirety.

Table 3, first row, reports for each DS the true number of composing entries and the true distance (in number of bins) between the first and the last event for the first SSE occurrence (*dx*) and for the second SSE occurrence (*dy*). The second row shows the corresponding values

Table 3. Diagonal structures of active synfire chains. The top row shows in bold the true number of repeating synchronous events forming the four pairs of repeating SSEs present in the data. The second row shows the number of entries composing each of the four associated DSs as found by the analysis, and in brackets the distance (in number of bins) between the first and the last event of the first SSE occurrence (*dx*) and the second SSE occurrence (*dy*). The other rows in the table show the median value of the entries in the intersection matrix (*second row*), in the probability matrix (*third row*) and in the joint probability matrix (*fourth row*) corresponding to each DS.

	DS 1	DS 2	DS 3	DS 4
true nr. entries (*dx*/*dy*)	**20**	**13**	**13**	**3**
nr. entries (*dx*/*dy*)	23 (14/15)	13 (9/9)	18 (10/10)	3 (2/2)
median i-mat value	93	88	53	14
median p-mat value	$< 10^{-22}$	$< 10^{-22}$	$< 10^{-22}$	$\sim 6.6 \cdot 10^{-8}$
median j-mat value	$< 10^{-22}$	$< 10^{-22}$	$< 10^{-22}$	$\sim 3.9 \cdot 10^{-14}$

doi:10.1371/journal.pcbi.1004939.t003

found by the analysis. Due to a larger bin width (3 ms) than the inter-link propagation delay (~ 2 ms), the found SSEs are shorter than the number of links in the chain: some successive events fall in the same time bin and are therefore merged into a single event. Nevertheless the composing spikes are correctly retrieved, as well as the associated units. Rows 3 to 5 in the table show the median values of matrix entries corresponding to the DS in the intersection matrix, the probability matrix and the joint probability matrix, respectively. As a comparison, the median values of the full matrices were 0, 1 and 1 respectively, indicating the high statistical significance of the entries forming the found DSs compared to the other entries in the matrices. Note that the kernel length employed (length $l_K = 5$) is shorter than the number of entries composing the DSs associated to the first SFC (≥ 9). Nevertheless, these DSs are successfully retrieved.

These results taken together demonstrate the ability of the method to retrieve repeated SSE activity, here generated by repeatedly active SFCs, in massively (2000) parallel spike trains. The method yields high significance values even when employing sub-optimal parameters, such as a larger bin size than the inter-link transmission delay orsignificantly smaller kernel length than the DS length.

Computational performance

The computational cost of the algorithm is almost entirely determined by the time required to evaluate the joint probabilities defined in Eq 6. The expression involves a nested sum of several terms. The number of terms grows with the number n of matrix entries covered by the kernel (determined by the kernel length l_K and the kernel width w_K) and with the number d of largest neighbors among which the joint significance is computed. These are free parameters of the analysis, whereas features of the data, such as the number of neurons or their firing rates, do not influence this step of the computation. As an estimate, for the values we employed in the manuscript ($l_K = 5$, $w_K = 5$, $d = 5$) the evaluation of a single entry took about 10 ms on a single core of a dual AMD 12-core Opteron 6174 machine with 64GB RAM using the Python code provided with this manuscript. Thus, the evaluation of a full matrix J of 200×200 entries took on average less than 7 minutes. However, the fact that single entries are evaluated independently may be easily exploited by parallelizing the analysis on multi-core machines or computer clusters, where each worker process is assigned to perform the computation for a subset of the matrix entries (see [37]).

Discussion

Temporal sequences of synchronous spike events (SSEs) have been postulated as a working mechanism of activity propagation in the cortex [38–40]. The present manuscript introduces a novel statistical method for the detection of SSEs in massively parallel spike train data, named ASSET (Analysis of Synchronous Spike EvenTs). The method is inspired by a visual technique first proposed in [12], which represents the repeated occurrence of an SSE as a sequence of large entries along the diagonals of an intersection matrix (diagonal structure, or DS) that indicates for any two time bins the number of neurons firing in both bins. ASSET automatizes the detection process of the original visual technique, assesses the statistical significance of SSEs by exploiting the multiple evidence of its events to derive their joint significance, and determines the structure and neuronal composition of the identified SSEs. In evaluating the null distribution, the method accounts for the temporal profiles of the firing rates of the observed neurons. As such, it detects SSEs which cannot be explained on the basis of rate coding mechanisms, and thus arise from spike correlations on a shorter temporal scale. Rate correlation is understood as a conceptually different mechanism of information processing than spike synchrony (as for SSEs), because it corresponds to stochastic (probability <1) rather than reliable (probability 1) neuron activation [29, 41–43].

We assessed the performance of ASSET in terms of false positive (FP) DSs found in various types of stochastic models which mimicked typical features of neuronal spike trains, such as variable firing rates in time or across neurons, different inter-spike interval distributions, and correlation structures differing from SSEs. We then additionally injected repeated SSE activity in the data to assess the power of the method in terms of true positive (TP) detections. The analysis performs two statistical tests on each pair of time bins, which amounts to tens of thousands of tests for a stochastic simulation of 1 s binned at 5 ms. In addition, entries passing the two tests have to lie close to each other in the matrix in order to be clustered into a common DS. To avoid FPs, the statistical threshold needs to be set to low values (here, 10^{-5}). This is possible without incurring into large levels of FNs because the joint tail probabilities associated to the second test are very low (typically $<10^{-5}$, and as low as 10^{-12}) for the entries corresponding to the embedded SSEs. Indeed, the method shows high performance, i.e. FP and FN rates both close to 0. We did not need to further correct the statistical thresholds by the amount of tests performed (e.g. by Bonferroni or FDR correction), because already the set of combined requirements that need to be fulfilled to identify a DS (success of the two tests for matrix entries, and proximity of these entries in order to be clustered into a DS) makes the analysis very conservative.

The underlying null hypothesis assumes independent Poisson spike trains, which enables an analytical formulation of the test statistics. The method proves to be robust to deviations from Poissonianity, such as a higher regularity of the inter-spike intervals, which may be observed in experimental data [22, 44]. It is also selectively sensitive to SSEs, but not to other models of spike correlation, such as synchronous events not organized in a temporal sequence. Nevertheless, anticipating scenarios where strong correlations not forming an SSE might indeed bias the statistics, we proposed a Monte-Carlo approach to account for these correlations. To this end, we constructed the null hypothesis by estimating the probability to find a certain degree of pattern overlap from the repeated generation of surrogate intersection matrices. In our test data the Monte-Carlo approach yielded results comparable to the analytical approach, yet at a considerably higher computational cost.

Furthermore, the method was able to distinguish SSEs from repeated precise temporal sequences of sharp, local rate transients from one group of neurons to another (rate propagation). Rate peaks increased the probability of the involved neurons to spike, which remained nevertheless a stochastic event, in contrast to SSE activity. Propagation of rate transients thus generated spike patterns that were different in composition (due to the stochastic activation of neurons) and less precisely timed than SSEs, but more closely resembled the latter as the rate modulation became higher and faster. It is likely that, for extremely high firing rates and very short rate jumps, ASSET would not distinguish the two models. The distinction between rate correlation and synchrony correlation (see e.g. [29]) has been formally questioned in [30]. Users of ASSET may want to identify waves of co-modulating rates like those defined in our model 6 as SSEs, rather than rejecting them. This is possible by decreasing the statistical thresholds α_1 and α_2 to less strict values than those used in this manuscript.

ASSET critically relies on the estimation of the firing rate profile of each neuron to compute the expected overlap of neuron IDs in the intersection matrix and thus to estimate the significance of the observed overlap. Estimating firing rate profiles on the basis of single trial spike data typically requires some kind of sliding window approach, such as convolution of each spike with a temporal kernel [16]. Temporal averaging smears out peaks of the underlying original firing rate that occur on a shorter time scale than the window width, and creates artificial peaks if the window width is excessively short. Single-trial rate estimates obtained by kernel convolution in the presence of time-stationary firing rates yielded high performance of ASSET, but led to impaired performance when the rates were non-stationary in time on a fast time scale, in particular if the rate excursion was coherent across neurons. The reason is that smoothing by convolution underestimates positive rate peaks and thus the expected overlap, yielding FPs. Estimating the firing rates on the basis of trial averages solved the problem. We here tested three such approaches, namely the peri-stimulus time histogram (PSTH, [15]), a trial-averaged kernel convolution with fixed kernel width [16] and a trial-averaged kernel convolution with an optimized kernel width [18]. Already a small number of trials reduced the FP rate considerably for all three methods, although best performance was achieved for the optimized-width kernel convolution. Importantly, cross-trial rate estimation works under the assumption of identical rate profiles across trials. Deviations from this assumption lead to a wrong estimation of the rate in single trials, that is required to calculate the probability matrix, and thereby enhances the FPs. This bias is amplified if neurons exhibit cross-trial variability in a coherent manner [19]. Latency variability is a special instance of cross-trial non-stationarity which causes a mis-estimation around the rate onset. In some cases the onset variability of rates can be corrected for by choosing a more proper alignment of trials, e.g. to the stimulus or behavioral event related to the rate change [45]. If this is not possible, we suggest to generate the probability matrix P on the basis of surrogate spike data, e.g. by spike train shifting or spike dithering [20]. The details of such an approach, however, still need to be explored.

We further investigated how the performance of ASSET relates to various other parameters of the SSEs, such as the number of its sequential synchronous events composing the SSE, the number of neurons in each synchronous event, and the total number of observed neurons. SSEs were statistically more significant and therefore easier to detect when they involved a larger fraction of the total neurons. However, they could be retrieved even when employing sub-optimal parameters such as a kernel length smaller than the length of the DS, given that the SSE involved enough neurons. In contrast, the method did not detect spatio-temporal patterns, i.e. a special case of SSE where each event is composed by a single spike only. Spatio-temporal patterns in a more general sense (with different time intervals between spikes) were suggested as signatures of synfire chain activity in data of low numbers of simultaneously recorded neurons [40, 46, 47].

A less constrained model of cortical processing (synfire braid) was proposed in [48], which incorporates synchronous input to individual neurons as in the synfire chain model, however transferred by connections of different temporal delays compensated by differences in their activation times. This model was further analyzed in [11] and termed polychronization. Since spike synchrony occurs with a temporal lag in this framework, one expects that, although ASSET is not designed to capture this type of coordinated spiking activity, it may still detect signatures of such activity by choosing a correspondingly larger bin width. We aim to explore such a scenario and other applications of ASSET to data from different types of network simulations that exhibit correlations on a fine temporal scale to study the potential of our method in identifying features of the underlying network model.

Importantly, increasing the number of neurons to be analyzed does not increase the computational cost of the method. Indeed, in order to evaluate significance values, the method relies on expressions involving a mere sum of the firing rates of individual neurons, which are virtually instantaneous to evaluate. Rapid advances in electrophysiology will soon enable the simultaneous recording of thousands of neurons [49]. These data promise to expose concerted mechanisms of neuronal coding that remained invisible so far. Our method is designed to keep up with these advances, and to be applicable to the next generation of large-scale recordings of spike data.

If an SSE occurs more than two times, it generates multiple DSs in the intersection matrix, each corresponding to one pair of occurrences. In [32] it was suggested to compute the overlap between triplets (or n-tuples) of bins rather than pairs, and generate a corresponding n-dimensional intersection matrix to visualize DSs in three (or n) dimensions. It is possible to extend this approach to ASSET and to exploit this higher-order evidence to increase the power of the method. This extension will be considered in future work.

Finally, we demonstrated the efficacy of ASSET on data of large-scale simulations of a balanced random network with embedded synfire chains, which were previously generated and analyzed in [12] with the original visual method. ASSET fully reconstructed the synfire chains active in the considered time period and did not return additional FPs. Differently from the original technique in [12], SSEs were here determined on the basis of their statistical significance and the results were obtained by an automated analysis workflow.

When analyzing real data, some parameters of the analysis such as the statistical thresholds α_1 and α_2 should be chosen optimally to minimize the risk of FPs, while at the same time not being excessively strict and thus missing true SSEs. Optimal values for the statistical thresholds can be inferred from independent surrogates of the original data, which can be created by one of several approaches, such as spike train shifting or spike dithering [50]. Such surrogate data contain slightly displaced spikes as compared to the original data, such that correlations in the original data (and in particular SSEs) are intentionally destroyed while other features of the data (e.g. firing rates or inter-spike interval regularities) are preserved. These uncorrelated data can be used to determine the expected value of entries in the probability and joint probability matrices under independence, and therefore to determine lower bounds for the thresholds α_1 and α_2 which ensure avoidance of FPs. Taking the least conservative of such values (the lower bounds) also minimizes the risk for FNs. Suitable values for other analysis parameters can be determined analogously. As illustrated in "Methods", some of these parameters are tightly associated to putative features of the searched SSEs (for instance, the kernel length and kernel width to the temporal span and the wiggliness of the SSE, respectively), and may therefore be tuned in order to optimize the detection of SSE swith specific characteristics.

Taken together, these results demonstrate that ASSET is a reliable and effective tool to detect and identify repeated sequences of synchronous spiking activity in massively parallel spike train data. Whether synchrony propagation constitutes a mechanism of information processing in the neuronal circuitry still remains an open question, that belongs to the more general debate about the role of fine temporal coding versus rate coding. Convincing theoretical arguments as well as experimental evidence have been provided for both processing mechanisms (see e.g. [2, 6, 51] vs [26, 52, 53]). However, observing fine-scale temporal correlations requires the simultaneous observation and analysis of sufficiently large portions of the involved neuronal circuitry: the severe subsampling of such circuitry so far characterizing most available data prevents this analysis. Next-generation recording technology promises to expose this concerted activity, whose analysis may finally resolve the long-standing question about the role of millisecond-precise spike correlations in cognitive processing. Our work gives a contribution in this direction by providing, to the best of our knowledge, the first tool for a statistical analysis of synchrony propagation applicable to data of hundreds or thousands of simultaneously recorded neurons. The ASSET analysis method is available as part of the Electrophysiology Analysis Toolkit (Elephant; http://neuralensemble.org/elephant/). In future work we plan to employ the method to investigate the presence of SSE activity in electrophysiological recordings from awake animals and study SSE occurrence in relation to behavior.

Acknowledgments

The authors wish to thank Tom Tetzlaff for stimulating discussions in particular in a very early phase of this project during a scientific stay of George Gerstein at the INM-6 in Jülich in spring

2012, Junji Ito for ideas on how to efficiently generate surrogates, and Markus Diesmann for many discussions on embedding this work into the modeling context.

Author Contributions

Conceived and designed the experiments: ET MD SG GG. Performed the experiments: ET CC. Analyzed the data: ET CC. Contributed reagents/materials/analysis tools: ET MD SG MH GG CC. Wrote the paper: ET SG MH MD CC.

References

1. Abeles M. Role of cortical neuron: integrator or coincidence detector? Israel Journal of Medical Sciences. 1982; 18:83–92. PMID: 6279540
2. Roy SA, Alloway KD. Coincidence detection or temporal integration? What the neurons in somatosensory cortex are doing. The Journal of Neuroscience. 2001; 21(7):2462–2473. PMID: 11264320
3. Bender VA, Bender KJ, Brasier DJ, Feldman DE. Two Coincidence Detectors for Spike Timing-Dependent Plasticity in Somatosensory Cortex. The Journal of Neuroscience. 2006; 26(16):4166–77.
4. Fino E, Paille V, Cui Y, Morera-Herreras T, Deniau JM, Venance L. Distinct coincidence detectors govern the corticostriatal spike timing-dependent plasticity. Journal of Physiology. 2010; 588(16):3045–3062. doi: 10.1113/jphysiol.2010.188466 PMID: 20603333
5. Perez-Orive J, Bazhenov M, Laurent G. Intrinsic and circuit properties favor coincidence detection for decoding oscillatory input. The Journal of Neuroscience. 2004; 24(26):6037–47. doi: 10.1523/JNEUROSCI.1084-04.2004 PMID: 15229251
6. Hong S, Ratté S, Prescott SA, De Schutter E. Single neuron firing properties impact correlation-based population coding. The Journal of Neuroscience. 2012; 32(4):1413–1428. doi: 10.1523/JNEUROSCI.3735-11.2012 PMID: 22279226
7. Braitenberg V, Schüz A. Anatomy of the Cortex: Statistics and Geometry. Berlin, Heidelberg, New York: Springer-Verlag; 1991.
8. Abeles M. Corticonics: Neural Circuits of the Cerebral Cortex. 1st ed. Cambridge: Cambridge University Press; 1991.
9. Diesmann M, Gewaltig MO, Aertsen A. Characterization of synfire activity by propagating 'pulse packets'. In: Bower JM, editor. Computational Neuroscience: Trends in Research. San Diego: Academic Press; 1996. p. 59–64.
10. Diesmann M, Gewaltig MO, Aertsen A. Stable propagation of synchronous spiking in cortical neural networks. Nature. 1999; 402(6761):529–533. doi: 10.1038/990101 PMID: 10591212
11. Izhikevich EM. Polychronization: Computation with Spikes. Neural Computation. 2006; 18:245–282. doi: 10.1162/089976606775093882 PMID: 16378515
12. Schrader S, Grün S, Diesmann M, Gerstein G. Detecting synfire chain activity using massively parallel spike train recording. Journal of Neurophysiology. 2008; 100:2165–2176. doi: 10.1152/jn.01245.2007 PMID: 18632888
13. Le Cam LM. On the distribution of Sums of Independent Random Variables. In: Neyman J, Cam LML, editors. Bernoulli, Bayes, Laplace: Proceeding of an International Research Seminar. New York: Springer-Verlag; 1963. p. 179–202.
14. Ester M, Kriegel HP, Sander J, Xu X. A density-based algorithm for discovering clusters in large spatial databases with noise. AAAI Press; 1996. p. 226–231.
15. Perkel DH, Gerstein GL, Moore GP. Neuronal Spike Trains and Stochastic Point Processes. I. The Single Spike Train. Biophysical Journal. 1967; 7(4):391–418.
16. Nawrot M, Aertsen A, Rotter S. Single-trial estimation of neuronal firing rates: from single-neuron spike trains to population activity. Journal of Neuroscience Methods. 1999; 1(94):81–92. doi: 10.1016/S0165-0270(99)00127-2
17. Shimazaki H, Shinomoto S. A method for selecting the bin size of a time histogram. Neural Computation. 2007; 19(6):1503–1527. doi: 10.1162/neco.2007.19.6.1503 PMID: 17444758
18. Shimazaki H, Shinomoto S. Kernel bandwidth optimization in spike rate estimation. Journal of Computational Neuroscience. 2010; 29:171–182. doi: 10.1007/s10827-009-0180-4 PMID: 19655238
19. Grün S, Riehle A, Diesmann M. Effect of cross-trial nonstationarity on joint-spike events. Biological Cybernetics. 2003; 88(5):335–351. doi: 10.1007/s00422-002-0386-2 PMID: 12750896

20. Grün S. Data-driven significance estimation of precise spike correlation. Journal of Neurophysiology. 2009;(101):1126–1140. doi: 10.1152/jn.00093.2008 PMID: 19129298
21. Grün S, Rotter S, editors. Analysis of Parallel Spike Trains. Springer; 2010.
22. Shinomoto S, Sakai Y, Funahashi S. The Ornstein-Uhlenbeck process does not reproduce spiking statistics of neurons in prefrontal cortex. Neural Computation. 1999; 11(4):935–951. doi: 10.1162/089976699300016511 PMID: 10226190
23. Farkhooi F, Strube-Bloss MF, Nawrot MP. Serial correlation in neural spike trains: Experimental evidence, stochastic modeling, and single neuron variability. Physical Review E. 2009; 79(2):021905. doi: 10.1103/PhysRevE.79.021905
24. Farkhooi F, Muller E, Nawrot MP. Adaptation reduces variability of the neuronal population code. Physical Review E. 2011; 83(5):050905. doi: 10.1103/PhysRevE.83.050905
25. Pipa G, Grün S, van Vreeswijk C. Impact of spike train autostructure on probability distribution of joint spike events. Neural Computation. 2013; 25(5):1123–63. doi: 10.1162/NECO_a_00432 PMID: 23470124
26. Shadlen MN, Newsome WT. Noise, neural codes and cortical organization. Current Opinion in Neurobiology. 1994; 4(4):569–579. doi: 10.1016/0959-4388(94)90059-0 PMID: 7812147
27. Softky WR. Simple codes versus efficient codes. Current Opininon in Neurobiology. 1995; 5:239–247. doi: 10.1016/0959-4388(95)80032-8
28. Shadlen MN, Newsome WT. Is there a signal in the noise? Current Opinion in Neurobiology. 1995; 5:248–250. doi: 10.1016/0959-4388(95)80033-6
29. Staude B, Rotter S, Grün S. Can Spike Coordination be differentiated from Rate Covariation? Neural Computation. 2008; 20:1973–1999. doi: 10.1162/neco.2008.06-07-550 PMID: 18336077
30. Amarasingham A, Geman S, Harrison MT. Ambiguity and non identifiability in the statistical analysis of neural codes. Proceedings of the National Academy of Sciences. 2015; 112(20):6455–6460. doi: 10.1073/pnas.1506400112
31. Helias M, Tetzlaff T, Diesmann M. Echoes in correlated neural systems. New Journal of Physics. 2013; 15:023002. doi: 10.1088/1367-2630/15/2/023002
32. Gerstein GL, Williams ER, Diesmann M, Grün S, Trengove C. Detecting synfire chains in parallel spike data. Journal of Neuroscience Methods. 2012; 206:54–64. doi: 10.1016/j.jneumeth.2012.02.003 PMID: 22361572
33. Staude B, Rotter S, Grün S. CuBIC: cumulant based inference of higher-order correlations in massively parallel spike trains. Journal of Computational Neuroscience. 2010; 29:327–350. doi: 10.1007/s10827-009-0195-x PMID: 19862611
34. Staude B, Grün S, Rotter S. Higher-order correlations in non-stationary parallel spike trains: statistical modeling and inference. Frontiers in Computational Neuroscience. 2010; 4. doi: 10.3389/fncom.2010.00016 PMID: 20725510
35. Grün S, Abeles M, Diesmann M. Impact of higher-order correlations on coincidence distributions of massively parallel data. In: Lecture Notes in Computer Science, 'Dynamic Brain—from Neural Spikes to Behaviors'. vol. 5286; 2008. p. 96–114.
36. Kuhn A, Aertsen A, Rotter S. Higher-order statistics of input ensembles and the response of simple model neurons. Neural Computation. 2003; 15:67–101. doi: 10.1162/089976603321043702 PMID: 12590820
37. Denker M, Wiebelt B, Fliegner D, Diesmann M, Morrison A. Practically Trivial Parallel Data Processing in a Neuroscience Laboratory. In: Grün S, Rotter S, editors. Analysis of Parallel Spike Trains. Springer; 2010.
38. Abeles M, Gerstein GL. Detecting Spatiotemporal Firing Patterns Among Simultaneously Recorded Single Neurons. Journal of Neurophysiology. 1988; 60(3):909–924. PMID: 3171666
39. Prut Y, Vaadia E, Bergman H, Haalman I, Hamutal S, Abeles M. Spatiotemporal Structure of Cortical Activity: Properties and Behavioral Relevance. Journal of Neurophysiology. 1998; 79(6):2857–2874. PMID: 9636092
40. Ikegaya Y, Aaron G, Cossart R, Aronov D, Lampl I, Ferster D, et al. Synfire chains and cortical songs: temporal modules of cortical activity. Science. 2004; 5670(304):559–564. doi: 10.1126/science.1093173
41. De la Rocha J, Doiron B, Shea-Brown E, Kresimir J, Reyes A. Correlation between neural spike trains increases with firing rate. Nature. 2007; 448(16):802–807. doi: 10.1038/nature06028 PMID: 17700699
42. Shea-Brown E, Josic K, de la Rocha J, Doiron B. Correlation and synchrony transfer in integrate-and-fire neurons: basic properties and consequences for coding. Physical Review Letters. 2008; 100:108102. doi: 10.1103/PhysRevLett.100.108102 PMID: 18352234

43. Schultze-Kraft M, Diesmann M, Gruen S, Helias M. Noise Suppression and Surplus Synchrony by Coincidence Detection. PLoS Computational Biology. 2013; 9(4):e1002904. doi: 10.1371/journal.pcbi.1002904 PMID: 23592953

44. Nawrot MP, Riehle A, Aertsen A, Rotter S. Spike Count Variability in Motor Cortical Neurons. In: European Journal of Neuroscience. vol. 12, Supl. 11; 2000. p. 506.

45. Grün S, Diesmann M, Aertsen A. 'Unitary Events' in Multiple Single-Neuron Spiking Activity. II. Non-Stationary Data. Neural Computation. 2002; 14(1):81–119. doi: 10.1162/089976602753284464 PMID: 11747535

46. Nadasdy Z, Hirase H, Czurko A, Csicsvari J, Buzsaki G. Replay and Time Compression of Recurring Spike Sequences in the Hippocampus. The Journal of Neuroscience. 1999; 19(21):9497–9507. PMID: 10531452

47. Shmiel T, Drori R, Shmiel O, Ben-Shaul Y, Nadasdy Z, Shemesh M, et al. Temporally precise cortical firing patterns are associated with distinct action segments. Journal of Neurophysiology. 2006; 96(5):2645–2652. doi: 10.1152/jn.00798.2005 PMID: 16885517

48. Bienenstock E. A model of neocortex. Network. 1995; 6:179–224. doi: 10.1088/0954-898X_6_2_004

49. Schwarz DA, Lebedev MA, Hanson TL, Dimitrov DF, Lehew G, Meloy J, et al. Chronic, wireless recordings of large-scale brain activity in freely moving rhesus monkeys. Nature Methods. 2014; 11:670–676. doi: 10.1038/nmeth.2936 PMID: 24776634

50. Louis S, Gerstein GL, Grün S, Diesmann M. Surrogate spike train generation through dithering in operational time. Front Comput Neurosci. 2010; 4(127). doi: 10.3389/fncom.2010.00127 PMID: 21060802

51. Ohiorhenuan IE, Mechler F, Purpura KP, Schmid AM, Hu Q, Victor JD. Sparse coding and high-order correlations in fine-scale cortical networks. Nature. 2010; 466:617–621. doi: 10.1038/nature09178 PMID: 20601940

52. Shadlen MN, Movshon AJ. Synchrony Unbound: A Critical Evaluation of the Temporal Binding Hypothesis. Neuron. 1999; 24:67–77.

53. London M, Roth A, Beeren L, Häusser M, Latham PE. Sensitivity to perturbations in vivo implies high noise and suggests rate coding in cortex. Nature Reviews Neuroscience. 2010; 466(1):123–128.

Nencki Award for George L. Gerstein

A. Aertsen et al. (eds.), *Introducing Computation to Neuroscience*, Springer Series in Computational Neuroscience, https://doi.org/10.1007/978-3-030-87447-6

Acta Neurobiol Exp 2009, 69: iv–vi

NENCKI AWARD FOR GEORGE L. GERSTEIN

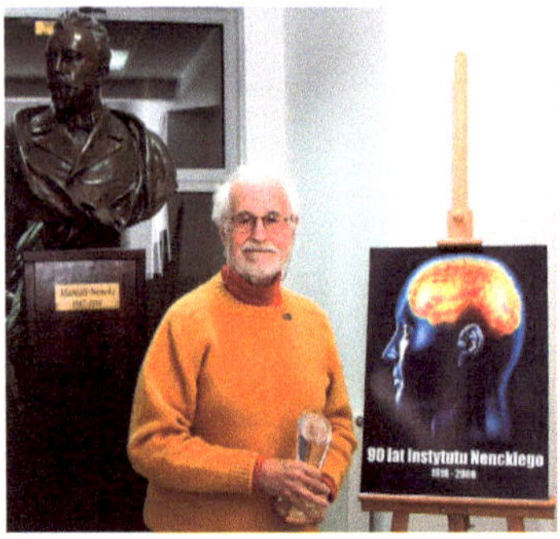

George L. Gerstein
First scientist to win the Nencki Award

In December 2008, the Nencki Institute of Experimental Biology is celebrating its 90th anniversary. On this occasion a prize has been established to honor scientists from all over the world who have contributed significantly to the scientific development of the Institute. The first winner of the Nencki Award is George L. Gerstein, Emeritus Professor of Neuroscience from the University of Pennsylvania School of Medicine

George L. Gerstein was born in Berlin 1933. His family (mother from Grodno and father from Riga) emigrated in 1938 to USA. George finished primary and high school in New York. He then moved to Harvard for college and graduate school (1948–1958) specializing in nuclear physics.

In 1958 Dr. Gerstein decided to switch his field of interest to neurobiology and he became a postdoctoral fellow at MIT, Cambridge. In the next three years he published five papers on computer analysis of electrophysiological signals starting with histogram averaging of single neurons (Gerstein 1960). In 1961–64 he was a junior faculty at MIT and continued his work on application of mathematical methods to physiological data together with Kiang, Fetz, Rodieck and Mandelbrot.

In 1964, George L. Gerstein moved to the University of Pennsylvania as Assistant Professor of Biophysics, built his own laboratory, started teaching graduate students and postdocs. He continued his work on spike train analysis, and published two now classic papers with Perkel and Moore on neuronal trains and stochastic point processes (Perkel et al. 1967). Working over the years at the same University, he changed positions to Professor of Physiology and Professor of Neuroscience but continued the same line of research. Among other achievements, he proposed the theoretical background for studying simultaneous activity of functional neuronal networks (Gerstein and Perkel 1969, Kristan and Gerstein 1970, Michalski et al. 1983) and worked out technology for recording single neurons from cerebral cortex of awake cats (Talwar et al. 2001). He also set up and worked with the nervous system of Aplysia and sensorimotor reflexes of crayfish claw and plasticity of networks controlling the underlying mechanisms. Throughout the years he published about 130 papers on neuroinformatics and sensory neurophysiology with several milestones in both areas (Kristan and Gerstein 1970, Emerson and Gerstein 1977, Gerstein et al. 1985, Abeles and Gerstein 1988, Sillito et al. 1994, Xing and Gerstein 1996, Maldonado and Gerstein 1996, Lindsey et al. 1997, Kisley and Gerstein 1999, 2001, Talwar and Gerstein 2001).

In the 1970s, the University of Pennsylvania received a grant from the Arthur P. Sloane Foundation (arranged by professor Eliot Stellar) to bring over postdocs from the Nencki Institute. In George Gerstein's laboratory several postdoctoral fellows from the Nencki Institute were offered their first research training abroad, which was otherwise difficult to organize that time (A. Wróbel, J. Czarkowska-Bauch, A. Michalski, M. Sarna). Some other researchers from Nencki Institute visiting Penn University at that time collaborated with the team of George Gerstein (R. Tarnecki and K. Turlejski). After that grant ran out, dr. Gerstein secured additional postdoctoral training for P. Musiał (Talwar et al. 2001). From 1970s professor Gerstein has visited the Nencki Institute to take part in several conferences organized by the Department of Neurophysiology. In 1974 he lectured at the "Brain and Behavior" symposium in Jabłonna (Gerstein 1974), and in 2004 he delivered the opening lecture at the workshop "Electrophysiological Semiotics of the Neuronal Systems" supported by a grant establishing a European Center of Excellence at the Nencki Institute (Gerstein 2004). He became a member of the ANE's Editorial Board in 1981. Just the following year, due to the difficulties in printing the Journal during the Martial Low in Poland ANE's lost its position on the ISI list and George Gerstein was among those western scientists who wrote letters in favor of bringing it back, and the action was ended successfully.

Professor Gerstein went on sabbatical leave and research stays in many academic institutions throughout the world: Moscow, Prague, Warsaw, Jerusalem, Tuebingen, London, and more recently Cambridge UK, Freiburg, Newcastle UK, Tel-Aviv, Tokyo. For his full sabbatical year at the Nencki Institute (1978–79) professor Gerstein brought over two laboratory mini-computers, integrated them with laboratory equipment and taught a group of researchers to use and program them to both control and analyze neurophysiological experiments. Although these computers were rather primitive machines by American standards (Z-80) he had to overcome a lot of bureaucratic obstacles in the US before he could export them to the Nencki. The training he offered to the Nencki staff started a new era of computational neuroscience in our Institute. He lived (with his children Mark, then age 14 and Eva, age 10) in the Ursynów district not long after it was built – still with raw countryside and forest nearby. It was a year of heavy snowstorms, and he still recalls that at one point the trains were so blocked by snow that the power plants almost ran out of coal. That year, however, he brought his warm and friendly personality to the atmosphere at the Nencki Institute, although the time was difficult and the situation was swelling for big changes of the Solidarity years, 1980–81.

Andrzej Wróbel

Selected papers of George L. Gerstein

Gerstein GL (1960) Analysis of firing patterns in single neurons Science 131: 1811–1812.

Perkel DH, Gerstein GL, Moore GP (1967) Neuronal spike trains and stochastic point processes. I. The single spike train. II. Simultaneous spike trains. Biophys J 7: 391–418, 419–440.

Gerstein GL, Perkel DH (1969) Simultaneously recorded trains of action potentials: analysis and functional interpretation. Science 164: 828–830.

Kristan WB, Gerstein GL (1970) Plasticity of synchronous activity in a small neural net. Science 169: 1336–1339.

Gerstein GL (1974) Plasticity in small neuronal assemblies. Acta Neurogiol Exp (Wars) 34: 81–92.

Emerson RC, Gerstein GL (1977) Simple striate neurons in the cat. II. Mechanisms underlying directional asymmetry and directional selectivity. J Neurophysiol 40: 136–155.

Michalski A, Gerstein GL, Czarkowska J, Tarnecki R (1983) Interaction between cat striate cortex neurons. Exp Brain Res 51: 97–107.

Gerstein GL, Perkel DH, Dayhoff JE (1985) Cooperative firing activity in simultaneously recorded populations of neurons: Detection and measurement. J Neurosci 5: 881–889.

Abeles M, Gerstein GL (1988) Detecting spatio-temporal firing patterns among simultaneously recorded single neurons. J Neurophysiol 60: 909–924.

Salganicoff M, Sarna M, Sax L, Gerstein GL (1988) Unsupervised waveform classification for multi-neuron recordings: a real-time, software-based system. I. Algorithms and implementation. J Neurosci Methods 25: 181–187.

Sarna MF, Gochin P, Kaltenbach J, Salganicoff M, Gerstein GL (1988) Unsupervised waveform classification for multi-neuron recordings: a real-time, software-based system. II. Performance comparison to other sorters. J Neurosci Methods 25: 189–196.

Sillito AM, Jones HE, Gerstein GL, West DC (1994) Corticofugal feedback in the visual system produces stimulus dependent correlation. Nature 369: 479–482.

Xing J, Gerstein GL (1996) Networks with lateral connectivity. II. Development of neuronal grouping and corresponding receptive field changes. J Neurophysiol 75: 200–216.

Maldonado PE, Gerstein GL (1996) Reorganization in the auditory cortex of the rat induced by intracortical microstimulation: A multiple single-unit study. Exp Brain Res 112: 420–430.

Musial P, Panecki S, Gerstein GL, Wróbel A (1996) Nonlinearities within the cat LGN cell receptive fields in simulated network with recurrent inhibition. Acta Neurobiol Exp (Wars) 56: 927–942.

Lindsey BG, Morris KF, Shannon R, Gerstein GL (1997) Repeated patterns of distributed synchrony in neuronal assemblies. J Neurophysiol 78: 1714–1719.

Kisley MA, Gerstein GL (1999) Trial to trial variability and state-dependent modulation of auditory evoked responses in cortex. J. Neuroscience 19: 10451–10460.

Talwar SK, Gerstein GL (2001) Reorganization in awake rat auditory cortex by local microstimulation and its effect on frequency discrimination behavior. J Neurophysiol 86: 1555–1572.

Kisley MA, Gerstein GL (2001) Daily variation and appetitive conditioning-induced plasticity of auditory cortex receptive fields. Eur J Neurosci 13: 1993–2003.

Talwar SK, Musial PG, Gerstein GL (2001) The role of mammalian auditory cortex in the perception of elementary sound properties. J. Neurophysiology 85: 2350–2358.

Musial PG, Baker SN, Gerstein GL, King EA, Keating JG (2002) Signal-to-noise ratio improvement in multiple electrode recording. J Neurosci Methods115: 29–43.

Gerstein GL, Kirkland KL, Musial PG, Talwar SK (2002) Recordings, behaviour and models related to corticothalamic feedback. Philos Trans R Soc Lond B Biol Sci 357: 1835–1841.

Gerstein GL (2004) Searching for significance in spatiotemporal firing patterns. Acta Neurobiol Exp (Wars) 64: 203–207.

Epilogue

George Gerstein: Obituary (1933–2018)

Ad Aertsen and Moshe Abeles

On March 28, 2018, our colleague and friend George Gerstein passed away.

George Gerstein was a great scientist, a dedicated teacher, and a true friend—in short, a real character, one of a kind. George had many facets: in science, in teaching, in interacting with students and colleagues, and in private. We will try to capture some of these aspects in the following paragraphs, being aware that we can only scratch the surface, based on our personal interactions with George. Fortunately, we could also draw from a manuscript by George's Ukrainian friend Dr. Valya (Valentina) Zelinskaya, and from our private correspondence with George's former wife Linda and their children Mark and Eva.

George Gerstein as a Neuroscientist

George was a truly unique person, in many ways. He profoundly influenced the development of scientific methodology in the neurosciences. His mission was as straightforward as it was ambitious, as formulated already in the abstract of

A. Aertsen
Bernstein Center Freiburg and Faculty of Biology, University of Freiburg, Freiburg, Germany
e-mail: aertsen@biologie.uni-freiburg.de

M. Abeles
Hebrew University, Jerusalem

Bar-Ilan University, Tel Aviv, Israel
e-mail: Moshe.Abeles@biu.ac.il

A. Aertsen et al. (eds.), *Introducing Computation to Neuroscience*, Springer Series in Computational Neuroscience, https://doi.org/10.1007/978-3-030-87447-6

his seminal paper with Benoit Mandelbrot in 1964, one of the very first papers ever to develop a mathematical model of single-neuron spiking activity (Gerstein and Mandelbrot 1964): *"Quantitative methods for the study of the statistical properties of spontaneously occurring spike trains from single neurons have recently been presented. Such measurements suggest a number of descriptive mathematical models. One of these, based on a random walk towards an absorbing barrier, can describe a wide range of neuronal activity in terms of two parameters. These parameters are readily associated with known physiological mechanisms."*

Here, in this short abstract, we meet several terms that characterize George Gerstein's approach to neuroscience, inherited from his solid background in physics: quantitative methods, statistical properties, mathematical models, model parameters, clearly linked to physiological mechanisms.

These various notions continued to be prominently present, in one way or the other, in all of George Gerstein's scientific contributions, throughout his long and productive academic career. From his very first papers in the early 1960s until his final ones in the late 2010s, all his contributions focused on a single, unifying scientific goal: to measure and quantify single and multiple neurons' spiking activities from experimental recordings in relation to behavior and, from the results of such analyses, to gain insight into these neurons' contributions to the functional role of the networks they participated in, preferably in the form of an appropriate mathematical model. Meanwhile, it is almost impossible to imagine neuroscience research without the many fruitful quantitative methods and measures developed by George Gerstein over the years.

Fortunately, we do not need to go through the many individual instances of George's work to document and illustrate these contributions. The various chapters in this Volume, introducing and reflecting on George Gerstein's scientific contributions to a wide range of topics in neuroscience research, written by a selection of his former students and close associates over the years, document and illustrate these contributions more than convincingly.

George Gerstein as a Teacher

George Gerstein was an influential teacher. Starting in his early years at MIT, throughout his long career at UPenn, and via his many extended research periods at Universities and scientific institutes all over the world, George was able to attract many students and postdocs into his unique way of thinking science. Among others, this is reflected in the large number of George's former students making their way successfully into academia and in the type of neuroscience research they contributed ever since, as is also wonderfully illustrated by their contributions in the various chapters of this Volume.

George took a great pride in contributing to the training of neuroscience students from Eastern European countries, especially from Poland. Thus, he spent a sabbatical year (1978–1979) at the Nencki Institute in Warsaw (Poland), together with his wife Linda and their children Mark and Eva. Moreover, he hosted an impressive series of Postdoctoral students from the Nencki Institute in his lab at UPenn over the years. When going to Poland in 1978, George managed to bring two Z-80 minicomputers to the Nencki Institute (not a trivial task at the time) and got them to work. These were the very first minicomputers available for neuroscience research in Poland at the time. The spike train cross-correlograms computed on these machines were henceforth, quite appropriately, called "Georgeograms." Hence, very appropriately, the very first Nencki Award was awarded to George Gerstein in 2008 (see 'Nencki Award for George L. Gerstein' this Volume).

Such activities did not go unnoticed in the USA at the time. George told me (AA) years later that some time after his stay in Poland, he received an unannounced visitor, dressed in a long trench coat and with a fedora hat, who insisted on interviewing him on which advanced technologies and scientific insights he had transferred to the "enemy," Needless to say that George threw him out and continued his dedicated contribution to training neuroscience students and postdocs, both from within and beyond the USA.

George Gerstein as a Person

Here, George's Ukrainian friend Dr. Valya Zelinskaya writes:

"George Gerstein was born in Berlin in 1933 to an intellectual Jewish family, which had emigrated from Riga a year before the October Revolt in 1917. In 1938, when he was 5 years old, the family left Nazi Germany, literally by 'the last train,' for the United States. There, at home in New York, according to his stories, there was everything necessary to satisfy the intellectual requests of his family, but never anything excessive. Thus he remained an ascetic for the rest of his life. In everyday life he lived minimally. In his opinion, for instance, a car requires 4 wheels to drive, but does not need any frills. Also, he was extremely modest in clothing and food. "I'm not posh" he often said. However, one luxury he didn't reject was books, art and travel. He had a highly developed feeling of beauty. He found it and admired it not only in the arts, but also in everything that surrounded him—a beautiful flower, an autumnal leaf."

"George's school years passed in New York. During these years George had read every book, from A to Z, available in the public library. Incredible, but he told me about this as we passed by the library, walking around Manhattan one day. At age 15, he moved to Harvard to continue his education. As he no longer lived with his family anymore, he did not speak Russian—sometimes for years, and he always wondered, how he had retained this language for life. He enjoyed speaking Russian, and liked when we called him by his Russian name '*Yura*.'"

"George liked to travel the world and share his scientific expertise/inspiration with others. He spent Sabbatical leaves in the Soviet Union, Poland, Israel, Germany, and England. In 1968 he came to a conference in Kiev (then still part of the Soviet Union). There he teamed up with Dr. Valentina Zelinskaya (alias Valya) and her husband who worked on data recorded by Valya in the Basal Ganglia of normal and 'Parkinsonian' rats. A couple of years later he spent half a Sabbatical year in Moscow and visited Kiev again. That half year was used by his, then, spouse Linda to study Churches in the region. Later, George spent many hours organizing the multitude of pictures she had made in a well-organized data base."

In Israel, George spent 5 extended research periods with Moshe Abeles at the Hebrew University in Jerusalem. A year before his third visit, Valya and her family (spouse, children, and grandchildren) emigrated to Israel. Valya came to work in the lab of Moshe Abeles, who did not know about her acquaintance with George. Imagine her and George's surprise and joy, when meeting on the first day he came to the lab. During this visit, he joined the retreat of students and teachers of the ICNC at the Hebrew University in Kibbutz Ein Gedi. Moshe Abeles remembers:

"In these retreats I always organized a half-day hike in the Judean desert. This breaks the barriers between students and professors and allows for honest discussions of problems. At this particular retreat, we hiked along a gorge of a dry river-bed running down towards the Dead Sea. All of a sudden, I saw Valya and George depart from the group and walking towards a paved path running on the elevated bank of the gorge. I caught up with them to inquire what had happened. Valya told me that George had eye problems that made it difficult for him to walk over the boulders and the uneven river bed. Prior to this, I did not notice any problems, but she was by far more sensitive than I was. This was the first sign of the progressive deterioration of George's eye sight. Later on, he had eye surgery that improved his eye sight a bit. He could read and write, work on the computer, developing software for analyzing data, and carry on his data analyses. In his next two extended visits to Israel, together with his new companion Janet, they both enjoyed the science and the country tremendously."

George Gerstein as a Colleague and Friend

For us, writing this obituary is a bit like riding a roller coaster, oscillating between reliving fond memories of our many constructive interactions with George and feelings of sorrow, due to George's deteriorating health and growing isolation in spite of his clear mind in his final years. We decided to focus on and rejoice our fond memories in the following and keep the sorrow to ourselves.

Interacting with George on neuroscience was a true pleasure and a headache at the same time. The pleasure came from his ceaseless creativity in suggesting new measures for quantifying neuronal spiking in relation to behavioral tasks, in relation to other neurons' spiking, and in relation to neuronal population activity measures such as LFP and EEG. The headache came from his stubborn insistence on truly understanding these measures and his highly detailed approach of calibrating them vis-a-vis a comprehensive set of possible neuronal interaction scenarios, each one implemented in a detailed neuronal network simulation scheme. Not unexpectedly, this typically took forever and, hence, application of his new measures to experimental data took considerably longer than the experimentalists had hoped for. Nevertheless, whenever these measures finally emerged from these hard testing grounds, they invariably proved to be extremely worthwhile and insightful.

George was not religious, but he was keenly aware of his Jewish ancestry. When he spent his sabbatical year in Tübingen in 1986–1987, he told me (AA) that he had had a very hard time to decide spending longer time in Germany, the nation that had erased all his relatives, except for his own small private family that managed to escape Germany in the very last moment. What ultimately convinced him was that he would be visiting and working with two non-Germans in Germany: Valentino Braitenberg (Fig. 1, from Italy) and myself (from NL). On several hikes we made on the Swabian Alb or elsewhere in the vicinity of Tübingen, George would invariably ask: "Just look at this country, it's just so beautiful! Why on earth did they leave this wonderful place and march east to claim more 'Lebensraum'?"

Several years later, in 1998, George joined me (AA) for the FENS Forum in Berlin. After enjoying a late afternoon coffee with us in a street café in downtown Berlin, George silently left, without notifying any of us. It was only 2 hours later that he returned. He had gone to re-visit the house where he had been born in 1933 and had spent a happy youth until the age of five. Remarkably, he said, he remembered it all—and nothing much had changed since then.

Fig. 1 George Gerstein (left) and Valentino Braitenberg at the MPI for Biological Cybernetics in Tübingen in summer, 1987

Over the years, George was drawn again and again to spend sabbatical (and otherwise) time in Israel, to be together with his Israeli colleagues and friends, and to practice his Hebrew and to enjoy speaking his Russian. Together with one of us (MA), he spent many hours looking for repeating spike patterns (Fig. 2). Until the present day, we do not know if they found any—that is, above chance level. But in some instances, that does not even matter—when the ultimate charm is in the search, not in the finding (Fig. 3).

Fig. 2 Moshe Abeles (left) and George Gerstein in the Judean Desert in the late 1990s

Fig. 3 Moshe Abeles, George Gerstein, and Ad Aertsen. Picture taken by our colleague Bill Newsome at Moshe's home, on the occasion of the "Being on Time" Symposium for Moshe Abeles at the Hebrew University in Jerusalem, January 3–5, 2007. We thank Bill for making this picture available to us for this purpose

Apart from science, George loved the arts, especially music. He loved to play the piano and, even more so, his clavichord that he kept in a "wohltemperiertes Zimmer" in his beloved home at Whitemarsh Road in Ardmore, PA. There, with spikes, amplifiers, computers, and math out of the way, he loved to spend time with his dearest friend, Johann Sebastian Bach, enjoying his preludes and fugues. After his last sabbatical leave in Freiburg in 2010, he wrote: "The social and cultural aspects of my stay in Freiburg were delightful. I was able to attend a good many concerts, particularly a series of organ recitals and various interesting performances at the Hochschule für Musik. And the Augustiner Museum provided a fascinating look at medieval art and architecture."

George Gerstein presented a rare and unique combination of human qualities. On behalf of those of us who had the privilege to work with him, to interact with him, and to learn from him, we thank him for this privilege. His passing away is a great loss, for all of us—but his being among us was a great experience and a true pleasure. His legacy as a scientist, a person, a colleague, and a friend will always remain with us.

Reference

Gerstein GL, Mandelbrot B (1964) Random walk models for the spike activity of a single neuron. Biophys J 4(1, Part 1):41–68. https://doi.org/10.1016/S0006-3495(64)86768-0

Publications of George L. Gerstein

Gerstein G, Niederer J, Strauch K (1957) Elastic scattering of 96-Mev protons. Phys Rev 108(2):427–432. https://doi.org/10.1103/PhysRev.108.427

Gerstein GL (1960) Analysis of firing patterns in single neurons. Science 131(3416):1811–1812. https://doi.org/10.1126/science.131.3416.1811

Gerstein GL, Kiang NS (1960) An approach to the quantitative analysis of electrophysiological data from single neurons. Biophys J 1(1):15–28. https://doi.org/10.1016/s0006-3495(60)86872-5

Geisler CD, Gerstein GL (1961) The surface EEG in relation to its sources. Electroencephalogr Clin Neurophysiol 13(6):927–934. https://doi.org/10.1016/0013-4694(61)90199-7

Gerstein GL (1961) Neuron firing patterns and the slow potentials. Electroencephalogr Clin Neurophysiol 20(Supplement):68–71

Gerstein G (1962) Mathematical models for the all-or-none activity of some neurons. IRE Trans Inf Theory 8(5):137–143. https://doi.org/10.1109/TIT.1962.1057761

Rodieck RW, Kiang NYS, Gerstein GL (1962) Some quantitative methods for the study of spontaneous activity of single neurons. Biophys J 2(4):351–368. https://doi.org/10.1016/S0006-3495(62)86860-X

Fetz EE, Gerstein G (1963) An RC model for spontaneous activity of single neurons. In: Quarterly Progress Report #71. Research Laboratory of Electronics. MIT, New York

Gerstein G (1963) Problems in the study of sensory coding in the nervous system. In: Northeast electronics research and engineering meeting (NEREM); Boston, Mass, vol 5. Winner, New York

Gerstein GL, Clark WA (1964) Simultaneous studies of firing patterns in several neurons. Science 143(3612):1325–1327. https://doi.org/10.1126/science.143.3612.1325

Gerstein GL, Kiang NYs (1964) Responses of single units in the auditory cortex. Exp Neurol 10(1):1–18. https://doi.org/10.1016/0014-4886(64)90083-4

Gerstein GL, Mandelbrot B (1964) Random walk models for the spike activity of a single neuron. Biophys J 4(1, Part 1):41–68. https://doi.org/10.1016/S0006-3495(64)86768-0

Gross CG, Schiller PH, Wells C, Gerstein GL (1967) Single-unit activity in temporal association cortex of the monkey. J Neurophysiol 30(4):833–843. https://doi.org/10.1152/jn.1967.30.4.833

Perkel DH, Gerstein GL, Moore GP (1967a) Neuronal spike trains and stochastic point processes. I. The single spike train. Biophys J 7(4):391–418. https://doi.org/10.1016/S0006-3495(67)86596-2

Perkel DH, Gerstein GL, Moore GP (1967b) Neuronal spike trains and stochastic point processes. II. Simultaneous spike trains. Biophys J 7(4):419–440. https://doi.org/10.1016/s0006-3495(67)86597-4

A. Aertsen et al. (eds.), *Introducing Computation to Neuroscience*, Springer Series in Computational Neuroscience, https://doi.org/10.1007/978-3-030-87447-6

Erulkar SD, Butler RA, Gerstein GL (1968) Excitation and inhibition in cochlear nucleus. II. Frequency-modulated tones. J Neurophysiol 31(4):537–548. https://doi.org/10.1152/jn.1968.31.4.537

Gerstein GL, Butler RA, Erulkar SD (1968a) Excitation and inhibition in cochlear nucleus. I. Tone-burst stimulation. J Neurophysiol 31(4):526–536. https://doi.org/10.1152/jn.1968.31.4.526

Gerstein GL, Gross CG, Weinstein M (1968b) Inferotemporal evoked potentials during visual discrimination performance by monkeys. J Comp Physiol Psychol 65(3, Part 1):526–528. https://doi.org/10.1037/h0025813

Gerstein GL, Perkel DH (1969) Simultaneously recorded trains of action potentials: analysis and functional interpretation. Science 164(3881):828. https://doi.org/10.1126/science.164.3881.828

Gerstein GL (1970) Functional association of neurons: Detection and interpretation. In: Schmitt FO (ed.) The Neurosciences, Second Study Program. Rockefeller University Press, Rockefeller, pp 648–661

Kristan Jr WB, Gerstein GL (1970) Plasticity of synchronous activity in a small neural net. Science 169(3952):1336–1339. https://doi.org/10.1126/science.169.3952.1336

Fernald RD, Gerstein GL (1972) Response of cat cochlear nucleus neurons to frequency and amplitude modulated tones. Brain Res 45(2):417–435. https://doi.org/10.1016/0006-8993(72)90472-6

Gerstein GL, Perkel DH (1972) Mutual temporal relationships among neuronal spike trains: Statistical techniques for display and analysis. Biophys J 12(5):453–473. https://doi.org/10.1016/S0006-3495(72)86097-1

Gerstein GL, Schuster RH, Wiens TJ (1972) Neural activity and behavioral plasticity in the crayfish claw. In: Kiger J (ed.) Biology of behavior. Oregon State University Press, New York, pp 67–82

Voronin LL, YA GG, Ioffe SV, Kudriashov IE (1973) A rapidly elaborated conditioned reflex with simultaneous recording of neuronal activity. Pavlov J Higher Nerv Activity 23:636–639 (In Russian)

Dickson JW, Gerstein GL (1974) Interactions between neurons in auditory cortex of the cat. J Neurophysiol 37(6):1239–1261. https://doi.org/10.1152/jn.1974.37.6.1239

Gerstein GL (1974) Plasticity in small neuronal assemblies. Acta Neurobiologiae Exp (Wars) 34(1):81–92

Perkel DH, Gerstein GL, Smith MS, Tatton WG (1975) Nerve-impulse patterns: A quantitative display technique for three neurons. Brain Res 100(2):271–296. https://doi.org/10.1016/0006-8993(75)90483-7

Tatton WG, Forner SD, Gerstein GL, Chambers WW, Liu CN (1975) The effect of postcentral cortical lesions on motor responses to sudden upper limb displacements in monkeys. Brain Res 96(1):108–113. https://doi.org/10.1016/0006-8993(75)90580-6

Voronin LL, Gerstein GL, Kudryashov IE, Ioffe SV (1975a) Elaboration of a conditioned reflex in a single experiment with simultaneous recording of neural activity. Brain Res 92(3):385–403. https://doi.org/10.1016/0006-8993(75)90324-8

Voronin LL, Gerstein GY, Ioffe SV, Kudriashov IE, Lekhtman GA (1975b) Conditioned reflex convenient for neurophysiological analysis. In: Adrianov O (ed.) XXII. Conference on problems of higher nervous activity (Abstracts), vol 2. Gorky, Moscow, pp 59–60 (In Russian)

Wiens TJ, Gerstein GL (1975) Cross connections among crayfish claw efferents. J Neurophysiol 38(4):909–921. https://doi.org/10.1152/jn.1975.38.4.909

Stevens JK, Gerstein GL (1976a) Spatiotemporal organization of cat lateral geniculate receptive fields. J Neurophysiol 39(2):213–238. https://doi.org/10.1152/jn.1976.39.2.213

Stevens JK, Gerstein GL (1976b) Interactions between cat lateral geniculate neurons. J Neurophysiol 39(2):239–256. https://doi.org/10.1152/jn.1976.39.2.239

Stevens JK, Emerson RC, Gerstein GL, Kallos T, Neufeld GR, Nichols CW, Rosenquist AC (1976) Paralysis of the awake human: visual perceptions. Vision Res 16(1):93–98. https://doi.org/10.1016/0042-6989(76)90082-1

Wiens TJ, Gerstein GL (1976) Reflex pathways of the crayfish claw. J Comp Physiol 107(3):309–326. https://doi.org/10.1007/BF00656740

Emerson RC, Gerstein GL (1977a) Simple striate neurons in the cat. I. Comparison of responses to moving and stationary stimuli. J Neurophysiol 40(1):119–135. https://doi.org/10.1152/jn.1977.40.1.119

Emerson RC, Gerstein GL (1977b) Simple striate neurons in the cat. II. Mechanisms underlying directional asymmetry and directional selectivity. J Neurophysiol 40(1):136–155. https://doi.org/10.1152/jn.1977.40.1.136

Lindsey BG, Gerstein GL (1977) Reflex control of a crayfish claw motor neuron during imposed dactylopodite movements. Brain Res 130(2):348–353. https://doi.org/10.1016/0006-8993(77)90281-5

Stafstrom CE, Gerstein GL (1977) A paradigm for position learning in the crayfish claw. Brain Res 134(1):185–190. https://doi.org/10.1016/0006-8993(77)90939-8

Gerstein GL, Perkel DH, Subramanian KN (1978) Identification of functionally related neural assemblies. Brain Res 140(1):43–62. https://doi.org/10.1016/0006-8993(78)90237-8

Gross CG, Bender DB, Gerstein GL (1979) Activity of inferior temporal neurons in behaving monkeys. Neuropsychologia 17(2):215–229. https://doi.org/10.1016/0028-3932(79)90012-5

Lindsey BG, Gerstein GL (1979a) Interactions among an ensemble of chordotonal organ receptors and motor neurons of the crayfish claw. J Neurophysiol 42(2):383–399. https://doi.org/10.1152/jn.1979.42.2.383

Lindsey BG, Gerstein GL (1979b) Proprioceptive fields of crayfish claw motor neurons. J Neurophysiol 42(2):368–382. https://doi.org/10.1152/jn.1979.42.2.368

Gerstein GL, Renfrew DH, Pubols BH (1980) Two behavioral paradigms for study of rapid changes in functional grouping of neurons. Acta Neurobiologiae Exp (Wars) 40(1):149–160

Gerstein GL, Michalski A (1981) Firing synchrony in a neural group: Putative sensory code. In: Székely G, L'abos E, Damjanovich S (eds.) Advances in physiological sciences. Neural Communication and control, vol 30. Pergamon Press, Akademiai Kiado, Debrecen/Budapest, pp 93–102. https://doi.org/10.1016/b978-0-08-027351-8.50012-3

Dayhoff JE, Gerstein GL (1983a) Favored patterns of spike trains. I. Detection. J Neurophysiol 49(6):1334–1348. https://doi.org/10.1152/jn.1983.49.6.1334

Dayhoff JE, Gerstein GL (1983b) Favored patterns of spike trains. II. Application. J Neurophysiol 49(6):1349–1362. https://doi.org/10.1152/jn.1983.49.6.1349

Gerstein GL, Bloom MJ, Espinosa IE, Evanczuk S, Turner MR (1983) Design of a laboratory for multineuron studies. IEEE Trans Syst Man Cybern SMC-B(5):668–676. https://doi.org/10.1109/TSMC.1983.6313059

Michalski A, Gerstein GL, Czarkowska J, Tarnecki R (1983) Interactions between cat striate cortex neurons. Exp Brain Res 51(1):97–107. https://doi.org/10.1007/BF00236807

Aertsen AMHJ, Gerstein GL (1985) Evaluation of neuronal connectivity: sensitivity of cross-correlation. Brain Res 340(2):341–354. https://doi.org/10.1016/0006-8993(85)90931-X

Gerstein GL, Aertsen AM (1985) Representation of cooperative firing among simultaneously recorded neurons. J Neurophysiol 54(6):1513–1528. https://doi.org/10.1152/jn.1985.54.6.1513

Gerstein GL, Aertsen A, Bloom M, Espinosa I, Evanczuk S, Turner M (1985a) Multi-neuron experiments: observation of state in neural nets. In: Haken H (ed.) Complex systems—operational approaches. Springer, Berlin, pp 58–70

Gerstein GL, Perkel H, Dayhoff E (1985b) Cooperative firing activity in simultaneously recorded populations of neurons: detection and measurement. J Neurosci 5(4):881–889. https://doi.org/10.1523/JNEUROSCI.05-04-00881.1985

Aertsen A, Gerstein GL, Johannesma P (1986) From neuron to assembly: neuronal organization and stimulus representation. In: Palm G, Aertsen A (eds.) Brain Theory. Springer, Berlin. https://doi.org/10.1007/978-3-642-70911-1_3

Kaltenbach JA, Gerstein GL (1986) A rapid method for production of sharp tips on preinsulated microwires. J Neurosci Methods 16(4):283–288. https://doi.org/10.1016/0165-0270(86)90053-1

Gerstein GL (1987) Information flow in cortical networks: interpreting multi-neuron experiments. In: von Seelen W, Shaw G, Leinhos U (eds.) Organization of neural networks. Weinheim VCH, Verlagsgesellschaft, pp 51–57

Abeles M, Gerstein GL (1988) Detecting spatiotemporal firing patterns among simultaneously recorded single neurons. J Neurophysiol 60(3):909–924. https://doi.org/10.1152/jn.1988.60.3.909

Espinosa IE, Gerstein GL (1988) Cortical auditory neuron interactions during presentation of 3-tone sequences: Effective connectivity. Brain Res 450(1–2):39–50. https://doi.org/10.1016/0006-8993(88)91542-9

Palm G, Aertsen AMHJ, Gerstein GL (1988) On the significance of correlations among neuronal spike trains. Biol Cybern 59(1):1–11. https://doi.org/10.1007/BF00336885

Salganicoff M, Sarna M, Sax L, Gerstein GL (1988) Unsupervised waveform classification for multi-neuron recordings: a real-time, software-based system. I. Algorithms and implementation. J Neurosci Methods 25(3):181–187. https://doi.org/10.1016/0165-0270(88)90132-X

Sarna MF, Gochin P, Kaltenbach J, Salganicoff M, Gerstein GL (1988) Unsupervised waveform classification for multi-neuron recordings: a real-time, software-based system. II. Performance comparison to other sorters. J Neurosci Methods 25(3):189–196. https://doi.org/10.1016/0165-0270(88)90133-1

Aertsen AM, Gerstein GL, Habib MK, Palm G (1989) Dynamics of neuronal firing correlation: modulation of "effective connectivity". J Neurophysiol 61(5):900–917. https://doi.org/10.1152/jn.1989.61.5.900

Gerstein GL, Bedenbaugh P, Aertsen AMHJ (1989) Neuronal assemblies. IEEE Trans Biomed Eng 36(1):4–14. https://doi.org/10.1109/10.16444

Gochin PM, Kaltenbach JA, Gerstein GL (1989) Coordinated activity of neuron pairs in anesthetized rat dorsal cochlear nucleus. Brain Res 497(1):1–11. https://doi.org/10.1016/0006-8993(89)90963-3

Kaltenbach JA, Gochin P, Gerstein GL (1989) Spectral time-course analysis of firing patterns in the dorsal cochlear nucleus. Hear Res 41(2-3):107–114. https://doi.org/10.1016/0378-5955(89)90004-X

Kreiter AK, Aertsen AMHJ, Gerstein GL (1989) A low-cost single-board solution for real-time, unsupervised waveform classification of multineuron recordings. J Neurosci Methods 30(1):59–69. https://doi.org/10.1016/0165-0270(89)90075-7

Lindsey BG, Shannon R, Gerstein GL (1989) Gravitational representation of simultaneously recorded brainstem respiratory neuron spike trains. Brain Res 483(2):373–378. https://doi.org/10.1016/0006-8993(89)90183-2

Gerstein GL, Turner MR (1990) Neural assemblies as building blocks of cortical computation. In: Schwatz EL (ed.) Computational neuroscience. MIT Press, Cambridge, pp 179–191

Gochin PM, Gerstein GL, Kaltenbach JA (1990) Dynamic temporal properties of effective connections in rat dorsal cochlear nucleus. Brain Res 510(2):195–202. https://doi.org/10.1016/0006-8993(90)91367-P

Aertsen AMHJ, Gerstein GL (1991) Dynamic aspects of neuronal cooperativity: fast stimulus-locked modulations of 'effective connectivity'. In: Krüger J (ed.) Neuronal Cooperativity, vol 49. Springer Series in Synergetics, Berlin, pp 52–67. https://doi.org/10.1007/978-3-642-84301-3_4

Gochin PM, Miller EK, Gross CG, Gerstein GL (1991) Functional interactions among neurons in inferior temporal cortex of the awake macaque. Exp Brain Res 84(3):505–516. https://doi.org/10.1007/BF00230962

MacGregor RJ, Gerstein GL (1991) Cross-talk theory of memory capacity in neural networks. Biol Cybern 65(5):351–355. https://doi.org/10.1007/BF00216968

Turner MR, Gerstein GL, Bajcsy R (1991) Underestimation of visual texture slant by human observers: a model. Biol Cybern 65(4):215–226. https://doi.org/10.1007/BF00206219

Gerstein GL, Gochin P (1992) Neuronal population coding and the elephant. In: Aertsen A, Braitenberg V (eds.) Information processing in the cortex: experiments and theory. Springer, Berlin, pp 139–160

Gochin PM, Bedenbaugh P, Gelfand JJ, Gross CG, Gerstein GL (1992) Intrinsic signal optical imaging in the forepaw area of rat somatosensory cortex. Proc Natl Acad Sci 89(17):8381–8383. https://doi.org/10.1073/pnas.89.17.8381

Lindsey BG, Hernandez YM, Morris KF, Shannon R, Gerstein GL (1992a) Respiratory-related neural assemblies in the brain stem midline. J Neurophysiol 67(4):905–922. https://doi.org/10.1152/jn.1992.67.4.905

Lindsey BG, Hernandez YM, Morris KF, Shannon R, Gerstein GL (1992b) Dynamic reconfiguration of brain stem neural assemblies: respiratory phase-dependent synchrony versus modulation of firing rates. J Neurophysiol 67(4):923–930. https://doi.org/10.1152/jn.1992.67.4.923

Gerstein GL, Turner MR (1993) Neural assemblies as building blocks of cortical computation. In: Schwartz EL (ed) Computational Neuroscience. MIT Press, Cambridge, pp 179–191

Xing J, Gerstein GL (1993) A neural network model for texture discrimination. Biol Cybern 69(2):97–108. https://doi.org/10.1007/BF00226193

Bedenbaugh P, Gerstein GL (1994) Rectification of correlation by a sigmoid nonlinearity. Biol Cybern 70(3):219–225. https://doi.org/10.1007/BF00197602

Gochin PM, Colombo M, Dorfman GA, Gerstein GL, Gross CG (1994) Neural ensemble coding in inferior temporal cortex. J Neurophysiol 71(6):2325–2337. https://doi.org/10.1152/jn.1994.71.6.2325

Sillito AM, Jones HE, Gerstein GL, West DC (1994) Feature-linked synchronization of thalamic relay firing induced by feedback from the visual cortex. Nature 369(6480):479–482. https://doi.org/10.1038/369479a0

Xing J, Gerstein GL (1994) Simulation of dynamic receptive fields in primary visual cortex. Vision Res 34(14):1901–1911. https://doi.org/10.1016/0042-6989(94)90314-X

Maldonado PE, Gerstein GL (1996a) Reorganization in the auditory cortex of the rat induced by intracortical microstimulation: A multiple single-unit study. Exp Brain Res 112(3):420–430. https://doi.org/10.1007/BF00227948

Maldonado PE, Gerstein GL (1996b) Neuronal assembly dynamics in the rat auditory cortex during reorganization induced by intracortical microstimulation. Exp Brain Res 112(3):431–441. https://doi.org/10.1007/BF00227949

Musial P, Panecki S, Gerstein GL, Wróbel A (1996) Nonlinearities within the cat LGN cell receptive fields in simulated network with recurrent inhibition. Acta Neurobiologiae Exp 56(4):927–942

Xing J, Gerstein GL (1996a) Networks with lateral connectivity. I. Dynamic properties mediated by the balance of intrinsic excitation and inhibition. J Neurophysiol 75(1):184–199. https://doi.org/10.1152/jn.1996.75.1.184

Xing J, Gerstein GL (1996b) Networks with lateral connectivity. II. Development of neuronal grouping and corresponding receptive field changes. J Neurophysiol 75(1):200–216. https://doi.org/10.1152/jn.1996.75.1.200

Xing J, Gerstein GL (1996c) Networks with lateral connectivity. III. Plasticity and reorganization of somatosensory cortex. J Neurophysiol 75(1):217–232. https://doi.org/10.1152/jn.1996.75.1.217

Bedenbaugh P, Gerstein GL (1997) Multiunit normalized cross correlation differs from the average single-unit normalized correlation. Neural Comput 9(6):1265–1275. https://doi.org/10.1162/neco.1997.9.6.1265

Lindsey BG, Morris KF, Shannon R, Gerstein GL (1997) Repeated patterns of distributed synchrony in neuronal assemblies. J Neurophysiol 78(3):1714–1719. https://doi.org/10.1152/jn.1997.78.3.1714

Kirkland KL, Gerstein GL (1998) A model of cortically induced synchronization in the lateral geniculate nucleus of the cat: a role for low-threshold calcium channels. Vision Res 38(13):2007–2022. https://doi.org/10.1016/S0042-6989(97)00385-4

Maldonado PE, Altman JA, Gerstein GL (1998) Neuron discharges in the rat auditory cortex during electrical intracortical stimulation. Neurosci Behav Physiol 28(1):48–59. https://doi.org/10.1007/BF02461912

Talwar SK, Gerstein GL (1998) Auditory frequency discrimination in the white rat. Hearing Res 126(1-2):135–150. https://doi.org/10.1016/S0378-5955(98)00162-2

Kirkland KL, Gerstein GL (1999) A feedback model of attention and context dependence in visual cortical networks. J Comput Neurosci 7(3):255–267. https://doi.org/10.1023/A:1008923203424

Kisley MA, Gerstein GL (1999a) The continuum of operating modes for a passive model neuron. Neural Comput 11(5):1139–1154. https://doi.org/10.1162/089976699300016386

Kisley MA, Gerstein GL (1999b) Trial-to-trial variability and state-dependent modulation of auditory-evoked responses in cortex. J Neurosci 19(23):10451–10460. https://doi.org/10.1523/jneurosci.19-23-10451.1999

Talwar SK, Gerstein GL (1999) A signal detection analysis of auditory-frequency discrimination in the rat. J Acoust Soc Am 105(3):1784–1800. https://doi.org/10.1121/1.426716

Baker SN, Gerstein GL (2000) Improvements to the sensitivity of gravitational clustering for multiple neuron recordings. Neural Comput 12(11):2597–2620. https://doi.org/10.1162/089976600300014863

Gerstein GL (2000) Cross-correlation measures of unresolved multi-neuron recordings. J Neurosci Methods 100(1–2):41–51. https://doi.org/10.1016/S0165-0270(00)00226-0

Kirkland KL, Sillito AM, Jones HE, West DC, Gerstein GL (2000) Oscillations and long-lasting correlations in a model of the lateral geniculate nucleus and visual cortex. J Neurophysiol 84(4):1863–1868. https://doi.org/10.1152/jn.2000.84.4.1863

Baker SN, Gerstein GL (2001) Determination of response latency and its application to normalization of cross-correlation measures. Neural Comput 13(6):1351–1377. https://doi.org/10.1162/08997660152002889

Gerstein GL, Kirkland KL (2001) Neural assemblies: technical issues, analysis, and modeling. Neural Netw 14(6–7):589–598. https://doi.org/10.1016/S0893-6080(01)00042-9

Kisley MA, Gerstein GL (2001) Daily variation and appetitive conditioning-induced plasticity of auditory cortex receptive fields. Eur J Neurosci 13(10):1993–2003. https://doi.org/10.1046/j.0953-816X.2001.01568.x

Talwar SK, Gerstein GL (2001) Reorganization in awake rat auditory cortex by local microstimulation and its effect on frequency-discrimination behavior. J Neurophysiol 86(4):1555–1572. https://doi.org/10.1152/jn.2001.86.4.1555

Talwar SK, Musial PG, Gerstein GL (2001) Role of mammalian auditory cortex in the perception of elementary sound properties. J Neurophysiol 85(6):2350–2358. https://doi.org/10.1152/jn.2001.85.6.2350

Gerstein GL, Kirkland KL, Musial PG, Talwar SK (2002) Recordings, behaviour and models related to corticothalamic feedback. Philos Trans R Soc London, Ser B Biol Sci 357(1428):1835–1841. https://doi.org/10.1098/rstb.2002.1166

Keating JG, Gerstein GL (2002) A chronic multi-electrode microdrive for small animals. J Neurosci Methods 117(2):201–206. https://doi.org/10.1016/S0165-0270(02)00115-2

Musial PG, Baker SN, Gerstein GL, King EA, Keating JG (2002) Signal-to-noise ratio improvement in multiple electrode recording. J Neurosci Methods 115(1):29–43. https://doi.org/10.1016/S0165-0270(01)00516-7

Gerstein GL (2004) Searching for significance in spatio-temporal firing patterns. Acta Neurobiologiae Exp (Wars) 64(2):203–207

Davies RM, Gerstein GL, Baker SN (2006) Measurement of time-dependent changes in the irregularity of neural spiking. J Neurophysiol 96(2):906–918. https://doi.org/10.1152/jn.01030.2005

Lindsey BG, Gerstein GL (2006) Two enhancements of the gravity algorithm for multiple spike train analysis. J Neurosci Methods 150(1):116–127. https://doi.org/10.1016/j.jneumeth.2005.06.019

Schrader S, Grün S, Diesmann M, Gerstein GL (2008) Detecting synfire chain activity using massively parallel spike train recording. J Neurophysiol 100(4):2165–2176. https://doi.org/10.1152/jn.01245.2007

Gerstein GL (2010) Gravitational Clustering. In: Grün S, Rotter S (eds.) Analysis of parallel spike trains. Springer series in computational neuroscience. Springer, Berlin, pp 157–172. https://doi.org/10.1007/978-1-4419-5675-0

Louis S, Gerstein GL, Grün S, Diesmann M (2010) Surrogate spike train generation through dithering in operational time. Front Comput Neurosci 4:1–16. https://doi.org/10.3389/fncom.2010.00127

Gerstein GL, Williams ER, Diesmann M, Grün S, Trengove C (2012) Detecting synfire chains in parallel spike data. J Neurosci Methods 206(1):54–64. https://doi.org/10.1016/j.jneumeth.2012.02.003

Picado-Muiño D, Borgelt C, Berger D, Gerstein G, Grün S (2013) Finding neural assemblies with frequent item set mining. Front Neuroinf 7:9. https://doi.org/10.3389/fninf.2013.00009

Gerstein GL (2015) Joint-Peri Stimulus Time Histogram (JPSTH). In: Jaeger D, Jung R (eds.) Encyclopedia of computational neuroscience. Springer, New York. https://doi.org/10.1007/978-1-4614-6675-8

Torre E, Canova C, Denker M, Gerstein G, Helias M, Grün S (2016) ASSET: analysis of sequences of synchronous events in massively parallel spike trains. PLoS Comput Biol 12(7):e1004939. https://doi.org/10.1371/journal.pcbi.1004939

Backside

This book brings together a selection of papers by George Gerstein, representing his long-term endeavor of making neuroscience into a real science—inspired by physics, where he had his roots. In this, George was many years ahead of the field, consistently striving for quantitative analyses, mechanistic models, conceptual clarity, and, foremost, real understanding. In doing so, George Gerstein pioneered Computational Neuroscience, many years before the term was even conceived.

In this book, we collected a selection of George Gerstein's many seminal contributions to neuroscience, be they experimental, theoretical, or computational, into a single, comprehensive volume. We also intend to provide the readers, especially the younger ones who may not even be aware that there was ever a time in neuroscience without the PSTH, the cross-correlation approach, or stochastic neuron models, with the fresh introduction of these various concepts in the original literature. Thus, in this sense, this book also serves as an introduction to various fields of neuroscience, experimental, theoretical, and computational.

We organized the material in a series of chapters by subject, ordered in time, each one containing a few reprints of George Gerstein's papers. Several of George's former students and collaborators kindly contributed an Introduction to these chapters, putting the reprinted papers in their historical and present context and adding the personal touch of their involvement in generating them. Taken together, we hope that this book may help the reader appreciate how much our current thinking on brain function owes to George Gerstein's research and the insights gained from it. George was not only an innovative thinker, in theory, experiment, and analysis methodology, but he also inspired by his mode of thinking science. This book is both a product of this inspiration and a tribute to it.

A. Aertsen et al. (eds.), *Introducing Computation to Neuroscience*, Springer Series in Computational Neuroscience, https://doi.org/10.1007/978-3-030-87447-6

Ad Aertsen is an Emeritus Professor of Neurobiology and Biophysics at the Faculty of Biology at the University of Freiburg and Founding Director of the Bernstein Center Freiburg. He pursued his postdoctoral studies in physiology with George Gerstein at the University of Pennsylvania, Philadelphia, USA, in 1983–1984. His main interest was and is the recording, analysis, and modeling of neuronal assembly activity in cortical and subcortical networks.

Sonja Grün is the Professor for Theoretical Systems Neurobiology at RWTH Aachen University, Germany, and the Director of the Institute Neuroscience and Medicine INM-6 and INM-10 at the Research Centre Jülich, Germany. Her research interests are (a) dynamical interactions in the brain network, (b) development of statistical analysis tools for activity data, (c) development of collaborative, reproducible workflows, and (d) closing the loop between neural network models and experimental data.

Pedro E. Maldonado earned a Bachelor's and a Master's degree at the Faculty of Sciences of the Universidad de Chile. He obtained a Ph.D. degree in physiology as a student of George L. Gerstein, at the University of Pennsylvania. He is now a Full Professor and the Chairman of the Department of Neuroscience at the Faculty of Medicine, Universidad de Chile, and an Associate Researcher at the Millennium Institute of Biomedical Neuroscience (BNI).

Günther Palm is the retired director of the Institute of Neural Information Processing at Ulm University. After studying mathematics, he worked in theoretical neuroscience and neuro-informatics. His main research interests are neural networks, associative memory, and Hebbian cell assemblies.

MIX
Papier aus verantwortungsvollen Quellen
Paper from responsible sources
FSC® C105338

If you have any concerns about our products,
you can contact us on
ProductSafety@springernature.com

In case Publisher is established outside the EU,
the EU authorized representative is:
Springer Nature Customer Service Center GmbH
Europaplatz 3, 69115 Heidelberg, Germany

Printed by Libri Plureos GmbH
in Hamburg, Germany